Science of Earth Systems

Stephen D. Butz

THOMSON
DELMAR LEARNING

Australia Canada Mexico Singapore Spain United Kingdom United States

Science of Earth Systems
Stephen D. Butz

Business Unit Executive Director:
Susan L. Simpfenderfer

Acquisitions Editor:
Zina M. Lawrence

Developmental Editor:
Andrea Edwards

Editorial Assistant:
Elizabeth Gallagher

Executive Production Manager:
Wendy A. Troeger

Production Manager:
Carolyn Miller

Production Editor:
Kathryn B. Kucharek

Executive Marketing Manager:
Donna J. Lewis

Channel Manager:
Nigar Hale

Cover Image:
PhotoDisc

Cover Design:
TDB Publishing Services

COPYRIGHT © 2004 by Delmar Learning, a division of Thomson Learning, Inc. Thomson Learning™ is a trademark used herein under license.

Printed in the United States
2 3 4 5 XXX 06 05 04 03

For more information contact Delmar Learning, 5 Maxwell Drive, Clifton Park, NY 12065-2919.

Or you can visit our internet site at http://www.delmarlearning.com

ALL RIGHTS RESERVED. No part of this work covered by the copyright hereon may be reproduced or used in any form or by any means—graphic, electronic, or mechanical, including photocopying, recording, taping, Web distribution or information storage and retrieval systems—without written permission of the publisher.

For permission to use material from this text or product, contact us by
Tel (800) 730-2214
Fax (800) 730-2215
http://www.thomsonrights.com

Library of Congress
Cataloging-in-Publication Data

Butz, Stephen D.
Science of earth systems /
Stephen D. Butz.
p. cm.
Includes index.
ISBN 0-7668-3391-7
1. Earth sciences. I. Title.

QE28 B967 2002
550—dc21

2002017421

NOTICE TO THE READER

Publisher does not warrant or guarantee any of the products described herein or perform any independent analysis in connection with any of the product information contained herein. Publisher does not assume, and expressly disclaims, any obligation to obtain and include information other than that provided to it by the manufacturer.

The reader is expressly warned to consider and adopt all safety precautions that might be indicated by the activities herein and to avoid all potential hazards. By following the instructions contained herein, the reader willingly assumes all risks in connection with such instructions.

The Publisher makes no representation or warranties of any kind, including but not limited to, the warranties of fitness for particular purpose or merchantability, nor are any such representations implied with respect to the material set forth herein, and the publisher takes no responsibility with respect to such material. The Publisher shall not be liable for any special, consequential, or exemplary damages resulting, in whole or part, from the readers' use of, or reliance upon, this material.

Preface
Science of Earth Systems

For much of the twentieth century, science has been divided into unique, highly specialized disciplines that have attempted to unlock the workings of nature. Until recently, each of these branches of science have remained virtually isolated from one another. Not only has this scientific isolation existed at the higher academic levels, but it is also standard practice in much of the secondary school science curriculum. Today, because of the changes that are being made by humankind on the Earth, scientists in these separate academic disciplines are finding it necessary to collaborate with one another on a unprecedented scale. It has now become necessary for them to share their research to try to understand the Earth's complex systems, and how those systems are being affected by human activity. This new area of collaborative research is called Earth system science, which is a combination of all the scientific disciplines into one body of knowledge. *Science of Earth Systems* is a textbook designed to introduce this new scientific discipline to secondary school students. This textbook brings together in one volume, the interactions that occur in the living and nonliving world. Whether students are studying the Earth sciences, agriculture, or environmental science, this text provides an understanding of the physical and biological processes that exist on our planet. *Science of Earth Systems* was written to meet the National Science Education Standards developed by the National Academy of Sciences for the Earth and space sciences, and can be easily aligned with Earth science-related standards and curriculum taught throughout the United States. This text provides an overview of all of the principle physical, chemical, and biological systems that occur on the Earth, and how they are being altered by human activity. Major topics include the role of science and technology in society, matter, energy, astronomy, geology, meteorology, oceanography, biology, environmental science, and much more. *Science of Earth Systems* introduces students to the processes that occur on our Earth by dividing the planet into five unique spheres, the exosphere, lithosphere, atmosphere, hydrosphere, and biosphere. Together these five spheres cover all the principal interactions between the Earth's physical and biological properties that make our planet unique. Any educator who needs to teach secondary students about the basic functions of the Earth and how those functions interact with the living world will find this textbook extremely useful. *Science of Earth Systems* begins with an introduction to the foundations of science and technology and an overview of the basic concepts of matter and energy. The textbook then presents the five individual spheres that together make up all of the Earth's systems.

Each unit in the textbook begins with a list of topics to be covered, and is divided into individual chapters that cover each specific subject to be learned. All of the chapters in this textbook begin with clear educational **objectives** to be learned by the student in the reading, and a list of important **terms to know.**

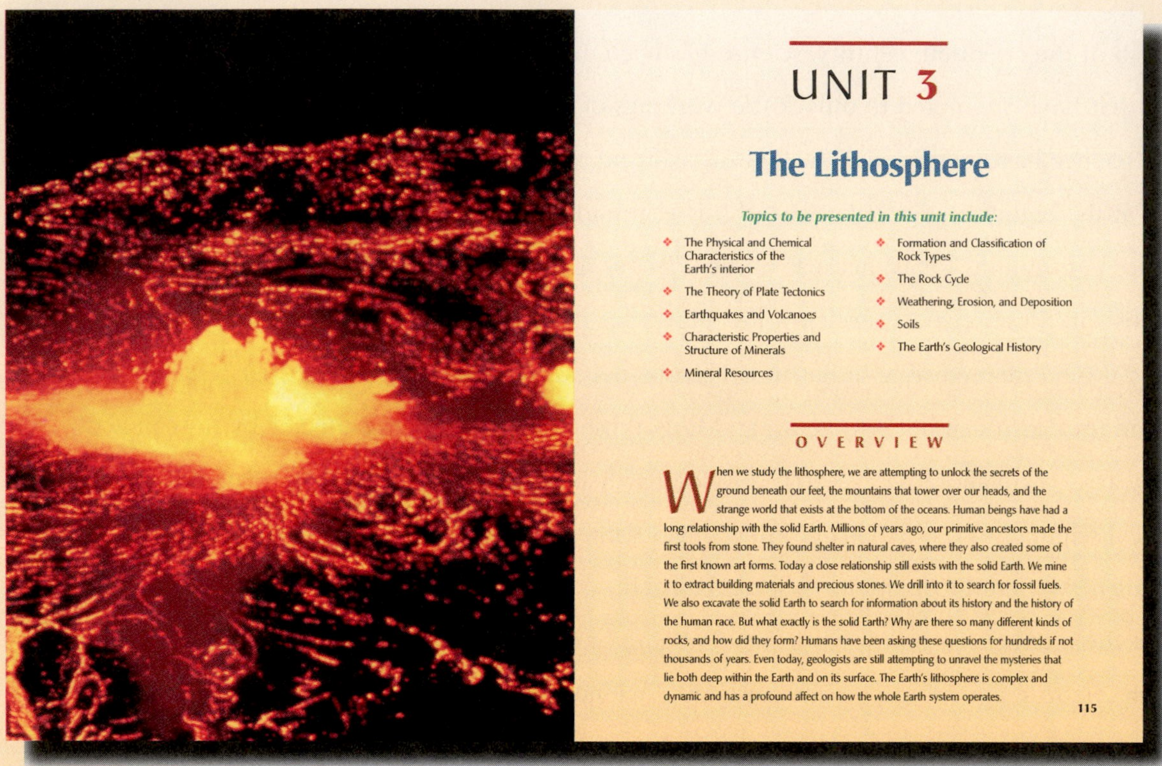

UNIT 3

The Lithosphere

Topics to be presented in this unit include:

- The Physical and Chemical Characteristics of the Earth's interior
- The Theory of Plate Tectonics
- Earthquakes and Volcanoes
- Characteristic Properties and Structure of Minerals
- Mineral Resources
- Formation and Classification of Rock Types
- The Rock Cycle
- Weathering, Erosion, and Deposition
- Soils
- The Earth's Geological History

OVERVIEW

When we study the lithosphere, we are attempting to unlock the secrets of the ground beneath our feet, the mountains that tower over our heads, and the strange world that exists at the bottom of the oceans. Human beings have had a long relationship with the solid Earth. Millions of years ago, our primitive ancestors made the first tools from stone. They found shelter in natural caves, where they also created some of the first known art forms. Today a close relationship still exists with the solid Earth. We mine it to extract building materials and precious stones. We drill into it to search for fossil fuels. We also excavate the solid Earth to search for information about its history and the history of the human race. But what exactly is the solid Earth? Why are there so many different kinds of rocks, and how did they form? Humans have been asking these questions for hundreds if not thousands of years. Even today, geologists are still attempting to unravel the mysteries that lie both deep within the Earth and on its surface. The Earth's lithosphere is complex and dynamic and has a profound affect on how the whole Earth system operates.

115

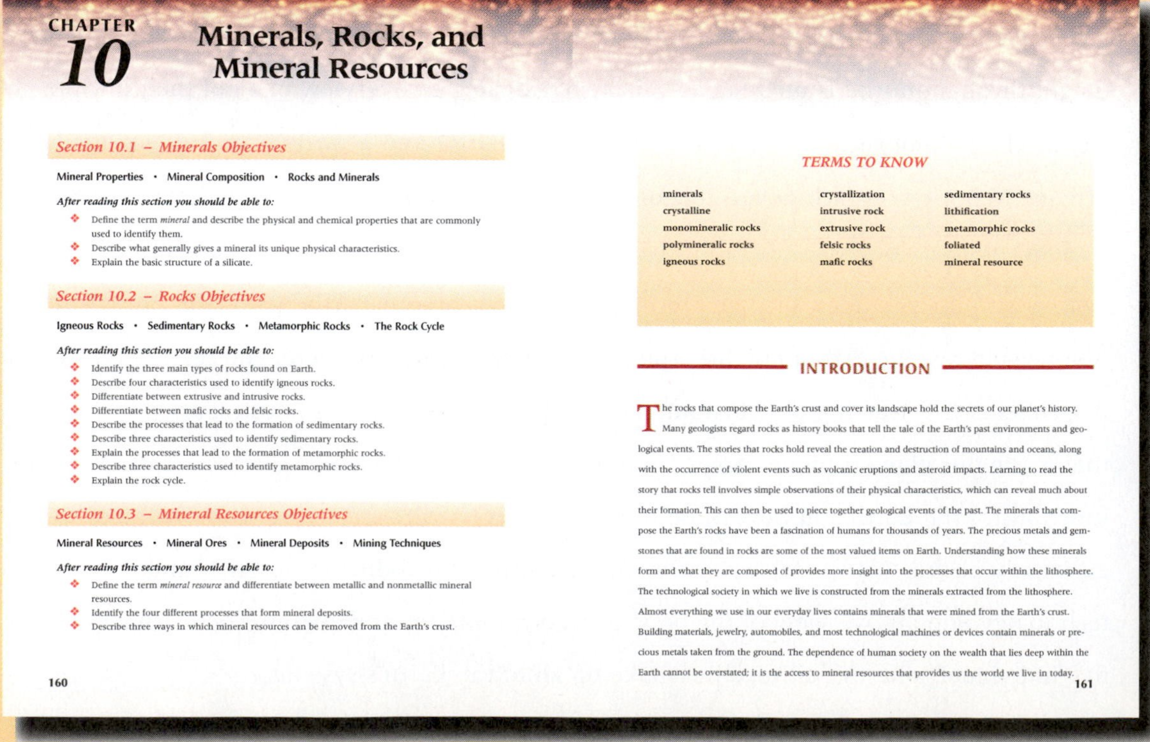

CHAPTER 10 Minerals, Rocks, and Mineral Resources

Section 10.1 – Minerals Objectives

Mineral Properties • Mineral Composition • Rocks and Minerals

After reading this section you should be able to:

- Define the term *mineral* and describe the physical and chemical properties that are commonly used to identify them.
- Describe what generally gives a mineral its unique physical characteristics.
- Explain the basic structure of a silicate.

Section 10.2 – Rocks Objectives

Igneous Rocks • Sedimentary Rocks • Metamorphic Rocks • The Rock Cycle

After reading this section you should be able to:

- Identify the three main types of rocks found on Earth.
- Describe four characteristics used to identify igneous rocks.
- Differentiate between extrusive and intrusive rocks.
- Differentiate between mafic rocks and felsic rocks.
- Describe the processes that lead to the formation of sedimentary rocks.
- Describe three characteristics used to identify sedimentary rocks.
- Explain the processes that lead to the formation of metamorphic rocks.
- Describe three characteristics used to identify metamorphic rocks.
- Explain the rock cycle.

Section 10.3 – Mineral Resources Objectives

Mineral Resources • Mineral Ores • Mineral Deposits • Mining Techniques

After reading this section you should be able to:

- Define the term *mineral resource* and differentiate between metallic and nonmetallic mineral resources.
- Identify the four different processes that form mineral deposits.
- Describe three ways in which mineral resources can be removed from the Earth's crust.

160

TERMS TO KNOW

minerals	crystallization	sedimentary rocks
crystalline	intrusive rock	lithification
monomineralic rocks	extrusive rock	metamorphic rocks
polymineralic rocks	felsic rocks	foliated
igneous rocks	mafic rocks	mineral resource

INTRODUCTION

The rocks that compose the Earth's crust and cover its landscape hold the secrets of our planet's history. Many geologists regard rocks as history books that tell the tale of the Earth's past environments and geological events. The stories that rocks hold reveal the creation and destruction of mountains and oceans, along with the occurrence of violent events such as volcanic eruptions and asteroid impacts. Learning to read the story that rocks tell involves simple observations of their physical characteristics, which can reveal much about their formation. This can then be used to piece together geological events of the past. The minerals that compose the Earth's rocks have been a fascination of humans for thousands of years. The precious metals and gemstones that are found in rocks are some of the most valued items on Earth. Understanding how these minerals form and what they are composed of provides more insight into the processes that occur within the lithosphere. The technological society in which we live is constructed from the minerals extracted from the lithosphere. Almost everything we use in our everyday lives contains minerals that were mined from the Earth's crust. Building materials, jewelry, automobiles, and most technological machines or devices contain minerals or precious metals taken from the ground. The dependence of human society on the wealth that lies deep within the Earth cannot be overstated; it is the access to mineral resources that provides us the world we live in today.

161

Each chapter contains **key vocabulary words** highlighted in bold, **Career Connections,** and many colorful **pictures** and **diagrams** that help illustrate the concepts presented.

30 ❖ UNIT 1 The Foundations of Science

Figure 2–6 Matter in the liquid state is fluid, which can be poured, and takes on the shape of its container. *(Courtesy of PhotoDisc.)*

more fluid property that has no definite shape (Figure 2–6). The arrangement of atoms in a liquid tend to be in long chains or clumps, rather than the orderly crystal patterns in a solid (Figure 2–7). The movement of atoms in a liquid is also less restrictive than in a solid. When a solid turns into a liquid, it is referred to as a change in state, or phase change. Phase changes usually depend on a certain temperature and pressure to which the matter is exposed. The temperature at which a solid changes into a liquid at a specific pressure is called its melting point. When a liquid changes into a solid it is called freezing.

The gaseous state of matter provides atoms with the highest degree of movement. Atoms in a gas move at high rates and undergo constant collisions with other atoms (Figure 2–8). The high degree of movement that the atoms experience in a gaseous form causes them to have no defined shape. When a gas is enclosed in a solid container, the number of collisions the freely moving molecules have with the walls of the container is known as pressure. The higher the number of collisions, the greater the pressure of the gas. When a liquid changes phase into a gas, it is called **vaporization** (Figure 2–9). The heat of vaporization is the specific temperature required to change a liquid into a gas, also known as its **boiling point.** The change from a gas to a liquid is called **condensation.** On some occasions a solid can change directly into a gas; this is called **sublimation.**

The fourth state of matter is called plasma, which is not as common as the other three. Plasma forms when the atoms that compose a gas become exposed to such high energy that they begin to ionize, or lose their electrons (Figure 2–10). The electrons are stripped

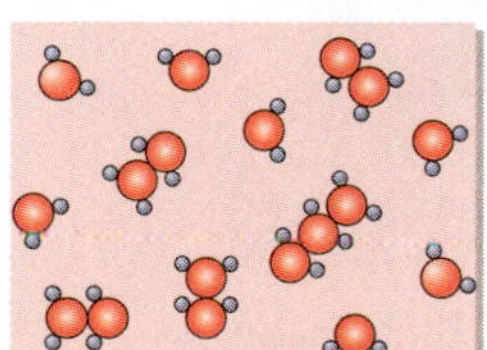

Figure 2–7 The chainlike configuration of molecules in a liquid.

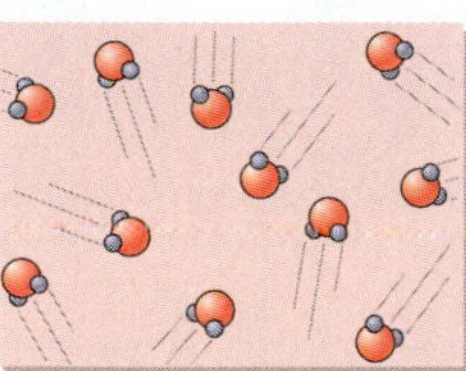

Figure 2–8 The rapidly moving molecules of a gas.

32 ❖ UNIT 1 The Foundations of Science

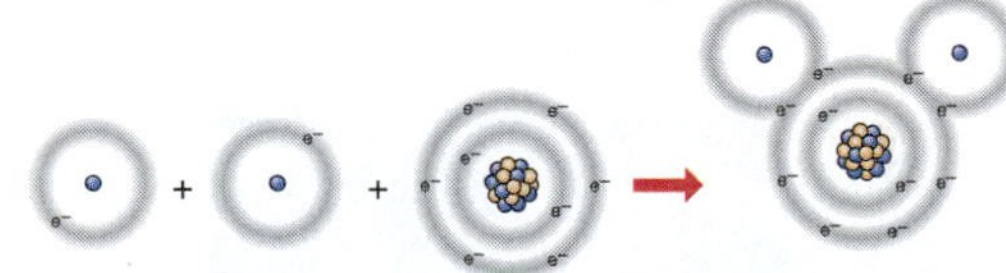

Figure 2–11 The formation of a water molecule by combining two atoms of hydrogen with one atom of oxygen.

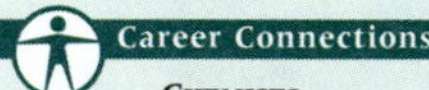

Career Connections

CHEMISTS

Chemistry is one of the most important scientific careers in the modern world. This vast career path involves the study of materials and compounds and how they interact with one another. Chemists analyze existing chemical compounds and perform experiments to discover new chemicals and materials. Chemists also often work with other scientists and engineers when conducting research. Branches of chemistry include analytical chemistry, organic chemistry, physical chemistry, and biochemistry. Analytical chemists identify the composition, structure, and properties of chemical substances. Organic chemists study the chemistry of living things, especially carbon-containing compounds, along with the synthesis of new drugs and commercial materials. Physical chemists study the properties of atoms and molecules and how they interact with one another, and biochemists study the chemistry of the human body and other living organisms. Chemistry courses in high school provide a good opportunity to experience what a career in chemistry might include. There are many jobs available for chemists once they gain a 4-year college degree. A related occupation is the chemistry technician, who works with chemists, helping them to perform their daily tasks. This type of work requires a less formal education.

the most important molecules on Earth is the water molecule. Water is a compound that is composed of two atoms of hydrogen with one atom of oxygen (Figure 2–11). When two or more compounds are mixed together and still maintain their unique physical and chemical properties, it is called a mixture. Air is a mixture of many gases, including nitrogen, oxygen, and carbon dioxide. When two substances are mixed to form a uniform substance, it is called a solution. Seawater is a solution of salt and other minerals in water.

SECTION REVIEW

1. What is matter?
2. Describe the three subatomic particles that make up the atom.
3. What is the difference between the atomic mass and atomic number of an atom?
4. What are isotopes and ions?
5. Describe the four states of matter.
6. What is a phase change?
7. What is an element, and how many have been identified on Earth?
8. Describe the difference between an atom and a compound.
9. Who was John Dalton?

In addition, each chapter contains **current research topics** and **internet sites.** Each chapter also highlights the lives and achievements of various **Earth System Scientists** throughout history.

CHAPTER 2 Matter and Energy ❖ 33

EARTH SYSTEM SCIENTISTS *JOHN DALTON*

John Dalton was born in England in 1766. He started his scientific career as a meteorologist, which began a lifelong dedication to weather observation in the area surrounding Manchester, England. One of his first scientific achievements was proposing that precipitation was formed by a change in atmospheric temperature, not by a change in air pressure. He also proposed the theory of the partial pressure of gases that today is known as Dalton's law of partial pressure. Later in his life he switched his scientific focus from meteorology and the study of the atmosphere to chemistry. In 1803 he developed his famous atomic theory, which stated that all matter was composed of small indivisible particles called atoms. He furthered hypothesized that atoms of a given element possess unique characteristics and weight and that three types of atoms exist: elements, compounds, and complex molecules. He published this theory in his book *New Systems of Chemical Philosophy* in 1808. Because of this theory, he is regarded by many to be the father of modern chemistry.

EARTH MATH

1) DETERMINE THE TOTAL ATOMIC MASS OF A MOLECULE OF WATER USING THE FOLLOWING INFORMATION: HYDROGEN HAS AN ATOMIC MASS OF 1.0079, AND OXYGEN HAS AN ATOMIC MASS OF 15.9994.

For more information go to this web link:
<http://www.stcms.si.edu/pom/pom_student.htm>

2.2 *Energy*

The relationship between matter and energy in the universe is the fundamental cause for the existence of the living and nonliving world. To fully understand the properties of matter, a full understanding of the properties of energy is equally important. In his world-famous equation $E = mc^2$, Albert Einstein revealed that energy and mass have a mutual relationship. But what exactly is energy, and how is it related to mass?

Energy

Energy is defined as the ability to do work or cause change. Matter that is exposed to energy is said to be in motion. Energy is classified in two basic forms: kinetic energy and potential energy. Kinetic energy is the energy of motion. When mass is in motion it is experiencing kinetic energy. The movement of the Earth around the Sun is kinetic energy. The movement of an athlete on a playing field is kinetic energy. The other type of energy is potential energy. Potential energy is stored energy. The gasoline in the tank of your car is a type of potential energy. It has the potential of powering your motor for a specific period. After riding a ski lift to the top of a mountain, you have a form of potential energy. The gravitational pull of the Earth will cause your

54 ❖ UNIT 2 The Exosphere

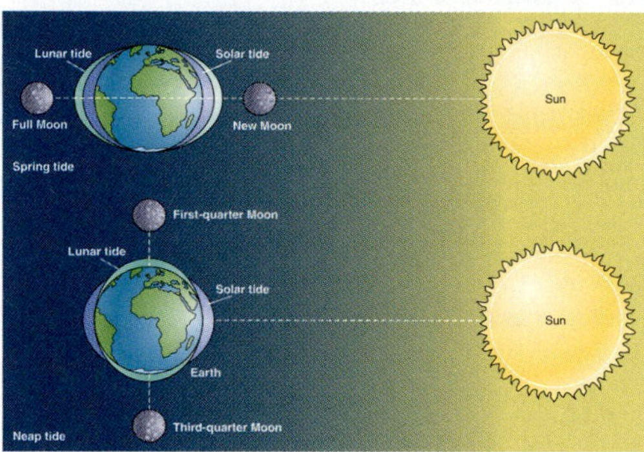

Figure 3–6 The effects of the Sun and Moon on the formation of spring and neap tides on the Earth.

system, which helps them to map celestial objects and also track their movement. The horizon can be divided into 360 degrees to make a circle that measures the horizontal location of a celestial object with respect to the Earth. The degrees on the horizon that marks a specific location is called the azimuth. The altitude of a celestial object is the angle in degrees above the horizon of a celestial object between the horizon and the zenith. The **zenith** is the point in the sky that is directly overhead, or 90 degrees from the horizon. Using these two coordinates, the azimuth and the altitude, an observer on the Earth can mark the location of a celestial object (Figure 3–7). Astronomers also use a more exact form of coordinates called declination and right ascension. This system is more complex and works much like lines of latitude and longitude, which is the coordinate system used on the surface of the Earth.

The Earth's Coordinate System: Latitude and Longitude

A coordinate system is a method of locating an exact location on a two-dimensional surface. Although the Earth's surface is not truly two dimensional, a coordinate system that uses intersecting horizontal and vertical lines is used to mark locations on the Earth. The Earth's coordinate system uses an imaginary horizontal line that divides the Earth in half, called the **equator.** This divides the planet into two hemispheres. The region north of the equator all the way to the North Pole is called the Northern Hemisphere. The region south of the equator all the way to the South Pole is called the Southern Hemisphere. Each hemisphere is further divided by horizontal lines that mark the location north or south of the equator. These parallel lines are called

84 ❖ UNIT 2 The Exosphere

Figure 5–13 The aurora borealis, also called the northern lights, light up the sky in the Northern Hemisphere as a result of the solar wind interacting with the magnetosphere and atmosphere. *(Courtesy of PhotoDisc.)*

Current Research

Scientists working for the National Aeronautics and Space Administration (NASA) who have been studying the Sun have discovered a unique feature never before seen on our nearest star. Recent images taken of the Sun's surface have revealed explosions that create earthquakelike ripples on the Sun's surface. The explosions on the Sun's surface that cause these ripples are being called sunquakes. The power of these sunquakes would measure 11.1 on the Richter scale if they occurred on Earth. The explosions that cause sunquakes are connected to the Sun's increased activity. Activity on the Sun is marked by increases in the occurrence of solar storms that cause sunspots and solar flares. An increase in the number of solar storms that cause these sunquakes occurs during the Sun's solar maximum on an average of every 11 years. Scientists are interested in these solar storms because of the effect that they might have on the Earth. Much of the current research that is discovering new information about the nature of the Sun is being done with observations taken through NASA's Solar and Heliospheric Observatory, or SOHO satellite. This satellite is designed to observe the Sun continually and orbits around the Sun approximately 930,000 miles away from the Earth.

sun at approximately 18,000 miles per hour. The Earth is shielded from this stream of particles by its magnetic field; however, at the North and South Poles, the solar wind collides with the atmosphere and lights up the nighttime sky with brilliant colors. This is known as the northern lights, or the aurora borealis, in the Northern Hemisphere and the australis borealis in the Southern Hemisphere (Figure 5–13). Solar flares that leap off the surface of the Sun can also cause a temporary increase in the intensity of the solar wind.

The Sun's Life Cycle

The Sun, like any star, must go through the stages of a star's life cycle. Because the Sun is believed to be approximately 4.5 to 4.7 billion years old, it is estimated that it will take another 5 to 6 billion years to use up all of its fuel. When this happens the Sun will begin to enter the red giant phase of its life (Figure 5-14). When the sun becomes a red giant, it will be approximately 100 times larger than it is now. This will cause the hydrosphere on Earth to boil away and the atmosphere to be

The end of each chapter is followed by a series of **review questions.** These questions include short answer, **Earth Math,** multiple choice, matching, and critical thinking. A **Chapter Summary** also highlights the topics that have been presented.

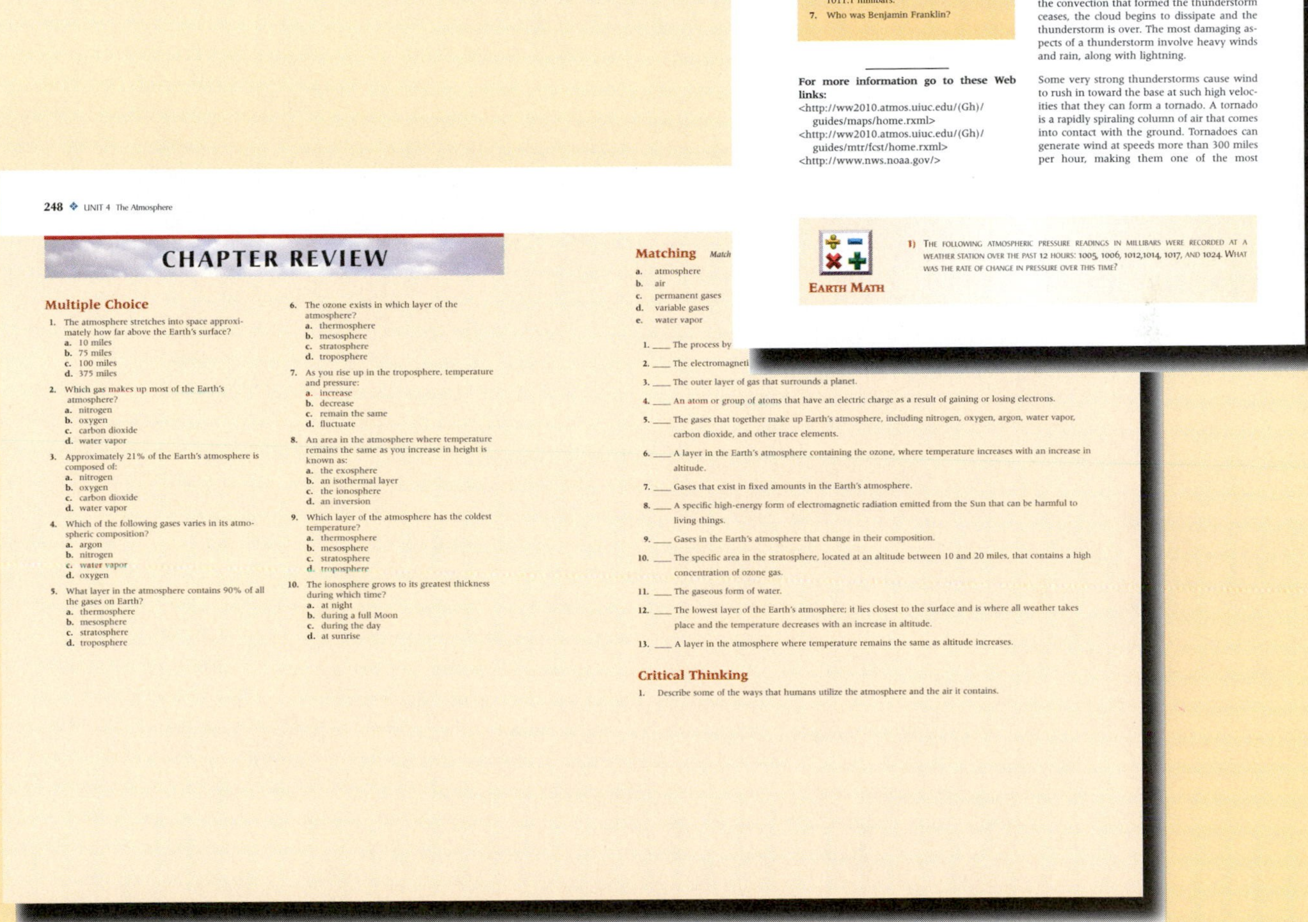

CHAPTER 19 Storms and Weather Forecasting ❖ 339

proaching warm front bringing extended precipitation to the region. Clearing night skies during winter can also signal an extreme drop in temperature.

SECTION REVIEW

1. What are the three main processes used to forecast the weather?
2. What weather information is commonly recorded by weather stations?
3. What are the common weather maps that depict the current state of the weather?
4. Describe how radar and satellites are used to forecast the weather.
5. What are some common changes that occur in the local conditions that can be used to forecast the weather?
6. Draw a station model for the following weather observation: temperature 24° Fahrenheit, dew point 20° Fahrenheit, 50% cloud cover, winds from the southwest at 25 knots, atmospheric pressure 1011.1 millibars.
7. Who was Benjamin Franklin?

For more information go to these Web links:
<http://ww2010.atmos.uiuc.edu/(Gh)/guides/maps/home.rxml>
<http://ww2010.atmos.uiuc.edu/(Gh)/guides/mtr/fcst/home.rxml>
<http://www.nws.noaa.gov/>

CHAPTER SUMMARY

Storms in the atmosphere can form dangerous conditions, leading to the damage of property and loss of life. There are three main types of potentially dangerous storms on the Earth: thunderstorms, tornadoes, and hurricanes. Thunderstorms form when the Earth's surface is heated, causing warm air to rise rapidly. This causes the formation of a small-scale convection cell that leads to the creation of a cumulonimbus cloud. Eventually the cloud grows large enough to create heavy precipitation and generate strong winds. The vertical winds inside a thunderstorm are called updrafts and downdrafts. These winds cause the separation of lighter ice crystals near the top of the cloud and heavier hail near the cloud base. The two different forms of precipitation build up opposite electrical charges within the cloud, eventually forming lightning. Lightning can strike from cloud to cloud, cloud to air, or cloud to ground. When the convection that formed the thunderstorm ceases, the cloud begins to dissipate and the thunderstorm is over. The most damaging aspects of a thunderstorm involve heavy winds and rain, along with lightning.

Some very strong thunderstorms cause wind to rush in toward the base at such high velocities that they can form a tornado. A tornado is a rapidly spiraling column of air that comes into contact with the ground. Tornadoes can generate wind at speeds more than 300 miles per hour, making them one of the most

EARTH MATH

1) THE FOLLOWING ATMOSPHERIC PRESSURE READINGS IN MILLIBARS WERE RECORDED AT A WEATHER STATION OVER THE PAST 12 HOURS: 1005, 1006, 1012,1014, 1017, AND 1024. WHAT WAS THE RATE OF CHANGE IN PRESSURE OVER THIS TIME?

248 ❖ UNIT 4 The Atmosphere

CHAPTER REVIEW

Multiple Choice

1. The atmosphere stretches into space approximately how far above the Earth's surface?
   a. 10 miles
   b. 75 miles
   c. 100 miles
   d. 375 miles
2. Which gas makes up most of the Earth's atmosphere?
   a. nitrogen
   b. oxygen
   c. carbon dioxide
   d. water vapor
3. Approximately 21% of the Earth's atmosphere is composed of:
   a. nitrogen
   b. oxygen
   c. carbon dioxide
   d. water vapor
4. Which of the following gases varies in its atmospheric composition?
   a. argon
   b. nitrogen
   c. water vapor
   d. oxygen
5. What layer in the atmosphere contains 90% of all the gases on Earth?
   a. thermosphere
   b. mesosphere
   c. stratosphere
   d. troposphere
6. The ozone exists in which layer of the atmosphere?
   a. thermosphere
   b. mesosphere
   c. stratosphere
   d. troposphere
7. As you rise up in the troposphere, temperature and pressure:
   a. increase
   b. decrease
   c. remain the same
   d. fluctuate
8. An area in the atmosphere where temperature remains the same as you increase in height is known as:
   a. the exosphere
   b. an isothermal layer
   c. the ionosphere
   d. an inversion
9. Which layer of the atmosphere has the coldest temperature?
   a. thermosphere
   b. mesosphere
   c. stratosphere
   d. troposphere
10. The ionosphere grows to its greatest thickness during which time?
   a. at night
   b. during a full Moon
   c. during the day
   d. at sunrise

Matching *Match*

a. atmosphere
b. air
c. permanent gases
d. variable gases
e. water vapor

1. ____ The process by
2. ____ The electromagneti
3. ____ The outer layer of gas that surrounds a planet.
4. ____ An atom or group of atoms that have an electric charge as a result of gaining or losing electrons.
5. ____ The gases that together make up Earth's atmosphere, including nitrogen, oxygen, argon, water vapor, carbon dioxide, and other trace elements.
6. ____ A layer in the Earth's atmosphere containing the ozone, where temperature increases with an increase in altitude.
7. ____ Gases that exist in fixed amounts in the Earth's atmosphere.
8. ____ A specific high-energy form of electromagnetic radiation emitted from the Sun that can be harmful to living things.
9. ____ Gases in the Earth's atmosphere that change in their composition.
10. ____ The specific area in the stratosphere, located at an altitude between 10 and 20 miles, that contains a high concentration of ozone gas.
11. ____ The gaseous form of water.
12. ____ The lowest layer of the Earth's atmosphere; it lies closest to the surface and is where all weather takes place and the temperature decreases with an increase in altitude.
13. ____ A layer in the atmosphere where temperature remains the same as altitude increases.

Critical Thinking

1. Describe some of the ways that humans utilize the atmosphere and the air it contains.

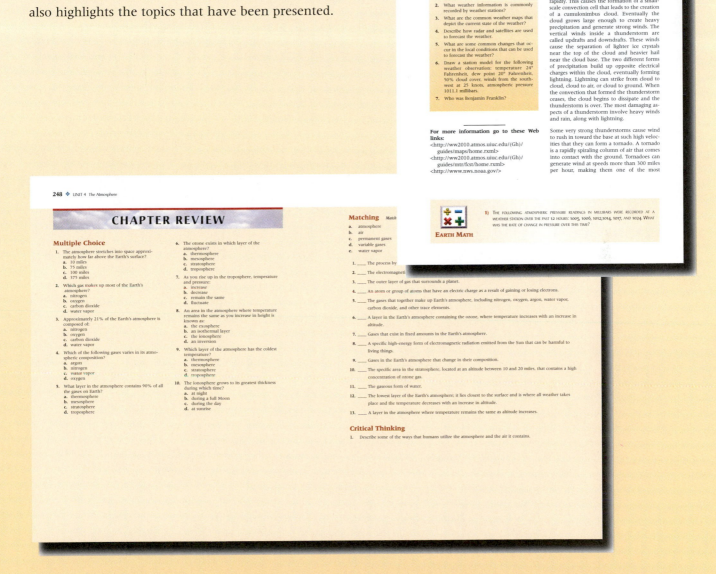

Also available with this textbook is a comprehensive **Instructor's Guide** and **Classmaster CD-ROM** that contains lesson planning suggestions, transparency masters for many of the informative illustrations and diagrams within the textbook, and a test bank. *Science of Earth Systems* also offers a companion **Laboratory Manual** that includes 45 laboratory experiments and activities that help reinforce the material learned in each chapter and a corresponding **Lab Manual Instructor's Guide.** Overall, *Science of Earth Systems* is a textbook with ancillary materials that offers the complete instructional materials needed for introducing students to the physical systems of the Earth and the interactions that exist between its living and nonliving components.

TABLE OF CONTENTS

Preface v

Unit 1: The Foundations of Science xvi

Chapter 1: Technology, Science, and Scientific Measurement 2

Section 1.1 - Technology 4
- Human Origins 4
- Stone Age Technology 6
- The Agricultural Revolution 6
- Bronze Age and Iron Age Technology 6
- The Industrial Revolution 7
- Technological Systems 9

Section 1.2 - The Foundations of Science 11
- The Birth of Science 11
- The Scientific Method 12
- Scientific Experiments 12
- Science, Technology, and Engineering 14

Section 1.3 - Scientific Measurement 15
- Observation and Measurement 15
- Units of Measurement 16
- Density 18
- Scientific Notation 19
- Percent Error 20

Chapter 2: Matter and Energy 24

Section 2.1 - The Properties of Matter 26
- Atoms 26
- The Elements 26
- States of Matter 29
- Compounds and Molecules 31

Section 2.2 - Energy 33
- Energy 33
- The Law of the Conservation of Energy 34
- Electromagnetic Radiation 34
- Energy Transfer 36
- Heat and Temperature 39

Unit 2: The Exosphere 44

Chapter 3: The Earth as a Planet 46
- The Earth's Composition 48
- The Earth's Shape 48
- The Size of the Earth 49
- The Earth's Rotation 50
- The Earth's Tilted Axis 51
- The Earth's Orbit 51
- Tides 52
- Apparent Motion of Celestial Objects 53
- The Earth's Coordinate System: Latitude and Longitude 54
- Topographic Maps 57

Chapter 4: The Moon, Earth's Closest Neighbor 64
- The Moon's Formation 66
- The Composition of the Moon 66
- The Moon's Surface 67
- The Moon's Orbit 68
- The Phases of the Moon 68
- Eclipses 68

Chapter 5: The Life of Stars 74

Section 5.1 - Life Cycle of Stars 76
- Star Formation and Life Cycle 76
- Classification of Stars 76
- Dwarf Stars 76
- Giant and Supergiant Stars 78
- Neutron Stars 79
- Black Holes 80

Section 5.2 - The Sun 81
- Formation of the Sun 81
- Composition of the Sun 81
- The Sun's Life Cycle 84

Chapter 6: The Solar System and its Place in the Universe 90

Section 6.1 - The Solar System **92**
The Formation of the Solar System 92
The Terrestrial Planets 92
The Gaseous Planets 96
Asteroids 99
Meteoroids 101
Comets 102

Section 6.2 - The Earth's Place in the Universe **104**
The Age and Size of the Universe 104
Galaxies 105
Quasars 107
Life in the Universe 109

Unit 3: The Lithosphere 114

Chapter 7: The Earth's Interior 116
The Earth's Core 118
The Mantle 118
The Earth's Crust 118
The Model of Earth's Interior 120

Chapter 8: Plate Tectonics 124
The Dynamic Crust 126
Evidence of Crustal Movement 126
Continental Drift 129
Sea Floor Spreading 131
Tectonic Plates 132
Plate Boundaries 132
Mantle Convection 136

Chapter 9: Earthquakes and Volcanoes 142

Section 9.1 - Earthquakes **144**
Causes of Earthquakes 144
Seismic Waves 144
Epicenter Location 145
Earthquake Measurement 147

Section 9.2 - Volcanoes **149**
Formation of a Volcano 149
Quiet Eruption Volcanoes 149
Explosive Eruption Volcanoes 150
Volcanic Hazards 153

Chapter 10: Minerals, Rocks, and Mineral Resources 160

Section 10.1 - Minerals **162**
Mineral Properties 162
Mineral Composition 166
Rocks and Minerals 166

Section 10.2 - Rocks **167**
Igneous Rocks 168
Sedimentary Rocks 169
Metamorphic Rocks 173
The Rock Cycle 175

Section 10.3 - Mineral Resources **177**
Mineral Resources 177
Mineral Ores 177
Mineral Deposits 178
Mining Techniques 179

Chapter 11: Weathering, Erosion, and Deposition 186
Physical and Chemical Weathering 188
The Process of Erosion 190
Agents of Erosion 192
The Process of Deposition 193

Chapter 12: Soils 200
Soil Minerals 202
Soil Organic Material 202
Soil Water and Air 203
Soil Organisms 204
Soil Structure 204
Parent Material 205
Soil Horizons 206
Soil Classification 207

Chapter 13: The Earth's Geologic History 214
Geologic Principles 216
Radiometric Dating 217
The Geologic Time Scale 218
The Archean Eon 218
The Proterozoic Eon 223
The Paleozoic Era 224
The Mesozoic Era 228
The Cenozoic Era 230

Unit 4: The Atmosphere 238

Chapter 14: Structure and Composition of the Atmosphere 240

- Atmospheric Composition 242
- Atmospheric Structure 243

Chapter 15: Insolation 250

- The Sun's Radiation 252
- Insolation and the Atmosphere 252
- Insolation and the Earth's Surface 253
- Angle of Insolation 254
- Duration of Insolation 256
- Heating the Atmosphere 256

Chapter 16: Atmospheric Temperature and Pressure 262

Section 16.1 - Atmospheric Temperature 264

- Temperature in the Atmosphere 264
- Distribution of Heat on Earth 264
- Radiative Cooling 267
- The Greenhouse Effect 268

Section 16.2 - Atmospheric Pressure 270

- Pressure in the Atmosphere 270
- Measuring Atmospheric Pressure 270
- High and Low Atmospheric Pressure 271
- Atmospheric Pressure and Moisture 273

Chapter 17: Humidity, Clouds, and Precipitation 280

Section 17.1 - Humidity 282

- Atmospheric Moisture 282
- Sources of Atmospheric Moisture 282
- Relative Humidity 283
- Dew Point Temperature 284

Section 17.2 - Clouds and Precipitation 287

- Cloud Formation 287
- Types of Clouds 290
- Formation of Precipitation 292
- Types of Precipitation 292
- Orographic Precipitation 295

Chapter 18: Wind, Air Masses, and Fronts 300

Section 18.1 - Wind 302

- Pressure Gradient 302
- Planetary Winds 302
- Pressure Systems 306
- Mesoscale Winds 307
- Local Winds 309
- The Jet Stream 309
- Wind Measurement 311

Section 18.2 - Air Masses and Fronts 312

- Air Mass Formation 312
- Source Regions and Classification of Air Masses 312
- Fronts 314
- Mid-latitude Cyclones 318

Chapter 19: Storms and Weather Forecasting 324

Section 19.1 - Storms 326

- Thunderstorms 326
- Tornadoes 328
- Hurricanes 330

Section 19.2 - Weather Forecasting 334

- Weather Data Collection 334
- Synoptic Weather Maps 335
- Weather Forecasts 335
- Weather Radar and Satellites 336

Chapter 20: Global Climate Change 344

- Revealing the Earth's Past Climate 346
- Ice Ages and Glaciations 347
- The Milankovitch Cycle 347
- Hot House Climates 351
- Goldilocks Syndrome 351
- Humans and Global Climate Change 352

Chapter 21: Acid Precipitation and Deposition **360**
The pH of Precipitation 362
Formation of Acid Precipitation 362
Anthropogenic Gases 362
Long Distance Transport of Acid-causing Pollutants 364
Effects of Acid Deposition on Aquatic Ecosystems 364
Effects of Acid Precipitation on Terrestrial Ecosystems 366
Effects of Acid Deposition on Human Beings and Building Materials 367
Control of Acid Deposition 369

Chapter 22: Ozone Depletion **374**
Ozone Gas 376
The Ozone Layer 376
Measuring Stratospheric Ozone 377
Effects of Ozone Depletion 379
The Ozone Hole 379
Reducing Ozone Depletion 380

Unit 5: The Hydrosphere 386

Chapter 23: Earth, The Water Planet **388**
Section 23.1 - The Blue Marble 390
The Earth's Oceans 390
Distribution of Freshwater on Earth 391
Section 23.2 - The Amazing Water Molecule 393
The Water Molecule 393
Adhesion and Cohesion 393
Heat Capacity of Water 394
Properties of Ice 395
Water as a Solvent 396

Chapter 24: The Hydrologic Cycle **402**
Evaporation 404
Water Vapor and Condensation 404
Precipitation and Surface Water 405
Runoff 406
Evapotranspiration 406
Infiltration and Groundwater 407

Chapter 25: Oceanography **412**
Seawater 414
Ocean Currents 414
Deep Ocean Circulation 416
Life Zones in the Ocean 416
Continental Shelves 419
Intertidal Zones 420

Chapter 26: Fresh Surface Water and Groundwater **426**
Section 26.1 - Fresh Surface Water 428
Lakes 428
Lake Productivity 428
Life Zones in Lakes 430
Watersheds and Rivers 431
Stream Features 434
Floodplains 435
Life Cycle of Rivers 436
Section 26.2 - Groundwater 438
Groundwater Recharge and the Water Table 438
Groundwater Flow 439
Aquifers and Groundwater Discharge 441
Groundwater Pollution 441

Chapter 27: Glaciers **448**
Anatomy of a Glacier 450
Glacial Movement and Moraines 451
Types of Glaciers 452
Glaciers and Global Climate 455

Chapter 28: Pollution of the Hydrosphere **460**
Sediment Pollution 462
Nutrient Pollution and Eutrophication 463
Toxic Organic Compounds 464
Toxic Inorganic Compounds 465
Disease-causing Agents 466
Thermal Pollution 468

Chapter 29: El Niño and the Southern Oscillation **474**
- The South Equatorial Current and Upwelling 476
- The Southern Oscillation 476
- The Effects of an El Niño Event 479
- La Niña Events 481
- Monitoring the Pacific Ocean 482

Chapter 30: Coral Bleaching **486**
- Coral and Coral Reef Systems 488
- The Bleaching of Corals 488
- Occurrence of Coral Bleaching 489
- Causes of Coral Bleaching 489

Unit 6: The Biosphere 496

Chapter 31: Ecological Systems **498**
- Ecology 500
- Habitats 500
- Populations 501
- Communities 503
- Ecosystems 504

Chapter 32: World Biomes and Marine Ecosystems **510**

Section 32.1 - World Biomes 512
- Biomes 512
- Tundra 512
- Coniferous Forests 512
- Temperate Forests 514
- Grasslands 516
- Savannas 516
- Deserts 518
- Tropical Rain Forests 520
- Chaparral 520
- Mountains 521

Section 32.2 - Marine Ecosystems 523
- Coastal Wetlands 523
- The Neritic and Intertidal Zones 523
- The Oceanic Zone 525
- The Benthic Zone 528
- Hydrothermal Vent Communities 528

Chapter 33: The Flow of Energy and Matter Through Ecosystems **534**

Section 33.1 - Energy Flow Within Living Systems 536
- Photosynthesis and Chemosynthesis 536
- Autotrophs and Heterotrophs 537
- Primary Production 537
- Primary and Secondary Consumers 538
- Food Chains and Webs 540
- The Energy Pyramid 540

Section 33.2 - Biogeochemical Cycling 544
- Biogeochemical Cycling 544
- Carbon Cycling 544
- Oxygen Cycling 547
- Nitrogen Cycling 547
- Phosphorus Cycling 550

Chapter 34: Biological Succession **556**
- Biological Succession 558
- Primary Succession and Pioneer Communities 558
- Secondary Succession 560

Chapter 35: Classification of the Living World **566**
- Taxonomy 568
- The Kingdom Monera 568
- The Kingdom Protista 570
- The Kingdom Fungi 571
- The Kingdom Plantae 572
- The Kingdom Animalia 574
- Invertebrate Animals 574
- Vertebrate Animals 580

Glossary **596**

Index **630**

UNIT 1

The Foundations of Science

Topics to be presented in this unit include:

- The Foundations of Technology
- The Foundations of Science
- Scientific Observation and Measurement
- The Properties of Matter
- The Properties of Energy

OVERVIEW

From the beginnings of humankind, ancient people were at one with their world. Their understanding of the processes of nature were vital for their own survival. It was only a natural development for early humans to want to learn about the world in which they lived. Other creatures who shared the Earth with early humans utilized their own special adaptations to survive. Humans, on the other hand, utilized the power of thought to help them survive in the often harsh conditions that exist in nature. From these early beginnings, humans were learning about the environment in which they lived, and experimenting with new ways to improve their situation. This was certainly the beginning of what we know today as science and technology. Gathering information about the natural world, and then putting that knowledge to use in a practical form has been a human trait for tens of thousands of years. This has resulted in a lasting relationship between science and technology, which continues to advance the human condition.

CHAPTER 1 Technology, Science, and Scientific Measurement

Section 1.1 – Technology Objectives

Human Origins • Stone Age Technology • The Agricultural Revolution • Bronze Age and Iron Age Technology • The Industrial Revolution • Technological Systems

After reading this section you should be able to:

- Define the term *technology,* and describe its importance to human beings.
- Identify the progressive evolutionary developments of human technology.
- Explain the systems approach to understanding how technologies solve problems.
- Identify the different parts of a technological systems model.

Section 1.2 – The Foundations of Science Objectives

The Birth of Science • The Scientific Method • Scientific Experiments • Science, Technology, and Engineering

After reading this section you should be able to:

- Define the term *science.*
- Explain the three main processes that together make up the scientific method.
- Discuss the importance of performing experiments in the process of scientific investigation.
- Explain the differences between science and engineering.

Section 1.3 – Scientific Measurement Objectives

Observation and Measurement • Units of Measurement • Density • Scientific Notation • Percent Error

After reading this section you should be able to:

- Define the term *observation.*
- Explain why scientists utilize scientific instruments.
- Identify the fundamental units of measurement.
- Differentiate between the standard and metric systems of measurement.
- Define the term *density,* and explain how it affects a substance's physical properties.
- Explain why scientists use scientific notation.
- Explain why calculating percent error is an important part of scientific measurement.

TERMS TO KNOW

technology
technological systems model
science
scientific method
observation
experiment
theory
hypothesis
inference
length
mass
weight
time
energy
density

INTRODUCTION

Try to imagine a world without science and technology. What would it be like, and more importantly, would you want to live in that world? For most of us the answer would certainly be no, for science and technology affect every aspect of our lives. The use of tools, materials, and technological processes are as important to human beings as wings are to a bird or claws are to a bear. These animals rely on these adaptations for their own survival, as humans rely on science and technology for our own survival. The link between human beings and technology is inseparable, because technology is an important extension of the human mind that allows us to do things and live like no other organism on the Earth. Equally important to humans is our search for knowledge and understanding of the world in which we live. This forms the base of what we call science, which has changed our understanding of the Earth and its place in the universe. But what exactly are science and technology and how have they changed our lives?

1.1 *Technology*

Human Origins

Humans possess the ability to extend their own capabilities by using tools. Since ancient times, people have devised ways to make specialized devices to improve their situation. This was the beginning of technology. The appearance of a new type of primate occurred on the African continent approximately 4 million years ago. This new species belonged to a family of what anthropologists called the hominid. Although this creature would hardly have looked like a modern-day human being, it possessed two very unique adaptations. The first was the ability to walk upright on two legs. The second was the ability to make stone tools. Walking upright is called bipedalism, which literally means "two feet." Probably the most significant aspect of the ability to stand and walk upright is that it allows for the arms and hands to be utilized in other ways. These ancestors of modern humans are called Homo habilis ("clever human") by anthropologists and are the first known hominid to create customized tools. Over the next 4 million years, the hominid family began to disperse and form new species, which eventually led to the development of modern humans (Figure 1–1).

Our early ancestors' ability to use their hands provided them with the opportunity to explore new ways in which to put them to use. Eventually this unique feature led to the development of what we call technology.

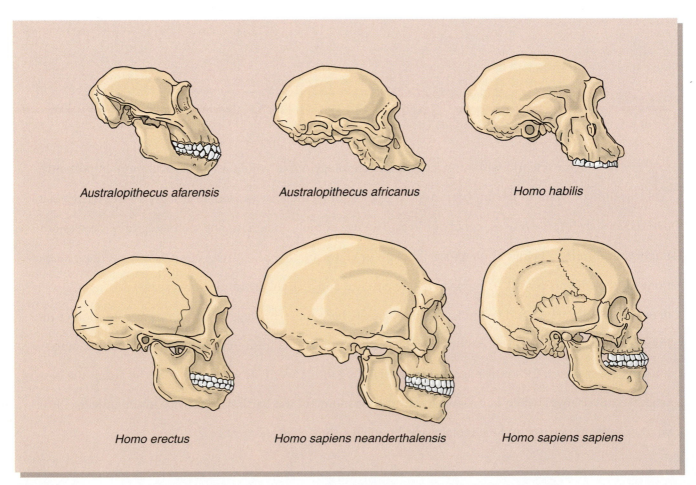

Australopithecus afarensis
Australopithecus africanus
Homo habilis
Homo erectus
Homo sapiens neanderthalensis
Homo sapiens sapiens

Figure 1–1 Comparison of hominid skull fossils traces the evolutionary development of human beings.

Major Breakthrough

The fossilized remains of the hominid known as Lucy, who is believed to have lived more than 3 million years ago in Africa, were unearthed from the Hadar region of Africa by anthropologist Don Johanson in 1974. This partial skeleton was one the oldest hominid remains ever found, until a recent discovery made in Kenya, Africa. French and African anthropologists recently found the skeletal remains of an early hominid that may be 6 million years old. The Millennium Man, as this early human ancestor is known, may extend the human family tree back by almost 3 million years!

Modern humans, or *Homo sapiens* ("intelligent human"), are not the only hominid to have used tools. There is sufficient archeological evidence to suggest that many of our hominid ancestors also used primitive tools. The oldest tools discovered so far are approximately 2 million years old. These consist of crude stone implements called choppers. Examples of these early stone tools are shown in Figure 1–2. Today even modern chimpanzees and

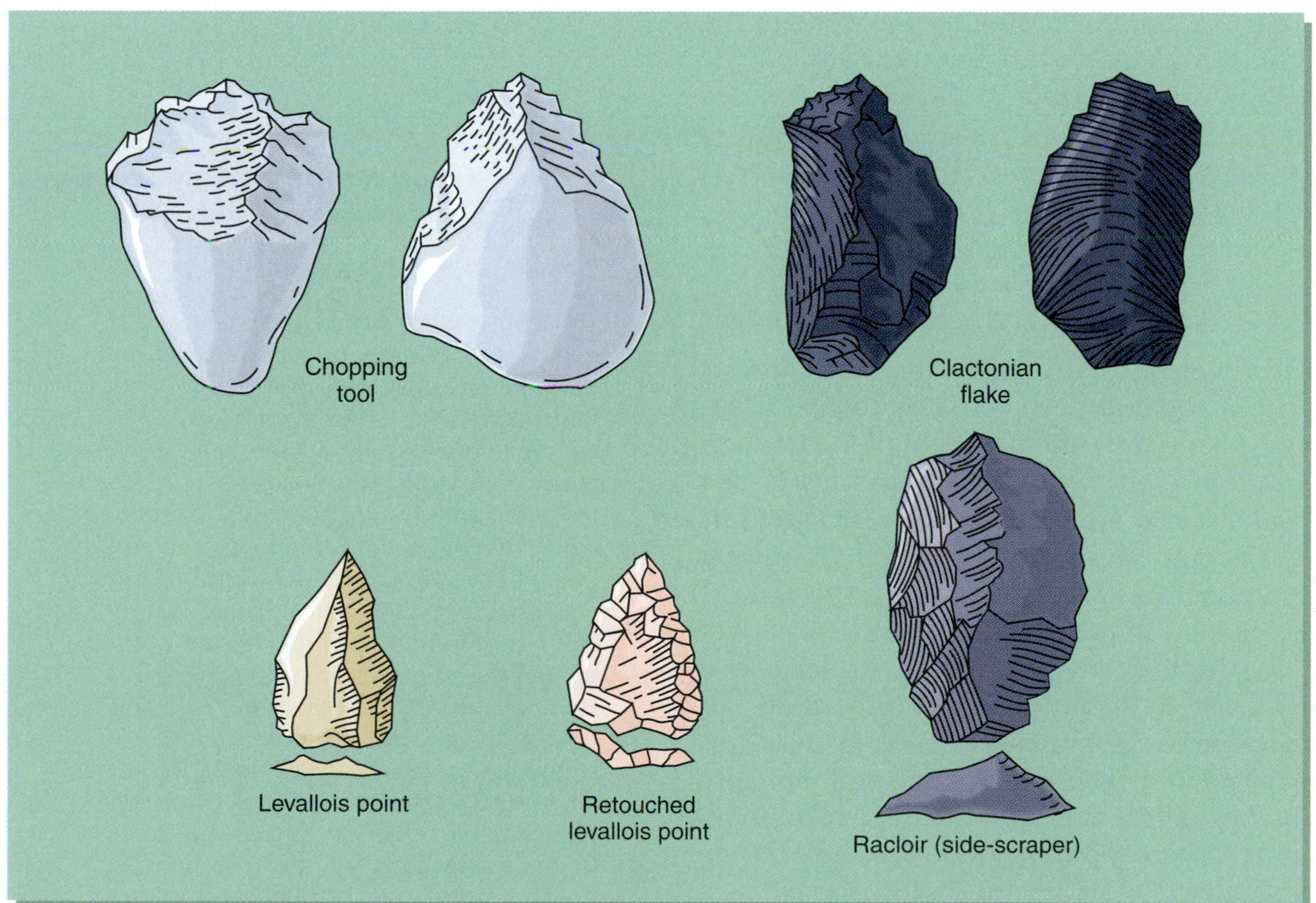

Figure 1–2 Examples of stone tools used by early humans for chopping and cutting.

other animals like sea otters use primitive tools like these.

Stone Age Technology

It is the use of tools by our early ancestors that began what we call **technology.** Technology ("the science of skill") is defined as all the ways that human beings extend their capabilities to meet their wants and needs and to solve problems. Our early hominid ancestors used stone tool technology to extend their capabilities for more than 2 million years. The period of time covering the use of stone tools is often called the Stone Age. The Stone Age is divided into two periods; the first is called the Paleolithic ("old stone") period. During the Paleolithic period, hominids used crude stones mainly for chopping and for breaking open animal bones to expose the nutritious marrow. Stone tools were also used for carving wood and for digging up plant roots. The later period of the Stone Age is called the Neolithic ("new stone"). During this period our ancestors began to manipulate stones to form customized sharp cutting tools for use in farming the land.

The Agricultural Revolution

The appearance of modern humans approximately 100,000 years ago led to major advancements in technology. Modern humans began to craft very specialized tools in the form of spears and knives. About 35,000 years ago, people began creating pieces of art and primitive musical instruments. Over time, technology improved and became a part of every day life. After fire and the wheel, one of the greatest (if not the most important) advancements in human technology was human domestication of plants and animals. This is called the Agricultural Revolution, which began approximately 10,000 years ago in southwest Asia (Figure 1–3).

During the Agricultural Revolution, humans created many customized tools made from stone, wood, and animal bones for the purpose of growing and harvesting food crops. Until then, people hunted and scavenged off the land to meet their food needs. Agriculture created a whole new, better way of life for humans. No longer did they need to move around the land in search of food; now they could settle in one area and grow their own food. One of the tools that helped tremendously in this endeavor was the wooden plow, which humans built approximately 7000 years ago (Figure 1–4). This development allowed humans to farm larger areas of land and grow more food.

Bronze Age and Iron Age Technology

As the human population grew, so did the number of technological advancements. One of these was the use of naturally occurring metal ore deposits such as copper

Career Connections

ANTHROPOLOGY

Scientists who study human beings, their culture, and their origins are called anthropologists. Anthropology is divided into four major fields: Physical anthropology, cultural anthropology, archaeology, and linguistics. Physical anthropologists, also called biological anthropologists, study the unique biological adaptations of humans and also the origin of the human species. Cultural anthropologists study the social behavior and specific cultures of humans throughout the world both in the past and today. Archaeologists study the history of humans by searching for the remains of ancient cultures. Linguists study the origins and development of human language. Careers in anthropology usually require a 4-year college degree and can lead to jobs in academic fields, corporations, nonprofit organizations, and state and federal governments.

Figure 1–3 The shift by early humans from a hunter-gather society to harvesting crops they have grown marked the beginning of the Agricultural Revolution more than 10,000 years ago.

and tin to create new, stronger tools. This began more than 5000 years ago and marked the beginning of the Bronze Age. Bronze is an alloy, or mixture, of copper and tin. Humans learned that different metals could be mixed together to form strong new tools. Eventually they found that iron was better than bronze for making tools. Iron was much stronger than bronze and quickly improved human tool-making abilities (Figure 1–5). This led to what is called the Iron Age.

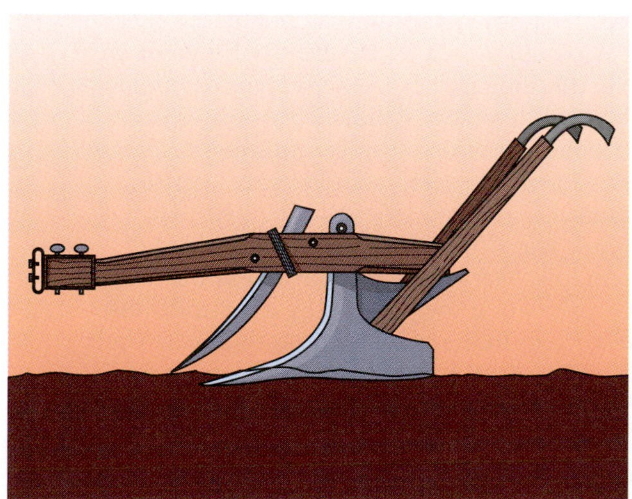

Figure 1–4 The moldboard plow enabled humans to efficiently prepare the land for the growth of crops.

The Industrial Revolution

As time passed, human beings began to manipulate their world with many different tools and processes that improved the

Figure 1–5 Smelting iron ore at high temperatures creates molten metal, which can then be molded into tools. *(Courtesy of PhotoDisc.)*

quality of life. Never before had the Earth seen such resourcefulness as in the human species. Human beings began to build cities, conquer the oceans, and construct complicated machinery. The culmination of human ingenuity began a new era called the Technological Revolution (or the Industrial Revolution), which started in the late 1800s (Figure 1–6). Human beings now possessed the ability to build machines and use industrial processes to perform work that would have been unthinkable only 100 years earlier.

The technology of stone tools has long been surpassed; however, it is important to realize that new technologies are usually the result of manipulating or combining older technological processes. Today complicated machines and tools perform the same tasks that were once performed by primitive stone tools. Many profound technological advancements have occurred over the course of human history, including the harnessing and manipulation of fire, the development of spears and the bow and arrow, the invention of the agricultural plow, use of the wheel, the development of shipbuilding, the invention of the steam engine and the railroad, the development of the internal combustion en-

EARTH SYSTEM SCIENTISTS *THE LEAKEY FAMILY*

No other family in the science of anthropology has made such an impact on our understanding of human origins as the Leakey family. Louis Leakey was born in Kenya, Africa, in 1903, where he became interested in the history of the human race. As a boy he became fascinated by the prehistoric stone tools he found while exploring the African landscape. He eventually went on to pursue an education in anthropology and began his life-long career in the search for human origins. Louis Leakey met his future wife, Mary Nicols, while on an archaeological dig in England. Mary Nicols was born into a family of archaeologists and specialized in the prehistory of human beings. In 1937 the newly married Leakeys returned to Kenya and began to raise a family. During this time they made some very important fossil discoveries that helped piece together our human ancestry. Most of their famous work was done in the Olduvai Gorge in Africa, where they unearthed many hominid fossil remains. Mary Leakey's most significant discovery was the unearthing of a series of fossilized footprints that were made by a 3.5-million-year-old human ancestor. This find proved that these ancient humans walked on two feet. Their son, Richard Leakey, who was born in 1944, also joined the family business. His work in anthropology continues to this day. Throughout the twentieth century the Leakey family made significant discoveries in tracing human origins, and they have influenced other anthropologists and archaeologists to pursue their line of work.

Technology Timeline

Paleolithic Period	• First use of stone tools by Homo habilis~2.5 million years ago • Emergence of modern humans (Homo sapiens)~100,000 years ago
Neolith Period	• Use of stone, wood, and animal bone tools • Emergence of art and musical instruments~35,000 years ago • Beginning of agricultural revolution~10,000 years ago • First known use of wooden plow in southwest Asia~7000 years ago • First known use of the wheel~5500 years ago
Bronze Age	• Emergence of copper and bronze in tool making~5000 years ago • Beginning of textile production in Mesopotamia~5000 years ago
Iron Age	• Smelting of iron in Africa~3000 years ago • Use of iron in Europe and China~2000 years ago • Construction of the Roman aqueducts~300 B.C.
Industrial Age	• Invention of the steam engine–1720 A.D. • Invention of the steamboat–1807 A.D. • First airplane flight–1903 A.D. • First moon landing–1969 A.D.
Information Age (the present)	

Figure 1–6 The technology timeline is divided into different historical periods based on advances in technology.

gine, the invention of the light bulb, the development of the telephone, and the development of modern computers (Table 1–1). The list could go on, but the point is quite clear: Technology has been used in countless ways to perform countless tasks to improve the human condition.

Technological Systems

Human problems and needs are now being addressed and resolved by complex technological systems that began with simple ideas and tools. A technological system is a system through which a process combines resources to provide a desired output. These systems are then studied to see how problems can be solved or how specific needs can be met. **Technological systems models** can be used to analyze specific systems and to understand how they operate. The basic technological systems model is composed of a command input, comparison device, adjustment, resource inputs, process, output, monitor, and feedback loop (Figure 1–7). This model helps to illustrate how a technological process works and how it can be improved. Today people are moving into a new period of technological innovation, often referred to as the Information Age. The use of computer technology is changing our way of life in ways that our ancestors could not have imagined.

TABLE 1–1 Important technological inventions throughout history

Woven cloth, 5000 B.C.	Typewriter, 1867 A.D.
Wheeled vehicles, 3500 B.C.	Bicycle, 1867 A.D.
Ox Drawn plow, 2500 B.C.	Four-stroke engine, 1876 A.D.
Maps, 510 B.C.	Telephones, 1876 A.D.
Lever, 250 B.C.	Light bulb, 1879 A.D.
Paper, 100 A.D.	Gasoline automobile, 1885 A.D.
Gunpowder, 850 A.D.	Radio, 1901 A.D.
Eyeglasses, 1249 A.D.	Airplane, 1903 A.D.
Mechanical clock, 1360 A.D.	SONAR, 1917 A.D.
Printing press, 1454 A.D.	Refrigerator, 1918 A.D.
Screwdriver, 1550 A.D.	Jet engine, 1936 A.D.
Graphite pencil, 1565 A.D.	Photocopier, 1938 A.D.
Thermometer, 1592 A.D.	Atomic bomb, 1945 A.D.
Telescope, 1608 A.D.	Transistor, 1948 A.D.
Barometer, 1643 A.D.	Satellite, 1957 A.D.
Microscope, 1673 A.D.	Argon laser, 1960 A.D.
Steam engine, 1712 A.D.	Manned spaceflight, 1961 A.D.
Hot air ballon, 1782 A.D.	Word processor, 1965 A.D.
Steam locomotive, 1804 A.D.	Videotape cassette, 1970 A.D.
Photography, 1822 A.D.	Computer floppy disk, 1970 A.D.
Electric motor, 1831 A.D.	Laser printer, 1980 A.D.
Steam shovel, 1836 A.D.	CD-ROM, 1984 A.D.
Telegraph, 1844 A.D.	Gene altered bacteria, 1987 A.D.
Sewing machine, 1846 A.D.	

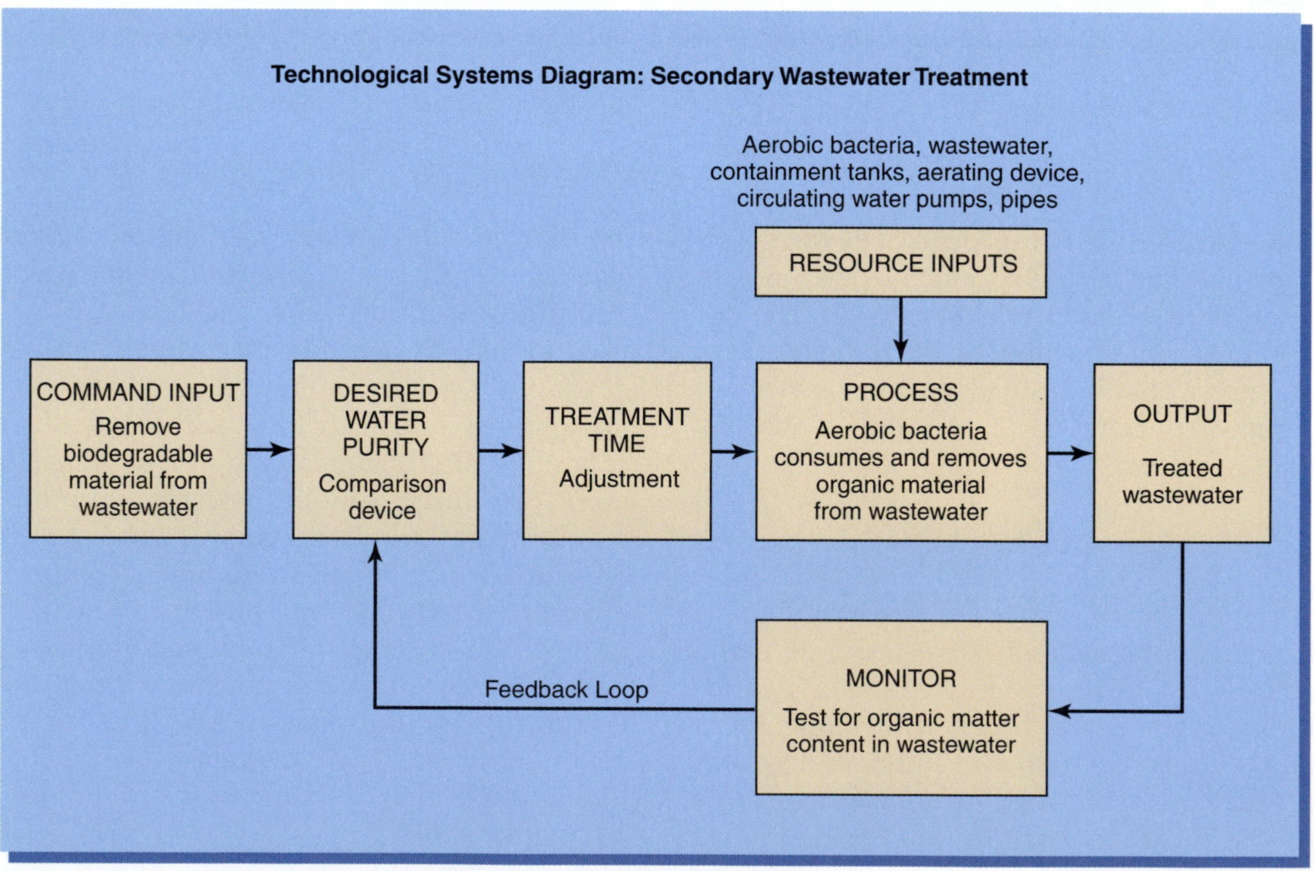

Figure 1–7 A technological systems diagram is used to model the technological process that is used to treat wastewater.

Current Research

Composite technology is creating the building materials of the future. A composite material is formed from combining two or more materials with unique properties. Composites are composed of a binder and a reinforcer. A binder completely surrounds and binds together the reinforcing material. Common bonding agents today include plastic resins. The reinforcer is a stronger material that provides the composite with strength and stability. First developed in the 1940s, composites have quickly become more widespread as building materials. The first successful composite material was fiberglass. Fiberglass composites are still widely used today in car bodies and boats. New composite materials are being developed that use plastic polymers and strong synthetic fibers. Some of the reinforcing agents used in composites include carbon fiber, or Kevlar. These materials are extremely lightweight but provide great strength and stability. Newer composite materials are currently under development that use new binders and reinforcers. The advantage of using a composite material involves strength, light weight, and the ability to be molded into any shape.

SECTION REVIEW

1. Define the term *technology,* and describe its importance to human beings.
2. List and describe the evolution of technology throughout human history.
3. Explain how the technological system model can be used to represent a technological process.
4. Who was Louis Leakey?

For more information go to these web links:
<http://www.mnh.si.edu/anthro/humanorigins/>
<http://echo.gmu.edu/cenyter/index.html>

1.2 *The Foundations of Science*

The Birth of Science

Although it is difficult to determine when humans actually began to practice science, it can be assumed from our ancient beginnings that we have always wondered at how our world works. This is the core of understanding what science is. **Science** is the search for knowledge about how the natural world operates. The actual practice of science involves observation, investigation, description, and explanation. The birth of formal science most likely had its roots in the study of celestial objects (Figure 1–8). Some of the world's first scientists were fascinated with the motion of the stars and planets in space. Careful observation and measurement of astronomical phenomena lead to the birth of astronomy and to the birth of modern science. Many of the world's most well known scientists were and are astronomers. Famous names like Aristotle, Newton, Kepler, Galileo, Copernicus, Einstein, Hubble, Hawking, and Sagan are all famous astronomers who helped to revolutionize science. Today science continues to improve our world and increase our understanding of how it works. To do this, scientists use a formal

Earth Math

1) If the Paleolithic period began 2.5 million years ago and ended approximately 100,000 years ago, how many years did the Paleolithic period last?
2) If the Information Age began in 1990, how long did the Industrial Age last?

Figure 1–8 Stonehenge in England is believed to be an ancient astronomical calendar used to track the movement of the heavens that was built more than 2000 years ago. *(Courtesy of PhotoDisc.)*

method of inquiry to attempt to unlock the secrets of nature; this is called the scientific method.

The Scientific Method

The **scientific method** of inquiry is based on three main concepts: observation, experimentation, and the development of theories or natural laws (Table 1–2). The first step in the scientific method is the actual observation and recording of facts. Much of the work of a scientist involves **observation** and the collection of data. This helps scientists to gain as much information as they can about the natural phenomena they are studying and then record that information in an organized way. Observation also involves conducting experiments. **Experiments** are controlled observations that help to answer questions about what scientists are trying to discover (Figure 1–9). The next step in the scientific method is the formulation of a **theory** that might explain how or why the natural phenomenon that is being studied is occurring. This is also called a **hypothesis,** which is an explanation that is supported by a set of facts. The final step in the scientific process is the formulation of a natural law that explains the phenomenon that is being studied. The formulation of a natural or a physical law helps to explain how certain aspects of the natural world operate and, more importantly, how they can be used to make predictions. Scientists often use observations they have made in the past to make inferences about what might occur in the future. An **inference** is a prediction or conclusion that is made about a future event based on previous scientific observations. The scientific method is a formal and organized procedure that scientists around the world use to make accurate investigations of the natural world.

Scientific Experiments

One of the most important aspects of the scientific method is experimentation. An experiment allows scientists to prove or disprove a hypothesis. Experiments are also an important way for students of science to learn about the natural world and gain knowledge by actually

EARTH SYSTEM SCIENTISTS *THALES OF MILETUS*

Thales was born in ancient Greece in approximately 620 B.C. He is considered by many historians to be the founder of scientific inquiry. His main achievements centered around the notion that natural events were the work of observable properties and processes, not the work of the gods. His achievements involved geography, engineering, mathematics, science, and politics. He is best known for his development of the scientific method and for his practical search for knowledge. He also did extensive work in astronomy and the prediction of eclipses. Thales' study of philosophy and science influenced many Greek philosophers, including Aristotle, Plato, Socrates, Pythagoras, and Diogenes.

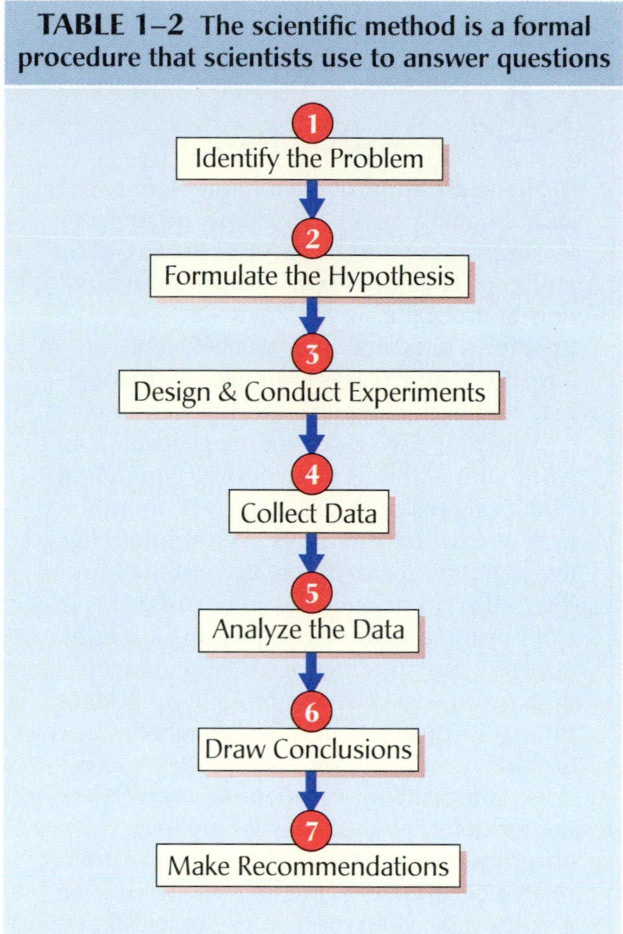

TABLE 1–2 The scientific method is a formal procedure that scientists use to answer questions

practicing science. Conducting experiments also follows an organized pattern, which includes stating the purpose of the experiment, creating a hypothesis, writing out step-by-step procedures, collecting and analyzing data, and formulating a conclusion. All experiments begin by stating the purpose for performing the experiment. The purpose explains exactly what you are trying to determine in your experiment. For example, if you wanted to know at what temperature water boiled, you could determine it by performing a simple experiment. Your purpose is simple: to discover the exact temperature at which water boils.

The next step in experimentation is formulating a hypothesis that might explain your experiment. A hypothesis is an explanation of how or why the phenomenon you are studying is occurring. A hypothesis is usually based on any previous knowledge you have about what you are studying. Using the boiling water example, you may already have an idea at what temperature water boils, so your hypothesis would state this. Or if you have no idea about the boiling temperature of water, you may state in your hypothesis that you believe water always boils at the same temperature, which you are trying to discover.

Writing out the procedures that you took to perform your experiment is another important aspect of experimentation. This allows other scientists or students to understand how the experiment is to be conducted. The procedure is also important because it allows for your experiment to be recreated by other scientists. A procedure is just a step-by-step explanation of what you did to conduct your experiment. It also may include special safety concerns and specific materials or instruments required to conduct your experiment. For example, if you are trying to determine the temperature of boiling water, you must take special precautions to prevent yourself from being burned by the water.

Collecting and recording data from your experiment is the next step in experimentation. Data collection is a precise way of making

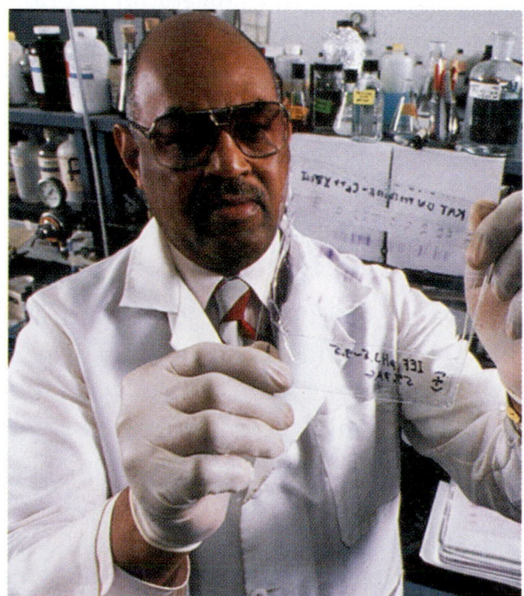

Figure 1–9 Scientists commonly conduct experiments to prove or disprove their theories. *(Courtesy of USDA/ARS #K1968-13.)*

accurate observations during your experiment. Often experiments are conducted more than once to collect more than one set of data, which allows for more accuracy. For this portion of your experiment, you may choose to record the temperature of the water every 30 seconds until it boils and then repeat the experiment again to gain a more accurate result. You must also consider what units of measurement you will use to record your data. In most scientific experiments the metric system is used, and temperature is often recorded in degrees Celsius. Calculating the results of your experiment involves the analyzation and organization of the data you have collected. Creating a graph or chart to analyze your data often is an effective way to organize your results. A graph is a way to visually display numbers and data that you have collected during an experiment. Graphs are an important tool in the scientific method because they can reveal trends that are occurring in your data.

The final portion of an experiment is stating a conclusion. The conclusion of your experiment should summarize the results of your experiment, which will either support or disprove your hypothesis. If your experiment revealed that the boiling point of water was 100° Celsius, then this should be stated in your conclusion. Carefully conducting an experiment in an organized way ensures that scientific discovery can be well documented and recreated.

Science, Technology, and Engineering

There has always been an important link between science and technology. The general goal of science has been to try to understand how the natural world works. The goal of technology is to take the knowledge gained by scientific inquiry and apply it in a practical way. Today many of the technologies that our society uses are the direct result of scientific inquiry and research. These include technologies like the airplane, the telephone, radio, computers, satellites, the light bulb, lasers,

Career Connections

ENGINEERS

Engineers are professional problem solvers who apply math and science to improve the human condition. There are many different types of engineers who work on many new and unique ways to benefit society. Engineers need a 4-year college degree in one of the four major branches of engineering: electrical, chemical, mechanical, or civil engineering. Electrical engineers develop and work with all kinds of electronic equipment. Chemical engineers use chemistry to produce new or existing chemicals. Mechanical engineers design, research, create, test, and produce all types of mechanical devices. This includes simple tools or complex machines like airplanes and automobiles. Civil engineers design and supervise the building of bridges, railroads, tunnels, roads, water supplies, sewers, and airports. Once out of college, engineers can make an excellent salary by working for industry, business, the military, government, education, or research. Another related occupation is engineering technician, a technician who works in the practical applications of engineering. This includes jobs in all the engineering fields but requires a less formal education. Some of the many engineering fields include the following:

- automotive
- aerospace (aero or astronautical)
- agricultural
- architectural
- bioengineering (biomedical, biomechanical, biochemical)
- ceramic
- computer
- environmental
- fire protection
- geological
- geothermal
- heating, ventilating, air-conditioning, and refrigeration
- industrial
- manufacturing
- materials
- metallurgy and materials
- mineral and mining
- naval
- nuclear
- ocean
- optical
- petroleum
- plant
- plastics
- robotics and automated systems
- safety
- software
- transportation

EARTH MATH

1) IN 1543 A.D., NICOLAS COPERNICUS HYPOTHESIZED THAT THE EARTH AND OTHER PLANETS REVOLVE AROUND THE SUN. HOW OLD IS THIS FUNDAMENTAL THEORY OF ASTRONOMY?

and countless more inventions. People who use scientific principles and apply them in a practical way are called engineers. Today engineers use math and science in a variety of ways to improve our way of life. Structural engineers, also called civil engineers, design buildings, bridges, tunnels, roads, airports, water supplies, and sewers that we depend on every day. Electrical engineers design all the electrical devices we use, including computers, communication equipment, entertainment devices, electrical power creation and distribution, appliances, and lighting. Chemical engineers create chemical substances that are used by both science and industry. Agricultural engineers help design new and efficient ways of producing food. One of the fastest growing fields of engineering today involves the use of biotechnology, also called bioengineering. Biotechnology is the use of living organisms to produce a usable product, or accomplish a specific task. The role of science and technology will play an important role in the future of the human race as more knowledge is gained about the world in which we live and, more importantly, how this knowledge can be used to improve the human condition.

For more information go to this web link:
<http://echo.gmu.edu/center/index.html>

SECTION REVIEW

1. What are the three main processes that together make up the scientific method?
2. Describe the series of steps involved in conducting an experiment.
3. What is the difference between a scientist and an engineer?
4. Who was Thales?

1.3 *Scientific Measurement*

Observation and Measurement

Much of a scientist's job involves making observations of the natural world. An **observation** can be as simple as using one of the five human senses. These are sight, touch, taste, smelling, and hearing. Today scientists extend their own senses by using scientific instruments. Instruments are devices that are designed to extend the human senses and make accurate measurements of the natural world. Scientific instruments can be as simple as a scale or meter stick or as complex as a satellite or radio telescope (Figure 1–10). Measurement is an important part of science, for it allows observations to be made more precisely.

Accurate measurements are also important for identifying unique physical or chemical properties of the Earth. Today scientists use four fundamental measurements to help observe the natural world: length, mass, time, and energy. **Length** measures the distance between two fixed points. **Mass** is a measure of the amount of matter in an object. On Earth the mass of an object is measured by determining its **weight.** Although weight can be used to measure mass, mass and weight are not always equal, because

Figure 1–10 Scientific instruments, such as a balance, rulers, satellites, telescopes, and compasses, are used to extend the human senses. *(Courtesy of PhotoDisc.)*

weight is just a measure of the pull of gravity on an object's mass. Therefore an object's weight can be different on the Moon than on the Earth; however, its mass remains constant. **Time** is forward movement, or specific interval between two events. **Energy** is measured as the electric charge of matter, which can be either positive or negative. Energy is also a measure of the movement of matter, called its kinetic energy. Common energy measurements include the heat or temperature of a substance.

Units of Measurement

All scientific measurements consist of a numerical quantity and a specific unit. A unit is a specific measurement that has a fixed value. One of the oldest length units of measurement still in use today is the foot. The Romans first introduced the foot as a unit of length that divided the distance from a person's heel to the tip of their big toe into 12 equal parts. They also introduced the mile as a measure of length that equaled 1000 paces. A pace was the distance a person could travel in two walking steps. Another ancient unit of measuring length is the yard, which was the distance between the tip of a person's nose and the thumb when the arm was fully extended. Today in the United States we recognize two systems of measurement. One is called the standard system of measurement, which includes units like the inch, foot, mile, pound, ounce, and yard. In the 1780s the French government developed a new system of measurement called the international system, also known as the metric system. Today the entire scientific community around the world uses the metric system for observation and measurement (Table 1–3). The metric system is

TABLE 1–3 Common units of measurement used in science

Length

Metric

1 kilometer (km) = 1,000 meters (m)
1 meter (m) = 100 centimeters (cm)
1 meter (m) = 1,000 millimeters (mm)
1 centimeter (cm) = 0.01 meter (m)
1 millimeter (mm) = 0.001 meter (m)

English

1 foot (ft) = 12 inches (in)
1 yard (yd) = 3 feet (ft)
1 mile (mi) = 5,280 feet (ft)
1 nautical mile = 1.15 miles

Metric-English

1 kilometer (km) = 0.621 mile (mi)
1 meter (m) = 39.4 inches (in)
1 inch (in) = 2.54 centimeters (cm)
1 foot (ft) = 0.305 meter (m)
1 yard (yd) = 0.914 meter (m)
1 nautical mile = 1.85 kilometers (km)

Area

Metric

1 square kilometer (km^2) = 1,000,000 square meters (m^2)
1 square meter (m^2) = 1,000,000 square millimeters (mm^2)
1 hectare (ha) = 10,000 square meters (m^2)
1 hectare (ha) = 0.01 square kilometer (km^2)

English

1 square foot (ft^2) = 144 square inches (in^2)
1 square yard (yd^2) = 9 square feet (ft^2)
1 square mile (mi^2) = 27,880,000 square feet (ft^2)
1 acre (ac) = 43,560 square feet (ft^2)

Metric-English

1 hectare (ha) = 2.471 acres (ac)
1 square kilometer (km^2) = 0.386 square mile (mi^2)
1 square meter (m^2) = 1.196 square yards (yd^2)
1 square meter (m^2) = 10.76 square feet (ft^2)
1 square centimeter (cm^2) = 0.155 square inch (in^2)

Volume

Metric

1 cubic kilometer (km^3) = 1,000,000,000 cubic meters (m^3)
1 cubic meter (m^3) = 1,000,000 cubic centimeters (cm^3)
1 liter (L) = 1,000 milliliters (mL) = 1,000 cubic centimeters (cm^3)
1 milliliter (mL) = 0.001 liter (L)
1 milliliter (mL) = 1 cubic centimeter (cm^3)

English

1 gallon (gal) = 4 quarts (qt)
1 quart (qt) = 2 pints (pt)

Metric-English

1 liter (L) = 0.265 gallon (gal)
1 liter (L) = 1.06 quarts (qt)
1 liter (L) = 0.0353 cubic foot (ft^3)
1 cubic meter (m^3) = 35.3 cubic feet (ft^3)
1 cubic meter (m^3) = 1.30 cubic yard (yd^3)
1 cubic kilometer (km^3) = 0.24 cubic mile (mi^3)
1 barrel (bbl) = 159 liters (L)
1 barrel (bbl) = 42 U.S. gallons (gal)

Mass

Metric

1 kilogram (kg) = 1,000 grams (g)
1 gram (g) = 1,000 milligrams (mg)
1 gram (g) = 1,000,000 micrograms (μg)
1 milligram (mg) = 0.001 gram (g)
1 metric ton (mt) = 1,000 kilograms (kg)

English

1 ton (t) = 2,000 pounds (lb)
1 pound (lb) = 16 ounces (oz)

Metric-English

1 metric ton (mt) = 2,200 pounds (lb) = 1.1 tons (t)
1 kilogram (kg) = 2.20 pounds (lb)
1 pound (lb) = 454 grams (g)
1 gram (g) = 0.035 ounce (oz)

Energy and Power

Metric

1 kilojoule (kJ) = 1,000 joules (J)
1 kilocalorie (kcal) = 1,000 calories (cal)
1 calorie (cal) = 4,184 joules (J)

Metric-English

1 kilojoule (kJ) = 0.949 British thermal unit (Btu)
1 kilojoule (kJ) = 0.000278 kilowatt-hour (kW-h)
1 kilocalorie (kcal) = 3.97 British thermal units (Btu)
1 kilocalorie (kcal) = 0.00116 kilowatt-hour (kW-h)
1 kilowatt-hour (kW-h) = 860 kilocalories (kcal)
1 kilowatt-hour (kW-h) = 3,400 British thermal units (Btu)
1 quad (Q) = 1,050,000,000,000,000 kilojoules (kJ)
1 quad (Q) = 2,930,000,000,000 kilowatt-hours (kW-h)

Fahrenheit (°F) to Celsius (°C): $°C = \dfrac{(°F - 32.0)}{1.80}$

Celsius (°C) to Fahrenheit (°F): $°F = (°C \times 1.80) + 32.0$

THE INVENTION OF NUMBERS

The invention of numbers that we use every day has played an important part in the development human civilization. Almost all ancient societies, such as the Babylonians, Egyptians, Mayans, Hindus, and Arabs, developed methods of counting. The base 10 system, which we still use today, is based on the digits 1, 2, 3, 4, 5, 6, 7, 8, and 9. This most likely developed as a result of humans' having 10 fingers to count with. The writing of these numbers was first used in India during the third century B.C. and are known as the Brahmi numbers. This system of numbers was adapted by other civilizations and eventually became known as the Hindu-Arabic numeral system (Figure 1–11). This system, which we still use today, is more than 2000 years old!

A similar system is the Roman numeral system, which uses hash marks to represent numbers from 1 to 9. The concept of the zero in number systems and mathematics came much later. The first recorded rules using a zero in mathematical expressions were written in a book called *The Opening of the Universe,* written in 630 A.D. by the Indian mathematician Brahmagupta. The significance of the zero in math and counting continues to this day. Even with our sophisticated computer technology, the concept of the zero, when applied to recording the year, created a worldwide scare and cost billions of dollars to fix. People around the world nervously waited to see how computers would be affected by the year 2000!

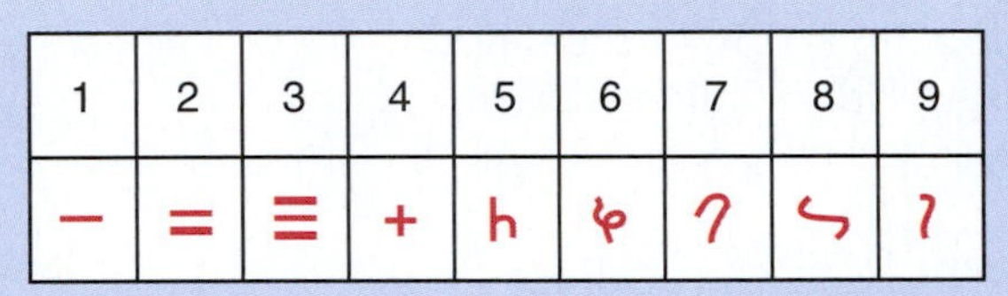

1	2	3	4	5	6	7	8	9
—	=	≡	+	h	[illegible]	[illegible]	[illegible]	[illegible]

Figure 1–11 An early numbering system developed, which forms the base of our modern numbers.

based on the fundamental unit of length called the meter. The meter is further divided into divisions of 100 called centimeters. Another fundamental unit of mass in the metric system is called the gram. The gram is equal to the mass of 1 cubic centimeter of liquid water.

Many measurable quantities of the natural world include two or more fundamental quantities. These are called derived units; for example, the volume of a substance is measured by determining its length, width, and height. Common units of volume in science include the cubic centimeter and milliliter. Another derived unit often used in observation is speed or acceleration. Speed is a measure of a specific distance traveled in a specific amount of time. Common speed or acceleration units of measurement include kilometers per hour and miles per hour.

Density

Density is an important derived unit of measurement that is used in science to determine the specific amount of mass per unit volume a substance has. Therefore density measures the amount of atoms and molecules that occupy a specific space. Density is determined by the following formula: density = mass/volume. The density of an object is important because it effects many of the physical properties of a substance (Figure 1–12). There is an important relationship between density and the three states of matter. For all objects, excluding water, the solid form is the most dense, liquid being less dense, and the gaseous form being the least dense form of matter. Objects that are less dense tend to float on more dense objects. Observe the results when you mix oil and water. The less

Career Connections

Surveyors and Surveying Technicians

Surveyors and their technicians operate instruments that are used to measure and map all parts of our world. Surveying involves many aspects of the modern world, including the accurate measurement of land for the sale of property. Surveyors also help to measure and lay out plans for the construction of roads, airports, bridges, and all other construction sites. Accurate surveys also help to generate many different maps of our world. Surveying work involves a lot of time outdoors, where accurate measurements are made using precise instruments. Surveyors usually must pass state tests to become certified and often require a 4-year college degree. Surveying technicians can learn their trade on the job or in a 2-year college program. Many vocational schools also offer surveying instruction.

dense oil floats on top of the more dense water. The temperature of an object also has a relationship to density. Generally, as you increase the temperature of a substance, its density decreases; therefore hot objects tend to be less dense than cooler objects. Pressure also has an effect on the density of a substance. When you increase the pressure on a substance, its density increases. Water is a unique substance that has unusual properties associated with density. The most dense form of water is its liquid form, at 4° Celsius. The reason that the liquid form of water is more dense than the solid form has to do with the crystal structure of ice. When liquid water freezes to form ice, the water molecules link together in an organized crystal pattern. This crystal pattern has a larger volume per unit mass than the liquid form of water, which causes ice to be less dense. The density of ice is 0.91 grams per cubic centimeter, and the density of liquid water is 1.0 grams per cubic centimeter. The lower density of ice compared with liquid water causes ice to float. In all other substances on Earth, the solid form of matter is always denser than the liquid form, causing it to sink (Table 1–4).

Scientific Notation

Scientific measurements are often recorded in a form called scientific notation, or exponential notation. This allows extremely large or very small numbers to be written much more easily. Scientific notation is written in the following way: $N \times 10^e$. N represents a number between 1 and 10, and e is the positive or negative exponent power of 10. The exponent represents the number of places to the right or left of the decimal point. If the exponent is a positive number, then it represents

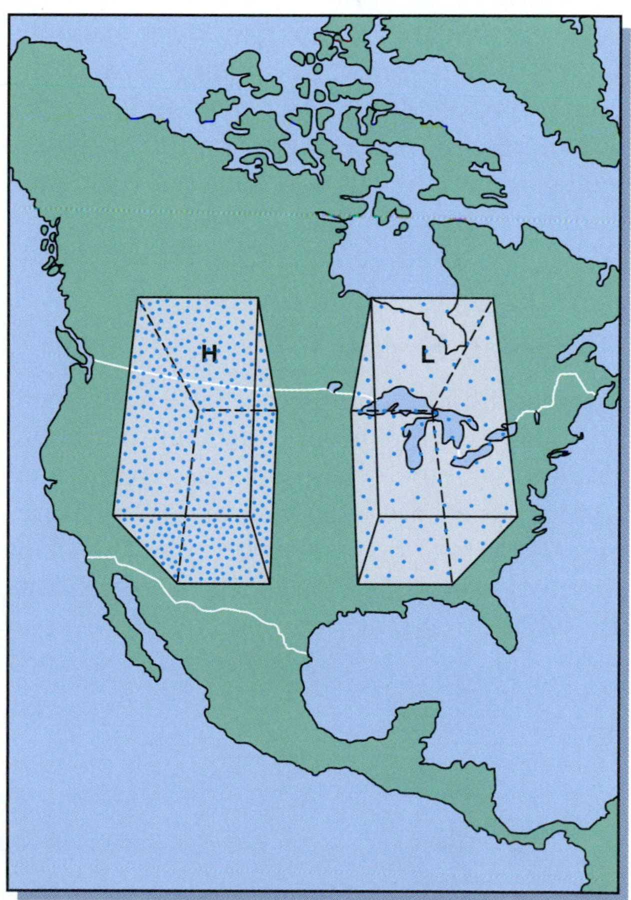

Figure 1–12 Two parcels of air with the same volume and different masses have different densities.

TABLE 1–4 Densities of common substances

Solid	Density (g/cm^3, 20°C)	Liquid	Density (g/cm^3, 20°C)	Gas	Density (g/L, 0°C*)
Gold	19.3	Water	1.00	Air	1.29
Lead	11.3	Gasoline	0.67	Oxygen	1.43
Copper	8.92	Milk	1.03	Hydrogen	0.090
Iron	7.86	Sea water	1.03	Helium	0.178
Aluminum	2.70	Blood	1.06	Carbon dioxide	1.96
Salt	2.16	Mercury	13.6		
Paper	0.70	Olive oil	0.92		
Balsa	0.20	Alcohol	0.79		
Redwood	0.44	Vinegar	1.01		
Rubber	1.1	Ether	0.70		
Ice	0.92	Carbon tetrachloride	1.59		

the number of decimal places to the right of the number N. This represents very large numbers that are greater than 1. For example, 4.5×10^6 is the scientific notation form of the number 4,500,000. If the exponent is a negative number, it represents the number of decimal places to the left of N. This is a very small number that is less than 1. For example, 2×10^{-3} is the scientific notation form of 0.002.

Percent Error

Another important aspect of scientific measurement is the concept of percent deviation, also known as percent error. The percent deviation is a measure of how accurate a measurement is to that of an accepted value. This helps scientists gauge how good their measurements are when they make an observation. Human error or faulty scientific instruments are a constant variable that can create inaccurate measurements. Calculating percent deviation involves the use of the following formula: percent deviation = ((the difference between the accepted value and the measured value/the accepted value) × 100). For example, if you determine the mass of a mineral sample to be 35 grams, and the accepted mass is actually 40 grams, then your percent error calculation is determined by the following calculation, ((40 grams − 35 grams)/ 40 grams) × 100 = 12.5%. No measurement in the science classroom is completely accurate; therefore calculating the percent error helps to reveal how good your measurement is. Common devices used for measuring, such as the speedometer of your car, have percent errors of approximately 3%. This means that the speed of your

EARTH MATH

1) WRITE THE FOLLOWING NUMBERS IN THEIR LONG FORM: 1.2×10^9 AND 6×10^{-5}
2) WRITE THE FOLLOWING NUMBERS IN SCIENTIFIC NOTATION: 91,000,000 AND 0.00003
3) DETERMINE THE PERCENT DEVIATION OF A MEASUREMENT OF 12 CENTIMETERS, WHEN THE ACCEPTED VALUE IS 11.5 CENTIMETERS.

SECTION REVIEW

1. Describe how a scientific observation is made.
2. What are the four fundamental units of measurement?
3. What system of measurement do scientists use?
4. What is a derived unit? Provide two examples of derived units.
5. Define density.
6. What are three examples of how the density of an object can be affected?
7. What is the densest form of water?
8. Approximately how old is our modern system of numbers, and who was the first to use them?

car is actually 3% more or less than the actual speed. An accurate measurement usually has a percent error of less than 3%.

For more information go to this web link:
<http://www.omnis.demon.co.uk>

CHAPTER SUMMARY

Throughout history, human beings have used technology to improve their lives by applying their knowledge to solve problems or perform tasks. This has resulted in a wealth of materials, tools, and processes that impact all aspects of our lives. Humans have often increased their knowledge of the world by careful investigation, which led to the development of what we know of today as science. Science advances human knowledge by a systematic process called the scientific method. This formal process consists of observation, experimentation, and the development of theories. Observation often involves the use of scientific instruments that extend the use of the human senses and help scientists collect accurate data. Performing experiments is probably the most important part of the scientific method because it allows scientists to test their theories and learn more about the world. There will always be an important relationship between science and technology because many technologies are often developed as a result of a scientific investigation. Practicing science often involves the recording of precise measurements. The fundamental units of measurement include length, mass, weight, time, and energy. Two systems of measurement are commonly used in the United States for accurate measurement. These are the standard system and metric system. Often two or more fundamental units of measurement are combined to form what is called a derived unit, such as density. Density is a measure of the amount of mass per unit volume of a substance, which is an important physical characteristic of many substances on the Earth. Other important aspects of scientific measurement include percent error calculation, which reveals the accuracy of a measurement, and scientific notation, which is used to record very large or very small measurements in an abbreviated form.

CHAPTER REVIEW

Multiple Choice

1. How old are the oldest stone tools discovered that represent early human technology?
   a. 10,000 years
   b. 100,000 years
   c. 500,000 years
   d. 2,000,000 years

2. The period that began when humans first started to grow their own food is called:
   a. the Stone Age
   b. the Agricultural Revolution
   c. the Industrial Revolution
   d. the Information Age

3. The application of scientific knowledge for a practical purpose is called:
   a. science
   b. a technological systems model
   c. technology
   d. an instrument

4. The search for knowledge about how the natural world operates is called:
   a. science
   b. a technological systems model
   c. technology
   d. an instrument

5. The scientific method consists of:
   a. proof, facts, and experiments
   b. purpose, procedure, and conclusion
   c. input, process, and output
   d. observation, hypothesis, and natural law

6. The part of an experiment that relates the step-by-step directions of what was done is called the:
   a. purpose
   b. hypothesis
   c. procedure
   d. conclusion

7. To make an observation, a person must use:
   a. experiments
   b. the senses
   c. mathematical calculations
   d. proportions

8. An interpretation based on previous observations is:
   a. an inference
   b. a fact
   c. a classification
   d. a measurement

9. Which unit of measurement defines the amount of matter an object contains?
   a. weight
   b. volume
   c. mass
   d. volume

10. Which term best describes the amount of space a substance occupies?
    a. density
    b. volume
    c. mass
    d. weight

11. In which state of matter do most materials on the Earth have their greatest density?
    a. gas
    b. liquid
    c. solid
    d. plasma

12. Compared with the density of liquid water, the density of ice is:
    a. always less
    b. always greater
    c. always the same
    d. sometime less and sometimes more

13. What is the density of a rock with a mass of 35 grams and a volume of 7.0 cubic centimeters?
    a. 42 g/cm^3
    b. 28 g/cm^3
    c. 0.2 g/cm^3
    d. 5.0 g/cm^3

14. How many centimeters are there in 1 meter?
    a. 1
    b. 10
    c. 100
    d. 1000

15. The mass of a rock is measured to be 51 grams, but its actual mass is 60 grams. What is the percent error of the rock's mass?
    a. 7%
    b. 9%
    c. 15%
    d. 18%

16. The circumference of the Earth is approximately 4.0 × 104 kilometers. This value is equal to:
    a. 400 km
    b. 4000 km
    c. 40,000 km
    d. 400,000 km

17. The distance from the Earth to the Sun is approximately 93,000,000 miles. This value is equal to:
   a. 9.3×10^7 miles
   b. 9.3×10^6 miles
   c. 93×10^7 miles
   d. 93×10^6 miles

Matching *Match the terms with the correct definitions.*

a. technology
b. technological systems model
c. science
d. scientific method
e. observation
f. experiment
g. theory
h. hypothesis
i. inference
j. length
k. mass
l. weight
m. time
n. energy
o. density

1. ____ A statement, or statements used describe a phenomena.
2. ____ The application of human knowledge to solve problems or perform tasks.
3. ____ An explanation based on a set of facts that can be tested.
4. ____ A model used to illustrate the steps in a technological process.
5. ____ A conclusion based on a set of observed facts.
6. ____ The practice of observing, identifying, describing, and explaining natural phenomena.
7. ____ A fundamental form of measurement that measures horizontal distance.
8. ____ The specific set of procedures that scientists utilize to gain knowledge.
9. ____ A fundamental unit of measurement that measures the amount of matter a substance contains.
10. ____ The direct perception of something by use of one of the five human senses.
11. ____ The force that results from the gravitational attraction of the Earth on an object, which is dependent on its mass (commonly known as how heavy something is).
12. ____ A controlled test to prove or disprove a hypothesis.
13. ____ A fundamental unit of measurement that records the specific interval that separates events.
14. ____ The mass per unit volume of a substance, usually expressed in grams per cubic centimeter.
15. ____ The ability to cause change, or perform work.

Critical Thinking

1. Briefly discuss some positive and negative effects of science and technology on humans and the Earth.

CHAPTER 2 Matter and Energy

Section 2.1 – The Properties of Matter Objectives

Atoms • The Elements • States of Matter • Compounds and Molecules

After reading this section you should be able to:

- Define the term *matter*.
- Describe the three particles that make up an atom.
- Identify the four states of matter.
- Explain what a phase change is.
- Describe the difference between an element and a compound.

Section 2.2 – Energy Objectives

Energy • The Law of the Conservation of Energy • Electromagnetic Radiation • Energy Transfer • Heat and Temperature

After reading this section you should be able to:

- Define the term *energy*.
- Discuss the difference between potential energy and kinetic energy.
- Identify the eight different forms of energy.
- Describe the relationship between wavelength and energy on the electromagnetic spectrum.
- Identify the five ways in which electromagnetic energy can interact with the environment.
- Differentiate between heat and temperature.
- Explain the three fundamental ways in which energy is transferred on the Earth.
- Define the term *latent heat*.

TERMS TO KNOW

matter
elements
vaporization
boiling point
condensation
sublimation
energy
electromagnetic radiation
electromagnetic spectrum
conduction
radiation
convection
temperature
heat
latent heat

INTRODUCTION

What do you, cars, buildings, trees, flowers, your family pet, computers, and television all have in common? They are all made up of a complex interaction of matter and energy. All things in the universe as we know it are made up of matter and energy. These two fundamental items enable all the wonders of our world to exist. We rely on their interactions everyday to feed us, keep us safe, provide us with knowledge, and entertain us. Understanding the basic concepts of matter and energy provides us with the ability to unlock the secrets of the world around us.

2.1 *The Properties of Matter*

Atoms

All substances that exist on the Earth consist of matter. **Matter** is something that occupies space and has mass. The amount of space that matter occupies is known as its volume. The mass of the matter is a measure of the amount of atoms that compose it. All matter is made up of fundamental particles called atoms. Atoms are extremely small particles that possess unique physical and chemical properties. The Greek philosopher Democritus first proposed the concept of an atom more than 2000 years ago. Democritus believed that all things were composed of tiny particles that could not be further divided. He called these particles *atomos*, which in Greek means "indivisible," or unable to divide. All atoms are composed of three subatomic particles: protons, electrons, and neutrons. Protons and neutrons form an atom's nucleus, around which revolves electrons. Because the atom is so small and cannot be seen with the human eye, many different models have been developed to represent how the atom might look (Figure 2–1). In 1913 physicist Niels Bohr proposed the Bohr model of the atom. His model showed the atomic nucleus being orbited by its electrons at specific levels. The Bohr model has become the most popular model of the atom; however, today it is believed the atomic nucleus is surrounded by an electron "cloud," which is composed of electrons possessing different energy levels. This electron cloud model is a variation of the original Bohr model.

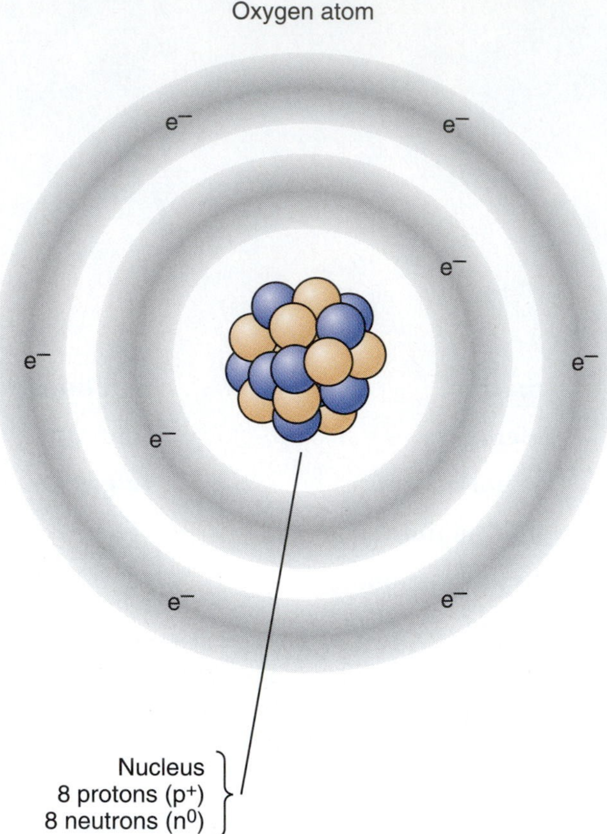

Figure 2–1 A model of an atom of oxygen, which shows the nucleus surrounded by an electron cloud.

The Elements

The subatomic particles that together form the atom provide matter with its unique properties. Protons are a subatomic particle that have a positive electric charge, whereas the electron possesses a negative electric charge. The third subatomic particle, the neutron, has no electric charge. Recent research has demonstrated that these subatomic particles are made up of even smaller basic particles called quarks, which form protons and neutrons when combined. Together the mass of the particular number of protons and neutrons in an atom's nucleus creates its atomic mass. Although the atomic mass is an important physical characteristic of an atom, atoms are classified by their atomic number. The atomic number is the number of protons in an atom's nucleus. The atomic number of an atom helps to classify the different atoms that exist, which are also known as **elements.** Together there are 109 different atoms or elements that have been identified based on their atomic number. Each element possesses its own unique chemical and physical properties. The element hydrogen has an atomic number of 1, which means that it has one proton. The 109 elements are arranged in an organized chart called the periodic table of elements, shown in Figure 2–2.

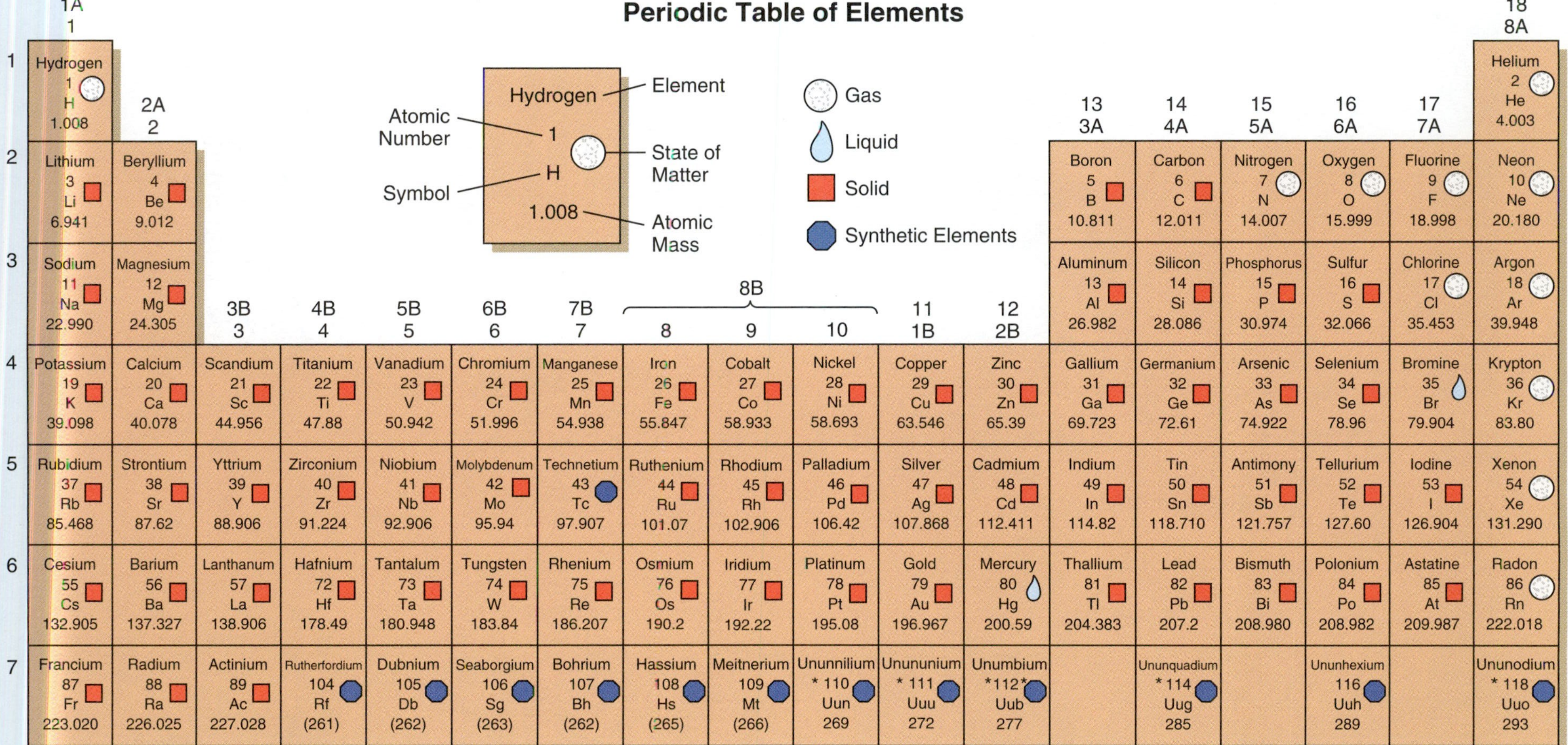

Figure 2–2 The periodic table of the elements.

TABLE 2–1 Common ions and their electric charges

Cations		
1+	*2+*	*3+*
Ammonium (NH_4^+)	Calcium (Ca^{2+})	Aluminum (Al^{3+})
Copper(I) (Cu^+)	Copper(II) (Cu^{2+})	Iron(III) (Fe^{3+})
Potassium (K^+)	Iron(II) (Fe^{2+})	
Silver (Ag^+)	Lead(II) (Pb^{2+})	
Sodium (Na^+)	Magnesium (Mg^{2+})	
	Mercury(I) (Hg_2^{2+})	
	Nickel (Ni^{2+})	
	Zinc (Zn^{2+})	
Anions		
1−	*2−*	*3−*
Chloride (Cl^-)	Carbonate (CO_3^{2-})	Phosphate (PO_4^{3-})
Cyanide (CN^-)	Oxide (O^{2-})	
Fluoride (F^-)	Peroxide (O_2^{2-})	
Hydrogen carbonate or bicarbonate (HCO_3^-)	Sulfate (SO_4^{2-})	
Hydrogen sulfide (HS^-)	Sulfide (S^{2-})	
Hydroxide (OH^-)		
Nitrate (NO_3^-)		

Elements usually have the same number of electrons and protons. For example, the element hydrogen, with an atomic number of 1, has one proton and one electron. Sometimes elements lose electrons, creating an imbalance in their electric charge. These atoms are called ions (Table 2–1). Ions with a positive electric charge are called cations. Ions with a negative electric charge are called anions. Common table salt (sodium chloride) is made up of the anion sodium (Na^+) and the cation chloride (Cl^-) (Figure 2–3). A variety of a specific element can exist when the number of neutrons differs in the nucleus of the same element; this is called an isotope. Many elements have different isotopes. Isotopes usually have the same chemical properties as atoms, but they possess a different atomic mass.

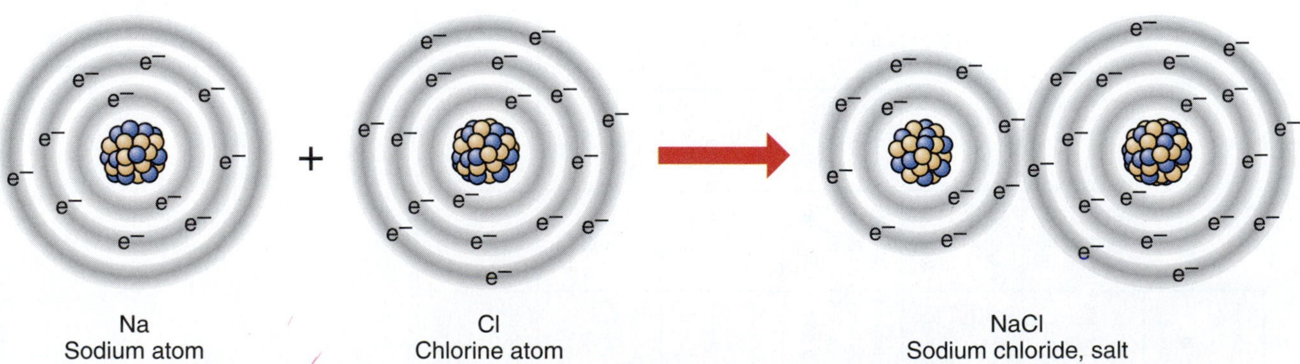

Figure 2–3 The sodium ion and chloride ion combine to form the compound known as table salt.

States of Matter

All matter in the universe exists in four distinct states that possess unique physical properties. These are known as the four states of matter: solids, liquids, gases, and plasma. The solid form of matter is the state in which atoms are most tightly packed together and are most restricted in their movement. Many solids are composed of atoms organized into a crystal pattern, which is an orderly, reoccurring arrangement of atoms (Figure 2–4). Atoms in a solid state possess the least amount of atomic movement (Figure 2–5). Because of the energy that is contained within an atom, they are always in constant motion. Atoms in a solid state of matter have movement that is limited to vibration because of their relative fixed position. Matter in the solid state also tends to have a definite shape.

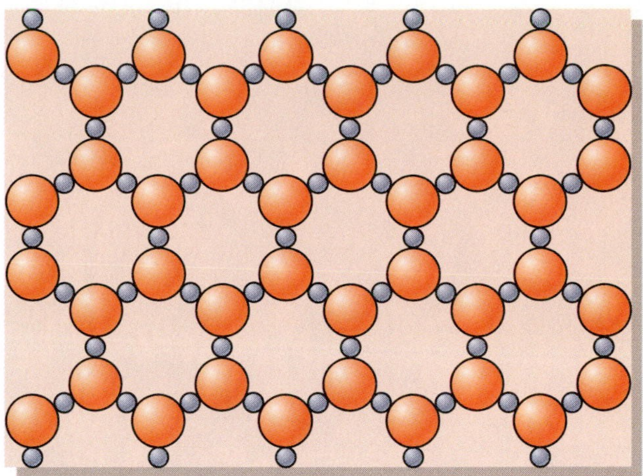

Figure 2–4 The orderly arrangement of molecules forming a crystal structure in a solid.

The next state of matter is liquid. Liquids are composed of atoms that are more loosely arranged. The atoms in a liquid have more freedom of movement, which gives liquid a

Figure 2–5 Examples of substances in the solid phase of matter.

Figure 2–6 Matter in the liquid state is fluid, which can be poured, and takes on the shape of its container. *(Courtesy of PhotoDisc.)*

more fluid property that has no definite shape (Figure 2–6). The arrangement of atoms in a liquid tend to be in long chains or clumps, rather than the orderly crystal patterns in a solid (Figure 2–7). The movement of atoms in a liquid is also less restrictive than in a solid. When a solid turns into a liquid, it is referred to as a change in state, or phase change. Phase changes usually depend on a certain temperature and pressure to which the matter is exposed. The temperature at which a solid changes into a liquid at a specific pressure is called its melting point. When a liquid changes into a solid it is called freezing.

The gaseous state of matter provides atoms with the highest degree of movement. Atoms in a gas move at high rates and undergo constant collisions with other atoms (Figure 2–8). The high degree of movement that the atoms experience in a gaseous form causes them to have no defined shape. When a gas is enclosed in a solid container, the number of collisions the freely moving molecules have with the walls of the container is known as pressure. The higher the number of collisions, the greater the pressure of the gas. When a liquid changes phase into a gas, it is called **vaporization** (Figure 2–9). The heat of vaporization is the specific temperature required to change a liquid into a gas, also known as its **boiling point.** The change from a gas to a liquid is called **condensation.** On some occasions a solid can change directly into a gas; this is called **sublimation.**

The fourth state of matter is called plasma, which is not as common as the other three. Plasma forms when the atoms that compose a gas become exposed to such high energy that they begin to ionize, or lose their electrons (Figure 2–10). The electrons are stripped

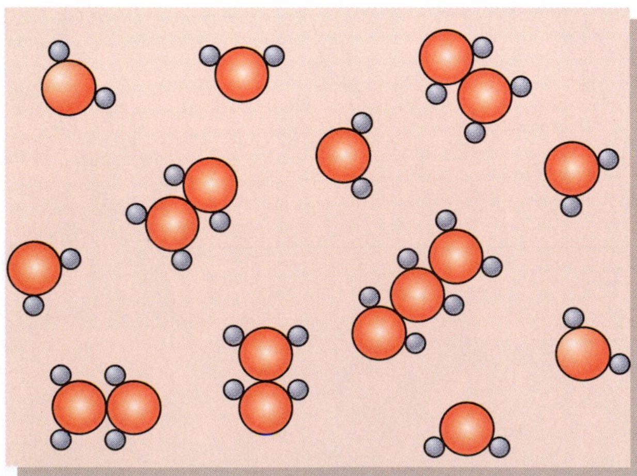

Figure 2–7 The chainlike configuration of molecules in a liquid.

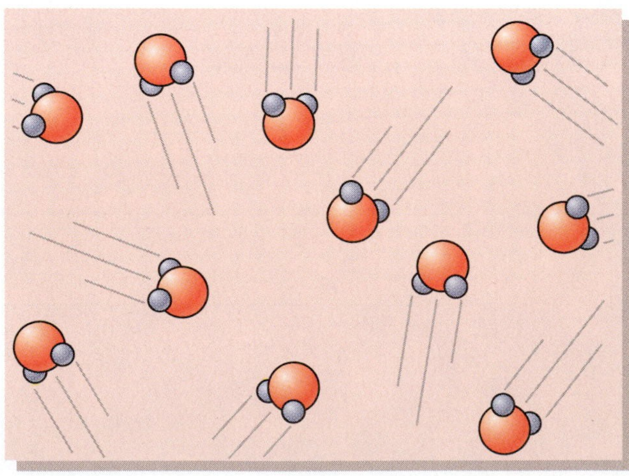

Figure 2–8 The rapidly moving molecules of a gas.

Figure 2–9 Steam rising from a volcano is the gaseous form of water. *(Courtesy of PhotoDisc.)*

Figure 2–10 The Sun is composed of plasma. *(Courtesy of PhotoDisc.)*

away from the atoms, creating a high-energy fluid gas mixture composed of ions and electrons. Stars are composed of plasma.

Compounds and Molecules

Although the basic building blocks of matter are the 109 different atoms that make up the Earth's elements, combinations of these atoms really form the physical aspects of the Earth. Combinations of atoms that are chemically joined are called compounds (Table 2–2). A compound is a substance composed of two or more elements chemically combined in specific proportions. Compounds are also known as molecules. The Earth is composed of more than 10 million different compounds that together form unique substances. One of

TABLE 2–2 Common compounds and the elements they are composed of

Common Name	Formula	Chemical Substance
Aspirin	$CH_3CO_2C_6H_4COOH$	Acetylsalicylic acid
Beet and cane sugars	$C_{12}H_{22}O_{11}$	Sucrose
Bleaching powder	$CaOCl_2$	Calcium oxychloride
Charcoal	$Ca_3(PO_4)_2 + C$	Calcium phosphate plus carbon
Clay	$H_2Al_2(SiO_4)_2 \cdot H_2O$	Hydrated ferric oxide
Common glass	$CaSiO_3 + Na_2SiO_3$	Calcium and other silicates
Diamond, graphite, fullerene (C_{60})	C	Carbon
Dry ice	CO_2	Frozen carbon dioxide
Fool's gold, pyrite	FeS_2	Iron disulfide
Grain alcohol	C_2H_5OH	Ethyl alcohol
Laughing gas	N_2O	Nitrous oxide
Limestone	$CaCO_3$	Calcium carbonate
Moth balls	$C_{10}H_8$	Naphthalene
Natural gas	CH_4	Impure methane
Nitroglycerin	$C_3H_5(NO_3)_3$	Glyceryl trinitrate
Quartz, agate, flint, chert	SiO_2	Silicon dioxide
Ruby, sapphire	Al_2O_3	Aluminum oxide
Rust	$(Fe_2O_3)_3 \cdot H_2O$	Hydrated ferric oxide
Soap lye	NaOH	Sodium hydroxide
Table and rock salts	NaCl	Sodium chloride

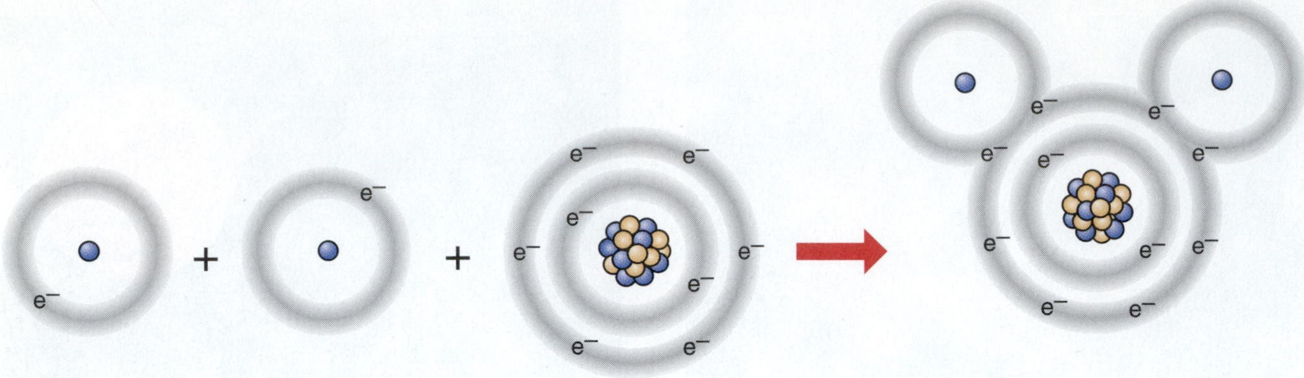

Figure 2–11 The formation of a water molecule by combining two atoms of hydrogen with one atom of oxygen.

Career Connections

CHEMISTS

Chemistry is one of the most important scientific careers in the modern world. This vast career path involves the study of materials and compounds and how they interact with one another. Chemists analyze existing chemical compounds and perform experiments to discover new chemicals and materials. Chemists also often work with other scientists and engineers when conducting research. Branches of chemistry include analytical chemistry, organic chemistry, physical chemistry, and biochemistry. Analytical chemists identify the composition, structure, and properties of chemical substances. Organic chemists study the chemistry of living things, especially carbon-containing compounds, along with the synthesis of new drugs and commercial materials. Physical chemists study the properties of atoms and molecules and how they interact with one another, and biochemists study the chemistry of the human body and other living organisms. Chemistry courses in high school provide a good opportunity to experience what a career in chemistry might include. There are many jobs available for chemists once they gain a 4-year college degree. A related occupation is the chemistry technician, who works with chemists, helping them to perform their daily tasks. This type of work requires a less formal education.

the most important molecules on Earth is the water molecule. Water is a compound that is composed of two atoms of hydrogen with one atom of oxygen (Figure 2–11). When two or more compounds are mixed together and still maintain their unique physical and chemical properties, it is called a mixture. Air is a mixture of many gases, including nitrogen, oxygen, and carbon dioxide. When two substances are mixed to form a uniform substance, it is called a solution. Seawater is a solution of salt and other minerals in water.

SECTION REVIEW

1. What is matter?
2. Describe the three subatomic particles that make up the atom.
3. What is the difference between the atomic mass and atomic number of an atom?
4. What are isotopes and ions?
5. Describe the four states of matter.
6. What is a phase change?
7. What is an element, and how many have been identified on Earth?
8. Describe the difference between an atom and a compound.
9. Who was John Dalton?

EARTH SYSTEM SCIENTISTS *JOHN DALTON*

John Dalton was born in England in 1766. He started his scientific career as a meteorologist, which began a lifelong dedication to weather observation in the area surrounding Manchester, England. One of his first scientific achievements was proposing that precipitation was formed by a change in atmospheric temperature, not by a change in air pressure. He also proposed the theory of the partial pressure of gases that today is known as Dalton's law of partial pressure. Later in his life he switched his scientific focus from meteorology and the study of the atmosphere to chemistry. In 1803 he developed his famous atomic theory, which stated that all matter was composed of small indivisible particles called atoms. He furthered hypothesized that atoms of a given element possess unique characteristics and weight and that three types of atoms exist: elements, compounds, and complex molecules. He published this theory in his book *New Systems of Chemical Philosophy* in 1808. Because of this theory, he is regarded by many to be the father of modern chemistry.

EARTH MATH

1) DETERMINE THE TOTAL ATOMIC MASS OF A MOLECULE OF WATER USING THE FOLLOWING INFORMATION: HYDROGEN HAS AN ATOMIC MASS OF 1.0079, AND OXYGEN HAS AN ATOMIC MASS OF 15.9994.

For more information go to this web link:
<http://www.stcms.si.edu/pom/pom_student.htm>

2.2 *Energy*

The relationship between matter and energy in the universe is the fundamental cause for the existence of the living and nonliving world. To fully understand the properties of matter, a full understanding of the properties of energy is equally important. In his world-famous equation $E = mc^2$, Albert Einstein revealed that energy and mass have a mutual relationship. But what exactly is energy, and how is it related to mass?

Energy

Energy is defined as the ability to do work or cause change. Matter that is exposed to energy is said to be in motion. Energy is classified in two basic forms: kinetic energy and potential energy. Kinetic energy is the energy of motion. When mass is in motion it is experiencing kinetic energy. The movement of the Earth around the Sun is kinetic energy. The movement of an athlete on a playing field is kinetic energy. The other type of energy is potential energy. Potential energy is stored energy. The gasoline in the tank of your car is a type of potential energy. It has the potential of powering your motor for a specific period. After riding a ski lift to the top of a mountain, you have a form of potential energy. The gravitational pull of the Earth will cause your

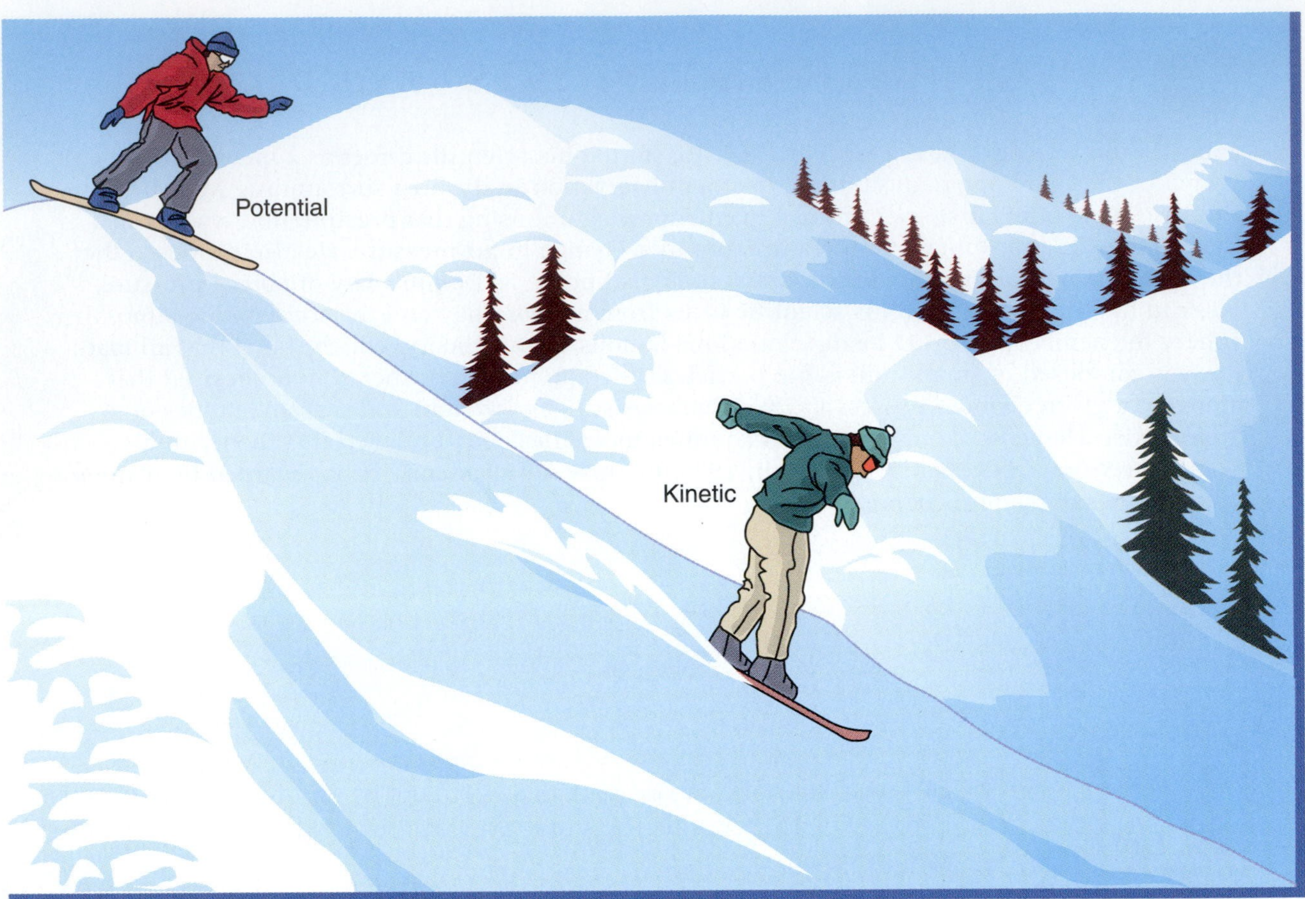

Figure 2–12 The snowboarder at the top of a mountain is an example of potential energy, whereas the snowboarder rapidly moving down the mountain illustrates kinetic energy.

body to move downhill on your snowboard (Figure 2–12). The sandwich in your lunch bag is also potential energy in the form of stored chemical energy—it will power your body for part of the school day.

The Law of the Conservation of Energy

Another important aspect of understanding energy is the law of the conservation of energy. This fundamental scientific law states that energy cannot be created or destroyed, but only changes form. What this law means is that energy can constantly change from one form to another, with no real gain or loss. The different forms in which energy can exist include mechanical, gravitational, radiant, thermal, electrical, magnetic, nuclear, and chemical. The light bulb that illuminates your classroom is an example of the law of the conservation of energy. The electricity that powers the light might have been generated by burning coal, which is a form of potential chemical energy. The burning coal produced heat energy that created steam, which turned a turbine generator by mechanical energy. The generator converted the mechanical energy into electrical energy that traveled to the light bulb through a network of wires. The light bulb converted the electrical energy into both radiant energy (light) and thermal energy (heat). So in the end no energy was really lost; it just changed forms.

Electromagnetic Radiation

Einstein's energy equivalency of mass equation ($E = mc^2$) means that all mass possesses energy. This type of energy is known as **electromagnetic radiation** (Figure 2–13). Electromagnetic radiation is the kinetic energy of movement or vibration given off by individual atoms or subatomic particles. All matter gives off electromagnetic radiation. Electromagnetic radiation travels in the form of a wave. Higher-energy electromagnetic radiation has smaller wavelengths than lower-energy electromagnetic waves. The **electromagnetic spectrum** indicates the energy and wavelength of the different forms of electromagnetic energy (Figure 2–14). The highest-energy electromagnetic waves are called gamma rays, which have a wavelength of approximately 1 billionth of a centimeter. The lowest-energy electromagnetic waves are radio waves, which have wavelengths that are as large as 10 kilometers. Visible light is also a form of electromagnetic radiation, which our eyes use to see. All electromagnetic waves travel at the speed of light, which is 186,000 miles per second.

The way in which electromagnetic radiation interacts with the environment plays an important role in many Earth processes. There are four basic ways that electromagnetic energy can interact with matter (Figure 2–15). Refraction causes electromagnetic waves to change direction or bend when they interact with matter. When a stick is placed halfway in water, it appears to be bent because of light being refracted by the water. Reflection is the bouncing of electromagnetic waves off a substance, like an image reflecting off a mirror. Scattering occurs when electromagnetic waves are refracted or reflected when they pass through a material. Some sunlight entering the atmosphere is scattered, which causes the sky to appear blue. Electromagnetic radiation can also be absorbed when matter takes in energy. The blacktop of a parking lot absorbs a great amount of solar energy, which makes it very hot in summer. Substances that are good at absorbing electromagnetic radiation also tend to be good at emitting electromagnetic radiation. Usually, materials that absorb high-energy, smaller-wavelength electromagnetic radiation reradiate it at longer, lower-energy wavelengths. The final way in which electromagnetic radiation interacts with matter is called transmission. Transmission is when electromagnetic radiation passes through matter without interacting with it, like light passing through a window.

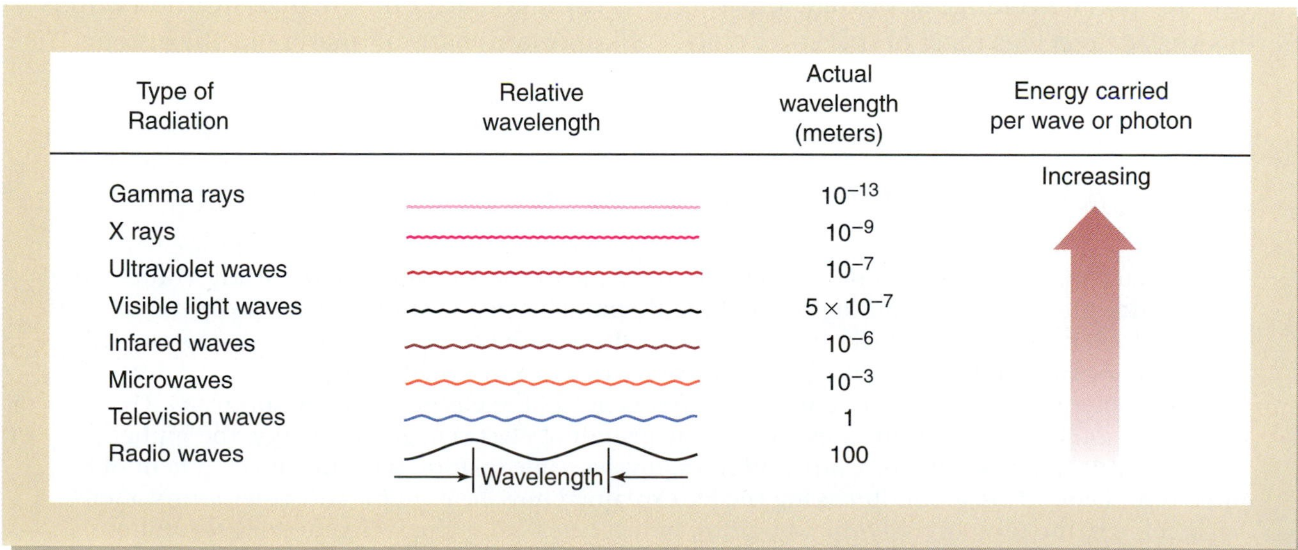

Type of Radiation
Relative wavelength
Actual wavelength (meters)
Energy carried per wave or photon
Gamma rays
X rays
Ultraviolet waves
Visible light waves
Infared waves
Microwaves
Television waves
Radio waves
10^{-13}
10^{-9}
10^{-7}
5×10^{-7}
10^{-6}
10^{-3}
1
100
Increasing
Wavelength

Figure 2–13 A chart showing the different types of radiation that make up the electromagnetic spectrum, their wavelengths, and energy levels.

Electromagnetic Spectrun

cm 0.000, 000, 000, 1 | 0.000, 000, 001 | 0.000, 000, 01 | 0.000, 000, 1 | 0.000, 001 | 0.000, 01 | 0.000, 1 | 0.001 | 0.01 | 0.1 | 1.0 | 10 | 100 | 1, 000

cm 10^{-10} 10^{-9} 10^{-8} 10^{-7} 10^{-6} 10^{-5} 10^{-4} 10^{-3} 10^{-2} 10^{-1} 10^{0} 10^{1} 10^{2} 10^{3}

Gamma rays
X rays
Ultraviolet
Visible
Infrared
Microwaves
Radio waves
Decreasing wavelength
Increasing wavelength

Violet | Blue | Green | Yellow | Orange | Red

4.0×10^{-5} 4.3×10^{-5} 4.9×10^{-5} 5.3×10^{-5} 5.8×10^{-5} 6.3×10^{-5} 7.0×10^{-5}

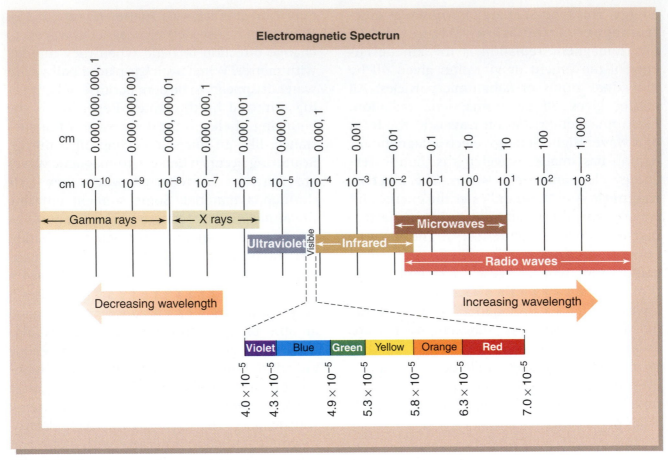

Figure 2–14 The electromagnetic spectrum.

Energy Transfer

All electromagnetic energy on Earth comes from the Sun. The Sun emits electromagnetic energy in all wavelengths. However, the Earth's surface receives most of the Sun's electromagnetic energy in the visible light range, with smaller amounts in the ultraviolet and infrared wavelengths. The interaction of energy and matter in, on, and around Earth is an important part of many Earth systems. The

EARTH SYSTEM SCIENTISTS *MAX PLANCK*

Max Planck was born in 1858 in Germany. Most of his life was devoted to the study of theoretical physics. He began his work by researching thermodynamics and the nature of electromagnetic radiation. His early contributions to physics dealt with the relationship of electromagnetic radiation to wavelengths and energy. He derived the Planck constant, which is used to determine the energy associated with a particular frequency of the electromagnetic spectrum. This breakthrough revealed that the shorter the wavelength of electromagnetic energy, the higher the energy level. His most famous contribution to physics came in 1900 with his development of quantum theory. This groundbreaking theory explained how heat, light, and other forms of energy come in the form of individual bundles, which he called quanta. His theories revolutionized modern physics. In his later life he helped apply his knowledge to support Einstein's famous theory of relativity.

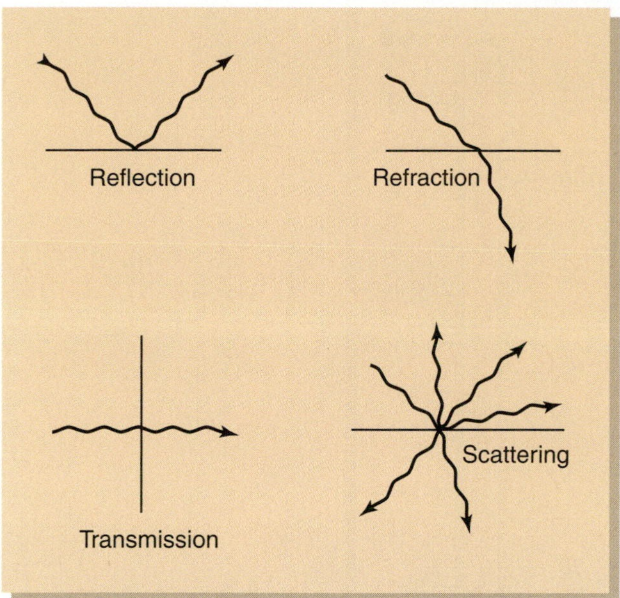

Figure 2–15 Four ways in which electromagnetic energy interacts with matter.

transfer of energy through the Earth takes place by three fundamental processes. **Conduction** is the transfer of kinetic energy by the direct contact of atoms and molecules. When you place a metal skillet on a fire, the heat of the fire is conducted through the metal and up the handle. Conduction is an important way that heat is transferred from the solid Earth to the atmosphere. Another way that heat is transferred throughout the Earth is **radiation.** Radiation is the movement of energy in empty space by electromagnetic waves. Feeling the heat of the Sun on your face on a hot summer day is sensing energy in the form of radiation. Radiation from the Sun is the main source of energy on Earth. The last method of energy transfer on Earth is **convection.** Convection is the circular movement of heat in a gas or liquid, which is caused by differences in temperature and density (Figure 2–16). This form of energy transfer causes hot air to rise up toward the ceiling of your house and cooler air to sink toward the floor. Convection of heat in the Earth's mantle and atmosphere is an important way that heat is distributed around the Earth.

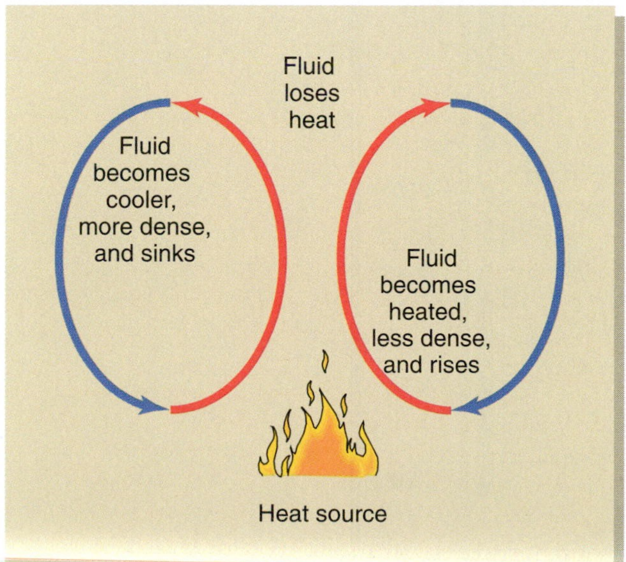

Figure 2–16 The formation of a convection cell in a fluid as result of differences in temperature and density.

Career Connections

PHYSICISTS

Physicists work to discover the basic principles of matter and energy and how they interact with one another. Their research leads to improved materials, better power generation, new electronics, and knowledge about the universe in which we live. There are many different branches of physics, including nuclear physics, elementary particle physics, molecular physics, acoustics, astrophysics, optics, biophysics, and fluid physics. Nuclear physicists study the processes that occur in nuclear reactions. Elementary particle physicists study the basic particles that compose matter and how they interact. Molecular physicists study the behavior of molecules and atoms. Acoustical physics includes the study of sound. Astrophysicists study the structure and movement of all objects in the universe. Optical physicists study the properties of light energy. Biophysicists research the physics associated with the human body and other living organisms, and fluid physics studies the properties of fluids. Physics requires a good knowledge of mathematics and requires at least a 4-year college degree. High school physics courses provide a good introduction to the type of work that physicists do.

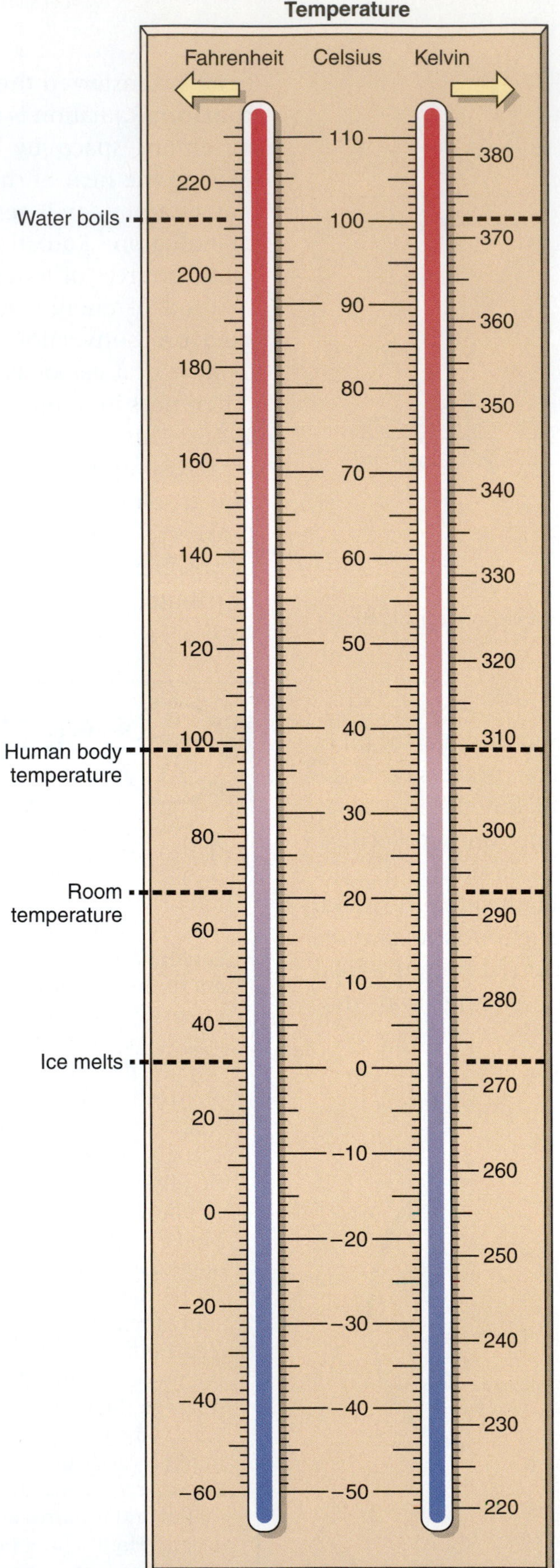

Figure 2–17 A comparison of the three scales used to measure temperature.

Heat and Temperature

Energy is measured on the Earth in the form of both heat and temperature. **Temperature** is the average kinetic energy of the molecules and atoms of a substance. Remember that all atoms are in motion; this motion can be in the form of vibrations, such as in a solid, or moving over long distances and rapidly colliding, such as what occurs in gases. The greater the amount of motion of the atoms in a substance, the higher the temperature.

Temperature on Earth is measured using three different scales (Figure 2–17). Degrees Fahrenheit and Celsius measure the various temperatures associated with the physical properties of the water molecule. The freezing point of water is 32° Fahrenheit (0° Celsius). This is the temperature at which liquid water becomes a solid. Water vaporizes at 212° Fahrenheit (100° Celsius). This is the temperature at which liquid water becomes water vapor, or the gaseous form of water. The third temperature scale is Kelvin, which measures the average kinetic energy of atoms or molecules. Zero Kelvin is the theoretical temperature at which all particle motion stops, causing the particles to emit no electromagnetic radiation. This temperature has never been reached on Earth. Zero Kelvin equals −459.67° Fahrenheit and −273.16° Celsius.

TABLE 2–3 The specific heat of common materials

Material		Specific Heat (calories/gram · C°)
Water	solid	0.5
	liquid	1.0
	gas	0.5
Dry air		0.24
Basalt		0.20
Granite		0.19
Iron		0.11
Copper		0.09
Lead		0.03

Heat is a measure of the flow of kinetic energy from one material to another, or the total amount of kinetic energy in a system. All heat flows from substances with high kinetic energy, or a source, to substances with low kinetic energy, called a sink. Therefore the energy flow in a system moves from a source to a sink. We sense this change in kinetic energy as heat. Heat is measured by using the calorie, which is a unit that denotes the amount of energy it takes to raise the temperature of 1 gram of water 1° Celsius. The specific heat of a substance is a measure of how much heat is required to raise 1 gram of the substance to 1° Celsius (Table 2–3). Specific heat is also called heat capacity. Liquid water has the highest specific heat of any natural substance on Earth. The specific heat of water is 1 calorie per gram. Other common substances on Earth have a lower heat capacity than water. The average heat capacity of rocks on Earth is about 0.2 calories per gram. Therefore water can hold much more heat energy than rock. This is why rocks heat and cool faster than water. The high heat capacity of water is an important property that helps the oceans regulate the climate of the Earth. Another important aspect of heat and its relationship to matter is that when matter changes state, heat energy is either absorbed or released. The potential energy that is released when a gas condenses into a liquid is called the **latent heat of condensation.** This means that heat is given off to the environment when a substance changes from a gas to a liquid. When 1 gram of water vapor condenses into liquid, 840 calories are released into the environment. The latent heat of freezing liquid water releases less energy into the environment, only 80 calories. When a solid changes phase into a liquid, or a liquid into a gas, energy is absorbed by a substance. When liquid water vaporizes into water vapor, it absorbs 840 calories. This is called the latent heat of vaporization. Latent heat is an important source of energy in the atmosphere, which is either absorbed or released when water vapor condenses or when liquid water vaporizes.

EARTH MATH

1) HOW MANY CALORIES ARE RELEASED INTO THE ATMOSPHERE WHEN 3 GRAMS OF WATER VAPOR CONDENSES TO FORM A CLOUD?

SECTION REVIEW

1. What is the definition of energy?
2. Describe the difference between kinetic energy and potential energy.
3. What is electromagnetic energy?
4. What is the relationship between wavelength and energy on the electromagnetic spectrum?
5. What is the source of all electromagnetic energy on Earth?
6. Which wavelengths of electromagnetic energy does the Earth's surface receive the most of?
7. List the five ways that electromagnetic energy can interact with the environment.
8. What is the difference between heat and temperature?
9. What are three ways by which heat can be transferred on Earth?
10. What is the specific heat of a substance?
11. Define the term *latent heat.*
12. Who was Max Planck?

For more information go to this web link:
<http://www.energy.gov/school/index.html>

CHAPTER SUMMARY

The universe is made up of two fundamental things: matter and energy. Matter is anything that occupies space and has mass. Matter is made up of individual particles called atoms. Atoms are composed of the subatomic particles called protons, neutrons, and electrons. Differing amounts of these subatomic particles make up unique atoms called elements. There are 109 different elements that have been identified on the Earth. Elements can be chemically combined to form complex substances called compounds. Energy is the ability to do work or cause change. There are two general types of energy: kinetic and potential. Kinetic energy is the energy of movement, and potential energy is stored energy. The law of the conservation of energy states that energy cannot be created or destroyed; it just changes forms. The different forms of energy include radiant energy, thermal energy, electrical energy, chemical energy, mechanical energy, nuclear energy, and magnetic energy. Electromagnetic energy, also called radiation, is a type of energy that travels in the form of a wave. The different types of electromagnetic energy are indicated by the electromagnetic spectrum, which classifies electromagnetic energy according to its wavelength. In general, short-wave radiation is higher in energy than long-wave radiation. The Sun emits all wavelengths of electromagnetic energy, but the surface of the Earth receives mostly the visible light form. Energy can be transferred in the environment by the processes of conduction, convection, or radiation. All these processes play an important role in the movement of energy through the Earth's systems. Energy is often measured in the form of heat or temperature. The temperature of a substance is a measure of the average kinetic energy of its atoms, which is commonly expressed as degrees Fahrenheit or Celsius.

CHAPTER REVIEW

Multiple Choice

1. Something that occupies space and has mass is called:
   a. energy
   b. weight
   c. matter
   d. volume

2. A particle that is composed of a nucleus surrounded by an electron is called:
   a. a proton
   b. an atom
   c. a neutron
   d. molecule

3. The state of matter that allows atoms to move freely is known as a:
   a. solid
   b. liquid
   c. gas
   d. plasma

4. Vaporization describes the phase change from a:
   a. solid to liquid
   b. liquid to gas
   c. gas to liquid
   d. liquid to solid

5. Condensation describes the phase change from a:
   a. solid to liquid
   b. liquid to gas
   c. gas to liquid
   d. liquid to solid

6. The process by which a solid changes phase into gas is called:
   a. boiling
   b. freezing
   c. melting
   d. sublimation

7. Two or more elements that are chemically combined create a:
   a. compound
   b. ion
   c. isotope
   d. atom

8. The energy that is associated with movement is called:
   a. radiant energy
   b. kinetic energy
   c. potential energy
   d. electromagnetic energy

9. The stored energy in gasoline, or in a sandwich, is classified as:
   a. radiant energy
   b. kinetic energy
   c. potential energy
   d. electromagnetic energy

10. High-energy electromagnetic radiation has:
    a. short wavelengths
    b. tall waves
    c. long wavelengths
    d. no waves

11. Hot air rising and cool air sinking in a room is an example of:
    a. conduction
    b. radiation
    c. convection
    d. reflection

12. Burning your hand on the handle of a pot on a stove is an example of heat transfer by:
    a. conduction
    b. radiation
    c. convection
    d. reflection

13. Which type of energy transfer travels in the form of a wave?
    a. conduction
    b. radiation
    c. convection
    d. reflection

14. The average kinetic energy of the atoms in a substance is known as:
    a. refraction
    b. heat
    c. latent heat
    d. temperature

15. Which type of energy is absorbed or released when a substance changes phase?
    a. heat
    b. latent heat
    c. temperature
    d. radiation

continued

Matching *Match the terms with the correct definitions.*

a. matter
b. elements
c. vaporization
d. boiling point
e. condensation
f. sublimation
g. energy
h. electromagnetic radiation
i. electromagnetic spectrum
j. conduction
k. radiation
l. convection
m. heat
n. temperature
o. latent heat

1. ____ The phase change from a solid to a gas.

2. ____ Something that occupies space and has mass.

3. ____ The ability to cause change or to perform work.

4. ____ The 109 identified atoms that have a definite number of protons, neutrons, and electrons.

5. ____ A type of energy that travels in the form of a wave and needs no medium for transfer.

6. ____ The phase change that occurs when a liquid changes into a gas.

7. ____ The range of specific wavelengths and frequencies that identify the specific forms of electromagnetic energy.

8. ____ The specific temperature at which a substance begins to change its phase from a liquid to a gas.

9. ____ The transfer of heat energy by direct molecular contact.

10. ____ The change in phase from a gas to a liquid.

11. ____ Waves or particles that are emitted from a substance.

12. ____ Heat energy that is either absorbed or released during a phase change.

13. ____ The transfer of heat energy in a fluid as a result of a change in density associated with a change in temperature.

14. ____ The average amount of kinetic energy of the atoms and molecules in substance.

15. ____ The measurable or perceived effect of energy that is transferred between two objects that have different temperatures.

Critical Thinking

1. Construct a flow chart that depicts all the energy transfers that occur from the Sun to the milk in your cereal.
2. Construct a list of all the elements you can identify that are found in common substances around your home or in school.

UNIT 2

The Exosphere

Topics to be presented in this unit include:

- The Earth as a Planet
- The Moon, Earth's Closest Neighbor
- The Life of Stars
- The Sun
- The Solar System
- The Earth's Place in the Universe

OVERVIEW

The **exosphere** is the area that is located outside of Earth's atmosphere, also known as outer space. Understanding the Earth's place in the exosphere and the specific regions of outer space is important to understanding the systems of the Earth. Many processes that occur on the Earth every day are the direct result of the processes that occur in the exosphere. Even in ancient times, thousands of years before humans left the Earth and began to explore space, people were fascinated with the world outside our own. The length of the day, the changing seasons, tides, and time are all influenced by the exosphere. This has in turn influenced every living thing on the Earth. Unlocking the secrets of outer space has helped us to learn more about our unique planet and how it functions.

CHAPTER

3 The Earth as a Planet

Objectives

The Earth's Composition • The Earth's Shape • The Size of the Earth • The Earth's Rotation • The Earth's Tilted Axis • The Earth's Orbit • Tides • The Apparent Motion of Celestial Objects • The Earth's Coordinate System: Latitude and Longitude • Topographical Maps

After reading this chapter you should be able to:

- Identify the four basic parts of the Earth's interior.
- Describe the true shape of the Earth, and explain three observations that prove its shape.
- Differentiate between the Earth's rotation and revolution.
- Describe the period needed to complete one rotation of the Earth and one revolution of the Earth.
- Define the terms *aphelion* and *perihelion*.
- Describe the position of the Earth's axis relative to the Sun during the equinoxes, summer solstice, and winter solstice.
- Explain how tides are formed on the Earth.
- Explain the concept of the apparent motion of celestial objects.
- Describe how azimuth and altitude are used to locate celestial objects.
- Explain how lines of latitude and longitude are used to locate an object on the Earth's surface.
- Describe how you can locate your latitude in the Northern Hemisphere.
- Explain what information you need to determine your longitude location on the Earth.
- Describe the type of information that can be found on a topographical map.
- Define the terms *contour line* and *contour interval*.

TERMS TO KNOW

terrestrial planet
exosphere
axis
rotation
revolution
ellipse
perihelion
aphelion
spring tides
neap tides
zenith
equator
latitude
longitude
contour lines
contour interval

INTRODUCTION

It is hard to imagine that we are all standing on a huge ball that is spinning like a top at more than 1000 miles per hour near the equator and that is also being propelled through space at speeds of approximately 66,000 miles per hour! The Earth's place in the solar system and its interactions with the Sun affect many of the Earth's systems. The length of the day, changing seasons of the year, and rise and fall of the tides are all examples of these complex interactions. It has taken human beings hundreds of years of careful observations to fully understand our planet's place in the solar system. As a result, human knowledge now rises out of the atmosphere and into the universe, revealing much about the planet we call home.

The Earth's Composition

The Earth is a **terrestrial planet.** A planet is any large body that orbits around a star. The term *terrestrial* means that it is composed mainly of rock and metal. Because the Earth is composed of rock, it has a relatively high density, approximately 5.5 grams per cubic centimeter. The interior of the Earth is made up of a dense combination of molten iron and nickel, which is surrounded by a plastic-like, less dense rock called the mantle (Figure 3–1). The movements of the liquid interior of the Earth generate a strong magnetic field that surrounds the planet. This causes the Earth to act much like a large magnet, with the poles of the magnet located near the poles of the Earth. This magnetic field stretches out through the atmosphere and acts as a protective barrier to deadly, high-energy solar radiation. This portion of the Earth is called the magnetosphere. The Earth's magnetic field allows us to use compasses to locate direction on the planet. A small magnetic needle that is allowed to spin freely aligns itself with the Earth's magnetic field and points to the magnetic North Pole. The outer portion of the Earth, also called the crust, is made up of two basic types of solid rock: granites and basalts. Granite rocks make up most of the Earth's continents, and basalts compose most of the ocean floor. Although the Earth is classified as a terrestrial planet, it also is surrounded by a thin envelope of gas called the atmosphere. Approximately 70% of the planet is covered in oceans, which have an average depth of 2 miles.

The Earth's Shape

The shape of the Earth is nearly spherical, which means it resembles a round ball. Long

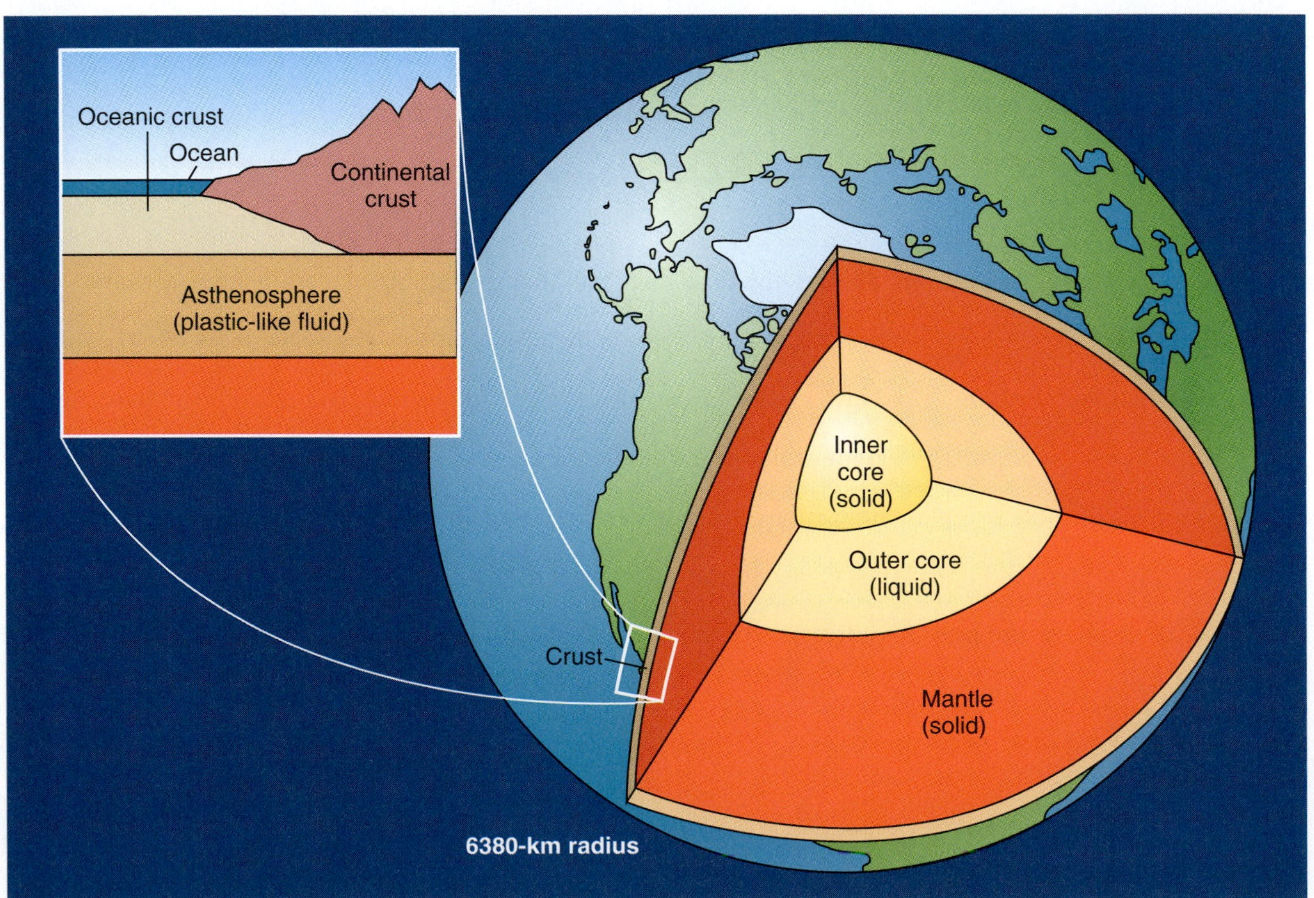

Figure 3–1 This cross section of the Earth reveals its internal structure, which is made up of the inner and outer core, mantle, and crust.

before satellites or spacecraft were able to photograph the Earth from space and reveal its spherical form, humans theorized about its shape. Three observations made hundreds of years ago hinted at the Earth's true shape. The first was the disappearance of ships as they moved over the horizon. This simple observation could only be explained by the ships' traveling over a curved surface. Ancient astronomers also observed a curved shadow move across the moon during a lunar eclipse. Another observation included the altitude of the North Star in the night sky. The North Star, also called Polaris, is a bright star that lies directly over the North Pole. Early astronomers noticed that the altitude of Polaris—that is, the height of the star above the horizon—changed as they moved farther south or north. As an observer traveled north, toward the North Pole, the altitude of Polaris increased until it was exactly 90 degrees above the horizon directly at the North Pole. Then, as an observer traveled south toward the equator, the altitude of Polaris decreased. Eventually, as the observer moves past the equator and into the Southern Hemisphere, the North Star dips below the horizon and can no longer be seen. The only explanation for this apparent change in altitude is the Earth's shape being round. The actual shape of the Earth, however, is not perfectly round, but is referred to as an oblate spheroid. The term *oblate* means that it is slightly flattened. The slight flattening of the Earth near the equator is caused by its rapid rotation. Because the Earth is slightly oblate, its diameter is slightly larger around the equator than around the poles. The equatorial diameter of the Earth is approximately 7928 miles, whereas the diameter of the Earth between the North and South Poles is 7900 miles. As a result of this slight bulging at the equator, if you were to fly around the world at the equator, you would have to travel 24,900 miles. This is 43 miles more than if you flew around the Earth starting at the North Pole. Because of this relatively small bulge at the equator, it can be assumed that the Earth is nearly a perfect sphere when viewed from space (Figure 3–2).

Figure 3-2 The Earth viewed from space is nearly a perfect sphere. *(Courtesy of PhotoDisc.)*

The Size of the Earth

The total surface area of the Earth is approximately 197 million square miles, of which more than 70% is covered by ocean. This is why the planet Earth appears blue when seen from space. Long before precise instruments were able to measure the size of the Earth, a Greek scientist by the name of Eratosthenes, who lived more than 2000 years ago, accurately measured the circumference of the Earth. He observed that the Sun shone directly down at the bottom of a well in the city of Syene, Egypt. He then measured the length of a shadow cast by an obelisk at the same time in the city of Alexandria. Finally, Eratosthenes used his knowledge of geometry and the distance between Syene and Alexandria to calculate the circumference of the Earth (Figure 3–3). His calculations revealed the circumference of Earth to be approximately 25,054 miles, which is only 175 miles off from the Earth's actual average circumference.

Sun ray striking top of wall in Alexandria
Sun ray entering well in Syene
7.2 angle measurement
wall
Alexandria
Syene
well

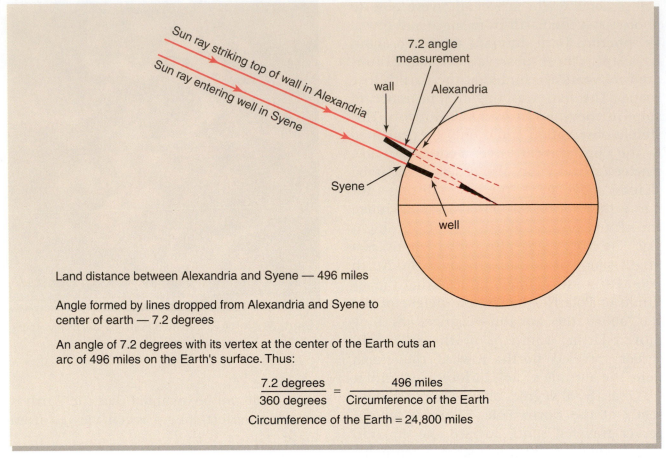

Land distance between Alexandria and Syene — 496 miles

Angle formed by lines dropped from Alexandria and Syene to center of earth — 7.2 degrees

An angle of 7.2 degrees with its vertex at the center of the Earth cuts an arc of 496 miles on the Earth's surface. Thus:

$$\frac{7.2 \text{ degrees}}{360 \text{ degrees}} = \frac{496 \text{ miles}}{\text{Circumference of the Earth}}$$

Circumference of the Earth = 24,800 miles

Figure 3–3 Eratosthenes' method for determining the circumference of the Earth, which he used more than 2000 years ago.

The Earth's Rotation

The Earth, like all planets that orbit the Sun, rotates on its axis. The **axis** of the Earth is an imaginary line through the center of the Earth that connects the North and South Poles. Early observations of the movement of stars, planets, and the Sun around the Earth caused ancient astronomers to believe that either the heavens were spinning around the Earth or the Earth was spinning on its axis.

EARTH SYSTEM SCIENTISTS *ERATOSTHENES*

Eratosthenes was a Greek scientist and philosopher who lived during the third century B.C. He also served as the librarian in the greatest library of the ancient world, located in Alexandria, Egypt. Eratosthenes accomplished many things during his 80-year life. He was the first person to accurately measure the circumference of the Earth, and he calculated the distance of the Sun from the Earth. He studied the importance of prime numbers in mathematics. Eratosthenes created star charts that cataloged more than 500 stars, and he developed a calendar that included leap years. He created accurate maps and charts of the Nile River and was the first to suggest that heavy rains in the mountains and lakes above the river were the cause of floods downriver. He also wrote an epic poem titled *Hermes*, which explains the fundamentals of the science of astronomy at the time.

It was not until 1851 that a French scientist by the name of Jean Foucault proved that the Earth was spinning by performing an original experiment. He suspended a pendulum from the ceiling and let it swing back and forth across the room. The pendulum would continue to swing in the same direction if the Earth were not spinning. Much to the onlookers' amazement, the pendulum slowly changed direction as Foucault had predicted. The pendulum altered its course because the Earth was rotating underneath it. Today copies of Foucault's pendulum swing to and fro in many science museums across the world. The changing direction of the pendulums knock over objects to prove the Earth is indeed spinning on its axis.

The speed at which the Earth is rotating depends on how far north or south of the equator you are. The greatest velocity is at the equator, and the least is located near the North and South Poles. This is due to the fact that the Earth is spherical. For example, if a line of roller skaters were skating around in a circle, the skater on the outside of the circle would have to skate the fastest to keep up with the other skaters in the line. The skater near the center of the circle would skate the slowest because he or she would be closest to the axis of rotation. The equator is the farthest point on the Earth from the axis of rotation and acts like the skater on the outside of the circle moving at the greatest speed. The North and South Poles are closest to the axis of rotation and act like the skaters near the center of the circle moving the slowest. If you are standing on the equator, the Earth's rotational speed is approximately 1037 mile per hour. At this speed the Earth makes one full **rotation,** or one rotational period, on its axis every 23 hours, 56 minutes, and 4.1 seconds. We round off this period to 24 hours, or 1 day. The rotation of the Earth causes half the world to be exposed to sunlight while the other half is bathed in darkness. This is why we experience day and night on Earth. This change in sunlight exposure causes the Earth to receive unequal heating, which helps define the Earth's weather and climate regions.

The Earth's Tilted Axis

The axis of the Earth is tilted approximately 23.5 degrees toward the plane of the ecliptic. The plane of the ecliptic is an imaginary line drawn from the center of the Earth to the center of the Sun. The tilt of the Earth's axis causes parts of the Earth to receive more sunlight than other parts of the Earth during the year. This causes the Earth's four seasons. When the Earth's axis is pointing towards the Sun, the Northern Hemisphere experiences the summer season, during which the day length increases (Figure 3–4). The longest days during the summer are between 14 and 15 hours long. The longest day falls around June 22, which is called the summer solstice. The North Pole experiences 24 hours of light for a time during the summer season. When the equator on Earth lines up with the plane of the ecliptic, the planet experiences the fall and spring seasons. These are also known as the equinoxes, which means that the length of day and night are equal, or 12 hours in length. The autumnal equinox (fall) occurs around September 23, and the vernal equinox (spring) occurs around March 21. When the Southern Hemisphere is tilted toward the Sun, the winter season has arrived in the Northern Hemisphere. During this time of year, the Northern Hemisphere experiences short days that are approximately 9 to 10 hours long. The shortest day is called the winter solstice, which occurs around December 21. The cycle of seasons repeats itself every year, which is the time it takes for the Earth to make one complete revolution around the Sun.

The Earth's Orbit

As the Earth spins on its axis, it is also traveling around the Sun. One complete movement of the Earth around the Sun is known as one **revolution.** The Earth revolves around the Sun at a speed of approximately 66,000 miles per hour. This speed varies as a result of the Earth's orbit around the Sun. One complete revolution of the Earth around the Sun takes approximately 365.25 days. An orbit is the

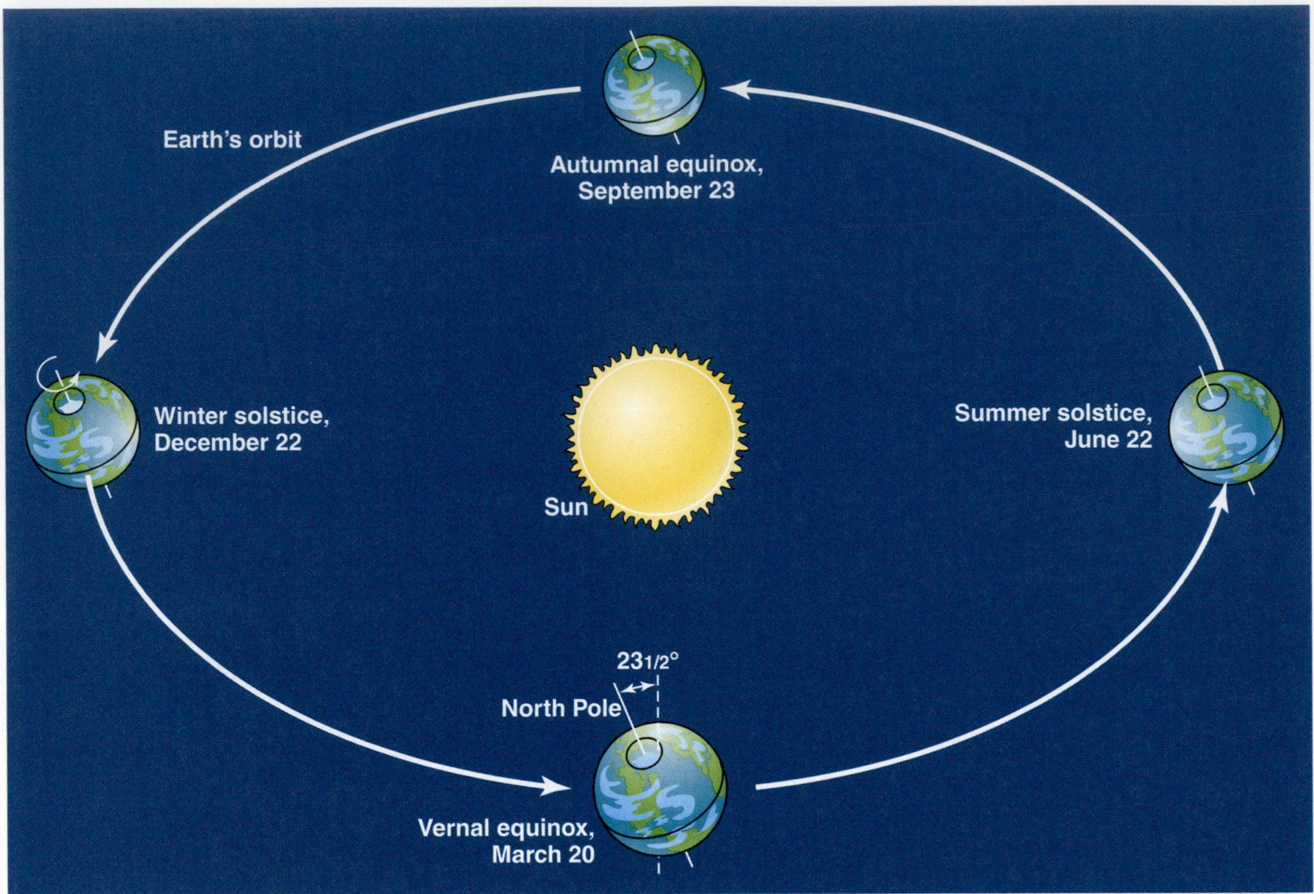

Figure 3–4 The Earth's tilted axis as it revolves around the Sun, causing the seasons.

specific path that an object takes around another object. The orbit of the Earth around the Sun is not a perfect circle, but is called an ellipse (Figure 3–5). An **ellipse** is a flattened circle. Although the Earth's orbit is elliptical, it is very nearly a perfect circle. The elliptical orbit of the Earth is caused by the gravitational attraction between the Earth and the Sun. Gravity is an attractive force that is created between two objects. The force of gravity depends on the mass of the two objects and their distance apart from each other. Sir Isaac Newton discovered the universal law of gravitation, which describes the properties of gravity. The gravitational force between the Sun and the Earth causes the Earth to be closer to the Sun during one portion of its orbit and farther from the Sun during another portion of its orbit. The point of the Earth's orbit when it is closest to the Sun is called the **perihelion.** During the perihelion, the Earth is approximately 91,349,000 miles from the Sun; this occurs sometime around the first of January every year. This is also the point in the Earth's orbit when it is traveling at its greatest speed. The **aphelion** is the point in the Earth's orbit when it is farthest from the Sun. This occurs during the first week of July every year, when the Earth is approximately 94,454,000 miles from the Sun, and marks the point in its orbit when is traveling the slowest.

Tides

The influence of gravity on the Earth, the Sun, and the Moon creates a noticeable effect on the Earth's oceans; this effect is called the tide. Tides are bulges that are created on the ocean's surface by the gravitational attraction of the Moon and Sun. These bulges cause a rise in sea level near the coastlines of land.

Because the Moon is closer to the Earth than the Sun, it has a greater influence on the formation of tides than the Sun. As the Moon moves around the Earth, the tidal bulge of ocean water follows it along, creating distinct times when high and low tides occur. The local geography of the coastlines of land masses also influences tides. This causes some places on Earth to experience one low and one high tide each day. These are called diurnal tides, and are common in the Caribbean. Semi-diurnal tides occur when there are two high tides and two low tides each day. This is also the result of the local geography; semi-diurnal tides are common along much of the coast of the Northeastern United States. During the times of the year when the Sun and the Moon are on the same side or on the opposite sides of the Earth, **spring tides** are created. A spring tide creates the highest and lowest tides of the year. The difference between the height of the high tide and the height of the low tide is called the tidal range. When the Sun and Moon are at right angles to one another relative to the Earth, a **neap tide** occurs (Figure 3–6). The gravitational attractions of the Sun and Moon cancel out each other and create the lowest tidal range of the year. The place on the Earth that experiences the greatest tidal range is the Bay of Fundy in Nova Scotia. The tidal range there is more than 47 feet each day! Tides have a great influence on the shipping industry and the currents associated with waterways. Many tidal currents can be extremely strong and very dangerous to swimmers or boaters.

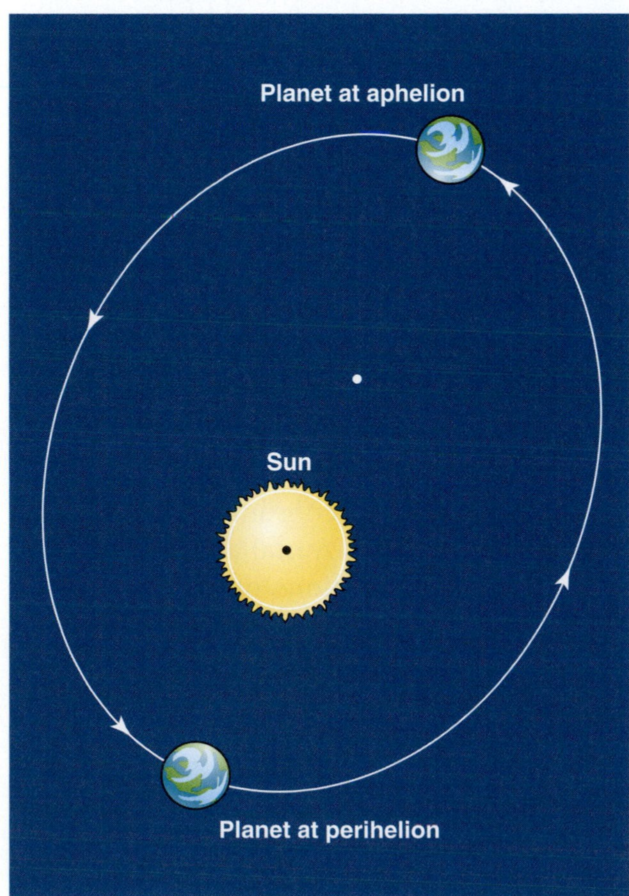

Figure 3–5 An exaggerated view of the Earth's elliptical orbit around the Sun, showing its aphelion and perihelion locations.

Apparent Motion of Celestial Objects

The movement of the Earth both on its axis and around the Sun creates a unique view of outer space. Because the Earth is spinning, objects in the night-time sky appear to move in regular motions. These objects are called celestial objects, and they include planets, moons, stars, comets, asteroids, and any other object located outside the Earth's atmosphere. The motion of celestial objects is called apparent motion. This apparent motion travels from east to west across the sky. The speed at which apparent motion travels is measured in degrees, with the sky representing 180 degrees from horizon to horizon. The apparent motion of celestial objects is approximately 15 degrees of sky per hour. Celestial objects appear to move at this rate because it takes 24 hours for the Earth to make one rotation. If you divide the 360 degrees that makes a complete circle around the Earth by the 24 hours it takes to make one complete rotation, you come up with 15 degrees of movement per hour. Celestial objects are not really moving; it is the spinning of the Earth that makes them appear to move. This is why it is called apparent motion. Astronomers have created a coordinate

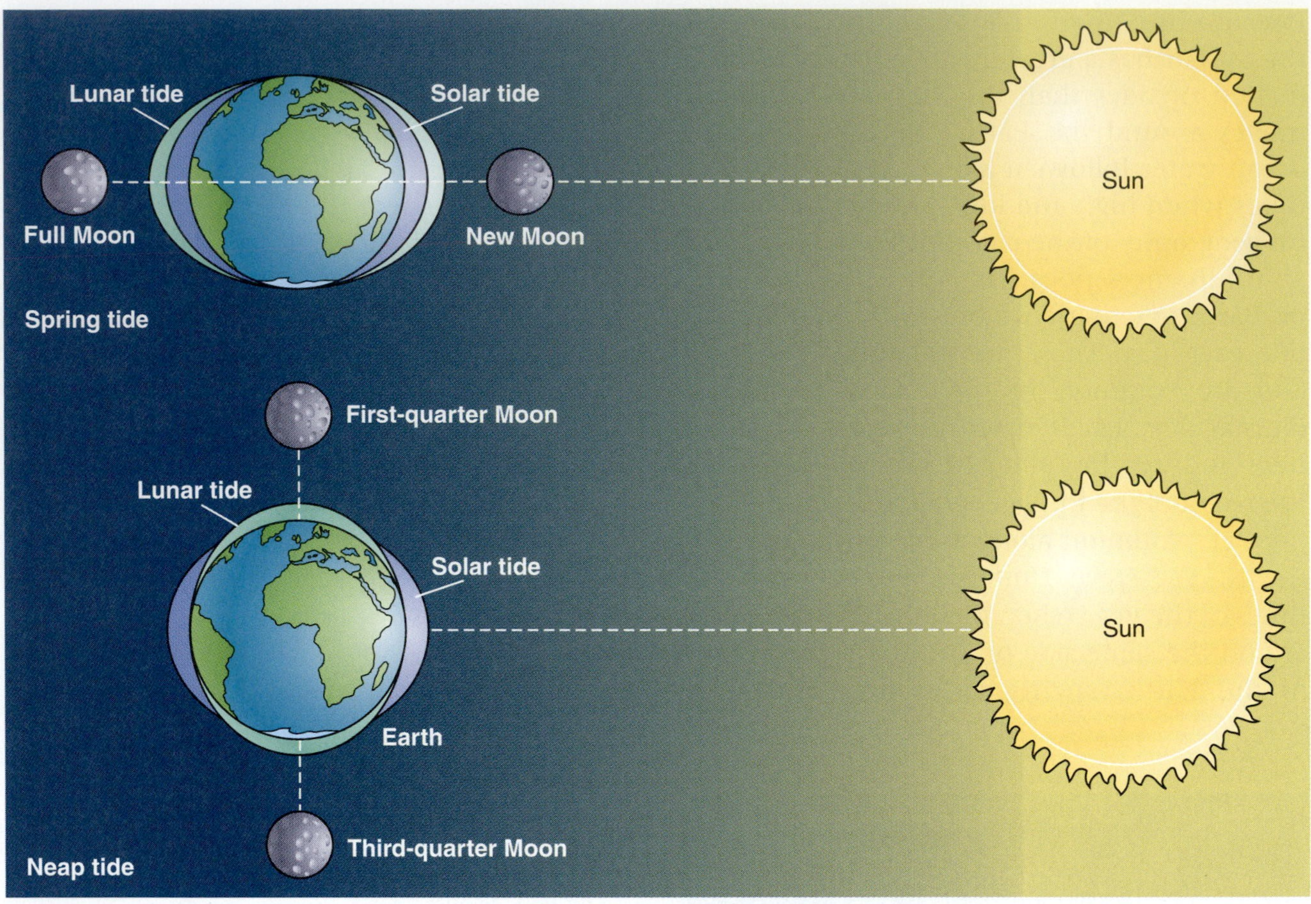

Figure 3–6 The effects of the Sun and Moon on the formation of spring and neap tides on the Earth.

system, which helps them to map celestial objects and also track their movement. The horizon can be divided into 360 degrees to make a circle that measures the horizontal location of a celestial object with respect to the Earth. The degrees on the horizon that marks a specific location is called the azimuth. The altitude of a celestial object is the angle in degrees above the horizon of a celestial object between the horizon and the zenith. The **zenith** is the point in the sky that is directly overhead, or 90 degrees from the horizon. Using these two coordinates, the azimuth and the altitude, an observer on the Earth can mark the location of a celestial object (Figure 3–7). Astronomers also use a more exact form of coordinates called declination and right ascension. This system is more complex and works much like lines of latitude and longitude, which is the coordinate system used on the surface of the Earth.

The Earth's Coordinate System: Latitude and Longitude

A coordinate system is a method of locating an exact location on a two-dimensional surface. Although the Earth's surface is not truly two dimensional, a coordinate system that uses intersecting horizontal and vertical lines is used to mark locations on the Earth. The Earth's coordinate system uses an imaginary horizontal line that divides the Earth in half, called the **equator.** This divides the planet into two hemispheres. The region north of the equator all the way to the North Pole is called the Northern Hemisphere. The region south of the equator all the way to the South Pole is called the Southern Hemisphere. Each hemisphere is further divided by horizontal lines that mark the location north or south of the equator. These parallel lines are called lines of **latitude** and are marked in degrees. The equator represents 0 degrees latitude, the North Pole is 90 degrees north latitude, and the South Pole represents 90 degrees south latitude. Each degree of latitude is further divided into 60 minutes, and 1 minute of latitude equals 60 seconds. This enables you to precisely measure your location on Earth either north or south of the equator (Figure 3–8). You can determine your latitude in the Northern Hemisphere by using the North Star, also called Polaris. The North Star is a bright star that is located directly above the Earth's axis at the North Pole. At the North Pole, Polaris is exactly 90 degrees above the horizon. As you begin to move south toward the equator, the altitude of Polaris in the nighttime sky begins to decrease. Therefore the angle at which the North Star is above the horizon equals your latitude location. For example, if you measured the altitude of the North Star above the horizon to be 45 degrees, your latitude location is 45 degrees north of the equator. Because the North Star is above the North Pole, it can only be used to determine your latitude north of the equator. Once you reach the equator, the North Star is no longer visible because it dips below the horizon. In the Southern Hemisphere, you can use the brightest star located in the constellation called the Southern Cross to estimate your latitude. The Southern Cross is located almost directly over the South Pole and acts like the Northern Hemisphere's Polaris.

Lines of latitude are only useful for determining an exact location either north or south of the equator. Another set of lines that run perpendicular to latitude lines are needed to

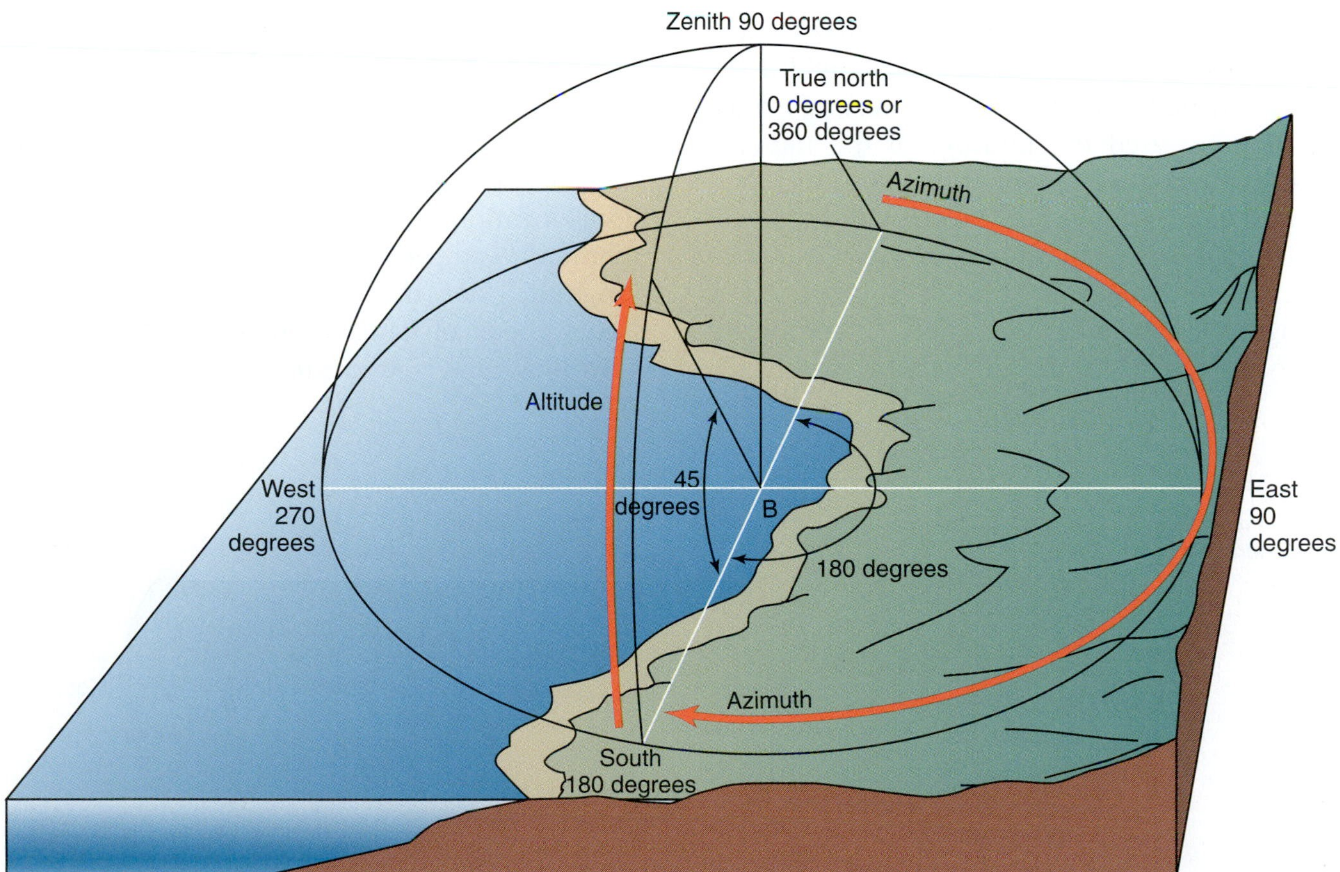

Figure 3–7 The location of the zenith and the horizon on the Earth and the use of altitude and azimuth to locate celestial objects.

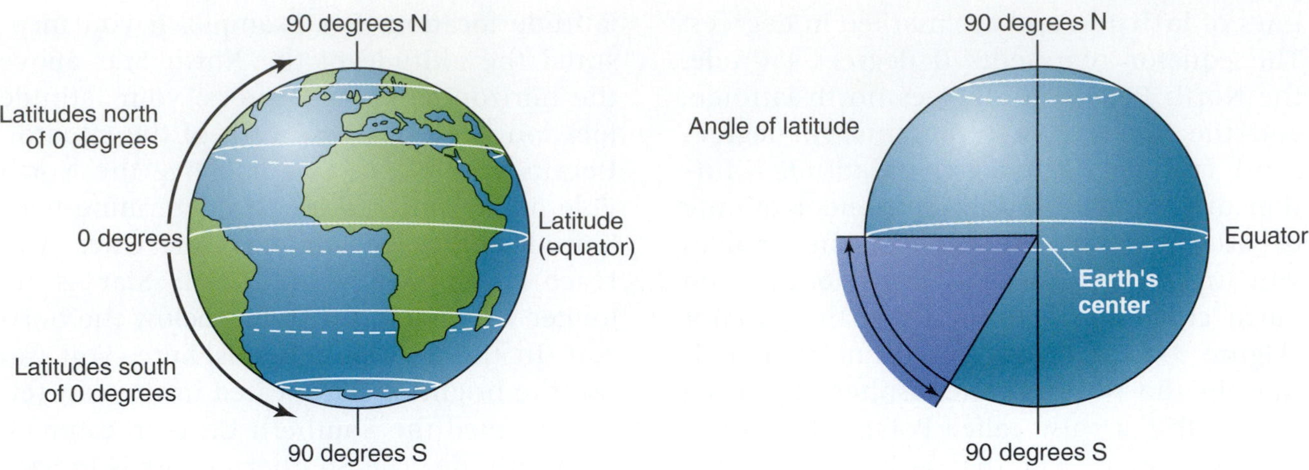

Figure 3–8 Lines of latitude on the Earth mark positions north and south of the equator.

mark a precise location on Earth. These lines are called lines of **longitude,** which are vertical lines that run from the North Pole to the South Pole. Lines of longitude also use degrees, minutes, and seconds as units of measurement. Unlike lines of latitude, which use the exact halfway point between the North and South Poles to represent 0 degrees, there is no natural halfway point that divides the Earth in half vertically. Therefore English astronomers used the location of the national observatory at Greenwich, England, to represent 0 degrees longitude. This imaginary line is called the prime meridian, or the Greenwich meridian. From this starting point, lines of longitude, also called meridians, mark a location on the Earth either west or east of the prime meridian (Figure 3–9). There are 180 degrees of longitude west of the prime meridian, and 180 degrees east of the prime meridian. The 180th meridian is also called the international date line. To determine a location in longitude, you must know the time difference between your location and the time at the prime meridian, also called Greenwich mean time. This is because the Earth is spinning on its axis from west to east. Every 15 degrees of longitude represents 1 hour of time; therefore the time difference between your location and the time at the prime meridian can be used to determine your longitude. For example, if your time is 3 hours ahead of Greenwich mean time, then your longitude is 45 degrees east of the prime

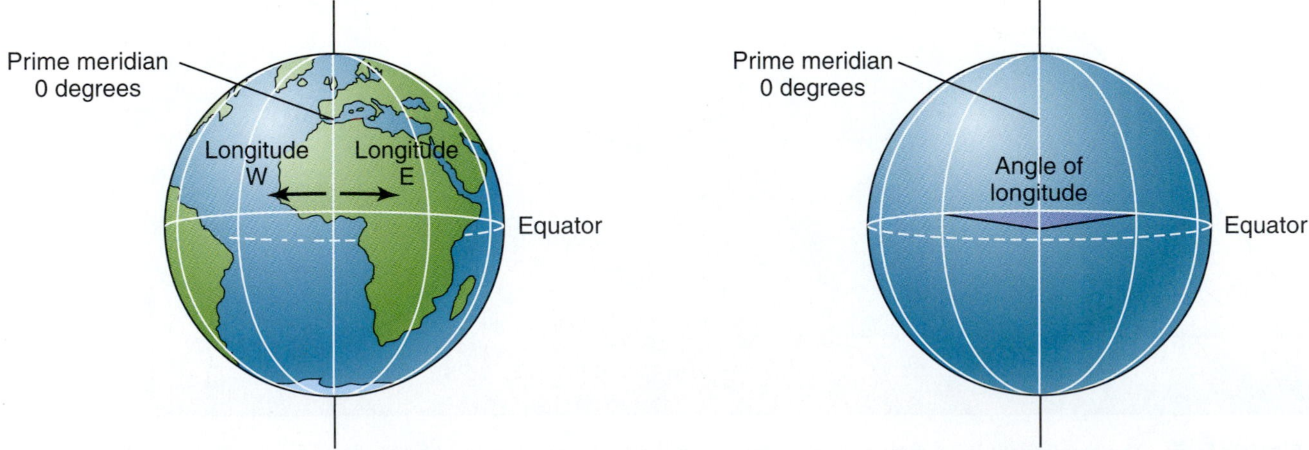

Figure 3–9 Lines of longitude on the Earth mark positions east and west of the prime meridian.

Career Connections

CARTOGRAPHER

Cartographers use accurate measurements of the Earth to produce precision maps of its surface. They gather many different types of information to make accurate maps. Geographers collect data on specific latitude and longitude locations of specific places on the Earth. They also collect geographical data and political information about a specific region. Cartographers use information about population, transportation routes, and local vegetation to create high-quality maps. Today, many cartographers use computers to help them in the creation of maps. Cartographers must be good at performing precision work, with attention to detail. Almost all cartography-related careers require a 4-year college education.

meridian. Because longitude requires an accurate knowledge of time, it wasn't possible to calculate longitude at sea until an accurate portable clock that could be used on board a ship was invented. This was achieved in 1728 by English clockmaker John Harrison. By using the latitude and longitude coordinate system, it is possible to exactly locate any point on the Earth.

Topographical Maps

Maps that are used to represent the three-dimensional surface of the Earth are called topographical maps; they show the true shape of the Earth's surface. Topographical maps represent changes in elevation on the Earth's surface by using contour lines (Figure 3–10). **Contour lines** are drawn on a map to represent a specific elevation of the land surface

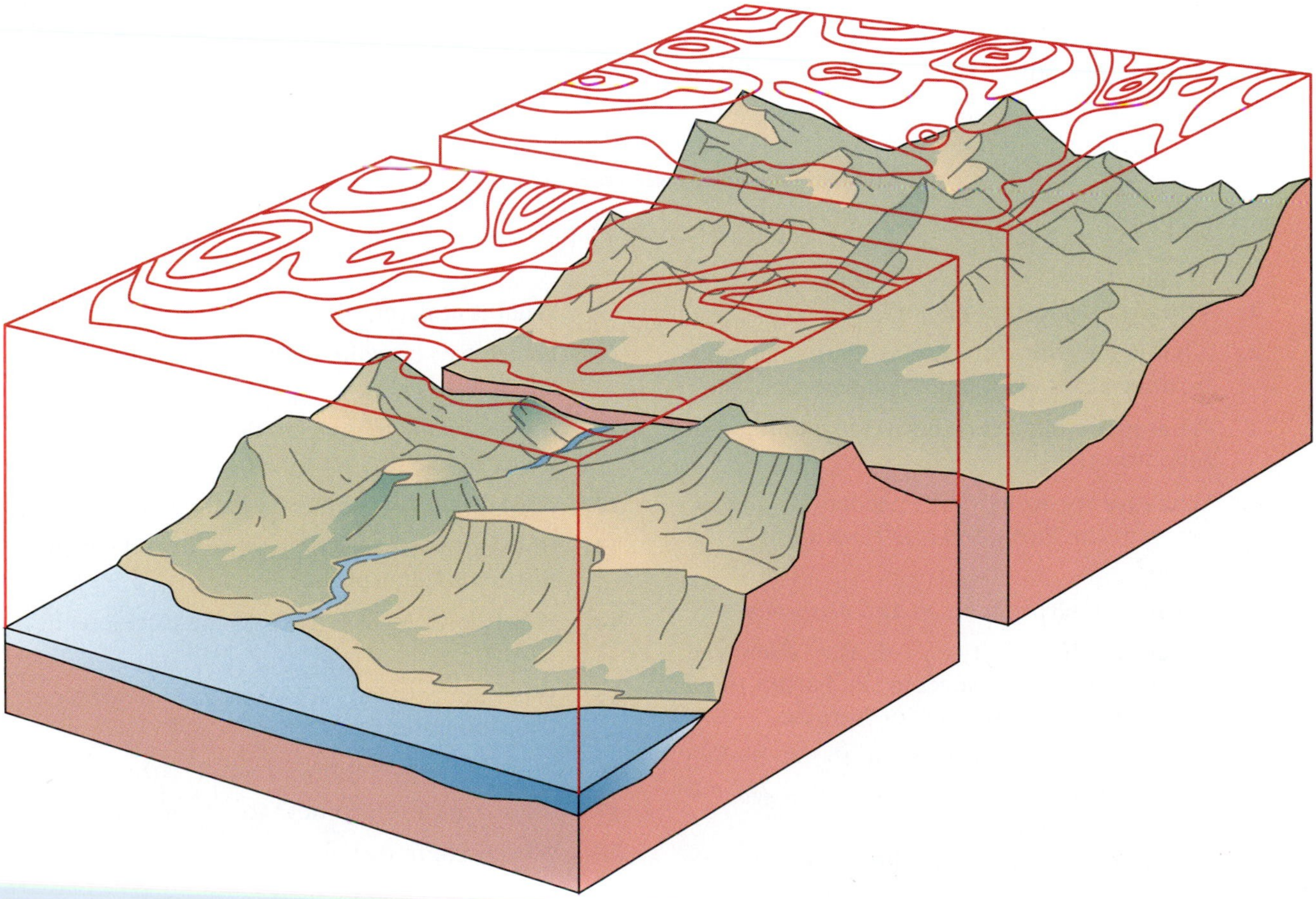

Figure 3–10 Contour lines are used to represent changes in elevation on a topographical map.

EARTH MATH

1) DETERMINE THE EQUATORIAL RADIUS IF THE EARTH'S EQUATORIAL DIAMETER IS 7928 MILES.
2) IF THE EARTH IS REVOLVING AROUND THE SUN AT APPROXIMATELY 66,000 MILES PER HOUR, HOW MANY MILES DOES IT TRAVEL IN SPACE EVERY DAY AND EVERY YEAR?
3) IF A CELESTIAL OBJECT TRAVELED 90 DEGREES OF ITS ARC ACROSS THE SKY AT MIDNIGHT, HOW FAR WOULD IT HAVE TRAVELED IN ITS ARC BY 3:00 A.M.?
4) WHAT IS YOUR APPROXIMATE LONGITUDE IF IT IS NOON AT THE PRIME MERIDIAN AND 5:00 P.M. AT YOUR LOCATION?
5) WHAT IS YOUR APPROXIMATE LONGITUDE IF IT IS MIDNIGHT AT YOUR LOCATION AND 6:00 A.M. AT THE PRIME MERIDIAN?

above sea level. By adding contour lines to a map, it is possible to see the true shape of the land. A **contour interval** is the specific division in height above sea level that each line represents. A common contour interval used on topographical maps in the United States is 20 feet. This means that the change in elevation between each contour line equals 20 feet. Mountains and valleys can be accurately represented on a two-dimensional map by adding contour lines. All of the United States and much of the world's land surface have been surveyed, and accurate topographical maps have been produced. Topographical maps show detailed features of the Earth's surface such as the locations of rivers, roads, swamps, lakes, ponds, railroads and other interesting features (Figure 3–11). These maps are useful to a variety of people and are easily available.

REVIEW

1. What is a terrestrial planet?
2. What is the true shape of the Earth, and what are three pieces of evidence that prove its shape?
3. What is the difference between rotation and revolution?
4. What are the exact times needed for one rotation of the Earth and one revolution of the Earth?
5. Define the terms *aphelion* and *perihelion.*
6. Describe the positions of the Northern Hemisphere with respect to the Earth's tilted axis and the four seasons of the year.
7. What is the significance of the following dates: December 21, March 21, June 22, and September 23?
8. How are tides formed on the Earth?
9. What is apparent motion?
10. Which two coordinates are used to locate celestial objects?
11. Approximately how far do celestial objects travel in the sky per hour?
12. What do lines of latitude measure on the Earth?
13. How can you determine your latitude in the Northern Hemisphere?
14. What do lines of longitude measure on the Earth?
15. What information do you need to determine your longitude?
16. What are topographical maps?
17. What are contour lines?
18. Who was Eratosthenes?

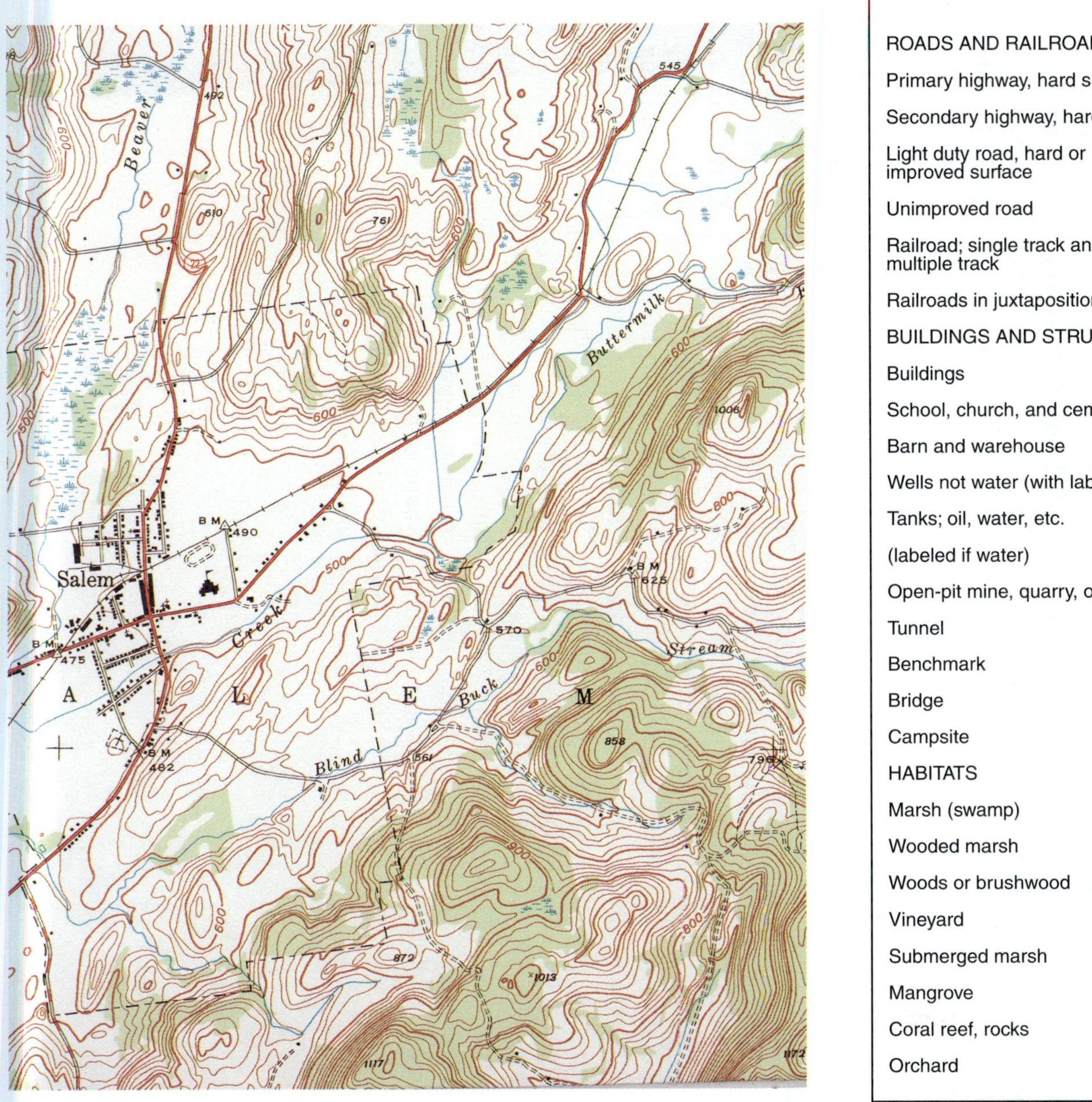

Figure 3–11 Topographical maps contain detailed information about the Earth's surface. *(Map courtesy of USGS.)*

For more information go to these web links:
<http://www.planetpals.com/planet.html>
<http://observe.arc.nasa.gov/nasa/earth/earth_index.shtml.html>
<http://mapping.usgs.gov>
<http://www.geog.ouc.bc.ca/physgeog/contents/chapter2.html>

CHAPTER SUMMARY

The Earth is a terrestrial planet that is composed mostly of a thin outer layer of rock, which surrounds a hot, molten mantle and dense iron core. Approximately 70% of the surface of the Earth is covered in water, and the planet is surrounded by an atmosphere. The Earth's shape is considered an oblate spheroid because it is slightly bulged at the equator. Evidence to support the Earth's nearly spherical shape includes photographs from space, the Earth's curved shadow on the Moon during a lunar eclipse, and the changing altitude of Polaris as you move southward from the North Pole toward the Equator. The Earth is spinning on its axis, which is called rotation, at approximately 1000 miles per hour near the equator. The time it takes for our planet to make one complete rotation is approximately 24 hours. This is commonly known as 1 day. The Earth's axis of rotation is tilted 23.5 degrees. This tilt causes the different seasons of the year. In addition to spinning on its axis, the Earth is also traveling around the Sun. This is known as a revolution. One complete revolution of the Earth around the Sun takes approximately 365 days, also known as 1 year. The path of the Earth moving around the Sun is called an orbit. The Earth's orbit is elliptical. An ellipse is a slightly flattened circle. The point in the Earth's orbit at which it is closest to the Sun is known as the perihelion. The point in its orbit at which it is farthest from the Sun is called the aphelion. The Sun and Moon's gravitational attraction have an effect on the Earth's oceans, known as tides. Tides are formed when the gravity of the Sun and Moon form large bulges in the ocean. This causes changes in the level of ocean water, called high and low tides. Observations of celestial objects from the Earth's surface reveal that they are moving in an arc across the sky. This movement is called apparent motion, because it is actually caused by the Earth's rotation. This causes stars, planets, and the Sun to appear to be moving, when it is really the Earth that is moving relative to them. Precise locations are identified on the Earth using the latitude and longitude coordinate system. Lines of latitude run parallel across the Earth's surface, marking locations north and south of the equator. Lines of longitude run from the North Pole to the South Pole, marking locations east and west of the prime meridian. The use of topographical maps enables the three-dimensional surface of the Earth to be displayed on a two-dimensional map. Contour lines are used to mark areas of equal elevation on a topographical map, which reveals the shape of the land. Topographical maps contain detailed information about the Earth's surface.

CHAPTER REVIEW

Multiple Choice

1. The true shape of the Earth is best described as a:
   a. perfect sphere
   b. perfect ellipse
   c. slightly oblate sphere
   d. highly eccentric ellipse

2. At sea level, which location is closest to the Earth:
   a. 45 degrees south latitude
   b. the equator
   c. 23 degrees north latitude
   d. the North Pole

3. The best evidence that the Earth has a spherical shape is provided by:
   a. the spherical shape of the Sun
   b. the change in time of sunrise and sunset through the year
   c. photographs from space
   d. viewing the stars at night

4. The spinning of an object on its axis is called:
   a. revolution
   b. rotation
   c. an ellipse
   d. an orbit

5. The tilted axis of the Earth points the Northern Hemisphere toward the Sun at which time?
   a. the autumnal equinox
   b. the winter solstice
   c. the vernal equinox
   d. the summer solstice

6. The movement of the Earth around the Sun is known as:
   a. a revolution
   b. a rotation
   c. an ellipse
   d. an axis

7. The orbit of the Earth around the Sun is best described as:
   a. a perfect circle
   b. slightly elliptical
   c. oblate
   d. extremely elliptical

8. The point in the Earth's orbit when it is closest to the sun is known as the:
   a. aphelion
   b. axis
   c. perihelion
   d. apparent motion

9. During which season of the year is the Earth farthest from the Sun?
   a. spring
   b. summer
   c. fall
   d. winter

10. The high tides that occur when the Moon and Sun are on the same side or opposite sides of the Earth are called:
    a. flood tides
    b. neap tides
    c. spring tides
    d. ebb tides

11. A star viewed at night appears to move at a rate of:
    a. 15 degrees per hour
    b. 30 degrees per hour
    c. 360 degrees per hour
    d. does not move at all

12. At which latitude is Polaris directly overhead?
    a. 0 degrees
    b. 23.5 degrees north
    c. 90 degrees south
    d. 90 degrees north

13. What is your latitude location if you have determined the altitude of Polaris to be 37 degrees above the horizon?
    a. 37 degrees south
    b. 53 degrees north
    c. 37 degrees north
    d. 90 degrees north

14. As a ship crosses the prime meridian and the altitude of Polaris is sighted as 45 degrees above the horizon, what is the ship's location?
    a. 45 degrees south latitude, 0 degrees longitude
    b. 45 degrees north latitude, 0 degrees longitude
    c. 0 degrees latitude, 45 degrees west longitude
    d. 0 degrees latitude, 45 degrees east longitude

15. What is your longitude location if your local time is 11:00 A.M. and the time at the prime meridian is 12:00 P.M.?
    a. 0 degrees longitude
    b. 15 degrees east longitude
    c. 15 degrees west longitude
    d. 30 degrees west longitude

continued

16. Contour lines showing elevations in feet above sea level on a topographical map are labeled in the following order: 20, 40, 60, 80, 100. What is the contour interval used on this map?
    a. 20 feet
    b. 10 feet
    c. 80 feet
    d. 1 foot

17. Blue lines on a contour map indicate:
    a. The location of streams
    b. Contour lines
    c. The locations of trees
    d. Town lines

Matching *Match the terms with the correct definitions.*

a. exosphere
b. axis
c. rotation
d. revolution
e. ellipse
f. perihelion
g. aphelion
h. spring tides
i. neap tides
j. zenith
k. equator
l. latitude
m. longitude
n. contour lines
o. contour interval

1. ____ The time of maximum tides that occurs during a Full or New Moon.
2. ____ The zone outside the Earth's atmosphere that is commonly known as outer space.
3. ____ The least amount of tidal activity that occurs when the Moon is at its first and third quarter positions in its orbit.
4. ____ A straight line around which an object rotates.
5. ____ The point in the sky that is directly above the observer, or 90 degrees above the horizon.
6. ____ The circular movement of a body around a central point called an axis.
7. ____ The imaginary line, also known as 0 degrees latitude, that divides the Earth into the northern and southern hemispheres.
8. ____ The movement of an object in an orbit around another object.
9. ____ Parallel lines that run east and west across the Earth's surface that measure location north or south of the equator.
10. ____ The flattened circular path of the orbits of most celestial objects around two foci, one of which is the Sun.
11. ____ Coordinate lines used on the Earth's surface that run north and south from pole to pole and measure a location east or west of the prime meridian.
12. ____ The point in a planet's orbit around the Sun when it is closest to the Sun.
13. ____ Lines that mark areas of equal elevation on a topographical map.
14. ____ The point in the orbit of a planet when it is farthest from the Sun.
15. ____ The specific change in elevation associated with each contour line on a topographical map.

Critical Thinking

1. Describe the possible effects on the Earth if the Earth was tilted 30 degrees on its axis.

CHAPTER

4

The Moon: Earth's Closest Neighbor

Objectives

The Moon's Formation • The Composition of the Moon • The Moon's Surface • The Moon's Orbit • The Phases of the Moon • Eclipses

After reading this chapter you should be able to:

- ❖ Identify the approximate age of the Moon.
- ❖ Describe how the Moon is believed to have formed and explain some evidence used to support these theories.
- ❖ Explain the composition of the Moon.
- ❖ Identify the various features found on the Moon's surface.
- ❖ Define the term *lunar month.*
- ❖ Identify the eight phases of the Moon and how they appear as viewed from the Earth's surface.
- ❖ Describe the positions of the Moon in its orbit around the Earth for each phase of the Moon.
- ❖ Define the term *eclipse.*
- ❖ Explain the difference between a lunar eclipse and a solar eclipse.

TERMS TO KNOW

silicate
deficient
volcanic rocks
basalt
mares
impact craters
Moon phases
lunar month
eclipse
umbra
penumbra
lunar eclipse
solar eclipse

INTRODUCTION

On July 20, 1969, human beings realized an age-old dream of landing on the Moon. The flight of Apollo 11 breached the previously uncrossable gap of space and placed the first human being on another object in our solar system other than Earth. Not only was this a great feat of modern technology, but it also revealed much about the composition of the Moon and the origins of the solar system. The Moon has fascinated human beings from our early beginnings. It is a source of wonder and inspiration for all who gaze at it. The Moon also influences the natural rhythms of life on Earth. Our calendar is set by the phases of the Moon, along with the functions of the human body. The influence of the Moon's gravity affects the oceans by creating tides and causes periodic eclipses of the Sun. Understanding the origin and composition of the Moon and role that the Moon plays in the Earth's system has helped us to understand much about our own planet.

The Moon's Formation

Rock samples brought back from the Moon reveal that our closest neighbor in space is approximately 4.5 billion years old. This is just as old as the Earth, which suggests that the Earth and the Moon formed at about the same time. The origin of the Moon is still being debated. One theory proposes that the Moon formed elsewhere in the solar system and was captured by the Earth's gravity as it passed close to our planet. Another theory suggests that debris orbiting the Earth that was left over from the formation of the planet eventually came together to form the Moon. The latest theory of how the Moon formed is gaining the most popularity. This theory states that a large object, perhaps as large as the planet Mars, struck the Earth and sent a large amount of the Earth's crust into space (Figure 4–1). This material eventually cooled and formed the Moon. Because of the age of the Moon, this event probably occurred early in the Earth's formation.

Evidence to support this theory comes from rock samples recovered from the Moon. They reveal that the Moon is mainly composed of **silicate** minerals and is **deficient** in iron. The lack of iron on the Moon, as compared with Earth, suggests that it was formed from the outer crust of the Earth, which contains much less iron than is located in its core. The average density of the Moon is approximately 3.3 grams per cubic centimeter, which is much less than Earth's density of 5.5 grams per cubic centimeter.

The Composition of the Moon

The rocks that were brought back from the Moon are very similar to **volcanic rocks** produced on the Earth. The volcanic eruptions that formed these rocks on the Moon occurred early in the Moon's history, between 4 and 2.5 billion years ago. These ancient eruptions created large plains of **basalt** lava that are called **mares.** The term *mare* comes from the Latin word for "sea," which is what early astronomers thought these flat regions on the Moon's surface to be. Since the formation of the mares, the Moon has had no known volcanic eruptions. The most recognizable feature of the Moon's surface are its **impact craters** (Figure 4–2). Impact craters are the bowl-like remains of celestial objects such as asteroids that collided with the Moon. Dating of the craters has revealed that the majority of the Moon's craters were formed more than 4 billion years ago and that the number of impacts of celestial bodies with the Moon surface has declined ever since. Scientists believe

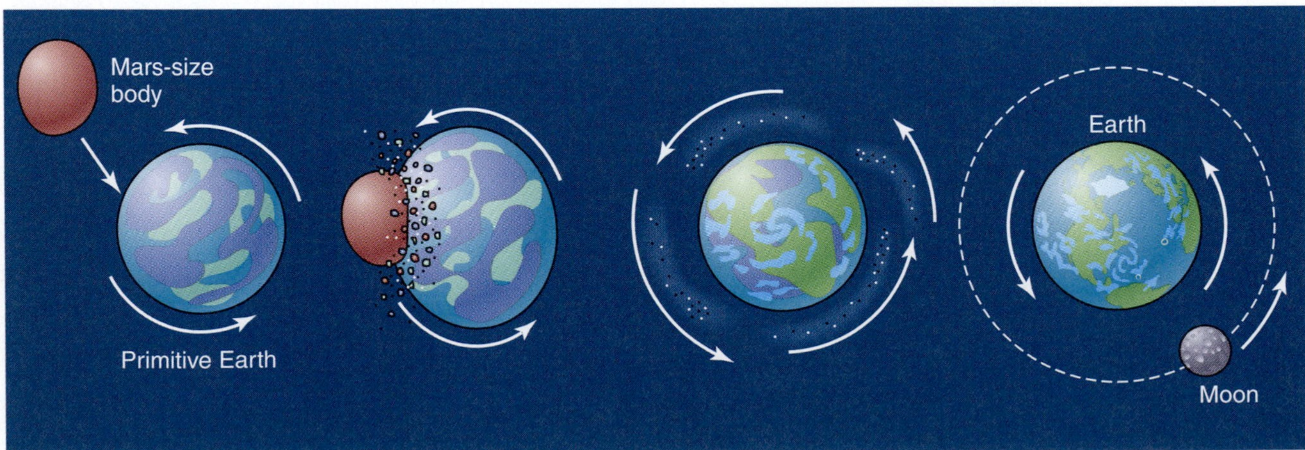

Figure 4–1 The theory of the Moon's formation by the impact of a Mars-sized object colliding with the Earth is known as the impact theory.

EARTH SYSTEM SCIENTISTS GALILEO GALILEI

Galileo was born in Pisa, Italy, in 1564. He began his life in the pursuit of science by studying mathematics. In 1583 he discovered the properties of pendulums. Later he experimented with the physics of motion and demonstrated how two objects of two different masses fall at the same rate. This famous experiment was reportedly enacted at the famous leaning tower of Pisa. In 1970 Galileo's famous experiment was demonstrated on the Moon, when an astronaut dropped a feather and a hammer, which fell at the same rate toward the Moon's surface. In 1593 Galileo began to study the properties of water and invented the thermometer. Galileo next turned his attentions to working with telescopes and used them to make the first accurate observations of the Moon surface. He also discovered the moons of Jupiter using the telescopes that he himself designed, including the Galilean refracting telescope. In 1632 Galileo published a book titled *Dialogue Concerning the Two Chief World Systems.* This book supported his belief that the Earth and other planets orbited around the Sun. This was known as the heliocentric theory, which was first proposed by Nicolaus Copernicus. This controversial theory was in conflict with the geocentric theory, which proposed that all planets and the Sun revolve around the Earth. The geocentric theory was supported by the church in Rome, which in 1633 charged Galileo with heresy. He was then placed under house arrest for the rest of his life. Galileo's life work helped form the foundation for modern physics and astronomy.

that the early Earth was also bombarded by celestial objects at the same time as the Moon; however, the Earth's craters have been long lost as a result of weathering and erosion.

Figure 4–2 The surface of the Moon is covered in dark, flat regions called mares and by large impact craters. *(Courtesy of PhotoDisc.)*

The Moon's Surface

Astronauts who landed on the Moon discovered that the surface is covered with a fine volcanic dust, which was formed by the impact craters. The Moon's surface has no atmosphere, and its gravity is 17% of Earth's gravity. The average surface temperature of the moon varies greatly during the day. At noon the temperature on the Moon can reach almost 200° F, and during the lunar night the temperature plummets to less than −250° F! Until recently it was believed that the Moon had no water, but explorations of the Moon's polar regions suggest that there may be some frozen water located on the Moon. The Moon's diameter is approximately 2160 miles, which is roughly 27% of the Earth's diameter. The Moon is approximately 238,866 miles from the Earth and completes one orbit around the Earth in 27 days and 7 hours. The period of the Moon's revolution around the Earth and the Moon's rotation on its own axis are the same; therefore the same side of the Moon always faces the Earth.

Career Connections

Lunar Geologist

A lunar geologist is a highly specialized geologist and astronomer who studies the geology of the Moon. Lunar geologists have analyzed the Moon rocks that were collected and brought back to the Earth during the Apollo missions. These scientists have unlocked the composition of the Moon and have pieced together its geological history. Today lunar geologists continue to study the Moon as a possible site for the future mining of mineral resources. Many space scientists believe that the Moon may someday be mined for minerals that can be used on Earth or for long-range space flight. Lunar geologists require a college education in both geology and space science and can be employed by NASA, academic institutions, or private industry.

The Moon's Orbit

As the Moon orbits around the Earth, one side of it is always illuminated by the Sun. This causes the different Moon phases as viewed from the Earth's surface. Different phases of the Moon occur at different points in its revolution around the Earth. The cycle of **Moon phases,** also called the **lunar month,** is approximately 29 days and 12 hours long. This is slightly longer than the time it takes the Moon to complete one orbit around the Earth. The difference in time between the Moon's orbital period and the lunar month is caused by the orbit of the Earth around the Sun. By the time the Moon has completed one full orbit, the Earth also has moved relative to the Sun. This makes the time between two New Moons 29.5 days.

The Phases of the Moon

The lunar month begins when the dark side of the Moon is facing Earth (Figure 4–3). This is called the New Moon, and because no sunlight is striking the Moon's surface, it appears like a dark disk in the sky. An eclipse of the Sun does not occur during every New Moon phase because the orbit of the Moon is tilted approximately 5 degrees around the Earth. The next phase of the Moon is called the Waxing Crescent phase; this phase occurs approximately 3 days after the New Moon. The Moon's surface is lit in a crescent shape and has completed one-eighth of its orbit around the Earth. The first Quarter Moon phase occurs approximately 7 days into the lunar cycle, when the Moon's surface as viewed from the Earth is half lit by sunlight. This is also called a Half Moon. The Waxing Gibbous phase occurs when three-quarters of the Moon's surface, as viewed from the Earth, is lit by sunlight. This occurs approximately 10 days into the lunar cycle. A Full Moon occurs when the side of the Moon facing the Earth is totally bathed in sunlight. This phase occurs about 14 days into the lunar cycle and marks the approximate halfway point of the Moon's orbit around the Earth. The next phase of the Moon is called the Waning Gibbous phase, which marks the point when three quarters of the Moon's surface is lit by the Sun, as viewed from the Earth. This occurs approximately 17 days into the lunar cycle. The third Quarter Moon phase takes place when the Moon has completed 75% of its orbit around the Earth. The Moon's surface facing the Earth is once again half bathed in sunlight, causing another Half Moon. This phase occurs approximately 20 days into the lunar cycle. The Waning Crescent phase is marked by one quarter of the moon's surface, as viewed from the Earth, being lit by the Sun. This occurs approximately 23 days into the lunar cycle. Finally, one lunar month is complete with the arrival of another New Moon. This occurs when the Moon has made one full lunar cycle, a full 29.5 days later.

Eclipses

Another important result of the Moon's orbit around the Earth is the occurrence of eclipses. An **eclipse** occurs when either the Moon or the Earth is shadowed from the Sun by the

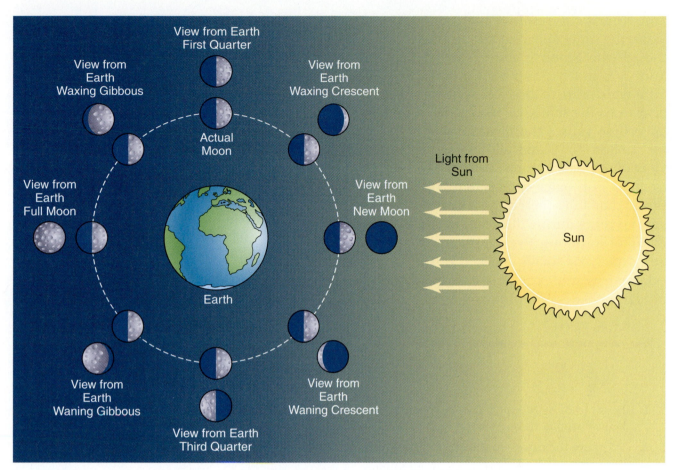

Figure 4–3 The eight different phases of the Moon as viewed from the Earth, and their locations in the Moon's orbit around the Earth.

other. The shadow of one celestial object that completely passes over another object is called the **umbra.** A partial shadow that occurs during an eclipse is called the **penumbra.** A **lunar eclipse** occurs when the Moon is shadowed from the Sun by the Earth (Figure 4–4). This type of eclipse can only occur during the Full Moon phase and can either be a partial eclipse or a total eclipse. A **solar eclipse** occurs when the Moon passes in front of the

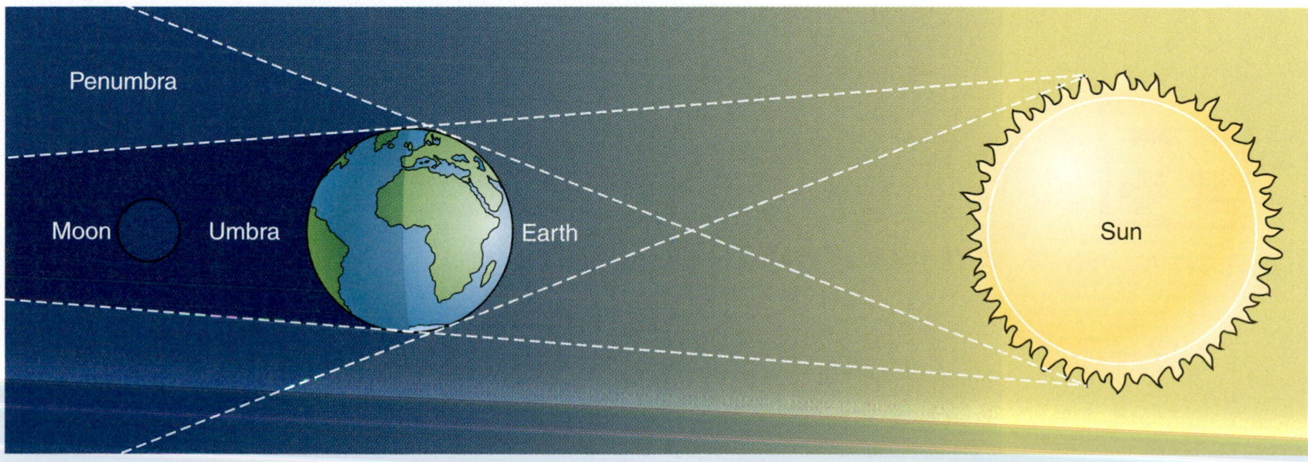

Figure 4–4 A lunar eclipse occurs when the Earth's shadow covers the Moon.

Sun. This is the result of the Sun and Moon being the same apparent size in the sky. Even though the Moon is much smaller than the Sun, it is much closer to the Earth than the Sun, which causes these celestial objects to appear to be the same size in the sky. A total solar eclipse causes the daytime sky to darken for a short time and allows for the unique view of the solar corona. The corona is the extremely hot outer atmosphere of the Sun. Both lunar and solar eclipses occur on Earth at various times during the year and are visible at different locations. Even though the Sun can be partially or totally blocked during a solar eclipse, it is still extremely dangerous to look directly at the Sun during any eclipse event.

REVIEW

1. How old is the Moon, and how do scientists believe it may have formed?
2. What are the mares on the Moon's surface?
3. What is the period of one orbit of the Moon around the Earth?
4. Why does the same side of the Moon face the Earth all the time?
5. What is a lunar month?
6. What are the eight main phases of the Moon?
7. What is a lunar eclipse?
8. Describe a solar eclipse.
9. Who was Galileo?

For more information go to this web link:
<http://seds.lpl.arizona.edu/nineplanets/luna.html>

CHAPTER SUMMARY

The Moon is a celestial object that is approximately one-quarter the size of the Earth. It is believed to have formed 4.5 billion years ago at the same time that the Earth was forming. Several theories explain the formation of the Moon. The most popular of these states that the Moon formed when an object roughly the size of Mars impacted the Earth. The resulting debris thrown up into space formed the Moon. The Moon's surface is made up of two distinct features: mares and impact craters. Mares are large, dark, flat areas on the Moon's surface formed by volcanic eruptions that occurred between 2.5 and 4.5 million years ago. Impact craters are large bowl-like impressions on the Moon's surface caused by asteroid and meteorite impacts. The surface of the Moon is also covered in a fine volcanic ash, which was revealed when NASA's Apollo missions sent humans to explore its surface. The rocks on the Moon are similar to volcanic rocks found on the Earth and are rich in silicates. It is now believed that frozen water may exist in the craters of the Moon near its polar regions. The Moon makes one complete orbit around the Earth in approximately 29 days, which is called a lunar

EARTH MATH

1) IF THE MOON'S PERIGEE (CLOSEST DISTANCE FROM THE EARTH) IS APPROXIMATELY 222,756 MILES AND THE MOON'S APOGEE (FARTHEST DISTANCE FROM THE EARTH) IS APPROXIMATELY 254,866 MILES, WHAT IS THE AVERAGE DISTANCE FROM THE EARTH TO THE MOON?
2) IF THE SUN'S DIAMETER IS APPROXIMATELY 400 TIMES THE DIAMETER OF THE MOON AND THE MOON'S DIAMETER IS APPROXIMATELY 2160 MILES, WHAT IS THE APPROXIMATE DIAMETER OF THE SUN?
3) IF THE MOON TRAVELS 13 DEGREES OF ARC IN SPACE, AS VIEWED FROM EARTH AS IT ORBITS THE PLANET EACH DAY, HOW MANY DAYS DOES IT TAKE TO TRAVEL 360 DEGREES OF ARC?

month. The same side of the Moon always faces toward the Earth. Different phases of the Moon are seen from the Earth depending on what point the Moon is at during its orbit. Moon phases are described by the amount of sunlight illuminating its surface as viewed from the Earth. Periodically the Moon passes in front of the Sun, which blocks it from the Earth. This is called a solar eclipse. When the Earth passes in between the Sun and the Moon and the Earth's shadow blocks the light illuminating the Moon, a lunar eclipse occurs.

CHAPTER REVIEW

Multiple Choice

1. The composition of the Moon most closely resembles:
   a. the Earth's core
   b. the Earth's mantle
   c. the Earth's crust
   d. the planet Mars

2. The craters on the Moon's surface were formed by:
   a. an impact with the Earth
   b. meteorites and asteroid impacts
   c. volcanic eruptions
   d. basaltic rocks

3. The age of the Moon is approximately:
   a. 4.5 billions years
   b. 2.5 billion years
   c. 15 billion years
   d. unknown

4. The periods of the Moon's rotation and revolution are equal. This results in:
   a. lunar eclipses
   b. the eight phases of the Moon
   c. neap tides
   d. the same side of the Moon facing the Earth

5. Which motion causes the Moon to show phases as viewed from the Earth:
   a. the rotation of the Moon on its axis
   b. the revolution of the Moon around the Earth
   c. the rotation of the Sun on its axis
   d. the revolution of the Sun around the Moon

6. As viewed from the Earth, the Moon's phases have shown which type of changes over the past fifty years:
   a. noncyclic and predictable
   b. noncyclic and unpredictable
   c. cyclic and predictable
   d. cyclic and unpredictable

7. Which phase of the Moon occurs halfway in the orbit of the Moon around the Earth?
   a. New Moon
   b. Waxing Gibbous
   c. Full Moon
   d. Waning Gibbous

8. The New Moon phase occurs when the Moon is positioned between the Earth and Sun, but these positions do not always cause an eclipse of the Sun because:
   a. the Moon's orbit is tilted relative to the Earth's orbit
   b. the New Moon phase is visible only at night
   c. the night side of the Moon faces the Earth
   d. the apparent diameter of the Moon is greatest during these phases

9. When the Moon is completely covered within the Earth's umbra, which occurs:
   a. a lunar eclipse
   b. a solar eclipse
   c. an annular eclipse
   d. no eclipse

10. When the Moon passes in front of the Sun, which occurs:
    a. a lunar eclipse
    b. a solar eclipse
    c. an annular eclipse
    d. no eclipse

continued

Matching *Match the terms with the correct definitions.*

a. silicate
b. deficient
c. volcanic rocks
d. basalt
e. mares
f. impact craters
g. Moon phases
h. lunar month
i. eclipse
j. umbra
k. penumbra
l. lunar eclipse
m. solar eclipse

1. ____ The series of different appearances of the Moon as observed from Earth, which results from the varying amount of light that illuminates the Moon at specific points in its orbit.
2. ____ A chemical compound that is composed of atoms of silicon and oxygen.
3. ____ The time it takes for the Moon to make one complete orbit around the Earth, which is approximately 29 days.
4. ____ Lacking something essential to life.
5. ____ The cutting off of all or part of the light of one celestial body by another.
6. ____ A common fine-grained volcanic rock that is dark in color, mafic, and dense.
7. ____ The area in shadow during an eclipse, which is totally blocked from the light.
8. ____ A Latin word for "seas" that is used to describe the flat, dark, plain-like areas that cover the surface of the Moon.
9. ____ The lighter area located next to the umbra, or darkened shadow, that occurs during an eclipse.
10. ____ Large bowl-like depressions that are left on the surface of a celestial object as the result of an impact by another celestial object, usually an asteroid or comet.
11. ____ The total or partial blocking of the Sun as viewed from the Earth when the Moon passes in front of it.
12. ____ Igneous rocks that form from cooled lava produced by a volcano.
13. ____ The total or partial blocking of sunlight striking the Moon's surface by the Earth as it moves directly in between the Moon and Sun.

Critical Thinking

1. One scientist once believed that the Moon should be destroyed, so that it no longer orbits our planet. How would the Earth be affected if the Moon did not exist?

CHAPTER 5 The Life of Stars

Section 5.1 – Life Cycle of Stars Objectives

Star Formation and Life Cycle • Classification of Stars • Dwarf Stars • Giant and Supergiant Stars • Neutron Stars • Black Holes

After reading this chapter you should be able to:

- Define the term *star.*
- Identify the four ways by which stars are classified.
- Explain the basic stages of a star's life cycle.
- Describe the characteristics of dwarf, giant, and supergiant stars.
- Explain the concept of a black hole.

Section 5.2 – The Sun Objectives

Formation of the Sun • Composition of the Sun • The Sun's Life Cycle

After reading this chapter you should be able to:

- Identify the approximate age of the Sun.
- Describe the composition of the Sun and its unique layers.
- Define the terms *solar cycle* and *solar wind* and explain how they affect the Earth.

TERMS TO KNOW

star
nuclear fusion
stellar nebula
luminosity
spectral class
red dwarf stars
white dwarf
red giants
blue supergiants
neutron star
pulsars
black hole
sunspots
solar flares
solar wind

INTRODUCTION

On a clear night, besides the Moon, the stars are by far the most visible of celestial objects. The grand display of twinkling starlight has captured the imagination of most all who have ever gazed up at the heavens. Ancient humans looked to the stars and grouped them into recognizable objects, like people or animals, called the constellations. Constellations are still used today to help locate specific stars. Stars also served as the first guides for navigation around the world. Studying the stars, learning how they form, and understanding their unique life cycles has enabled us to further understand our place in the universe. The closest star to the Earth, and the most important to life, is the Sun. All the energy that sustains life on our planet is derived from the Sun. The rising and setting of the Sun marks the beginning of our day, and our calendars and watches are all set to its rhythms. Almost all ancient peoples worshiped the life-giving properties of the Sun in their own unique ways. The Sun provides the mechanism for the movement of matter and energy for many of the Earth's systems. Human beings have created many myths and legends to explain what the Sun is, but it wasn't until the beginning of the nineteenth century that we began to fully understand this bright object that is so important to our lives.

5.1 Life Cycle of Stars

Star Formation and Life Cycle

A **star** is a large, hot, glowing ball of gas that is powered by **nuclear fusion.** Stars begin their lives as a large cloud of dust and gas, called a **stellar nebula** (Figure 5–1). Eventually the gravitational attraction of the atoms that compose the cloud cause it to begin to collapse. As the cloud collapses it begins to heat up, and its density and pressure increase. This stage in the life of a star is known as the proto star stage. Eventually, the cloud heats to extreme temperatures, and the pressure within is great enough to start a fusion reaction. The fusion reaction is caused by atoms of hydrogen joining together to form atoms of helium. The result is the release of an incredible amount of energy. This begins the stage in the life of a star known as the main sequence stage, when the star begins to shine. Eventually, after a star has used up all of its hydrogen, the nuclear reactions that have kept it burning begin to slow down. This causes the star to begin to expand and cool down, and eventually the star burns itself out. Not all stars are the same; in fact, they differ greatly in their size, color, and brightness.

Figure 5–1 This image shows the Orion Nebula among a background of stars. *(Courtesy of PhotoDisc.)*

Classification of Stars

Stars are classified by their size, temperature, color, and brightness. This classification is usually based on a star's age or stage of life. Two astronomers, Ejnar Hertzsprung and Henry Russell, developed a key that is used to classify stars. This is called the Hertzsprung-Russell diagram, or HR diagram. The HR diagram is used to classify stars by their color, temperature, and **luminosity** (Figure 5–2). The color of a star, also known as its **spectral class,** can range from blue to white, yellow, and red. The temperature of a star can be classified in either Kelvins or degrees Celsius. The coolest stars have temperatures around 2500° Celsius, and the hottest stars can have temperatures of more than 20,000° Celsius. The luminosity of a star measures how bright a star is or how much radiation it gives off relative to that of our Sun. The Sun has a luminosity of 1. The dimmest stars can have a luminosity of only 0.0001, and the brightest stars can have a luminosity of up to 1,000,00. These extremely bright stars are 1 million times brighter than our Sun.

Dwarf Stars

The coolest stars with the lowest luminosity are called **red dwarf stars.** These have temperatures between 2500° and 3000° Celsius. The color of these stars is red because of their relatively low temperatures (Figure 5–3). An example of a red dwarf star is Barnard's star, which is approximately 5.8 light years from Earth. A light year is the distance that light travels in 1 year. Because light travels at about 186,000 miles per second, this extremely far distance is equal to approximately 5.8 trillion miles. Amazingly, Barnard's star is the third closest star to the Earth!

The **white dwarf** is another star classification. These stars are in the final stages of their life cycles. Although they have used up all their fuel, they still shine brightly in the sky. This is the result of the leftover heat that the core of the star is still radiating into space (Figure 5–4).

Luminosity and Temperature of Stars
(Name on italics refers to star shown by a ⊕)

Luminosity (relative to the Sun): 1.000.000 Massive stars; 10,000; 100; 1; 0.01; Small stars 0.0001

Blue Supergiants
Rigel ⊕
Supergiants
Betelgeuse ⊕
Red Giants
Polaris ⊕
⊕ Aldebaran
Main Sequence
⊕ Sinus
⊕ Alpha Centauri
Sun ⊕
White Dwarfs
⊕ Procyon B
Red Dwarfs
Barnards Star ⊕

20,000 10,000 5,000 2,500
Temperature (°C)
Blue stars White stars Yellow stars Red stars
Color

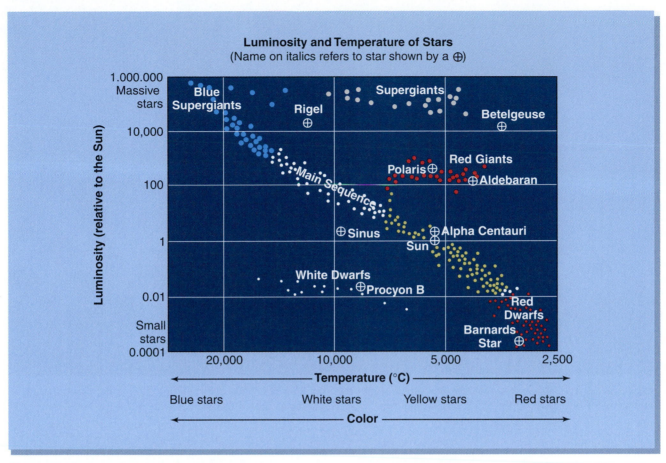

Figure 5–2 The HR diagram is used to classify stars by their color, temperature, luminosity, and size.

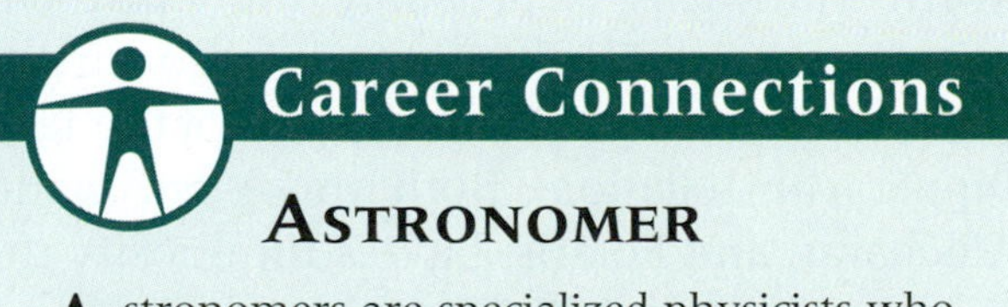

Astronomer

Astronomers are specialized physicists who observe and study the universe, galaxies, stars, planets, moons, and other celestial objects. They are also interested in utilizing their knowledge for practical applications such as space flight, communication, and navigation. Many astronomers also work to develop new instruments that can be used to study objects in space. These include specialized optical and radio telescopes, remote-operated spacecraft, and satellites. Astronomers often work conducting research for universities or with NASA. Some astronomers are employed by private industry or museums. To be an astronomer you must be interested in physics, mathematics, and computers, as well as celestial phenomena. All astronomers must have a college education.

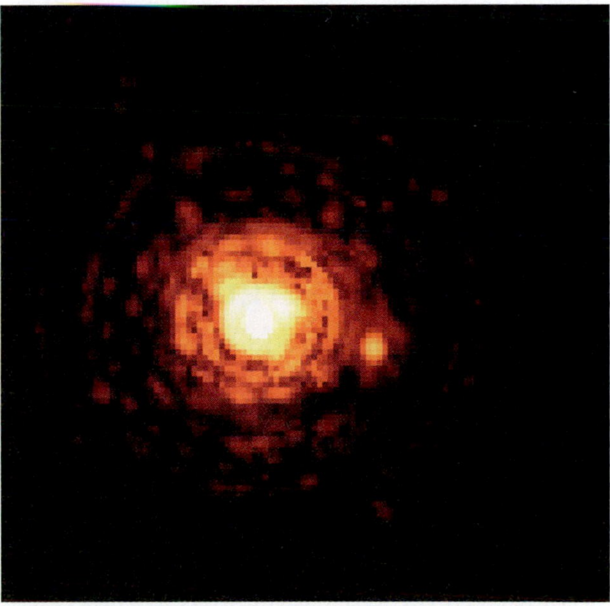

Figure 5–3 The small red dwarf star on the right, Gliese 623b, is 60,000 times dimmer and 10 times smaller than the Sun. The larger red dwarf star at center is its companion, Gliese 623a. *(Courtesy of C. Barbieri [University of Padua], NASA, ESA.)*

Figure 5–4 This image of the Red Spider Nebula shows the formation of a white dwarf at the center of the cloud of gas that has just been blown off the star's surface. *(Courtesy of Garrelt Mellema [Leiden University] et. al., HST, ESA, NASA.)*

The color of white dwarf stars can vary from light blue to true white or pale yellow. These stars have temperatures that range from 6000° to 16,000° Celsius. The luminosity of these stars is approximately 0.01 times that of our Sun. The star Procyon B is an example of a white dwarf; it is approximately 11.3 light years from the Earth and is located in the constellation Canis Minor.

Giant and Supergiant Stars

Very large stars that are in the later stages of their lives are called **red giants.** These stars form when the aging star begins to expand as it uses up its fuel. This causes the star to enlarge and to shine more brightly. These massive stars have a luminosity that is more than 100 times that of our Sun, which is the result of their extreme size. Red giants also have relatively moderate temperatures between 4000° and 7500° Celsius. An example of a red giant star is Aldebaran, which is located in the constellation Taurus. This red giant is located approximately 68 light years from Earth. Another red giant is Polaris, also called the North Star. This is probably the most well known star on Earth and is located directly over the North Pole. Polaris is part of the Little Dipper constellation. Both these red giants, Aldebaran and Polaris, have a luminosity that is approximately 100 times that of our Sun.

Supergiant stars are the largest of all stars. These stars can either be red supergiants or **blue supergiants.** These extremely large stars are also the brightest of all stars, with luminosities between 10,000 and 1,000,000 times brighter than our Sun. An example of a red supergiant star is Betelgeuse, which has a temperature of approximately 3000° Celsius. Betelgeuse is located in the constellation of Orion at an extreme distance of approximately 650 light years from the Earth (Figure 5–5). The blue supergiant Rigel is an extremely hot star with a temperature that is more than 10,000° Celsius. Rigel is also located in the constellation of

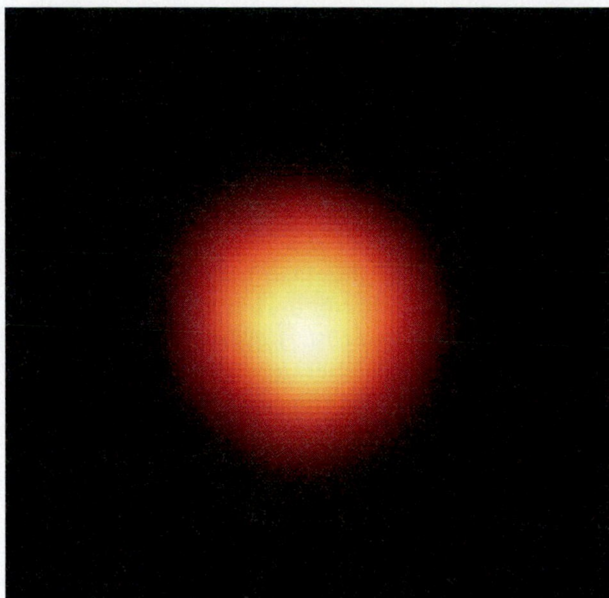

Figure 5–5 The red supergiant star Betelgeuse, which is 1000 times larger than the Sun but has a much lower temperature. *(Courtesy of A. Dupree [CFA], R. Gilliland [STScI], NASA.)*

Figure 5–6 Stars in the constellation of Orion are photographed using a technique that reveals their unique colors. At the lower right is the blue supergiant Rigel. *(© David Malin.)*

Orion and is approximately 815 light years from Earth (Figure 5–6).

Neutron Stars

Almost all stars end their lives as white dwarfs; however, some dying stars can result in the formation of a few unique objects. A **neutron star** forms when a star has burned up all its fuel and collapses under its own gravity. Some neutron stars emit radio waves at regular intervals as they spin on their axis. These neutron stars are also known as **pulsars.** Acting much like a lighthouse, the rapidly rotating pulsar emits radio waves out into space. Because the star is rapidly spinning, the radio waves it sends out are received on Earth at regular intervals. Some neutron stars are formed when the dying star explodes violently, causing a bright flash of light. This rare event is called a supernova (Figure 5–7). A supernova explosion was recorded by ancient astronomers more than 1000 years ago. Today the remains of this supernova form the Crab Nebula in the constellation Taurus.

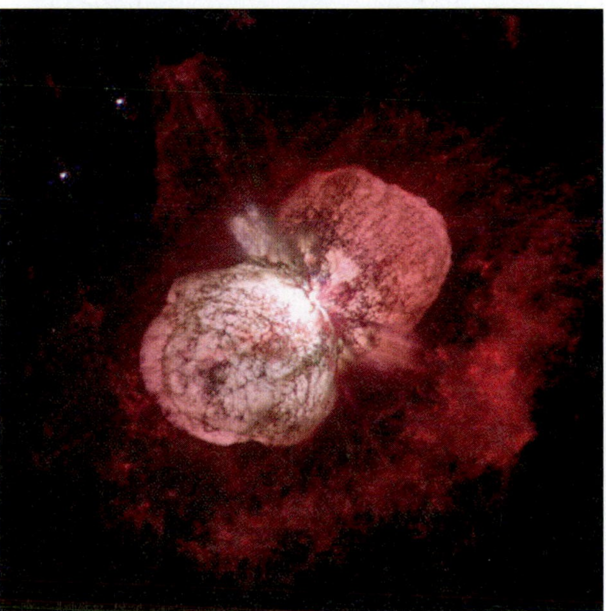

Figure 5–7 Gas and dust billows outward as a result of a supernova explosion of the star Eta Carinae, which exploded violently approximately 150 years ago. *(Courtesy of Jon Morse [University of Colorado] and NASA.)*

EARTH SYSTEM SCIENTISTS *STEPHEN HAWKING*

Astronomer Stephen Hawking was born in England in 1942. His early work in physics and mathematics involved the concepts of space and time. While in graduate school, Hawking was diagnosed with a degenerative disease that eventually left him in a wheelchair and unable to speak. However, this disability did not prevent him from achieving his goal as a respected mathematician and astrophysicist. In 1970 Hawking began to research the most bizarre of celestial phenomena, black holes. His work centered on the interaction of matter and energy within and around a black hole. In 1988 he published his famous book, *A Brief History of Time,* which explained the theories of space, time, relativity, and black holes to the general public. Today he still continues his work and holds the highly esteemed position of Lucasian professor of mathematics at Cambridge University in England, a position once held by Sir Isaac Newton.

Black Holes

Although almost all stars end their lives as white dwarfs and neutron stars, some extremely massive stars can form an object known as a **black hole.** A black hole forms when a massive star ends its life and begins to collapse. The extremely dense collapsing star creates a gravitational pull so strong that even light cannot escape from the star. This creates the object known as a black hole. The concept of a black hole is theoretical because technically they cannot be seen; however, they may be detected by their effects on the light of surrounding stars and by the presence of strong x-ray emissions. A black hole may exist in the constellation of Cygnus, and possibly at the center of the Milky Way galaxy. One of the most distant stars ever detected, Deneb, is also located in the constellation of Cygnus. This supergiant star is approximately 1600 light years from the Earth. The closest star to our planet, besides our own Sun, is the Centauri star system, which consists of three medium-sized stars that are similar to our Sun. These stars are approximately 4.4 light years from the Earth.

SECTION REVIEW

1. What is a star?
2. How are stars classified?
3. Describe how a star forms.
4. Describe the characteristics of dwarf stars.
5. What are the characteristics of the giant and supergiant stars?
6. What is a neutron star?
7. What is a pulsar?
8. What is a black hole?
9. Who is Stephen Hawking?

For more information go to this Web link:
<http://astron.berkeley.edu/~bmwndez/ay10/cycle/cycle.html>

EARTH MATH

1) IF 1 LIGHT YEAR IS EQUAL TO APPROXIMATELY 5.8 TRILLION MILES, HOW MANY MILES FROM EARTH IS THE CLOSEST STAR SYSTEM?

2) ANOTHER WAY TO MEASURE THE DISTANCE TO A STAR IS BY USING THE UNIT CALLED THE PARSEC. ONE PARSEC IS EQUAL TO APPROXIMATELY 3.26 LIGHT YEARS. KNOWING THIS, DETERMINE HOW MANY PARSECS THE NEAREST STAR SYSTEM IS TO THE EARTH.

5.2 *The Sun*

Formation of the Sun

The Sun is a main sequence star of medium size that radiates energy into space at all wavelengths of the electromagnetic spectrum. It lies at the center of the solar system, approximately 93 million miles from the Earth. At this distance the energy from the Sun takes approximately 8 minutes to reach the Earth. The Sun is believed to have formed from a swirling cloud of dust and gas called the Solar Nebula approximately 4.5 to 4.7 billion years ago (Figure 5–8). This nebula eventually collapsed under its own gravity, and the process of nuclear fusion began, which caused the Sun to shine.

Composition of the Sun

The Sun is composed of approximately 74% hydrogen, 24% helium, and trace amounts of other elements, such as carbon, oxygen, silicon, and iron. The Sun is made up of four principal layers that surround a central core (Figure 5–9). The core of the Sun is where the thermonuclear reactions occur that power the Sun. There hydrogen atoms are fused together to create helium and extreme amounts of energy. The pressure inside the Sun's core is approximately 7 trillion pounds per square inch. The pressure at the Earth's surface is only 14.7 pounds per square inch! The temperature in the Sun's core is approximately 27 million° Fahrenheit. The layer that surrounds the Sun's core is called the convective zone. The convective zone is where the heat and light energy created in the Sun's core are transferred outward by convection currents.

Figure 5–8 This image of the Sun, taken by NASA's Solar and Heliospheric Observatory (SOHO) satellite, reveals its surface. *(Courtesy of NASA.)*

The next layer of the Sun is called the photosphere. This is the visible surface of the Sun, which is composed of hot gases that act like a cloud surrounding the inner core and convection zone. The photosphere has a diameter of approximately 864,000 miles, which is approximately 109 times larger than the Earth. The photosphere is the region where hot gases rise and cooler gases descend back toward the Sun's center. These regions are known as granules, which are approximately 620 miles across. These granules give the photosphere a rough surface (Figure 5–10).

The average temperature of the photosphere is approximately 9700° Fahrenheit. Darker regions of cooler gas appear on the surface of the photosphere; these are called **sunspots** (Figure 5–11). Sunspot regions are about 3000° Fahrenheit cooler than the rest of the photosphere. Sunspots are also related to the occurrence of **solar flares,** which are hot regions of gas that leap off the Sun, sending out bursts of solar energy (Figure 5–12).

The number of sunspots on the Sun's surface changes from year to year, creating what is called a solar cycle. A solar cycle, or sunspot cycle, is the time from a period of high sunspot activity to a time when there are very few sunspots. The time of high sunspot activity is called the maxima, and the time of low sunspot activity is called the minima. One complete solar cycle, which is the time it takes to go from maxima to maxima, is 11.1 years.

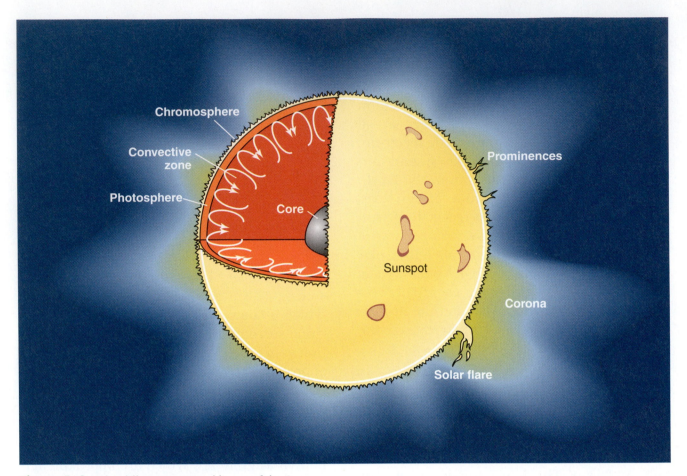

Figure 5–9 The different parts and layers of the Sun.

Scientists have related sunspot activity to periodic climate changes on the Earth. For example, periodic droughts that affect the Midwestern United States often occur on the same solar cycle as sunspot activity. Scientists are also concerned with the occurrence of solar flares, because these violent eruptions on the Sun can cause disruptions to satellite communications and electrical power distribution on Earth. These are caused by solar flares emitting large bursts of radiation that eventually strike the Earth's atmosphere.

EARTH SYSTEM SCIENTISTS *HANS BETHE*

Hans Bethe was born in 1906 in Germany. He earned his doctorate in physics in 1928 at the University of Munich and became a professor of physics. He fled Germany in 1933, when Hitler gained power, and moved to England. His major contribution to science was unraveling the mechanics behind the energy production in stars. He was the first to propose that stars gain their energy from a fusion reaction that combined two atoms of hydrogen to form one atom of helium, along with a great amount of energy. In 1975 he moved to the United States and became a professor of physics at Cornell University. His work on the energy of stars eventually won him the Nobel Prize for Physics in 1976. He also became the chief of theoretical physics at the Los Alamos laboratory in New Mexico during World War II. This work was part of the Manhattan Project, which developed the first atomic bombs. For much of his life he was also involved in ensuring that scientists maintain a social responsibility for their work.

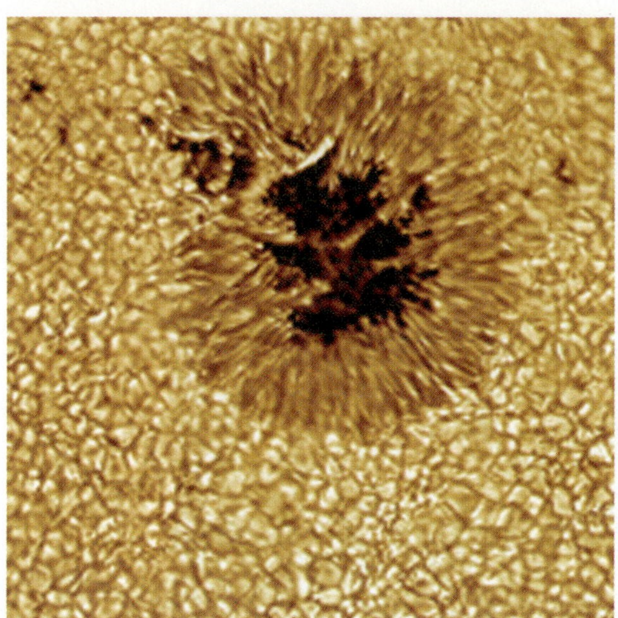

Figure 5–10 Close-up view of a sunspot surrounded by granules on the Sun's surface. *(Courtesy of T. Rimmele, M. Hannal/NOAO/AURA/NSF.)*

The layer of the Sun above the photosphere is called the chromosphere. This invisible layer of hot gas is approximately 2000 miles thick and has temperatures ranging from 7000° to 90,000° Fahrenheit. The chromosphere merges with the final layer of the Sun, called the corona. The corona is only visible during a solar eclipse and reaches approximately one million miles into space.

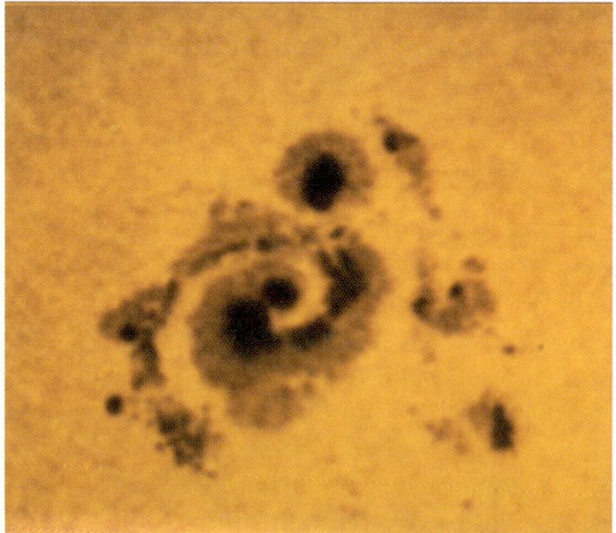

Figure 5–11 Image of a group of dark sunspots on the surface of the Sun. *(Courtesy of NOAO/AUEA/NSF.)*

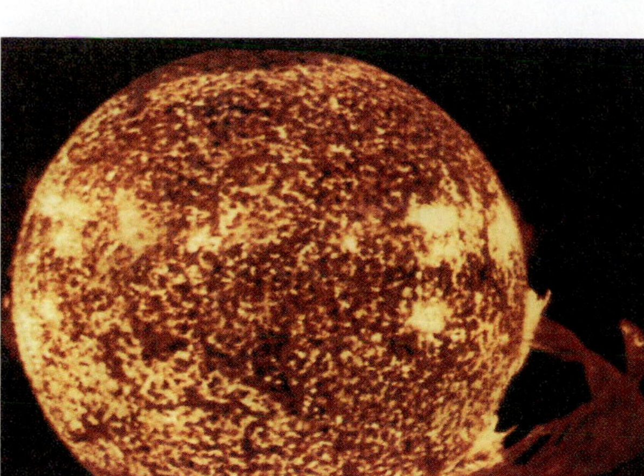

Figure 5–12 A large solar flare leaps off the surface of the Sun (lower right). *(Courtesy of NASA.)*

The corona can have temperatures greater than 3 million degrees Fahrenheit. The corona produces the **solar wind,** which is a stream of charged particles that move outward from the

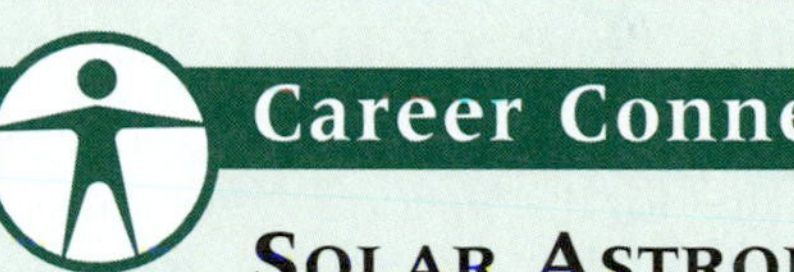

Career Connections

Solar Astronomer

A solar astronomer studies all aspects of our Sun. The Sun is the closest star to the Earth and has an important relationship with life on Earth. Solar astronomers study the Sun to better understand how it works and, more importantly, how it impacts life on Earth. This specialized aspect of astronomy is also known as "space weather." Space weather is the change in space as a result of changes that occur with the Sun. Solar astronomers study these changes that occur within the Sun and how they affect our technology and life on the planet. Predicting periodic solar flares and other solar storms is one of the important jobs of a solar astronomer. Solar storms can interrupt power distribution and communication on Earth, so it is the job of the solar astronomer to study these events and determine how they may affect the Earth. Solar astronomers use a number of specialized telescopes, spacecraft, and other scientific instruments to study the Sun. Solar astronomers can work in the academic fields, for the government, or for private industries like power companies.

Figure 5–13 The aurora borealis, also called the northern lights, light up the sky in the Northern Hemisphere as a result of the solar wind interacting with the magnetosphere and atmosphere. *(Courtesy of PhotoDisc.)*

Current Research

Scientists working for the National Aeronautics and Space Administration (NASA) who have been studying the Sun have discovered a unique feature never before seen on our nearest star. Recent images taken of the Sun's surface have revealed explosions that create earthquakelike ripples on the Sun's surface. The explosions on the Sun's surface that cause these ripples are being called sunquakes. The power of these sunquakes would measure 11.1 on the Richter scale if they occurred on Earth. The explosions that cause sunquakes are connected to the Sun's increased activity. Activity on the Sun is marked by increases in the occurrence of solar storms that cause sunspots and solar flares. An increase in the number of solar storms that cause these sunquakes occurs during the Sun's solar maximum on an average of every 11 years. Scientists are interested in these solar storms because of the effect that they might have on the Earth. Much of the current research that is discovering new information about the nature of the Sun is being done with observations taken through NASA's Solar and Heliospheric Observatory, or SOHO satellite. This satellite is designed to observe the Sun continually and orbits around the Sun approximately 930,000 miles away from the Earth.

sun at approximately 18,000 miles per hour. The Earth is shielded from this stream of particles by its magnetic field; however, at the North and South Poles, the solar wind collides with the atmosphere and lights up the nighttime sky with brilliant colors. This is known as the northern lights, or the aurora borealis, in the Northern Hemisphere and the australis borealis in the Southern Hemisphere (Figure 5–13). Solar flares that leap off the surface of the Sun can also cause a temporary increase in the intensity of the solar wind.

The Sun's Life Cycle

The Sun, like any star, must go through the stages of a star's life cycle. Because the Sun is believed to be approximately 4.5 to 4.7 billion years old, it is estimated that it will take another 5 to 6 billion years to use up all of its fuel. When this happens the Sun will begin to enter the red giant phase of its life (Figure 5–14). When the sun becomes a red giant, it will be approximately 100 times larger than it is now. This will cause the hydrosphere on Earth to boil away and the atmosphere to be

Figure 5–14 The Sun's mass will increase drastically when it becomes a red giant in approximately 5 billion years, which will incinerate many of the planets of the solar system. *(Courtesy of James Gitlin/STScI AVL.)*

blown off, which will end life on Earth as we know it.

The outer planets, such as Saturn and Jupiter, will also have their atmospheres blown off during this phase of the Sun's life. Eventually the Sun will expel its outer layers of hot gas and develop into a white dwarf. The Sun during this time will be about the size of the Earth but will have an extreme density of more than 1000 grams per cubic centimeter. Although this scenario means certain doom to our solar system, we should not worry, for this will not begin to occur for approximately 5 billion more years!

For more information go to this Web link:
<http://sohowww.nascom.nasa.gov/>

SECTION REVIEW

1. Approximately how old is the Sun?
2. What is the Sun made of?
3. Describe the different layers that together make up the Sun.
4. What is a solar cycle, and why is it important to the Earth?
5. What is the solar wind, and how does it affect the Earth?
6. Who is Hans Bethe?

CHAPTER SUMMARY

Stars are luminous balls of gas that are powered by nuclear fusion. There are billions of stars that exist in the universe. Stars are classified by their particular color, temperature, size, and luminosity. This is often displayed on the HR diagram. All stars begin their lives in the form of a stellar nebula, which is a large cloud of collapsing dust and gas. Eventually the nebula collapses to the point where nuclear fusion begins to occur, causing the star to shine. This is called the main sequence stage of a star's life cycle. Eventually a star uses up all its fuel and begins to expand and cool. During this stage of a star's life, it is called a red giant. The red giant is a very large star that has a red appearance and is cooler than a main sequence star. The outer layer of gas of a red giant eventually is blown off into

EARTH MATH

1) If the diameter of the Earth is approximately 7918 miles and the Sun is 109 times the diameter of the Earth, what is the approximate diameter of the Sun?

2) Using the diameter you calculated from question 1, if the chromosphere of the Sun is 2000 miles thick, what percentage of the total size of the Sun does this layer occupy?

space, leaving a small, extremely hot, dense object called a white dwarf. The white dwarf then cools to form a neutron star, and the life cycle is complete. Many neutron stars emit radio waves at regular intervals and are known as pulsars. Some massive stars can also explode violently during their red giant phase, causing what is called a supernova. Another type of object, called a black hole, can form from a massive star. A black hole is created from the remnants of a star; it has a gravitational attraction so great it prevents light from leaving its surface.

The closest star to the planet Earth is the Sun. The Sun is a medium-aged, main sequence star. Like all stars, the Sun is powered by the process of nuclear fusion. The Sun's central core is surrounded by four unique layers: the convective zone, photosphere, chromosphere, and corona. The photosphere is the visible surface of the Sun that we see from Earth. The surface of the photosphere has darker, cooler regions called sunspots. The number of sunspots on the photosphere changes periodically in what is called a solar cycle. The point in the solar cycle when there are a high number of sunspots is called the solar maxima. The point when there are very few sunspots is known as the solar minima. The average period between solar maxima and minima is 11 years. Increased solar activity also coincides with an increasing number of solar flares. A solar flare is a large eruption of hot gas off the surface of the Sun. Solar flares and increased sunspot activity can affect the Earth in a number of ways. The sun emits a stream of particles into space known as the solar wind. The interaction of the solar wind with the Earth's magnetic field creates the phenomenon known as the northern lights. The Sun is estimated to be approximately 5 billion years old, which is in the middle of its life cycle. Scientists believe the Sun will enter its red giant phase in approximately 5 billion years.

CHAPTER REVIEW

Multiple Choice

1. Which process causes a star to shine?
   a. nuclear fission
   b. nuclear fusion
   c. radioactive decay
   d. sunspots

2. Which is the correct sequence of events in a star's life cycle?
   a. white dwarf, red giant, main sequence, nebula
   b. red giant, nebula, white dwarf, main sequence
   c. nebula, main sequence, red giant, white dwarf
   d. main sequence, white dwarf, red giant, nebula

3. Which star classification causes a star's unique color?
   a. luminosity
   b. brightness
   c. size
   d. temperature

4. During which of the following stages of a star's life is it hottest?
   a. proto star
   b. red giant
   c. red dwarf
   d. white dwarf

5. Which of the following is the hottest type of star?
   a. blue supergiant
   b. white dwarf
   c. red giants
   d. red dwarf

6. Which type of star acts like a "lighthouse" that periodically emits radio waves into space?
   a. black hole
   b. neutron star
   c. pulsar
   d. supernova

7. Which of the following stars is closest to the Earth?
   a. Rigel
   b. Polaris
   c. Betelgeuse
   d. Alpha Centauri

8. What stage of a star's life is the Sun in now?
   a. proto star
   b. main sequence
   c. red giant
   d. white dwarf

9. What is the approximate age of the Sun?
   a. 2.5 billion years old
   b. 5 million years old
   c. 5 billion years old
   d. 10 billion years old

10. Which layer of the Sun is visible only during a solar eclipse?
    a. convective layer
    b. photosphere
    c. chromosphere
    d. corona

11. On which layer of the Sun do sunspots occur?
    a. convective layer
    b. photosphere
    c. chromosphere
    d. corona

12. Approximately how long is a solar cycle?
    a. 11 years
    b. 22 years
    c. 5 billion years
    d. 10 billion years

13. Climate change, power interruption, communication problems, and the northern lights are all associated with:
    a. granules
    b. solar flares
    c. nuclear fusion
    d. supernovas

14. The Sun is approximately how many times bigger than the Earth?
    a. 10 times
    b. 20 times
    c. 50 times
    d. 100 times

15. Scientists estimate the Sun will enter the red giant phase in approximately:
    a. 11 years
    b. 11 billion years
    c. 5 billion years
    d. never

continued

Matching *Match the terms with the correct definitions.*

a. star
b. nuclear fusion
c. stellar nebula
d. luminosity
e. spectral class
f. red dwarf stars
g. white dwarf
h. red giant
i. blue supergiant
j. neutron star
k. pulsars
l. black hole
m. sunspots
n. solar flares
o. solar wind

1. ____ A classification of a star that is dimmer than the Sun, white, small, and extremely hot.

2. ____ A large, shining, spherical celestial object that is held together by its own gravity and is undergoing nuclear fusion.

3. ____ A classification for a dim star that is small, red, and relatively cool.

4. ____ A nuclear reaction that is caused by combining, or fusing, two elements, which results in the creation of a great amount of energy.

5. ____ A classification used to describe a star's unique spectrum.

6. ____ The stream of particles and electromagnetic radiation that is emitted by the Sun and travels out into space in all directions.

7. ____ A classification for a star that is very bright, large, red, and relatively cool.

8. ____ Large flamelike emissions of hot plasma and radiation that leap off the surface of the Sun.

9. ____ A classification of massive stars that are blue, bright, and extremely hot.

10. ____ Dark spots that appear on the surface of the Sun, which are believed to be cooler areas on its surface.

11. ____ A classification for a star that is extremely dense and mostly composed of neutrons.

12. ____ A theoretical celestial object with a strong gravitational attraction that prevents light from escaping its surface.

13. ____ A type of neutron star that regularly emits periodic radio signals.

14. ____ A large mass of collapsing gas and dust that forms stars and planets.

15. ____ A measure of the rate at which stars radiate electromagnetic energy into space.

Critical Thinking

1. Explain what you think humans will need to do on Earth to survive when the Sun begins to enter the red giant phase.

CHAPTER

6 The Solar System and its Place in the Universe

Section 6.1 – The Solar System Objectives

The Formation of the Solar System • The Terrestrial Planets • The Gaseous Planets • Asteroids • Meteoroids • Comets

After reading this section you should be able to:

- Identify all the celestial objects that together make up the solar system.
- Differentiate between a terrestrial planet and a gas giant.
- Identify the nine planets in the solar system and some of their unique features.
- Define the term *asteroid* and identify the location of the asteroid belt within the solar system.
- Differentiate among a meteoroid, meteor, and meteorite.
- Define the terms *comet* and *Oort cloud.*

Section 6.2 – The Earth's Place in the Universe Objectives

The Age and Size of the Universe • Galaxies • Quasars • Life in the Universe

After reading this section you should be able to:

- Explain the approximate age of the universe and how it is believed to have begun.
- Define the term *galaxy* and identify the three different types of galaxies.
- Identify the approximate location of the solar system within the Milky Way galaxy.
- Define the term *quasar.*

TERMS TO KNOW

planets
terrestrial planets
gas giants
meteor
moon
asteroid belt
asteroids
meteoroids
meteors
meteorites
meteor showers
comets
Oort cloud
galaxy
Milky Way
quasars

INTRODUCTION

When you are gazing up at the sky on a clear night, you are sharing a view that has been seen by countless people for thousands of years. Technology may have changed our world, but the one thing that has remained the same is the view of the heavens. This creates a unique connection between all people who are living and have lived on the Earth. Because the universe is so vast and contains billions of star systems, it is possible that we are not the only ones who are trying to unlock its secrets. The awesome display of the celestial sphere has inspired generations of humans to ponder their place in the universe. Over time our explanations of the heavens have changed, revealing much about the vastness of space and all the wonders it contains. Each time a scientific discovery is made that reveals new information about the universe, further questions are raised and the universe becomes more fascinating to us.

6.1 The Solar System

The Formation of the Solar System

The solar system is a group of celestial objects that exist in a region centered by our sun. The celestial objects that make up the solar system include planets, moons, asteroids, meteors, comets, and the Sun. Our solar system is believed to have been formed from a swirling cloud of dust and gas called the Solar Nebula about 4.5 to 4.7 billion years ago (Figure 6–1). Eventually the inner portions of this nebula created the Sun, and the outer portions formed the **planets;** their moons, also called satellites; and other interplanetary material found in our solar system. The largest objects in the solar system, excluding the Sun, are known as planets. The solar system has nine planets that have been positively identified. A planet is a large body composed of rock and gas that orbits a star. In our solar system we divide the nine planets into two main groups, the inner **terrestrial planets** and the outer gaseous planets known as **gas giants.**

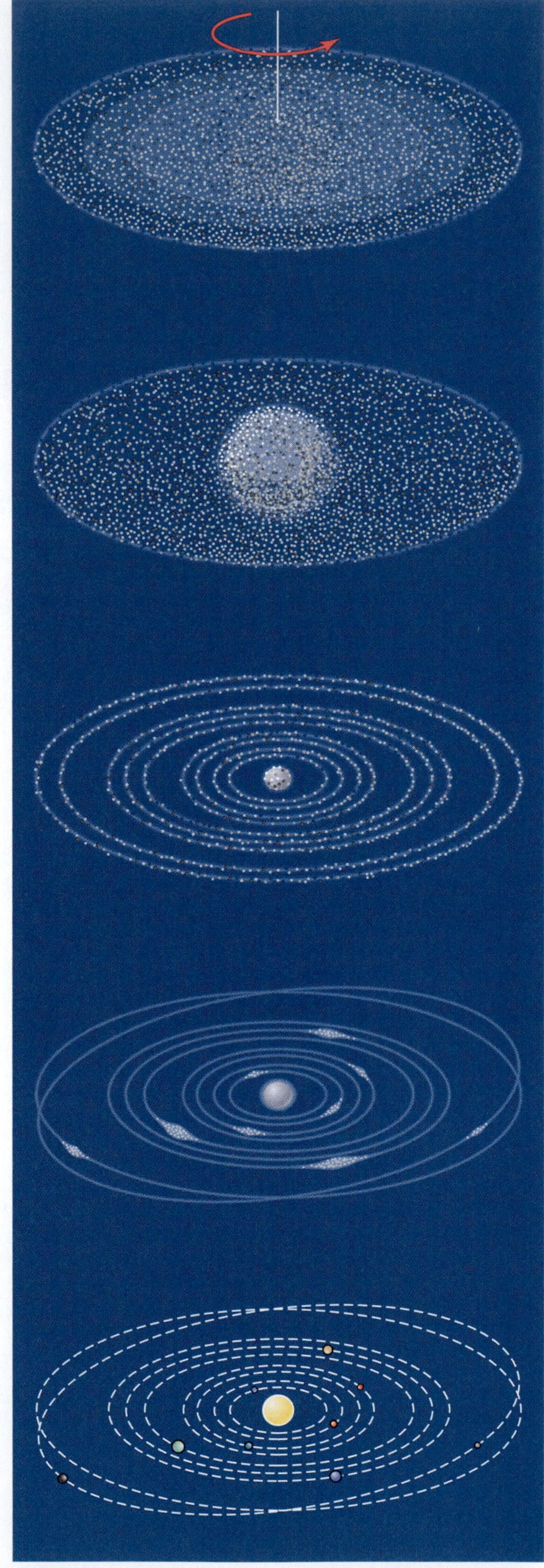

Figure 6–1 The formation of the solar system from the Solar Nebula.

The Terrestrial Planets

The inner terrestrial planets are planets that are relatively close to the Sun and are mainly composed of rock. These include Mercury, Venus, Earth, and Mars. The outer gaseous planets, also called the gas giants, lie further out in space from the Sun and mainly are composed of gases, although some may have rocky cores. The gaseous planets include Jupiter, Saturn, Uranus, Neptune, and Pluto. The distance that is used to measure how far a planet is from the Sun is called the astronomical unit, or AU. One astronomical unit is equal to the distance of the Earth from the Sun, which is approximately 93 million miles.

Mercury is the planet that is closest to the Sun, at a distance of 0.387 AU, or approximately 36 million miles. Mercury is about one-third the size of Earth and has no atmosphere. This tiny planet is also the fastest of all planets,

EARTH SYSTEM SCIENTISTS

NICOLAS COPERNICUS

Nicolas Copernicus was born in Poland in 1473. During his life, he devoted much of his time to the science of astronomy. Using mostly the observations of other astronomers of his time, he developed his own theory about the Earth's place in the universe. Since 200 A.D. scholars and the Roman Catholic Church believed that the Earth was the center of the universe, around which all other celestial objects revolved. This was known as the geocentric model of the universe, which was developed by a philosopher named Ptolemy. Copernicus's theory proposed that the Earth was not the center of the solar system or the universe. His observations suggested that the Earth, along with the other planets, revolved around the Sun, which also moved around the universe. He also proposed that the Earth's axis was tilted at approximately 23.5 degrees, which accounted for the apparent motion of celestial objects. Copernicus went on to explain that the Earth's atmosphere rotated with the Earth. His theory is known as the heliocentric theory and was first published in 1513 as an anonymous text. Copernicus also calculated the distance of the planets from the Earth using the astronomical unit, which is the distance of the Earth from the Sun. Because of the controversial nature of his theories, which went against the views of the church in Rome, Copernicus did not publish his full work on heliocentric theory until he was on his deathbed in 1543. Ironically, he dedicated his controversial book, De Revolutionibus, to Pope Paul III. Copernicus's groundbreaking book, which formed the foundations of modern astronomy, was on the Catholic Church's forbidden book list until 1835.

moving at approximately 29 miles per second around the Sun. Because Mercury has no atmosphere, it experiences wide temperature variations between day and night. At night the surface of Mercury can be as cold as −300° Fahrenheit, and during the day the surface heats up to more than 700° Fahrenheit. This is the greatest temperature fluctuation of any planet in the solar system. The surface of Mercury is covered with impact craters much like our own Moon (Figure 6–2). Mercury has a slow rotation rate of almost 59 days and takes more than 87 days to orbit around the Sun.

The planet Venus is located 0.723 AU from the Sun, or approximately 67 million miles. Venus is roughly the same size as the Earth and has an atmosphere that mainly consists of carbon dioxide gas and sulfuric acid clouds (Figure 6–3). Venus's clouds also reflect sunlight from the planet, which makes it the brightest planet in the sky. Venus is usually visible shortly after sunset, resembling a very bright star. The surface of Venus is shrouded by thick clouds of sulfuric acid that are most likely the result of volcanic eruptions. These clouds exist high above the planet's surface, where temperatures are approximately 70° Fahrenheit. This moderate temperature allows liquid water to mix with sulfur compounds and form sulfuric acid. Far down

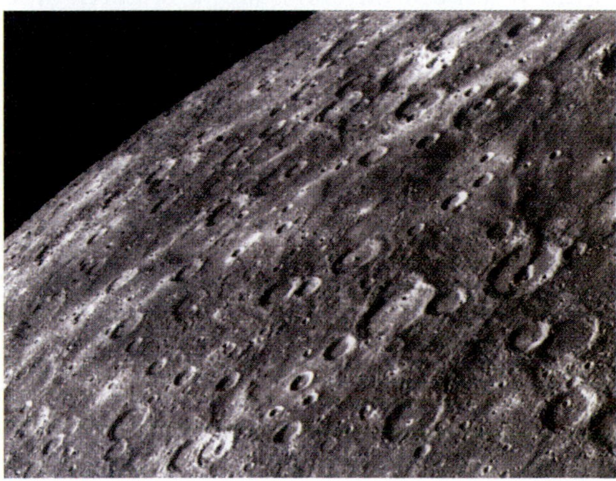

Figure 6–2 An image of the cratered surface of Mercury. *(Courtesy of NASA/NSSDC.)*

Figure 6–3 An image of Venus, which is roughly the same size as Earth. *(Courtesy of PhotoDisc.)*

below these clouds, however, the average surface temperature of Venus is approximately 800° Fahrenheit, with a surface pressure 90 times that on Earth. The extreme surface temperatures on Venus are believed to be the result of a runaway greenhouse effect. This is caused by the great amount of carbon dioxide in the planet's atmosphere, which traps heat on the surface. It takes more than 224 days for Venus to orbit around the Sun and 243 days to make one rotation on its axis. Venus's rotation on its axis is opposite to that of Earth's rotation. This is called retrograde motion, and it causes the Sun to rise in the west and set in the east on Venus!

Earth is the third planet from the Sun, at a distance of 1 AU, or 93 million miles. Earth is the only planet in the solar system that has liquid water and supports life. The Earth takes 365 days to orbit the Sun and 24 hours to make one full rotation on its axis. Our home planet also is orbited by one moon. The atmosphere on Earth is composed mainly of nitrogen and oxygen gas.

Mars is the fourth planet from the Sun and is the last of the four terrestrial planets. Mars is roughly half the size of the Earth and is located 1.52 AU from the Sun, or 141.5 million miles (Figure 6–4). Mars has a very thin atmosphere that mainly is composed of carbon dioxide gas. This causes wide temperature fluctuations on the surface of Mars. During the Martian day, the surface temperature can be more than −27° Fahrenheit, and at night the temperature can plummet to less than −117° Fahrenheit. Although the Martian atmosphere is thin, it can develop winds at speeds exceeding 62 miles per hour. Winds can create dust storms that cover large parts of the planet. The Martian atmosphere can also contain clouds. Martian clouds are composed of frozen water or dry ice (frozen carbon dioxide).

The surface of Mars has large canyons and channels that look much like they were formed from flowing liquid water. Because of this evidence, scientists believe that at some

Career Connections

PLANETARY SCIENTIST

Planetary scientists study the origin, evolution, and environments of planets in both the solar system and around other stars. To do this, planetary scientists must have a wide-ranging knowledge of meteorology, geology, biology, chemistry, and physics. They must then apply this knowledge to the study of planets beyond Earth. Because it is not easy to send scientists to other planets, planetary scientists must use long-range observations to infer what a planet's environment is like. This requires the use of many scientific instruments to gather data to learn more about planets. Remote sensing spacecraft, telescopes, and powerful computers aid planetary scientists in their attempts to learn more about planets. Although early astronomers made discoveries about the properties of planets in the solar system, it has been the work of planetary scientists over the past 20 to 30 years that has given us detailed knowledge of other worlds. These specialized astronomers are also interested in discovering how humans may someday live on other planets and if any life may exist somewhere other than Earth. Planetary scientists require a college education and usually conduct research in the academic fields or with NASA.

Figure 6–4 An image of Mars. *(Courtesy of PhotoDisc.)*

time in Mars's past, the surface may have contained liquid water. The channels that may have been excavated by flowing water are approximately 3.9 billion years old. Since that time, the Martian climate has been much too cold to contain liquid water. Some researchers believe that the evidence of once-flowing water on the Martian surface suggests that it also once contained life.

Others believe that there may still be living organisms on Mars. Scientists have recently discovered unique forms of bacteria that live in the ice and rock of Antarctica and in the extreme temperatures of hot springs on the Earth. This points to the possibility of some type of hardy life form that may also exist on Mars. Future missions to the planet will search for possible fossil evidence of extinct Martian life, as well as existing life forms. A meteor found in Antarctica recently is believed to be a piece of Mars that was blasted into space by an asteroid impact with the Martian surface. This meteor has been studied, and microscopic features that look very similar to fossilized bacteria on Earth were found in the meteor. Some scientists believe that this may be the first sign of a life form on a planet other than Earth. Mars is also home to the solar system's largest volcano, called Olympus Mons (Figure 6–5). This volcano is more than 78,000 feet high and 310 miles wide.

Mars also contains two polar ice caps. These are not composed of water, however, but of frozen carbon dioxide (Figure 6–6). Mars orbits the Sun in about 686 days and takes 1 day to make a complete rotation on its axis. Mars also is orbited by two **moons,** Phobos and

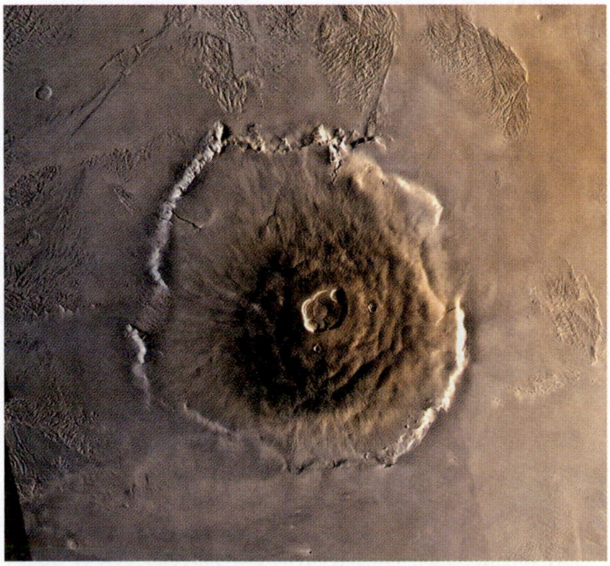

Figure 6–5 An image of the Olympus Mons on Mars, which is the largest known volcano in the solar system at more than 78,000 feet high and 310 miles wide. *(Courtesy of NASA/NSSDC.)*

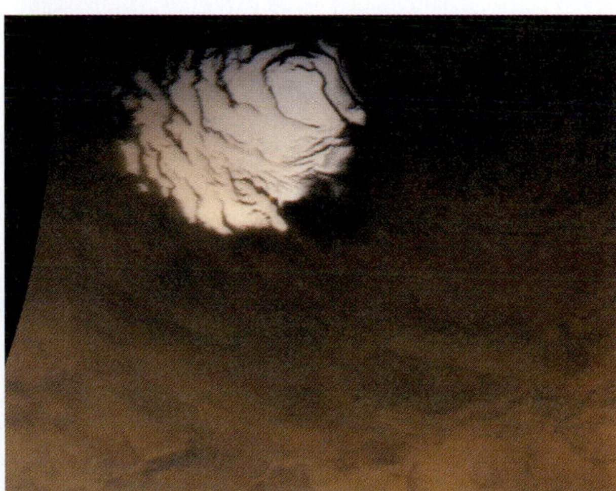

Figure 6–6 An image of the southern polar ice cap on Mars, which is believed to be composed of mostly carbon dioxide. The northern polar ice cap on Mars may contain a greater amount of frozen water. *(Courtesy of NASA/NSSDC.)*

Demos. Neither of the Martian moons possesses an atmosphere.

The Gaseous Planets

The first of the five gaseous planets, Jupiter, is located 5.2 AU from the Earth, or 483 million miles. Jupiter is the largest planet in the solar system, with a mass that is only one-thousandth that of the Sun's mass (Figure 6–7). Jupiter's diameter is 88,730 miles, and it takes a little more than 11 years to orbit the Sun. Unlike the inner terrestrial planets, Jupiter is composed mainly of hydrogen, helium, ammonia, and methane gas. Jupiter's cloudy surface is a variety of colors, including brown, red, orange, and white. These clouds are mostly composed of ammonia and form turbulent swirls around the planet in unique bands. These bands often result in large storms that can be seen on the planet. The most famous of Jupiter's storms is the great red spot. This storm resembles a hurricane on Earth but is large enough to hold two Earths. The great red spot is a storm that has been raging on Jupiter for more than 300 years. Below the thick atmosphere of Jupiter is believed to be layers of liquid hydrogen. There may also be a slushy liquid core located deep within the planet. Jupiter is also surrounded by three ring systems. Jupiter's rings are not as visible as Saturn's rings because they are made up of tiny dark particles. These particles are most likely the debris that was kicked up into space when Jupiter's moons were impacted by asteroids.

Figure 6–7 An image of Jupiter, the largest of the gas giants. *(Courtesy of PhotoDisc.)*

Figure 6–8 An image of Ganymede, one of Jupiter's moons, which is the most heavily cratered object in the solar system. *(Courtesy of NASA/NSSDC.)*

Jupiter is surrounded by at least 39 moons. The first moons of Jupiter were discovered in the year 1610 A.D. by Galileo using his newly invented telescope. These include Io, Europa, Ganymede, and Callisto. The remainder of Jupiter's moons are designated by a number, such as J6, J7, and so on. Ganymede is the largest moon in the solar system, with a diameter of approximately 3274 miles (Figure 6–8). This is about the size of the planet Mercury. This moon is composed of rock and water ice. Ganymede's surface also contains the most impact craters of any object in the solar system.

The moon Callisto is also composed of water ice and rock but appears very different from Ganymede (Figure 6–9). Its surface has been shaped by large outflows of water from its interior. This has created lines of frozen water that radiate outward from the impact craters.

The two inner moons of Jupiter include Io and Europa. Io is the closest moon to Jupiter and is also one of the solar system's most active moons. The surface of Io is covered with active volcanoes and has a thin atmosphere composed of sulfur dioxide. At one time Io was observed to have eight volcanic eruptions occurring at once (Figure 6–10). Io's volcanoes spew out molten rock, which spreads out onto the planet's surface, and gaseous sulfur compounds that are blasted more than 190 miles into the atmosphere. These sulfur compounds then condense in the cold atmosphere into solid particles that float down to the surface like snow. This sulfur "snow" covers the surface of the planet with a fine dust of white and orange sulfur compounds. The extreme volcanic activity that occurs on Io is believed to be caused by tidal forces that squeeze the moon and generate internal heat that powers the volcanoes. This is much like the heat that is generated when you twist a piece of metal wire. The surface temperature of Io is approximately −225° Fahrenheit, although some volcanic regions can have temperatures as high as 70° Fahrenheit. In these regions, large pools of liquid sulfur are believed to exist.

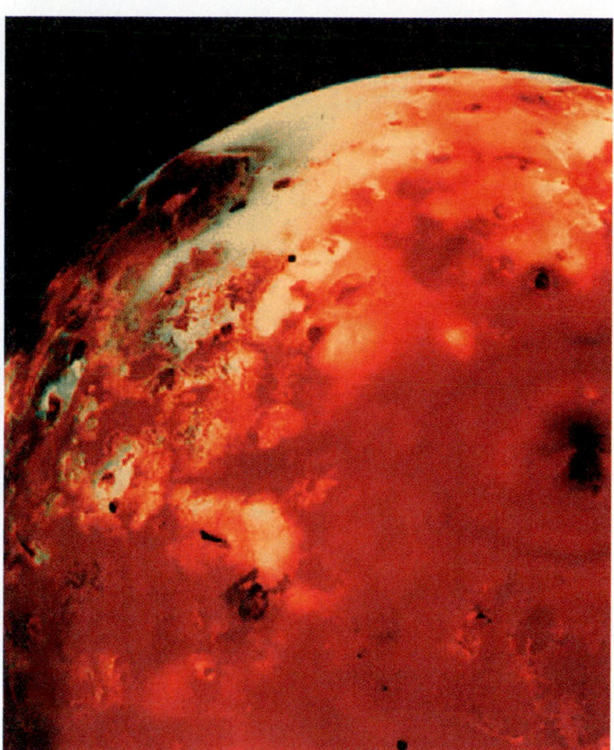

Figure 6–10 An image of the colorful surface of Io, one of Jupiter's moons, which is covered in many active volcanoes that spew out sulfur compounds. *(Courtesy of NASA/NSSDC.)*

Figure 6–9 An image of Callisto, one of Jupiter's moons, which has a surface covered in rock and water ice. *(Courtesy of NASA/NSSDC.)*

Jupiter's second closest moon to its surface is Europa. This moon is unique in that its surface is completely covered with frozen water. Some scientists believe that below this frozen water there may exist liquid water. Evidence of this lies on the surface of Europa, which is covered with large cracks. Some of these cracks resemble the cracked ice fields off the coast of Antarctica on Earth. Future missions planned by the National Aeronautic and Space Administration (NASA) will try to locate a possible liquid ocean beneath the ice of Europa (Figure 6–11). If liquid water does exist on Europa, it may be the most likely place in the solar system, besides the Earth, to support life in some form.

The second gaseous planet in the solar system is Saturn. Saturn is probably the most visually stunning of all the planets because of its colorful rings. Saturn is located 9.6 AU from the Sun, or 886.2 million miles. Saturn is also composed mainly of hydrogen and helium

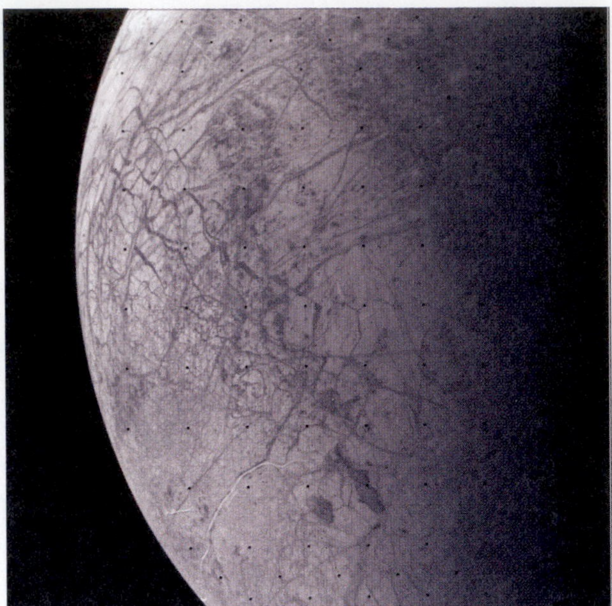

Figure 6–11 An image of the cracked ice on the surface of Europa, one of Jupiter's moons, which is believed to contain a liquid ocean beneath the ice. *(Courtesy of NASA/NSSDC.)*

gas, with clouds made of ammonia and methane. These clouds are not as spectacular as Jupiter's clouds and form regular bands around the planet. The bands move in alternating east-west directions at speeds of more than 1000 miles per hour. The most striking attribute of the planet Saturn is its unique rings (Figure 6–12). The planet is orbited by seven main ring systems, which are composed of billions of ice particles. These particles range in size from tiny grains of sand to giant boulders. This debris reflects different wavelengths of light giving each ring its unique color. If we could enter the ring systems of Saturn, it would appear as if we were in a great blizzard. Saturn also is orbited by at least 22 moons. The largest of Saturn's moons, Titan, is the only moon in the solar system that contains a substantial atmosphere. Titan is larger than the planets Mercury and Pluto and has an atmosphere that is mainly composed of nitrogen gas. Researchers believe that Titan may have an atmosphere that is similar to that of the primitive Earth. Below Titan's thick atmosphere, the surface temperature is −289° Fahrenheit. At this temperature there may be large lakes composed of ethane and methane covering portions of the surface. In 2004 the Cassini-Huygens spacecraft will launch a probe into Titan's atmosphere to study this fascinating moon more closely.

Uranus is the seventh planet from the sun, located 19.2 AU, or 1.783 billion miles, from the center of the solar system (Figure 6–13). Uranus is also a gaseous planet that is about four times larger than the Earth. It takes 84 years for Uranus to orbit the Sun and more than 17 hours to make one full rotation. Uranus is unique among the planets in that its axis of rotation is tilted 90 degrees, so it appears to be spinning on its side as compared with the rotation of the other planets. The planet is composed mostly of ammonia and methane, with an atmosphere of hydrogen and helium. The methane and ammonia give Uranus its blue-green color. Uranus also is surrounded by a ring system and at least 18 moons.

Neptune is the eighth planet in the solar system and is located 30.1 AU, or 2.794 billion miles, from the Sun. Neptune is also a gaseous planet composed of hydrogen, helium, and methane. The methane gives Neptune its blue color (Figure 6–14). Neptune orbits the Sun in approximately 168 years and makes one

Figure 6–12 An image of Saturn, showing its distinctive rings. *(Courtesy of PhotoDisc.)*

Figure 6–13 An image of Uranus. *(Courtesy of NASA/NSSDC.)*

complete rotation on its axis every 19 hours. Neptune also has a small ring system and eight moons. One of Neptune's moons, Triton, is the coldest object in space so far recorded. The surface temperature of this moon is −390° Fahrenheit! Astronomers believe that this moon will eventually collide with Neptune in about 10 million to 100 million years.

Figure 6–14 An image of Neptune, which appears blue as a result of the high percentage of methane in its atmosphere. *(Courtesy of PhotoDisc.)*

The ninth and final planet in the solar system is Pluto. Pluto is located 39.4 AU, or 3.666 billion miles, from the Sun. Pluto is a very small planet, with a diameter of only 1430 miles. Pluto is a gaseous planet composed of frozen methane, nitrogen, and carbon monoxide. It takes Pluto more than 247 years to orbit the Sun, and it completes one rotation on its axis in 6 days. Pluto has one moon, Charon, which is half its size (Figure 6–15). This cold moon is composed mostly of frozen water. Table 6–1 summarizes information about the nine planets of our solar system.

Asteroids

Lying between Mars and Jupiter is a region of our solar system known as the **asteroid belt.** This area is filled with a debris field of rocky **asteroids** that also orbit the Sun between a distance of 2.2 and 3.3 AU from the center of the solar system. Asteroids are the rocky remains of the solar system's formation (Figure 6–16). There may be thousands of asteroids in the asteroid belt that range in size from a small car to more than 600 miles in diameter. The largest asteroid so far discovered in the asteroid belt is called Ceres, which has a diameter of more than 633 miles. Other large belt asteroids include Vesta, Pallas, and Hygeia.

Not all asteroids lie within the asteroid belt. Some are located inside the orbit of the Earth around the Sun; these are called Aten asteroids. One of the largest asteroids ever discovered, Chiron, orbits the Sun near the planet Jupiter. This large asteroid is estimated to be more than 248 miles in diameter. Some asteroids are classified by their ability to cross the Earth's orbit. These are a concern to astronomers because they have the potential to strike the Earth's surface. This class of asteroids is known as Apollo asteroids. Researchers estimate that there may be between 300 and 700 of this type of asteroid. Most Apollo asteroids

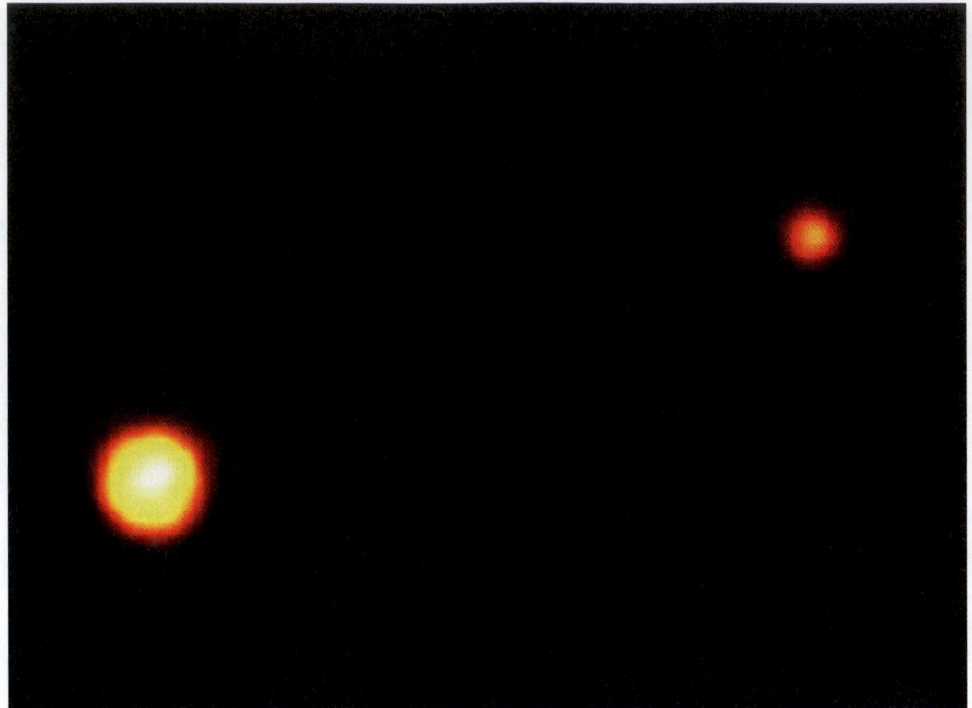

Figure 6–15 An image of Pluto and its moon, Charon. *(Courtesy of NASA/NSSDC.)*

TABLE 6–1 Solar system data table

Object	Mean Distance from Sun (millions of km)	Period of Revolution	Period of Rotation	Eccentricity of Orbit	Equatorial Diameter (km)	Mass (Earth = 1)	Density (g/cm³)	Number of Moons
Sun	–	–	27 days	–	1,392,000	333,000.00	1.4	–
Mercury	57.9	88 days	59 days	0.206	4,880	0.553	5.4	0
Venus	108.2	224.7 days	243 days	0.007	12,104	0.815	5.2	0
Earth	149.6	365.26 days	23 hr 56 min 4 sec	0.017	12,756	1.00	5.5	1
Mars	227.9	687 days	24 hr 37 min 23 sec	0.093	6,787	0.1074	3.9	2
Jupiter	778.3	11.86 years	9 hr 50 min 30 sec	0.048	142,800	317.896	1.3	39
Saturn	1,427	29.46 years	10 hr 14 min	0.056	120,000	95.185	0.7	18
Uranus	2,869	84.0 years	17 hr 14 min	0.047	51,800	14.537	1.2	21
Neptune	4,496	164.8 years	16 hr	0.009	49,500	17.151	1.7	8
Pluto	5,900	247.7 years	6 days 9 hr	0.250	2,300	0.0025	2.0	1
Earth's Moon	149.6 (0.386 from Earth)	27.3 days	27 days 8 hr	0.055	3,476	0.0123	3.3	–

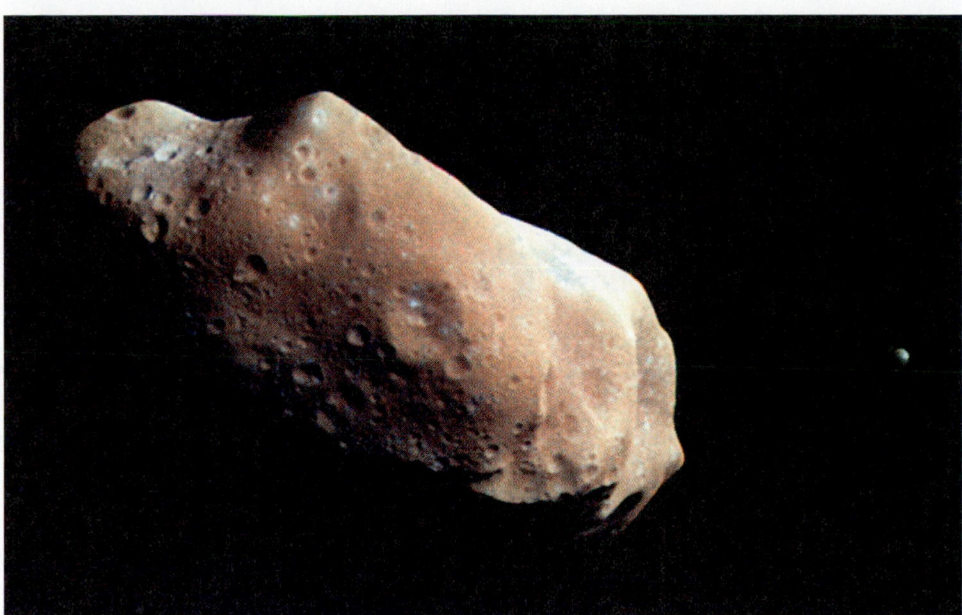

Figure 6–16 An image of the cratered surface of an asteroid. *(Courtesy of NASA/NSSDC.)*

are no larger than 3000 feet in diameter. In 1989 Asteroid 1989 FC passed within 430,000 miles of the Earth, which is the closest an asteroid has come to the Earth so far. Astronomers estimate that an asteroid strikes the Earth once every 2000 years. The last recorded strike by a celestial object occurred in 1908 over Northern Siberia in Asia. This event, known as the Tunguska blast, happened in an extremely remote region of Russia. The object exploded above the Earth's surface and leveled thousands of acres of trees. This explosion did not leave an impact crater behind and may have been a comet instead of an asteroid.

Meteoroids

Smaller chunks of rock that are located in the solar system are called **meteoroids, meteors,** and **meteorites.** Meteoroids are fast-moving chunks of rock that travel through space at high velocities. These rocks are the remains of the early solar system, comets, and fragments of our own Moon and nearby planets. These objects range in size from tiny grains of sand to large boulders. When a meteoroid's orbit crosses that of Earth, it may burn up in the atmosphere, creating a meteor, also known as shooting star (Figure 6–17). It is estimated that more than 1000 tons of meteoroid material rains down on the Earth every day! Meteors that make it all the way to the Earth's surface are called meteorites. Meteorites are grouped into three categories based on their composition. Stony meteorites are composed of silicate rock material. Iron meteorites are composed of an iron-nickel alloy and are very dense. Stony-iron meteorites are composed of a mixture of silicate rock and iron.

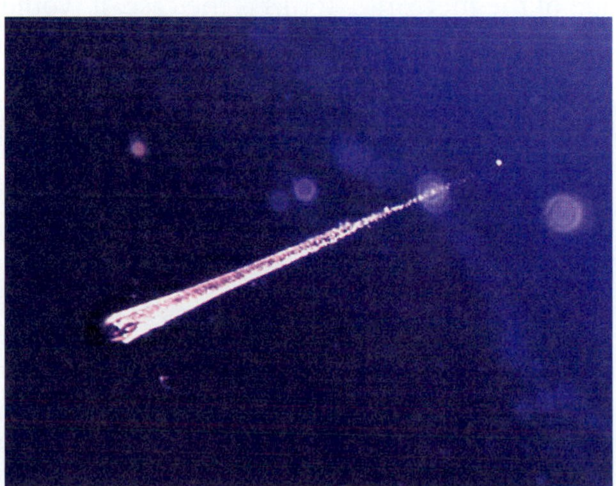

Figure 6–17 An image of a meteor burning up as it enters the Earth's atmosphere. *(Courtesy of ESA, NASA.)*

During certain times of the year, many meteors enter the Earth's atmosphere at once. These events are called **meteor showers.** Many meteor showers are associated with the debris left behind by a comet as it orbits around the Sun. The Perseid meteor shower occurs around the middle of August every year, when the Earth's orbit passes through the debris of comet 1862 III. During the height of this periodic meteor shower, on average a meteor enters the atmosphere once every minute. The largest iron meteorite found on the Earth fell in Southwest Africa and weighed more than 119,000 pounds. The largest stony meteorite fell in Kansas in the United States and weighed approximately 2200 pounds. Many meteorites are found on the ice fields of Antarctica, where they are easily visible.

In 1984 a team of meteorite hunters found a stony meteorite that probably came from the planet Mars. After careful analysis, this meteorite, called ALH 840001, was determined to have been a piece of the Martian surface that was launched into space by an asteroid impact with the planet approximately 16 million years ago. This Martian rock then entered the Earth's atmosphere and landed on Antarctica as a meteorite approximately 13,000 years ago. What is most fascinating about this meteorite is what was found inside it. Scientists have found what they believe are the fossil remains of bacteria inside of the meteorite. If this is the case, it would be the first known example of life existing elsewhere in the solar system.

Comets

Comets are celestial objects that are composed of rock and ice that orbit the Sun. Once referred to as dirty snowballs, these objects are unique in their appearance among the heavens. A comet consists of three parts: the nucleus, coma, and tail (Figure 6–18). A comet's nucleus is the rock and ice portion of the comet, which can range in size from half a mile to 6 miles in diameter. The nucleus of a comet is composed of rock and icy chunks of water, methane, carbon dioxide, and ammonia. The coma surrounds the nucleus and is composed of the dust and gas that was once held within the nucleus. The coma can be extremely large, sometimes reaching the size of the planet Jupiter. Behind the comet lies the comet's tail. This is a stream of ionized gas blown away from the coma by the solar wind. The tail always points away from the Sun, as the solar wind interacts with the gas and dust of the comet. Comets are believed to be another remnant of the solar system's formation. Most comets come from a region at the edge of the solar system called the **Oort cloud.** The Oort cloud is located about 100,000 AU from the Sun and may contain as many as 1 trillion comets. Comets probably leave the Oort cloud as a result of the gravity

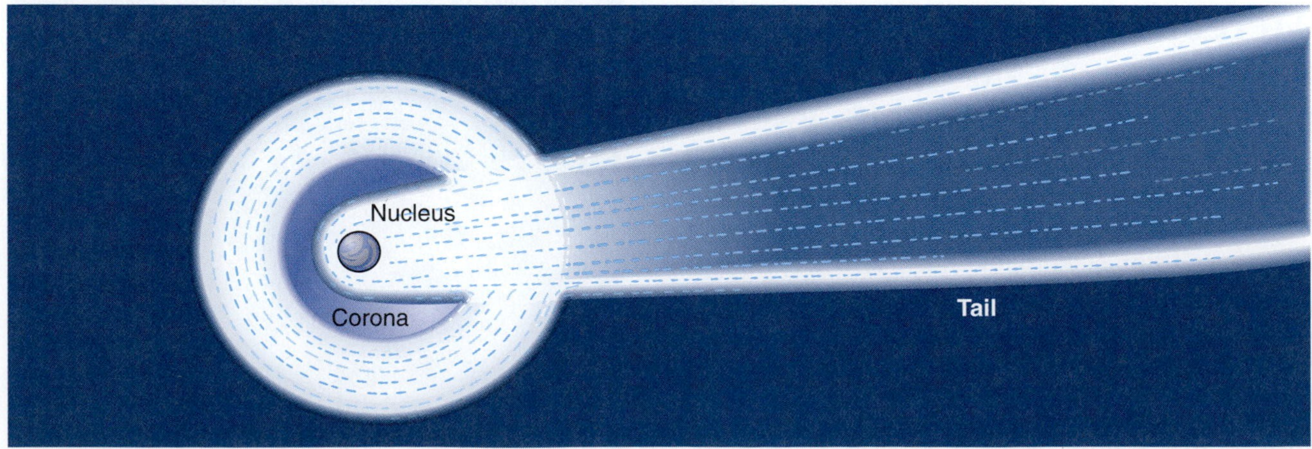

Figure 6–18 The parts of a comet.

from nearby stars and galaxies disturbing their orbits. Another source region for comets in the solar system may be located just outside Pluto's orbit. This area is known as the Kuiper belt and may contain billions of comets.

There are two main types of comets, which are classified based on their period of revolution around the Sun. Long-period comets take an extremely long time to orbit the Sun, such as 30 million years. Short-period comets orbit the sun in periods less than 200 years. The famous Halley's comet takes 76 years to make one complete orbit around the Sun. This comet was last seen in 1986 and will be visible again on Earth in the year 2061. Some short-period comets move at much faster orbits around the Sun. Many of these comets take only 3 to 14 years to complete one orbit. In July 1994 the comet Shoemaker-Levy 9 collided with Jupiter, creating an incredible opportunity to see the effects of a comet impact with the massive planet.

Current Research

A British astronomer studying the orbits of comets around the Sun has proposed that the solar system may contain a tenth planet. Professor John Murray believes that the orbits of long-period comets that he has been observing are being influenced by a massive object outside the orbit of Pluto. Other astronomers believe that this massive planet may be one day discovered to be a brown dwarf. A brown dwarf is an object that is three to four times as massive as Jupiter but is not large enough to become a star. Murray also believes that this planet is orbiting the Sun in an opposite direction to that of the other planets in the solar system. Astronomers hypothesize that this massive object may be a companion to the Sun and is likely orbiting between 20,000 and 50,000 AU from the Sun. Murray suggests that this mystery planet is probably going to be very difficult to detect because it is most likely 10 times dimmer than the dimmest stars.

SECTION REVIEW

1. What is the solar system, and how long ago did it form?
2. What are terrestrial planets, and how many are there in the solar system?
3. List all the terrestrial planets.
4. What are the gaseous planets, and how many are there in the solar system?
5. List all the gaseous planets.
6. Which two moons in the solar system are scientists eager to study, and why?
7. What is an asteroid, and where do most of them originate?
8. What is the difference between a meteoroid, a meteor, and a meteorite?
9. What are the three types of meteorites?
10. What is a comet, and where do they originate?
11. Who was Copernicus?

For more information go to these Web links:

<http://seds.lpl.arizona.edu/nineplanets/nineplanets/nineplanets.html#toc>

<http://csep10.phys.utk.edu/astr161/lect/index.html>

<http://www.nationalgeographic.com/solarsystem/>

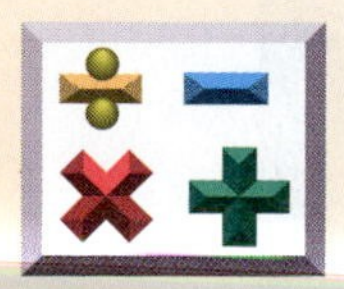

EARTH MATH

1) How many miles away is the Oort cloud located from the Sun?
2) If Jupiter has a diameter that is approximately 11.21 times greater than the Earth and the Earth's diameter is approximately 7926 miles, determine the approximate diameter of Jupiter.

6.2 The Earth's Place in the Universe

The Age and Size of the Universe

The universe is the total amount of volume in which all energy and matter exist. This includes all the stars in the sky, the solar system, and galaxies. The universe is constantly expanding as everything races away from a central point. Astronomers have discovered that the universe is expanding by analyzing the red shift associated with distant galaxies. Red shift is the shifting of wavelengths of light toward the red end of the visible portion of the electromagnetic spectrum. It is caused by an increase in the wavelength of light emitted from an object that is moving away from an observer. After careful analysis of light from many galaxies, astronomers have concluded that the universe is presently expanding.

The present size of the known universe is difficult to estimate; however, the most distant objects located so far are approximately 10 to 13 billion light years from Earth. The universe is believed to be between 14 to 20 billion years old. In comparison, Earth is approximately 4.5 billion years old. The birth of the universe began with an event known as the big bang. This event released all the known energy in the universe, which began to spread out from a central point. As the energy created from the big bang began to spread wider, it cooled to form clumps of matter. This was in the form of elements such as hydrogen and helium. Eventually, as energy and matter continued to spread apart and the universe grew larger, random clumps of dust and gas

Figure 6–19 An image of Orion A, a giant molecular cloud. *(Courtesy of T.A. Rector, B. Wolpa, G. Jacoby, AURA, NOAO, NSF.)*

Figure 6–20 An image showing a cluster of galaxies in the constellation of Hydra. *(© Anglo-Australian Observatory, Photograph by David Malin.)*

called giant molecular clouds began to form (Figure 6–19).

Galaxies

Portions of these clouds began to collapse to form the millions of stars that together make up galaxies. A **galaxy** is a grouping of millions or billions of individual stars (Figure 6–20). Each one of these stars sprang from the giant molecular clouds known as stellar nebulae. Galaxies appear as faint clouds of light when viewed through a telescope, unlike the bright points of light given off by stars. When viewed

EARTH SYSTEM SCIENTISTS *EDWIN HUBBLE*

Edwin Hubble was an American astronomer who was born in 1889. He began his career as an astronomer by attending the University of Chicago. After serving in the United States Army during World War I, he took up a position as the astronomer for the Mount Wilson Observatory in Pasadena, California, where he spent the rest of his career. It was there that he began to research stellar nebulae and other objects outside our own galaxy. His work led to the classification of galaxies, the study of their speed, and their particular direction of movement. This led him to develop Hubble's law, which stated that the more distant a galaxy, the greater its speed. Hubble also developed the Hubble constant, which was the ratio of a galaxy's speed to its distance. His work led to the understanding that all objects in the universe are moving away from a central point. Using Hubble's observations, astronomers can now estimate the approximate age of the universe. Hubble died in 1953. He will forever be linked to the advancement of our knowledge of the universe and to the famous space telescope that today bears his name.

with powerful telescopes, galaxies are revealed to be amazing objects that contain billions of stars.

There are three main types of galaxies: spiral, elliptical, and irregular. A spiral galaxy appears much like a rotating pin wheel. At the center of a spiral galaxy is the nucleus, which is surrounded by many spiral arms (Figure 6–21). The nucleus of a spiral galaxy appears like a large bulge of light, also called a halo, that is composed of millions of stars. Large groups of stars that surround the halo are called a globular cluster. The outlying spiral arms are also composed of stars that all rotate around the central nucleus.

The solar system is located in a spiral galaxy called the **Milky Way.** The solar system's location in the Milky Way galaxy is in one of the spiral arms, about halfway from the galactic center. The solar system revolves around the center of the Milky Way galaxy at approximately 136 miles per second. The Milky Way galaxy is approximately 100,000 light years in diameter and contains billions of individual stars. Many of these stars are similar to our Sun. All the stars that fill the nighttime sky are located within the Milky Way galaxy (Figure 6–22).

Many astronomers believe there is a black hole at the center of our galaxy. The Milky Way is part of a group of galaxies known as the Local Group. The Local Group contains approximately 27 galaxies, including a spiral galaxy called M31 and the Andromeda galaxy, which is the closest galaxy to Earth. This galaxy is located approximately 2 million light years from Earth.

Figure 6–21 An image of the M51 spiral galaxy. *(Courtesy of T.A. Rector and Monica Ramirez/NOAO/AURA/NSF.)*

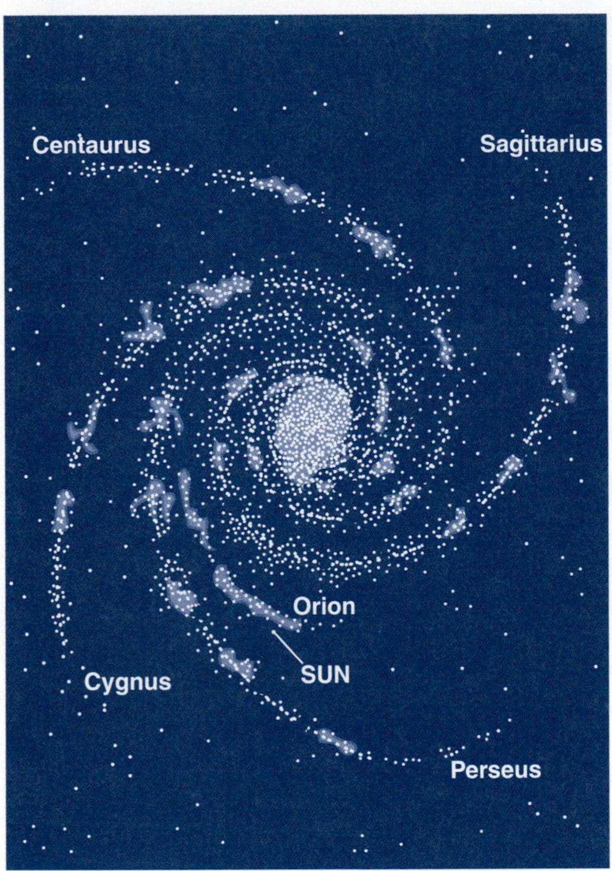

Figure 6–22 The location of the solar system within the Milky Way galaxy, near the center of one of its spiral arms.

An elliptical galaxy has a nucleus and a halo of stars but no spiral arms (Figure 6–23). These galaxies can range in size from 300,000 light years or more in diameter to only a few thousand light years across. The smaller elliptical galaxies are known as dwarf ellipticals. Astronomers believe that elliptical galaxies are the most common type of galaxy in the universe.

The third type of galaxy is the irregular galaxy. These galaxies have no well-defined shape. An example of an irregular galaxy is the Large Magellanic Cloud, which is the closest galaxy to the solar system (Figure 6–24). The Large Magellanic Cloud is approximately 175,000 light years away from Earth. Galaxies are believed to be well distributed throughout the universe. Astronomers estimate that there may be a billion galaxies in the universe, with each galaxy made up of more than 1 billion stars.

Quasars

Besides galaxies, another type of structure is found on the edges of the universe. These

Figure 6–23 An image of the elliptical galaxy M87. *(Courtesy of NOAO/AURA/NSF.)*

Figure 6–24 An image of an irregular galaxy called the Large Magellanic Cloud. *(Courtesy of NOAO/AURA/NSF.)*

objects, called **quasars,** are faint clouds of light racing away from the center of the universe at incredible speeds. These are the most distant objects ever discovered. Quasars get their name from the term *quasi-stellar radio sources,* because many of them have been found to emit radio waves. Some quasars are moving away from the Earth at the incredible rate of almost 90% of the speed of light! These are indeed the fastest objects yet dis-

EARTH SYSTEM SCIENTISTS *CARL SAGAN*

Carl Sagan was an American astronomer who devoted his life to the study of astronomy and to sharing his love and knowledge of the universe with the general public. Sagan began his career in astronomy by gaining his doctorate from the University of Chicago in 1960. In 1968 he took up the position of Director of Planetary Studies and eventually became the David Duncan Professor of Astronomy and Space Science at Cornell University. His early work involved research of the planet Venus, which revealed its hostile surface temperatures and physical features. His work also showed that Venus was experiencing a runaway greenhouse effect, which caused its high surface temperatures and was the result of increased levels of carbon dioxide gas in its atmosphere. Sagan was also the co-founder of the Planetary Society, which has become the largest space interest group in the world. In 1980 he produced and starred in the famous television series Cosmos, which helped present the science of astronomy to the general public. His work also included the search for the origins of life on Earth and the possibility of life elsewhere in the universe. This began the science of exobiology. In the 1970s he helped to design a series of plaques and records that were attached to the Pioneer 10 and Voyager spacecraft containing a message from Earth to possible extraterrestrial beings. During his life he published eight books, for one of which he received the Pulitzer Prize. Sagan died in 1996, leaving behind him a legacy of a life devoted to the search for understanding the universe and humanity's place within it.

covered in the universe. Quasars also emit incredible amounts of energy, more than 100 times the amount of energy released by the entire Milky Way galaxy. Astronomers believe that quasars may get their energy from extremely large black holes located at their centers. To create the energy emitted from a quasar, these black holes would have to be 1 billion times larger than our own Sun!

Life in the Universe

The brilliant objects that make up the universe—stars, galaxies, and quasars—are amazing to say the least. Although these objects have been studied, many questions still must be answered about the nature of the universe. Probably the most pressing of these is whether there are other planets in the universe like our own Earth. Recent discoveries have revealed that some stars in our own galaxy are being orbited by planets. This discovery suggests that all stars may indeed have their own planetary systems. If this is true, could some of these planets support life? The size and scale of the universe suggests that there may be millions of planets like our own Earth. If so, then it is possible that some of these planets do harbor life. As one astronomer said, "If the Earth were the only place in the universe to support life, what a great waste of space!" The distances between stars, however, are so great that we may never be able to visit planets outside our own solar system to see if there is life. Some researchers are searching the universe for radio signals or laser pulses that may indicate intelligent life on another planet. This may be the only way to truly discover if life does exist outside of Earth.

Career Connections

Exobiologist

An exobiologist studies the possibilities of life existing somewhere else in the universe other than on Earth. Exobiology closely resembles traditional biology but applies its knowledge of life processes to other planets, moons, or in space. Today an exobiologist studies how life might exist on other worlds and what types of life might be found there. A branch of exobiology also studies how living organisms from Earth survive in the microgravity of space. Many exobiologists are interested in studying life on Earth that can withstand extreme environments. Organisms like bacteria found living in hot springs and deep in the ice and rock of Antarctica are of special interest to exobiologists. If these organisms can survive in these harsh environments on Earth, then similar organisms might exist elsewhere in our solar system. Someday in the near future exobiologists may travel to Mars or to Jupiter's moon Europa to search for life. Exobiologists require a college education in biology and planetary science and can find work in the academic fields or with NASA.

For more information go to these Web links:

<http://csep10.phys.utk.edu/astr162/lect/index.html>

<http://csep10.phys.utk.edu/astr161/lect/index.html>

Earth Math

1) If the fastest quasar is moving at 90% of the speed of light and the speed of light is 186,000 miles per second, determine how fast this quasar is moving away from the center of the universe.

2) Using the speed you determined for the fastest quasar in question 1, how long would it take for a quasar to travel from the Earth to the Sun? The distance from the Earth to the Sun is approximately 93 million miles.

SECTION REVIEW

1. How old is the universe estimated to be, and what event began the universe as we know it?
2. What is a galaxy, and what are the three types of galaxies?
3. What is the name of the galaxy in which the solar system is located, and what type of galaxy is it?
4. What is the approximate size of our galaxy?
5. How many galaxies are believed to exist in the universe?
6. Which galaxy is closest to our own galaxy, and how far away is it?
7. What is a quasar?
8. Who were Edwin Hubble and Carl Sagan?

CHAPTER SUMMARY

The solar system is a group of celestial objects that all orbit around a central star called the Sun. A celestial object includes any object that is outside the Earth's atmosphere. The celestial objects that together make up the solar system include planets, moons, asteroids, comets, and meteoroids. There are nine planets in the solar system, which are divided into two categories, the terrestrial planets and the gas giants. Terrestrial planets are composed mainly of rock and orbit closest to the Sun. These include Mercury, Venus, Earth, and Mars. The gas giants are large planets composed mainly of gases. These planets orbit farther from the Sun and include Jupiter, Saturn, Uranus, Neptune, and Pluto. Each planet has its own unique physical features; Earth is the only known planet in the Solar system that supports life. Many planets are orbited by a moon or moons.

Asteroids are large chunks of rock that orbit the Sun. Many asteroids are believed to originate in the asteroid belt, which is located between the orbits of Mars and Jupiter. Meteoroids are smaller chunks of rock traveling through space at high speeds that occasionally enter the Earth's atmosphere. These are commonly known as shooting stars. Comets also exist in the solar system; they are composed of frozen compounds that surround a rocky core. As a comet orbits around the Sun, the solar wind vaporizes its frozen compounds and sends them trailing off into space, forming their characteristic tail.

The solar system is part of a large group of stars that together form the Milky Way galaxy. A galaxy is an extremely large cluster of stars. There are millions of galaxies that exist in the universe. The universe is believed to have begun approximately 15 billion years ago with an event called the big bang. All the stars and galaxies in the universe appear to be traveling away from one central point as a result of the big bang. The most distant objects ever detected in space are bright, high-energy quasars. These celestial objects are believed to be traveling near the speed of light.

CHAPTER REVIEW

Multiple Choice

1. Which of the following is not part of the solar system:
   a. the Sun
   b. the Milky Way
   c. the Moon
   d. comets

2. The planets that orbit closest to the Sun are also known as:
   a. gas giants
   b. elliptical
   c. irregular
   d. terrestrial

3. The planets that orbit farthest from the Sun are also known as:
   a. gas giants
   b. elliptical
   c. irregular
   d. terrestrial

4. Which of the following planets is one third the size of the Earth and has no atmosphere?
   a. Jupiter
   b. Mars
   c. Venus
   d. Mercury

5. Which of the following planets is approximately the same size of the Earth and has an atmosphere containing mostly carbon dioxide?
   a. Jupiter
   b. Mars
   c. Venus
   d. Mercury

6. Which of the following planets is believed to have once contained flowing liquid water?
   a. Jupiter
   b. Mars
   c. Venus
   d. Mercury

7. Which of the following planets contains a giant hurricane the size of three Earths?
   a. Neptune
   b. Pluto
   c. Jupiter
   d. Saturn

8. Which of the following planets contains a highly visible ring system?
   a. Uranus
   b. Pluto
   c. Jupiter
   d. Saturn

9. Which of the following planets has an axis of rotation that is tilted 90 degrees?
   a. Uranus
   b. Pluto
   c. Jupiter
   d. Saturn

10. Which of the following is located between the orbits of Mars and Jupiter?
    a. Oort cloud
    b. Halley's comet
    c. asteroid belt
    d. Perseid showers

11. What celestial object is also called a "dirty snowball"?
    a. asteroid
    b. comet
    c. meteoroid
    d. gas giant

12. Approximately how long ago do astronomers believe the big bang occurred?
    a. 15 billions years ago
    b. 10 billion years ago
    c. 5 billion years ago
    d. 4.5 billion years ago

13. What type of galaxy is the Milky Way?
    a. irregular
    b. spiral
    c. elliptical
    d. circular

14. The solar system is located in what part of the Milky Way galaxy?
    a. near the center
    b. in the middle of one of its arms
    c. between two arms
    d. on the outside edge

15. What are the fastest-moving celestial objects discovered so far in the universe?
    a. quasars
    b. galaxies
    c. meteoroids
    d. pulsars

continued

Matching *Match the terms with the correct definitions.*

a. planet
b. terrestrial planet
c. gas giant
d. meteor
e. moon
f. asteroid belt
g. asteroid
h. meteoroids
i. meteorites
j. meteor shower
k. comet
l. Oort cloud
m. galaxy
n. Milky Way
o. quasars

1. ____ A very high energy celestial object believed to be a type of galaxy, which is rapidly moving away from the center of the universe.

2. ____ An object that is orbiting the Sun, is smaller than a planet, and has no atmosphere.

3. ____ Small chunks of rock, no larger than a few feet in diameter, that travel through space.

4. ____ The name for the galaxy of stars in which the solar system is located.

5. ____ A meteor that does not burn up in the atmosphere and strikes the Earth's surface.

6. ____ A large grouping of stars.

7. ____ An event that describes a group of meteors entering into and burning up in the Earth's atmosphere.

8. ____ A hypothetical area that is located approximately 100,000 AU from the orbit of Pluto, where comets are believed to originate.

9. ____ A mixture of frozen compounds and rock that orbits the Sun; its distinct tail, composed of vaporized gas and dust, always points away from the Sun.

10. ____ A region in the solar system located between the orbits of Mars and Jupiter, where there are a high number of asteroids.

11. ____ A large celestial object that orbits around a star.

12. ____ The name for any large celestial body that orbits around a planet.

13. ____ A planet that is mostly composed of rock.

14. ____ A small chunk of rock, no larger than a few feet in diameter, that is traveling through space and enters the Earth's atmosphere, commonly known as a shooting star.

15. ____ A classification of planets that are extremely large and are composed mainly of gases.

Critical Thinking

1. Most all astronomers believe that life may exist elsewhere in the universe. Why do you think they have come to this conclusion?

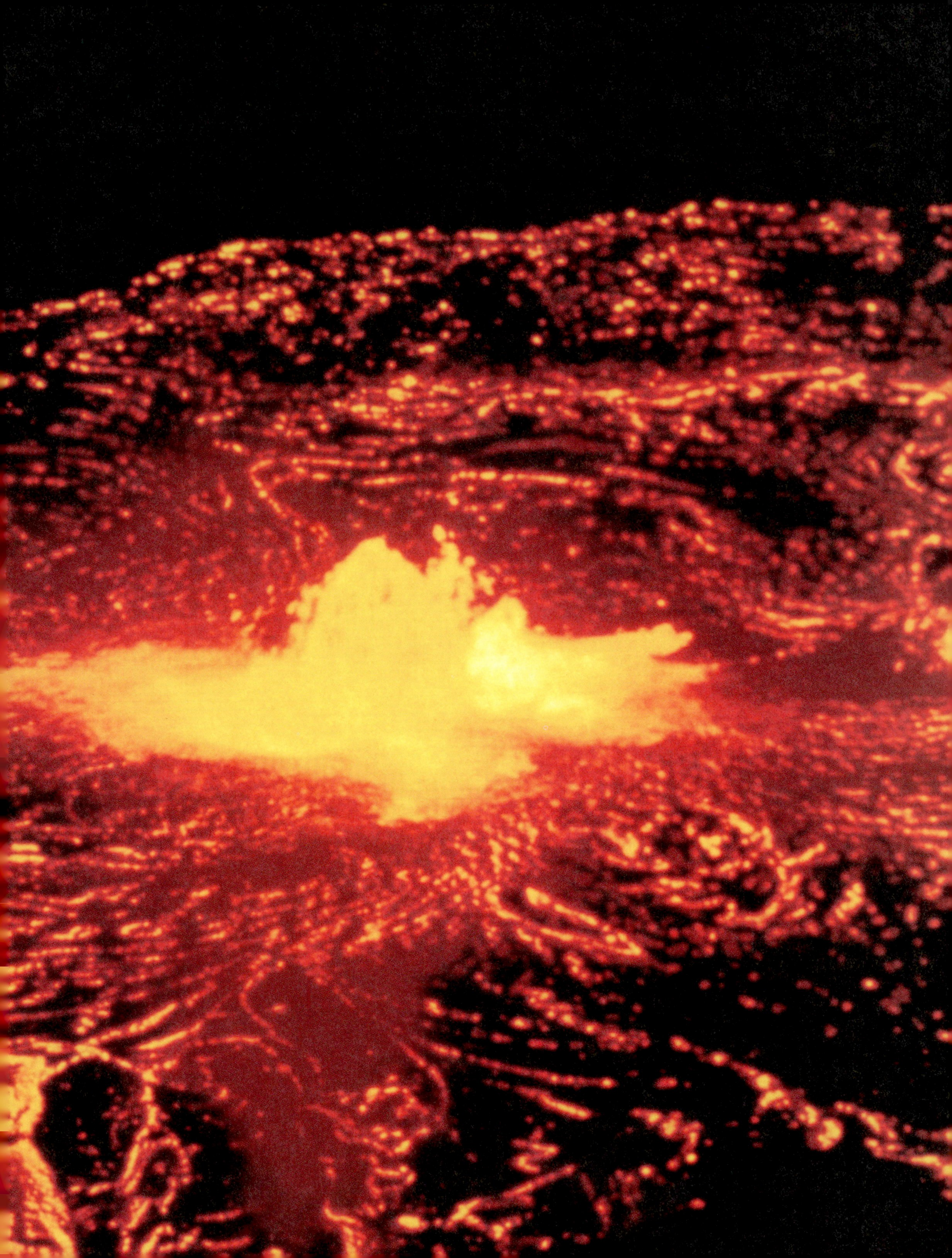

UNIT 3

The Lithosphere

Topics to be presented in this unit include:

- The Physical and Chemical Characteristics of the Earth's interior
- The Theory of Plate Tectonics
- Earthquakes and Volcanoes
- Characteristic Properties and Structure of Minerals
- Mineral Resources
- Formation and Classification of Rock Types
- The Rock Cycle
- Weathering, Erosion, and Deposition
- Soils
- The Earth's Geological History

OVERVIEW

When we study the lithosphere, we are attempting to unlock the secrets of the ground beneath our feet, the mountains that tower over our heads, and the strange world that exists at the bottom of the oceans. Human beings have had a long relationship with the solid Earth. Millions of years ago, our primitive ancestors made the first tools from stone. They found shelter in natural caves, where they also created some of the first known art forms. Today a close relationship still exists with the solid Earth. We mine it to extract building materials and precious stones. We drill into it to search for fossil fuels. We also excavate the solid Earth to search for information about its history and the history of the human race. But what exactly is the solid Earth? Why are there so many different kinds of rocks, and how did they form? Humans have been asking these questions for hundreds if not thousands of years. Even today, geologists are still attempting to unravel the mysteries that lie both deep within the Earth and on its surface. The Earth's lithosphere is complex and dynamic and has a profound affect on how the whole Earth system operates.

CHAPTER

7 The Earth's Interior

Objectives

The Earth's Core • The Mantle • The Earth's Crust • The Model of Earth's Interior

After reading this chapter you should be able to:

- Describe the composition of the Earth's inner and outer cores.
- Identify the main features of the Earth's mantle.
- Define the term *lithosphere.*
- Differentiate between oceanic and continental crust.
- Define the term *asthenosphere.*
- Explain how geologists were able to construct a model of the Earth's interior.

TERMS TO KNOW

lithosphere
mantle
Earth's core
convection cells
radioactive decay
friction
residual heat
crust
oceanic crust
continental crust
asthenosphere
inference

INTRODUCTION

The nineteenth-century novel *A Journey to the Center of the Earth,* written by Jules Verne, tells of an incredible scientific journey that went deep into the Earth. The story follows an expedition that discovered a fascinating underground world where strange seas and bizarre creatures existed. This classic work of fiction drew on the human fascination with what lies below the surface of the Earth. Many religions describe the underground world as a hot, hellish place where demons and monsters reside. Ancient peoples held funeral ceremonies deep in natural caves and caverns, which were believed to be the entrance to the underground world. This lead to the term *underworld,* which soon became synonymous with death and the afterlife. Some early scientists actually believed they could dig through the Earth from one side to the other. They even developed plans for special drilling vehicles designed to take up the task. The Earth's interior has been part of myth and wonder throughout the ages. The ground beneath our feet provides stability for us to walk on, build on, and drive on. The solid rock that lies below the surface is extremely hard and dense, as compared with most other substances on the Earth. But is the Earth totally composed of solid rock? Certainly the first time human beings gazed at the hot, molten rock spewing out from a volcano, they must have wondered what really lies below the surface. The advancement of science and technology in the twentieth century has finally given us a realistic view of what lies below us and has opened the door to further understanding of the Earth's systems.

The Earth's Core

The **lithosphere** is the solid outer crust of the Earth, which is made up of solid rock and the hot plasticlike upper **mantle.** If we could take the Earth and cut it in half, we would discover that it is made up of three major parts or layers (Figure 7–1). **Earth's core** is the very center of the Earth; it is divided into two parts, an outer liquid portion and an inner solid portion. The composition of both parts of the core are believed to be iron and nickel. You would have to travel more than 1800 miles to reach the Earth's outer core and another 1300 miles to reach the solid core. Geologists have used information gathered from the composition of meteorites to theorize that the Earth's core is made up of iron and nickel. These very dense elements are believed to have settled toward the center of the Earth during its formation. The Earth's core is extremely hot, reaching temperatures of more than 8000° Fahrenheit.

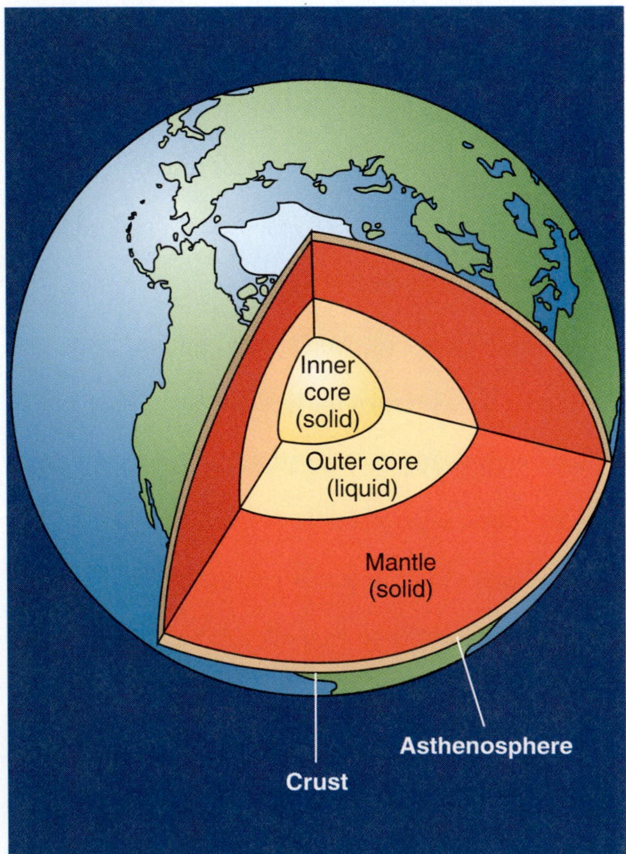

Figure 7–1 The Earth's interior, which consists of the solid inner core, liquid outer core, mantle, and crust.

The movement of seismic waves through the Earth, generated by earthquakes, have revealed that the Earth's core is divided into a solid inner portion and a liquid outer portion. This is due to the different way in which these waves interact with liquids and solids. Geologists have used techniques like this to study the Earth's interior, because it is extremely hot and under high pressure, making it impossible to observe directly.

The Mantle

The next layer of the Earth's interior is called the mantle. The mantle makes up most of the Earth's total volume and is mostly composed of a hot solid material consisting of the elements silicon, oxygen, iron, aluminum, and magnesium. The mantle is an approximately 1800-mile thick region that is also exposed to extremely high pressures. It is believed that heat from the Earth's hot core moves upward through the mantle, forming large **convection cells,** much like in air or water. The extreme heat that is generated deep within the core and mantle comes from **radioactive decay, friction,** and **residual heat** left over from the Earth's formation. The upper part of the mantle is made up of a plasticlike material called the asthenosphere, which flows like thick syrup. It is estimated that the Earth's heat increases approximately 1° Fahrenheit for every 50 feet in depth.

The Earth's Crust

The outer part of the Earth is composed of solid rock called the **crust.** The Earth's crust varies in its thickness from more than 40 miles underneath high mountains to only 3 miles underlying some parts of the ocean. The crust is relatively thin compared with the other parts that make up the Earth's interior. The Earth's crust varies in its density depending on its location around the planet. The crust that lies under the world's oceans, also called **oceanic crust,** has an average density of 3 grams per cubic centimeter. The crust that makes up the

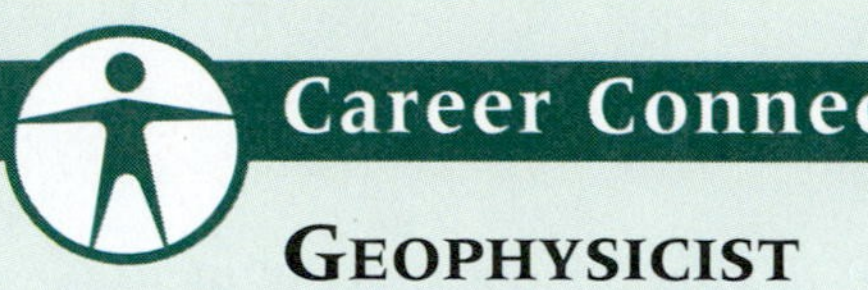

Career Connections

GEOPHYSICIST

A geophysicist utilizes a knowledge of physics to study the Earth's interior, along with its gravitational and magnetic fields. These scientists are trying to unlock the secrets of the mechanisms that exist deep within the Earth, how they affect life at the surface, and our planet's magnetic field. Geophysicists study the heat that is generated deep within the Earth and how it is distributed around the globe. This work is an important aspect of understanding plate tectonics. Geophysicists are also interested in the stress that builds up in the Earth's plates, which eventually leads to earthquakes. These scientists use a wide array of computers and sensing instruments to study the Earth's interior. His theories were published in 1795 in his book titled Theory of the Earth.

world's continents, also known as the **continental crust,** has an average density of 2.7 grams per cubic centimeter and is composed mainly of granitic rock. The oceanic crust is slightly denser than continental crust because it forms under the intense pressure of the deep ocean and is composed mainly of basalt rock. The solid crust of the Earth floats on top of the plasticlike upper mantle, much like a cracker floating on hot soup. Only seven elements make up nearly 99% of the Earth's crust. Oxygen and silicon are the most abundant elements, composing approximately 72% of the rocks in the crust. The remaining elements that make up most of the Earth's crust include aluminum, iron, calcium, magnesium, and sodium. Together, the lower part of the crust and the upper mantle form a unique layer called the **asthenosphere.** The asthenosphere is the plasticlike layer of rock on which the solid crust "floats" (Figure 7–2).

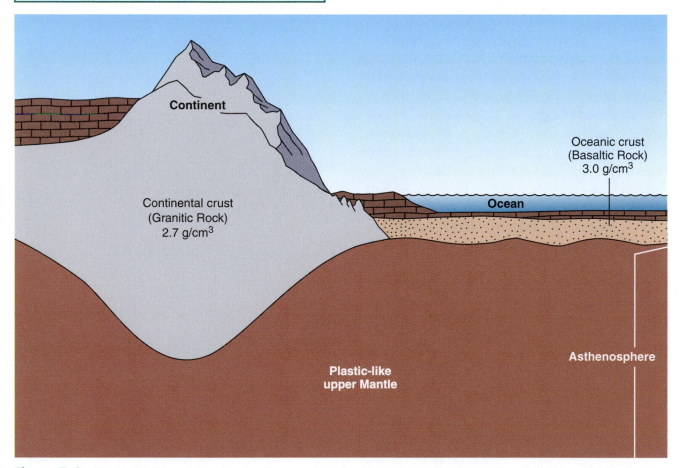

Figure 7–2 The Earth's lithosphere, showing the oceanic and continental crust floating on the plasticlike upper mantle called the asthenosphere.

EARTH SYSTEM SCIENTISTS *JAMES HUTTON*

James Hutton was born in Edinburgh, Scotland, in 1726. After receiving his doctorate of medicine in 1749, he spent the next 20 years farming the land and traveling extensively in his native Scotland. His travels and his growing interest in geology gave him unique insight into the geological history of the Earth. He was the first scientist to propose that the Earth was continually undergoing a slow process of erosion and creation of new landforms. His greatest contribution to modern geology included the theory of uniformitarianism, which stated that the same geological processes that are occurring on Earth today have been occurring throughout its history. This fundamental geological concept forms the base of modern geology. He also proposed that the erosion of the continents led to the formation of sedimentary rocks on the floor of the ocean. These rocks were then thrust up to form new mountain chains by the internal heat of the Earth. He also pointed out the difference between igneous rock formation and sedimentary rock formation. His theories were published in his book titled *Theory of the Earth.*

The Model of Earth's Interior

Amazingly, geologists have developed a model of the Earth's interior by studying the way that seismic waves, which are produced by earthquakes, travel through the Earth. They have also studied the composition of the interior of meteorites to help them better understand the Earth's interior. The deepest that humans have penetrated into the Earth by drilling is only 9 miles, and the deepest mine has only reached a little more than 2 miles deep. Therefore the model of the Earth's interior is constructed by inference, because we are unable to actually observe what exists deep below the surface. An inference is a conclusion based on gathered evidence. Geologists can only infer about the Earth's interior because it is impossible for humans to travel there. Many theories in geology are based on inference because the scale of the Earth is so large and much of the lithosphere cannot be studied by direct observation, such as the oceans or the atmosphere.

REVIEW

1. What is the composition of the Earth's inner and outer core?
2. Approximately how thick is the mantle region of Earth's interior?
3. Describe the effect on temperature and pressure of increasing depth into the Earth's interior.
4. What elements make up 99% of the Earth's crust?
5. How do geologists know what the Earth's interior is like?
6. Who was James Hutton?

EARTH MATH

1) IF THE RADIUS OF THE EARTH IS APPROXIMATELY 4000 MILES, DETERMINE THE PERCENTAGE THAT EACH PORTION OF THE EARTH'S INTERIOR OCCUPIES (CRUST, MANTLE, OUTER CORE, INNER CORE). ASSUME THAT THE AVERAGE THICKNESS OF THE CRUST IS 20 MILES AND THE AVERAGE THICKNESS OF THE MANTLE IS 1800 MILES.

For more information go to these Web links:

<http://geollab.jmu.edu/Fichter/PlateTect/erthstru.html>

<http://www.solarviews.com/eng/earthint.htm>

<http://interactive2.usgs.gov/learningweb/students/homework_geology.asp>

CHAPTER SUMMARY

The Earth is divided into three distinct layers: the core, mantle, and lithosphere. The Earth's core is divided into two parts, the liquid outer core and the solid inner core. Both are believed to be composed of iron and nickel. The mantle is the next layer, which is located between the core and the lithosphere. The mantle makes up most of the total volume of the Earth and is extremely hot and dense. The extreme heat of the inner portions of the Earth is derived mainly from radioactive decay, friction, and residual heat left over from when the planet formed. The outer layer of the Earth is called the lithosphere. The lithosphere is further subdivided into the solid outer layer of rock called the crust and the hot, plasticlike upper mantle called the asthenosphere. There are two basic types of crust: oceanic and continental. Oceanic crust forms below the oceans, is composed of basalt rock, and is denser than continental crust. Continental crust makes up the continents, is composed of granite rock, and is less dense than oceanic crust. Both types of crust float on the hot, plasticlike asthenosphere. Because humans cannot actually explore the Earth's interior, they have constructed a model of what they believe to exist below the surface by analyzing the way seismic waves travel through the Earth.

CHAPTER REVIEW

Multiple Choice

1. The inner core of the Earth is believed to contain:
   a. liquid iron
   b. solid iron
   c. liquid iron and nickel
   d. solid iron and nickel

2. Which objects have geologists studied to learn about the composition of the Earth's core:
   a. comets
   b. asteroids
   c. planets
   d. meteorites

3. Which of the following portions of the Earth's interior make up most of the total volume of the planet:
   a. crust
   b. mantle
   c. outer core
   d. inner core

4. As compared with oceanic crust, continental wcrust is:
   a. less dense and composed of granite
   b. more dense and composed of granite
   c. less dense and composed of basalt
   d. more dense and composed of basalt

5. As compared with continental crust, oceanic crust is:
   a. less dense and composed of granite
   b. more dense and composed of granite
   c. less dense and composed of basalt
   d. more dense and composed of basalt

6. The thin outer crust of rock is also called the:
   a. asthenosphere
   b. mantle
   c. continents
   d. lithosphere

7. The hot, plasticlike layer of rock on which the Earth's crust floats is known as the:
   a. asthenosphere
   b. mantle
   c. continents
   d. lithosphere

8. Geologists have constructed a model of the Earth's interior by:
   a. direct observation
   b. inference
   c. guessing
   d. tunneling

9. Which of the following is used to reveal the composition of the Earth's interior:
   a. satellite images
   b. seismic waves
   c. maps
   d. drills

10. Which elements compose most of the Earth's crust:
    a. hydrogen and oxygen
    b. aluminum and magnesium
    c. silicon and oxygen
    d. iron and nickel

Matching *Match the terms with the correct definitions.*

a. lithosphere
b. Earth's core
c. mantle
d. convection cells
e. radioactive decay
f. friction
g. residual heat
h. crust
i. oceanic crust
j. continental crust
k. asthenosphere
l. inference

1. ____ The extremely hot and dense center of the Earth that is believed to be composed of iron and nickel.

2. ____ An area of flowing, plasticlike molten rock located directly below the Earth's crust.

3. ____ The solid outer layer of the Earth that is composed of rock.

4. ____ A conclusion based on a set of observed facts.

5. ____ The solid outer layer of the Earth that is composed of rock and soil.

6. ____ The specific portion of the Earth's outer crust that is relatively thick, lower in density, and composed primarily of granitic rock that makes up the continents.

7. ____ The extremely hot, dense inner layer of the Earth that makes up most of the planet's total volume.

8. ____ The portion of the Earth's crust that is formed and lies below the oceans is typically more dense than continental crust, and is composed of basaltic rock.

9. ____ The circular movement of a fluid caused by a change in temperature and density associated with the transfer of heat.

10. ____ Heat left over from something.

11. ____ The breakdown of one element by the release of subatomic particles, which form a new element.

12. ____ The rubbing of one surface or object against another.

Critical Thinking

1. What observations do you think lead humans to create various myths about the hot, dark underworld that lies below the Earth?

CHAPTER

8 Plate Tectonics

Objectives

The Dynamic Crust • Evidence of Crustal Movement • Continental Drift • Sea Floor Spreading • Tectonic Plates • Plate Boundaries • Mantle Convection

After reading this chapter you should be able to:

- Identify four observations that can be used to prove that the Earth's crust is dynamic.
- Explain the geological principle of original horizontality.
- Describe the formation of a geosynclines.
- Explain the theory of continental drift, and provide three pieces of evidence used to support it.
- Describe the concept of sea floor spreading, and provide three pieces of evidence used to support it.
- Explain the theory of plate tectonics.
- Describe the three types of tectonic plate boundaries.
- Identify the locations of frequent earthquakes and volcanoes on the Earth's surface.
- Explain the concept of mantle convection.
- Define the term *hot spot.*

TERMS TO KNOW

strata	**geosyncline**	**rift valleys**
lateral forces	**isostasy**	**subduction**
anticline	**mid-ocean ridges**	**volcanic arc**
syncline	**magnetic north**	**mantle convection**
fault	**plate boundaries**	**hot spot**

INTRODUCTION

The theory of plate tectonics is one of the most exciting scientific discoveries of our time. Rarely in science is there the development of one explanation for the cause of many phenomena. This is known as a unifying theory. Plate tectonics explains some of the most well known natural phenomena on the Earth, including earthquakes, volcanoes, continental drift, deep ocean trenches, mid-ocean ridges, mountains, and deformed rocks, to name a few. Today, plate tectonics is widely accepted, but it wasn't that long ago when a group of scientists were laughed at by their peers, when it was declared that the continents floated along the surface of the Earth like boats in a pond. The story of plate tectonics is also interesting because its development reveals the way that science progresses and how theories develop. It is a perfect illustration of the scientific method at work and the advancement of human knowledge, all of which is the result of the development of new ideas based on careful observation, experimentation, and use of technology.

The Dynamic Crust

One of the most fundamental concepts in geology, which also provides evidence to support the theory of plate tectonics, is the notion that the Earth's crust is constantly in motion. At first this may seem like an outlandish idea, for how could the solid rock that composes the landscape move? All it takes is a trip to the mountains or a drive down the highway where the road cuts through rock formations to reveal that the Earth's crust is moving. Sometimes the rocks exposed near a road cut appear to have been tilted or folded by some great amount of force. Evidence that the Earth is dynamic, or constantly in motion, can be found all around the globe.

Evidence of Crustal Movement

The appearance of deformed rock layers suggests that the Earth's crust is moving in some way. To understand how deformed rocks lead to the notion that the Earth's crust is in a state of motion, you must understand the concept of original horizontality. Original horizontality is a fundamental principle in geology that assumes that most all rock forms in horizontal layers on the Earth's surface. These horizontal layers are called **strata** (Figure 8–1).

Just like the layers of cake, strata reveals the original orientation of the rock when it formed. There are three categories of deformed rock. Folded deformed rock occurs when

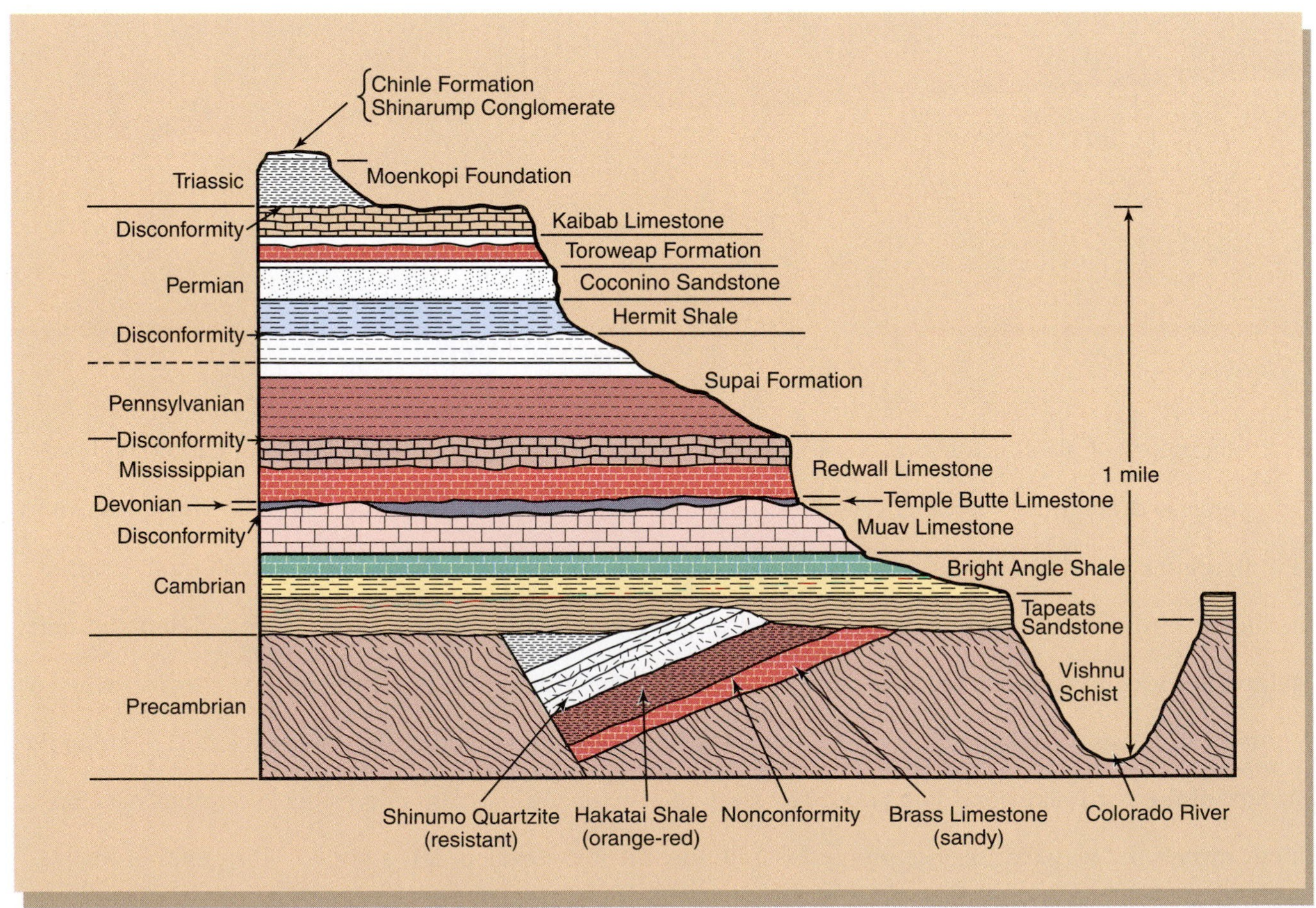

Figure 8–1 Diagram of the rock strata that makes up the Grand Canyon.

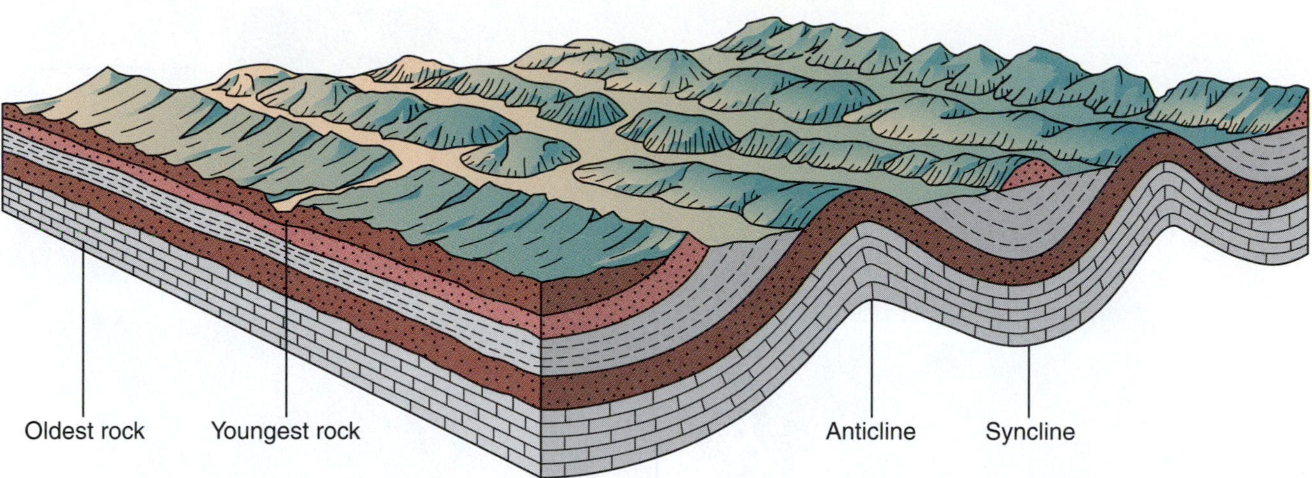

Figure 8–2 Deformed rock strata showing an anticline and syncline.

lateral forces on the Earth's crust push together and fold strata either upward or downward. Folded rock strata creating an archlike formation is called an **anticline.** Folded rock strata forming a bowl-like depression is called a **syncline** (Figure 8–2). The second type of deformed rock is called tilted rock. This occurs when rock strata tilts upward or downward at an angle as a result of crustal movement (Figure 8–3). The third type of deformed rock is called faulted rock. This occurs when a large crack occurs in the rock, also called a **fault,** and the strata moves upward or downward along the fault line (Figure 8–4).

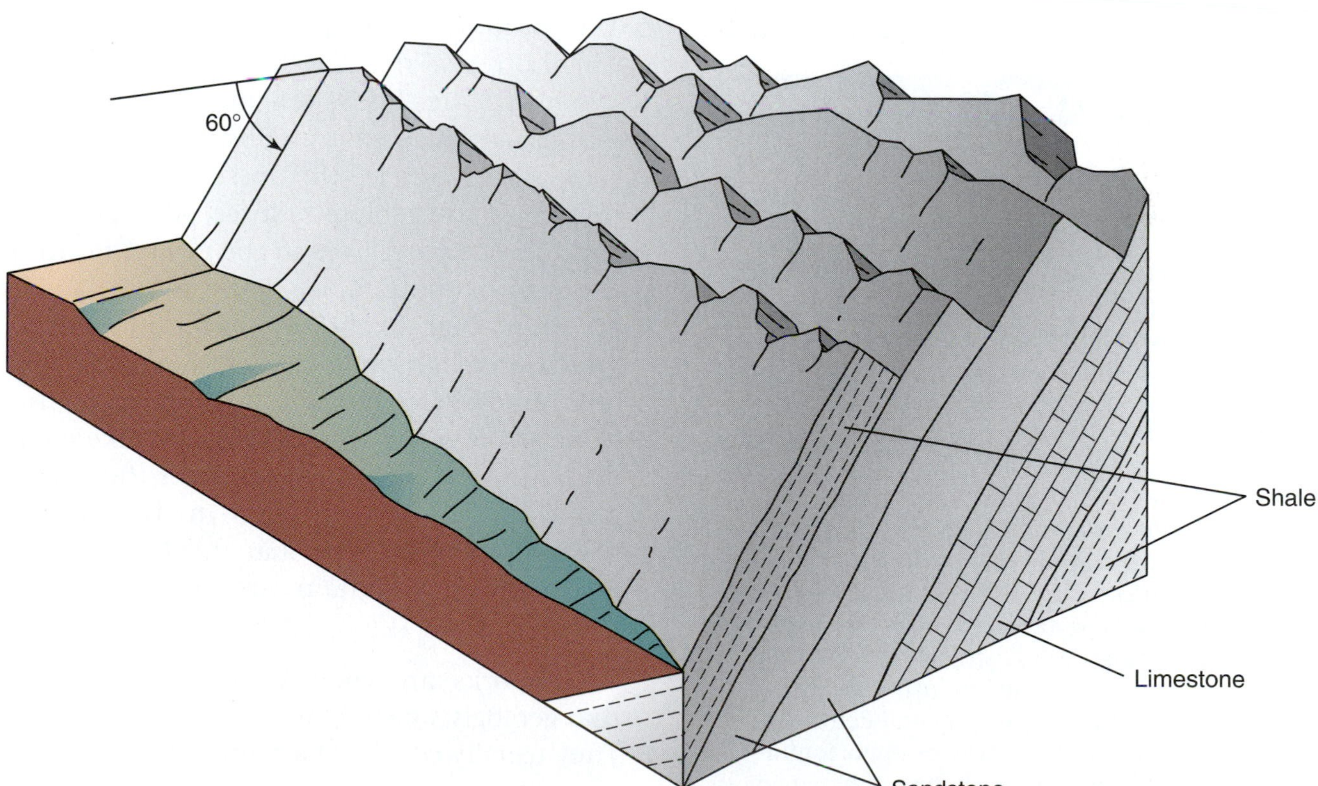

Figure 8–3 Rock strata that has been tilted 60 degrees to the horizon.

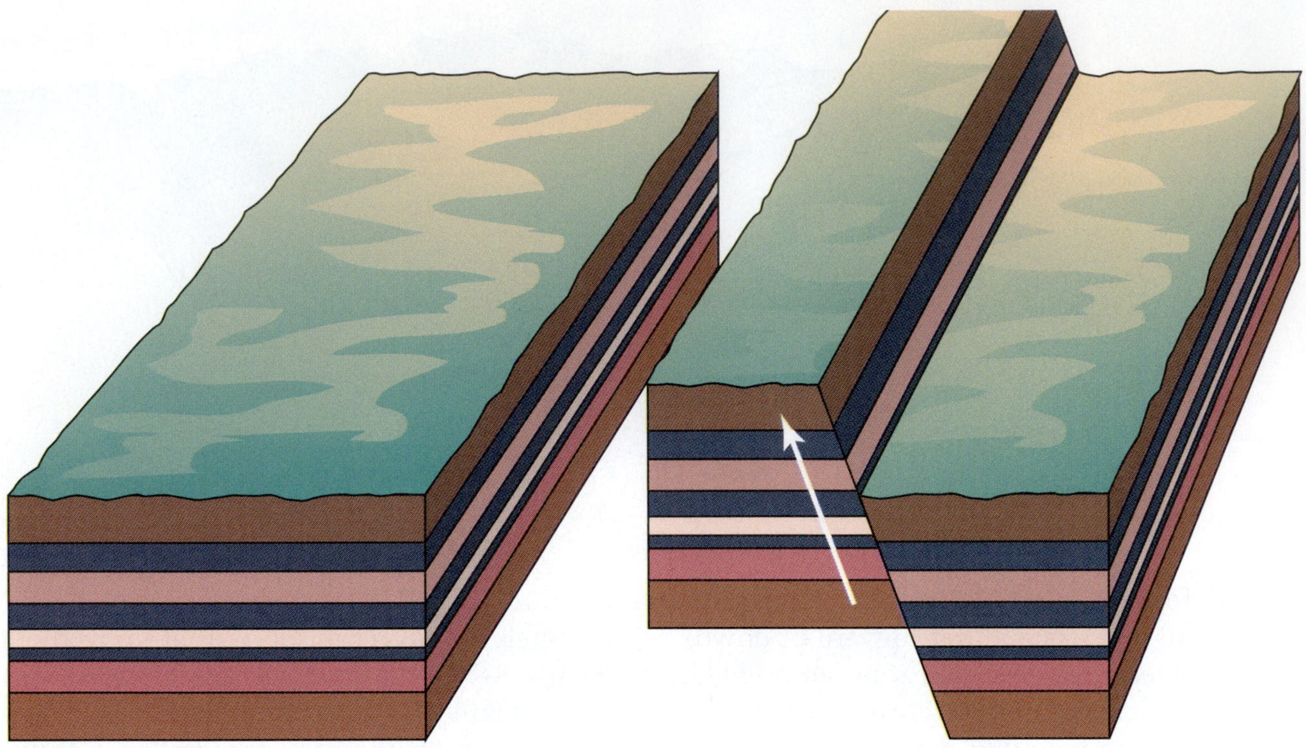

Figure 8–4 The development of a fault that displaces rock strata.

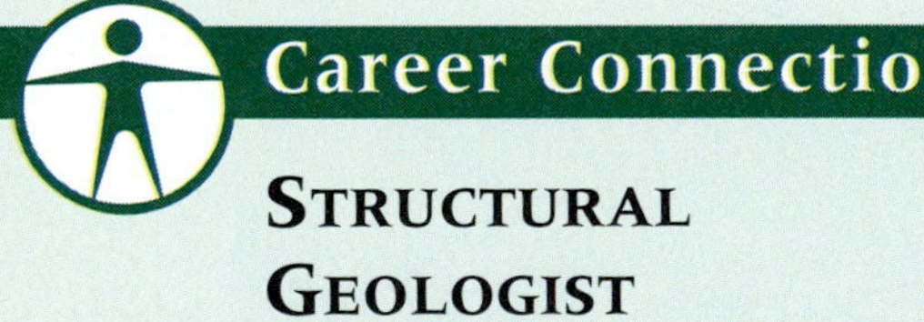

Structural Geologist

A structural geologist studies the processes that deform the Earth's crust. This includes researching the mechanisms that lead to faulting, fracturing, and folding of rock; tectonic stress; crust deformation; and mountain building. Structural geologists travel the world to study rock formations to piece together the processes that have shaped our planet. They also use sophisticated computers and sensors to study the Earth's crust and the forces that continue to alter it. A career in structural geology begins with a 4-year degree in geology. Many structural geologists are employed by the oil industry because of the relationship between oil and deformed rock formations. Other career opportunities include academic research and employment with a government agency.

The location of fossils in sedimentary rocks can also reveal that the Earth's crust is indeed dynamic. The fossils of marine organisms found high in the mountains suggest that the rocks that contain the fossils were moved there by some geological force. All marine organisms live their lives in the ocean, far from the tops of mountains. Many marine fossils are found high in the Himalayan mountains, which are thousands of feet above sea level and hundreds of miles from the ocean. Also, some shallow water marine fossils, like coral, are found deep at the bottom of the ocean. These phenomena are known as displaced fossils, which suggests that the Earth's crust and the rocks it contains are moving in some way (Figure 8–5).

Benchmarks are specific elevation markers that geologists have placed all over the world. They usually consist of a bronze disk on top of a cement pillar and are marked on topographical maps. Benchmarks are used to mark the exact latitude and longitude coordinates

Figure 8–5 The discovery of displaced marine fossils located high above sea level suggests that the Earth's crust has been moved. *(Courtesy of Coolrox.com.)*

and elevation of a specific point on the Earth's surface. Measurements taken over time at these benchmarks reveal that their positions are changing. This also provides evidence that the surface of the Earth is moving.

A **geosyncline** is a large, shallow ocean basin where sediments are accumulating near the edges of continents. The thickness of some of the sediment layers has been measured and reveals that the sediment layer is deeper than the water that it is in. This suggests that the accumulating sediments must be pushing down on the crust beneath the ocean, creating a large depression. This depression is called a geosyncline (Figure 8–6). The weight of the sediments in a geosyncline also causes the nearby continent to rise upward in elevation as it becomes lighter from the lost sediments. The concept of the lighter, less dense continents floating higher on the Earth's mantle and the more dense ocean basins sinking into the mantle is called **isostasy.** Isostasy is the theory that states that the Earth's crust is always in a state of equilibrium. This means that as weight is added to one part of the Earth's crust, it will cause it to become heavier and sink into the mantle. The part of the crust that is losing weight becomes lighter and floats higher on the mantle, causing it to rise, much like what occurs to a lighter person sitting on the mattress of a bed when a heavier person sits down on the mattress. The heavier person sinks down into the mattress and causes the lighter person to rise upward. Isostasy and the occurrence of geosynclines indicate the Earth's crust is indeed dynamic.

Continental Drift

From the time when the first maps were being drawn showing the accurate shapes of the world's continents, scientists have wondered at the jigsaw puzzle–like shapes of the land. One cannot help but notice the way that the South American and African continents appear to fit together. Geologists began to wonder if this was a coincidence or if at some time during the Earth's past they were once joined as one landmass. The first theories of how the Earth formed were developed around the idea that the planet was once composed of liquid rock.

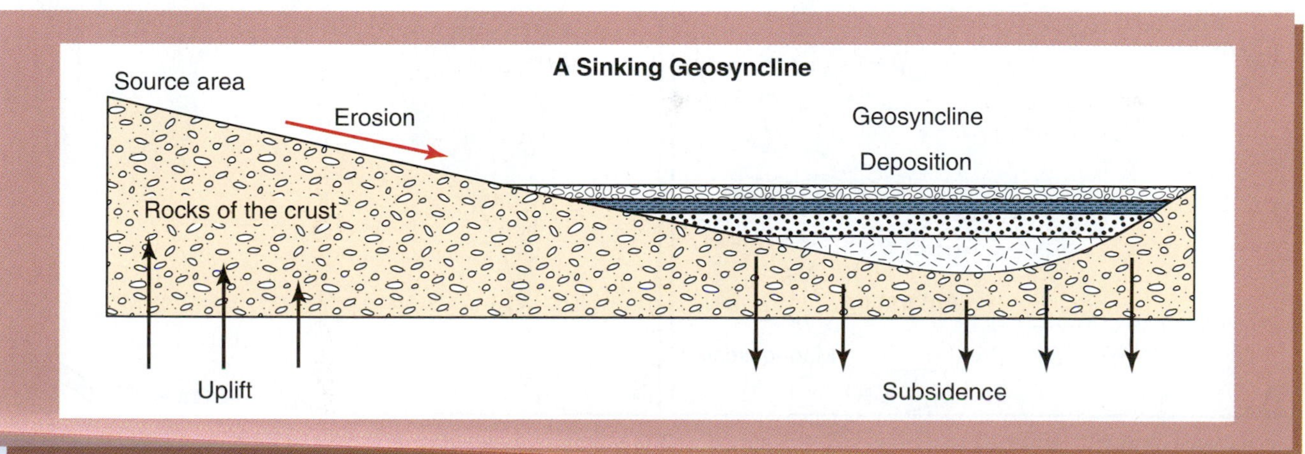

Figure 8–6 The formation of a geosyncline.

Once this rock began to cool, the outer surface of the Earth began to wrinkle and shrink. This was the explanation that described the formation of mountains and valleys, that the Earth's crust was only capable of vertical movement. It meant that the Earth's surface could only move upward to form mountains or downward to form ocean basins.

When scientists studied fossils located on the western coast of Africa and the Eastern coast of South America, they found many of them to be of the same species of organism. To explain this phenomenon, scientists theorized that ancient land bridges once connected the two continents but had long ago disappeared. It wasn't until 1912 that a meteorologist by the name of Alfred Wegner proposed a new theory that explained the jigsaw puzzle-like appearance of the continents and the similarities of fossils found there. He called it the theory of continental drift, which explained that the continents had once been joined together and have slowly drifted apart by some unknown mechanism. He supported his theory by continuing to compare fossil samples on either side of the continents, which revealed that many of them were of the same species. He also compared rock types on the different continents and showed that they were similar in their composition. The last piece of supporting evidence for his theory was the similarities that existed in the way glaciers left scars in rocks on the separate continents. He showed that the direction of the glacial movement lined up perfectly if the separate continents were rejoined (Figure 8–7). Wegner's fellow scientists at the time rejected his theory for two reasons. Their first argument was that Wegner was a meteorologist, not geologist, and could not know about the Earth's continents. Secondly, Wegner failed to provide a

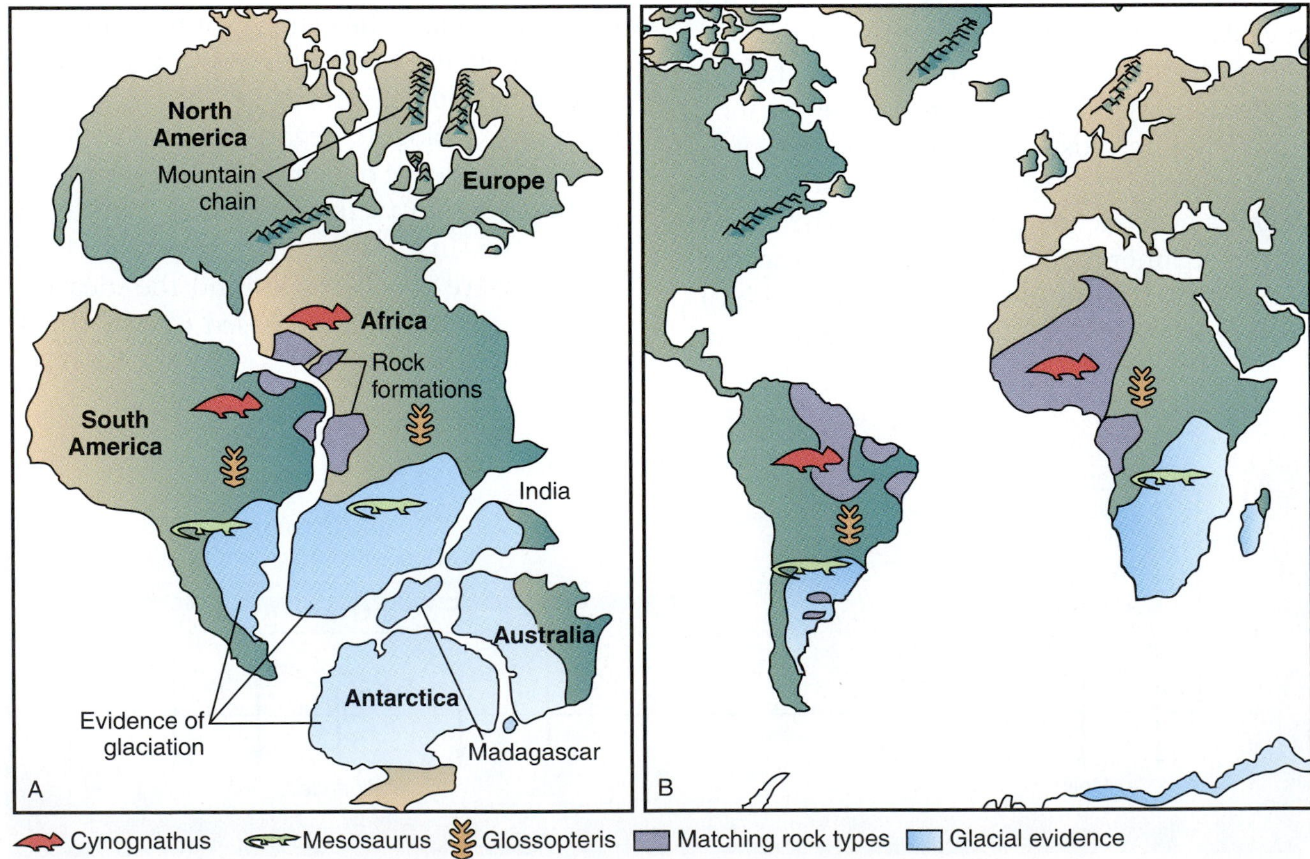

Figure 8–7 The fossil and rock formation correlations used to support the theory of continental drift.

EARTH SYSTEM SCIENTISTS *ALFRED WEGNER*

Alfred Wegner was born in Germany in 1880 and began his career in science as a meteorologist. While traveling and studying the various climates of the world, he began to notice that the shapes of the continents, along with their regional geology, were very similar. He observed how some of the continents appeared to be large pieces of a jigsaw puzzle. This lead him to begin formulating his theory of continental drift, which proposed that the continents were joined at some time in the Earth's past and had somehow drifted apart. To support this theory he traveled the world comparing fossils and rock formations in areas around Brazil and West Africa. He published his theory in his 1929 book, titled *The Origins of Continents and Oceans.* His theory was widely rejected by geologists at the time because Wegner was a meteorologist, not a geologist. They also rejected his theory because of the lack of an explanation of the mechanisms causing the continents to drift. Through much of his life, Wegner also continued to pursue meteorology, especially in the polar regions of the world. He died at the age of 50 during an expedition in Greenland. It took more than 30 years after the death of Wegner for his theory of continental drift, now known as plate tectonics, to become validated by the scientific community.

mechanism for how the massive continents were able to drift across the Earth's surface. Today geologists have gathered enough evidence from around the world to prove continental drift and have pieced together a timeline of how the continents have moved relative to one another in the past. At one time all the continents were joined together to form one supercontinent called Pangea. Pangea formed about 290 million years ago.

Sea Floor Spreading

During the period after World War II, detailed maps that were being constructed of the ocean floor revealed large underwater mountain ranges called **mid-ocean ridges.** One of these underwater mountain chains runs down the center of the Atlantic Ocean and is known as the Mid-Atlantic Ridge. During the 1950s and 1960s, scientists who studied the mid-ocean ridges began to put together theories on how new crust was being formed at the center of these ridges and was moving outward from the ridge center. This was called sea floor spreading and helped to explain how the continents might drift apart (Figure 8–8).

Further evidence to support sea floor spreading was gathered by studying how the crust on either side of the mid-ocean ridges contained alternating bands of magnetic crystals. Scientists had known that the Earth's magnetic field has periodically flipped at different times in the past. This meant that the locations of the **magnetic north** and south would periodically change from time to time. When liquid rock containing iron cools, the resulting iron crystals that form align themselves toward magnetic north. Maps made of the magnetic orientation of rocks on either side of the ridge appeared like a mirror image. After studying these maps, scientists then concluded that new crust was being formed at the mid-ocean ridge, which was then being pushed outward and away on both sides from the ridge center by the formation of newer rock (Figure 8–9).

Scientists also dated the rock samples around the mid-ocean ridges, which showed that the farther the rock was located from the ridge center, the older it was. Further analysis of heat around the mid-ocean ridges revealed that the ocean crust became cooler as you moved away from the ridge center. This pointed to the possibility that large convection cells in the underlying mantle may be the force causing the Earth's crust to move around on the Earth's surface. These discoveries were

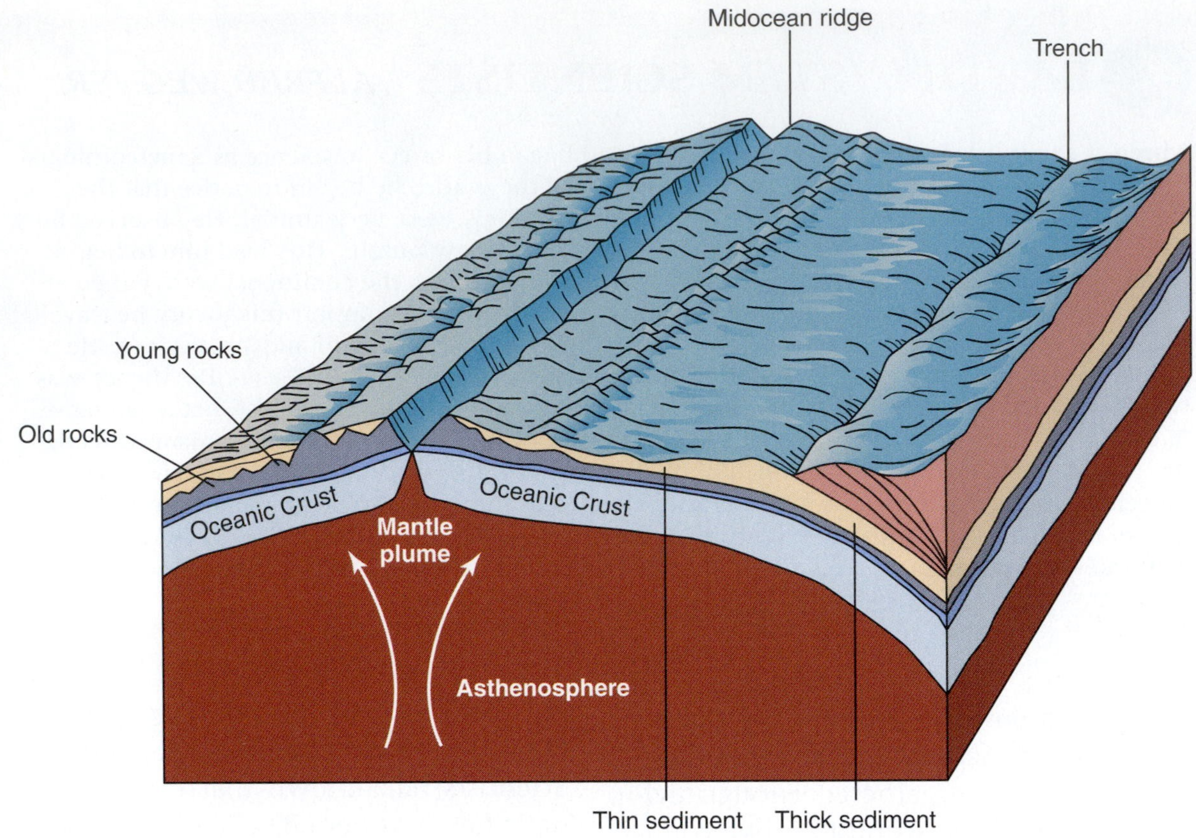

Figure 8–8 A cross section of a mid-ocean ridge showing the formation of new ocean crust at the ridge center, causing the sea floor to spread apart.

providing Wegner's continental drift theory with the mechanism needed to explain what caused the continents to move.

Tectonic Plates

As the theory of continental drift progressed, scientists were beginning to put together the idea that the Earth's surface is composed of a series of large, moving plates. This new theory, called plate tectonics, was built on the foundations of Alfred Wegner's theory of continental drift. Plate tectonics describes the Earth's surface as being divided up into distinct plates that move relative to one another. These solid plates float on the semiliquid upper mantle, making the Earth's surface resemble a broken egg shell. It was also observed that edges of the tectonic plates, known as **plate boundaries,** are areas where volcanoes are located and also where frequent earthquakes occur. These areas, known as zones of earthquakes and volcanoes, mark the locations of plate boundaries. It is the movement of plates and the buildup of tectonic stress along plate boundaries that cause earthquakes (Figure 8–10). There have been 14 major plates identified on the Earth that together form the solid outer crust. All the plates float on the underlying asthenosphere and move relative to one another. Tectonic plates move at varying rates, with some moving at only 0.5 inches per year and others moving more rapidly, up to 4 inches per year.

Plate Boundaries

The point where two tectonic plates meet is called a plate boundary. There are three main types of plate boundaries on the Earth (Figure 8–11). The first type is called a divergent plate boundary. This is the boundary where two plates are moving away from one another.

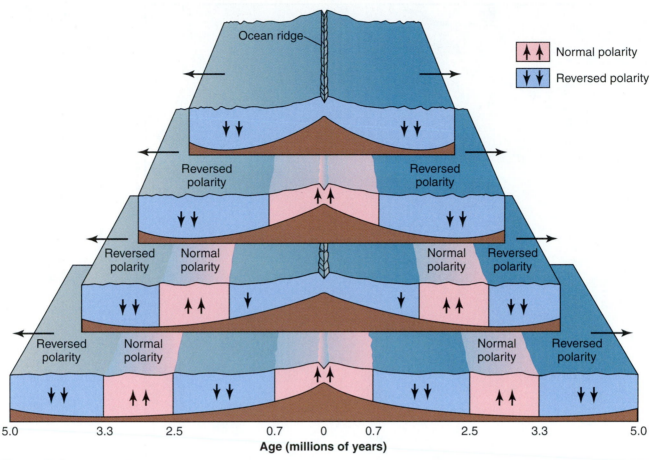

Figure 8–9 A map of the magnetic reversals associated with the Mid-Atlantic Ridge, which supports the concept of sea floor spreading.

EARTH SYSTEM SCIENTISTS *HARRY HESS*

Harry Hess was born in New York in 1906. After attending Yale and Princeton, where he studied electrical engineering and geology, Hess began his research in geophysics and oceanography. In 1931 he began his lifelong study of the ocean floor. He conducted his research by taking soundings and samples of the bottom of the ocean from all parts of the world. His research lead him to observe that certain parts of the ocean floor were younger than other portions. He eventually developed his theory of sea floor spreading in 1962. This ground-breaking theory proposed that new portions of the ocean floor were being formed at the mid-ocean ridges from rising currents of molten mantle. The new rock formed at the ridges and was then pushed outward as new rock rose up from the Earth's mantle and cooled, much like a conveyor belt. Hess's theory of sea floor spreading gave the struggling theory of continental drift the mechanism it needed to be proven. He then further hypothesized that the crust would be pushed along until it collided or ran under other tectonic plates. Harry Hess's work revolutionized the theory of plate tectonics and our knowledge of the Earth's systems.

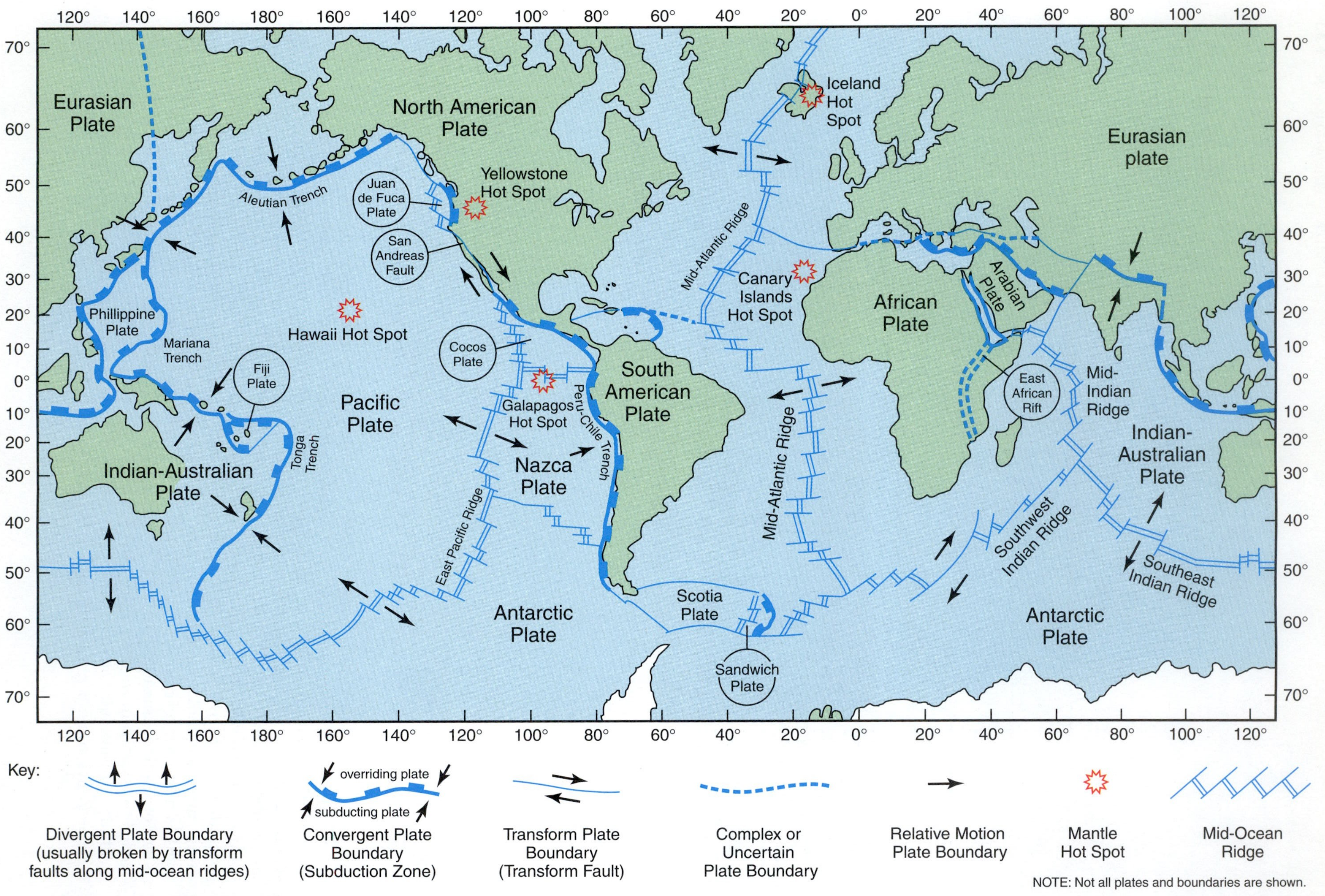

Figure 8–10 A map of the tectonic plates showing their relative movement and the different plate boundaries.

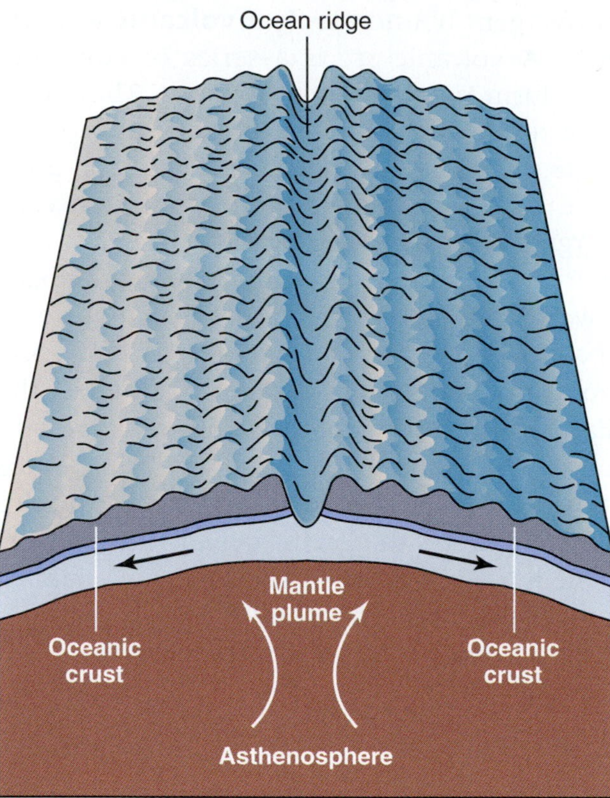

Figure 8–11 A mid-ocean ridge is an example of a divergent plate boundary.

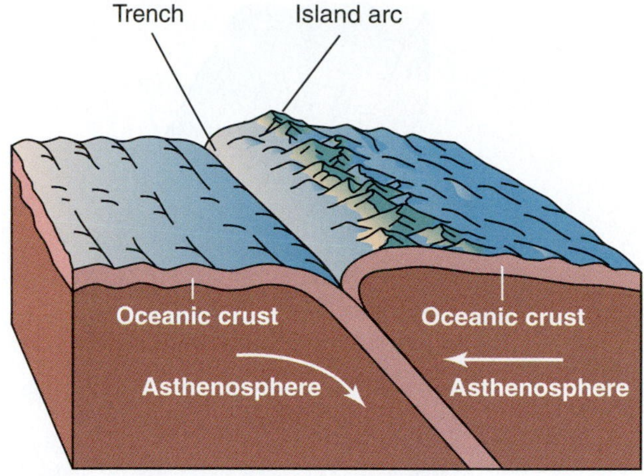

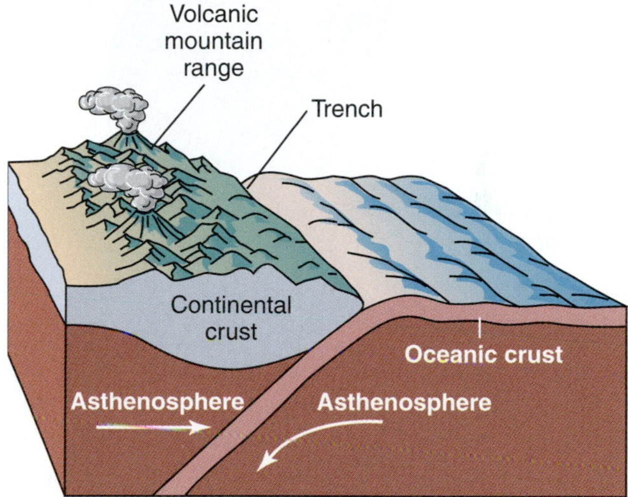

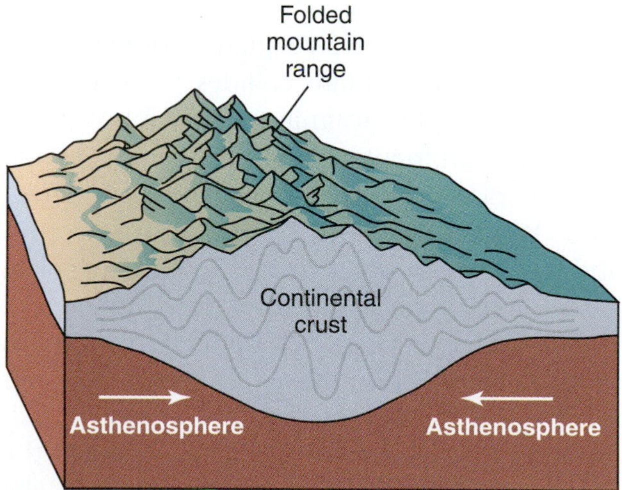

Figure 8–12 Three different types of convergent plate boundaries.

These are usually areas where new crust is formed from magma. Magma is hot, liquid rock that flows beneath the Earth's surface. Divergent plate boundaries are usually associated with mid-ocean ridges, like the Mid-Atlantic Ridge that runs down the middle of the Atlantic Ocean. Divergent plate boundaries on land form **rift valleys,** like the great African Rift Valley located near the eastern side of the African continent.

Areas where two plates come together and collide into one another are called convergent plate boundaries (Figure 8–12). Two things can occur when plates converge. First, if the two plates are both composed of less dense continental crust, they will collide, buckle, and rise upward in elevation to form mountains. This is what is occurring today as India continues to collide into Asia to form the Himalayan mountains. The Himalayas are continuing to rise in elevation as a result of

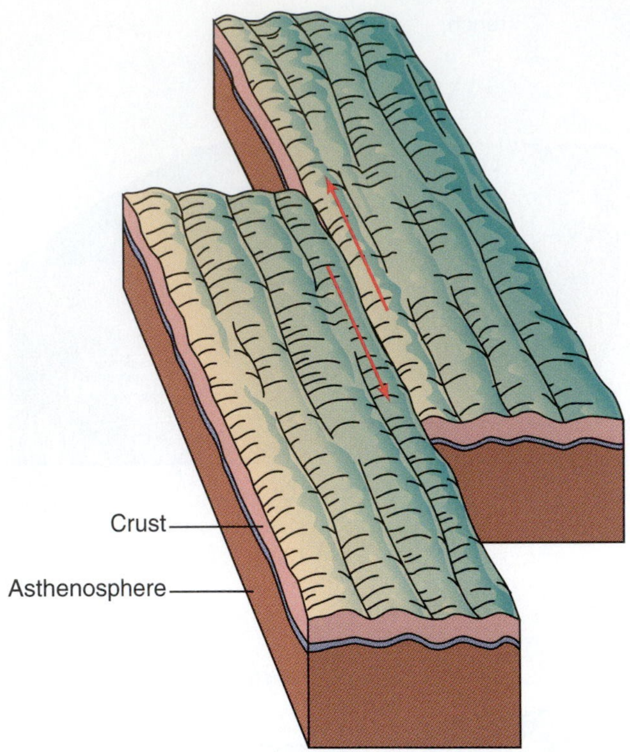

Figure 8–13 An example of a transform fault plate boundary.

the collision of these two plates. The Appalachian mountain range, located along the east coast of North America, formed when the African continent collided with North America approximately 290 million years ago. This important geological event also formed the supercontinent called Pangea.

The other result of two plates converging is when an oceanic plate collides with a continental plate. The oceanic crust is more dense than the continental crust, which causes the oceanic plate to slide under the continental plate. When this occurs, it is called **subduction** and results in the solid crust of the oceanic plate descending downward into the mantle, where it melts. The area where subduction occurs is called a subduction zone. Subduction zones form some of the deepest ocean trenches in the world. The Marianis trench, located at the bottom of the Pacific Ocean near Japan, is the deepest point in the ocean at more than 7 miles deep. Another feature associated with convergent boundaries is a **volcanic arc** system. A volcanic arc is a series of volcanoes that form near subduction zones. They form as a result of the sinking crust of an oceanic plate, which melts under great heat and pressure as it slides beneath the plate it is converging with. The sinking crust brings with it a great amount of sea water, which rises back toward the Earth's surface as a mixture of magma and steam in a network of faults and volcanic vents. It eventually reaches the Earth's surface to form a chain of active volcanoes. A subduction zone like this exists along the western edge of South America where it meets the Pacific ocean. This convergent boundary formed the volcanic chain of mountains that runs along the coast of South America known as the Andes mountains.

The third type of plate boundary occurs when two tectonic plates slide along one another. This is called a transform fault plate boundary (Figure 8–13). Also known as a lateral fault, these areas can build up a great amount of tension between the two plates as a result of friction. Eventually the plates slip past one another and cause earthquakes. The famous San Andreas fault in California is an example of a transform fault plate boundary.

Mantle Convection

Mantle convection is the driving force that causes the tectonic plates to move around the Earth's surface. Convection is the way that the Earth's extremely hot interior dissipates heat toward the surface. Hot plumes of mantle rise upward toward the crust from deep within the asthenosphere, forming large convection cells. The specific area where hot magma rises up from the Earth's interior and breaks through the crust is called a **hot spot.** Hot spots are areas where magma breaks through the earth's surface, forming lava that solidifies to form new crust. Mid-ocean ridges are located over hot spots, as well as some volcanic island chains.

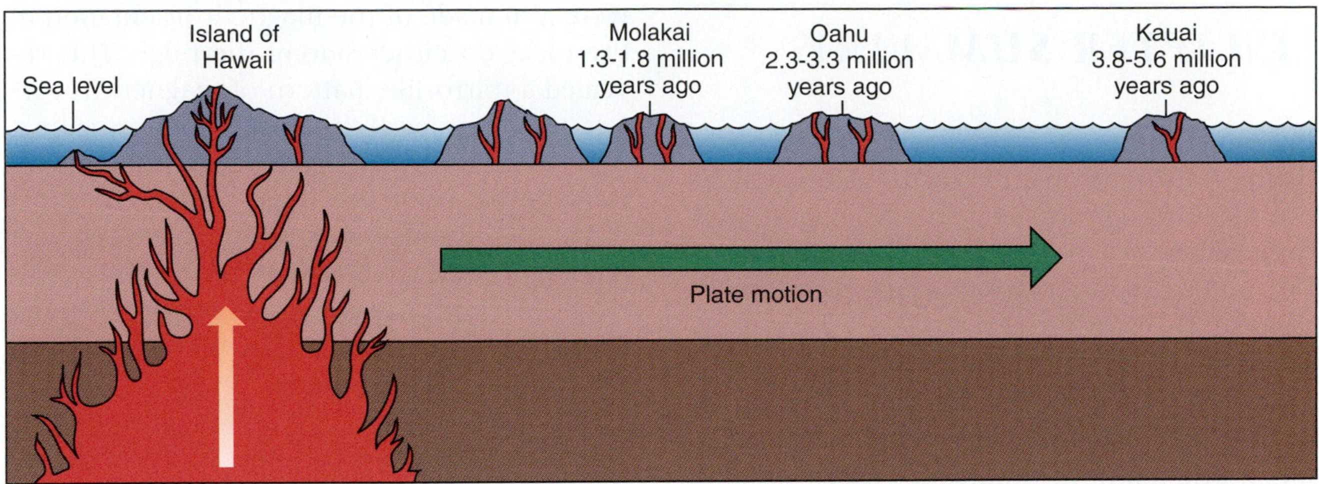

Figure 8–14 The formation of the Hawaiian Islands as the Pacific plate moves over a hot spot.

The Hawaiian islands in the middle of the Pacific Ocean were formed as the Pacific plate moved over a hot spot (Figure 8–14). The result is a chain of active volcanic islands, with the oldest islands being the least active. As the Pacific plate moves over the hot spot, the older islands, which formed earlier, slowly move away from the hot spot and begin to cool. Mantle convection and hot spots act like giant conveyor belts on which the tectonic plates move. Once the mantle dissipates its heat to the crust, the cooler mantle sinks downward back into the Earth and the whole convection cycle begins again. The result of mantle convection and plate tectonics is an active crust that is constantly changing as new rock is produced and older rock is returned to the mantle.

REVIEW

1. What are the three type of deformed rock formations?
2. List four things that might suggest that the Earth's crust is dynamic.
3. What are three pieces of evidence that Wegner used to prove his theory of continental drift?
4. What are the three tectonic plate boundaries?
5. Explain how sea floor spreading is related to the theory of plate tectonics.
6. What is the mechanism by which the tectonic plates are forced to move?
7. Who were Alfred Wegner and Harry Hess?

For more information go to these Web links:

<http://observe.arc.nasa.gov/nasa/earth/tectonics/Tectonics1.html>

<http://www.cotf.edu/ete/modules/msese/earthsysflr/plates1.html>

<http://geollab.jmu.edu/Fichter/PlateTect/index.html>

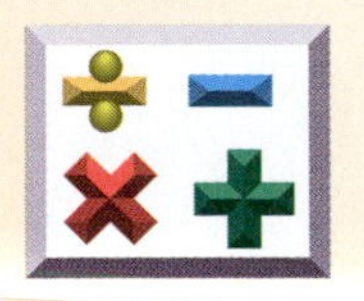

EARTH MATH

1) If a rock sample is located 50 miles to the east of the Mid-Atlantic Ridge and moves at approximately 1 inch per year, how many years ago did the rock form?

CHAPTER SUMMARY

Plate tectonics is a unifying theory that explains the occurrence of earthquakes, volcanoes, deformed rocks, mountains, continental drift, deep ocean trenches, and mid-ocean ridges. Simple observations of the rock formations on the Earth reveal that they have been moved and deformed by geological forces. This implies that the Earth's crust is dynamic, or in a state of motion. The geological principle of original horizontality is used to support the dynamic nature of the Earth's crust. Original horizontality states that all sedimentary rock layers were laid down horizontally on the Earth's surface. Evidence to support the notion that the crust is dynamic includes displaced fossils, tilted or folded rock formations, moving benchmarks, the presence of geosynclines, and continental drift.

The continental drift theory was first proposed by Alfred Wegner in 1912. It stated that the continents were joined together at one time in the past and have now drifted apart. Wegner used the jigsaw puzzle–like appearance of the coasts of Africa and South America, along with similar fossils and rock formations found on both continents, to prove that they were once joined. The lack of a mechanism to explain how the massive continents were moved made Wegner's theory unacceptable to many geologists at the time. Shortly after World War II, scientists began to make detailed maps of the sea floor. These maps revealed the existence of the Mid-Atlantic Ridge, which is a long chain of mountains that runs directly down the center of the Atlantic Ocean. Further investigation of this ridge system revealed that the age of the rocks increases as you move away from the ridge center. Studies were also made of the magnetic orientation of the rocks on either side of the ridge. This revealed a mirrorlike pattern of magnetic reversals on both sides of the ridge. Monitoring heat flow along the ridge has revealed that as you move away from the ridge center, the crust cools. All this evidence suggests that new crust is being formed at the center of the mid-ocean ridges, which is causing the sea floor to spread apart. This is the mechanism that can explain the continental drift theory.

The theory of plate tectonics states that the Earth's crust is divided into distinct plates that move relative to one another on the Earth's surface. Areas where plates interact form what are called plate boundaries. These areas are also zones where frequent earthquakes and volcanoes occur. It is the movement of the plates and buildup of tectonic stress that cause earthquakes. There are three different types of plate boundaries: divergent, convergent, and transform. Divergent boundaries occur where two plates are moving apart, like at the mid-ocean ridges. Convergent boundaries occur where two plates come together. This results in the formation of mountain ranges and volcanic island arcs. When one plate slides underneath another plate, it is called subduction. Transform fault plate boundaries occur where two plates slide alongside one another, such as at the famous San Andreas fault in California. The mechanism that moves tectonic plates is caused by convection currents in the Earth's mantle. These form as the heat from the Earth's core is dissipated outward toward the crust. Mantle convection acts like large conveyor belts on which the tectonic plates move. Hot spots occur in the Earth's crust where mantle plumes break through the surface, forming volcanoes and mid-ocean ridges.

CHAPTER REVIEW

Multiple Choice

1. An observer discovers shallow-water marine fossils at an elevation of 15,000 feet above sea level. What is the best explanation for this observation?
   a. the level of the ocean was once 15,000 feet higher
   b. violent earthquakes caused the crust to collapse
   c. marine organisms evolved into land organisms
   d. the crust was lifted up by some force

2. A sandstone layer is found tilted 75 degrees from the horizon. What most likely caused this tilt?
   a. the sediments that formed the sandstone were laid down at a 75-degree angle
   b. the sandstone layer has changed positions because of crustal movement
   c. the sandstone has recrystallized in that orientation.
   d. all sandstone is deposited by the wind

3. Which is the best evidence of crustal movement?
   a. molten rock in the Earth's outer core
   b. tilted sedimentary rock layers
   c. sediment found on top of bedrock
   d. marine fossils found below sea level

4. A sequence of thickly bedded rock containing shallow marine fossils was most likely formed by:
   a. an ocean trench
   b. a meandering stream
   c. a geosyncline
   d. a glacial lake

5. The term *isostasy* refers to a:
   a. line of equal air pressure
   b. series of anticlines
   c. deflection of the winds by the Earth's rotation
   d. condition of balance between segments of the Earth's crust

6. Which statement best supports the theory that all the continents were once a single landmass?
   a. rocks of the ocean ridges are older than the surrounding sea floor
   b. rock and fossil correlations can be made where the continents appear to fit together
   c. marine fossils can be found at high elevations on all continents
   d. great thicknesses of shallow water sediments are found in the middle of continents

7. Which is the best evidence supporting the concept of sea floor spreading?
   a. earthquakes occur at great depth beneath the sea floor
   b. sandstones and limestones can be found in both North America and Africa
   c. volcanoes appear at random within the ocean crust
   d. rock along the mid-ocean ridges are younger than those farther from the ridge

8. Which best describes a major characteristic of earthquakes and volcanoes?
   a. they are centered at the poles
   b. they are located at the same geographical areas
   c. they are related to the formation of glaciers
   d. they are restricted to the Southern Hemisphere

9. Approximately how many tectonic plates make up the Earth's crust?
   a. 5
   b. 9
   c. 14
   d. 20

10. The area where two tectonic plates come together is known as a:
    a. convergent boundary
    b. divergent boundary
    c. transform boundary
    d. hot spot

11. The area where two tectonic plates move apart from each other is called:
    a. convergent boundary
    b. divergent boundary
    c. transform boundary
    d. hot spot

12. The area where two tectonic plates slide along one another is known as a:
    a. convergent boundary
    b. divergent boundary
    c. transform boundary
    d. hot spot

13. Evidence of subduction occurs between the:
    a. African and South American plates
    b. Australian and Antarctic plates
    c. Pacific and Antarctic plates
    d. Nazca and South American plates

continued

14. Heat detected below the mid-ocean ridges that gradually decreases as you move away from the ridge center is evidence that:
    a. subduction is occurring
    b. new crust is formed at the ridge center
    c. the Earth's magnetic poles are reversed
    d. earthquakes occur there

15. What is the mechanism that drives plate tectonics?
    a. the Earth's rotation
    b. sea floor spreading
    c. hot spots
    d. mantle convection

Matching *Match the terms with the correct definitions.*

a.	strata	f.	geosynclines	k.	rift valley
b.	lateral forces	g.	isostasy	l.	subduction
c.	anticline	h.	mid-ocean ridge	m.	volcanic arc
d.	syncline	i.	magnetic north	n.	mantle convection
e.	fault	j.	plate boundaries	o.	hot spot

1. ____ The direction on the Earth where a magnetic needle points north.
2. ____ Horizontal layers of sedimentary rocks.
3. ____ A type of divergent tectonic plate boundary located on the ocean floor where new crust is formed that pushes on the two plates, causing them to spread apart.
4. ____ Forces that push in on something from both sides.
5. ____ Large convection cells that are believed to exist in the Earth's upper mantle.
6. ____ A type of fold in rock strata that forms an archlike shape.
7. ____ A chain of volcanic islands that forms near a convergent plate boundary located below the ocean.
8. ____ A bowl-like fold or depression in rock strata.
9. ____ The movement of one tectonic plate underneath another at a convergent plate boundary.
10. ____ A large break or crack in a rock mass formed by tectonic stress that results in the displacement of rock strata.
11. ____ A valley that forms along a divergent plate boundary, where two tectonic plates are spreading apart.
12. ____ A bowl-like depression in the Earth's crust formed from the deposition of large amounts of sediments.
13. ____ Specific areas on the Earth's crust where two or more tectonic plates interact with one another.
14. ____ The theory that explains how the Earth's crust is in balance, causing the continents to float at different levels on the asthenosphere below.
15. ____ A term used to describe a specific point located near the middle of a tectonic plate that experiences volcanic activity.

Critical Thinking

1. If Africa and North America continue to move apart, what geological events will most likely occur and where?

CHAPTER

9

Earthquakes and Volcanoes

Section 9.1 – Earthquakes Objectives

Causes of Earthquakes • Seismic Waves • Epicenter Location • Earthquake Measurement

After reading this section you should be able to:

- Define the term *earthquake* and explain how they are caused.
- Define the term *seismic wave* and identify the characteristics associated with P-waves and S-waves.
- Differentiate between the focus and epicenter of an earthquake.
- Describe the process by which an earthquake's epicenter can be located.
- Differentiate between the Richter and Mercalli scales used to measure earthquakes.

Section 9.2 – Volcanoes Objectives

Formation of a Volcano • Quiet Eruption Volcanoes • Explosive Eruption Volcanoes • Volcano Hazards

After reading this section you should be able to:

- Define the terms *volcano, lava,* and *eruption.*
- Differentiate between quiet eruptions and violent eruptions.
- Identify the six different types of volcanoes found on Earth.
- Describe the specific hazards that are associated with volcanoes.

TERMS TO KNOW

earthquake
seismic waves
seismograph
P-wave
S-wave
surface wave
focus
epicenter
volcano
lava
magma chambers
lava vents
eruptions
caldera
pyroclastic flow
lahar

INTRODUCTION

In 79 A.D. a violent volcanic eruption of Mount Vesuvius buried the Roman cities of Pompeii and Herculaneum. Almost 2000 years later, archaeologists working in Italy continue to excavate these ancient cities and are learning more about the culture of ancient Roman society. Geologists too are interested in cities like Pompeii and Herculaneum because they reveal the violent nature of volcanic eruptions and how they can affect human civilization. These cities illustrate the longstanding relationship that humans have had with natural disasters such as earthquakes and volcanoes. This relationship continues today; millions of people live in areas that may potentially be affected by these natural disasters. Understanding what causes these often violent events and the potential damage to human life and property they can cause is the best way to minimize their threat. As a result of intensive research, geologists and engineers are now enabling human society to coexist with the dynamic nature of the Earth, which causes volcanoes and earthquakes (Figure 9–1). Although we may never be able to control these massive events, it is becoming increasingly possible to warn people when they may occur and also to create methods to prevent the loss of life and destruction they have caused too often in the past.

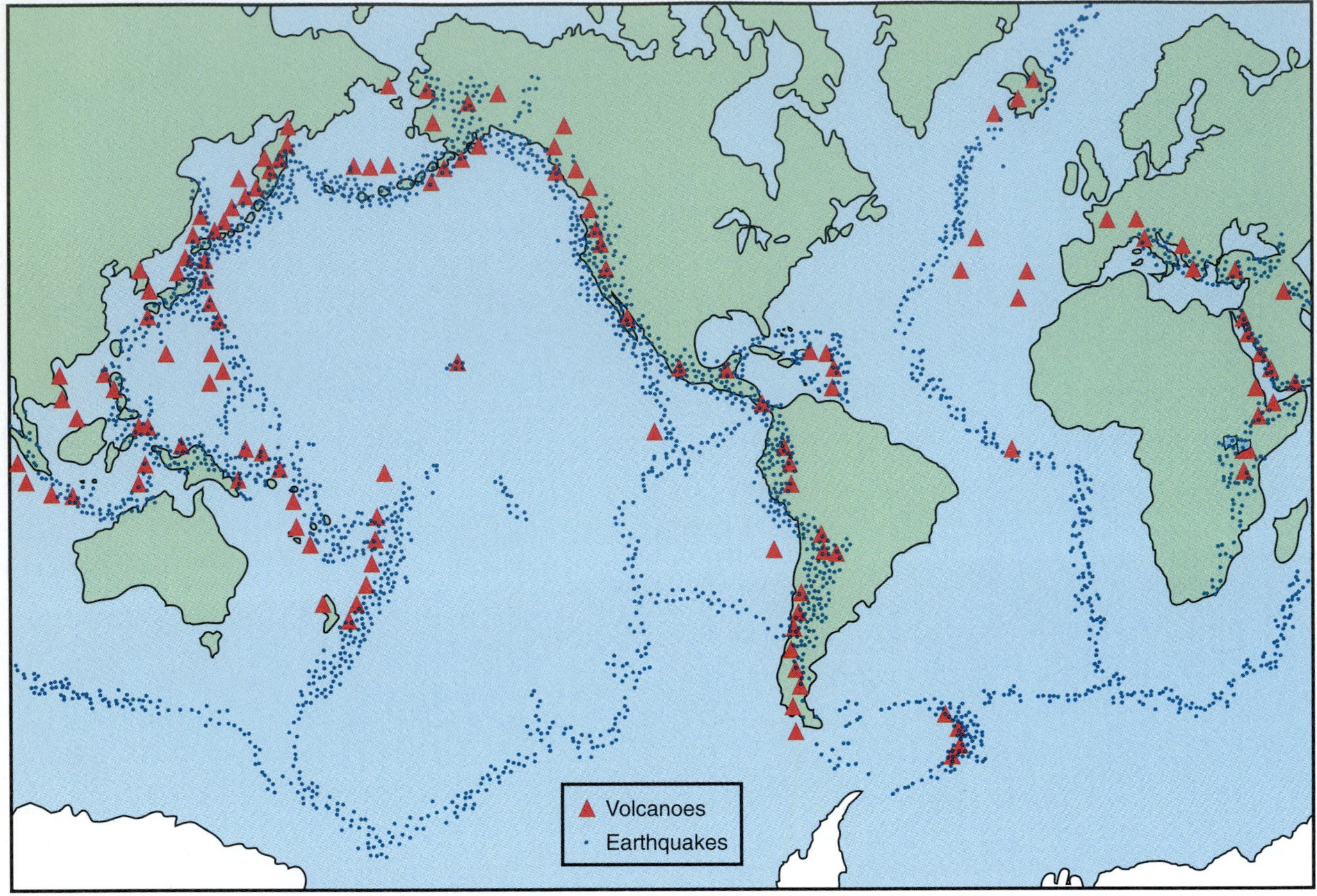

Figure 9–1 A map showing the locations of earthquakes and volcanoes on the Earth reveals a definite pattern, which geologists have linked to interactions occurring between tectonic plates.

9.1 *Earthquakes*

Causes of Earthquakes

One of the most destructive forces on Earth is the **earthquake.** An earthquake is caused by a natural rapid shaking of the ground, which is the result of displaced rocks (Figure 9–2). Displaced rocks are rocks that have been moved by tectonic forces. These forces cause strain to build up in the rigid rocks that form the Earth's crust. Eventually the rocks can no longer take the strain, and they break, much like bending a wooden ruler. If you gently apply force to both ends of the ruler, at first it will bend; however, if you increase the force at the ends of the ruler, the strain becomes too great and the ruler breaks. When this occurs in rock, it causes energy to dissipate outward from where the rock mass has fractured. The energy travels through the Earth in **seismic waves.** When seismic waves reach the Earth's surface, they move the ground and cause earthquakes. Some earthquakes occur on the ocean floor and can cause undersea landslides or displacement of the sea floor, which can form tidal waves, also known as tsunamis.

Seismic Waves

The seismic waves produced by earthquakes can be detected using a **seismograph** (Figure 9–3). A seismograph uses a series of

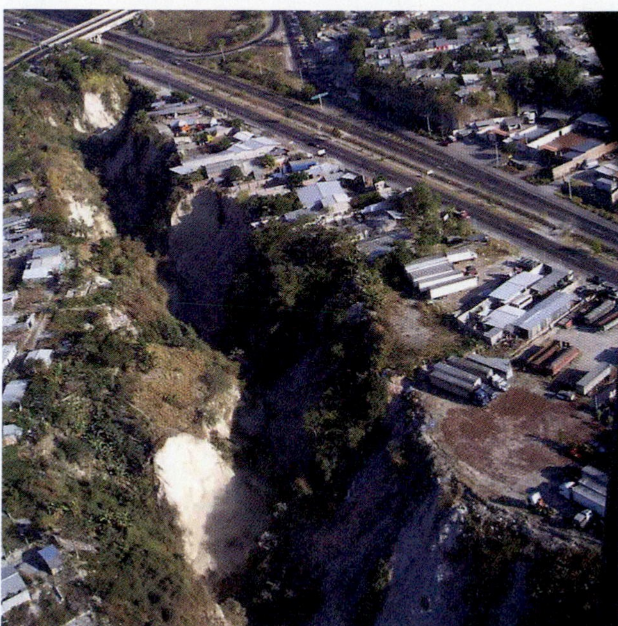

Figure 9–2 The massive displacement of the Earth's surface, caused by an earthquake. *(Photograph by Edwin L. Harp/ U.S. Geological Survey.)*

springs and weights to detect the waves produced when earthquakes occur. Earthquakes produce three distinct types of seismic wave. The first type of wave produced by an earthquake is called a compressional primary wave, also known as a **P-wave.** These waves travel through the Earth by compressing and expanding the material through which they pass. P-waves move like sound waves and vibrate the particles of rock in which they travel. They can therefore travel through solids, liquids, and gases. P-waves are the fastest type of seismic wave. The second type of seismic wave is the secondary wave, or **S-wave.** S-waves travel through the ground perpendicular to their forward motion, which is in an up-and-down pattern much like the waves that move through water. S-waves move much more slowly than P-waves and can only travel through solid material, not liquids or gases. The third type of seismic wave is a **surface wave.** Surface waves are produced by a complex interaction of P- and S-waves reaching the Earth's surface. The result is a rolling motion of the ground as the surface waves move along the surface of the Earth. Surface waves are the most destructive of all seismic waves. The source of seismic waves is the area in the Earth where rock has broken apart violently; this is called the **focus.** The area on the Earth's surface located directly above the focus is known as the earthquake's **epicenter.**

Epicenter Location

It is possible to discover the exact location of an earthquake's epicenter by recording the arrival of P- and S-waves on a seismograph.

EARTH SYSTEM SCIENTISTS *ROBERT MALLET*

Robert Mallet was born in Dublin, Ireland, in 1810. Shortly after attending college, he worked, along with his father, as a structural engineer. Together they helped design and build many large structures in Ireland, such as churches, railroads, and bridges. Mallet was also interested in the dynamics of earthquakes. He was one of the first scientists to conduct experiments to determine the speed and direction of movement of shock waves through rock and sand generated by blasts of dynamite. He also wrote the first research papers explaining the properties of earthquakes. In 1850 he moved to London, England, where he continued to practice as a consulting engineer while also writing scientific articles about the nature and location of earthquakes around the world.

Seismograms

Chicago

St. Petersburg

El Paso

All times corrected to Greenwich Mean Time

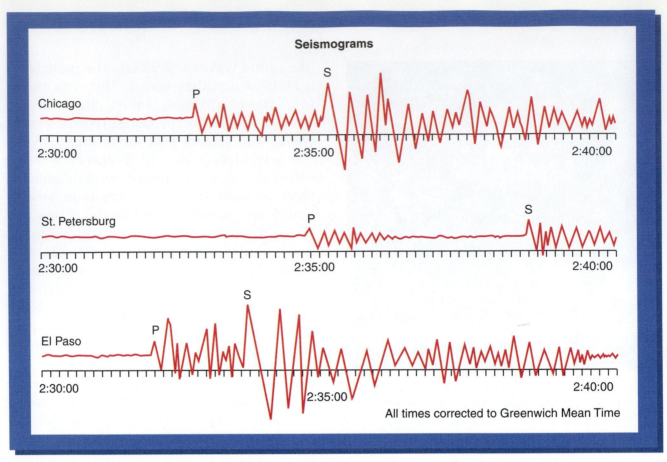

Figure 9–3 Seismographs record the arrival of seismic waves generated by an earthquake.

Epicenter Location Map

Delay between P and S waves (minutes)

Distance to epicenter in thousands of kilometers

Chicago 1400 km

El Paso 1000 km

St. Petersburg 2650 km

Kilometers

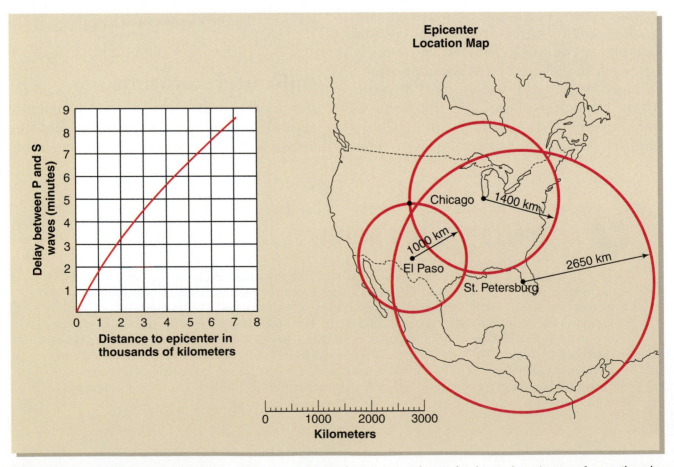

Figure 9–4 P- and S-wave arrival times at different seismograph stations can be used to locate the epicenter of an earthquake.

Career Connections

SEISMOLOGIST

Seismologists study the occurrence of earthquakes and the properties of the Earth's interior. The science of seismology uses shock waves to reveal the composition of the Earth's interior. By setting off controlled explosions, seismologists use sensors and other scientific instruments to study the movement of the shock waves through different parts of the Earth. This important research has revealed many underground features, such as faults. Seismologists also use these techniques to help locate oil and mineral resources. Studying earthquakes is another important aspect of seismology. Using seismographs, seismologists monitor and track earthquakes all over the world. Their research also involves earthquake prediction, which may someday prevent the loss of life associated with these natural disasters.

Because P-waves travel faster than S-waves, it is possible to use the difference in their arrival times to a seismograph station to locate the distance to the epicenter (Figure 9–4). This is easy if you know the speeds at which both P- and S-waves travel through the Earth. This information can be used to develop a chart that uses the difference in arrival times of P- and S-waves and distance to the epicenter. Because this chart reveals only distance and not direction, data from at least three different seismograph stations is required to accurately pinpoint the epicenter.

Earthquake Measurement

The strength of earthquakes are measured using two different types of scales (Table 9–1). The Richter scale measures the equivalent energy produced by an earthquake on a scale from 0 to 9. The Richter scale is a logarithmic scale, which means that for every increase in number, the earthquake gains a tenfold increase in magnitude. For example, an earthquake with a Richter magnitude of 5 has the equivalent energy of a medium-sized atomic bomb. Another method of measuring earthquakes is the modified Mercalli scale. This scale categorizes earthquakes on a scale from I to XII and measures the amount of destruction the earthquake can cause to society. For example, a level II earthquake is felt by few people, and suspended objects may swing. A level X earthquake would level most masonry buildings and bend railroad tracks.

Current Research

Satellite technology is now being applied to the monitoring of earthquakes. Researchers at the National Aeronautic and Space Administration's Jet Propulsion Laboratory have been using satellite images taken by the European Space Agency's Remote Sensing-2 satellite to map the precise movements of tectonic plates after an earthquake. The satellite photographs an image of a region on the Earth's surface from an angle to reveal the three-dimensional changes that have occurred as a result of an earthquake. Using images photographed before and after an earthquake occurs, researchers then combine data to produce a map that shows precise plate movements along fault lines. The data gained from these satellite images can then be applied to predicting future earthquakes. Often earthquake movement is unknown in remote regions that have not been studied at ground level. The satellite images provide researchers with data from all along the fault lines. By determining how much a plate has moved along a fault line, scientists can increase their ability to locate areas that may experience earthquakes in the near future.

TABLE 9-1 The Richter and modified Mercalli scales used to measure earthquakes

Level	Characteristic Effects in Populated Areas
I	Generally not felt; detectable by seismographs
II	Felt by few people; objects may swing if suspended
III	Felt by few people, mostly indoors; vibrations like a passing truck
IV	Felt by many people indoors but few outdoors; windows, dishes, and doors rattle
V	Felt by nearly everyone; sleepers awaken; small, unstable objects may fall and break; doors move
VI	Felt by everyone; some heavy furniture moves; people walk unsteadily; windows and dishes break; books fall from shelf; bushes and trees visibly shake
VII	Difficult to stand; moderate to heavy damage to poorly constructed buildings; plaster, loose bricks, titles, and stones fall; small landslides along slopes; water becomes turbid
VIII	Difficult to steer cars; damage to good unbraced masonry; chimneys, monuments, towers, and elevated tanks fall; tree branches break; steep slopes crack
IX	Extensive building damage; good masonry damaged seriously; foundations crack; serious damage to reservoirs; underground pipes break
X	Most masonry, frame structures, and foundations destroyed; numerous large landslides; water thrown on banks of rivers and lakes; railroad tracks bend slightly
XI	Few masonry buildings stand; railroad tracks bend severely; many bridges destroyed; underground pipelines completely inoperative
XII	Nearly total destruction; large rock masses displaced; objects thrown into the air

Comparison of Richter Magnitude and Energy Released

Richter Number	Approximate Energy Released (Amount of TNT)
1	170 grams
2	6 kilograms
3	179 kilograms
4	5 metric tons
5	179 metric tons
6	5,643 metric tons
7	179,100 metric tons
8	5,643,000 metric tons

SECTION REVIEW

1. Describe what happens in the Earth to cause earthquakes.
2. What are the three types of seismic waves generated by earthquakes? Describe their general characteristics.
3. What is the equivalent energy of a magnitude 8 earthquake?
4. Who was Robert Mallet?

For more information go to these Web links:

<http://www.crustal.ucsb.edu/ics/understanding/>

<http://www.seismo.unr.edu/ftp/pub/louie/class/100/plate-tectonics.html>

EARTH MATH

1) YOUR SEISMOGRAPH RECORDS THE ARRIVAL OF P-WAVES GENERATED BY AN EARTHQUAKE AT 2:03 P.M., AND THE ARRIVAL OF THE S-WAVES OCCURS AT 2:06 P.M. USING THE TRAVEL TIMES OF P- AND S-WAVES GRAPH ON PAGE 148, DETERMINE HOW FAR THE EPICENTER IS FROM YOUR LOCATION.

2) DETERMINE HOW MUCH MORE ENERGY IS RELEASED IN A MAGNITUDE 8 EARTHQUAKE THAN IN A MAGNITUDE 5 EARTHQUAKE.

9.2 *Volcanoes*

Formation of a Volcano

A **volcano** marks a point on the Earth's surface where hot molten rock, or **lava,** flows from beneath the crust. The location of volcanoes is often associated with convergent tectonic plate boundaries, where the solid crust of a subducting tectonic plate begins to descend back into the Earth's mantle and melt. The sinking plate brings with it a large amount of seawater, which mixes with the molten rock to make steam that expands in the rock, making it less dense. This causes the molten rock to rise upward toward the Earth's surface, where it forms large chambers called **magma chambers.** Magma chambers feed a small network of **lava vents** that lead to the Earth's surface (Figure 9–5). This is where the lava that forms volcanoes originates. Volcanoes can also form near divergent plate boundaries such as the Mid-Atlantic Ridge, or over hot spots such as the Hawaiian Islands.

Quiet Eruption Volcanoes

Volcanoes are often classified by the types of **eruption** they produce. An eruption is the release of lava, ash, steam, and gases from deep within the Earth. Quiet eruptions occur from volcanoes called shield cones. Shield cones are formed when lava gently flows up from a central vent. A vent is a pipelike crack in the Earth's crust through which lava flows. The lava cools at the surface and solidifies, forming a series of layers that resemble a shield lying on the ground. The volcanoes that formed the Hawaiian islands are classified as shield cones (Figure 9–6). Many volcanic islands in the ocean were formed from shield cones that slowly built up from layers of cooling lava, which eventually rose above sea level.

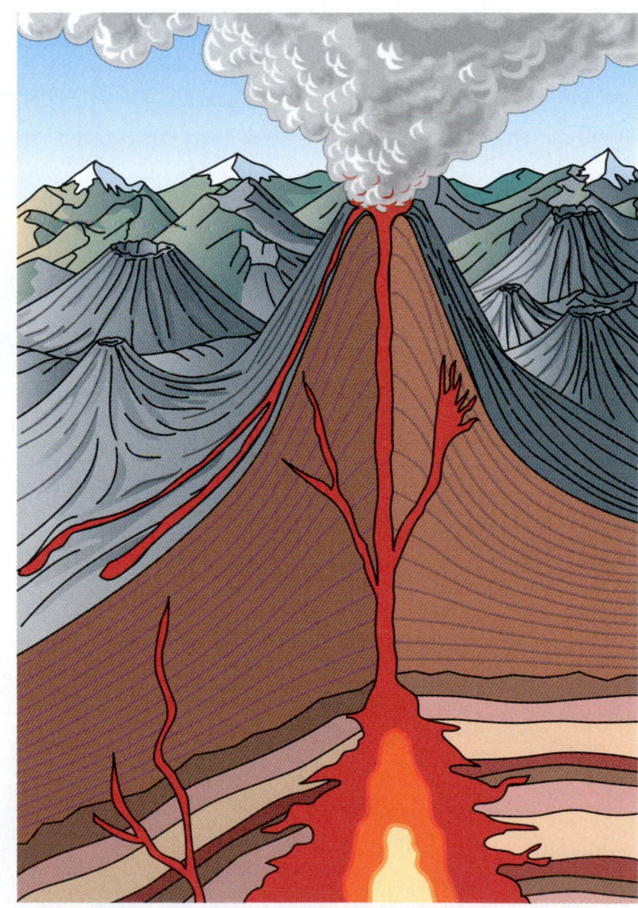

Figure 9–5 A cross section of a volcano showing the magma chambers and volcanic vents, which feed lava up to the surface.

Figure 9–6 The Mauna Loa shield cone volcano, located on the island of Hawaii. *(Courtesy of PhotoDisc.)*

Another type of quiet eruption volcano is called a fissure eruption. A fissure is a long crack in the Earth's crust from which lava flows (Figure 9–7). These types of eruptions can produce large amounts of lava that cool to form huge sheets, or plateaus. The Deccan lava flows in India are probably the world's largest lava flow produced from a fissure. This massive lava flow occurred approximately 65 million years ago and has been linked by some researchers with the extinction of the dinosaurs. A fissure eruption of this size may have spewed millions of tons of toxic gas and ash that could have altered the environment on a scale that may have lead to a mass extinction. Another large-scale fissure eruption formed the Columbia River plateau, which covers parts of Washington, Oregon, and Idaho in the United States. Both these lava flows cover thousands of square miles and are hundreds of feet thick. Fissure eruptions also frequently occur on the island of Iceland, which lies over the Mid-Atlantic Ridge.

Explosive Eruption Volcanoes

The other classification of volcanoes is the explosive eruption. These are much more vio-

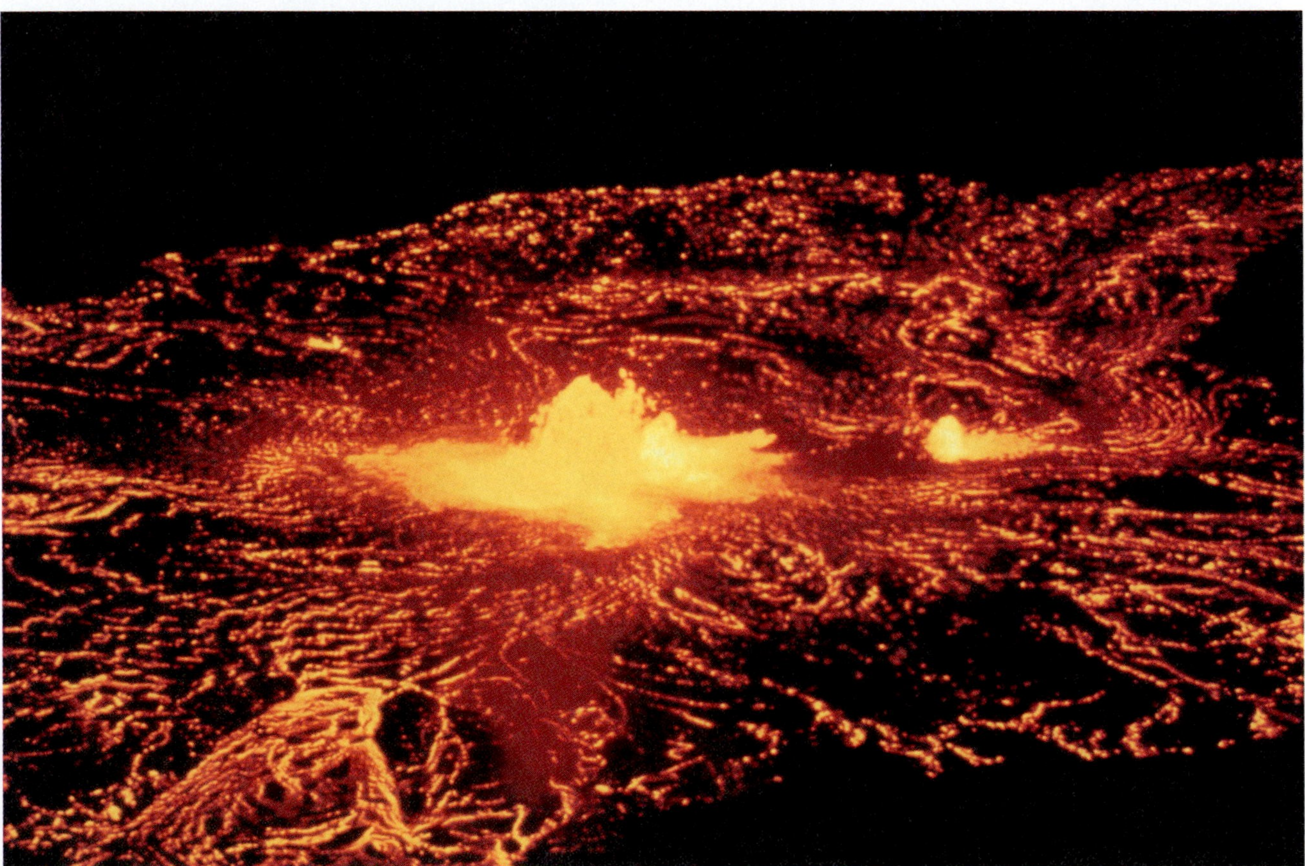

Figure 9–7 Lava pours from a fissure, creating a lava flow. *(Courtesy of PhotoDisc.)*

lent than quiet eruptions and pose the greatest threat to human life and property. One type of volcano that produces explosive exceptions is called a composite cone. Composite cone volcanoes are large, mountainlike structures that have formed from layers of lava and pyroclastic material (Figure 9–8). The term *pyroclastic* means "fire broken" and refers to rocks that were blasted apart as a result of an explosive eruption. Composite cones usually surround one central vent and are located on the continents near subduction zones associated with convergent plate boundaries. The hot magma that rises up in the central vent of a composite cone does not flow as easily as lava produced from fissures or shield cones. Therefore great amounts of pressure build up behind the "sticky" lava, until a great explosion occurs. The result is one of the most destructive occurrences on Earth. Some eruptions from composite cone volcanoes can be so destructive that the entire mountain may be blown apart. Examples of composite cone volcanoes include Mount Ranier in Washington State and the famous Mount Vesuvius in Italy (Figure 9–9).

Figure 9–8 Mount St. Helens in Washington State is an example of a composite cone volcano. *(Courtesy of PhotoDisc.)*

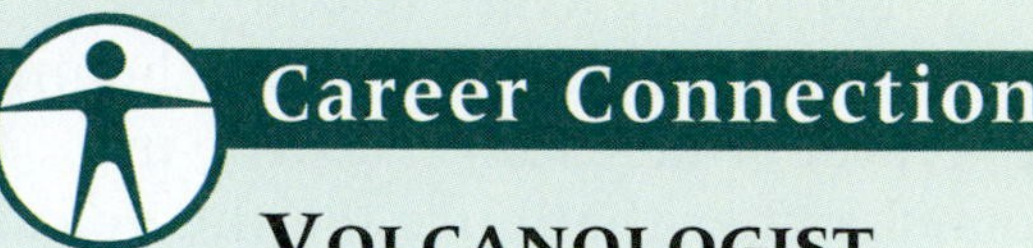

Volcanologist

A volcanologist studies all aspects of volcanoes. This often dangerous profession puts scientists in the path of one of nature's deadliest forces. The study of volcanoes includes researching the forces that lead to volcano formation and the characteristics of molten rock that form lava. Volcanologists are also interested in studying volcanic eruptions and their deadly consequences. Currently, volcanologists are researching ways to better predict the eruptions associated with volcanoes to prevent loss of life. Some volcanologists also study mineral resources that are associated with volcanic rocks. Careers in volcanology require a college degree in geology and the willingness to work in sometimes dangerous environments.

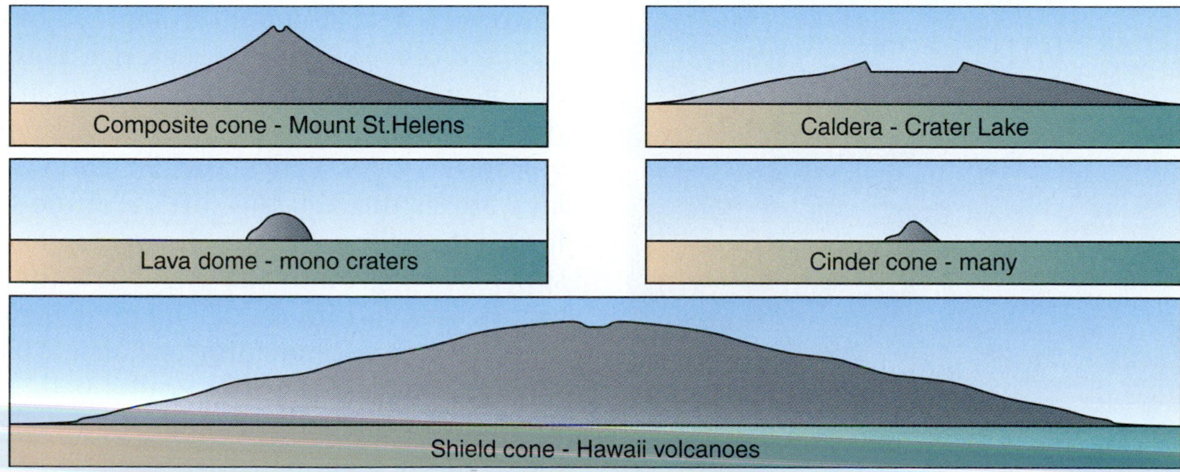

Figure 9–9 The size and shapes of different types of volcanoes.

When a composite cone is completely destroyed by an eruption, it leaves behind a large crater that sinks down into the magma chamber that once fed the volcano. This type of structure is called a **caldera,** or volcanic crater. When calderas eventually cool down, they can be filled with water to form large lakes. Crater Lake in Oregon is a large caldera (Figure 9–10). Many calderas can still produce explosive eruptions, although they appear to be inactive. Large calderas exist in Yellowstone National Park. The hot magma below the surface heats groundwater and produces the active geysers that attract millions of tourists to the park each year.

Another type of volcano that produces explosive eruptions is called a lava dome. Lava domes are smaller mounds of hardened lava that cover a lava vent. These structures can grow slowly as new lava forms underneath the dome and pushes outward. Lava domes are often found in the center of the craters of cinder cones. Often the lava vents that feed the lava dome become plugged. This results in a buildup of great pressure that may eventually cause an explosive eruption. After the famous Mount St. Helens volcanic eruption that occurred in Washington State in 1980, a small lava dome formed at the center of its crater (Figure 9–11).

Current Research

Researchers from England believe that supervolcanoes may cause drastic changes in the global climate. Unlike common eruptions from volcanic mountains, supervolcanoes erupt from calderas. Calderas are craterlike volcanoes that lie over large magma chambers. Over thousands of years, pressure builds up under these calderas, and eventually they explode with great force. The explosions that are caused from a caldera equal that of an asteroid or comet impact on the Earth, which is thousands of times more violent than common eruptions. The result may be a drastic change in global climate, caused by the gas and dust kicked up high into the atmosphere. Surprisingly, the research team points to the Yellowstone caldera in the United States as the next likely supervolcano to erupt. Professor Bill McGuire believes the Yellowstone caldera is due to explode in the year 2076. This is based on a periodic eruption rate of 600,000 years. The last time Yellowstone erupted was 640,000 years ago. A supervolcano located in Italy, which is much smaller than the one in Yellowstone, is also expected to erupt sometime in the future. McGuire points out that ice core data have shown that the explosions of supervolcanoes in the Earth's past have lowered global temperatures during spring and summer by as much as 20° Fahrenheit. In comparison, large-scale mountain volcanoes affect the global temperature by only 3° Fahrenheit.

Figure 9–10 The caldera that formed Crater Lake in Oregon. *(Courtesy of PhotoDisc.)*

The last type of volcano that may cause an explosive eruption is the cinder cone. A cinder cone is a small mound composed of lava, ash, and other pyroclastic material. This type of volcano is found in many places around the Earth. They are also the smallest type of volcano. Cinder cones usually form when lava breaks through the surface of the Earth for the first time. The resulting eruption can spew lava high into the air. Often the lava that is thrown into the air hardens before it hits the ground and forms what is called a volcanic bomb. Once the eruption that has formed the cinder cone stops, it usually remains inactive permanently.

Figure 9–11 A lava dome developing in the crater of Mount St. Helens. *(Photograph by Lyn Topinka/U.S. Geological Survey.)*

Volcanic Hazards

One of the most dangerous results of an erupting volcano is **pyroclastic flow.** Unlike lava, which moves at a fairly slow pace, pyroclastic flows can move at speeds of more than 100 miles per hour. These rapidly moving flows contain extremely hot gases, water, ash, and debris that rushes down the side of an erupting volcano, destroying everything in its path. The famous Roman cities of Pompeii and Herculaneum were destroyed by the immense pyroclastic flow created when Mount Vesuvius erupted in 79 A.D. Fine ash produced from volcanic eruptions, called volcanic ash, can be sent high into the atmosphere, where it can be transported over long distances. The result is the buildup of ash on the ground miles away from the erupting volcano. This ash collects like gray snow on the ground and can fill the air, making it difficult to breath. Airplanes that fly through these ash clouds can experience engine failure and crash. Ash sent into the atmosphere from the Mount St. Helens eruption in 1980 traveled all the way to New York (Table 9–2).

The 1991 Mount Pinatubo eruption in the Philippines created so much ash that it fell thousands of miles away (Figure 9–12). So much ash entered the Earth's atmosphere from this eruption that it is believed to have altered the climate of the entire planet. The ash buildup can become so heavy that the roofs of buildings may collapse. Heavy rains caused by the huge amount of steam sent into the atmosphere during an eruption can mix with the ash fall and cause dangerous mudflows. These rapidly moving flows of wet volcanic ash can completely destroy whole towns. The Roman city of Herculaneum was covered by mudflows that were more than 60 feet deep after the eruption of Mount Vesuvius in 79 A.D. The Mount Pinatubo eruption in 1991 created mudflows that were more than 650 feet deep.

EARTH SYSTEM SCIENTISTS *EDUARD SUESS*

Eduard Suess was born in London in 1831 and became a professor of geology at the University of Prague. He studied many aspects of geology, including paleontology, economic geology, and structural geology. In 1909 he published a book that presented many of his geologic theories, titled *The Face of the Earth.* Some of the research that Suess presented in his book included the history of the world's oceans, the nature of crustal movement, the structure of mountain chains around the globe, and the unique structural geology of the continents. He also proposed that there once was a large supercontinent on the Earth, which he called Gondwanaland. This marked the beginnings of the concept of continental drift, which was later developed by Alfred Wegner. Other research that Suess conducted during his life included work on seismology and the fossils of the Danube River Basin.

TABLE 9-2 Some of the world's most active volcanoes

Volcano Name	Location	Date of Last Activity
Africa and the Indian Ocean		
Lengai Ol Doinyo	Tanzania	1993
Nyamuragira	Zaire	1992
Piton de la Foumaise	Zaire	1992
Antarctica		
Mount Erebus	Ross Island	1990
Big Ben	Heard Island	1986
Deception Island	South Shetland Island	1970
Asia		
Aso	Japan	1993
Krakatau (Anak Krakatau)	Indonesia	1993
Mayon	Philippines	1993
Sakura-jima	Japan	1993
Sheveluch	Russia	1993
Central America and the Caribbean		
Arenal	Costa Rica	1994
Pacaya	Guatemala	1994
Santiaguito (Santa Maria) Dome	Guatemala	1993
Rincon de la Vieja	Costa Rica	1992
Poas	Costa Rica	1991
North America		
Mount St. Helens	Washington State	1980
Mauna Loa Kilauea	Hawaii	2001

Another product of volcanic eruptions can be the release of large amounts of toxic gas. Gases like sulfur dioxide can mix with steam to form sulfuric acid, which can rain down on the surrounding area. Volcanoes can also emit carbon dioxide. This colorless, odorless gas can flow down the sides of volcanoes and suffocate all the living things in its path. Debris flows, also known as **lahars,** are generated when melting snow that has collected on volcanoes with high elevations suddenly melts during an eruption (Figure 9–13). Lahars can move rapidly down the side of the volcano, taking with them trees, rock, mud, and water. The result is a deadly wall of rapidly moving debris that can easily destroy anything in its path. A lahar formed from the Nevado Del Ruiz volcano in South America was more than 120 feet high and traveled more than 30 miles away from the volcano. This lahar completely destroyed the village of Armero, killing more than 20,000 people.

For more information go to these Web links:

<http://wwwhvo.wr.usgs.gov/>

<http://volcano.und.nodak.edu/vw.html>

Figure 9–12 The massive eruption of the Mount Pinatubo volcano in the Philippine Islands during July 1991. *(Courtesy of PhotoDisc.)*

Figure 9–13 An immense lahar created by the 1982 eruption of Mount St. Helens. *(Photograph by U.S. Geological Survey.)*

SECTION REVIEW

1. What are the types of volcanoes associated with quiet eruptions?
2. What are the types of volcanoes associated with explosive eruptions?
3. Describe the direct effects that volcanoes can have on the surrounding area.
4. Who was Eduard Suess?

EARTH MATH

1) A huge fissure volcano poured lava onto the land surface in southeastern Washington State more than a million years ago. The lava covered an area of 30,000 square miles, with an average depth of 5000 feet. How many cubic feet of lava was produced by this eruption?

CHAPTER SUMMARY

An earthquake is the violent shaking of the ground caused by the energy released from fractured rock below the Earth's surface. The energy of an earthquake travels in the form of waves through the Earth's interior, called seismic waves. There are three type of seismic waves that cause earthquakes. P-waves are the fastest traveling seismic waves and can move through solids, liquids, and gases. S-waves move more slowly than P-waves and can only travel through solids. The third type of seismic wave is called a surface wave. Surface waves form at the Earth's surface by an interaction of P- and S-waves and are the most damaging of all seismic waves. The point in the Earth's interior where a rock mass is fractured, causing an earthquake, is called the focus. The area on the Earth's surface that is directly over the focus is called the epicenter. Earthquakes are often located near tectonic plate boundaries, where the Earth's crust is under constant strain. The epicenters of earthquakes can be located by using a seismograph, which is a scientific instrument that detects seismic waves. Because P- and S-waves travel at different rates through the Earth, the difference in their arrival time, recorded on a seismograph, can be used to calculate how far away the earthquake occurred. Using data from at least three different seismographs pinpoints the earthquake's epicenter location. Two scales are used to measure the intensity of earthquakes, the Richter scale and the modified Mercalli scale. The Richter scale measures the energy equivalency of an earthquake. The modified Mercalli scale measures the potential damage to structures caused by an earthquake.

Volcanoes occur where hot, molten rock called lava erupts at the Earth's surface. Volcanoes are often associated with convergent tectonic plate boundaries where one oceanic plate is subducting beneath another plate. This causes seawater to mix with the upper mantle, forming a less dense magma, which rises toward the surface, forming a volcano. Volcanoes are classified by the type of eruption they produce. Quiet eruptions occur when lava slowly flows from the ground, forming fissure volcanoes and shield cone volcanoes. Violent eruptions occur when gases, ash, rock, and lava are violently released from a volcano. These are associated with composite cones, calderas, lava domes, and cinder cones. Violent eruptions usually produce pyroclastic material, which is a mixture of hot gas, ash, lava, and rock. Two types of hazards are often associated with violent erupting volcanoes: lahars and mudflows. A lahar is a rapidly moving mixture of lava, gas, water, and ash that flows down the side of a volcano at speeds of more than 100 miles per hour. Mudflows are formed by melting snow on high-altitude volcanoes, which mixes with rock, ash, and other debris that races down the side of a volcano.

CHAPTER REVIEW

Multiple Choice

1. The sudden break of a rock mass beneath the Earth's surface causes:
   a. isostasy
   b. an earthquake
   c. erosion
   d. mantle convection

2. The locations of earthquakes are usually associated with:
   a. the center of continents
   b. the centers of landscape regions
   c. plate boundaries
   d. zones of erosion

3. The place in the Earth's crust where a rock mass has moved, causing an earthquake, is called the:
   a. epicenter
   b. zenith
   c. core
   d. focus

4. The point on the Earth's surface above where an earthquake originates is called the:
   a. epicenter
   b. zenith
   c. core
   d. focus

5. When the sea floor moves as a result of an earthquake, what can occur?
   a. seismic waves
   b. volcanoes
   c. a tsunami
   d. a focus

6. Which statement is true regarding the speed of P- and S-waves?
   a. S-waves travel faster then P-waves
   b. P-waves travel faster than S-waves
   c. They both travel at the same speed
   d. Their speed cannot be determined

7. Which statement is true regarding P-waves?
   a. They travel through solids, liquids, and gases
   b. They can only travel through solids
   c. They travel more slowly than S-waves
   d. They are the most damaging of all seismic waves

8. Which statement is true regarding S-waves?
   a. They travel through solids, liquids, and gases
   b. They can only travel through solids
   c. They travel more slowly than P-waves
   d. They are the most damaging of all seismic waves

9. Which statement is true regarding surface waves?
   a. They travel through solids, liquids, and gases
   b. They can only travel through solids
   c. They travel more slowly than S-waves
   d. They are the most damaging of all seismic waves

10. A seismograph station recorded the difference in arrival times between P- and S-waves to be 4 minutes. Using the P- and S-wave travel time diagram, how far away is the epicenter from this station?
    a. 1000 km
    b. 1900 km
    c. 2600 km
    d. 5200 km

11. Many volcanoes occur in specific regions associated with:
    a. subduction zones
    b. mountain ranges
    c. islands
    d. transform plate boundaries

12. Magma rises to the Earth's surface forming lava and volcanoes as a result of:
    a. earthquakes
    b. sliding rock along faults
    c. tsunamis
    d. mantle mixing with seawater

13. Shield cones and fissure eruptions cause:
    a. violent eruptions
    b. lahars
    c. quiet eruptions
    d. mudflows

14. Composite cones, lava domes, calderas, and cinder cones all cause:
    a. violent eruptions
    b. lava fields
    c. quiet eruptions
    d. plateaus

15. A rapidly moving mass of hot gas, ash, lava, and rock material is called a:
    a. mudflow
    b. pyroclastic flow
    c. lava flow
    d. lahar

16. A large mass of melted snow and debris that runs down the side of an erupting volcano is known as:
    a. mudflow
    b. pyroclastic flow
    c. lava flow
    d. lahar

continued

Matching *Match the terms with the correct definitions.*

a. earthquake
b. seismic waves
c. seismograph
d. P-wave
e. S-wave
f. surface wave
g. focus
h. epicenter
i. volcano
j. lava
k. magma chambers
l. eruption
m. caldera
n. pyroclastic flow
o. lahars

1. ____ The point on the Earth's surface directly above the focus of an earthquake.
2. ____ A rapid flow of mud and debris formed from the rapid melting of snow and ice associated with a volcanic eruption.
3. ____ The violent, rapid shaking of the Earth caused by a rupture in the crust.
4. ____ An opening in the Earth's crust through which gas, dust, lava, and other pyroclastic materials flow to the surface.
5. ____ The extremely hot gas, ash, and volcanic material that is ejected from a volcano during an eruption and rapidly moves downhill.
6. ____ Energy released by an earthquake that travels through the Earth in the form of waves.
7. ____ The point in the Earth's crust where a rock mass is broken or moved, causing an earthquake.
8. ____ A large crater (more than 1 mile in diameter) caused by a violent volcanic eruption.
9. ____ A seismic wave formed from the interaction of other seismic waves at the Earth's surface caused by an earthquake, which causes the ground to move in a wavelike rolling motion.
10. ____ A scientific instrument that is used to detect seismic waves generated by earthquakes.
11. ____ The sudden release of lava or pyroclastic material from a volcano.
12. ____ A seismic wave generated at the focus of an earthquake that travels in the form of a wave and can only pass through solids.
13. ____ Tubes, tunnels, or large cavities in the Earth's crust through which magma travels or collects.
14. ____ A seismic wave produced at the focus of an earthquake that is the fastest of all seismic waves and can travel through all states of matter.
15. ____ Hot, molten volcanic rock that flows freely on the Earth's surface.

Critical Thinking

1. Describe the reasons why you would or would not choose to live near a tectonic plate boundary.

CHAPTER

10 Minerals, Rocks, and Mineral Resources

Section 10.1 – Minerals Objectives

Mineral Properties • Mineral Composition • Rocks and Minerals

After reading this section you should be able to:

- ❖ Define the term *mineral* and describe the physical and chemical properties that are commonly used to identify them.
- ❖ Describe what generally gives a mineral its unique physical characteristics.
- ❖ Explain the basic structure of a silicate.

Section 10.2 – Rocks Objectives

Igneous Rocks • Sedimentary Rocks • Metamorphic Rocks • The Rock Cycle

After reading this section you should be able to:

- ❖ Identify the three main types of rocks found on Earth.
- ❖ Describe four characteristics used to identify igneous rocks.
- ❖ Differentiate between extrusive and intrusive rocks.
- ❖ Differentiate between mafic rocks and felsic rocks.
- ❖ Describe the processes that lead to the formation of sedimentary rocks.
- ❖ Describe three characteristics used to identify sedimentary rocks.
- ❖ Explain the processes that lead to the formation of metamorphic rocks.
- ❖ Describe three characteristics used to identify metamorphic rocks.
- ❖ Explain the rock cycle.

Section 10.3 – Mineral Resources Objectives

Mineral Resources • Mineral Ores • Mineral Deposits • Mining Techniques

After reading this section you should be able to:

- ❖ Define the term *mineral resource* and differentiate between metallic and nonmetallic mineral resources.
- ❖ Identify the four different processes that form mineral deposits.
- ❖ Describe three ways in which mineral resources can be removed from the Earth's crust.

TERMS TO KNOW

minerals
crystalline
monomineralic rocks
polymineralic rocks
igneous rocks
crystallization
intrusive rock
extrusive rock
felsic rocks
mafic rocks
sedimentary rocks
lithification
metamorphic rocks
foliated
mineral resource

INTRODUCTION

The rocks that compose the Earth's crust and cover its landscape hold the secrets of our planet's history. Many geologists regard rocks as history books that tell the tale of the Earth's past environments and geological events. The stories that rocks hold reveal the creation and destruction of mountains and oceans, along with the occurrence of violent events such as volcanic eruptions and asteroid impacts. Learning to read the story that rocks tell involves simple observations of their physical characteristics, which can reveal much about their formation. This can then be used to piece together geological events of the past. The minerals that compose the Earth's rocks have been a fascination of humans for thousands of years. The precious metals and gemstones that are found in rocks are some of the most valued items on Earth. Understanding how these minerals form and what they are composed of provides more insight into the processes that occur within the lithosphere. The technological society in which we live is constructed from the minerals extracted from the lithosphere. Almost everything we use in our everyday lives contains minerals that were mined from the Earth's crust. Building materials, jewelry, automobiles, and most technological machines or devices contain minerals or precious metals taken from the ground. The dependence of human society on the wealth that lies deep within the Earth cannot be overstated; it is the access to mineral resources that provides us the world we live in today.

10.1 Minerals

Mineral Properties

The Earth's crust is composed of **minerals.** Minerals are naturally occurring, crystalline, inorganic substances that have unique physical and chemical properties. The definition of a mineral is complex and therefore should be further explored to fully understand what a mineral is. The term *naturally occurring* refers to the process by which a material is formed naturally on the Earth. Minerals are not created by humans; they exist naturally in the Earth. Minerals are also **crystalline.** This means that they are made up from atoms and molecules that are arranged in definite patterns. The term *inorganic* means nonliving, or not formed from a living thing. All minerals are inorganic. Rocks such as amber, for example, are not formed from minerals because amber is the hardened sap from a tree that grew millions of years ago.

Minerals are identified by a series of specific physical and chemical properties that make them unique. The color of a mineral can be used to identify it because some minerals possess unique colors (Figure 10–1). Sulfur is a bright yellow mineral that can be easily identified by its color. Color alone, however, is not sufficient to identify all minerals. This is due to the fact that many minerals share the same color, such as pyroxine and olivine. These two minerals are both green, making it difficult to use color alone to identify them. Other minerals, such as quartz, can come in a range of colors. Rosy quartz is pink, smoky quartz is gray, and some quartz is clear, lacking any color at all.

Figure 10–1 The various colors of the mineral calcite. *(Courtesy of The Rockdoctor.)*

Figure 10–2 The mineral magnetite showing its metallic luster. *(Courtesy of Coolrox.com.)*

The *luster* of a mineral is the way that its surface reflects light. Luster is also an identifying characteristic for minerals. Some minerals possess a metallic luster that resembles polished metal. Magnetite has a metallic luster, as does galena (Figure 10–2). Other minerals can be classified as having a glassy luster. The minerals halite and quartz are examples of these because they both appear like transparent glass. Other luster categories include dull, pearly, greasy, and earthy.

The streak of a mineral is the small powder trail that is left behind when a mineral is rubbed against a rough surface (Figure 10–3). Often a white plate made from porcelain, called a streak plate, is used to determine a mineral's streak. Many minerals produce a streak that is different from their overall color.

Figure 10–3 The characteristic streaks for the minerals hematite and galena. *(Courtesy of The Rockdoctor.)*

The mineral fluorite, for example, has a light blue-green color, but it produces a white streak when rubbed against a porcelain plate. Some minerals are so hard that they do not produce a streak.

Hardness is the ability of a mineral to resist being scratched; it is often used to identify particular minerals. The Mohs scale of mineral hardness can be used to determine a mineral's hardness (Table 10–1). This measures mineral hardness on a scale from 1 to 10. A mineral hardness of 1 represents an extremely soft mineral, such as talc. Talc can easily be scratched by a fingernail. A harder mineral such as quartz has a hardness of 7 on the scale and can be used to scratch glass. The hardest of all minerals is represented by a 10 on the hardness scale. A diamond, with a hardness of 10, cannot be scratched by any other mineral.

Cleavage is the tendency for minerals to break apart along specific surfaces or planes.

TABLE 10-1 Mohs scale of mineral hardness

Mineral	Hardness	Relative Hardness
Graphite	0.7	Can be scratched by fingernail
Talc	1	
Gypsum	2	
Calcite	3	Can be scratched by copper penny
Flourite	4	Can be scratched by steel
Apatite	5	
Orthoclase	6	Can be scratched by glass
Quartz	7	
Topaz	8	Can be scratched by quartz
Corundum	9	Can be scratched by topaz
Diamond	10	Hardest of all minerals

EARTH SYSTEM SCIENTISTS *FRIEDRICH MOHS*

Friedrich Mohs was born in Germany in 1773 and gained his education at the Freiberg Mining Academy. He spent his life teaching and studying the unique properties of minerals. He continued to teach mining science in Vienna until his death in 1839. Mohs's work involved the development of a mineral classification system similar to the ones that were being developed for use in biology. To help classify minerals, he created the Mohs scale of hardness, which identifies unique minerals by their ability to be scratched by certain materials. His scale is still widely used today to identify specific minerals.

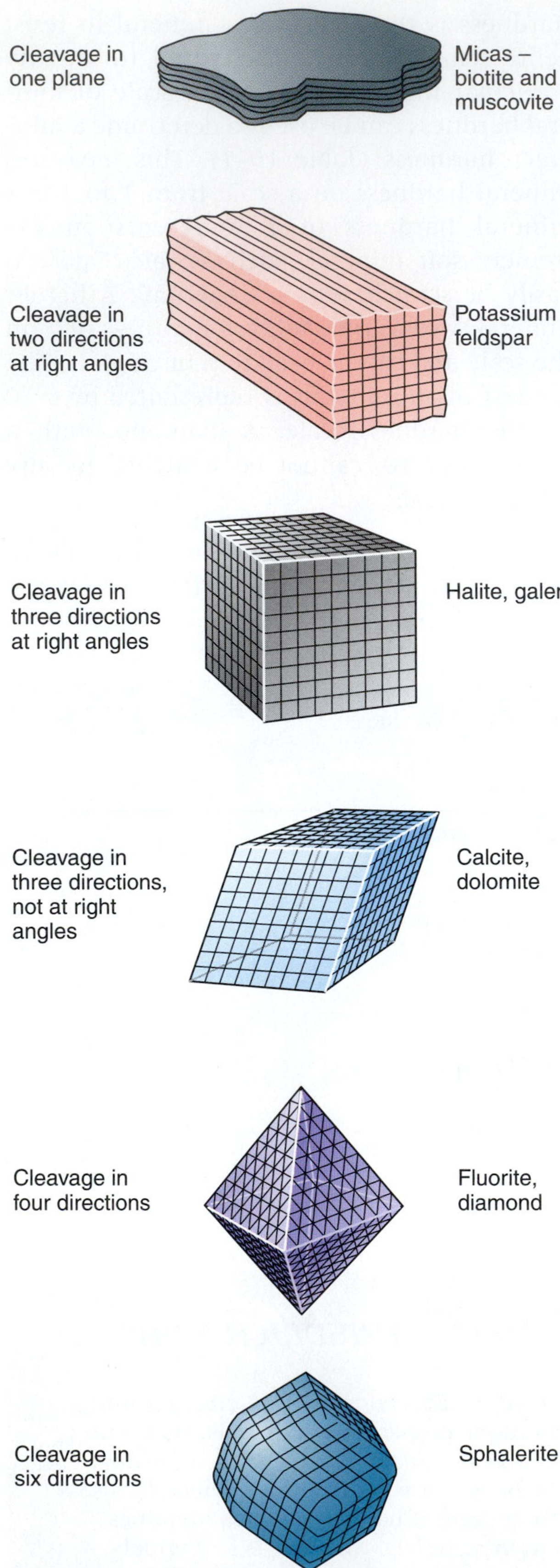

Figure 10–4 Diagram of the various types of mineral cleavage.

A mineral's crystalline structure determines how a mineral forms cleavage (Figure 10–4). Minerals such as mica produce cleavage in only one direction. This results in mica's breaking apart into thin sheets. Other minerals show different types of cleavage. Halite, the mineral that forms rock salt, cleaves in three directions or planes. This causes it to always break apart into cubes or rectangles. Calcite is a common mineral that always breaks apart along cleavage planes at 75-degree angles. Minerals that do not have particular cleavage patterns and break apart randomly are called fractured.

Specific gravity is a measure of a mineral's relative density. The density of a mineral is recorded in grams per cubic centimeter. The least dense minerals, such as calcite and talc, have a specific gravity of approximately 3. Hematite and magnetite are more dense and possess a specific gravity of approximately 5. The densest of all minerals is gold, which, in its pure form, has a specific gravity of 19.3. This means that gold is more than 19 times heavier than talc.

Some minerals can be identified by using specific chemical tests. Minerals that contain calcite, which is composed of calcium carbonate, can be identified by dropping a small amount of hydrochloric acid on them. The acid breaks apart the calcium carbonate molecule to form carbon dioxide gas. This causes the surface of the mineral to fizz. Geologists often use the acid test to determine if a rock contains calcium carbonate.

Minerals can possess special properties that are also used to identify them (Table 10-2). Some minerals are magnetic, such as magnetite, and can be easily identified by their magnetic properties. Other minerals are luminescent, meaning that they glow when exposed to ultraviolet light. Fluorite is a luminescent mineral. Other minerals, such as quartz, are piezoelectric, which means that they produce a weak electric current when exposed to increased pressure. A unique

TABLE 10-2 Characteristic properties of common minerals

Properties of Common Minerals

LUSTER	HARD-NESS	CLEAVAGE	FRACTURE	COMMON COLORS	DISTINGUISHING CHARACTERISTICS	USE(S)	MINERAL NAME	COMPOSITION*
Nonmetallic luster	1	★		White to green	Greasy feel	Talcum powder, soapstone	Talc	$Mg_3Si_4O_{10}(OH)_2$
	2		★	Yellow to amber	Easily melted, may smell	Vulcanize rubber, sulfuric acid	Sulfur	S
	2	★		White to pink or gray	Easily scratched by fingernail	Plaster of paris and drywall	Gypsum (Selenite)	$CaSO_4 \cdot 2H_2O$
	2–2.5	★		Colorless to yellow	Flexible in thin sheets	Electrical insulator	Muscovite Mica	$KAl_3Si_3O_{10}(OH_2)$
	2.5	★		Colorless to white	Cubic cleavage, salty taste	Food additive, melts ice	Halite	NaCl
	2.5–3	★		Black to dark brown	Flexible in thin sheets	Electrical insulator	Biotite Mica	$K(Mg,Fe)_3$ $AlSi_3O_{10}(OH)_2$
	3	★		Colorless or variable	Bubbles with acid	Cement, polarizing prisms	Calcite	$CaCO_3$
	3.5	★		Colorless or variable	Bubbles with acid when powdered	Source of magnesium	Dolomite	$CaMg(CO_3)_2$
	4	★		Colorless or variable	Cleaves in 4 directions	Hydrofluoric acid	Flourite	CaF_2
	5–6	★		Black to dark green	Cleaves in 2 directions at 90°	Mineral collections	Pyroxene (commonly Augite)	$(Ca,Na)(Mg,Fe,Al)$ $(Si,Al)_2O_6$
	5–5	★		Black to dark green	Cleaves at 56° and 124°	Mineral collections	Amphiboles (commonly Hornblende)	$Ca,Na(Mg,Fe)4(Al,Fe,Ti)3$ $Si_6O_{22}(O,OH)_2$
	6	★		White to pink	Cleaves in 2 directions at 90°	Ceramics and glass	Potassium Feldspar (Orthoclase)	$KAlSi_3O_8$
	6	★		White to gray	Cleaves in 2 directions, striations visible	Ceramics and glass	Potassium Feldspar (Na-Ca Feldspar)	$(Na,Ca)AlSi_3O_8$
	6.5		★	Green to gray or brown	Commonly light green and granular	Furnace bricks and jewelry	Olivine	$(Fe,Mg)_2SiO_4$
	7		★	Colorless or variable	Glassy luster, may form hexagonal crystals	Glass, jewelry, and electronics	Quartz	SiO_4
	7		★	Dark red to green	Glassy luster, often seen as red grains in NYS metamorphic rocks	Jewelry and abrasives	Garnet (commonly Almandine)	$Fe_3Al_2Si_3O_{12}$
Either	1–6.5		★	Metallic silver or earthy red	Red-brown streak	Ore or iron	Hematite	Fe_2O_3
Metallic luster	1–2	★		Silver to gray	Black streak, greasy feel	Pencil lead, lubricants	Graphite	C
	2.5	★		Metallic silver	Very dense (7.6 g/cm3), gray-black streak	Ore of lead	Galena	PbS
	5.5–6.5		★	Black to silver	Attracted by magnet black streak	Ore of iron	Magnetite	Fe_3O_4
	6.5		★	Brassy yellow	Green-black streak, cubic crystals	Ore of sulfur	Pyrite	FeS_2

★ = Chemical symbols

Al = aluminum	Cl = chlorine	H = hydrogen	Na = sodium	S = sulfur
C = carbon	F = fluorine	K = potassium	O = oxygen	Si = silicon
Ca = calcium	Fe = iron	Mg = magnesium	Pb = lead	Ti = titanium

Career Connections

MINERALOGIST

A mineralogist works to identify and classify all crystalline minerals. This includes the study of the elements that make up individual minerals and their structural arrangement. Mineralogists also determine the unique properties of individual minerals and the conditions by which they form in the Earth. This knowledge is then applied to the practical use of particular minerals. The work of a mineralogist involves the use of precise instruments to help identify specific minerals. These include powerful electron microscopes, x-ray diffraction machines, spectrometers, and computers. Many mining companies employ mineralogists to help identify regions where particular minerals may be found and how to process them. Other private industries employ mineralogists for use in industrial processes and commercial applications in which minerals are used. A career in mineralogy requires a college education.

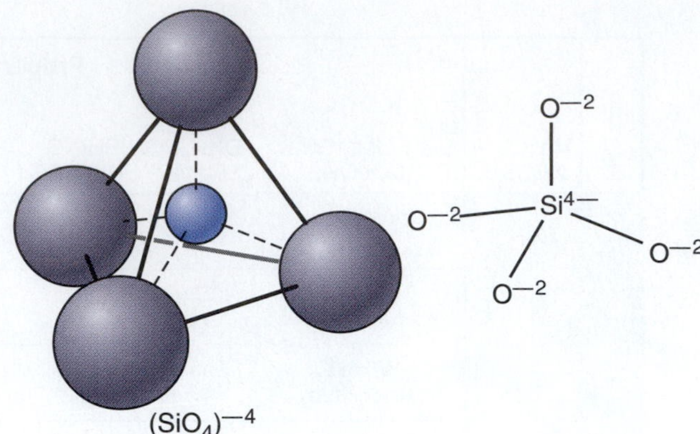

Figure 10–5 The structure of a silicate tetrahedron.

flame color produced when a mineral is exposed to fire is also a special property that can be used for identification.

Mineral Composition

Different minerals are composed of specific elements that together give them their unique properties. Even though many unique minerals exist, most are made up of only two elements, silicon and oxygen. Silicon and oxygen are by far the most abundant elements in the Earth's crust by mass. Molecules that are formed from atoms of oxygen and silicon are also called silicates. Oxygen atoms make up more than 46% of the Earth's crust by mass. Silicon composes more than 28% of the Earth's crust by mass. The other main elements that compose most minerals are aluminum, iron, calcium, sodium, magnesium, and potassium.

Because silicon and oxygen account for most minerals, it is important to understand how they combine to form a mineral's unique structure. The joining of four oxygen atoms with one atom of silicon results in the formation of a silicon-oxygen tetrahedron (Figure 10–5). A tetrahedron is a four-sided object that resembles a three-dimensional triangle. The four corners of the tetrahedron are composed of oxygen atoms, which surround a central silicon atom. The silicon-oxygen tetrahedron is an important structure that gives many minerals their unique properties. The way that these tetrahedrons are arranged within a mineral creates its unique crystalline structure. Quartz is a common mineral that is composed of only silicon and oxygen atoms.

Rocks and Minerals

Most all rocks are composed of minerals. Exceptions to this include coal and amber, which are rocks formed from the remains of once-living organisms. Rocks that are formed from only one specific mineral are called **monomineralic rocks.** Rock salt is a monomineralic rock because it contains only the mineral halite. Another monomineralic rock is limestone; it only contains the mineral calcite. Rocks that are formed from more than one mineral are called **polymineralic rocks.** Most rocks that form the Earth's crust are polymineralic. Approximately 2000 minerals

TABLE 10-3 Common rock-forming minerals

Mineral	Abundance in Crust, %	Rock in Which Found
Plagioclase*	39	Igneous rocks mostly
Quartz	12	Detrital sedimentary rocks, granites
Orthoclase	12	Granites, detrital sedimentary rocks
Pyroxenes	11	Dark-colored igneous rocks
Micas	5	All rock types as accessory minerals
Amphiboles	5	Granites and other igneous rocks
Clay minerals	5	Shales, slates, decomposed granites
Olivine	3	Iron-rich igneous rocks, basalt
Others	11	Rock salt, gypsum, limestone, etc.

*Feldspar group of minerals

have been identified in the Earth's crust, with only about 20 to 30 making up most of the rocks on Earth. The minerals that form the rocks in the Earth's crust are called the rock-forming minerals (Table 10–3). They include feldspar, quartz, talc, calcite, olivine, magnetite, pyrite, and mica.

For more information go to these Web links:

<http://mineral.galleries.com>

<http://geollab.jmu.edu/Fichter/Minerals/index.html>

<http://www.netspace.net.au/~mwoolley/top.htm>

SECTION REVIEW

1. What is the definition of a mineral?
2. List the chemical and physical properties that are used to identify minerals.
3. Draw a model of the silicon-oxygen tetrahedron.
4. Provide two examples of a monomineralic rock.
5. List five rock-forming minerals.
6. Who was Friedrich Mohs?

10.2 *Rocks*

Rocks are the naturally formed, solid material that makes up the Earth's crust. Most rocks are made up of one or more minerals, which are called the rock-forming minerals. Rocks on Earth are classified on the basis of their origin and formation. The three main categories of rocks on Earth are igneous, sedimentary, and metamorphic.

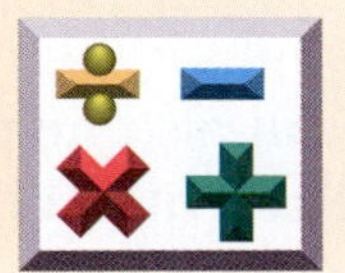

EARTH MATH

1) DETERMINE THE DENSITY FOR A MINERAL SAMPLE THAT HAS A MASS OF 104 GRAMS AND A VOLUME OF 20 CUBIC CENTIMETERS.

Igneous Rocks

Igneous rocks are rocks that have formed from the cooling and solidification of molten rock. Molten rock comes from the Earth's upper mantle and is called magma. When magma reaches the Earth's surface and comes into contact with air or water, it is called lava. When molten rock cools, it becomes a solid by forming crystals. The cooling and solidification of molten rock is also called **crystallization.** This process also occurs when liquid water reaches its freezing point. At 32° Fahrenheit, liquid water freezes and forms a network of crystals, which becomes ice. When magma cools to form igneous rock beneath the Earth's surface, these rocks are classified as **intrusive.** This is because the magma has intruded into the Earth's crust. The opposite of intrusive rock is **extrusive rock,** which forms when lava cools on the Earth's surface (Figure 10–6).

Figure 10–7 Coarse-textured granite formed from the slow cooling of intrusive magma below the Earth's surface. *(Photo by Pamela Gore, Georgia Perimeter College.)*

Igneous rocks are identified by their texture, color, density, and mineral composition. The texture of a rock is influenced by the size, shape, and arrangement of a rock's crystals. The texture of an igneous rock is defined by the time it took for the rock to cool and solidify. Rocks that cool slowly form very large crystals. These igneous rocks are usually formed from intrusive molten rock that has slowly cooled deep in the Earth's crust. Granite is an intrusive igneous rock that forms from slow cooling (Figure 10–7). The larger crystals that make up granite can be easily seen with the naked eye.

Figure 10–6 Extrusive rock forming from cooling lava at the Earth's surface. *(Courtesy of PhotoDisc.)*

Rocks that are cooled quickly form very small crystals or even no crystals at all. These rocks are extrusive and are the result of volcanic activity (Figure 10–8). Basalt is a common extrusive rock containing small crystals that is formed from volcanoes. Some extrusive rocks that are formed beneath the ocean cool so quickly that they develop a glassy texture. Obsidian, also called volcanic glass, is formed when lava cools rapidly. Some extrusive rock cools so quickly that air gets trapped inside, giving the rock a porous texture, which is referred to as vesicular. Pumice is a fine-grained extrusive rock with a porous texture that resembles a sponge.

The color of an igneous rock can also help to classify it. Color is usually referred to as being either light or dark. Lighter-colored igneous rocks tend to contain feldspar and silicate minerals, which are not very dense. These in-

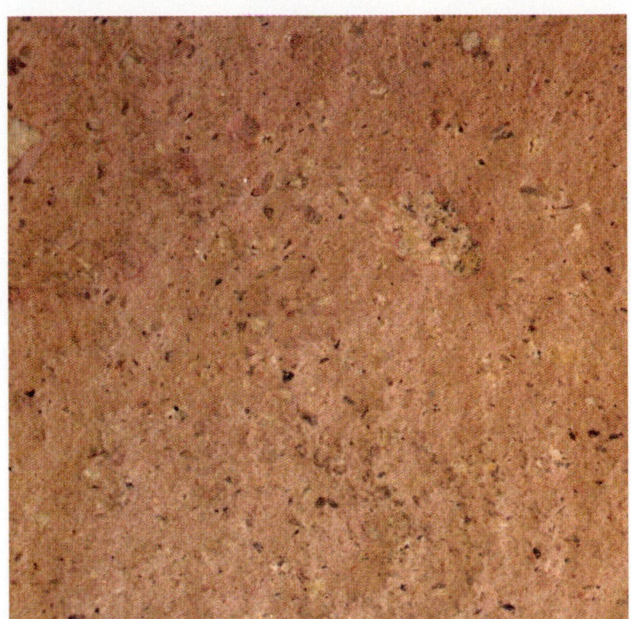

Figure 10–8 The fine texture of rhyolite formed from the rapid cooling of extrusive lava. *(Photo by Pamela Gore, Georgia Perimeter College.)*

clude rocks such as granite and pumice. Darker-colored igneous rocks are rich in iron and magnesium and are much more dense. Basalt and gabbro are examples of darker-colored igneous rocks. The composition of minerals in igneous rock can also help to identify them. Igneous rocks containing high percentages of quartz and potassium feldspar minerals are classified as being **felsic rocks.** These are rocks that are lighter in color and lower in density. The term *felsic* is derived from the words *feldspar* and *silicate.* Igneous rocks that contain high percentages of pyroxine, olivine, and plagioclase minerals are called **mafic rocks.** The term *mafic* is derived from the symbols for the chemical elements magnesium (Mg) and iron (Fe). These igneous rocks are darker in color and have a high density. Igneous rocks can be easily identified by using the scheme for igneous rock identification shown in Figure 10–9.

Sedimentary Rocks

Sedimentary rocks are formed from the accumulation of sediments, which are tiny rock particles that were weathered from, or broken off from, preexisting rock or organic material. These tiny rock particles are chemically or physically removed from their parent rock and then transported by wind, water, or glacial action to form sedimentary rock. The sediments that form these rocks are often classified by their unique texture. The texture of a sedimentary rock can be clastic and crystalline. The term *clastic* refers to the rock's being composed of individual rock fragments that have been bonded together. *Crystalline* texture refers to the rock's being composed of crystals. Another texture type, called *bioclastic,* refers to the rock's being formed from the remains of living organisms. Many sedimentary rocks form at the bottom of large bodies of water where large amounts of sediments have settled. Sedimentary rocks also usually contain horizontally arranged parallel layers called strata (Figure 10–10). These mark the different layers of accumulated sediments that formed the rock. Because sedimentary rocks are formed from accumulating sediment, they also may contain fossils. Fossils are the hardened impressions of once-living organisms that died long ago and were buried in the sediments that eventually turned into rock. Sedimentary rocks are the only type of rocks that contain fossils. The process by which accumulating sediment turns into a solid mass, or sedimentary rock, is called **lithification.** There are four main processes that cause accumulated sediments to undergo lithification.

Cementation is the lithification of sediments by binding them together with a substance such as iron oxide, silicates, or calcium carbonate. These substances act as cement to tightly bond the sediments together to form a solid mass (Figure 10–11). Iron oxide, commonly called rust, is an excellent binding agent. For example, old automobile engines can become seized when their parts are allowed to rust. This is the same process that binds together sediment. Concrete uses the same binding action to form a solid mass that is used in construction. Cement is a mixture

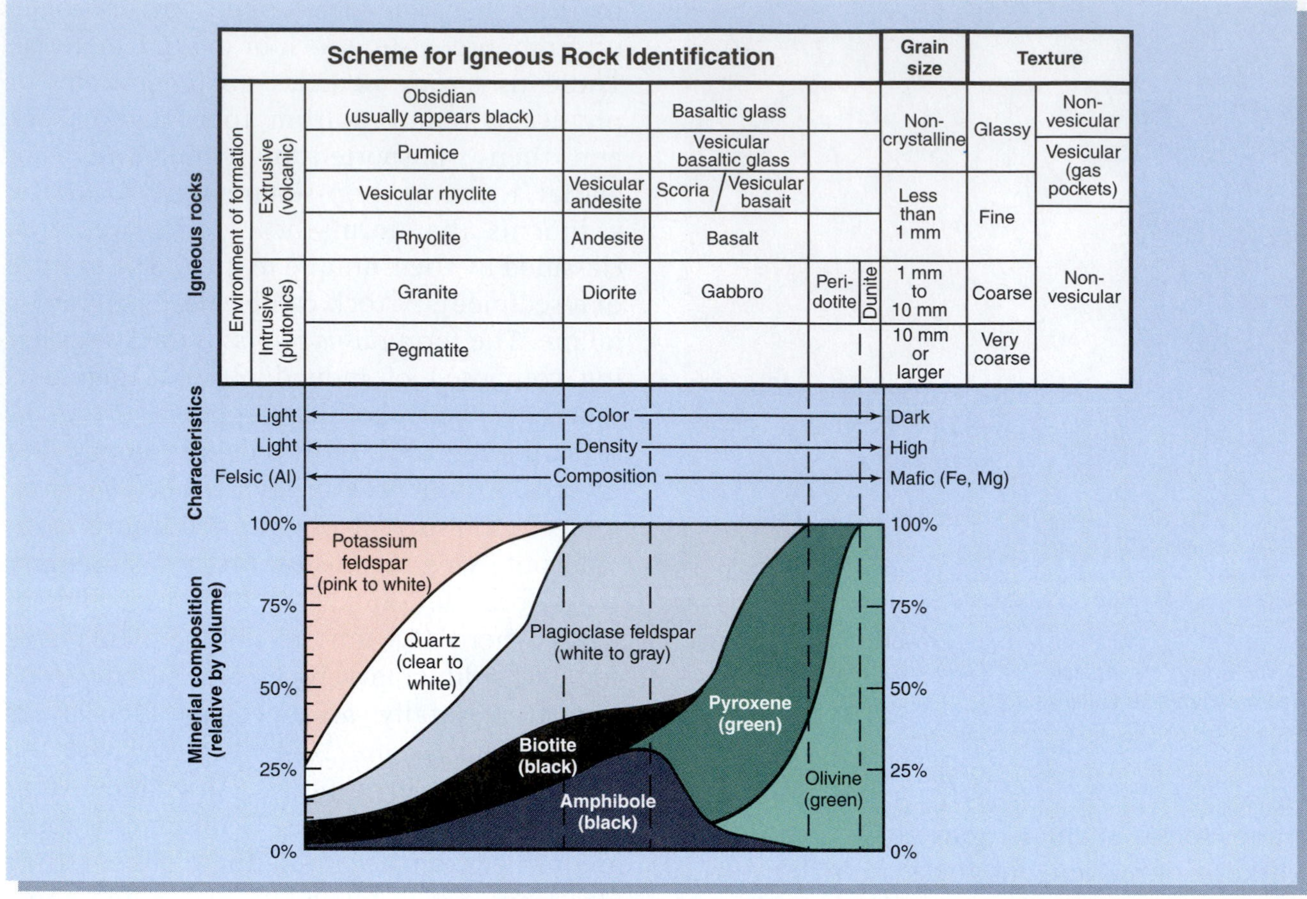

Figure 10–9 The scheme for igneous rock identification.

of iron, silica, and limestone (calcium carbonate) that binds together crushed stone. Common sedimentary rocks that are formed by the process of cementation are sandstone and conglomerate.

Compression and compaction is another way that sediments can become lithified to form sedimentary rock. As fine sediment settles to the bottom of a large lake or ocean, it begins to collect and is slowly compressed together. As newer sediments collect on top of older sediments, they increase the overlying weight and pressure, also known as overburden. Eventually the weight and pressure from the accumulating overburden becomes so great that the underlying sediments become compacted together to form a solid mass. Many sedimentary rocks form at the bottom of the ocean, where the extreme weight of the overlying water helps to compress and compact the sediment particles together. Shale is a sedimentary rock that is formed from the compression and compaction of clay and silt-sized sediment. Approximately 70% of all the sedimentary rocks on Earth are shale.

Chemical processes such as precipitation and evaporation can also form sedimentary rocks. Minerals such as calcite can become dissolved in water to form a solution of calcium carbonate and water. Eventually the calcium carbonate can precipitate out of solution to form small mineral particles of calcite. The term **precipitation,** when used in geology, means the separation of solid particles out of a solution. These minerals can precipitate and build up over time to form solid rock masses. This process occurs underground when slightly acidic groundwater dissolves the calcite. If the

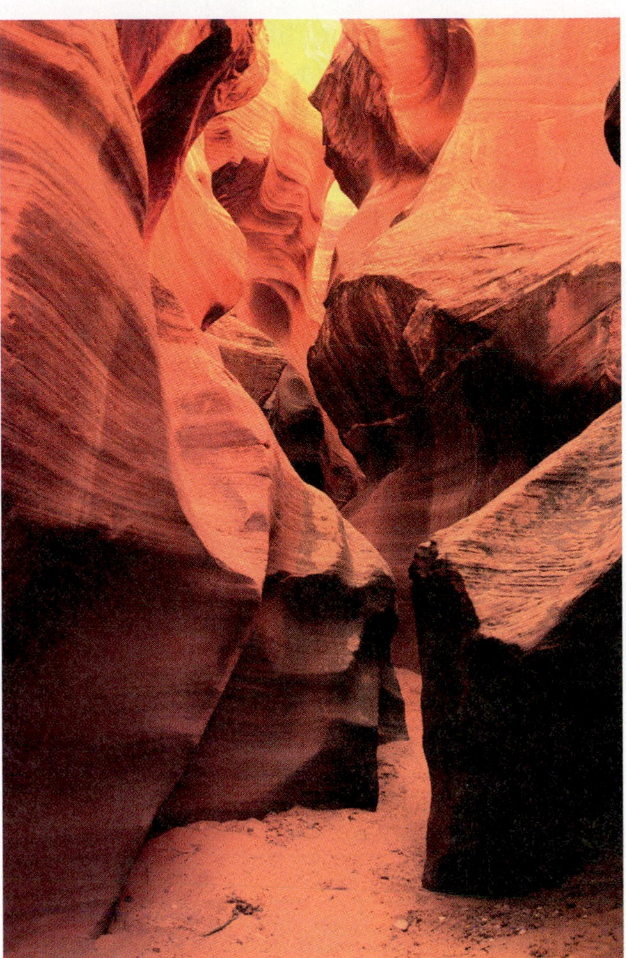

Figure 10–10 Horizontal layers, or strata, of sandstone, which is a typical characteristic of sedimentary rock formations. *(Courtesy of PhotoDisc.)*

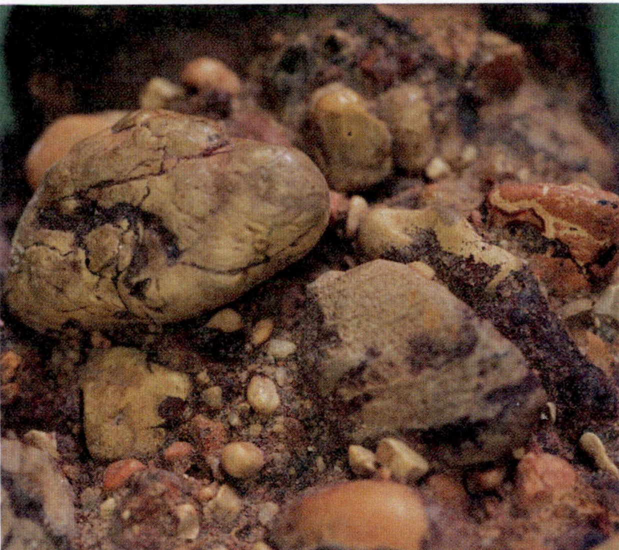

Figure 10–11 Large sediment particles that are cemented together to form conglomerate. *(Photo by Pamela Gore, Georgia Perimeter College.)*

groundwater penetrates into underground caverns and caves, the calcite precipitates out of solution and forms sedimentary rocks such as stalagmites and stalactites, which are also known as chemical limestone.

Another method of chemically forming sedimentary rocks involves evaporation. Seawater contains high amounts of dissolved sodium and chloride ions. If the seawater becomes trapped in some way and is exposed to warm temperatures for a long time, eventually all the water will evaporate. The sodium chloride that is left behind binds together to form halite. Together the halite makes up rock salt, also called an evaporate sedimentary rock, because it was formed as a result of evaporation. Large rock salt deposits can be found deep in the Earth's crust in the middle of continents far from the ocean. These deposits can be more than 10 feet thick in some places. This suggests that at one time the area was covered by a shallow sea that must have evaporated as a result of change in the environment. Sedimentary rocks that are formed from chemical action all tend to be monomineralic in composition. Rock salt only contains the mineral halite, and chemical limestone is composed of the mineral calcite.

The last method of sedimentary rock formation involves some type of living organism. In the ocean many microscopic organisms such as phytoplankton and zooplankton spend their lives floating freely through the water. Eventually these organisms die and settle at the bottom of the ocean. Over time a great amount of these once-living organisms can accumulate on the sea floor and become compacted into a solid mass. Many of the tiny organisms that collect on the sea floor build tiny shells composed of calcium. When these organisms collect on the sea floor and are lithified by compression and compaction, they form a solid mass called fossil limestone.

Another biological method of sedimentary rock formation involves the remains of ancient plants that once thrived around swamps. These plants died and fell to the bottom of the swamp, where they collected over time. Eventually the swamp dried up and became buried under accumulating sediments. The plant remains trapped deep in the Earth were then compressed and compacted together to form what is called bituminous coal, a biologically derived sedimentary rock. Although it appears to be black, and looks nothing like a plant, it is made from the fossilized remains of plants. Coal is one of the few rocks on the Earth that is not made up of minerals. It is made up of hydrocarbon, which is a compound formed mainly from atoms of carbon and hydrogen by living things. Because of the differences in the formation of sedimentary rocks, geologists divide them into two distinct categories (Figure 10–12). The first are the inorganic, land-derived sedimentary rocks. This category includes all sedimentary rocks that are formed from the weathering of preexisting rock. Examples of inorganic, land-derived sedimentary rocks include siltstone, conglomerate, shale, and sandstone. The other category of sedimentary rock is chemically or organically formed sedimentary rock. These are rocks that are formed from some type of chemical process such as precipitation or evaporation or are the remains of once-living organisms. Examples of these type of sedimentary rocks include chemical limestone, rock salt, gypsum, fossil limestone, and bituminous coal.

Scheme for Sedimentary Rock Identification

CHEMICALLY AND/OR ORGANICALLY FORMED SEDIMENTARY ROCKS					
Texture	**Grain Size**	**Composition**	**Comments**	**Rock Name**	**Map & Symbol**
CRYSTALLINE	Coarse to fine	Calcite	Crystals from chemical precipitates and evaporites	Chemical limestone	
	Varied	Halite		Rock salt	
	Varied	Gypsum		Rock Gypsum	
	Varied	Dolomite		Dolostone	
BIOCLASTIC	Microscoptic to coarse	Calcite	Cemented shell fragments or precipitates of biologic origin	Limestone	
	Varied	Carbon	Plant remains	Coal	
INORGANIC LAND-DERIVED SEDIMENTARY ROCKS					
Texture	**Grain Size**	**Composition**	**Comments**	**Rock Name**	**Map & Symbol**
CLASTIC (fragmental)	Pebbles, cobbles, and/or boulders embedded in sand, silt, and/or clay	Mostly quartz, feldspar, and clay minerals; may contain fragments of other rocks and minerals	Rounded fragments	Conglomerate	
			Angular fragments	Breccia	
	Sand (0.2 to 0.006 cm)		Fine to coarse	Sandstone	
	Silt (0.006 to 0.0004 cm)		Very fine grain	Siltstone	
	Clay (less than 0.0004 cm)		Compact: may split easily	Shale	

Figure 10–12 The scheme for sedimentary rock identification.

EARTH SYSTEM SCIENTISTS *WILLIAM SMITH*

William Smith was born in England in 1769. His primary occupation for most of his life was builder of canals. Although he had no formal education, Smith became interested in the geology of England. In 1799 he began to record the detailed arrangement of rock strata all across England. His research lead to the creation of some of the first detailed geological maps. In 1815 he published a map titled *A Delineation of the Strata of England and Wales,* which showed in detail and scale the unique rock formations of his native country. Later he published geological charts, maps, and descriptions that identified the unique fossil species that were found within particular rock formations in England. Smith's work formed the foundation for the creation of accurate geological maps that eventually were made for all parts of the world. Much of the work he accomplished during his life was unappreciated because of his lack of education and his work as a canal builder. It was not until 8 years before his death in 1839 that he was recognized for his achievements by the Geological Society of London.

Metamorphic Rocks

Metamorphic rocks form when preexisting rocks undergo a change as a result of exposure to intense heat and pressure. The rocks from which metamorphic rocks form are also known as parent rocks. Often these rocks are either sedimentary or igneous rocks. The formation of metamorphic rocks requires a great amount of heat and pressure; therefore metamorphic rocks are often associated with mountains. The forces that cause mountains to rise up from the Earth's surface are great enough to produce metamorphic rocks. The formation of metamorphic rocks usually involves the recrystallization of the minerals inside the parent rock. Recrystallization is the new arrangement of the atoms and molecules in the rock that gives it new properties. Recrystallization does not require the rock to melt. The only rocks that form from melted rock are igneous rocks. The recrystallization that occurs to form metamorphic rocks is the result of intense heat and pressure, not melting. Metamorphic rocks also become **foliated** when they are exposed to heat and pressure. Foliation is the formation of distinct layers in the rock. The more intense the heat and pressure that the rock is exposed to, the thicker the bands of foliation; however, not all metamorphic rocks become foliated (Figure 10–13).

Metamorphic rocks often show a distorted structure, such as folding or curving (Figure 10–14). This is the result of the intense heat and pressure to which the rock was exposed at one time. Metamorphic rocks also have a

Figure 10–13 Foliation of minerals in a gneiss. *(Photo by Pamela Gore, Georgia Perimeter College.)*

higher density than their parent rocks. The increased density is due to the extreme pressure they have experienced.

Metamorphic rocks are often located within an existing mass of either igneous or sedimentary rock. This is caused by the process of contact metamorphism, which occurs when magma intrudes into an existing rock layer and the heat of contact between the magma and the rock layer causes a metamorphic rock to form. An example of this is the intrusion of magma into an existing layer of limestone. The magma eventually cools to form an igneous rock, such as granite, within the limestone. The area of limestone that was exposed to the heat of contact with the magma may metamorphose to form marble. Marble is a metamorphic rock whose parent rock was limestone.

Metamorphic rocks are classified mainly by their mineral composition, foliation, and texture (Figure 10–15). Common metamorphic rocks include slate, which forms from shale; quartzite, whose parent rock is sandstone; gneiss, which forms from basalt; and anthracite coal, which forms from bituminous coal.

Career Connections

PETROLOGIST

A petrologist studies the formation, composition, distribution, and classification of rocks. This includes a thorough knowledge of the conditions that lead to the formation of igneous, metamorphic, and sedimentary rocks. Identifying particular rock types and their unique properties is another important part of the work done by a petrologist. Many petrologists work for the mining or materials industry, in which they help locate specific rock types required for commercial or industrial uses. Other research performed by a petrologist includes the dating of rocks. Determining how long ago a rock formed, and therefore how old it is, is an important part of putting together the geological time scale of the Earth's history. This aspect of petrology is called geochronology. A career in petrology requires a college education and time spent in both the outdoors, where rock samples are collected, and in the lab, where they can be analyzed and tested.

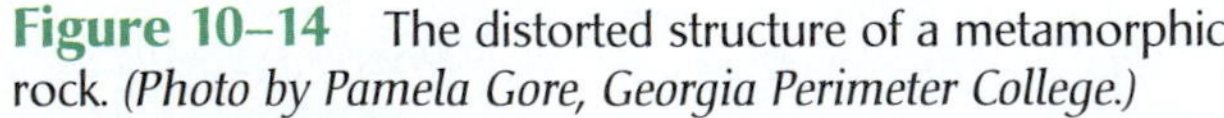

Figure 10–14 The distorted structure of a metamorphic rock. *(Photo by Pamela Gore, Georgia Perimeter College.)*

Scheme for Metamorphic Rock Identification

Texture		Grain Size	Composition	Type of Metamorphism	Comments	Rock Name	Map & Symbol
FOLIATED	MINERAL ALIGNMENT	Fine	MICA	Regional	Low-grade metamorphism of shale	Slate	
FOLIATED	MINERAL ALIGNMENT	Fine to medium	MICA, QUARTZ, FELDSPAR, AMPHIBOLE, GARNET	(Heat and pressure increase with depth) ↓	Foliation surfaces shiny from microscopic mica crystals	Phyllite	
FOLIATED	MINERAL ALIGNMENT	Fine to medium	MICA, QUARTZ, FELDSPAR, AMPHIBOLE, GARNET		Platy mica crystals visible from metamorphism of clay or feldspar	Schist	
FOLIATED	BAND-ING	Medium to coarse	MICA, QUARTZ, FELDSPAR, AMPHIBOLE, GARNET, PYROXENE		High-grade metamorphism; some mica changed to feldspar; segregated by mineral type into bands	Gneiss	
CRYSTALLINE		Fine	Variable	Contact (Heat)	Various rocks changed by heat from nearby magna/lava	Hornfels	
CRYSTALLINE		Fine to coarse	Quartz	Regional or Contact	Metamorphism of quartz sandstone	Quartzite	
CRYSTALLINE		Fine to coarse	Calcite and/or dolomite	Regional or Contact	Metamorphism of limestone or dolostone	Marble	
CRYSTALLINE		Coarse	Various minerals in particles and matrix	Regional or Contact	Pebbles may be distorted or stretched	Meta-conglomerate	

Figure 10–15 The scheme for metamorphic rock identification.

The Rock Cycle

Because of the dynamic nature of the Earth's crust, geologists have constructed a model that represents the cycling of rock material on Earth. This model is called the rock cycle, which is represented by a series of interconnecting processes that lead to the formation of all three main rock types (Figure 10–16). The rock cycle is based on a cyclic pattern because there is no real starting point for rock formation. However, billions of years ago when the Earth was forming, all rocks began as igneous rock. Today many different processes are occurring on and in the Earth's crust to form or re-form rock. These processes are illustrated in the rock cycle and include important terms such as *uplift, weathering and erosion, burial, heat, pressure, melting, compaction, sedimentation,* and *solidification.* All these terms refer to the specific processes that lead to rock formation and are important in understanding how rocks form on Earth.

EARTH MATH

1) IF LAVA AT THE EARTH'S SURFACE COOLED FROM A TEMPERATURE OF 450° CELSIUS TO 35° CELSIUS IN 30 MINUTES, WHAT WAS ITS COOLING RATE PER MINUTE?

2) IF A MAGMA INTRUSION COOLED FROM A TEMPERATURE OF 700° CELSIUS TO 90° CELSIUS IN 2 DAYS, WHAT WAS ITS COOLING RATE PER MINUTE?

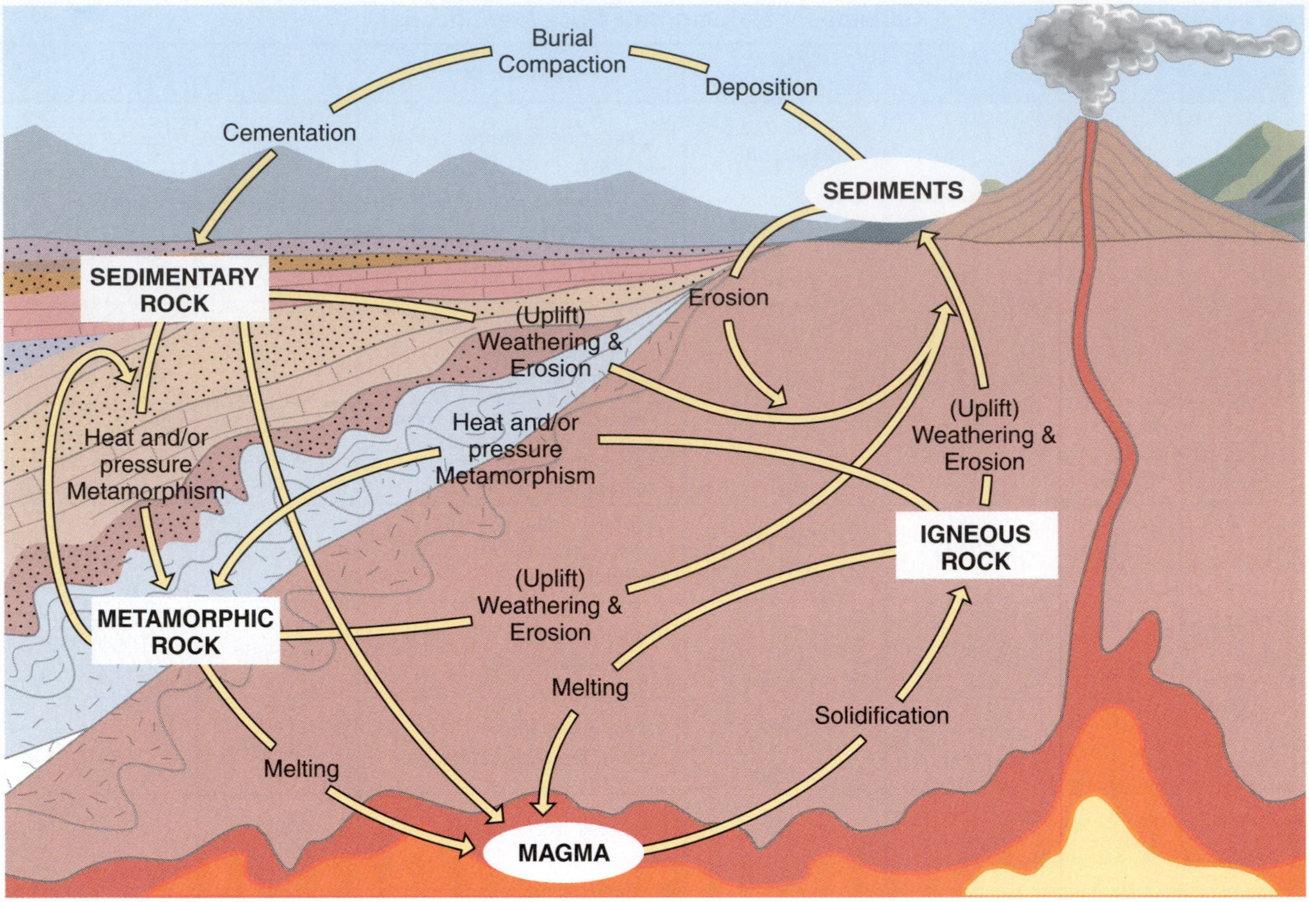

Figure 10–16 The rock cycle illustrates the processes that form rocks on the Earth.

For more information go to these Web links:

<http://geollab.jmu.edu/Fichter/IgnRx/IgHome.html>

<http://seis.natsci.csulb.edu/basicgeo/IGNEOUS_TOUR.html>

<http://www.geocities.com/RainForest/Canopy/1080/igneous.htm>

<http://www.dc.peachnet.edu/~pgore/geology/geo101/sedrx.htm>

<http://www.geocities.com/RainForest/Canopy/1080/sedimentary.htm>

<http://geollab.jmu.edu/Fichter/MetaRx/>

<http://www.geocities.com/RainForest/Canopy/1080/metamorphic.htm>

<http://www.dc.peachnet.edu/~pgore/geology/geo101/meta.htm>

EARTH SYSTEM SCIENTISTS *ABRAHAM WERNER*

Abraham Werner was a German geologist born in 1749. After receiving his education at the Freiberg Mining School, he went on to become a world-renowned geology instructor. His research involved the explanation of how the different rocks were formed on the Earth. He developed an early version of the rock cycle that explained the processes that lead to the formation of rocks he called primary, sedimentary, recent, and volcanic. Werner's theory proposed that most rocks on Earth were formed from water as the principal agent. At that time, other scientists disagreed with Werner and believed that rocks were formed from cooling molten rock. Today both these processes are considered responsible for rock formation on Earth.

SECTION REVIEW

1. Describe the difference between intrusive and extrusive igneous rocks.
2. How does the rate of cooling affect the size of crystals in igneous rocks?
3. What is the difference between mafic and felsic igneous rocks?
4. What are the four methods of lithification that form sedimentary rocks?
5. What are three examples of inorganic, land-derived sedimentary rocks?
6. What are three examples of chemically or organically formed sedimentary rocks?
7. What causes sedimentary or igneous rocks to metamorphose?
8. List three characteristics of metamorphic rocks.
9. What are four examples of metamorphic rocks? What are their parent rocks?
10. Using the rock cycle diagram, explain the path that a sedimentary rock can take when it becomes a metamorphic rock, then a sedimentary rock once again.
11. Who was William Smith?
12. What did Abraham Werner develop?

10.3 *Mineral Resources*

Mineral Resources

A **mineral resource** is any mineral, compound, or pure element that exists naturally in the Earth's crust and is used by humans in some way. Mineral resources are classified in two broad categories: metallic mineral resources and nonmetallic mineral resources (Table 10-4). Metallic mineral resources are the abundant metal elements that exist in the Earth in large quantities. These include iron, aluminum, magnesium, titanium, and manganese. Other metallic mineral resources that are not so abundant in the Earth's crust are called scarce metals; these include gold, silver, copper, zinc, lead, tin, and nickel. Nonmetallic mineral resources include construction materials such as gravel, sand, clay, sandstone, shale, and limestone. Limestone is a major component of cement. Other nonmetallic mineral resources include phosphates, sodium chloride, and sulfur. Phosphate mined from the Earth is used for agricultural fertilizers, and sodium chloride is also known as common salt. Some nonmetallic mineral resources are used for abrasives such as garnet, which is used for sandpaper, or for ceramics such as clay, pumice, and quartz.

Mineral Ores

Minerals are often found in the Earth in a form called ore. Mineral ores are natural rock that contain the desired mineral to be extracted. Mineral ores are often found in specific rock formations called mineral deposits. A mineral deposit usually contains rocks with a high percentage of the desired mineral resource. A mineral ore deposit that contains a relatively high concentration of the desired mineral resource is considered a high-concentration deposit. For example, a high concentration of aluminum ore, also called bauxite, contains about 35% aluminum. The remaining 65% is unusable

TABLE 10-4 Classification of mineral resources

Metallic Mineral Resources	Nonmetallic Mineral Resources
Abundant Metals Iron, aluminum, manganese, magnesium, titanium	**Minerals for Industrial and Agricultural Use** Phosphates, nitrates, carbonates, sodium chloride, fluorite, sulfur, borax
Scarce Metals Copper, lead, zinc, tin, gold, silver, platinum-group metals, molybdenum, uranium, mercury, tungsten, bismuth, chromium, nickel, cobalt, columbium	**Construction Materials** Sand, gravel, clay, gypsum, building stone, shale and limestone (for cement)
	Ceramics and Abrasives Feldspar, quartz, clay, corundum, garnet, pumice, diamond

SOURCE: James R. Craig, David L. Vaughan, and Brian J. Skinner, *Resources of the Earth* (Englewood Cliffs, N.J.: Prentice Hall, 1988).

rock. A high-concentration deposit of iron ore can contain almost 70% iron. Some high-concentration ores, such as copper, contain only 3% to 4% copper. A low-concentration deposit contains lower percentages of the desired mineral resource. A low-concentration deposit of iron contains only 20% iron.

Mineral Deposits

Mineral deposits originate from four different processes. Igneous mineral deposits occur as a result of the formation of igneous rocks that contain mineral resources. These mineral deposits form by volcanic activity, intrusive magma, near hydrothermal vents. Examples of igneous mineral deposits include copper,

EARTH SYSTEM SCIENTISTS — *JOHN WESLEY POWELL*

John Wesley Powell was born in 1834 in Illinois. He became interested in the study of natural history at a young age, especially with gathering specimens while exploring the outdoors. In 1850 he became the head of the Illinois Society of Natural History, where he continued his study of the Illinois landscapes. Powell's interest centered around the study of geology. After serving as a Union officer during the American Civil War, in which he lost his arm, he conducted a boating expedition down the Colorado River through the Grand Canyon. The results of his journey were published in his book titled *The Exploration of the Colorado River* in 1875. Shortly after his famous trip, Powell was appointed as the head of the U.S. Geological Survey. He continued his work in geomorphology by conducting a survey of the natural resources in Colorado, Utah, and Arizona. The results of his expeditions opened up many parts of the American Southwest, which had previously been unexplored and unmapped. His research also helped reveal the geological processes that led to the formations of much of the Southwest. In addition, Powell studied the processes of erosion, mountain formation, and volcanoes. Later in his life he began to work on plans to create dams and irrigation canals to supply water to the desert regions of the American Southwest.

Career Connections

SEDIMENTOLOGIST

A sedimentologist studies the geological processes that lead to the formation of sedimentary rocks. These scientists usually specialize in one of the three different types of sedimentary rock: clastic, carbonate, and precipitate rocks. Clastic rocks are formed from sand and clays. Carbonate rocks are formed from living organisms, and precipitate rocks form when elements precipitate from a solution. Each rock classification has different types of useful minerals associated with its formation that may be desired by the mining industry. Important natural resources such as zinc, iron, lead, and oil are found in many sedimentary rock formations. Sedimentologists can find work in the mining and petroleum industries, as well as in academic research. Their work involves outdoor exploration to search for new rock formations, as well as laboratory analysis. They also require skills in map reading and a knowledge of computer technology. A college education focusing on geology and environmental sciences is desired for this occupation.

diamonds, lead, and zinc. Sedimentary mineral deposits form by sedimentary processes such as precipitation and evaporation associated with the ocean. Typical sedimentary mineral resources include rock salt, manganese, and iron. Weathered deposits of mineral resources also occur as result of the leaching of minerals from soil. The aluminum ore bauxite is an example of a weathered mineral deposit.

Mining Techniques

Because mineral resources often exist deep within the Earth's crust, techniques have been developed to remove them and make them available for use. The removal of mineral resources from the lithosphere is called mining. Mining practices fall under two broad categories, underground mining and surface mining. Underground mining involves the digging of mine shafts directly into the Earth's crust (Figure 10–17). This is an extremely dangerous undertaking because the potential for collapse of the mine shaft is very real. Mine shafts are dug using a variety of methods that include rock drills and explosives. Once the ore deposit is drilled or blasted free, it must be transported out of the mine shaft. This involves a complex arrangement of elevators and railroad tracks that transport the heavy ore back to the surface. Today the deepest mine that has ever been excavated is in South Africa and burrows 2.3 miles into the Earth. Many ore deposits that lie deep within the Earth's crust can only be accessed by underground mines. One of the problems associated with underground mining involves the release of toxic heavy metals into the environment. Many deep underground mines become filled with groundwater, which must be pumped to the surface to keep the mine shafts from filling with water. Some of this water is naturally acidic and can dissolve toxic heavy metals exposed in the mine. When the water is discharged at the surface, it can pollute nearby surface waters.

Surface mining is the practice of removing mineral ore deposits from near the Earth's surface. There are many different types of surface

Figure 10–17 The shaft of a deep underground mine. *(Courtesy of Corbis.)*

mining operations. Open pit mining involves the removal of massive amounts of ore by some of the world's largest machines (Figure 10–18). The bucket excavators that extract ore in open pit mines can remove more than 100 tons in one single scoop (Figure 10–19). The Bingham Canyon open pit copper mine in Utah is the world's largest mine (Figure 10–20). It has created a hole in the Earth almost 3000 feet deep and has produced more than 3 billion tons of copper ore. Open pit mines cause extreme damage to the landscape as a result of massive erosion and runoff of water carrying toxic metals and sediments.

Another type of surface mining operation is strip mining. This involves the excavation of shallow strips along the Earth's surface. This is less disruptive to the environment because the mine is backfilled immediately after the ore is removed. Many strip mines are also replanted with vegetation to stabilize the landscape and prevent erosion. Hydraulic mining

Figure 10–18 Open pit mines remove massive amounts of ore from the earth's crust. *(Courtesy of PhotoDisc.)*

Figure 10–19 An open pit coal mining operation. *(Courtesy of Corbis.)*

Figure 10–20 The Bingham Canyon Copper Mine, located in Utah, is the world's largest open pit mine. *(Courtesy of EyeWire.)*

also removes ore from the surface. This is done by blasting high-pressure water at the rock containing the ore deposit. This method of mining is also damaging to the environment because the water used causes massive erosion and flooding to the surrounding landscape. Gold is mined in this way today in many parts of the world.

The last type of surface mining method is called dredging. Dredging is done when mineral ore deposits lie beneath a body of water. A large dredging bucket is used to remove the ore, which creates sediment plumes that can disrupt the aquatic environment. Today many mining practices are being improved to help lessen the impact that they have on the environment.

SECTION REVIEW

1. Define the term *mineral resource* and explain the two categories of mineral resources.
2. What are some examples of metallic mineral resources?
3. What are two examples of nonmetallic mineral resources?
4. What are the four ways that mineral ore deposits form on the Earth?
5. Describe three ways that mineral resources are mined from the Earth's crust.
6. Who was John Wesley Powell?

For more information go to this Web link:
<http://minerals.usgs.gov/>

CHAPTER SUMMARY

Minerals are the naturally occurring, inorganic, crystalline substances formed in the Earth's crust that have unique physical and chemical properties. Individual minerals are identified by their unique properties, which include color, luster, hardness, streak, and cleavage. Many minerals that make up the rocks in the Earth's crust are composed of oxygen and silicon. These elements combine to form the chemical compound known as a silicate. Silicates possess a unique structure that is formed from four atoms of oxygen surrounding one atom of silicon. This structure is known as a silicate tetrahedron. The internal arrangement of atoms in a mineral provides it with its unique physical and chemical properties. Although hundreds of different minerals exist in the Earth's crust, only a few make up most of the rocks in the crust. These are called the common rock-forming minerals, which include feldspar, quartz, talc, calcite, olivine, magnetite, pyrite, and mica. Rocks are formed from one or many types of minerals.

There are three types of rocks that form on the Earth: igneous, sedimentary, and metamorphic. Igneous rocks are formed from the crystallization of cooling lava or magma. These types of rocks are classified by their relative color, texture, composition, and density. Common igneous rocks include granite, gabbro, and basalt. Sedimentary rocks are formed from the lithification of accumulating sediments. The sediments are bonded together by different processes to form one solid mass of rock. These

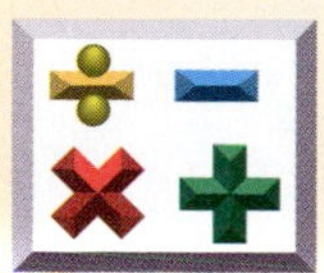

EARTH MATH

1) IF 3 BILLION TONS OF COPPER ORE HAS BEEN MINED FROM THE BINGHAM CANYON MINE IN UTAH AND THE ORE CONTAINS AN AVERAGE OF 6% COPPER, HOW MUCH COPPER HAS THE MINE PRODUCED?

processes include cementation, compression, compaction, precipitation, or evaporation. Sedimentary rocks are the only type of rocks that contain fossils and are commonly classified by their texture, grain size, and composition. Common sedimentary rocks include sandstone, shale, and limestone. The third type of rock that is found on the Earth is metamorphic rock. Metamorphic rock is formed from preexisting rock that has been exposed to extreme heat and pressure. This causes the minerals in the rock to recrystallize, forming a new type of rock. Metamorphic rocks can be foliated, which means that the minerals that they contain align themselves into unique layers. Other metamorphic rocks can be folded, bent, or distorted in some way. Common metamorphic rocks include slate, gneiss, and marble.

The processes that form rocks on the Earth are often displayed in a model called the rock cycle. This model shows the various pathways and processes by which rocks form and re-form on the Earth.

Many rocks in the Earth's crust contain valuable mineral resources. A mineral resource is a mineral that is useful to society. There are two basic types of mineral resources: metallic and nonmetallic. Metallic resources include iron, zinc, aluminum, gold, and silver. Nonmetallic resources are mostly used as building materials, such as gravel, limestone, slate, and granite. A concentration of a mineral resource in the Earth is called a mineral deposit. The raw form of the mineral that is extracted from the Earth is called mineral ore. Various techniques are used to extract, or mine, mineral resources from the ground. These can often be large-scale above-ground or below-ground operations, which may cause damage to the environment.

CHAPTER REVIEW

Multiple Choice

1. The physical properties of minerals are largely caused by:
   a. volume
   b. melting point
   c. organic composition
   d. internal arrangement of atoms

2. A student rubs a mineral sample on a porcelain plate. The student is trying to determine a mineral's:
   a. density
   b. luster
   c. hardness
   d. streak

3. A student scratches the mineral with a fingernail. The student is trying to determine a mineral's:
   a. density
   b. luster
   c. hardness
   d. streak

4. Which mineral is mostly made from silicate?
   a. quartz
   b. magnetite
   c. mica
   d. calcite

5. Which mineral fizzes when exposed to acid?
   a. quartz
   b. magnetite
   c. mica
   d. calcite

6. Which physical property is classified as metallic, glassy, earthy, or dull?
   a. density
   b. luster
   c. hardness
   d. streak

7. Which two processes result in the formation of igneous rocks?
   a. evaporation
   b. recrystallization
   c. crystallization
   d. cementation

8. Which property is common to mafic rocks?
   a. high density
   b. intrusive formation
   c. quartz composed
   d. light color

9. Which igneous rock cools most rapidly?
   a. granite
   b. gabbro
   c. basalt
   d. marble

10. Which of the following is a coarse-grained, intrusive, light-colored, low-density igneous rock?
    a. granite
    b. gabbro
    c. basalt
    d. marble

11. Which process most likely formed sandstone?
    a. evaporation
    b. recrystallization
    c. crystallization
    d. cementation

12. Which rock is most likely organic in origin?
    a. limestone
    b. sandstone
    c. basalt
    d. conglomerate

13. Which rock is most likely to contain fossils?
    a. basalt
    b. granite
    c. shale
    d. marble

14. Which rock has a clastic texture?
    a. rock salt
    b. gypsum
    c. marble
    d. sandstone

15. Metamorphic rocks result from the:
    a. erosion of rocks
    b. recrystallization of rocks
    c. crystallization of magma
    d. cementation of sediments

16. The metamorphism of preexisting rock most likely results in the rock's becoming:
    a. melted
    b. more dense
    c. fossilized
    d. eroded

17. The alignment of minerals forming bands in a rock caused by recrystallization is called:
    a. folding
    b. clastic
    c. foliation
    d. cementation

continued

18. Heat and pressure of a rock mass caused by an igneous intrusion is also known as:
   a. vertical sorting
   b. foliation
   c. contact metamorphism
   d. chemical evaporation

19. Which of the following is considered a nonmetallic mineral resource?
   a. gold
   b. limestone
   c. zinc
   d. copper

20. Which of the following mineral resources is classified as a sedimentary deposit?
   a. iron
   b. rock salt
   c. aluminum
   d. copper

Matching *Match the terms with the correct definitions.*

a. mineral
b. crystalline
c. monomineralic rocks
d. polymineralic rocks
e. igneous rocks
f. crystallization
g. intrusive rock
h. extrusive rock
i. felsic rocks
j. mafic rocks
k. sedimentary rocks
l. lithification
m. metamorphic rocks
n. foliated
o. mineral resource

1. ____ A type of igneous rock that is formed at the Earth's surface from the solidification of lava.
2. ____ The valuable minerals that are located in specific locations in the Earth's crust that can be mined.
3. ____ A classification of igneous rocks that are light colored, low density, and contain silicates and aluminum.
4. ____ A naturally occurring, inorganic, crystalline substance that has specific physical properties.
5. ____ The layered or wavy structure that forms in some metamorphic rocks.
6. ____ A substance or structure that is made up of crystals.
7. ____ A class of rocks that are formed when igneous or sedimentary rocks are changed into a new rock by exposure to intense heat and pressure.
8. ____ A specific type of rock that is made from only one mineral, such as rock salt or limestone.
9. ____ The process of converting sediments into one solid mass of rock.
10. ____ A type of rock formed on the Earth from solidifying magma or lava.
11. ____ Rocks that are composed of two or more different minerals.
12. ____ A type of rock that is formed from rock particles that are compacted or cemented together into one solid mass.
13. ____ A specific class of igneous rocks that are generally dark, dense, and contain iron and magnesium.
14. ____ Igneous rock that is formed from magma seeping into an existing rock mass.
15. ____ Rock changing phase from a liquid to a solid, also called solidification.

Critical Thinking

1. List the items in or around your home or at school that contains minerals, rocks, or mineral resources. Explain how your life would be changed if they did not exist.

CHAPTER

11 Weathering, Erosion, and Deposition

Objectives

Physical and Chemical Weathering • The Process of Erosion • Agents of Erosion • The Process of Deposition

After reading this chapter you should be able to:

- Define the term *physical weathering* and provide three examples of this type of weathering.
- Define the term *chemical weathering* and provide three examples of this type of weathering.
- Describe the factors that affect the rate of weathering.
- Define the terms *sediment* and *erosion.*
- Describe three agents of erosion.
- Identify the factors that lead to mass wasting.
- Explain the relationship between transported sediment size and velocity of water.
- Describe how a sediment's size, shape, and density affect its settling rate.
- Differentiate between horizontal sorting and graded bedding.
- Explain how glaciers erode and deposit sediments.

TERMS TO KNOW

physical weathering
chemical weathering
oxidation
carbonation
humid
erosion
runoff
stream erosion
suspension
mass wasting
deposition
graded bedding
horizontal sorting
moraines
glacial till

INTRODUCTION

One of the most dynamic systems on the Earth is the breakdown and movement of rock material on its surface. This three-part system consists of the weathering, erosion, and deposition of rocks and sediments. It involves larger rocks being reduced into smaller rock particles, transporting them over long distances, and then placing them in a new location. When children scoop up a shovel full of sand to make a sand castle at the beach, they are touching rock particles that have made a long journey from the mountains to the sea. The breakdown and movement of rock on the Earth is an important part of the rock cycle, which constantly recycles this material on the planet. Weathering, erosion, and deposition also impact many aspects of our modern society. Agriculture depends on the deposition of sediments to renew the fertility of the soil. Farmers are also concerned about erosion because they want to keep their soil in place. Our buildings, homes, and transportation systems are constantly being affected by these processes, which costs millions of dollars each year to combat. Learning about the powerful forces that drive weathering, erosion, and deposition is vital to a complete understanding of the Earth's systems.

Physical and Chemical Weathering

The weathering of rock at the Earth's surface is an important part of understanding the dynamic nature of the Earth's lithosphere system. Weathering is simply the breakdown of rocks into smaller rock particles known as sediment. The formation of sediment is an important part of the rock cycle and also important to life on Earth. Weathering on Earth is divided into two categories: physical weathering and chemical weathering. **Physical weathering,** also called mechanical weathering, is the process by which rock is broken down into smaller particles by a physical process, with no chemical changes occurring (Figure 11–1). An example of physical weathering includes freezing and thawing, also known as frost action. When water freezes it expands, and can apply a great amount of pressure on any surrounding rock. The pressure can actually break the rock apart into smaller pieces. Rocks exposed to climates that experience freezing and thawing can be weathered rapidly. Another example of physical weathering is heating and cooling. When rocks are heated, they expand as their molecules move farther apart. Then, when they cool rapidly, they contract as their molecules pack more tightly together. The action of expansion and contraction can break apart rocks and weather them over time. Physical weathering by expansion and contraction usually occurs in climates that experience severe temperature changes in a short time, such as in deserts. During the daylight hours in a desert climate, the temperature can rise rapidly, causing the rocks to heat up quickly. Then at night the temperature drops quickly, causing the rocks to rapidly cool. This causes the rocks to expand and contract continually day after day, which eventually breaks them apart.

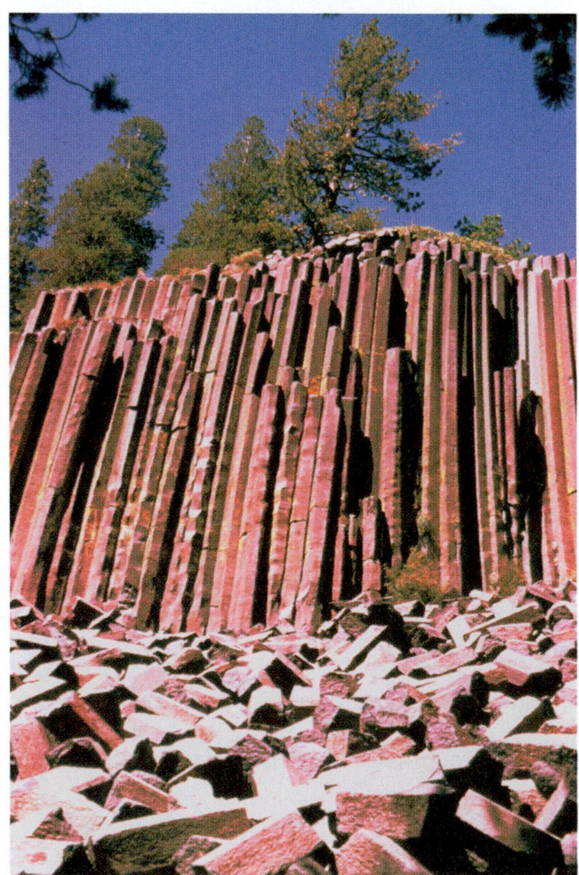

Figure 11–1 A combination of heating and cooling and frost action has physically weathered this rock outcrop, creating a pile of smaller rocks at its base, called talus. *(Courtesy of the National Park Service.)*

Abrasion is another means of physical weathering; this is the rubbing together of rock particles. Much like the way sandpaper breaks down wood, sediments in either water or wind can weather rock (Figure 11–2). The action of living organisms can also break down rock. The roots of trees can grow into the cracks of rocks and pry them apart. Burrowing organisms such as ants or rodents can break apart rocks over time. Some minerals in rock can absorb water, causing the rock to swell and expand. Eventually when it dries out, the outer layer of rock can peel away. This is called exfoliation, or the peeling away of layers of rock, which is an example of physical weathering. Finally, another physical weathering process that breaks rock down into particles or sediments is called pressure unloading. Pressure unloading occurs when rocks that have been buried deep in the ground and have been subject to extreme pressure are exposed at the surface. The release of the pressure causes the rocks to ex-

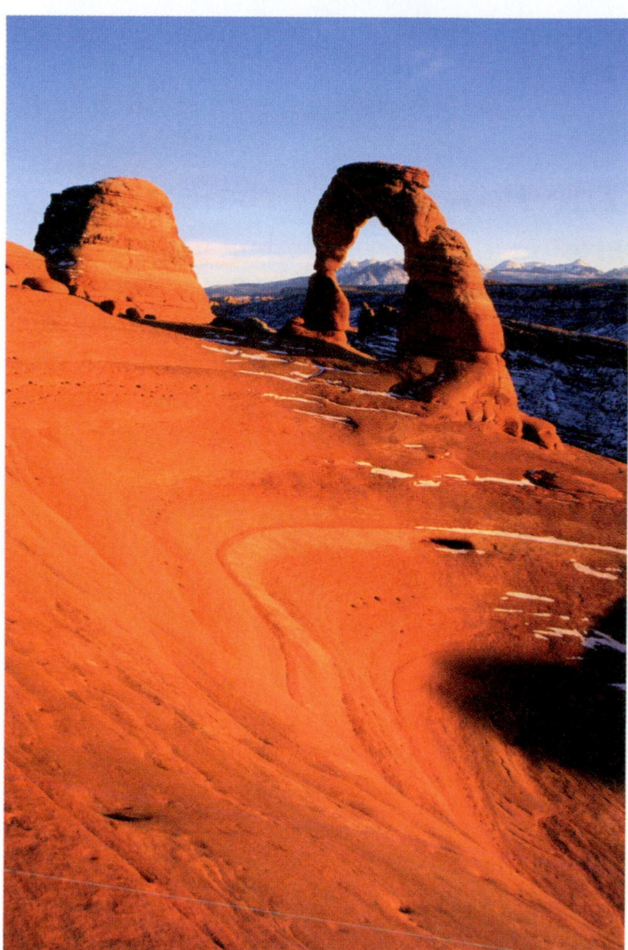

Figure 11–2 The unique shape of this rock formation is the result of the physical weathering process of abrasion by wind. *(Courtesy of PhotoDisc.)*

pand and break apart. This process affected many rocks that were covered by thick glaciers. The weight of the overlying ice compressed the rock. When the glaciers melted and retreated, the weight was removed, causing the rock to expand and crack. This physical weathering process is still occurring in parts of the northeastern United States.

Chemical weathering occurs when rock breaks down into smaller particles as result of a chemical process. **Oxidation** is a chemical weathering process that occurs when minerals in the rock react with atmospheric oxygen. Oxidation changes the physical properties of the rock, causing it to break down. The formation of iron oxide, commonly known as rust, occurs when iron-containing rocks combine with oxygen in the air to form iron oxide. This causes the rock to crumble and break apart. If you have ever observed the effect that rust has on automobiles, you can appreciate the power that chemical weathering has on solid objects. Another form of chemical weathering is called **carbonation.** This occurs when carbon dioxide mixes with water to form carbonic acid. Carbonic acid can then break down rock into smaller particles. This type of weathering can occur at the Earth's surface when atmospheric moisture mixes with carbon dioxide in the air to form carbonic acid. This weak acid then rains down on exposed rocks and chemically weathers them. Limestone rocks are chemically weathered by this process. Carbonation can also occur underground when groundwater containing carbon dioxide slowly dissolves surrounding rocks. This eventually leads to the formation of large underground caverns and caves. Many caves and caverns on the Earth are formed when rock formations containing limestone are slowly dissolved by ground water (Figure 11–3).

The last form of chemical weathering is called hydration. Hydration is the absorption of water by certain minerals in rock. Some minerals, such as feldspar, hornblende, and biotite, break down as a result of hydrolysis, or the splitting of water. Other minerals, such as halite and calcite, can also be dissolved by water.

The rate of weathering, or how long it takes for rock to be weathered, depends on a few specific factors. The climate in which the rock is located greatly affects the rate of weathering. Generally rock that is exposed to warm, **humid** climates weathers at a higher rate. The size of the rock particle that is being weathered is also an important factor in determining how long it takes a rock to weather. Smaller rock particles expose more surface area per unit volume than larger particles and therefore weather at a higher rate. The mineral composition of the particular rock that is being weathered is also an important factor for weathering rates. Some minerals are more

Figure 11–3 Spectacular caverns and caves like this one were formed by the chemical weathering of limestone by groundwater. *(Courtesy of PhotoDisc.)*

resistant to weathering than others. For example, quartz and orthoclase minerals are very weather resistant. Olivine and pyroxene minerals weather very easily. Probably the most important factor that determines weathering rates is time. The more time that a rock is exposed to the forces of weathering, the more it will break down into smaller particles.

Once a rock has been weathered, it usually ends up in one of two forms, sediments or soil. Sediments are the rock particles or fragments that are produced by the weathering of rock. Sediments are classified by their size; the smallest rock particles are clay, silt, and sand, and the larger fragments are pebbles, cobbles, and boulders. Soil is the other result of weathered rock, which is actually a mixture of mineral particles with organic material, air, and water. Soil formation and the classification of soils will be discussed in depth in Chapter 12.

The Process of Erosion

Erosion is another important process that affects the lithosphere system. Erosion is the process of transporting sediments from their place of origin and depositing them elsewhere. The principal force that transports sediments is gravity. The force of gravity along with some type of transporting material is called a transporting agent. Water is by far the most significant agent of erosion and is responsible for moving millions of tons of sediments around the world every year.

There are four main ways by which water erodes sediments on Earth. Raindrop and runoff erosion are caused when liquid precipitation, rain, impacts the Earth's surface and dislodges, or moves, sediment from its location. Heavy rains can cause a great amount of

Career Connections

GEOMORPHOLOGIST

A geomorphologist studies the geological processes that lead to the formation of unique landscapes on the Earth. This includes researching conditions that lead to the formation of mountains, valleys, plains, and canyons. They examine the effects of climate and geological processes in specific regions around the world. Geomorphologists study the short-term and long-term effects of erosion and deposition in specific regions. They also examine how human beings are altering landscapes. This information is then used to prevent environmental damage and for land use planning. Geomorphologists require a college education with an emphasis on developing skills in map making, geology, surveying, computers, and environmental science. They can find employment with state and federal government agencies and in academic fields.

sediment to be dislodged. Precipitation that does not infiltrate the ground collects on the surface and can flow down slopes, carrying with it dislodged sediments. This is called **runoff** and accounts for millions of tons of eroded sediments each year. Exposed farm fields and construction sites are extremely susceptible to erosion by raindrop impact and runoff. Once sediments enter into water, they become transported by **stream erosion.** Stream erosion transports sediments either by dissolving the sediments into solution, carrying the smaller sediments in **suspension,** or by the force of the moving water bouncing and rolling the sediments downstream. The transportation of sediments by stream erosion greatly depends on the velocity of the flowing water. The greater the velocity of the water, the greater the size of the sediment that is transported (Figure 11–4). Sediments that have been transported by moving water eventually become rounded by the abrasion that occurs as the particles bounce and roll downstream.

Another method of water transport of sediment involves the ocean. The coastal interface between the ocean and the land is an area that experiences massive erosion by water. The power of waves and ocean currents can both erode and transport rock particles over long distances. A sandy beach along the shoreline is constantly changing its shape as sediment, in the form of sand, is transported by the power of the ocean. When frozen water accumulates on land in great quantities over a long period, glaciers can form. Glaciers are large masses of ice that can move along the surface of the Earth. Glaciers are like great bulldozers that plow through rock and move it to new locations. Glaciers are also called "dirty snowballs" because there is a great amount of rock and sediment trapped in the ice. When glaciers melt and retreat, the rock particles that were trapped in the ice are deposited to form glacial sediments. The glacial sediments that were deposited to form New York's Long Island were transported all the way from Canada by glacial action (Figure 11–5).

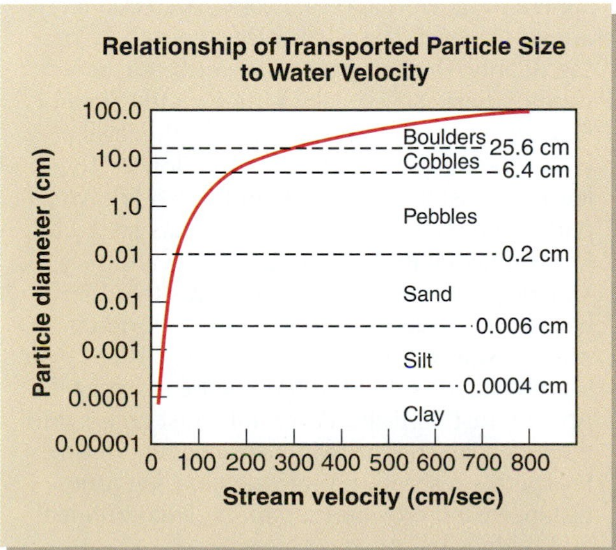

Figure 11–4 The relationship between sediment particle size and the velocity of the water needed to transport it.

Figure 11–5 A glacial moraine composed of unsorted, angular sediments. *(Courtesy of Duncan Heron.)*

WILLIAM MORRIS DAVIS

William Morris Davis was born in Philadelphia in 1850. He gained his college education at Harvard University, where he later taught meteorology, geology, and geomorphology. At the start of his career, he conducted many geographical surveys around the world. His major achievements in science included the study of the formation of landforms and landscapes. Because of this, he is regarded as the father of geomorphology. In 1889 he developed theories to explain the regular cycle of erosion and deposition. Davis developed a theory to explain the life cycle of rivers. This included a detailed explanation of the formation of steep river valleys, which signify young rivers, and the wide, flat meanders and floodplains of a mature river.

Agents of Erosion

Water is definitely the dominant agent of erosion on Earth, but not the only one. The action of air moving across the Earth's surface, or wind, can also transport sediments over long distances. Wind erosion often occurs in dry climates where sediments are dried out and exposed to the power of wind, which moves them. Some wind-deposited sediments found in the midwestern United States are more than 100 feet thick and have been built up over thousands of years. Wind erosion can also greatly affect agriculture if farm fields are left exposed to the atmosphere. In the 1930s a series of droughts and windstorms caused disastrous wind erosion in the midwest. This was called the dust bowl, and it resulted in a mass migration of farmers from the region because of the extreme effects of wind erosion. The constantly shifting sand dunes in the world's deserts illustrate the movement of sediments by the force of wind.

The last agent of erosion is called **mass wasting.** Mass wasting is the downhill movement of sediments by the force of gravity. As sediments are formed on slopes, they are exposed to the tug of gravity, which moves them downward. As the slope increases, the potential for the downward movement of sediments also increases. This is because the frictional forces that hold sediments in place on a slope lessen as slope increases; therefore land with a steeper slope also has a greater erosion rate. Rapid mass wasting occurs when sediments on a slope are forced to

Current Research

An atmospheric scientist from the University of Miami is studying the effects of the long-distance transport of dust from Africa on the citizens of Florida. Professor Joseph Prospero's research has revealed that dust kicked up from storms in West Africa rises up into the atmosphere to an elevation of 15,000 to 20,000 feet above sea level. This dust is then carried across the Atlantic Ocean by planetary-scale easterly winds, where it eventually settles on the state of Florida. What concerns Prospero is the level at which the dust enters the air over Florida. The levels of dust in the air in Florida, also known as particulate matter, are above the limits set by the Environmental Protection Agency for atmospheric pollutants. Prospero believes that the haze that is visible in many parts of Florida is often mistaken for human pollution. His research suggests that the haze is the result of the African dust particles. When this dust enters into the lungs, it can react with sensitive lung tissue. Prospero's research reveals the amazing long-distance wind erosion that affects two continents and bridges the Atlantic Ocean.

Figure 11–6 The result of the rapid mass wasting of a hillside. *(Photograph by U.S. Geological Survey.)*

move down slope suddenly (Figure 11–6). These type of events are called rock slides, mud slides, or avalanches.

Rapid mass wasting usually occurs when water accumulates on the slope, which greatly reduces the friction and increases the weight of the sediments. This causes the sediments to suddenly slide downhill. The opposite of rapid mass wasting is called creep. Creep occurs as sediments slowly move downhill over a long period. Evidence of creep can be seen in the leaning of fence posts pointing downhill as a result of the sediments that they are laid in slowly moving downward (Figure 11–7). Trees can also lean downhill as a result of creep.

The Process of Deposition

Once sediments have been transported by some type of agent, they are released and then settle in a new location. This is called **deposition.** Deposition of sediments is usually caused when the velocity of the transporting agent, usually wind or water, decreases. The decrease in velocity causes the rock particles to settle out of suspension, forming a sediment deposit. Factors that also affect the deposition of sediments include particles size, shape, and density. Smaller particles tend to settle at a slower rate than larger particles. The more spherical a sediment particle is, the faster its rate of settling; and flatter particles tend to settle at slower rates. In addition, the greater the density of a particle, the greater its settling rate. Taken together, these factors result in unique sediment deposits.

Figure 11–7 A line of fence posts points downhill, revealing evidence of creep on a hillside.

When the velocity of the transporting agent of sediments of mixed sizes is reduced quickly, they are deposited in a pattern called **graded bedding.** Graded bedding is a sediment deposit in which the size of the sediment increases with depth (Figure 11–8). This pattern results when the larger particles settle out first, and the smallest last. Graded bedding of sediments is often found near the area where streams enter into a large, deep lake.

Another type of sediment deposit is called **horizontal sorting** (Figure 11–9). This occurs as a result of the transporting agent's

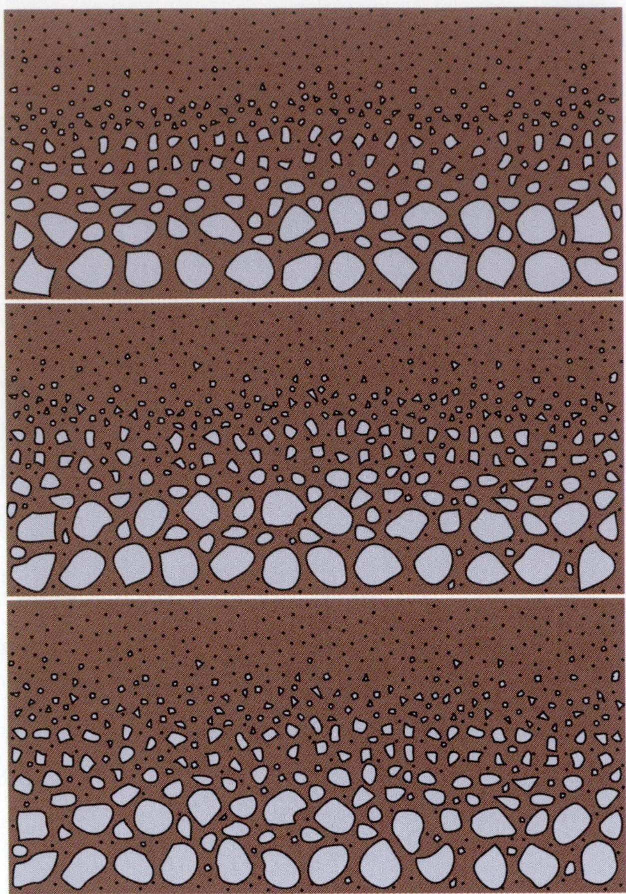

Figure 11–8 A series of vertically sorted sediment deposits known as graded bedding.

velocity decreasing at a slow rate over a long distance. Horizontal sorting results in a deposit in which larger particles settle first, and then the smaller particles settle farther out in the direction of the movement of the transporting agent. This type of deposition occurs where rivers empty into the ocean. Horizontal sorting causes sand deposits to settle nearer the shore and silt and clay deposits to form out in deeper water.

Another unique type of deposition is related to glaciers. When glaciers begin to melt, the sediments that they are carrying are left behind in large deposits called **moraines.** Moraines form large, unsorted sediment deposits also known as **glacial till.** In this type of deposition, sediment particles are randomly deposited. You can often differentiate between glacial deposits and deposits associated with moving liquid water by the shape and sorting of the sediment particles. Unlike sediments transported by liquid water, which become rounded because of abrasion, glacial sediments remain jagged and angular. This is because the particles are frozen in glacial ice and are not as heavily abraded. Also, glacial deposits are usually unsorted because the particles do not settle at different rates as a result of melting, unlike the sorted sediments associated with liquid water deposits. Almost all naturally occurring gravel pits are the remains of sediments that were deposited by either glaciers or moving water thousands of years ago. They often reveal much about what the environment was like when the sediments were deposited.

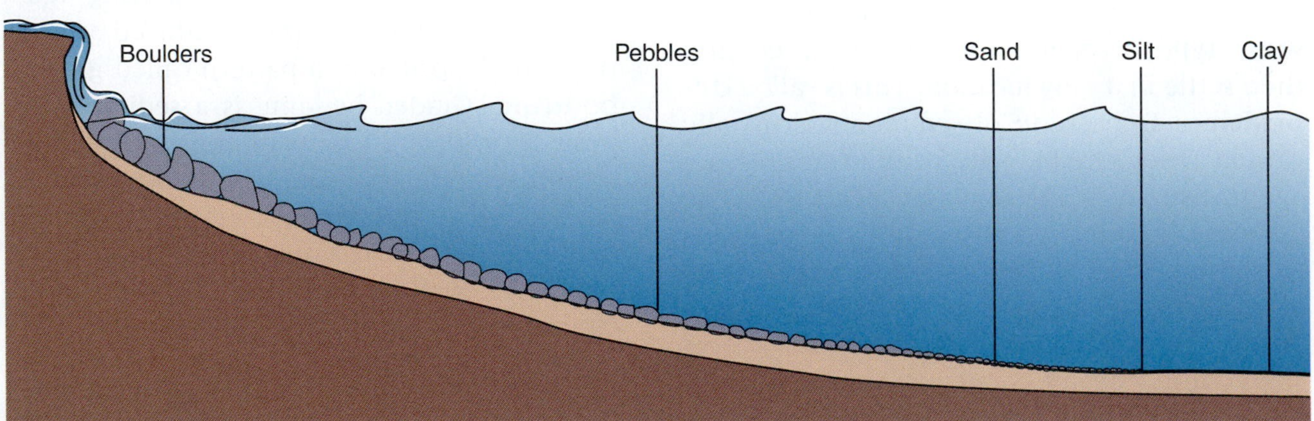

Figure 11–9 The horizontal sorting of sediments deposited by flowing water.

EARTH MATH

1) CALCULATE THE DENSITY OF THE FOLLOWING SEDIMENT PARTICLES, AND THEN DETERMINE WHICH ONE WILL SETTLE OUT OF WATER FIRST: SAMPLE A WEIGHS 50 GRAMS, AND ITS AREA IS DETERMINED TO BE 15 CUBIC CENTIMETERS. SAMPLE B WEIGHS 60 GRAMS, AND ITS AREA IS DETERMINED TO BE 20 CUBIC CENTIMETERS.

REVIEW

1. Define physical weathering and provide four examples.
2. Define chemical weathering and provide two examples.
3. What are some of the factors that affect the rate of weathering?
4. Describe two specific agents of erosion.
5. Describe three ways by which sediments can be deposited.
6. Who was William Morris Davis?

For more information go to this Web link:
<http://www.geog.ouc.bc.ca/physgeog/conents/11g.html>

CHAPTER SUMMARY

The breakdown and movement of rock material on the Earth involves three fundamental processes. These are weathering, erosion, and deposition. Weathering is the breakdown of large rocks into smaller rock particles. There are two types of weathering processes on the Earth: physical weathering and chemical weathering. Physical weathering involves the breakdown of rock by a physical process. Examples of physical weathering include freezing and thawing, heating and cooling, abrasion, and root action. Chemical weathering is the breakdown of rock as a result of a chemical change. Examples of chemical weathering include hydration, carbonation, and oxidation.

The product of weathering is called sediment. Sediment is classified by its size, smaller particles being clay, silt, and sand. Larger sediments include pebbles, cobbles, and boulders. Once sediments are created, they are eventually transported to a new location. This process is known as erosion. The specific way that sediments are transported is called an agent of erosion. The three principal agents of erosion are water, wind, and glaciers. The movement of sediment by water depends on two factors: the velocity of the water and the size of the sediment. Generally, the higher the velocity of the water, the larger the size of sediment that can be transported.

The erosion of large amounts of sediment and soil from land into a body of water is known as runoff. The rapid movement of sediment down a hillside is called mass wasting. Rapid mass wasting occurs very quickly and is known as a rockslide, mudslide, or avalanche. Slow mass wasting is called creep, which occurs over a long period. Once sediments are eroded, they eventually settle in a new location. This is called deposition. Deposition usually occurs when the transporting agent slows its velocity, causing the sediment particles to settle. The rate at which particles settle depends on the particles' size, shape, and density.

There are two main types of sediment deposits: graded bedding and horizontal sorting. Most sediment deposits are associated with liquid water, but glacial ice can also result in deposition. Glaciers scrape along rocks and gather sediments as they advance across the landscape. The sediments then become mixed in with the ice and carried along with the glacier. Eventually, when the glacial ice that contains the sediments melts, the sediments are deposited in large piles called moraines. Glacial deposits are usually composed of unsorted, angular sediments also known as glacial till.

CHAPTER REVIEW

Multiple Choice

1. Which is the best example of physical weathering?
- **a.** the cracking of a rock mass by the freezing and thawing of water
- **b.** the transportation of sediment in a stream
- **c.** the reaction of limestone with acids in rainwater
- **d.** the formation of a sandbar along the side of a stream

2. Which property of water makes frost action a common form of physical weathering?
- **a.** Water dissolves many Earth materials.
- **b.** Water expands when it freezes.
- **c.** Water cools the surrounding area when it evaporates.
- **d.** Water loses heat when it freezes.

3. The main cause of chemical weathering on the Earth is:
- **a.** rock abrasion
- **b.** heating and cooling of a rock
- **c.** reactions of rock material with air and water
- **d.** contraction of water when it freezes

4. Which type of climate causes the fastest chemical weathering?
- **a.** cool and dry
- **b.** cold and humid
- **c.** hot and dry
- **d.** hot and humid

5. On Earth the dominant agent of erosion is:
- **a.** wave action
- **b.** moving ice
- **c.** running water
- **d.** moving air

6. As the velocity of the water increases, the size of the sediment particle transported:
- **a.** decreases
- **b.** stays the same
- **c.** increases
- **d.** varies

7. Which of the following causes an increase in runoff?
- **a.** an increase in slope
- **b.** a decrease in rainfall
- **c.** an increase in biological activity
- **d.** a decrease in air temperature

8. A flowing body of water transports sediments by rolling and bouncing, in solution, and by:
- **a.** sublimation
- **b.** transpiration
- **c.** suspension
- **d.** evaporation

9. Generally, if the particle size of sediments increases from clay to sand, the settling time:
- **a.** decreases
- **b.** increases
- **c.** remains the same
- **d.** varies

10. As the velocity of a stream decreases, there will most likely be an increase in:
- **a.** erosion by the stream
- **b.** deposition by the stream
- **c.** the size of particles transported by the stream
- **d.** the amount of material transported by solution in the stream

11. A sediment deposit that contains larger particles on the bottom and increasingly smaller particles toward the top is known as:
- **a.** graded bedding
- **b.** glacial till
- **c.** moraine
- **d.** horizontal sorting

12. A sediment deposit that is formed as moving water slowly decreases its velocity over a long distance is called:
- **a.** graded bedding
- **b.** glacial till
- **c.** moraine
- **d.** horizontal sorting

13. A hill consisting of angular sediments that are unsorted was most likely deposited by:
- **a.** wind
- **b.** a glacier
- **c.** a river
- **d.** a landslide

14. A gravel pit containing rounded sediments that are horizontally sorted was most likely deposited by:
- **a.** wind
- **b.** a glacier
- **c.** a river
- **d.** a landslide

continued

15. Which sediment particle takes the least amount of time to settle: Particle A, a flat, low-density particle; or Particle B, a spherical, high-density particle?
    a. Particle A
    b. Particle B
    c. They would both settle at the same rate.
    d. They would both float.

16. Which rock would weather at a faster rate?
    a. a very large rock
    b. a very tiny rock
    c. a rock located in a dry climate
    d. a rock frozen in glacial ice

Matching *Match the terms with the correct definitions.*

a. physical weathering
b. chemical weathering
c. oxidation
d. carbonation
e. humid
f. erosion
g. runoff
h. stream erosion
i. suspension
j. mass wasting
k. deposition
l. graded bedding
m. horizontal sorting
n. moraines
o. glacial till

1. ____ The breakdown of rocks into smaller rock particles by a chemical process.
2. ____ Large amounts of unsorted glacial sediments that are deposited by a melting glacier.
3. ____ The process of breaking down rocks into smaller rock particles by a physical process in which no chemical changes take place.
4. ____ An accumulation of glacial sediments.
5. ____ The process of adding oxygen to a chemical compound.
6. ____ A specific type of sediment deposition, occurring in still water, which results in larger particles settling first, then progressively smaller particles settling on top of the larger ones.
7. ____ The addition of carbon dioxide gas to something.
8. ____ A specific form of deposition of sediments that results from the reduction of the velocity of water at the mouth of a river that enters into a body of water. This causes larger particles to settle closest to the mouth and decrease in size as you move farther into the still water.
9. ____ A term used to describe a region that has high atmospheric moisture.
10. ____ To put or place something down.
11. ____ The movement of rock particles or soil by wind, water, and the force of gravity.
12. ____ The rapid, down-slope movement of large masses of rock and soil.
13. ____ The rapid loss of soil, sediments, or other substances as a result of being washed away by rain or melting snow.
14. ____ Free-moving, solid particles that are hanging in a liquid.
15. ____ The movement of rock, soil, or sediments in a flowing body of water.

Critical Thinking

1. A gravel pit was discovered that contained horizontally sorted, rounded sediment particles at its base, covered by angular, unsorted sediments. Describe the series of events and changes in the environment that would result in this geological formation.

CHAPTER

12 Soils

Objectives

Soil Minerals • Soil Organic Material • Soil Water and Air • Soil Organisms • Soil Structure • Parent Material • Soil Horizons • Soil Classification

After reading this chapter you should be able to:

- Define the term *soil.*
- Describe the composition of a typical soil.
- Explain how minerals are gained by or lost from soil.
- Define the term *humus* and explain why it is an important part of healthy soil.
- Differentiate between fertile soil and infertile soil.
- Describe the three main states of soil moisture.
- Identify the organisms that are important for the formation of healthy soil.
- Define the term *loam.*
- Define the term *parent material* and describe the four different types of parent material.
- Describe the five different soil horizons.
- Identify one soil order and describe its unique characteristics.

TERMS TO KNOW

soil	**tilling**	**loess**
leaching	**nitrogen-fixing bacteria**	**soil profile**
organic material	**loam**	**soil horizon**
humus	**parent material**	**top soil**
soil moisture	**alluvial soils**	**subsoil**

INTRODUCTION

Probably the most important aspect of the lithosphere, with regard to sustaining life on Earth, is the existence and formation of soil. The green plants that grow in soil and depend on it to sustain their lives convert the sun's energy into stored chemical energy that other animals require to live. Without soils, life on land would not exist in its present form. Soils have also helped to shape civilization as we know it. This natural resource is of great importance to human beings. All the great societies that have flourished on the planet owe their success to fertile soils. Without a stable and plentiful food base grown in soil, advancement in civilization surely could not take place. This relationship continues today, as more people are added to the planet who depend on soils to sustain their food supply.

Soil Minerals

Soil is a complex arrangement of minerals and organic material mixed with air, water, and microorganisms. Approximately 45% of a typical soil is made up of minerals. This an important aspect of soil because plants require minerals for healthy growth. The minerals that make up the soil's inorganic material are classified by their size. From smallest to largest, they include clay, silt, sand, and gravel. Soils receive their minerals from the weathering of rock, decomposing organic material, and fertilizer created by human beings. Plants require 16 essential elements for healthy growth, making the mineral content of the soil extremely important for crops.

Soils lose minerals, as well as gain them (Figure 12–1). It is the loss of the essential elements for plant growth that can make a soil infertile, or unable to support the healthy growth of crops. Soil minerals are lost by erosion, or the removal of minerals by an agent of erosion. Soil minerals can also be lost by **leaching.** Leaching is the downward movement of minerals that are dissolved in water. Leaching moves essential minerals deep down into the soil where they cannot be accessed by plants. Minerals are also removed from soil by plant uptake. Over time plants remove essential minerals to build their bodies, which can create deficiencies in a soil. This is why farmers periodically spread fertilizers on their fields.

Soil Organic Material

The **organic material** in a soil composes approximately 5% of its total bulk. The organic matter in a soil is made from the decayed remains of living organisms. This dark organic

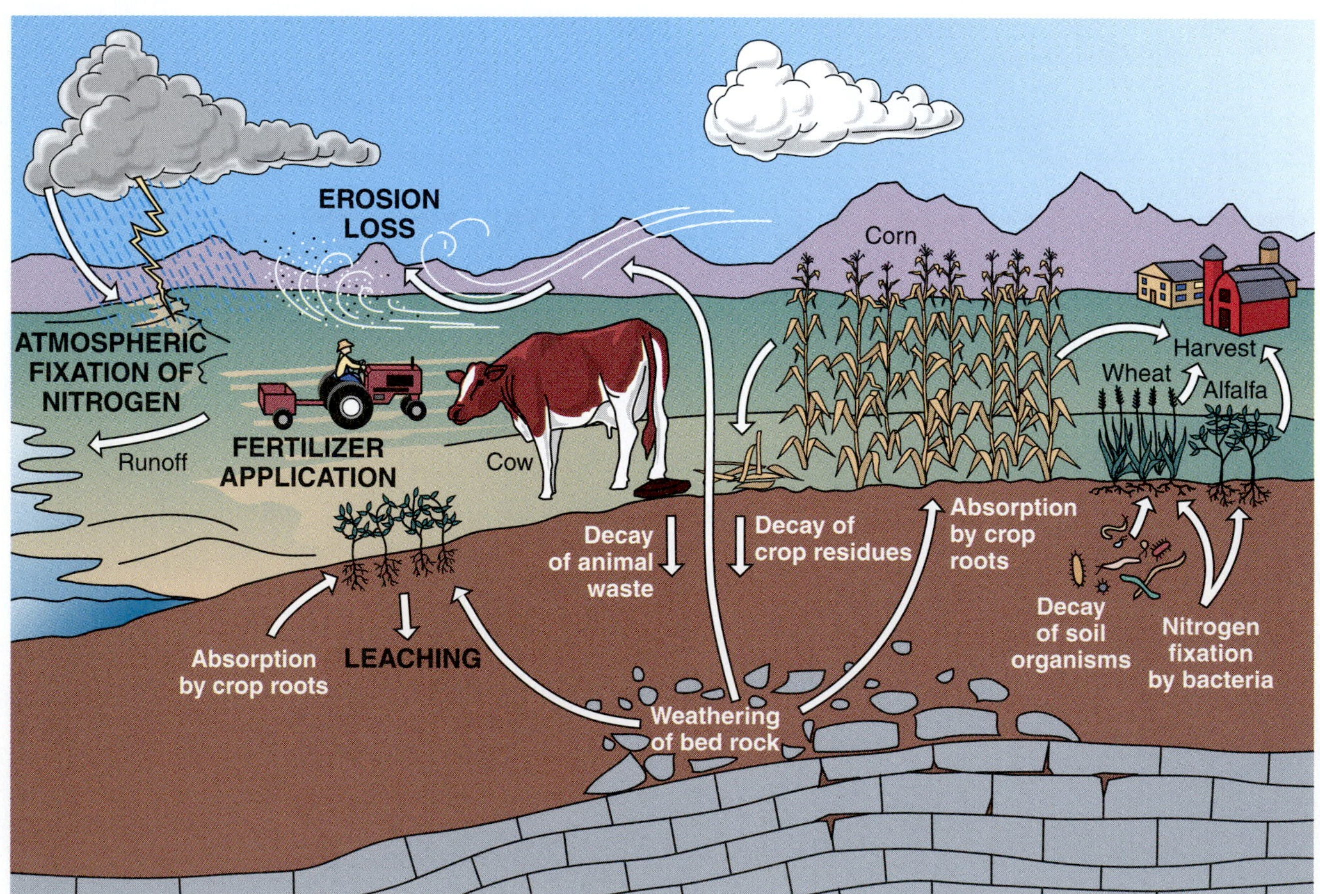

Figure 12–1 Soils can gain and lose essential minerals in a variety of ways.

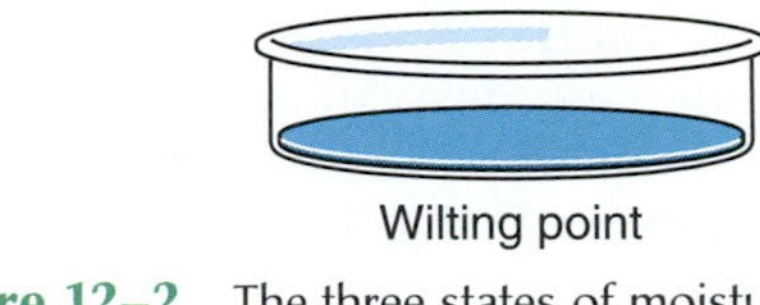

Figure 12–2 The three states of moisture in a soil's pore spaces are saturation, field capacity, and the wilting point.

material is also called **humus;** it is made up of decomposed leaves, roots, twigs, insects, animals, worms, and so on. Any living thing that dies, falls to the Earth's surface, and decays can form humus. Humus is also the result of composting. When composted grass, leaves, or vegetable matter turns into a dark material, it has formed into humus. Although humus only makes up 5% of soil, it is extremely important because it acts like glue to hold the mineral particles together in a soil. This enables soils to resist erosion. The humus in a soil also helps to form pore spaces. These little pockets that form in the soil hold water and air. A soil with a good amount of humus becomes spongy and provides a good home for the many beneficial bacteria that also live in the soil. It is important to add organic material to a soil to maintain its fertility. Farmers periodically spread animal manure or grow grass and other crops that they plow under into their fields. Both these practices help maintain organic material in the soil.

Soil Water and Air

The amount of water that is held in a soil's pore spaces is called **soil moisture** (Figure 12–2). Soil moisture is important for plant growth, but it also helps to bind together all the soil particles. Soil with good moisture levels resists erosion better than drier soil. Soil can also contain too much moisture and become saturated. Saturation of a soil occurs when the soil's pore spaces are completely filled with water. The optimum amount of moisture in a soil is called field capacity and occurs when the soil's pore spaces contain 50% water and 50% air. When a soil loses almost all its moisture, it is at its wilting point. This type of soil moisture is called the wilting point because the water levels in the soil are so low that plants begin to dry out and wilt. If this goes on for an extended period, the plants will eventually die.

A fertile soil also needs to contain air. The air in soil is similar to the air in the atmosphere because it contains the same amounts of nitrogen and oxygen and slightly higher amounts of carbon dioxide (Figure 12–3). Oxygen is especially important for a productive soil, because

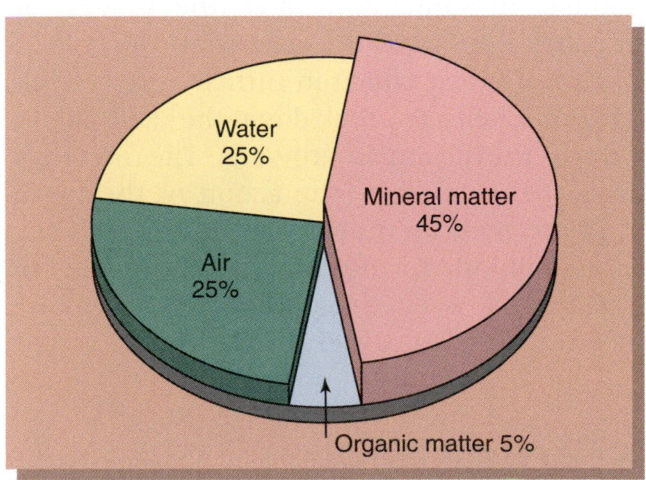

Figure 12–3 The composition of a typical soil that is good for crop growth.

the cells that make up a plant's root system require oxygen for healthy growth. Air in a soil is located in the pores, along with the water. Turning over the soil and mixing it, also called **tilling,** is how air is introduced into a soil. It is important to ensure that soils used for growing plants are well aerated by occasionally tilling them.

Soil Organisms

Living organisms are also an important part of a soil. The organism that is most associated with soil is probably the earthworm. Earthworms help to mix and aerate soils but are not the only organisms in soil. Bacteria are important soil organisms that help add nitrogen to the soil. Nitrogen is an essential element for plant growth and is not derived from soil minerals. **Nitrogen-fixing bacteria** that reside in soil convert atmospheric nitrogen into plant-available forms such as nitrates. Other organisms, such as fungi and insects, help to improve the structure of a soil and also add organic material. Rodents that burrow into soil also help to mix a soil by digging deep into the ground. Another important living organism that helps to form soil is lichen. Lichen are actually two organisms that live symbiotically. This means that they help each other survive. Lichen are composed of algae and fungi. The algae produce food from sunlight through the process of photosynthesis. The algae share the food with the fungi, which in turn secrete a weak acid that helps to break down the material on which the lichen is growing. The minerals that are released by the action of the weak acid are then used by the algae as nutrients. The symbiotic relationship of lichen helps to create soil over time by slowly breaking down rock to form soil minerals.

Career Connections

SOIL CONSERVATION TECHNICIAN

A soil conservation technician provides technical assistance to anyone who is concerned with the appropriate use of soil. The soil conservation technician works with farmers, foresters, landowners, state and local governments, and builders. The main goals of persons in this occupation are to prevent the loss of fertile soil, the pollution of local waterways, and the destruction of natural resources. Soil conservationists must have knowledge in soil science, erosion control, agricultural engineering, hydrology, and chemistry. They apply their knowledge to develop land use plans to get the most productivity from the land without creating environmental damage. Most soil conservation technicians work with the soil and water conservation services that exist throughout the country. College programs in natural resource conservation provide an excellent foundation for a career in soil conservation.

Soil Structure

The *structure* of a soil refers to the arrangement of the mineral particles and organic material in a specific soil (Figure 12–4). A well-structured soil tends to be spongy, with a high amount of humus and a mixture of mineral particle sizes. The mixture of mineral sizes can reveal a lot about how a soil will grow crops. A well-structured soil for crop growth is classified as **loam** and contains nearly equal portions of sand, silt, and clay. Soil that contains a higher percentage of sand is called sandy loam. Sandy loams tend to have excessive drainage and cannot maintain good soil moisture. The opposite of this is a soil that has a high percentage of clay. This reduces the amount of pore spaces in the soil and prevents water from infiltrating it. Most poor-structured soils are classified as dense. They do not support good crop growth and are highly erodible.

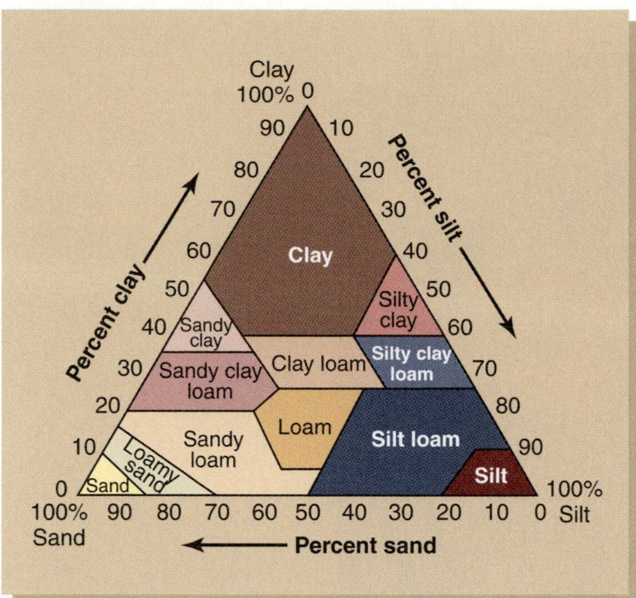

Figure 12–4 The soil textural triangle is used to classify a soil based on its mineral composition.

Parent Material

The **parent material** of a soil is the rock from which the soil minerals were derived. Rocks exposed to weathering break down to form mineral particles in a soil. The type of parent material that formed a particular soil greatly influences the fertility of that soil. For example, if the parent material for a particular soil is granite or sandstone, which are high in silicates, then the resulting soil tends to be nutrient deficient. Other types of parent rock, such as limestone, siltstone, and shale, are high in calcium and other essential elements that help to form fertile soils.

There are two main types of parent material: residual and transported. Residual soils are formed from parent rock that is located directly below them. These soils tend to be younger, thin, and not yet fully developed. Transported parent material falls into three main categories, depending on how the parent rock material was transported. Glacial transport soils are soils formed from the parent material created by glaciers. Glacial till is the sedimentary remains of glaciers and is composed of sediment in a variety of sizes. This makes glacial soils very rocky but also extremely fertile because they contain an assortment of ground-up minerals. Water-transported soils are formed from the deposition of sediments transported by liquid water. These are also called **alluvial soils.** The periodic flooding that occurs near rivers deposits deep layers of sediments that help form rich, fine-grained, fertile soils. These soils tend to be excellent for crop growth. The third type of transported parent material is a wind-transported soil, also known as a **loess.** Soils formed from loess occur in areas where the

EARTH SYSTEM SCIENTISTS *EDMUND RUFFIN*

Edmund Ruffin was born in Virginia in 1794 and became the United States' first soil researcher. After years of farming the land, Ruffin began to realize that techniques needed to be developed to revitalize the soil. A mostly self-educated man, Ruffin began to experiment with different techniques to return fertility to the soil and to understand the chemistry of the soil. He studied the effects of poor plowing and drainage on the soil and devised ways to increase the productivity of the soil. He used mixtures of clay, sand, and crushed seashells to return soil nutrients. In 1832 he published *An Essay on Calcareous Manures,* which presented techniques for improving the fertility of the soil. Ruffin also toured the southern United States giving lectures about the science of soil management. When the U.S. Civil War broke out, Ruffin became an outspoken Confederate and was one of the volunteers to fire the first cannon shots at Fort Sumter, which began the war in 1861.

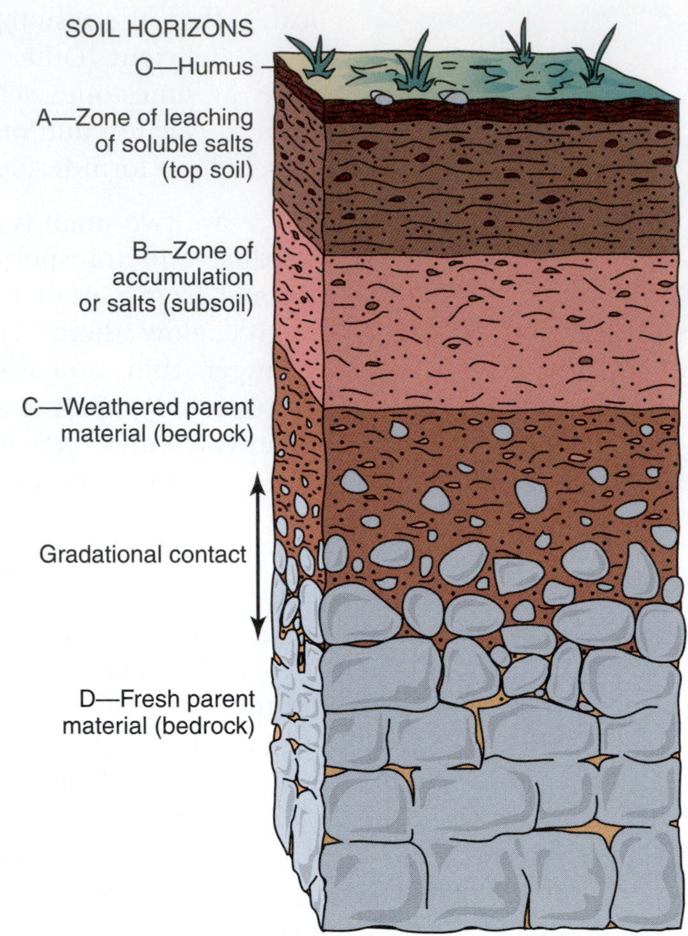

Figure 12–5 A soil profile reveals the different horizons that make up a soil.

wind has deposited thick layers of sediments. The American Midwest has excellent, fine-grained, loess-formed soil that is some of the most productive agricultural soil in the world.

Soil Horizons

A **soil profile** is a cross-sectional view of a soil that helps to classify it by identifying its unique horizons (Figure 12–5). A **soil horizon** is a unified layer in the soil that has similar physical and chemical features. Soils are usually divided into five horizons. The O-horizon, also called the organic layer, is the layer of decaying organic material that forms at the top of a soil. The leaf litter that collects on the forest floor is an example of an O-horizon. This is an important horizon that helps to add organic matter to a soil and also helps to increase soil moisture by preventing excess evaporation. The next horizon in a soil is the A-horizon, or top soil. This is the darker soil that is rich in minerals, air, and water; it is the zone in which most plant root systems grow and forms the base for all productive agriculture in the world. The B-horizon lies below the A-horizon and is known as the **subsoil.** This horizon is usually a light tan or reddish brown colored layer that is low in organic material and very dense. The subsoil may be high in iron content as a result of the leaching of minerals that occurs from the A-horizon to the B-horizon. The next layer in a typical soil is the C-horizon, or unconsolidated parent material. *Unconsolidated* means "broken up," which describes the fractured rock of the soil's parent material. The final layer in a soil profile is the D-horizon, also called the bedrock. This is the solid mass of rock on which a soil rests. The bedrock is either an igneous, metamorphic, or sedimentary rock mass. It is important to note that not all soil types have all soil horizons. Some soils lack an O-horizon because of the lack of available organic material, such as a desert soil. Other soils may not have A-horizons or C-horizons because of the unique geography, climate, and living organisms in the area where the soils are formed. The type of soil profile and its unique horizon arrangement help soil scientists classify the soils of the world.

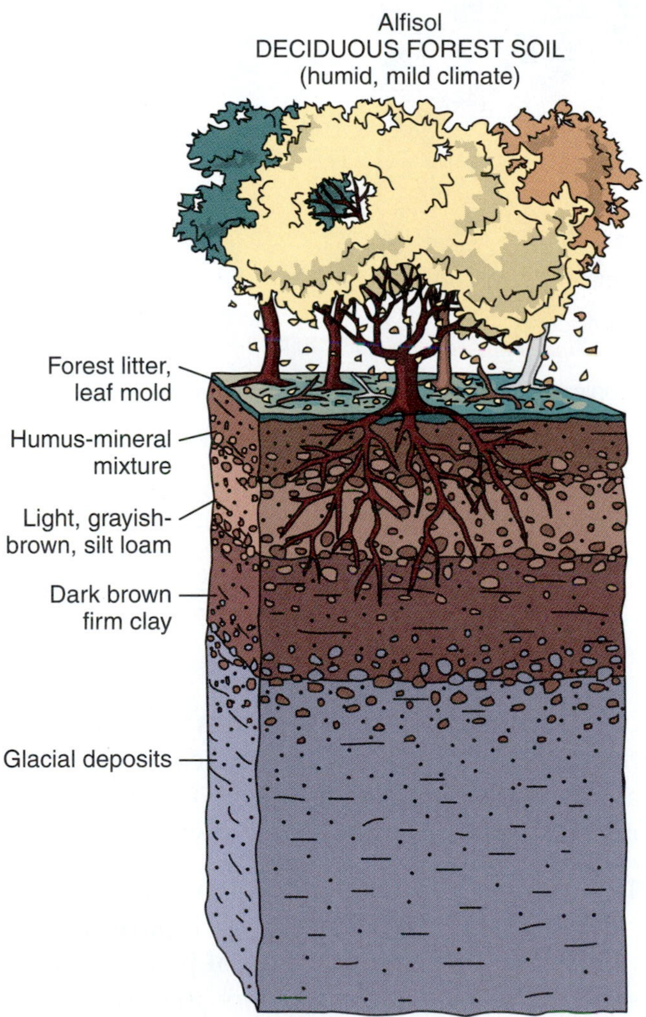

Figure 12–6 The profile of an Alfisol soil order associated with the deciduous forests of the northeastern United States reveals its unique horizon characteristics.

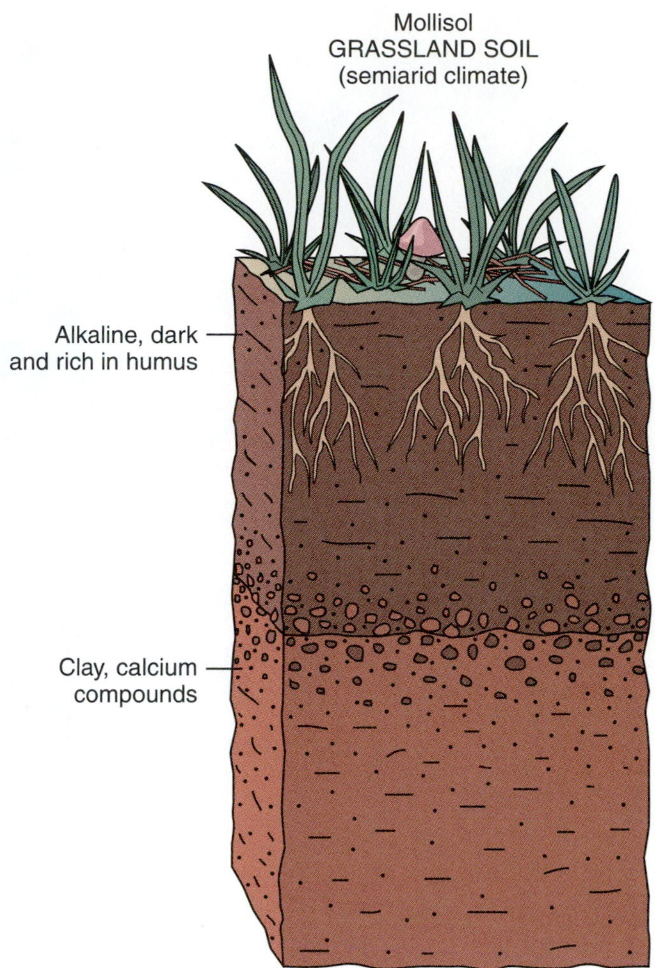

Figure 12–7 The profile of a Mollisol soil order associated with the grasslands of the midwestern United States reveals its unique horizon characteristics.

Soil Classification

Soils around the world is classified into groups called orders. There are a total of ten soil orders identified on the Earth. Alfisols are a soil order that describes soils that form in the deciduous forest of the northeastern United States (Figure 12–6). These soils have extensive O-horizons, dark rich A-horizons that support excellent plant growth, and C-horizons that are formed from glacial deposits.

Mollisols make up a soil order containing soils that form under the grasslands in the midwestern United States (Figure 12–7). These soils have no O-horizon, an extremely deep and rich A-horizon, and a C-horizon

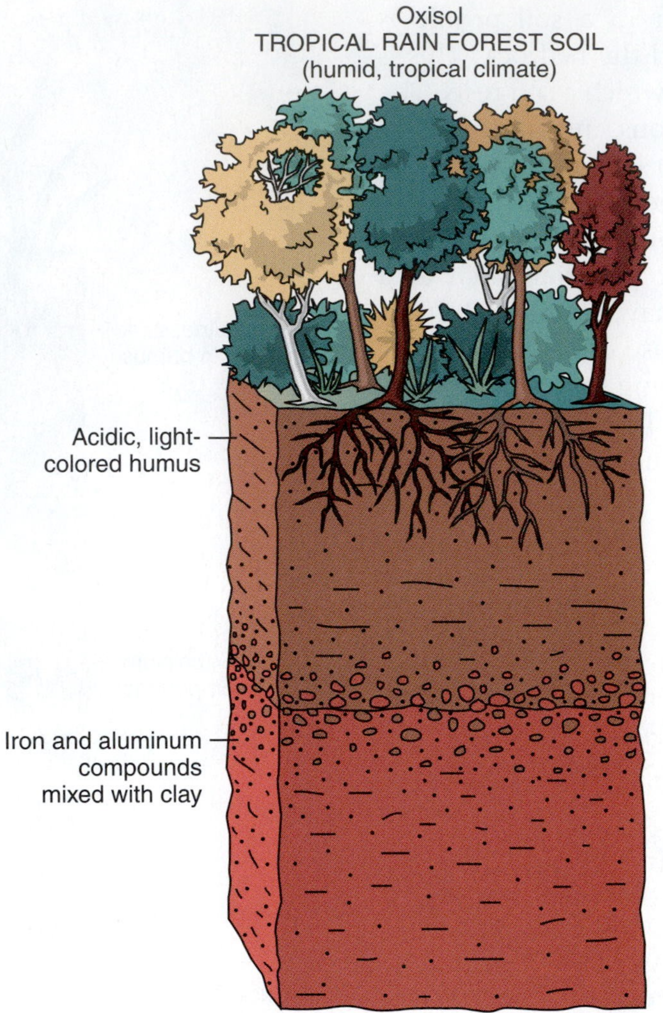

Figure 12–8 The profile of an Oxisol soil order associated with tropical rainforests reveals its unique horizon characteristics.

that is composed from loess, alluvial, or glacial till.

The Oxisol soil order describes the type of soil that forms in the tropical rainforest regions (Figure 12–8). This soil has a thin O-horizon, barely any A-horizon, and a B-horizon that is dense and high in iron. When the vegetation of tropical rainforests is cleared away, the B-horizon is exposed to heavy rains and the hot tropical sun, which bakes it into a hard, brick-like material. This type of soil is not good for crop growth. Many thousands of acres of tropical rainforests are cleared each year for agriculture, and unfortunately Oxisol soils do not support long-term crop growth.

The world's soil is an important natural resource on which our food supply relies (Figure 12–9). The formation of just 1 inch of soil may take a thousand years, making it important to practice sound soil conservation techniques that will protect and sustain this fragile part of the lithosphere.

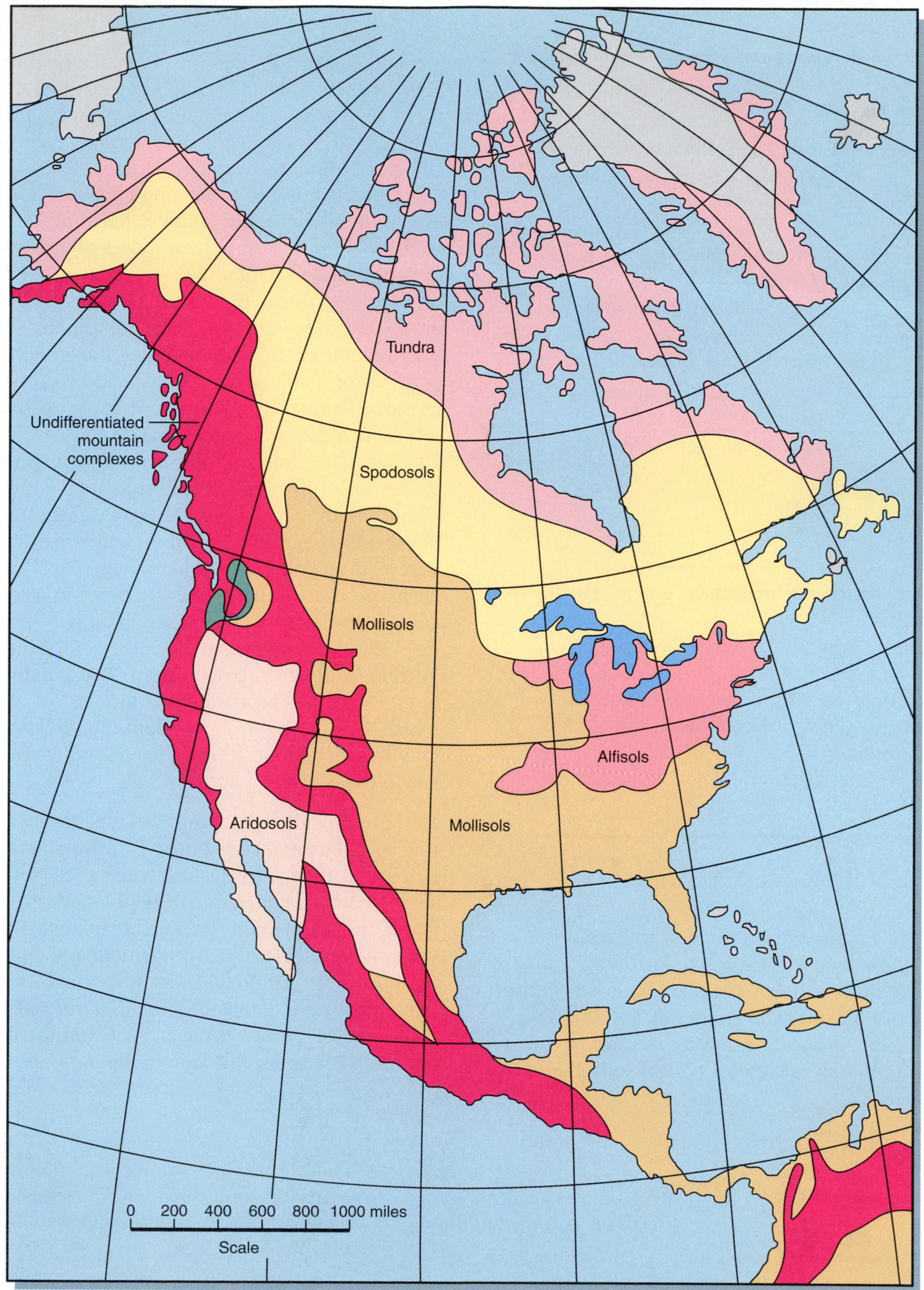

Figure 12–9 A map of North America showing the location of the continent's main soil orders.

REVIEW

1. What is the composition of a typical soil?
2. What is humus, and why is it important to soil?
3. What are three processes by which soil loses mineral nutrients?
4. What is the optimum amount of air and water content in a soil's pore spaces?
5. What are the three types of transported parent material that forms soil?
6. List the five soil horizons and describe their unique features.
7. Who was Edmund Ruffin?

For more information go to these Web links:

<http://ltpwww.gsfc.nasa.gov/globe/index.htm>
<http://www.fmnh.org/ua/default.htm>
<http://www.usda.gov/news/usdakids/index.html>

CHAPTER SUMMARY

Soil is an important natural resource on which many organisms, including humans, depend for their survival. A soil is a mixture of minerals, organic material, air, and water in specific proportions. Minerals make up approximately 45% of a typical soil. Plants that grow in soil utilize these minerals for healthy growth. Minerals in a soil are classified by their size, from the smallest to the largest, which include clay, silt, sand, and gravel. A soil gains minerals from decaying organic material, weathering of rocks, and fertilizer. Soils lose minerals by plant uptake, leaching, and runoff. Organic material composes about 5% of a typical soil. Organic material in a soil is often in the form of humus. Humus is the dark, moist, completely decayed remains of organic matter. It holds together the mineral particles of a soil, which helps a soil to resist erosion. Humus also forms pore spaces in a soil, which are tiny spaces that contain water and air. The amount of water in the pore spaces of a soil is called soil moisture, which composes approximately 25% of a typical soil. There are three types of soil moisture: saturation, wilting point, and field capacity. Field capacity occurs when 50% of the pores in a soil are filled with air and 50% are filled with water. This is the optimum amount of moisture needed for healthy crop growth. The remaining 25% of a typical soil is composed of air, which is also required by plants for healthy growth. The composition of air in a soil is similar to the air in the atmosphere.

There are also many organisms that reside in the soil. Earthworms burrow through soil, helping to turn it over and aerate it. Bacteria also live in the soil, some of which add nitrogen to the soil, which is needed by plants. The structure of a soil is the arrangement and composition of its minerals. Loam is a type of soil structure that is good for crop growth and is composed of nearly equal parts of sand, silt, and clay. The rocks that supply the minerals in

EARTH MATH

1) IF IT TAKES APPROXIMATELY 1000 YEARS FOR 1 INCH OF TOP SOIL TO DEVELOP IN A PARTICULAR PART OF THE WORLD, HOW DEEP IS THE SOIL AFTER THE GLACIER RETREATED 12,500 YEARS AGO?

a soil are known as parent material. There are four different types of parent material: alluvial, loess, glacial, and residual. Soils are classified by their unique profile, which is a cross-sectional view of a soil. Soil profiles reveal the unique layers of soil, called horizons. These horizons all have their own unique physical and chemical characteristics. The five soil horizons are the O-horizon, A-horizon, B-horizon, C-horizon, and D-horizon. Specific soils are classified into unique soil orders that are based on the characteristics of a soil's horizons.

CHAPTER REVIEW

Multiple Choice

1. A typical soil contains approximately what percentage of minerals?
   a. 5%
   b. 10%
   c. 25%
   d. 45%

2. The smallest mineral particle in a soil is called:
   a. clay
   b. silt
   c. sand
   d. gravel

3. A source of minerals in a soil is:
   a. leaching
   b. runoff
   c. plant uptake
   d. organic material

4. Organic material makes up approximately what percentage of a typical soil?
   a. 5%
   b. 10%
   c. 25%
   d. 45%

5. The dark, moist, decayed organic material that holds the mineral particles together in a soil is called:
   a. top soil
   b. subsoil
   c. humus
   d. parent material

6. When a soil's pore spaces contain 50% water and 50% air it is known as:
   a. saturation
   b. the wilting point
   c. full
   d. field capacity

7. Air and water compose approximately what percentage of a typical soil?
   a. 25% air, 25% water
   b. 50% air, 50% water
   c. 45% air, 5% water
   d. 5% air, 45% water

8. Which type of organism fixes nitrogen in the soil?
   a. earthworms
   b. lichen
   c. bacteria
   d. fungus

9. Which of the following parent material is transported by wind?
   a. loess
   b. alluvial
   c. glacial
   d. residual

10. When liquid water deposits parent material that forms soil it is known as:
    a. loess
    b. alluvial
    c. glacial
    d. residual

11. Which soil horizon is commonly called top soil?
    a. O-horizon
    b. A-horizon
    c. B-Horizon
    d. C-horizon

12. Which soil horizon is known as the subsoil?
    a. O-horizon
    b. A-horizon
    c. B-Horizon
    d. C-horizon

continued

13. Which soil horizon is mostly composed of decaying organic material?
    a. O-horizon
    b. A-horizon
    c. B-Horizon
    d. C-horizon

14. This type of soil order is commonly associated with the fertile grassland soils of the American Midwest:
    a. Alfisol
    b. Mollisol
    c. Oxisol
    d. Aridosol

15. The tropical rainforests that are being cleared for agriculture in South America grow in which type of soil?
    a. Alfisol
    b. Mollisol
    c. Oxisol
    d. Aridosol

Matching *Match the terms with the correct definitions.*

a. soil
b. leaching
c. organic material
d. humus
e. soil moisture
f. tilling
g. nitrogen-fixing bacteria
h. loam
i. parent material
j. alluvial soils
k. loess
l. soil profile
m. soil horizon
n. top soil
o. subsoil

1. ____ The movement of chemicals that are dissolved in water from a higher layer of soil downward into a lower level of soil or groundwater.
2. ____ The layer of soil, also known as the B-horizon, that lies directly below the top soil.
3. ____ A mixture of minerals, organic material, air, and water that forms at the surface of the Earth.
4. ____ The uppermost layer of soil that contains a high amount of organic material, also called the A-horizon.
5. ____ A term used to describe a soil that contains specific portions of sand, silt, and clay.
6. ____ A well-defined layer of soil that has specific characteristics.
7. ____ A specific type of bacteria that converts atmospheric nitrogen (N_2) to plant-usable forms of nitrogen such as nitrate (NO_3^-)
8. ____ The term used to describe the specific rocks from which soil minerals are derived.
9. ____ A term used to describe material that is derived from a living thing.
10. ____ Rich, fertile soil formed by the deposition of minerals by liquid water.
11. ____ The completely decomposed remains of organic debris that is an important part of a soil.
12. ____ The cross-sectional view of a particular soil that shows all the soil's horizons.
13. ____ The amount of water that is present in a soil.
14. ____ A type of soil found in the American Midwest that is formed from wind-transported parent material.
15. ____ The process of turning over, or plowing, the soil.

Critical Thinking

1. As the world's population grows, our dependence on soil used to grow crops increases. What specific practices do you think can ensure that our soils will continue to produce healthy crops in the future?

CHAPTER

13

The Earth's Geologic History

Objectives

After reading this chapter you should be able to:

- Describe the techniques used to relatively date rock formations.
- Define the term *index fossil* and explain how it is used to date rock formations.
- Explain how radiometric dating is used to determine the absolute age of rocks.
- Identify the four elements commonly used for radiometric dating and list their individual half-lives.
- Identify the three major eons that make up the geologic time scale and list their relative time periods.
- Define the term *mass extinction.*
- Use the geologic time scale to date Earth's important geologic events.
- Use the geologic time scale to identify the appearance or disappearance of specific living organisms on Earth.

TERMS TO KNOW

geologic time scale
principle of uniformity
principle of superposition
index fossils
half-life
radiometric dating
eon
stromatolites
Precambrian
eras
mass extinctions
Pangea
epochs

INTRODUCTION

Scientists believe that the Earth is approximately 4.6 billion years old. Discovering the events that occurred and the type of organisms that lived during this long period is like unraveling a great mystery. Scientists must act like detectives and gather clues using many different techniques of investigation and observation to help them reveal the mysteries of the Earth's past. These techniques have helped develop theories of how our planet formed and how it has changed over time, what life forms it has supported, and how those life forms have evolved during its long past.

Geologic Principles

Scientists have pieced together the history of the Earth using the **geologic time scale,** which divides the Earth's history into distinct, relative time periods. Geologic time is based on important geologic principles. The first is the **principle of uniformity,** which states that the processes occurring today on the Earth are the same processes that have occurred throughout the planet's history. This means that the conditions that lead to the formation of rocks today are the same as in the past. This important scientific concept was first proposed by Scottish farmer James Hutton in the late eighteenth century. Using this concept, geologists began to piece together other important geologic concepts such as the **principle of superposition,** which states that in undisturbed rock layers, the older rocks lie below younger rocks. Therefore it is possible to create a time scale showing the formation of rocks relative to one another. In addition, sedimentary rock layers can be identified and traced from one location on the Earth to another. This helps to correlate geologic history over wide geographical areas.

Other important clues that help to create relative dates include the identification of unique volcanic ash layers in rock layers. These layers act like time markers in rock strata that can be traced over wide areas. Geologic features such as igneous intrusions, which occur when hot molten rock flows into cracks and crevices of older rock layers and solidifies, can reveal much about the dates of rock. The intrusion is younger than the rock layers that it is in, which helps to determine the relative date of its formation. Rock features such as faults and volcanic rocks can also help piece together the Earth's history. Faults are younger than the rock formations in which they are found, and volcanic rock usually flows through and over older rock. Together all these relative dating techniques help to establish the series of events that have occurred in a rock formation (Figure 13–1).

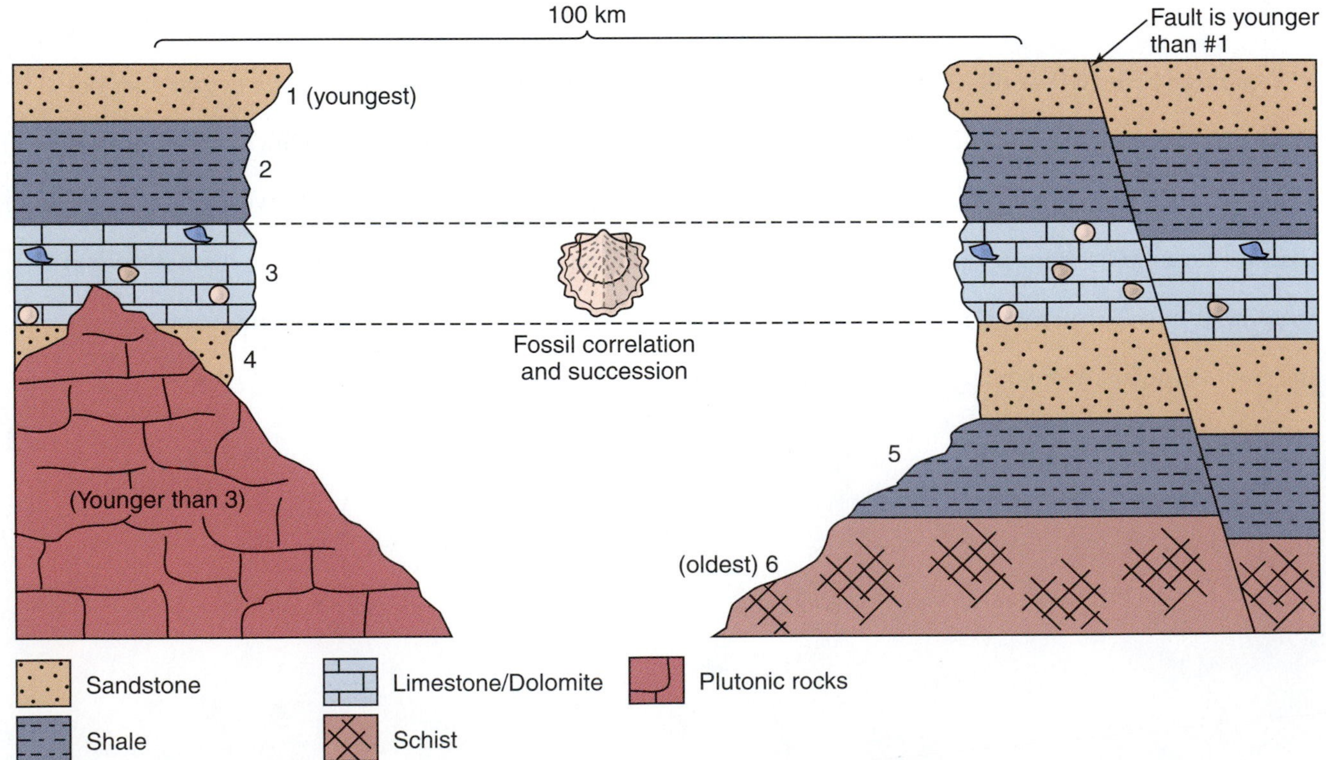

Figure 13–1 This geologic cross section shows how superposition, faults, igneous intrusions, index fossils, and rock correlations are used to relatively date rock formations.

Figure 13–2 Fossils of marine organisms found in rocks located high above sea level suggest that the environment was much different during the time these organisms were alive. *(Courtesy of PhotoDisc.)*

The discovery of the fossil remains of once living organisms add more information to the geologic time scale. Fossil evidence can reveal much about what the environment was like when these organisms lived. For example, the remains of fossilized fish suggest that these organisms lived in an aquatic environment (Figure 13–2).

Fossils that lived for a relatively short time on Earth but are found all over the globe can be used to establish relative dates of rock formations. These are known as **index fossils.** If an index fossil is identified as living during a specific period, the same species found in a new rock formation can be used to establish the rock formation's relative age. Using all these methods, geologists create a geologic history of the Earth, along with assumptions of what life was like during these periods.

Radiometric Dating

In 1895 French physicist Antoine Henri Becquerel discovered that uranium atoms undergo radioactive decay. His discovery revealed that uranium breaks down over time to form a new element, lead. The original element is called the parent element, and the product of its decay is known as the daughter element. Therefore uranium is the parent element, and lead is the resulting daughter element. Becquerel also discovered that radioactive decay occurs at a specific rate, called the **half-life.** The half-life of an element is the time it takes for half of a number of parent elements to decay into their daughter elements (Table 13–1). For example, if you start with 2 grams of uranium 238, 1 gram will remain after one half-life has elapsed. This technique is called **radiometric dating;** it allows geologists to estimate how old rocks were when they formed. By calculating the ratio of parent elements to daughter elements and using the half-life of the known element, it is possible to date rocks accurately and determine the absolute age of a rock. For example, the half-life of uranium 238 is 4.5 billion years. This means that it will take 4.5 billion years for half, or 50%, of the uranium in the

TABLE 13–1 The half-lives and daughter elements of radioactive isotopes commonly used for radiometric dating

Radioactive Isotope	Approximate Half-Life	Decay Product
Rubidium 87	50 billion years	Strontium 87
Potassium 40	1.3 billion years	Argon 40
Uranium 238	4.5 billion years	Lead 206
Uranium 235	700 million years	Lead 207
Carbon 14	5730 years	Nitrogen 14

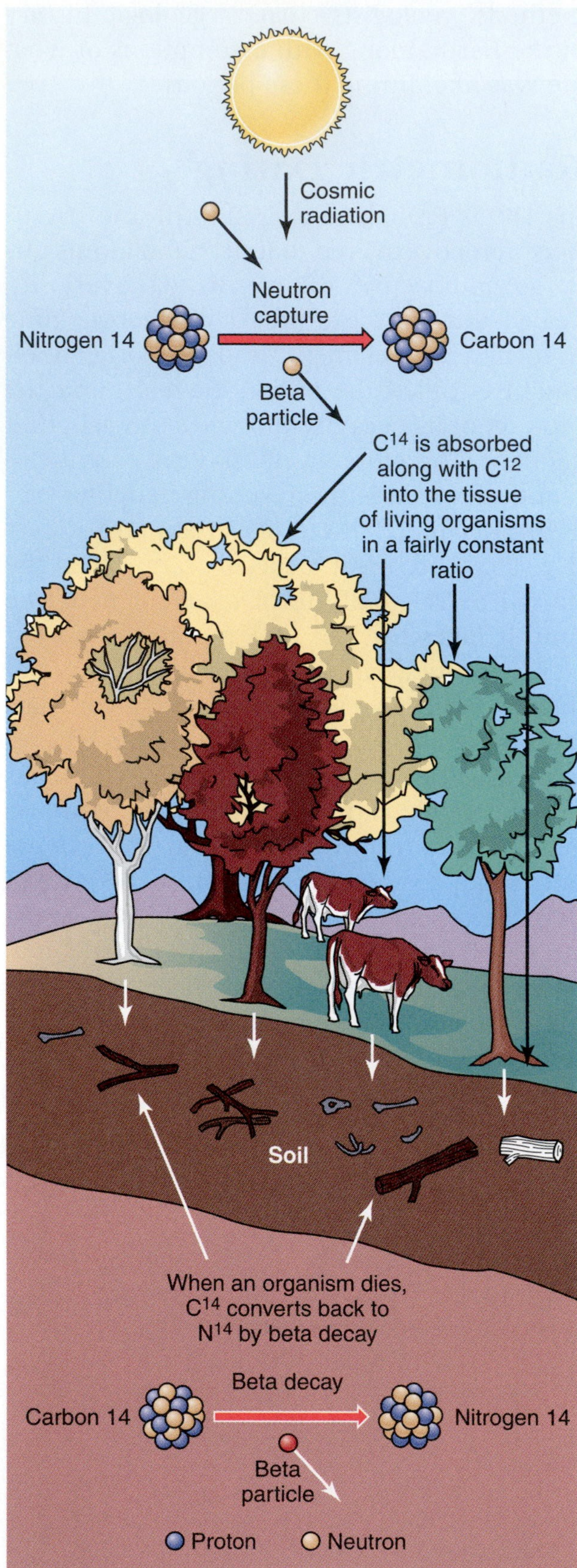

Figure 13–3 The process by which carbon 14 enters living organisms. Determining the ratio of carbon 14 to its daughter element, nitrogen 14, can reveal how long ago the organism died.

rock to break down into lead. Therefore if you found a rock that contained 50% lead and 50% uranium, you could estimate that it formed 4.5 billion years ago, because half the element has decayed. The element uranium is only good for dating extremely old rocks, but other elements that have shorter half-lives can be used to date younger rocks. These elements include potassium 40, which has a half-life of 1.3 billion years; uranium 235, with a half-life of 700 million years; and carbon 14, with a half-life of 5730 years. Because all living things contain carbon, the isotope carbon 14 can be used to radiometrically date the remains of living things (Figure 13–3).

Current exploration of objects in the solar system has also helped to piece together the Earth's history. When astronauts brought back rock samples from the Moon, scientists were able to accurately date the rocks' age. The age of many meteorites has also been determined. The dating of rocks by radioactive decay, the existence of fossil evidence, and the well-established principles of geology combine to create the fascinating story of our planet's history. Our knowledge of this history is constantly being expanded as more scientific discoveries are made about the geology of Earth.

The Geologic Time Scale

The geologic time scale is used to represent the history of the Earth (Figure 13–4). The largest period represented in the geologic time scale is the **eon.** Three eons make up the Earth's history: the Archean, Proterozoic, and Phanerozoic eons.

The Archean Eon

The Archean eon began approximately 4.5 billion years ago, when scientists believe the Earth formed (Figure 13–5). The beginning of the Archean eon was a violent time in the Earth's history. The planet was exposed to constant bombardment by meteorites and asteroids, as evidenced by the surface of the Moon.

EARTH SYSTEM SCIENTISTS *GEORGES CUVIER*

Born in 1769, Georges Cuvier was a comparative anatomist who studied the similarities in all forms of animals and their unique physical characteristics. He continued the classification work begun by Carolus Linnaeus and helped to group organisms into unique categories called phyla. In 1812 his worked turned to the classification of the fossil remains of organisms. He was the first scientist to determine that fossil bones were the remains of once living organisms. Cuvier also theorized that catastrophic events at different times during the Earth's past wiped out many species on the Earth, which caused new species to become more widespread. This was the first suggestion of the mass extinction, which is an important aspect of the geologic time scale. His work also involved the reconstruction of many fossil skeletons and the identification of extinct organisms. Because of this, he is often considered the father of paleontology.

EARTH SYSTEM SCIENTISTS *ARTHUR HOLMES*

Arthur Holmes was a geologist who in 1913 first proposed a geologic time scale that described the history of the Earth. Using the new science of radioactive dating, he theorized that the age of the Earth was approximately 4.5 billion years old. He also helped to date the different eras that together make up the geologic time scale. Later in his career he hypothesized that the mechanism behind plate tectonics might be large-scale convection cells in the Earth's mantle. This hypothesis was given without any research to back it up, however, and was not truly proven until 30 years later. Arthur Holmes continued his work in geology until his death in 1965.

During this time, the Earth was extremely hot, and much of the surface was covered in molten rock. As the Earth began to cool, it expelled gases from its interior, which is called outgassing. These gases were mainly carbon dioxide, water vapor, and hydrogen. The lighter element hydrogen was released into space, leaving carbon dioxide and water vapor to make up the Earth's early atmosphere.

Eventually the Earth cooled to the point that the surface began to solidify and form the Earth's crust. This marked the time when rocks first formed on Earth. The oldest rocks found on the planet today are located in parts of Canada, Greenland, and Australia, and are between 4- and 3.8-billion-year-old metamorphic and sedimentary rocks. They reveal that the Earth's crust may have formed more than 4 billion years ago. During the formation of the Earth's crust, the water vapor that was being expelled from the Earth's interior was condensing in the atmosphere and raining down on the Earth's surface. This began the creation of today's oceans. The first billion years of Earth's history resulted in the formation of the Earth's crust and oceans. Eventually, as the planet's environment became more stable, scientist's the first life forms appeared.

The Earth was unique compared with other planets in the solar system because it contained a great amount of liquid water. It was in the oceans that the first organisms flourished. The oldest fossils discovered on Earth are the remains of **stromatolites** that were found in Australia (Figure 13–6). Stromatolites are colonies of single-celled cyanobacteria that form large, slimy mats in warm, shallow saltwater.

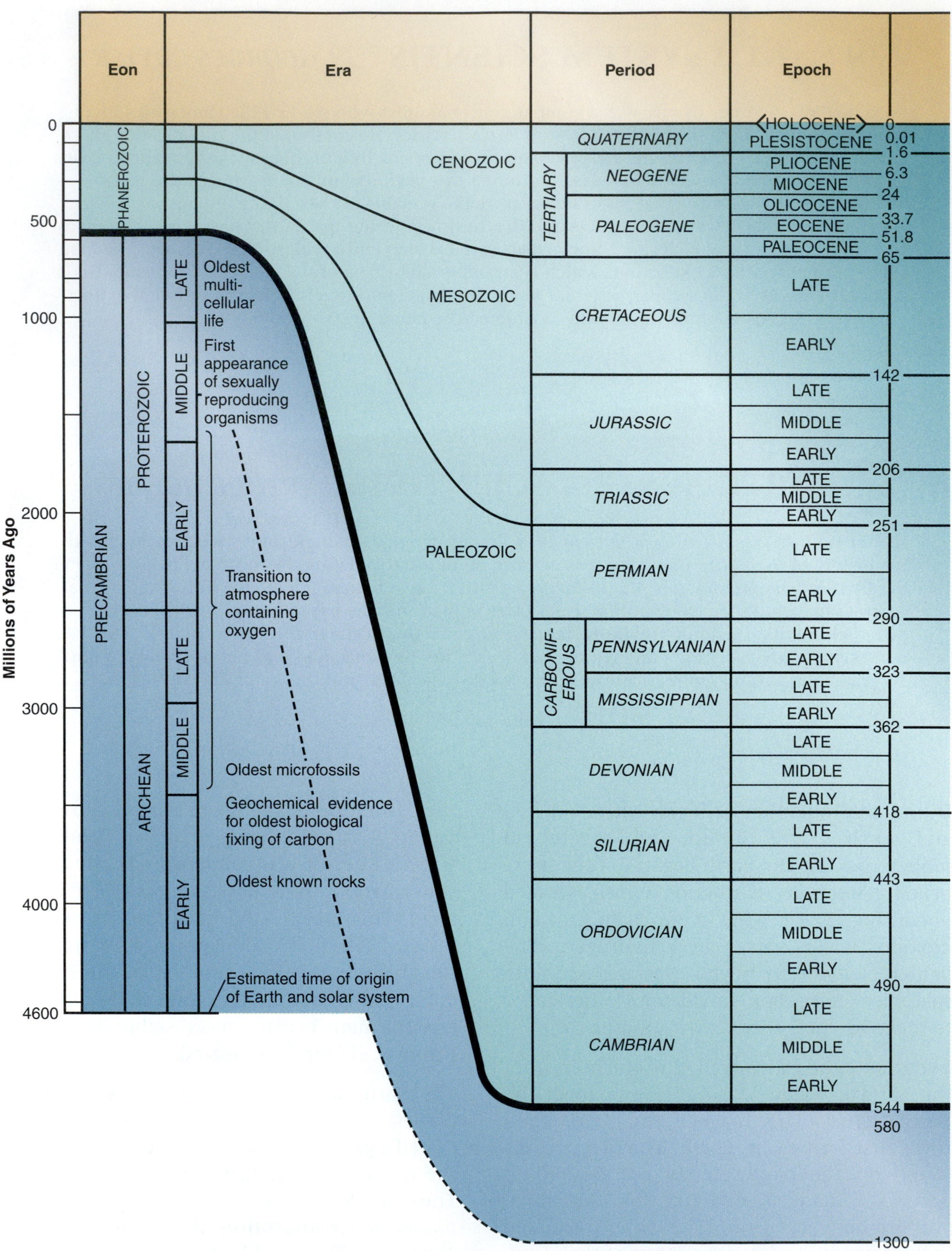

Figure 13–4 The geologic time scale.

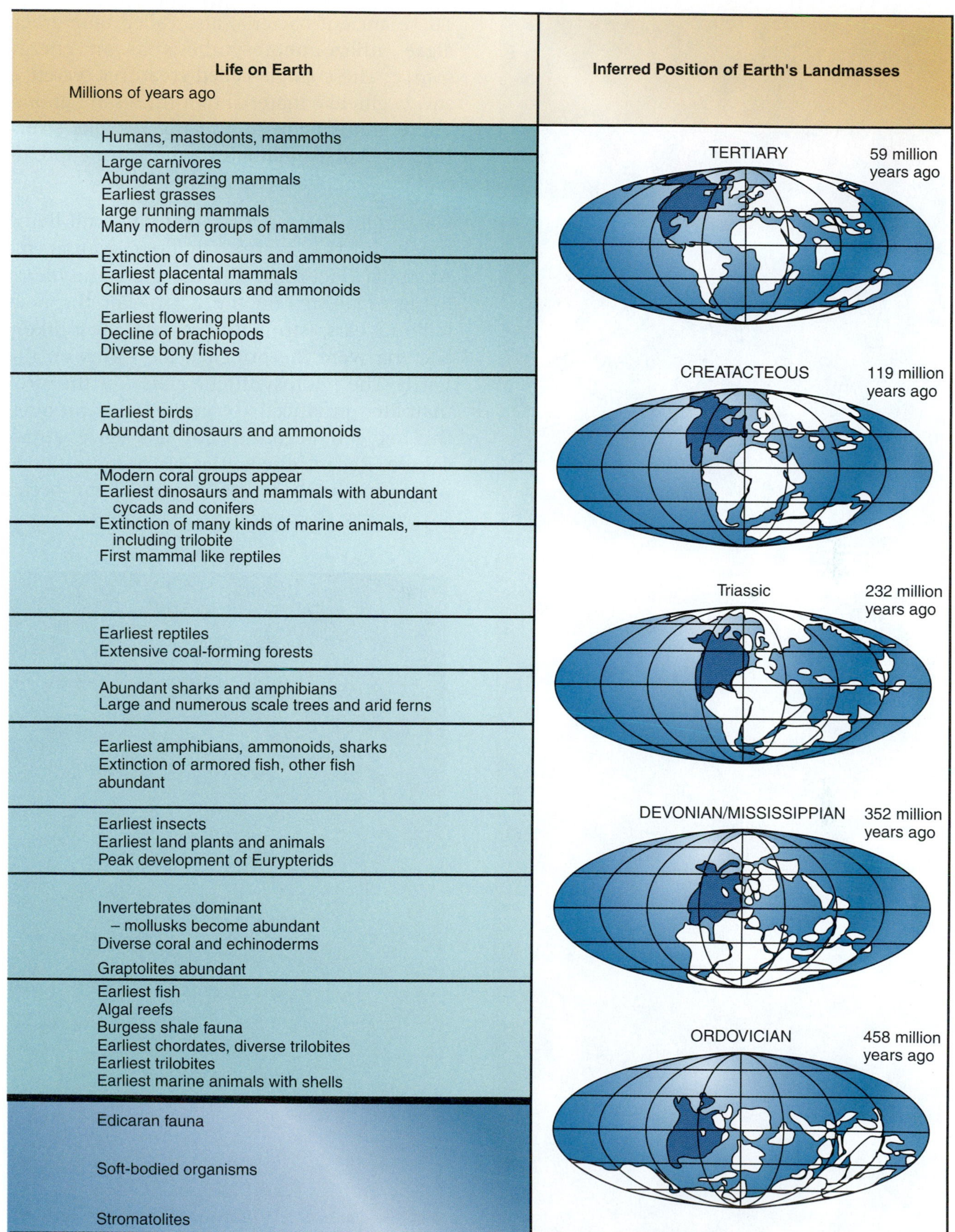

Life on Earth
Millions of years ago
Inferred Position of Earth's Landmasses
Humans, mastodonts, mammoths
Large carnivores
Abundant grazing mammals
Earliest grasses
large running mammals
Many modern groups of mammals
Extinction of dinosaurs and ammonoids
Earliest placental mammals
Climax of dinosaurs and ammonoids
Earliest flowering plants
Decline of brachiopods
Diverse bony fishes
Earliest birds
Abundant dinosaurs and ammonoids
Modern coral groups appear
Earliest dinosaurs and mammals with abundant cycads and conifers
Extinction of many kinds of marine animals, including trilobite
First mammal like reptiles
Earliest reptiles
Extensive coal-forming forests
Abundant sharks and amphibians
Large and numerous scale trees and arid ferns
Earliest amphibians, ammonoids, sharks
Extinction of armored fish, other fish abundant
Earliest insects
Earliest land plants and animals
Peak development of Eurypterids
Invertebrates dominant
– mollusks become abundant
Diverse coral and echinoderms
Graptolites abundant
Earliest fish
Algal reefs
Burgess shale fauna
Earliest chordates, diverse trilobites
Earliest trilobites
Earliest marine animals with shells
Edicaran fauna
Soft-bodied organisms
Stromatolites
TERTIARY
59 million years ago
CREATACTEOUS
119 million years ago
Triassic
232 million years ago
DEVONIAN/MISSISSIPPIAN
352 million years ago
ORDOVICIAN
458 million years ago

Figure 13–4, cont'd The geologic time scale.

Figure 13–5 The early Earth as it may have appeared during the Archean eon.

These ancient bacteria, also called blue-green algae, utilize photosynthesis as an energy source. These single-celled creatures secrete a sticky, gluelike material that traps sand grains. These grains help to form the rocklike structures called stromatolites, which are approximately 3.5 billion years old.

Amazingly, there are stromatolites still living today in the warm, shallow waters off Australia (Figure 13–7). These are the oldest living organisms on the Earth. For the next billion years, stromatolites and most likely bacteria were the principal life forms on the Earth. The photosynthetic action of the stromatolites produced oxygen as a byproduct; this oxygen began to fill the early atmosphere. Recent discoveries of communities of organisms living at the bottom of the ocean

Figure 13–6 Stromatolite fossils, which are approximately 3.5 billion years old, are some of the oldest living things discovered on the Earth. (*Courtesy of Joseph Deuel/Petrified Sea Gardens.*)

Figure 13–7 Today stromatolites still thrive in the shallow waters off the coast of Australia. *(Courtesy of Eva Boogaard/Lochman Transparencies.)*

near volcanic vents suggest that life also may have existed on the early Earth in a similar deep-water environment.

The Proterozoic Eon

The end of the Archean eon came approximately 2.5 billion years ago, and the Proterozoic eon began. Geologic evidence suggests that during this eon crustal plate movements began, which led to the formation of the first mountain ranges. Stromatolites were also widespread during the Proterozoic eon. The microscopic fossil remains of bacteria and other single-celled organisms were found in rocks near Lake Superior and are approximately 1.9 billion years old. Some of these bacteria are similar to species that are alive today. These organisms used respiration to gain energy, which suggests that the atmosphere contained a level of oxygen similar to that found in today's atmosphere. Later in the Proterozoic eon, multicelled creatures began to flourish in the Earth's oceans for the first time. The fossil remains of algae and soft-bodied aquatic creatures have been found to be 570 to 900 million years old.

There is also geologic evidence to suggest that the first glaciers began to appear during this time. Remains of glacial deposits formed when ice that contained rocks, also called glacial till, melted and was deposited at the bottom of either a glacial lake or coastal ocean. Rocks formed from glacial till are known as tillites. During the late Proterozoic eon, approximately 700 million years ago, tillite glacial deposits were found on many continents, suggesting that much of the land was covered by glaciers.

The period before 570 million years ago, which includes the Archean and Proterozoic eons, is also known as the **Precambrian** time. By the end of Precambrian time, much of the North American continent was formed. One of the oldest Precambrian rock formations in the world is called the Canadian Shield, which today makes up parts of eastern Canada, Greenland, and the Adirondack Mountains of New York. This ancient land mass is approximately 2 billion years old. The end of the Proterozoic eon occurred approximately 570 billion years ago, and the eon in which we live today, the Phanerozoic eon, began.

The Paleozoic Era

The Phanerozoic eon is subdivided into three different periods called **eras.** This division is based on the occurrence of **mass extinctions** of organisms. Each era is then further subdivided into specific periods. The Paleozoic, or "early life," era consists of six specific periods based on the principal life forms that were alive at that time. The early Paleozoic era consists of the Cambrian period and the Ordovician period (Figure 13–8).

The Cambrian period began 570 million years ago and lasted for about 65 million years. Many of the continents on the planet during this time were flooded by shallow inland seas. The beginning of the Cambrian period marked the emergence of shelled marine creatures such as trilobites. The Ordovician period began approximately 505 million years ago with the emergence of a variety of marine life forms, including crinoids, corals, snails, starfish, and segmented worms. There is some fossil evidence to suggest that during the late Ordovician period plants may have begun to inhabit land. Much of the globe was covered

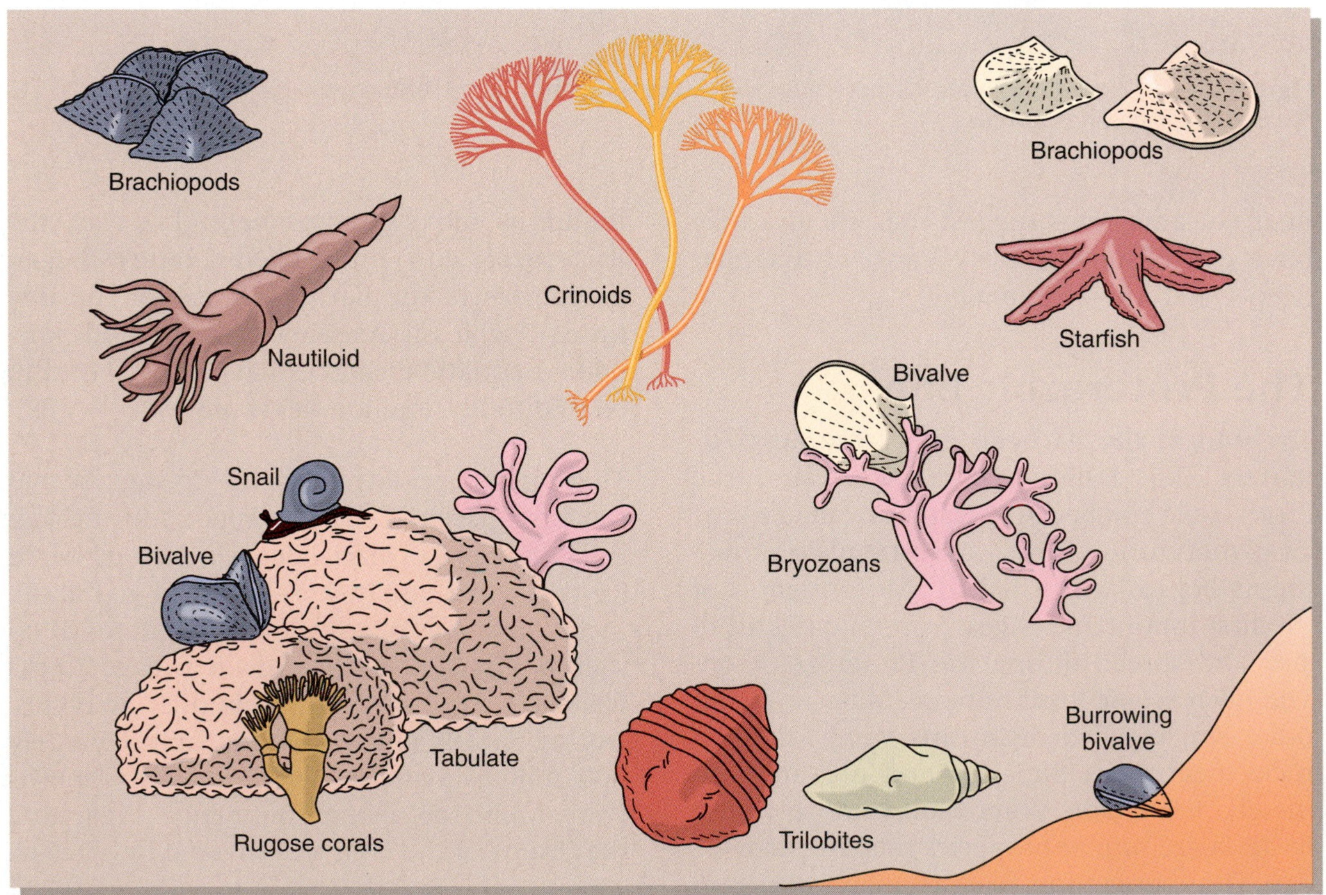

Figure 13–8 An example of marine organisms that lived during the Cambrian and Ordovician periods between 544 and 443 million years ago.

in water as in the Cambrian period, providing the perfect shallow sea habitats that are preserved in Ordovician sediments; these sediments contain some of the best-preserved fossils on the planet.

Some regions of the planet may have been covered in glaciers during the later part of the Ordovician period. The Taconic Mountains were formed on the eastern part of North America when the continent collided with a string of volcanic islands in the proto-Atlantic Ocean, called the Iapetus Ocean. The end of the Ordovician period was marked by the mass extinction of the trilobite species.

The middle Paleozoic era began 438 million years ago and is divided into two periods: the Silurian and the Devonian. During the Silurian period, complex coral reefs began to form and swimming organisms such as the eurypterid emerged (Figure 13–9). This sea creature, which resembled a large aquatic scorpion, swam along the bottom of shallow water in search of prey.

Figure 13–9 Eurypterid fossils from the Silurian period, between 443 and 418 million years ago. *(Photo by Pamela Gore, Courtesy of the Denver Museum of Nature and Science.)*

The first true land plants also emerged during the Silurian period. These semiaquatic plants resembled today's marsh vegetation but were much simpler in their design. During this time, much of North America was flooded by a large inland sea. During the late Silurian period, this sea began to evaporate and shrink, leaving behind large salt deposits that are still being mined today.

The Devonian period began about 408 million years ago with the emergence of jawed fish. These were the first fish that had scales and fins like today's modern freshwater and marine fish species. Some of the jawed fish that lived during the late Devonian period grew to more than 30 feet in length. Other strange fish also flourished during this time, including the armored fish (Figure 13–10). These heavily protected fish were covered in unique armor; some even had armor that covered parts of their eyes!

The armored fish became extinct during the end of the Devonian period. Also during the late Devonian period the first trees began to appear on Earth, leading to the development of the first forests. The once barren landscape of Earth was now populated by many different plants, which contributed to the development of fertile soils. The first animals to inhabit the land began to emerge during the late Devonian period. These animals were amphibians that periodically crawled out from the water to search for food on land. Many of these creatures were primitive lungfish, which had strong, footlike fins that may have helped them to move about on land. The end of the Devonian period was marked by the mass extinction of many tropical marine species. This extinction is believed to have been caused by the cooling of the planet because apparently only tropical species were affected. Geologic evidence from the late Devonian also supports that global cooling occurred during this time, because there appear to have been glaciers covering many land areas.

The middle Paleozoic era ended approximately 363 million years ago and the rise of

Figure 13–10 The fossilized remains of a placoderm, a type of armored fish that became extinct in the middle of the Devonian period approximately 400 million years ago. *(Photo by Pamela Gore, Georgia Perimeter College.)*

Figure 13–11 A paleontologist exposes a fossilized jaw bone as part of his laboratory work. *(Courtesy of PhotoDisc.)*

Career Connections

PALEONTOLOGIST

Paleontologists are specialized geologists who study the rock formations of the world in search of the fossilized remains of once living organisms. These scientists explore many parts of the globe to uncover knowledge about the Earth's living history. Paleontologists painstakingly uncover the remains of ancient plants and animals to piece together our planet's past (Figure 13–11). The most well known paleontologists search for dinosaurs and are attempting to reveal the lives of these amazing creatures. Much of the paleontologist's career is spent in the laboratory analyzing and restoring fossil remains extracted from the Earth. All paleontologists are highly specialized and must have a solid understanding of geology. It is also important for paleontologists to have knowledge about biology, so they can apply it to the ancient creatures that they are studying. Paleontologists require at least a 4-year college degree and often work within the academic fields. Many paleontologists become college professors or museum curators and travel to exotic places all over the world. There are also career opportunities for laboratory technicians who work with paleontologists. These jobs require less education but can be extremely rewarding, because they are often involved in the restoration and preparation of fossil remains.

large coal swamps began. This is known as the Carboniferous period because a great amount of hydrocarbons were produced and buried deep in the ground. Hydrocarbons are molecules that contain hydrogen and carbon, and eventually they were transformed into coal. Coal swamps were large areas of the Earth that were partially flooded with freshwater (Figure 13–12). Large trees, ferns, and other plants grew in abundance in and around these swamps. When these plants died, they fell into the swamps and became buried in sediments. Over time the remains of this decaying swamp vegetation were transformed by the heat and pressure of the Earth into coal.

Also during the Carboniferous period, large winged insects began to appear. These creatures resembled today's houseflies and dragonflies. The ancestor of the modern cockroach first appeared during the Carboniferous period. Although the first amphibians appeared during the late Devonian period, the Carboniferous period marked the true rise of land-dwelling amphibians, which ranged from mouse sized to more than 20 feet in length. Many of the shallow seas of the world during this time contained a variety of crinoids (Figure 13–13).

Crinoids are fernlike creatures that attach themselves to the sea floor with long stems and filter out tiny microorganisms with their fanlike arms. These widespread organisms covered the bottom of the ocean, forming underwater meadows. Ancestors of modern sharks also lived in the oceans during this time. The end of the Carboniferous period, about 290 million years ago, marked the appearance of two important organisms: the conifers and the reptiles. Conifers are cone-bearing trees often called evergreens. The ancestors of today's pine, spruce, and fir trees began to appear at the start of the Permian period 290 million years ago. The formation of the Appalachian Mountain chain in eastern North America, known as the Appalachian orogeny, began at this time. It occurred as a result of the ancient African continent colliding with North America, which helped to form the supercontinent called **Pangea** (Figure 13–14).

The first reptiles appeared during this time. Reptiles were the first creatures known to lay eggs that could survive on dry land. Many of these early reptiles resembled today's lizards and alligators; however, some had large fins on their backs. Much of the climate during the later part of the Permian period was very hot and dry. The end of the Permian was marked by a mass extinction of both aquatic

Figure 13–12 A Carboniferous-period coal swamp that existed between 323 and 290 million years ago.

and land-based organisms. This was probably the greatest mass extinction of species in the Earth's history. The cause of the extinction is not well understood, but it is believed that a drop in sea level along with global cooling might have been contributing factors. The end of the Permian period came about 245 million years ago and also marked the end of the Paleozoic era.

The Mesozoic Era

The Mesozoic era, which began approximately 245 million years ago, is known as the age of the dinosaurs. The beginning of the Mesozoic era, called the Triassic period, marked the rise of many new species of both aquatic and land-based organisms. The rise of large fish and swimming reptiles occurred during this period, along with many clams and other bivalve mollusks. The appearance of modern corals occurred during this time. On land the dominant plants were ferns; the first mammals also appeared. These mammals were small rodents similar to mice and rats.

The supercontinent called Pangea began to break apart during the Triassic period. This began the formation of the Atlantic Ocean, which formed as North America began to move apart from Africa, South America, and Eurasia. The first turtles and flying reptiles appeared during the Jurassic period. The pterosaur, or flying lizard, grew to approximately 2 feet in length. The rise of the first true dinosaurs came about 208 million years ago, at the start of the Jurassic period (Figure 13–15).

Figure 13–13 Crinoid fossils from the Mississippian period, between 362 and 232 million years ago. *(Courtesy of Coolrox.com.)*

During the Jurassic period, the sauropods appeared; these were immense land-dwelling dinosaurs that grew to more than 90 feet in length. Other smaller dinosaurs, such as the stegosaurus, which had large armored plates and a spiked tail, also lived during this period. Many marine reptiles flourished, including the plesiosaur, which had a long neck and four large fins and grew to more than 40 feet in length. Much of the land was covered by forests that contained fernlike trees called cycads, along with many cone-bearing trees and the ancestors of modern broad-leafed trees such as the ginkgo. The end of the Jurassic period came approximately 146 million years ago with the appearance of the first birdlike organisms, called archaeopteryx. The archaeopteryx was the first winged reptilelike creature to be covered in feathers.

Much of North America was covered by a large inland sea during the Jurassic period, which today is represented by sedimentary rock known as the Morrison Formation. Many excellent dinosaur fossils have been preserved in these rocks.

The Cretaceous period began approximately 146 million years ago with the continued domination of the dinosaurs. During this time flowering plants and broad-leafed trees began to dominate the land. The first crabs and snails began to emerge during the Cretaceous period, along with much of the marine plankton that we see today. Many well-known dinosaurs, such as the tyrannosaurus, the triceratops, and the duck-billed dinosaur, flourished at this time. Other smaller creatures, such as snakes, salamanders, frogs, and turtles, lived alongside the great dinosaurs of the Cretaceous period.

The breakup of Pangea occurred throughout the Cretaceous period, creating the continents that exist today. The climate of the time was very hot and humid, which may have been the result of global warming caused by extensive volcanic activity. The end of the Cretaceous period was marked by the most famous extinction in all the Earth's history, the extinction of the dinosaurs. The Cretaceous extinction occurred

Figure 13–14 The supercontinent of Pangea as it appeared 250 million years ago.

Figure 13–15 Dinosaurs that roamed Earth during the Jurassic period, between 206 and 142 million years ago.

approximately 65 million years ago and is the subject of great debate. Some scientists believe that the extinction of the dinosaurs was caused by the spread of disease, whereas others believe that a change in climate was the cause.

The most fascinating theory to explain the extinction of the dinosaurs came from Luis Alvarez, a scientist studying a unique layer of ash that was deposited around the world during the time of the great extinction. The ash layer contained an element called iridium, which is found in high levels in comets and asteroids. Alvarez hypothesized that a large asteroid or comet may have struck the Earth 65 million years ago causing massive damage and climate change. He believed that the climate change led to the extinction of the dinosaurs. Recently the remains of a large crater in the Gulf of Mexico, near Central America, was discovered. This crater has been dated to approximately 65 million years ago and may have been created by an asteroid impact that led to the extinction of the dinosaurs (Figure 13–16).

The Cenozoic Era

The extinction of the dinosaurs resulted in the emergence of mammals as the dominant life form on the land. The death of the dinosaurs allowed mammals to flourish on the Earth like never before. The end of the Cretaceous period 65 million years ago also marked the end of the Mesozoic era and began the Cenozoic era in which we live today. The Cenozoic era is further subdivided into the Tertiary and Quaternary periods, and the Tertiary period can be divided into the Paleogene and Neogene periods. This time began the emergence of

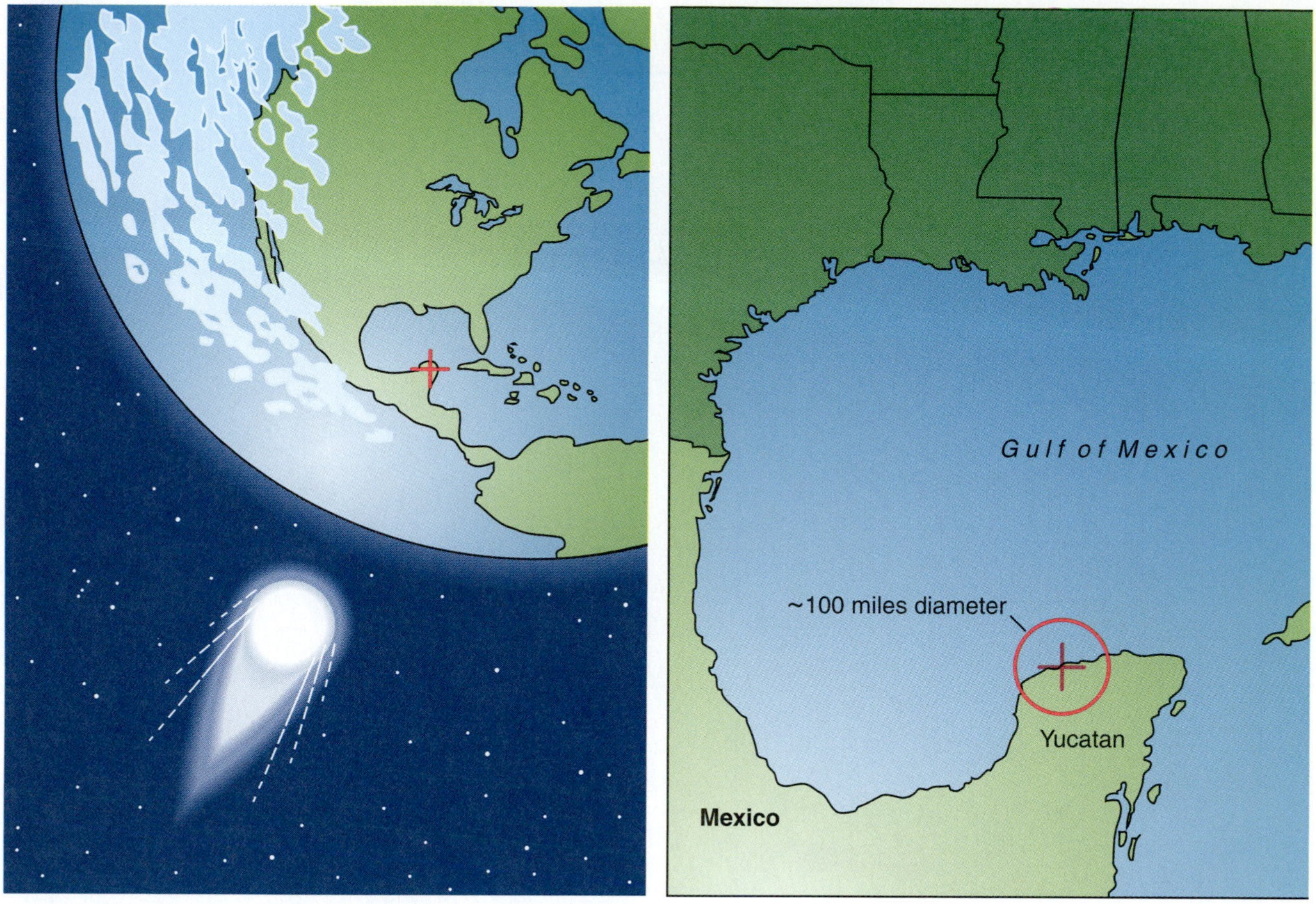

Figure 13–16 The asteroid impact theory suggests the extinction of the dinosaurs was caused by a 6-mile-wide asteroid that impacted the Earth near the present-day Yucatan Peninsula. The 100-mile-wide Chicxulub crater in the Gulf of Mexico is believed to have been caused by this asteroid impact 65 million years ago.

EARTH SYSTEM SCIENTISTS *LUIS ALVAREZ*

Luis Alvarez began his career as a physicist who studied light, radar, and the nucleus of the atom. During World War II, he worked on the Manhattan Project, the top secret development of the atomic bomb at Los Alamos. While there he developed a detonating trigger for the first atomic bombs. He also rode on the Enola Gay, the B-29 bomber that dropped the first atomic bomb on Japan. In 1968 he received the Nobel Prize for his work in high-energy physics. Alvarez served as a member of the Warren Commission, which was a panel of scientists and government officials assigned to investigate the assassination of President John F. Kennedy.

Luis Alvarez gained most of his fame late in his life, when he teamed up with his son, Walter, who was geologist. In 1980 Luis and Walter Alvarez theorized that the dinosaurs had become extinct as a result of an asteroid impact. Their theory was based on a thin layer of volcanic ash that appeared at about the same time the dinosaurs became extinct. This ash layer contained traces of a rare element, iridium, which is found in asteroids and comets. The Alvarez team gathered evidence from around the world to support their theory, which is still being debated. Luis Alvarez died in 1988.

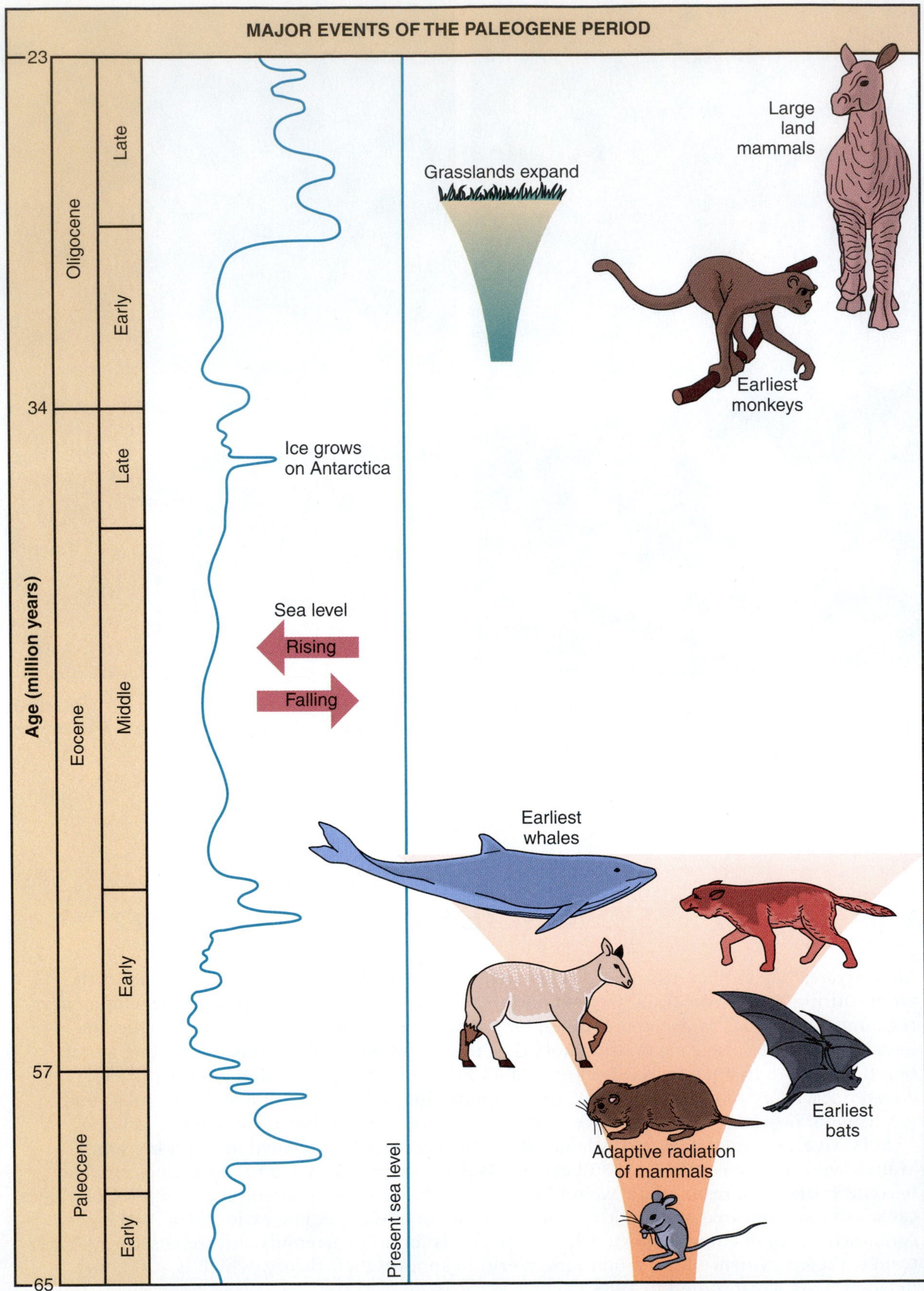

Figure 13–17 Animals that first appeared during the Tertiary period, 65 million years ago.

large modern mammals, many of which survive today, such as whales, tigers, lions, horses, monkeys, and wolves (Figure 13–17).

The Tertiary period began the appearance of grasses on the Earth, which led to the widespread development of grazing animals. The continents were in positions similar to where they are today, and many of the organisms that lived in the oceans and on the land would be easily recognizable. The Tertiary period lasted for about 63 million years and is further subdivided into distinct periods called **epochs.** The end of the Tertiary period came approximately 2 million years ago with the emergence of the hominids, which marked the beginning of human beings.

The emergence of our human ancestors approximately 2 million years ago brings us to the present period, the Quaternary. This period has included large-scale glaciation of much of the North American continent. It was during this cold time on Earth that humans began to flourish. The oldest fossil evidence for the existence of our own species, Homo sapiens, is dated to approximately 100,000 years ago. The glaciers in North America did not begin to melt until approximately 11,000 years ago. This glacial melting created the landscapes that we live in today throughout much of the world.

The geologic time scale is an important aspect of the Earth's history; it reveals the gradual changes that have occurred on our planet over time. Piecing together the history of the planet has been the result of exhausting research by countless numbers of scientists and researchers regarding rock formations and the fossils they contain. Although many pieces of Earth's fascinating history were not preserved and have been lost forever, it is still possible to tell a reasonable story of the Earth's past. There is one thing that the geologic record does show, however; it is the gradual advancement of life in many different forms over time, from its earliest beginnings to its most advanced forms. Understanding the Earth's geologic history is indeed understanding ourselves.

REVIEW

1. What is the principle of uniformity?
2. Describe the principle of superposition.
3. What is the half-life of an element?
4. What is radiometric dating?
5. What are the four principal elements used for radiometric dating?
6. Briefly describe some of the events that took place on Earth during the Archean eon.
7. Briefly describe some of the events that took place on Earth during the Proterozoic eon, also known as Precambrian time.
8. When did the Cambrian period begin, and what were some of the organisms alive during that time?
9. What were some of the highlights of the Ordovician period?
10. How long ago did the Devonian period begin? Describe some of the organisms that lived during that time.
11. What was the significance of the Carboniferous period?
12. Which mountains formed during the Permian period, and how long ago did they begin to form?
13. What organisms flourished on Earth during the Mesozoic era?
14. What event is believed to have ended the Cretaceous period?
15. During what geologic period did our human ancestors first appear, and how long ago did this occur?
16. How old is the first fossil evidence of our own species, Homo sapiens?
17. Who was Georges Cuvier?
18. Explain who Arthur Holmes was.
19. What were Luis Alvarez's contributions to science?

EARTH MATH

1) Determine approximately how many years the dinosaurs lived on the Earth.
2) What percentage of the Earth's total geologic history (4.5 billion years) did the Archean eon occupy?
3) What percentage of the Cenozoic era (65 million years) have humans occupied?
4) How much longer did the dinosaurs exist on the Earth compared with humans?

For more information go to these Web links:

<http://www.cotf.edu/ete/modules/msese/earthsysflr/geotime.html>

<http://www.ucmp.berkeley.edu/help/timeform.html>

CHAPTER SUMMARY

The history of the Earth stretches back over 4.5 billion years. This long span of time is divided into unique sections that together make up the geologic time scale. The geologic time scale divides the Earth's history into distinct periods based on specific geologic events and the appearance or disappearance of unique life forms. Geologists use the geologic principles of uniformity and superposition to relatively date the age of rock formations. Other relative dating techniques involve the use of unique volcanic ash layers in rocks, igneous intrusions, faults, and index fossils. Radioactive isotopes can also be used by geologists to determine the absolute age of a rock by using the known half-life of an element. This technique is known as radiometric dating. Elements that are commonly used for radiometric dating include uranium 238, potassium 40, and carbon 14.

Both relative and absolute dating techniques have been used to construct the geologic time scale, which is divided into three large spans of time called eons. The Archean eon began approximately 4.5 billion years ago with the formation of the Earth, and it ended 2.5 billion years ago. The Proterozoic eon began 2.5 billion years ago and ended approximately 544 million years ago. Together these two eons are also known as Precambrian time, which means "before life." This is because most of the life forms on the Earth appear in the fossil record beginning 544 million years ago. The current eon, known as the Phanerozoic, began 544 million years ago and continues today. This span of time is further subdivided into geologic eras, periods, and epochs based on the first appearance of a species of organisms in the fossil record and when they disappeared.

Geologic time is often marked by mass extinctions, which occurs when a large amount of the species alive on the Earth die suddenly. Mass extinctions have occurred many times during the Earth's past, including the one 65 million years ago that marked the extinction of the dinosaurs. As more evidence is revealed about the Earth's history, new pieces of information will be added to the geologic time scale that will provide more detail about the Earth's past.

CHAPTER REVIEW

Multiple Choice

1. Which geologic principle states that the same geologic processes occurring today have occurred throughout the Earth's history?
   a. superposition
   b. uniformity
   c. catastrophism
   d. radiometric

2. Unless a series of rock layers are overturned, the layers at the bottom are:
   a. the same age as the ones at the top
   b. the youngest rocks
   c. the oldest rocks
   d. the result of an intrusion

3. What is the relative age of a fault that cuts across many rock layers?
   a. the fault is the same age as the bottom layer it cuts across
   b. the fault is the same age as the top layer it cuts across
   c. the fault is older than all the rock layers it cuts across
   d. the fault is younger than all the layers it cuts across

4. An igneous intrusion is 50 million years old What is the most probable age of the rock surrounding the intrusion?
   a. 10 million years
   b. 25 million years
   c. 40 million years
   d. 60 million years

5. Which of the following elements is used for radiometric dating?
   a. cobalt 60
   b. plutonium 244
   c. potassium 40
   d. silicon 28

6. If a rock sample contains half the original amount of potassium 40, approximately how old is the rock?
   a. 2.6 billion years
   b. 4.5 billion years
   c. 1.3 billion years
   d. 0.7 billion years

7. The geologic time unit known as the era is based on the occurrence of:
   a. life
   b. asteroids
   c. mass extinctions
   d. rock formations

8. The geologic time scale has been subdivided into time units called periods based on:
   a. fossil evidence
   b. rock thickness
   c. rock types
   d. radiometric dating

9. During which geologic period did the first forms of life appear?
   a. Archean
   b. Proterozoic
   c. Phanerozoic
   d. Paleozoic

10. Which geologic period marked the first appearance of trilobites on Earth?
    a. Cretaceous
    b. Cambrian
    c. Ordovician
    d. Tertiary

11. How many years ago did the Permian period begin?
    a. 544 million
    b. 142 million
    c. 490 million
    d. 290 million

12. The end of what geologic period marked the extinction of the dinosaurs?
    a. Carboniferous
    b. Cretaceous
    c. Devonian
    d. Jurassic

13. Approximately how long ago did the supercontinent Pangea begin to break apart?
    a. 251 million years
    b. 490 million years
    c. 119 million years
    d. 65 million years

14. During what geologic period did the Appalachian Mountains begin to form?
    a. Cambrian
    b. Silurian
    c. Permian
    d. Quaternary

15. Approximately how long ago did the glaciers that covered much of North America begin to melt?
    a. 11,000 years
    b. 100,000 years
    c. 1 million years
    d. 4 million years

continued

Matching *Match the terms with the correct definitions.*

a. geologic time scale
b. principle of uniformity
c. principle of superposition
d. index fossils
e. half-life
f. radiometric dating
g. eon
h. stromatolites
i. Precambrian
j. era
k. mass extinction
l. Pangea
m. epoch

1. ____ The largest division of geologic time, measured in billions or hundreds of millions of years.

2. ____ The scale of time that divides the Earth's history into distinct periods based on geologic events and the appearance or disappearance of specific life forms.

3. ____ A large unit of geologic time when not much life flourished on the Earth; it started with the formation of the planet and ended approximately 544 million years ago.

4. ____ A natural law used in geology that states that the geologic processes that are currently shaping the Earth have been occurring throughout the Earth's history.

5. ____ Large moundlike layers of sediments that form when cyanobacteria traps sand in warm, shallow ocean water.

6. ____ A natural law used in geology that states that in undisturbed rock layers the oldest layers are located at the bottom and the youngest are at the top.

7. ____ A specific type of fossil organism that lived for a short time over a wide geographical area; it is used to identify specific rock formations and the period when they were formed.

8. ____ A method of dating objects using the known decay rate of certain radio isotopes.

9. ____ The smallest division of geologic time that is measured in millions or thousands of years.

10. ____ The time it takes for half of a mass of a radioactive isotope to decay into its daughter element.

11. ____ The name of a supercontinent that existed more than 250 million years ago, when all the present-day continents were joined together into one great landmass.

12. ____ The widespread disappearance of a great number of species on Earth in a short time.

13. ____ The second largest division of geologic time; it is measured in hundreds of millions or millions of years and marks the mass extinction of species on Earth.

Critical Thinking

1. The geologic record reveals that many extinctions have occurred on the Earth, and the planet's environment has constantly undergone change. What do these two things tell you about the nature of life on Earth?

UNIT 4

The Atmosphere

Topics to be presented in this unit include:

- Structure and Composition of the Atmosphere
- Insolation
- Atmospheric Temperature and Pressure
- Humidity, Clouds, and Precipitation
- Wind, Air Masses, and Fronts
- Storms and Weather Forecasting
- Global Climate Change
- Acid Precipitation and Deposition
- Ozone Depletion

OVERVIEW

When astronauts orbiting the Earth in the space shuttle gazed back at the surface of the planet, they noticed that the only thing that separated it from the vacuum of space was a thin blue line. This thin blue line that surrounds the Earth is called the atmosphere. If the whole Earth was represented by an apple, the atmosphere would be only as thin as the apple's skin. The atmosphere acts as a protective blanket around the Earth, which allows for life to flourish on its surface. This fragile part of the Earth, although mostly invisible, plays an important part in the Earth system in which we live.

CHAPTER 14

Structure and Composition of the Atmosphere

Objectives

After reading this chapter you should be able to:

- Define the terms *atmosphere* and *air.*
- Identify the main gases that make up the atmosphere.
- Identify the five layers of the atmosphere.
- Define the term *isothermal layer.*
- Explain where the ozone layer is located in the atmosphere and what function it performs.
- Describe the characteristics of the ionosphere.

TERMS TO KNOW

atmosphere
air
permanent gases
variable gases
water vapor
troposphere
isothermal layer
stratosphere
temperature inversion
ultraviolet radiation
ozone layer
ions
solar radiation

INTRODUCTION

The air in which we live is an invisible sea of molecules that surrounds us and sustains life. We sense this invisible sea every time the wind blows or we look up at clouds moving through the bright blue sky. Not only does air sustain life on Earth, it plays an important role in distributing heat and water around the planet. Air also acts as a barrier that surrounds the Earth and protects its surface from the harsh environment of space. Probably the most well known aspect of the air in which we live is how it forms weather. All these things point to the importance of understanding the Earth's atmosphere and its role in the Earth's systems.

Atmospheric Composition

The **atmosphere** is defined as the thin envelope of gas that surrounds the Earth (Figure 14–1). From the Earth's surface, the gases that surround the planet stretch approximately 375 miles above sea level. Past this point lies the near vacuum of space.

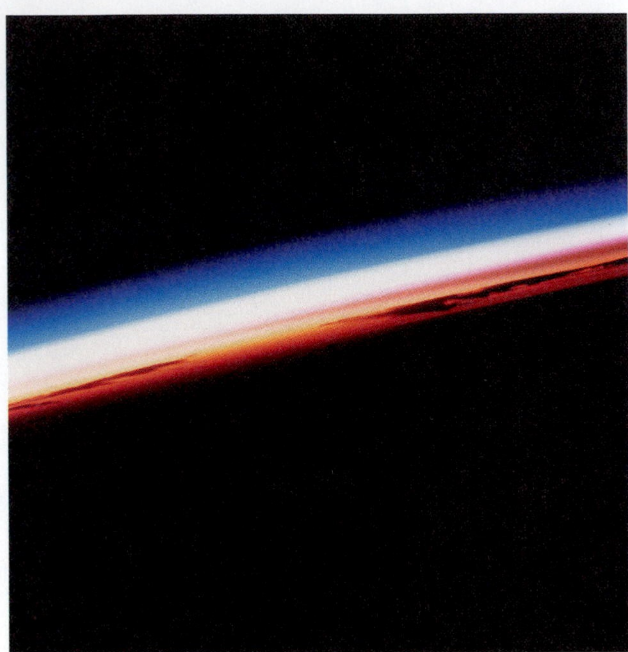

Figure 14–1 The atmosphere of the Earth as viewed from space appears to be a thin blue line. *(Courtesy of PhotoDisc.)*

The atmosphere is composed of gases that together are commonly known as **air.** These gases are divided into two main groups: permanent and variable (Table 14–1). **Permanent gases** are gases that maintain a constant level in the atmosphere. These include nitrogen, oxygen, argon, neon, helium, hydrogen, and xenon. Nitrogen gas is a diatomic molecule, meaning that it contains two atoms of nitrogen. It composes approximately 78% of the Earth's atmosphere. Oxygen gas in the atmosphere is also a diatomic molecule, made from two atoms of oxygen. Oxygen composes approximately 21% of the Earth's atmosphere and is vital to life on Earth.

Geological evidence suggests that more than 2 billion years ago the atmosphere was much different than today, containing no oxygen. The photosynthetic action of aquatic plants and algae eventually filled the atmosphere with oxygen to its present levels. Prior to this, much of the Earth's early atmosphere primarily was composed of water vapor and carbon dioxide. Eventually much of the carbon dioxide was absorbed by the oceans and by rocks. The water vapor in the Earth's early atmosphere condensed and rained down to the surface to collect and form the oceans and fresh surface water that cover the planet today. The remaining permanent gases that

TABLE 14–1 The permanent and variable gases that make up the Earth's atmosphere, and their composition

Permanent Gases			Variable Gases			
Gas	Symbol	Percent (by Volume) Dry Air	Gas (and Particles)	Symbol	Percent (by Volume)	Parts per Million (ppm)*
Nitrogen	N_2	78.08	Water vapor	H_2O	0 to 4	
Oxygen	O_2	20.95	Carbon dioxide	CO_2	0.035	355
Argon	Ar	0.93	Methane	CH_4	0.00017	1.7
Neon	Ne	0.0018	Nitrous oxide	N_2O	0.00003	0.3
Helium	He	0.0005	Ozone	O_3	0.000004	0.04
Hydrogen	H_2	0.00006	Particles (dust, soot, etc.)		0.000001	0.01
Xenon	Xe	0.000009	Chlorofluorocarbons (CFCs)		0.00000001	0.0001

*For CO_2, 355 parts per million means that out of every million air molecules, 355 are CO_2 molecules.

EARTH SYSTEM SCIENTISTS ARISTOTLE

Aristotle was born in Macedonia in 384 B.C. during the rise of ancient Greece. He studied under the great philosopher Plato and eventually became one the greatest thinkers of the ancient world. He was the tutor of the future king of Greece, later known as Alexander the Great. In 335 B.C. he established his own school in Athens, where he taught philosophy, natural history, mathematics, and astronomy. He also helped to establish the great library of Alexandria in Egypt. Aristotle's book *Meteorologica* was the first to describe the processes that lead to the formation of clouds, precipitation, wind, lightning, and other phenomena associated with weather. This book is where the term *meteorology* was derived and was widely accepted as scientific fact for almost 2000 years.

make up today's atmosphere are found at much smaller levels than nitrogen or oxygen. These are known as trace gases.

The **variable gases** that make up the Earth's atmosphere include water vapor, carbon dioxide, methane, ozone, nitrous oxide, and chlorofluorocarbons (CFCs). **Water vapor** is the gaseous form of water in the atmosphere, which varies in its composition around the globe. In humid tropical climates near the Earth's equator, water vapor content can be as high as 4% in the atmosphere. In the drier polar regions, water vapor is as low as 0.5%. Water vapor content in the atmosphere changes daily and is an important aspect of the local weather.

Carbon dioxide is another variable gas in the atmosphere, composing a small portion of the air with levels of 0.035%; however, as a result of industrial processes, levels of this gas are slowly increasing. Even though carbon dioxide composes a small portion of the total atmosphere, it plays an important role in trapping the Earth's heat. Because of this, it is known as a greenhouse gas; it traps heat much like the glass of a greenhouse. Methane, another variable gas in the atmosphere, is produced as a byproduct of anaerobic decomposition, which is the breakdown of organic material without oxygen. The main sources of methane in the Earth's atmosphere are from decomposing bacteria in livestock and rice paddies. The remaining variable gases that compose air are found in extremely small amounts but can play equally important roles in the atmosphere. CFCs are human-created gases that now exist in the stratosphere, where they are destroying the ozone layer.

Both permanent and variable gases have one thing in common: They are invisible. This is especially apparent on a clear night, when you can see the stars shining in space. Why then does the sky appear blue during the day? The blue appearance of the sky during the day is caused by the scattering of blue visible light by gas molecules and particles in the atmosphere. At sunrise and sunset, the sky often appears red or orange in color. This is also caused by the scattering of light by the atmosphere. The sky appears red or orange because the light has to travel through more of the atmosphere, therefore scattering more blue wavelengths of light and allowing only colors such as red or orange to pass through.

Atmospheric Structure

The atmosphere is divided into five distinct layers based on their unique characteristics. The **troposphere** is the layer that lies closest to the Earth's surface and is approximately 7 miles thick. All weather on Earth takes place in the troposphere, which also contains more than 90% of the atmosphere's gases. All the clouds that float through the skies form in the troposphere. The troposphere also contains all the oxygen we need to breathe. As you ascend into the troposphere, the temperature of the surrounding air decreases (Figure 14–2). The pressure of the air also decreases with height in the troposphere. This is apparent when you climb high mountains or visit areas that are located at higher elevations. Because these places are higher in the troposphere, the air is thinner, making it harder for you to breathe. You can also sense the change in pressure with height in the troposphere when your ears pop as you rise in elevation. This is caused by the air located in your inner ear equalizing with the air pressure outside your ear.

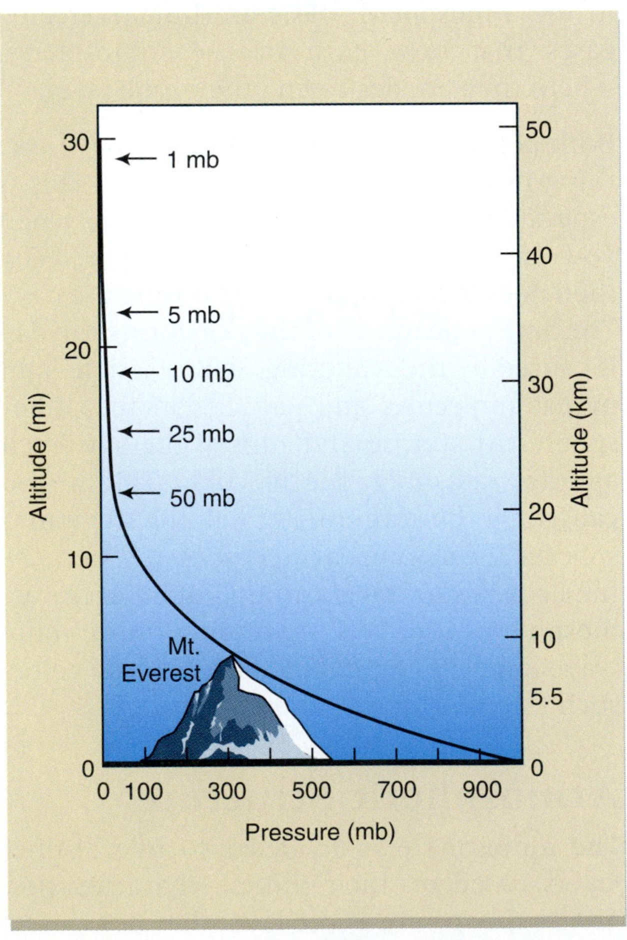

Figure 14–2 Air pressure decreases with height in the troposphere.

You may also notice a decline in temperature as you climb a high mountain. In some places on Earth it can be a sunny and warm day at the base of a mountain. When you climb the mountain and move higher in the troposphere, eventually it may be cold enough to form snow. The average decline in temperature in the troposphere is approximately 3.5° Fahrenheit for every 1000 feet in elevation. At the top of the troposphere, temperatures can be as low as −80° Fahrenheit. Eventually, as you rise toward the top of the troposphere, the air temperature begins to level off. The transitional layer that lies at the top of the troposphere is called the tropopause. In the tropopause, temperature remains stable as you increase in height. This is also called an **isothermal layer,** which is a layer that maintains the same temperature as you increase in altitude. The tropopause is approximately 6 miles thick, and eventually the air temperature begins to rise again. This marks the entry into the next layer of the atmosphere, called the stratosphere.

The base of the **stratosphere** is located approximately 7 to 10 miles above the Earth's surface and rises to a height of approximately 30 miles above the Earth. The stratosphere is the area of the atmosphere where temperature rises as you increase in height. When the air temperature increases with altitude in the atmosphere, it is known as a **temperature inversion.** Although the temperature increases with height, it is still very cold in the stratosphere. The highest temperature in the stratosphere is about 25° Fahrenheit. Approximately 99.9% of all the gases that together form the Earth's atmosphere lie below the stratosphere.

One of the most important aspects of the stratosphere is the region called the **ozone layer.** Ozone is an unstable molecule composed of three atoms of oxygen that help to block deadly **ultraviolet radiation** coming from the sun. The ozone layer is the main reason that the stratosphere experiences a temperature inversion. Heat energy is released into the air in the stratosphere when ultraviolet radiation strikes the ozone layer. The ozone layer lies approximately 15 miles above the Earth's surface. The air temperature at the top of the stratosphere eventually begins to level off, forming another isothermal transition layer called the stratopause.

Above the stratopause, approximately 35 miles above the Earth's surface, begins the next layer of the atmosphere, known as the mesosphere. In the mesosphere the air temperature begins to decrease again as you increase in height. The lowest temperature in the Earth's atmosphere, approximately −130° Fahrenheit, occurs in the mesosphere. This portion of the Earth's atmosphere rises to approximately 55 miles above the Earth. The air in the mesosphere is extremely thin and has very low pressure. The mesosphere is also exposed to high levels of ultraviolet radiation, which would burn a person's skin if it were exposed. Another transitional layer lies above the mesosphere, where the temperature again begins to level off with increasing height. This isothermal region is called the mesopause and marks the boundary between the mesosphere and the next layer of the atmosphere, called the thermosphere (Figure 14–3).

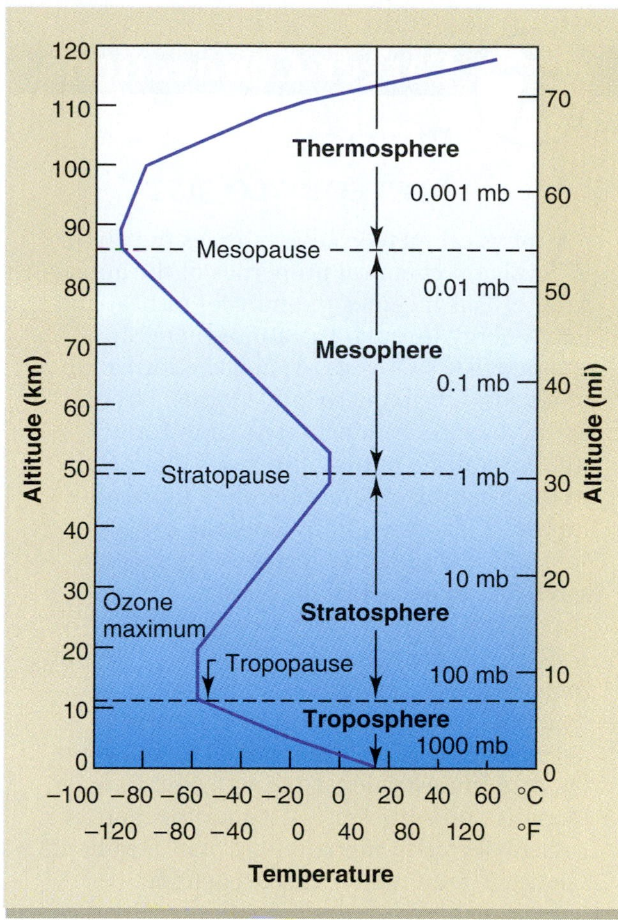

Figure 14–3 The layers of the atmosphere and their characteristic temperature, pressure, and altitude.

The thermosphere begins at approximately 60 miles above the Earth, where the air temperature begins to rise once again. Temperatures can be as high as 120° Fahrenheit in the thermosphere; however, there are so few air molecules at this altitude that the temperature cannot be compared with that of surface air temperatures. The temperature in the thermosphere is just a measure of how fast the air molecules are moving. The air is so thin in the thermosphere that an air molecule will collide with another air molecule only after it has traveled an approximate distance of more than 3 miles. In comparison, an air molecule traveling in the troposphere will strike another air molecule after it has traveled only one millionth of an inch!

The top of the thermosphere is located approximately 312 miles above the Earth's surface, where the final layer of the atmosphere, the exosphere, begins. The exosphere is the region where air molecules escape the gravitational pull of Earth and travel off into space. This marks the top of the Earth's atmosphere and is where the cold near-vacuum of space begins.

Another important aspect of the Earth's atmosphere is a region called the ionosphere, which is an area in the atmosphere that contains a high concentration of **ions.** An ion is an atom or molecule that has an electric charge. The ions that make up the ionosphere are formed by incoming **solar radiation** that strikes the air in the atmosphere. The solar

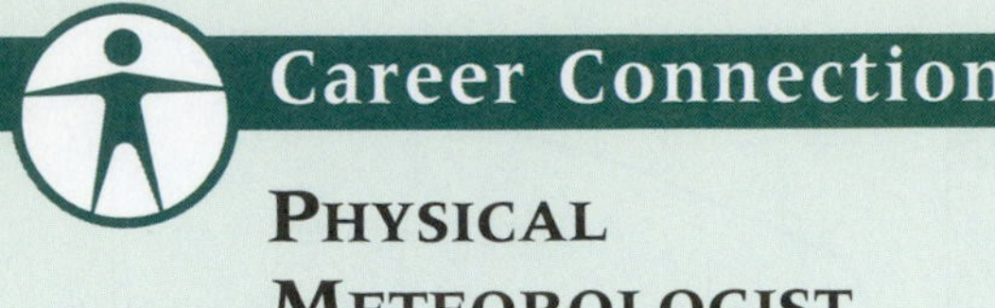

Physical Meteorologist

A physical meteorologist studies the physical and chemical properties of the atmosphere. This includes the transfer of heat and light energy through the atmosphere, the characteristics of its gases, and the formation of clouds, precipitation, and storms. Physical meteorologists conduct research to further understand the natural processes that occur in the atmosphere and how they impact life on Earth. Their work involves the use of sophisticated scientific instruments such as satellites, weather balloons, radar, and aircraft, along with the use of computers to collect and analyze atmospheric data. Physical meteorologists can find employment at colleges and universities or at government agencies such as the National Aeronautic and Space Administration (NASA) or the National Oceanic and Atmospheric Administration. A college degree in meteorology or atmospheric science is required for this occupation.

radiation strips away electrons from atoms in the atmosphere and forms ions. Because the formation of ions requires sunlight, the size of the ionosphere tends to be larger during the daytime and smaller at night. During the day, the ionosphere can form at approximately 37 miles above the Earth's surface, near the top of the stratosphere. At night the ionosphere shrinks and begins at a higher altitude of about 75 miles above the Earth, in the thermosphere. On average the ionosphere is approximately 200 miles thick.

The ionosphere plays an important role in radio communications on Earth. AM radio waves emitted from stations on the Earth's surface are reflected back when they strike the ionosphere. During the day, the sun prompts the ionosphere to thicken, causing the bottom layer to form closer to the Earth's surface. This causes the radio waves to bounce back toward the surface more quickly, forcing them to travel shorter distances. At night, when the ionosphere thins, its bottom layer rises higher in the atmosphere and radio waves take longer to bounce back. This causes them to travel extremely long distances at night. As a result of this, radio stations increase their transmission power in the daytime and decrease it at night to compensate for the changing effects of the ionosphere on the distance a radio wave travels (Figure 14–4). At night it is possible to pick up radio stations on your AM radio from far around the Earth.

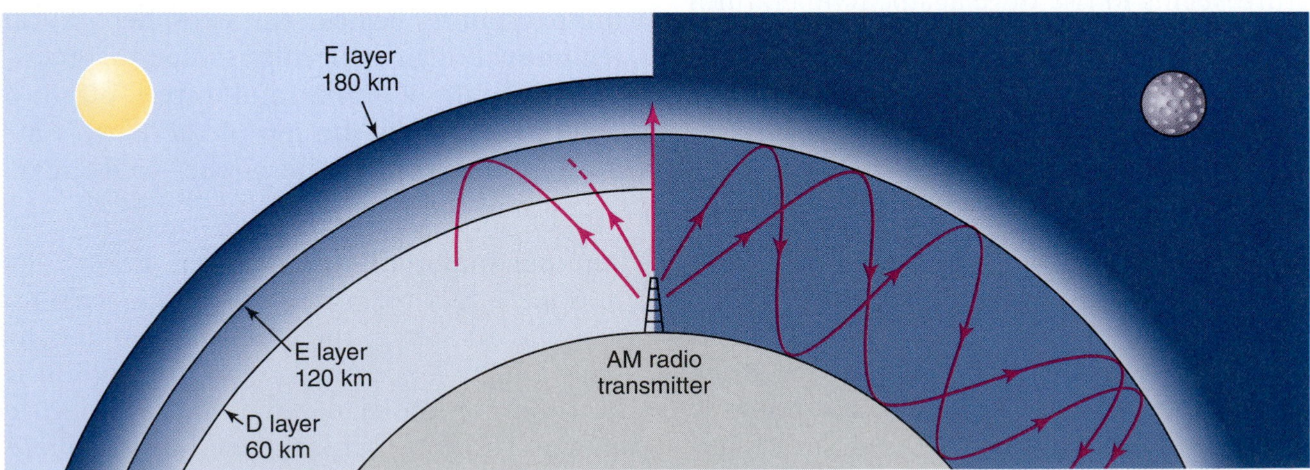

Figure 14–4 The effect of the changing size of the ionosphere on radio communications during the day and night.

EARTH MATH

1) USING THE INFORMATION PROVIDED IN THE CHAPTER, DETERMINE THE THICKNESS OF BOTH THE THERMOSPHERE AND THE MESOSPHERE.

2) IF THE TEMPERATURE IN THE TROPOSPHERE DECLINES 3.5° FAHRENHEIT FOR EVERY 1000 FEET IN ELEVATION, WHAT WOULD THE APPROXIMATE TEMPERATURE BE AT THE TOP OF MOUNT EVEREST (29,022 FEET) IF THE TEMPERATURE IS 70° FAHRENHEIT AT SEA LEVEL?

REVIEW

1. What is the atmosphere?
2. Describe what a permanent atmospheric gas is and list all the Earth's permanent gases.
3. Describe what a variable gas is and list all the Earth's variable gases.
4. Name the unique layers of the atmosphere.
5. Which layers of the atmosphere experience temperature inversions?
6. Which layers of the atmosphere are isothermal?
7. Describe the ionosphere and explain how it affects radio waves.
8. Who was Aristotle?

For more information go to these Web links:

<http://csep10.phys.utk.edu/astr161/lect/earth/atmosphere.html>

<http://www.earth.nasa.gov/science/Science_atmosphere.html>

CHAPTER SUMMARY

The atmosphere is the thin envelope of gases that surrounds the Earth. The atmosphere is commonly known as air, which is composed of 78% nitrogen, 21% oxygen, and other trace gases. The gases in the atmosphere are classified as permanent or variable. Permanent gases remain at fixed levels in the atmosphere and include nitrogen, oxygen, argon, neon, helium, hydrogen, and xenon. Variable gases change in their composition and include water vapor, carbon dioxide, methane, ozone, nitrous oxide, and chlorofluorocarbons (CFCs).

The atmosphere is divided into five layers based on their own unique characteristics. The troposphere is the layer of the atmosphere that lies closest to the Earth. This is the part of the atmosphere in which we all live and where all weather takes place. As you rise into the troposphere, both temperature and pressure decrease with height. Above the troposphere lies another layer of the atmosphere called the stratosphere. The stratosphere contains a layer of ozone gas called the ozone layer. The ozone layer acts as a shield that protects the Earth's surface from deadly ultraviolet radiation. The other layers of the atmosphere include the mesosphere and the thermosphere. Transitional layers located between all the layers of the atmosphere are called isothermal layers. These are parts of the atmosphere where temperature remains the same with an increase in height. Another part of the atmosphere is the ionosphere. It is located within the stratosphere and the thermosphere and is composed of ions created by incoming solar radiation.

CHAPTER REVIEW

Multiple Choice

1. The atmosphere stretches into space approximately how far above the Earth's surface?
   a. 10 miles
   b. 75 miles
   c. 100 miles
   d. 375 miles

2. Which gas makes up most of the Earth's atmosphere?
   a. nitrogen
   b. oxygen
   c. carbon dioxide
   d. water vapor

3. Approximately 21% of the Earth's atmosphere is composed of:
   a. nitrogen
   b. oxygen
   c. carbon dioxide
   d. water vapor

4. Which of the following gases varies in its atmospheric composition?
   a. argon
   b. nitrogen
   c. water vapor
   d. oxygen

5. What layer in the atmosphere contains 90% of all the gases on Earth?
   a. thermosphere
   b. mesosphere
   c. stratosphere
   d. troposphere

6. The ozone exists in which layer of the atmosphere?
   a. thermosphere
   b. mesosphere
   c. stratosphere
   d. troposphere

7. As you rise up in the troposphere, temperature and pressure:
   a. increase
   b. decrease
   c. remain the same
   d. fluctuate

8. An area in the atmosphere where temperature remains the same as you increase in height is known as:
   a. the exosphere
   b. an isothermal layer
   c. the ionosphere
   d. an inversion

9. Which layer of the atmosphere has the coldest temperature?
   a. thermosphere
   b. mesosphere
   c. stratosphere
   d. troposphere

10. The ionosphere grows to its greatest thickness during which time?
    a. at night
    b. during a full Moon
    c. during the day
    d. at sunrise

Matching *Match the terms with the correct definitions.*

a. atmosphere
b. air
c. permanent gases
d. variable gases
e. water vapor
f. troposphere
g. isothermal layer
h. stratosphere
i. temperature inversion
j. ozone layer
k. ultraviolet radiation
l. ions
m. solar radiation

1. ____ The process by which a warm layer of air overlies a cold layer of air.
2. ____ The electromagnetic radiation emitted from the Sun.
3. ____ The outer layer of gas that surrounds a planet.
4. ____ An atom or group of atoms that have an electric charge as a result of gaining or losing electrons.
5. ____ The gases that together make up Earth's atmosphere, including nitrogen, oxygen, argon, water vapor, carbon dioxide, and other trace elements.
6. ____ A layer in the Earth's atmosphere containing the ozone, where temperature increases with an increase in altitude.
7. ____ Gases that exist in fixed amounts in the Earth's atmosphere.
8. ____ A specific high-energy form of electromagnetic radiation emitted from the Sun that can be harmful to living things.
9. ____ Gases in the Earth's atmosphere that change in their composition.
10. ____ The specific area in the stratosphere, located at an altitude between 10 and 20 miles, that contains a high concentration of ozone gas.
11. ____ The gaseous form of water.
12. ____ The lowest layer of the Earth's atmosphere; it lies closest to the surface and is where all weather takes place and the temperature decreases with an increase in altitude.
13. ____ A layer in the atmosphere where temperature remains the same as altitude increases.

Critical Thinking

1. Describe some of the ways that humans utilize the atmosphere and the air it contains.

CHAPTER

15 Insolation

Objectives

After reading this chapter you should be able to:

- Define the term *insolation*.
- Differentiate between long-wave and short-wave radiation emitted from the Sun.
- Identify all the pathways insolation can take when it strikes the Earth's atmosphere.
- Explain what happens to insolation when it strikes the Earth's surface.
- Describe how water and land differ in their ability to absorb and release heat.
- Explain the relationship between the intensity of insolation and the angle at which it strikes the Earth's surface.
- Describe the relationship between the intensity and angle of insolation and latitude location on the Earth's surface.
- Explain what causes the duration of insolation striking the Earth's surface to change throughout the year.
- Identify four ways by which the atmosphere is heated.

TERMS TO KNOW

insolation
short-wave radiation
long-wave radiation
albedo
heat capacity
reradiate
angle of insolation
duration
terrestrial radiation
latent heat of condensation

INTRODUCTION

The warm sun on your face, a hot car seat, and a scalding parking lot beneath your feet on a sunny summer day all illustrate ways in which the Sun's energy can interact with our planet. Even though the Sun is 93 million miles away, the radiation it emits is so strong that it greatly affects many aspects of the Earth. Almost all the energy on our planet is derived from the Sun, which drives many processes on the Earth's surface. Knowing how solar energy interacts with the atmosphere, hydrosphere, and lithosphere is fundamental to a clear understanding of the Earth's systems.

The Sun's Radiation

Insolation is the incoming radiation from the Sun that is received by the Earth's atmosphere and surface. Insolation is the driving force that moves the atmosphere and creates weather on Earth. The Sun emits radiation in all wavelengths of the electromagnetic spectrum (Figure 15–1). Most of the Sun's radiation received by the Earth is in the form of **short-wave radiation.** This high-energy radiation makes up approximately 88% of the radiation that is received on Earth. Short-wave radiation emitted by the Sun includes ultraviolet, visible light, and near-infrared radiation. The remaining 12% of the radiation received on Earth from the Sun is low-energy **long-wave radiation.** This includes far-infrared and microwave radiation.

Insolation and the Atmosphere

Short-wave and long-wave radiation can take many different paths after they come into contact with the Earth's atmosphere (Figure 15–2). When the Sun's radiation strikes the atmosphere, much of it can be directly absorbed by the gases present in the air. This is what causes the thermosphere and the stratosphere to experience an increase in temperature with height, known as a temperature inversion. Some gases, such as stratospheric ozone, absorb a large amount of incoming ultraviolet short-wave radiation. On average, 19% of the insolation received on Earth is absorbed by the atmosphere. Some of the solar radiation that strikes the Earth's atmosphere is scattered or reflected back into space. The scattering of short-wavelength visible light by gas molecules and particles in the atmosphere gives the sky its blue appearance during the day. If there are clouds present in the atmosphere, they may also reflect radiation back into space. Approximately 26% of the insolation received by the Earth's atmosphere is scattered or reflected back into space by clouds.

The insolation that is not absorbed, scattered, or reflected back into space by the atmosphere strikes the land surface or the oceans. When short-wave radiation strikes the surface of the Earth, some of it is immediately reflected back into space. This often occurs on snow-covered landscapes or on glaciers. White objects such as snow reflect 95% of the radiation received from the sun. Calm water can reflect up to 10% of the solar radiation received. Overall, land or water on the Earth reflects back approximately 4% of the Sun's incoming radiation. Combining this with the 26% that is

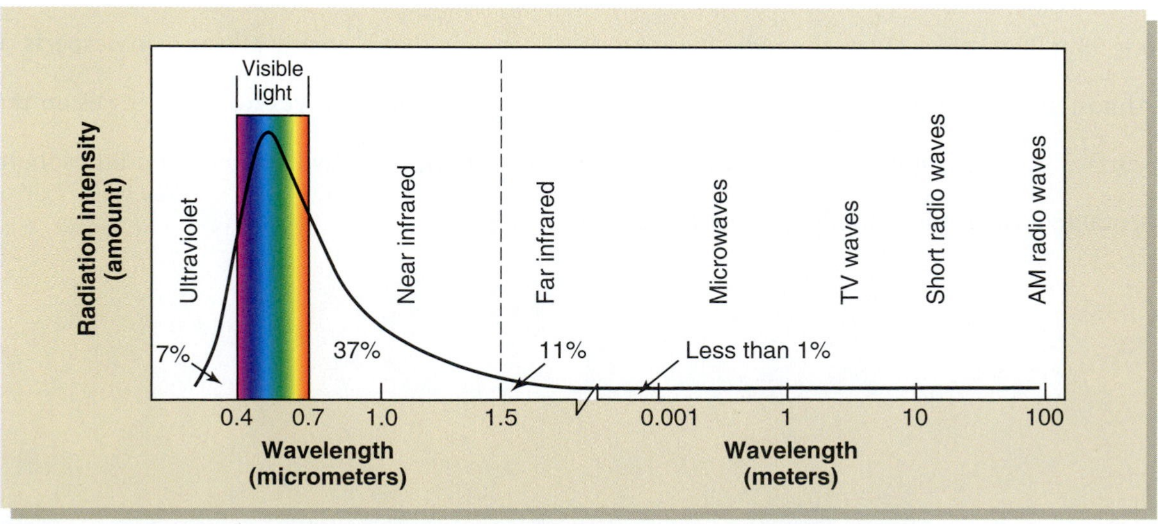

Figure 15–1 A portion of the amount and type of electromagnetic energy emitted by the Sun.

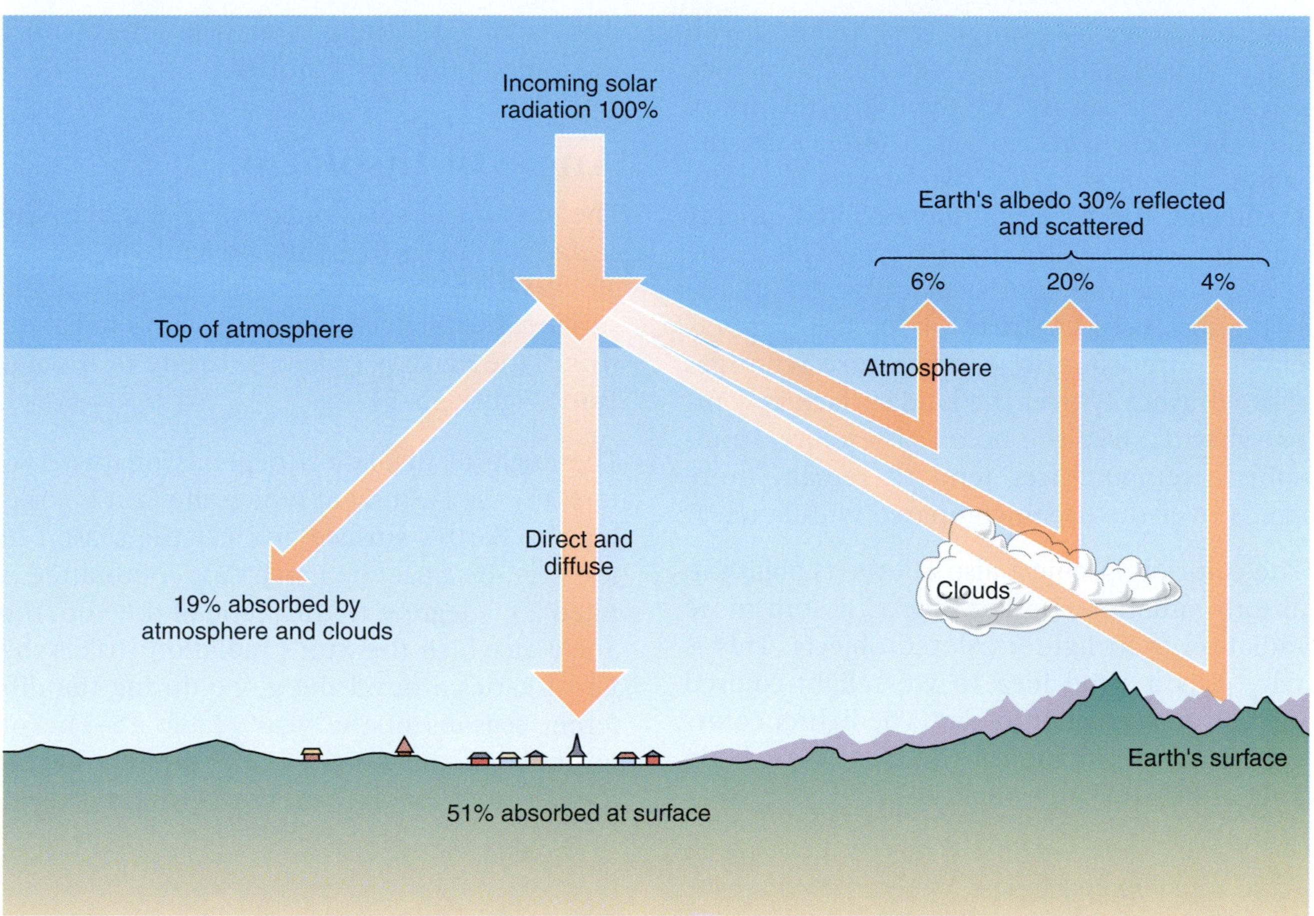

Figure 15–2 The general pathways that insolation takes when it strikes the Earth's atmosphere.

reflected or scattered by the atmosphere means that a total of 30% of the insolation received on Earth is returned to space.

The ability of an object in space to reflect radiation is also called its **albedo.** The albedo measures the reflectivity of objects in space. The Earth's albedo is therefore approximately 30%; that is, 30% of the radiation it receives is reflected back into space. Thus approximately 51% of the Earth's incoming solar radiation is absorbed at the surface by the world's oceans or land masses.

Insolation and the Earth's Surface

The land surface absorbs and releases heat from the Sun at a much higher rate than water because the rocks that make up the land surface have a lower **heat capacity** than water. Heat capacity is the ability of a substance to absorb or release heat and therefore alter its temperature. Water has a higher heat capacity than rock, which causes it to absorb a great amount of heat without a sudden change in temperature. This means that insolation absorbed by the ocean is stored for a longer time and released back into the atmosphere at a much slower rate. Rocks possess a lower heat capacity than water, which causes them to rapidly absorb heat.

Objects that are good absorbers of radiation also tend to be good radiators. As a result of this, rocks quickly **reradiate** their stored energy back into the atmosphere in the form of long-wave infrared radiation, which causes them to cool quickly. The difference between the heat capacity of water and rocks can be easily observed by a simple experiment. If you place a bucket of cool water next to a rock outside on a sunny day, you should notice that the rock heats up much faster than the water. If you leave the two objects out until nighttime, the rock will have cooled quickly and the water will remain warm. This is the result of the different heat capacities of the two substances. As a result of these differences, atmospheric temperature changes more drastically over the land surface as compared with over the ocean. Therefore atmospheric temperatures tend to remain more stable over the ocean compared with land.

The color of an object also affects its ability to absorb radiation. Darker objects absorb more radiation than lighter-colored objects. This is why it is a good idea to wear light-colored clothes on a hot, sunny day. The lighter colors reflect more radiation away from your body and therefore keep you cooler. If you wear dark-colored clothing, you will absorb more radiation and become hotter.

Angle of Insolation

The amount of solar radiation that strikes the surface of the Earth greatly depends on the angle at which the radiation is received on the surface. The angle at which the sun's radiation strikes the Earth is called the **angle of insolation** (Figure 15–3).

The angle of insolation depends on two factors: (1) the latitude at which the Sun is striking the Earth's surface and (2) the season of the year. Because the Earth's axis of rotation is tilted 23.5 degrees to the plane of the Sun, the angle at which the Sun's radiation strikes the Earth varies around the globe during the different seasons of the year (Table 15–1). For example, as the Northern Hemisphere begins to tilt away from the Sun when winter arrives,

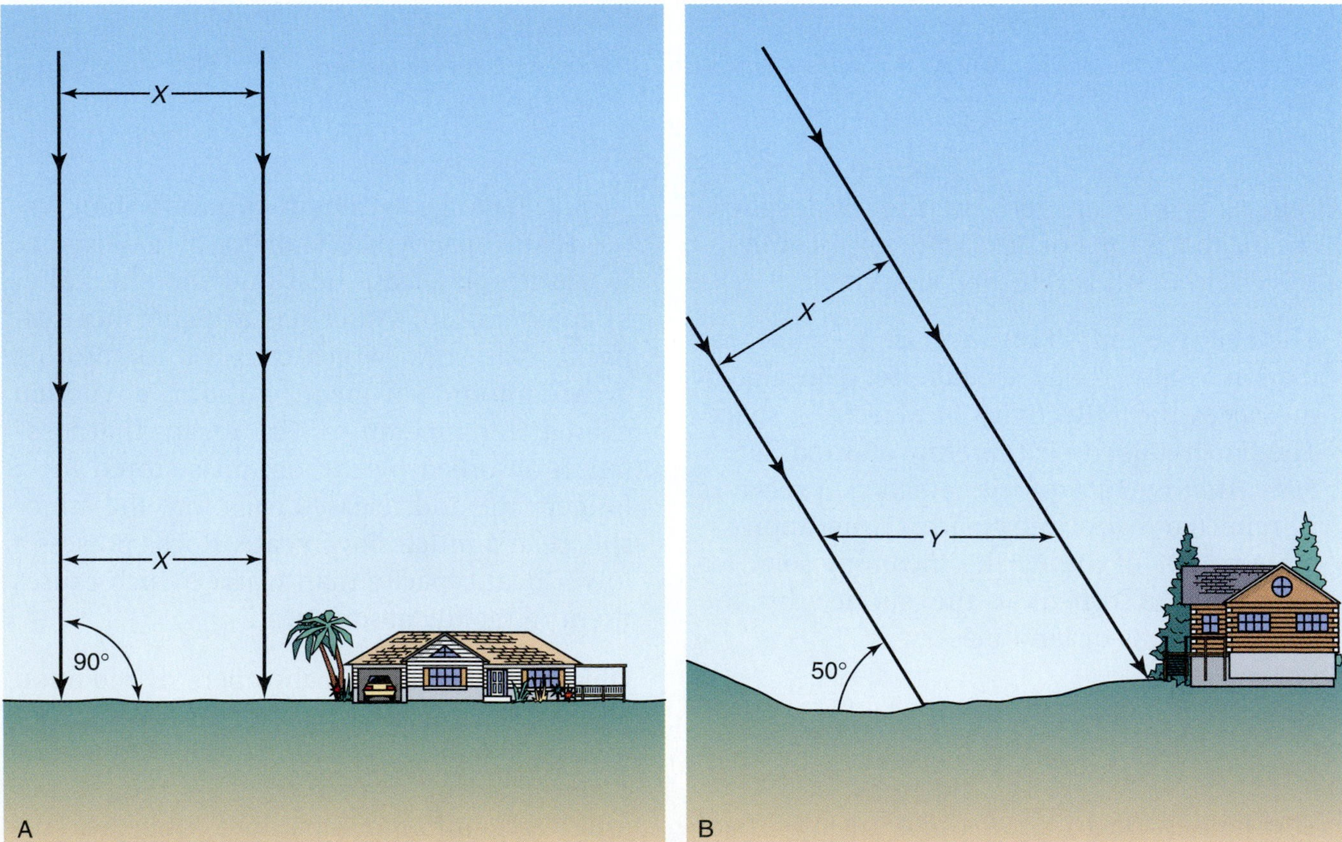

Figure 15–3 The intensity of insolation that strikes the Earth's surface is related to the angle at which it is received. The maximum insolation angle of 90 degrees has more intensity per unit area than an insolation angle of 50 degrees.

TABLE 15–1 Changing angle of insolation throughout the year for specific latitudes on Earth

Latitude	Summer Solstice June 21 — Angle of Insolation at Noon	Equinoxes March 21 September 23 — Angle of Insolation at Noon	Winter Solstice December 21 — Angle of Insolation at Noon
90° N	$23^1/_2°$	0°	–
80° N	$33^1/_2°$	10°	–
70° N	$43^1/_2°$	20°	–
60° N	$53^1/_2°$	30°	$6^1/_2°$
50° N	$63^1/_2°$	40°	$16^1/_2°$
40° N	$73^1/_2°$	50°	$26^1/_2°$
30° N	$83^1/_2°$	60°	$36^1/_2°$
20° N	$86^1/_2°$	70°	$46^1/_2°$
10° N	$76^1/_2°$	80°	$56^1/_2°$
0°	$66^1/_2°$	90°	$66^1/_2°$
10° S	$56^1/_2°$	80°	$76^1/_2°$
20° S	$46^1/_2°$	70°	$86^1/_2°$
30° S	$36^1/_2°$	60°	$83^1/_2°$
40° S	$26^1/_2°$	50°	$73^1/_2°$
50° S	$16^1/_2°$	40°	$63^1/_2°$
60° S	$6^1/_2°$	30°	$53^1/_2°$
70° S	–	20°	$43^1/_2°$
80° S	–	10°	$33^1/_2°$
90° S	–	0°	$23^1/_2°$

the angle of insolation striking the higher northern latitudes decreases. This causes the Northern Hemisphere to receive lower angles of insolation during winter, when it is tilted away from the Sun. During summer in the Northern Hemisphere, the angle of insolation is increased because the Earth is tilted toward the Sun. Therefore it receives a greater amount of insolation per unit area.

The nearly spherical shape of the Earth also affects the angle at which the Sun's energy strikes the Earth. This is because the curving surface of the Earth causes insolation to strike the higher latitudes at a lower angle. Generally the lower the angle of insolation, the lower the amount of solar energy received at the surface and therefore the lower the surface temperature. Areas located in the higher latitudes have a lower angle of insolation than areas closer to the equator. This results in a general decrease in surface temperature at latitudes farther north or south of the equator. Near the equator the Sun strikes the Earth at nearly a 90-degree angle all year. This concentrates a greater amount of solar radiation over a smaller area and creates high surface temperatures. Near the poles the angle of insolation is much lower, which results in cooler surface temperatures. The changing seasons on Earth reveal how the changing angle of insolation can affect the temperature throughout the year. In the United States during winter the angle of insolation is at its lowest; therefore the surface receives less insolation and experiences colder temperatures.

Career Connections

Climatologist

A climatologist collects information about the Earth's long-term temperature, insolation, wind, precipitation, and humidity in specific geographical regions. This information is then used to understand and predict the long-term weather patterns around the world. A climatologist's research helps other scientists and engineers to construct efficient heating and cooling systems for the construction industry. Their work also helps to improve agriculture and land use. Climatology involves the collection and organization of vast amounts of weather data from all around the world. This data is then analyzed using powerful computers and is used to determine specific climate regions that exist on our planet. Climatologists work for government agencies and in academic research, and they must have a college education specializing in meteorology.

During the summer, however, the Sun's angle of insolation is much greater, which concentrates more solar energy at the surface, therefore raising the temperature.

Duration of Insolation

Another factor that affects the amount of insolation received at the Earth's surface involves the **duration** that the surface is exposed to the Sun. Areas on the Earth that are located in the higher latitudes receive differing amounts of sunlight throughout the year. Generally in the United States the Sun shines for a longer period in summer than it does in winter. This is the result of the Earth's tilted axis of rotation, which causes greater amounts of insolation to be received by the Northern Hemisphere during summer and lesser amounts during winter. The difference in heating that results from the effect of the Sun's changing angle of insolation and duration throughout the year creates an unequal distribution of heat in the atmosphere (Figure 15–4).

Heating the Atmosphere

The Earth's atmosphere is heated by four processes. The first process through which the atmosphere is heated is from the direct absorption of radiation from the sun. As we have seen, gases in the atmosphere absorb both long-wave and short-wave radiation. The absorption of radiation by molecules in the atmosphere is then transferred into heat energy.

The second process through which the atmosphere is heated is from the reradiation of long-wave radiation from the Earth's surface. High-energy, short-wavelength radiation that strikes the Earth's surface is absorbed by the land surface. The land then reradiates long-

Length of Time from Sunrise to Sunset for Various Latitudes on Different Dates

Northern Hemisphere (read down)				
Latitude	March 20	June 21	Sep. 22	Dec. 21
0°	12 hr	12.0 hr	12 hr	12.0 hr
10°	12 hr	12.6 hr	12 hr	11.4 hr
20°	12 hr	13.2 hr	12 hr	10.8 hr
30°	12 hr	13.9 hr	12 hr	10.1 hr
40°	12 hr	14.9 hr	12 hr	9.1 hr
50°	12 hr	16.3 hr	12 hr	7.7 hr
60°	12 hr	18.4 hr	12 hr	5.6 hr
70°	12 hr	2 mo	12 hr	0 hr
80°	12 hr	4 mo	12 hr	0 hr
90°	12 hr	6 mo	12 hr	0 hr
Latitude	Sept. 22	Dec. 21	March 20	June 21
Southern Hemisphere (read up)				

Figure 15–4 The change in the duration of insolation for different latitudes during the different seasons of the year.

EARTH SYSTEM SCIENTISTS *EDMUND HALLEY*

Edmund Halley was born in England in 1656. He studied astronomy at Oxford University, which began a lifelong interest in celestial objects. From 1676 to 1678, using a telescope, he created the first chart of stars in the Southern Hemisphere. Later he became friends with the famous Sir Isaac Newton, whom he greatly admired. Halley traveled throughout Europe, often meeting many of the famous astronomers of the time. He also served as captain aboard a British Navy ship. During this time he formulated theories on the creation of the trade winds, atmospheric pressure, and insolation. He was the first person to propose that the unequal heating of the planet caused the formation of the trade winds. He also researched the change in air pressure that is associated with elevation. His work studying climate helped to reveal the distribution of heat and air around the globe. Halley is most remembered for his work with comets. He used mathematics to predict the orbital period of comets, especially the one that bears his name, Halley's comet.

wave, low-energy infrared radiation back into the atmosphere. This type of radiation is also called **terrestrial radiation.** Gases in the atmosphere absorb this infrared radiation and are heated.

The third way by which the atmosphere is heated is called conduction. Conduction is the transfer of heat by direct molecular contact. Not all the heat energy stored in the rocks at the Earth's surface is reradiated back into the atmosphere. Some of this energy is transferred by conduction through the direct contact of the hot rocks with the gas molecules in the atmosphere.

The fourth process that heats the atmosphere is called **latent heat of condensation.** This occurs when water vapor, which is the gaseous form of water, condenses to form liquid water. This change in phase results in the release of heat into the surrounding environment. When water vapor condenses, it gives off latent heat to the surrounding atmosphere. The absorption or release of heat energy when a change in phase occurs is known as latent heat. The heat energy that is released into the atmosphere when water condenses is gained by water molecules when they change phase from a liquid to a gas, known as evaporation. This change in phase removes heat from the surrounding environment, causing a cooling effect. This is what happens to your body when you sweat. The sweat on your skin evaporates and draws heat away from your body. This heat is then stored in water vapor as it rises into the atmosphere. Eventually,

EARTH MATH

1) WIEN'S LAW (WAVELENGTH IN MICROMETERS = 2897 MICROMETERS/THE SURFACE TEMPERATURE OF AN OBJECT IN DEGREES KELVIN) DESCRIBES THE CALCULATION THAT CONVERTS THE TEMPERATURE OF AN OBJECT IN DEGREES KELVIN INTO THE PEAK WAVELENGTH OF RADIATION IT EMITS IN MICROMETERS. USING WIEN'S LAW, AT WHAT PEAK WAVELENGTH DOES THE SUN EMIT ITS RADIATION IF ITS SURFACE TEMPERATURE IS 6000 DEGREES KELVIN?

2) USING WIEN'S LAW, DETERMINE THE PEAK WAVELENGTH OF RADIATION IN MICROMETERS THAT THE EARTH'S SURFACE EMITS WITH AN AVERAGE TEMPERATURE OF 300 DEGREES KELVIN.

when the water vapor condenses, the heat it contained is released to the surrounding environment as latent heat. Large clouds in the atmosphere that are formed by the condensation of water vapor give off large amounts of latent heat to the air. This helps to form massive storm systems. In this way the heat from your body that causes you to sweat is transferred into the atmosphere and helps form a cloud!

REVIEW

1. Define the term *insolation.*
2. Name the two main forms of radiation emitted from the Sun and provide one example of each form that strikes the Earth.
3. Approximately how much of the radiation received by the Earth is reflected back into space?
4. Approximately how much of the radiation received by the Earth is directly absorbed by the atmosphere?
5. What is the angle of insolation, and why does it vary around the globe?
6. List the four ways by which the atmosphere is heated.
7. Who was Edmund Halley?

For more information go to these Web links:

<http:www.noaa.gov/solar.html>

<http:www.noqa.gov/radiation.html>

CHAPTER SUMMARY

Insolation is the amount of incoming solar radiation received by the Earth. The Sun emits radiation in all wavelengths of the electromagnetic spectrum. Most insolation striking the Earth is high-energy, short-wave radiation in the ultraviolet and visible light range and low-energy, long-wave radiation in the infrared and microwave range. Insolation can interact with the atmosphere in a number of ways. Approximately 19% of the insolation that strikes the Earth is absorbed by the atmosphere, and 26% is reflected back into space by clouds or scattered throughout the atmosphere. The remaining 55% strikes the Earth's surface. Approximately 4% of this insolation is reflected back into space by snow, ice, or calm water. The remaining 51% of the insolation is then absorbed by objects on land or by water. Once this high-energy, short-wave radiation is absorbed at the surface, it eventually gets reradiated back into the atmosphere in a lower-energy, long-wave infrared form.

The angle at which insolation strikes the Earth's surface greatly affects the intensity of the radiation received. Generally the greater the angle of insolation, the greater the intensity of radiation received at the surface. Because the Earth is spherical, its curved surface receives different angles of insolation, which causes some parts of the planet to receive more intense insolation than others. This results in a decrease in the angle of insolation at higher latitudes; that is, it causes areas on the Earth that are located at the higher latitudes to receive insolation at a lower intensity. Areas near the equator receive insolation at the highest angle and therefore receive the most intense insolation on the planet.

The duration of insolation is also important; this refers to how long the Sun is shining on one spot on the Earth. Because the Earth's axis of rotation is tilted 23.5 degrees, different parts of the planet receive varying amounts of sunlight during different seasons of the year. When winter arrives in the Northern Hemisphere, the duration of insolation is very short. Six months later, during summer, the duration of insolation is greatest. Both latitude and season of the year greatly affect insolation, which in turn influences weather and climate on the Earth. This is due to an increase in surface temperatures as a result of the increasing angle or duration of insolation. Insolation heats the atmosphere in four principal ways. These include the direct absorption of short-wave radiation from the Sun; absorption of long-wave radiation reradiated from the Earth's surface; conduction; and latent heat of condensation.

CHAPTER REVIEW

Multiple Choice

1. Electromagnetic energy that reaches the Earth from the Sun is called:
   a. insolation
   b. conduction
   c. specific heat
   d. terrestrial radiation

2. Approximately how much insolation is reflected back into space?
   a. 19%
   b. 30%
   c. 51%
   d. 75%

3. Which type of electromagnetic energy radiated from the Sun does the Earth's surface receive most?
   a. infrared
   b. ultraviolet
   c. gamma rays
   d. visible light

4. Approximately what percentage of insolation is absorbed by the Earth's surface?
   a. 19%
   b. 30%
   c. 51%
   d. 75%

5. Rocks heat and cool quicker than water because they:
   a. are harder
   b. have a higher heat capacity
   c. have a lower heat capacity
   d. are less dense

6. Which substance would absorb the greatest amount of radiation in the shortest amount of time?
   a. a white rock
   b. a black rock
   c. a cup of water
   d. a glacier

7. During which time of the year is the angle of insolation greatest at 45 degrees north latitude?
   a. winter
   b. spring
   c. summer
   d. fall

8. Generally as latitude increases, the angle of insolation:
   a. decreases
   b. increases
   c. stays the same
   d. varies

9. As its angle decreases, the intensity of insolation:
   a. remains the same
   b. decreases
   c. varies
   d. increases

10. As latitude decreases, the angle of insolation:
    a. remains the same
    b. decreases
    c. varies
    d. increases

11. During what time of the year is the duration of insolation longest in the Northern Hemisphere?
    a. winter
    b. spring
    c. summer
    d. fall

12. Which latitude would generally receive the greatest amount of insolation?
    a. 90 degrees
    b. 45 degrees
    c. 23.5 degrees
    d. 0 degrees

13. As the angle of insolation decreases, the surface temperature generally:
    a. remains the same
    b. decreases
    c. varies
    d. increases

14. Long-wave radiation emitted from the Earth's surface is also called:
    a. insolation
    b. conduction
    c. specific heat
    d. terrestrial radiation

15. Which method of heating the Earth's atmosphere involves a phase change?
    a. conduction
    b. latent heat
    c. radiation
    d. absorption

16. Which method of heating the atmosphere involves direct molecular contact?
    a. conduction
    b. latent heat
    c. radiation
    d. absorption

Matching

Match the terms with the correct definitions.

a. insolation
b. short-wave radiation
c. long-wave radiation
d. albedo
e. heat capacity
f. reradiate
g. angle of insolation
h. duration
i. terrestrial radiation
j. latent heat of condensation

1. ____ The process by which an object takes in electromagnetic radiation and re-emits it into the atmosphere.
2. ____ Heat energy that is released into the atmosphere by the condensation of water vapor.
3. ____ The ability for a substance to absorb, contain, and release heat energy.
4. ____ The specific angle at which incoming solar radiation strikes the Earth's surface.
5. ____ The amount of incoming solar radiation on the Earth.
6. ____ The length of time that it takes for an event to occur.
7. ____ A high-energy, short-wavelength form of electromagnetic energy.
8. ____ Long-wavelength infrared radiation that is reradiated into the Earth's atmosphere from the land surface.
9. ____ A low-energy form of electromagnetic radiation such as infrared or radio waves.
10. ____ The reflective ability of an object or surface.

Critical Thinking

1. Some astronomers have proposed that the ice caps on Mars that contain both frozen water and carbon dioxide should be spray-painted black. Why do you think they believe this is a good idea?

CHAPTER 16

Atmospheric Temperature and Pressure

Section 16.1 – Atmospheric Temperature Objectives

Temperature in the Atmosphere • Distribution of Heat on Earth • Radiative Cooling • The Greenhouse Effect

After reading this section you should be able to:

- Define the term *temperature* and differentiate between the molecular motions associated with cool temperatures and warm temperatures.
- Identify the freezing and boiling points of water in degrees Fahrenheit and Celsius.
- Describe three factors that cause heat to be unequally distributed on the Earth.
- Explain the way that heat is distributed in a convection cell.
- Describe the process of radiative cooling.
- Explain the greenhouse effect and identify three greenhouse gases.

Section 16.2 – Atmospheric Pressure Objectives

Pressure in the Atmosphere • Measuring Atmospheric Pressure • High and Low Atmospheric Pressure • Atmospheric Pressure and Moisture

After reading this section you should be able to:

- Define the term *atmospheric pressure*.
- Identify the two units commonly used to measure atmospheric pressure.
- Describe the processes that form low and high atmospheric pressure on the Earth.
- Explain how atmospheric moisture affects air pressure.

TERMS TO KNOW

temperature	**convection cell**	**barometer**
thermometer	**radiative cooling**	**adiabatic cooling**
freezing point	**greenhouse effect**	**atmospheric moisture**
boiling point	**greenhouse gases**	
heat gradient	**atmospheric pressure**	

INTRODUCTION

The narrow temperature range that exists on Earth enables our planet to support water in all three states of matter. This is unique among all the planets of the solar system and allows life to flourish on the Earth. The changes in temperature and pressure that occur on Earth are important indicators of changing weather. Today, because of our understanding of the relationship between temperature and pressure in the atmosphere, it is possible to provide better short-term and long-term forecasts. As a result, human society is able to better adapt to our planet's dynamic environment.

16.1 *Atmospheric Temperature*

Temperature in the Atmosphere

The temperature on the Earth is greatly dependent on the atmosphere. **Temperature** is a measure of the average speed of atoms or molecules. The greater the speed of the molecules or atoms in a substance, the higher the temperature. When you go outside on a warm summer day, you are actually sensing the effects of rapidly moving atoms and molecules in the air. On a cold winter day, the movement of the molecules is much slower and you experience a lower temperature. The movement of the molecules is also known as kinetic energy, or the energy of movement. Generally the greater the kinetic energy, the higher the temperature. You sense heat as a result of the kinetic energy of the surrounding air. The faster the movement of the molecules, the higher the kinetic energy and therefore the greater the amount of heat generated.

Temperature is measured using an instrument called a **thermometer.** Today many electronic thermometers are used to measure the kinetic energy of a substance; however, the first widespread thermometers were made of glass and mercury. Mercury is a metal that is in its liquid form at temperatures above −35° Fahrenheit. In the original thermometers the mercury was kept in a small tube, and it expanded or contracted when it was exposed to different temperatures. The tube could then be calibrated using a scale, and temperatures could be recorded. In the United States, temperature is commonly recorded using the Fahrenheit scale, which was introduced in 1714 by the German scientist Gabriel Fahrenheit. This temperature scale is based on the **freezing point** of water being 32° and the **boiling point** of water being 212°. In 1742 a Swedish astronomer by the name of Anders Celsius derived a new type of temperature scale based on 0° marking the freezing point of water and 100° marking the boiling point of water. Today almost every country in the world uses the Celsius scale, although the United States still clings to the Fahrenheit scale. During the nineteenth century in England, Lord Kelvin created another temperature scale, which was based on the actual kinetic energy of the atoms or molecules. The Kelvin scale begins at absolute zero, which is the theoretical temperature at which all molecules stop moving. Using the Kelvin scale, the freezing point of water occurs at 273 Kelvins and the boiling point of water is 373 Kelvins (Figure 16–1).

Distribution of Heat on Earth

As we saw in Chapter 15, the Earth is unequally heated by differences in insolation received at the surface. This is mainly the result of the Earth's tilted axis and nearly spherical shape. The length of time that the Earth's surface is exposed to solar radiation also changes throughout the year, depending on the season. Another factor that affects the unequal distribution of solar radiation on the Earth is the rotation of the Earth itself. Half the Earth receives sunlight while the other half is bathed in darkness. All these factors lead to the unequal distribution of heat on the Earth, creating what is called a **heat gradient,** which is simply the change in temperature over a specific distance. A heat gradient exists on Earth between the poles, where temperatures are low, and the equator, where temperatures are high. A heat gradient also exists between the higher temperatures at the Earth's surface and the cooler temperatures high in the troposphere.

Temperature gradients on Earth result in the flow of heat from areas of high temperature to areas of low temperature. It is the flow of heat in a temperature gradient that leads to the distribution of heat by radiation, conduction, and convection. Recall that radiation is

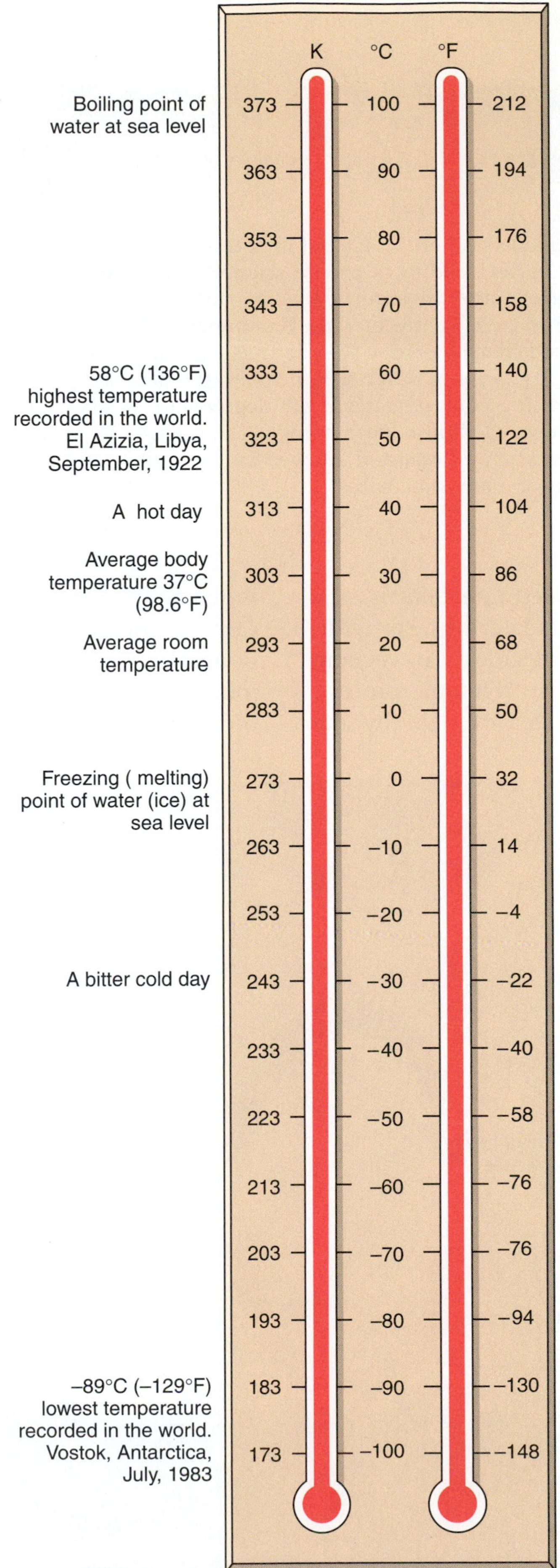

Figure 16–1 A comparison of the three principal temperature scales.

EARTH SYSTEM SCIENTISTS *DANIEL GABRIEL FAHRENHEIT*

Daniel Gabriel Fahrenheit was born in Poland in 1686 but lived most of his life in Amsterdam. He became interested in the building of precise scientific instruments. He traveled Europe for many years, learning from other instrument makers, and eventually returned to Amsterdam to produce his own instruments. He invented the first mercury glass thermometer in 1714. He then used his thermometer to create a precise scale for the accurate measurement of temperature. His scale marked the freezing point of water at 32 degrees, the heat of the body at approximately 90 degrees, and the boiling point of water at 212 degrees. Today his name still identifies the name of the scale he devised almost 300 years ago. The creation of an accurate thermometer was a great advancement to science because it allowed for the study of the effects of temperature on the physical properties of many substances.

the transfer of heat by electromagnetic waves, and conduction is the transfer of heat by direct contact between molecules. Convection, however, is a much more dynamic process. Convection is the transfer of heat by the movement of a fluid such as air or water. Convection begins when a portion of the fluid is heated in some way. The heat causes the molecules of the fluid to become less dense; therefore the fluid begins to rise. The heated,

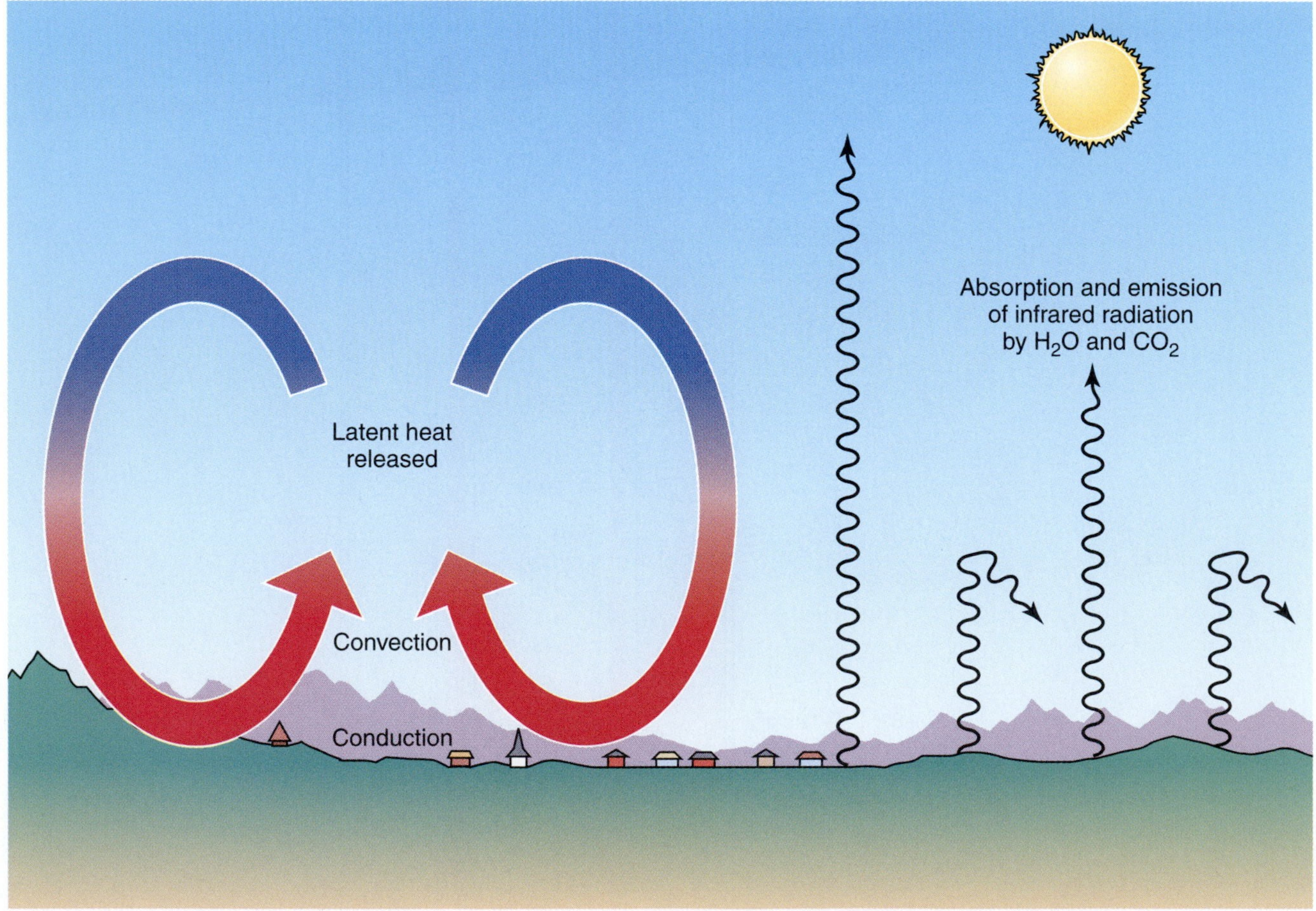

Figure 16–2 The formation of convection cells in the atmosphere helps to distribute heat around the planet.

Figure 16–3 Radiative cooling of the atmosphere occurs rapidly on a clear night as heat is lost to space. Cloudy nights help to trap heat in the atmosphere and keep it relatively warm.

rising fluid then begins to cool as its heat is dissipated. The cooling fluid becomes more dense and begins to sink downward again. The sinking fluid pushes under the warmer fluid and causes it to rise, and the process repeats itself. The result is a **convection cell,** which is formed by the circular movement of hot rising fluid and descending cooler fluid (Figure 16–2). Convection can occur in the Earth's air, water, and mantle. Convection that occurs in the troposphere is responsible for the distribution of heat around the Earth.

Radiative Cooling

Another process that affects the transfer of heat in the atmosphere is called **radiative cooling.** During the daytime, insolation received by the Earth's surface is absorbed, causing the temperature of the surface to rise. As insolation continues throughout the day, the surface is continually heated; the heat is then either conducted or radiated back into the atmosphere. This process continually heats the atmosphere during the day. As a result, the warmest temperatures during the day tend to occur around 3:00 P.M. At night, even though insolation stops, the Earth's surface continues to heat the atmosphere by radiating and conducting its stored heat. Eventually the surface cools and the heating of the atmosphere stops. The atmosphere then begins to cool as it radiates its heat back into space. Radiative cooling can result in the rapid decline of the temperature in the atmosphere. Extreme radiative cooling occurs on cloudless nights (Figure 16–3). Thus the coldest temperatures occur at night during winter. Cloudy nights can trap some of the heat that is being radiated back into space by

Career Connections

Atmospheric Scientist

An atmospheric scientist conducts specific research dealing with the unique aspects of the Earth's atmosphere. This includes the physical and chemical properties of the gases that surround the planet. Some atmospheric scientists also research the atmospheric characteristics of other planets within the solar system. Today much of the work of an atmospheric scientist involves the study of human-created pollutants on Earth. The effects of atmospheric pollution have begun to interrupt the natural balance of the Earth's ecosystems. Atmospheric scientists are researching the ways these pollutants are interacting with the atmosphere and how they are altering its chemistry. Major topics such as global warming and ozone destruction are an important part of the work of atmospheric scientists, who are searching for ways to lessen the impact of human beings on the atmosphere. Jobs in the atmospheric sciences can be found in government agencies, such as the National Oceanic and Atmospheric Administration (NOAA) or the National Aeronautics and Space Administration (NASA), and in academic research. Atmospheric scientists must have at least a 4-year college degree.

acting as a blanket that slows down the cooling of the atmosphere. Therefore cloudy nights are usually much warmer in winter than cloudless nights.

The Greenhouse Effect

High-energy, short-wave radiation penetrates the atmosphere to heat the Earth's surface. The surface then reradiates this energy back into the atmosphere in a low-energy, long-wave form, called infrared radiation. The infrared radiation can then be trapped by some of the gases in the atmosphere, causing its temperature to rise. This process that heats our planet is often referred to as the **greenhouse effect,** because the same process also occurs in a greenhouse (Figure 16–4). The short-wave infrared radiation that the Sun emits travels through the clear glass of a greenhouse and is absorbed by the solid floor and walls inside. These solid materials then reradiate long-wave (infrared) radiation back into the air of the greenhouse, which is trapped by the glass. The glass's trapping the long-wave infrared radiation can cause the air temperature inside the greenhouse to rise rapidly. The same process can also occur in a car. Short-wave radiation penetrates the windows of a car and is absorbed by the seats and

Earth Math

1) During the month of January the average temperature at the North Pole is −50 degree Fahrenheit and the average temperature at the equator is 80 degrees Fahrenheit, with a distance of 6250 miles between them. Determine the temperature gradient in the Northern Hemisphere using the following formula: gradient = change in temperature/change in distance.

2) A temperature reading of −52 degrees Fahrenheit is recorded at a height of 19 miles above sea level in the stratosphere, and at sea level the temperature is recorded at 52 degrees Fahrenheit. Using the gradient calculation from Question 1, determine the temperature gradient from the stratosphere to the Earth's surface.

3) Temperature values often need to be converted from one unit to another. Use the following formulas to convert 68 degrees Fahrenheit to Celsius and Kelvin and 37 degrees Celsius to Fahrenheit and Kelvin:

$F = 9/5 \times \text{Celsius} + 32$ $C = 5/9 \times (F-32)$ $K = 5/9 \times F + 255.36$

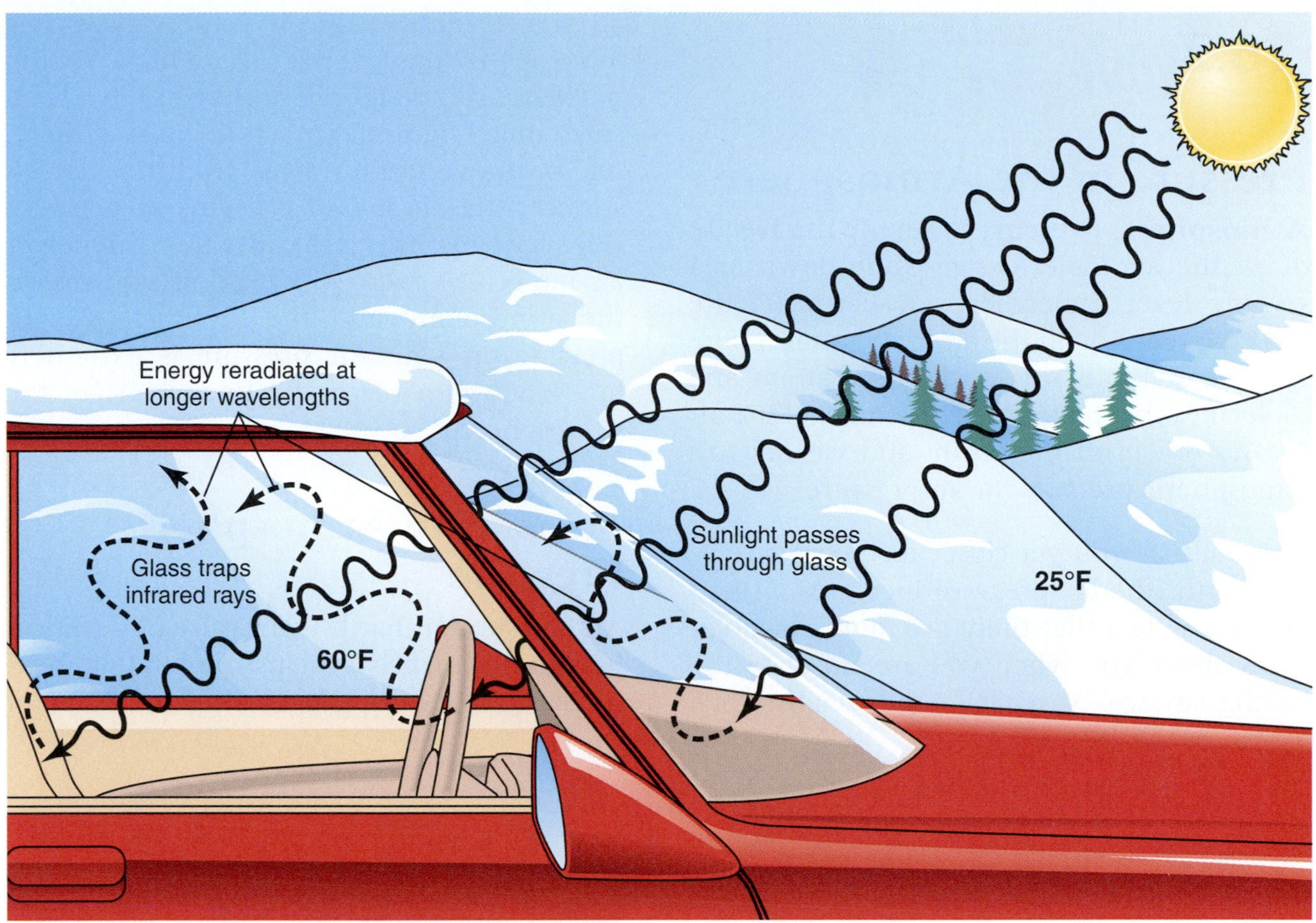

Figure 16–4 The greenhouse effect helps to heat the Earth in the same way that it heats the inside of a car.

dashboard. The seats and dashboard then reradiate long-wave radiation into the car and it is trapped by the glass, causing the air temperature inside the car to raise rapidly.

Gases in the atmosphere act like the glass in a greenhouse to help trap the long-wave radiation, therefore heating the planet. These type of gases are also known as **greenhouse gases.** Greenhouse gases in the Earth's atmosphere include water vapor, carbon dioxide, and methane. The concentrations of some of these gases are increasing in the atmosphere, which may be causing the temperature of the atmosphere to rise over time.

For more information go to this Web link:
<http://www.noa.gov/climate.html>

SECTION REVIEW

1. Define the term *temperature.*
2. What are the three scales used to measure temperature?
3. Describe the two types of temperature gradients that occur on Earth.
4. Briefly explain the process by which convection distributes heat.
5. What is radiative cooling?
6. Describe the greenhouse effect.
7. Who was Daniel Gabriel Fahrenheit?

16.2 *Atmospheric Pressure*

Pressure in the Atmosphere

Atmospheric pressure is simply the weight of all the air molecules pressing down at a specific level somewhere on Earth (Figure 16–5). Even though the air in the atmosphere appears invisible, it is made up of atoms and molecules that have mass. The mass of the air presses down on the Earth, and we call this atmospheric pressure, or air pressure.

Generally as you get closer to the surface of the Earth, the air pressure increases. This is due to the fact that there are more air molecules above you when you are closer to the Earth's surface. At sea level the weight of the air above you presses on your body with a force of 14.7 pounds per square inch. You do not sense this weight because you have been surrounded by it all your life, but it is there. As you move farther away from the Earth's surface, the air pressure begins to decrease (Figure 16–6). This is because there are fewer air molecules over your head as you rise in the atmosphere and therefore less weight pressing on you. Air pressure is extremely low when you reach the upper part of the stratosphere.

Measuring Atmospheric Pressure

Air pressure on Earth is measured by using a **barometer,** which is an instrument that measures the weight of the air at a specific point

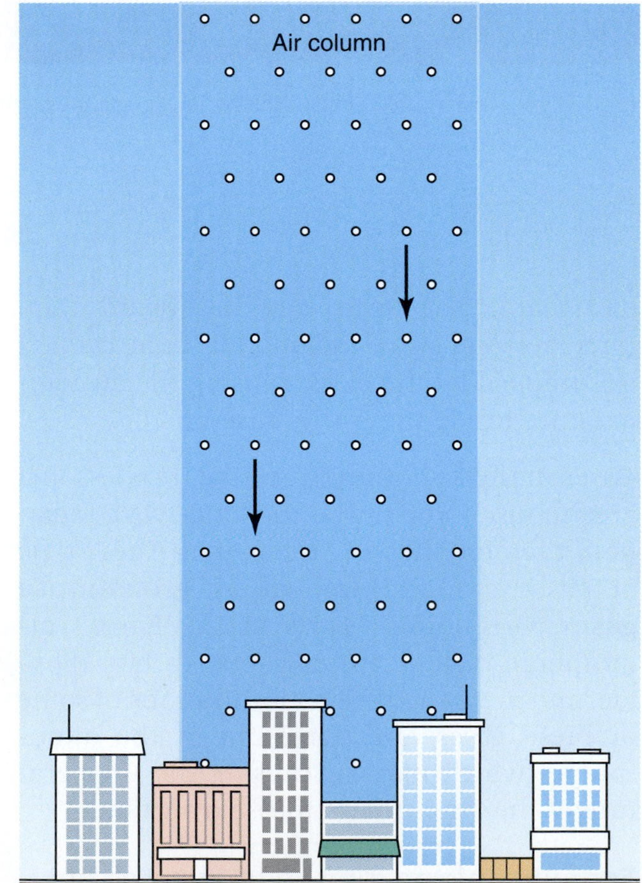

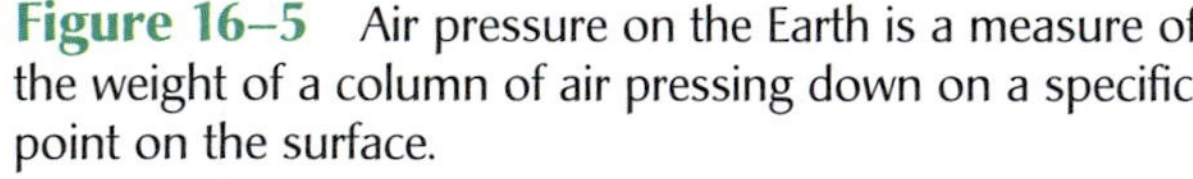
Figure 16–5 Air pressure on the Earth is a measure of the weight of a column of air pressing down on a specific point on the surface.

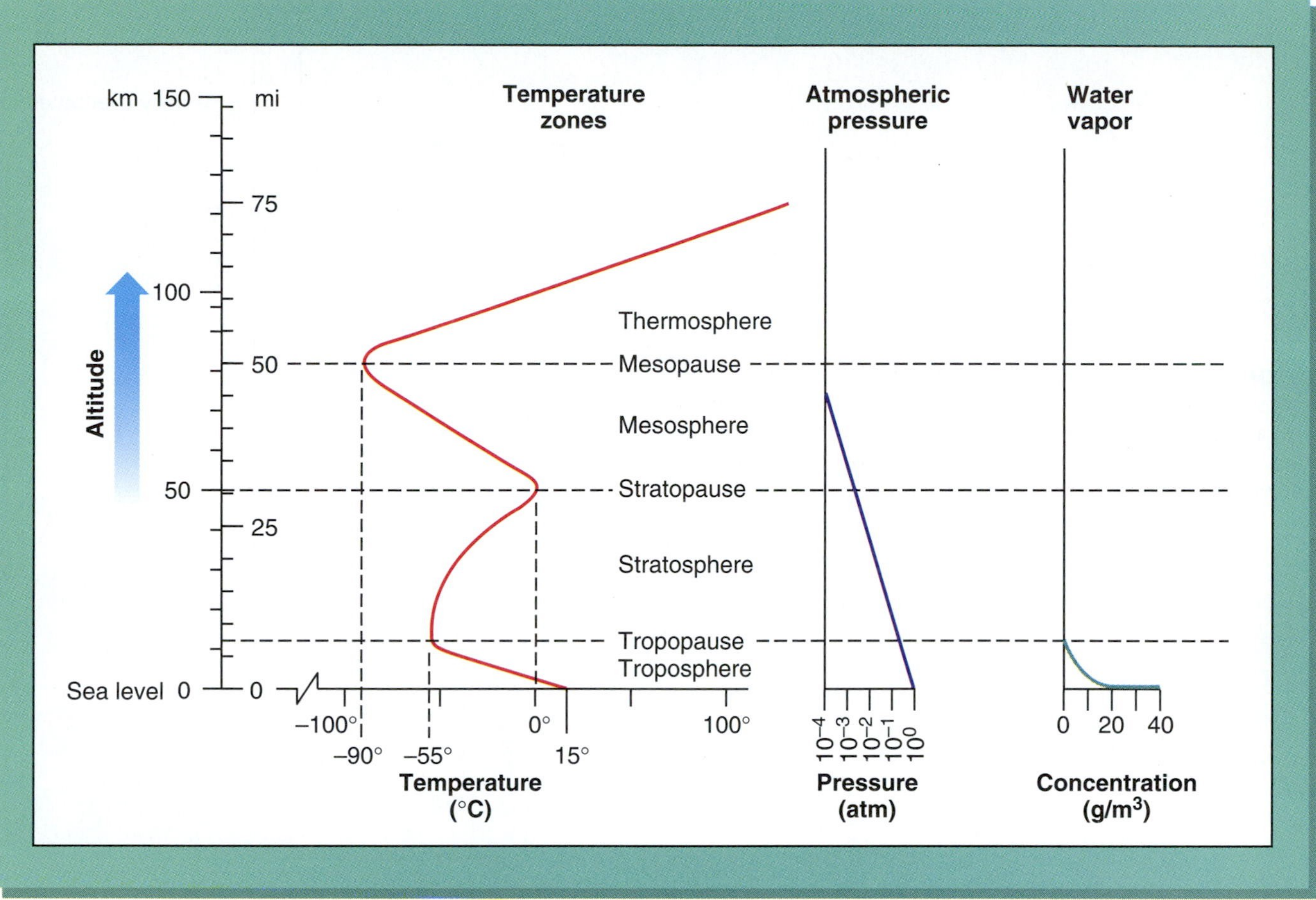

Figure 16–6 Air temperature, pressure, and water vapor decrease with an increase in height in the atmosphere.

on the Earth. Air pressure is also known as barometric pressure. Barometric pressure is measured using two types of units: inches of mercury and millibars. Inches of mercury is the oldest unit used to measure barometric pressure (Figure 16–7). It was derived by the use of a barometer that contained mercury to record the changes in air pressure. As the air pressure increased, it pressed down on a dish that held a pool of mercury inside a glass tube. The increased air pressure caused the mercury to rise in the tube, and the change in air pressure was recorded as the inches of mercury that the liquid metal occupied in the tube. The mercury barometer was invented in 1643 by an Italian scientist named Torricelli.

The average air pressure at sea level, also known as 1 standard atmosphere, is 29.92 inches of mercury. Today there are many different types of barometers that record atmospheric pressure without the need for mercury. The most common unit of measurement to record air pressure is the millibar. This is the form of barometric pressure used in meteorology and weather forecasting.

High and Low Atmospheric Pressure

Atmospheric pressure on Earth is often influenced by the air temperature. Warm air that is heated near the Earth's surface begins to

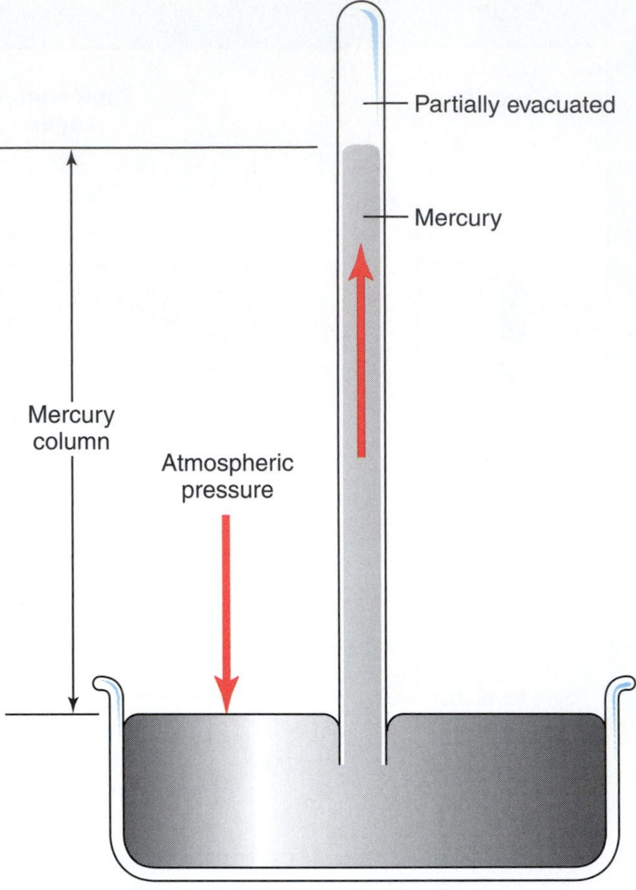

Figure 16–7 A mercury barometer measures the air pressure as a result of the weight of the air pushing down on a pool of mercury, causing it to rise or lower in the tube. This type of barometer was invented by Evangelista Torricelli in 1643.

EARTH SYSTEM SCIENTISTS

EVANGELISTA TORRICELLI

Evangelista Torricelli was born in Italy in 1608. He studied mathematics in Rome and eventually became the personal secretary to Galileo. After Galileo's death, he became a professor of mathematics in Florence, Italy. Torricelli began to work on the concept that the air that made up the atmosphere must have weight. In 1643 he invented a device to measure the weight of the air. He took a long glass tube that was closed on one end and filled it with mercury. He then inverted the tube into a pool of mercury and measured the height of the column of liquid metal. He theorized that the weight of the air was pushing on the pool of mercury, causing it to rise into the tube. This device eventually became known as the mercury barometer. Torricelli also noticed that the height of the column often changed from day to day. This observation led to the belief that air pressure also changed daily.

Career Connections

Biometeorologist

A biometeorologist studies the interactions between atmospheric phenomena and living things. This unique branch of atmospheric science attempts to understand the ways in which weather may influence organisms on Earth. Changes in atmospheric pressure and moisture can cause discomfort for many people in the form of aches and pains in their joints or by inducing headaches. Some animals are also very sensitive to changes in the atmosphere. A biometeorologist is interested in researching these phenomena to try to reverse their effects or to use them to better forecast the weather. Biometeorologists require a college education in both meteorology and biology. They often work in academic research.

rise in the atmosphere because warm air is less dense than cooler air. The lifting of warm air that is less dense than the surrounding air and that weighs less creates an area of low pressure at the surface. Low pressure at the surface is usually associated with warmer, lighter, less dense air. When the warm air rises into the atmosphere, it begins to expand and cool. This is called **adiabatic cooling,** which is the cooling of air as a result of expansion (Figure 16–8).

When the rising warm air begins to cool, the cooler air becomes more dense and begins to sink back toward the Earth's surface. The area on the Earth's surface where cool, dense air sinks downward experiences high atmospheric pressure. Therefore warm air is generally associated with low pressure and cooler air associated with high pressure (Figure 16–9). Meteorologists identify areas of low and high pressure on weather maps because they can be used to help forecast the weather. Areas of low pressure on a weather map are marked with a red capital L. Areas of high pressure on weather maps are marked with a blue capital H. Generally, low pressure indicates poor weather and high pressure indicates good weather.

Atmospheric Pressure and Moisture

Air pressure is also influenced by the amount of water vapor present in the air, also known as **atmospheric moisture.** Generally, the more moisture in the air, the less dense it is. The water molecule weighs much less than the molecules of nitrogen and oxygen that compose most of the atmosphere. When the

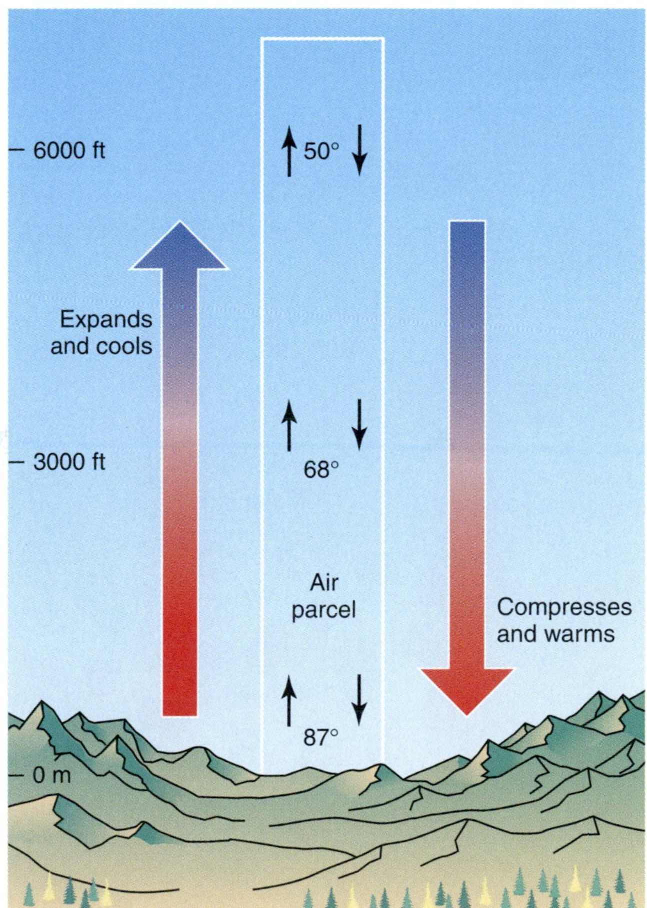

Figure 16–8 As air rises into the atmosphere, its pressure decreases, causing it to expand and cool. This is known as adiabatic cooling.

Falling air
Northern Hemisphere
Rising air
High Pressure
Low Pressure
Coriolis force effect

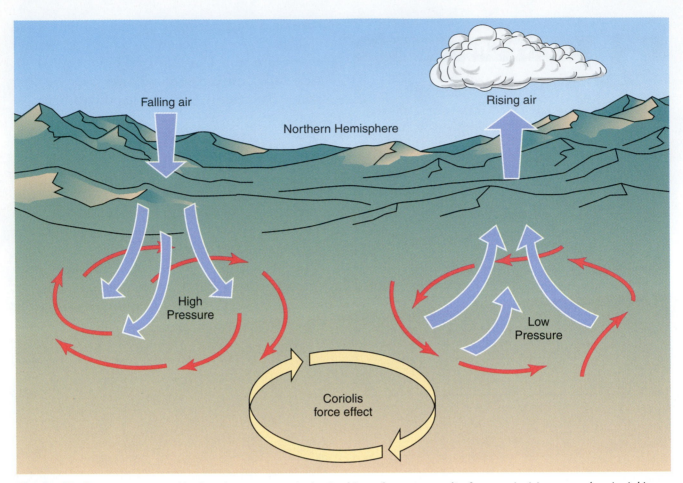

Figure 16–9 The creation of high or low pressure at the Earth's surface as a result of warm air rising or cooler air sinking.

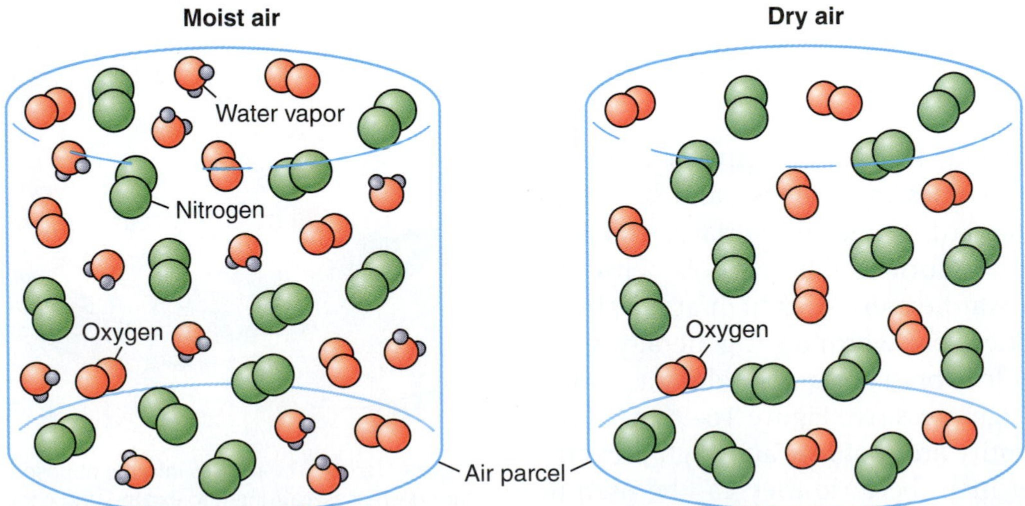

Figure 16–10 Moist air is lighter and lower in pressure than dry air as a result of the water molecules weighing less and displacing the heavier molecules of nitrogen and oxygen.

EARTH MATH

1) BAROMETRIC PRESSURE IS MEASURED IN BOTH MILLIBARS AND INCHES OF MERCURY. IF 1 INCH OF MERCURY EQUALS 33.865 MILLIBARS, HOW MANY MILLIBARS EQUAL 29.92 INCHES OF MERCURY?

air has a higher moisture content, the water molecules displace the nitrogen and oxygen and cause the air to become lighter and less dense (Figure 16–10). The lighter, less dense moist air rises away from the Earth's surface, resulting in lower atmospheric pressure. The opposite effect occurs with drier air, which contains less atmospheric moisture. Drier air weighs more and is denser than moist air, and therefore it tends to sink toward the Earth's surface. This forms areas of high atmospheric pressure.

SECTION REVIEW

1. Define the term atmospheric pressure.
2. How is atmospheric pressure measured?
3. Describe the characteristics of low-pressure air.
4. Describe the characteristics of high-pressure air.
5. What type of general weather conditions are associated with low pressure and with high pressure?
6. Who was Torricelli?

For more information go to this Web link:
<http://kids.earth.nasa.gov/archive/air_pressure/>

CHAPTER SUMMARY

The surface temperatures on the Earth depend on the atmosphere. Temperature is a measure of the average kinetic energy of a substance. Molecules of air that are moving relatively slowly have a cooler temperature. Rapidly moving molecules of air are higher in temperature. The specific temperature of a substance is measured using a thermometer. This scientific instrument records the temperature in degrees Fahrenheit, degrees Celsius, or Kelvins. The Earth's tilted axis, rotation, and spherical shape cause it to be unequally heated by the Sun. This results in specific heat gradients on the planet, which creates unique climate and weather. A heat gradient is the specific change in temperature over a specific distance.

Heat is distributed throughout the atmosphere in three ways: radiation, conduction, and convection. The unequal heating of the Earth helps to form large-scale convection cells, which distribute heat around the planet. Heat can also be lost from the atmosphere, which cools the planet. This is known as radiative cooling. On a clear night the atmosphere can radiate its heat into space, causing rapid cooling. The coldest nights on the Earth usually occur on clear nights as a result of radiative cooling.

Heat is trapped on the Earth by a mechanism known as the greenhouse effect. The atmosphere acts like the glass in a greenhouse, which traps the long-wave infrared radiation that is reradiated into the atmosphere from the land surface. Atmospheric gases that trap this heat are known as greenhouse gases. These include water vapor, carbon dioxide, and methane.

Air pressure on the Earth is caused by the weight of the molecules of air pressing down on the surface. This is also known as barometric pressure. Generally the weight of the air at the Earth's surface is approximately 14.7 pounds per square inch. The pressure in the atmosphere decreases with an increase in height, because there are fewer air molecules as you move away from the surface. The pressure of the atmosphere is measured by using a scientific instrument called a barometer. Atmospheric pressure is usually recorded in either inches of mercury or millibars.

The temperature of the air influences air pressure on the Earth. Warm air tends to be less dense and therefore rises away from the surface. This helps to form an area of low pressure at the surface. Cooler air is more dense and sinks toward the surface. This helps to form an area of high pressure at the surface. Changes in air pressure are good indicators of changes in the weather. Low air pressure is usually associated with poor weather and high pressure with fair weather. The amount of water in the air, also known as atmospheric moisture, can influence the air pressure. Moist air tends to be lower in air pressure than drier air.

CHAPTER REVIEW

Multiple Choice

1. As the temperature of the atmosphere increases, the average kinetic energy of the air molecules:
   a. decreases
   b. stays the same
   c. increases
   d. varies

2. At what temperature does water freeze in degrees Celsius?
   a. 0 degrees
   b. 32 degrees
   c. 100 degrees
   d. 212 degrees

3. What is the equivalent temperature in degrees Fahrenheit of 21° Celsius?
   a. −4 degrees
   b. 65 degrees
   c. 70 degrees
   d. 37 degrees

4. A weather balloon is released into the atmosphere. What changes will it record as it increases its altitude?
   a. increasing air temperature and pressure
   b. increasing air temperature and decreasing pressure
   c. decreasing air temperature and pressure
   d. decreasing air temperature and increasing pressure

5. The Earth's tilted axis, rotation, and curved surface all cause:
   a. tides on the Earth
   b. wind on the Earth
   c. convection cells on the Earth
   d. unequal heating of the Earth

6. Which type of radiation is trapped by greenhouse gases?
   a. short wave
   b. ultraviolet
   c. long wave
   d. visible light

7. Which of the following is not a greenhouse gas?
   a. water vapor
   b. oxygen
   c. methane
   d. carbon dioxide

8. The process by which the atmosphere cools on a clear night is called:
   a. conduction cooling
   b. convection cooling
   c. adiabatic cooling
   d. radiative cooling

9. The process of air expanding as it rises is called:
   a. conduction cooling
   b. convection cooling
   c. adiabatic cooling
   d. radiative cooling

10. As the temperature of the air increases, its density:
    a. decreases
    b. stays the same
    c. increases
    d. varies

11. An air pressure of 29.65 inches of mercury is equivalent to:
    a. 984.0 millibars
    b. 999.0 millibars
    c. 1001.0 millibars
    d. 1004.0 millibars

12. Which type of atmospheric pressure results from warm, rising air?
    a. low pressure
    b. adiabatic pressure
    c. high pressure
    d. radiative pressure

13. When air loses moisture, air pressure generally:
    a. increases
    b. remains the same
    c. decreases
    d. fluctuates rapidly

14. Which type of weather conditions are usually associated with low-pressure systems:
    a. fair weather
    b. high pressure and clear skies
    c. poor weather
    d. cool temperatures

continued

Matching *Match the terms with the correct definitions.*

a. temperature
b. thermometer
c. freezing point
d. boiling point
e. heat gradient
f. convection cell
g. radiative cooling
h. greenhouse effect
i. greenhouse gases
j. atmospheric pressure
k. barometer
l. adiabatic cooling
m. atmospheric moisture

1. ____ A scientific instrument used to measure temperature.

2. ____ The amount of water vapor present in the air.

3. ____ The specific temperature at which a substance begins to change its phase from a liquid to a gas, also known as the boiling point.

4. ____ The average amount of kinetic energy of the atoms and molecules in a substance, which is commonly expressed as the degree of hot or cold measured by a thermometer.

5. ____ The process by which rising air is cooled by expansion.

6. ____ The specific temperature at which a liquid changes phase into a solid.

7. ____ A term used to describe short-wave radiation passing through the atmosphere and being absorbed by the Earth's surface, which then reradiates the energy in a long-wave form that is trapped by gases in the atmosphere and heats the planet.

8. ____ An instrument used to measure atmospheric pressure, also called barometric pressure.

9. ____ The change in heat energy over a specific distance.

10. ____ The weight of a column of air at a specific point in the atmosphere, usually measured in millibars or inches of mercury.

11. ____ The circular movement of a fluid caused by a change in temperature and density associated with the transfer of heat.

12. ____ The cooling of an object as a result of its emitting electromagnetic radiation into the atmosphere or into space.

13. ____ Specific gases in the atmosphere that trap long-wave radiation, such as water vapor, methane, and carbon dioxide.

Critical Thinking

1. Explain why hot air balloons are launched early in the morning and not during the middle of the day during fall in the Northern Hemisphere.

CHAPTER 17 Humidity, Clouds, and Precipitation

Section 17.1 – Humidity Objectives

Atmospheric Moisture • Sources of Atmospheric Moisture • Relative Humidity • Dew Point Temperature

After reading this section you should be able to:

- Define the term *humidity* and describe its general relationship with air temperature.
- Describe the three ways by which moisture enters the atmosphere.
- Define the term *relative humidity.*
- Define the term *dew point* and describe how it can be determined.

Section 17.2 – Clouds and Precipitation Objectives

Cloud Formation • Types of Clouds • Formation of Precipitation • Types of Precipitation • Orographic Precipitation

After reading this section you should be able to:

- Define the term *cloud* and explain the process that leads to its formation.
- Describe four ways by which air can be uplifted to form clouds.
- Differentiate between low clouds, middle clouds, high clouds, and fog.
- Describe the process that leads to the formation of precipitation.
- Identify the six types of precipitation and how they are formed.
- Define the term *orographic precipitation.*

TERMS TO KNOW

humidity
water vapor
evapotranspiration
sublimation
dew point
cloud
condensation
condensation nuclei
lapse rate
uplift
convergence
fog
precipitation
updrafts
orographic precipitation

INTRODUCTION

Anyone who has experienced the moist air of a hot summer night has felt the direct effects of humidity. This often uncomfortable state of the atmosphere not only causes human discomfort, but greatly affects the weather. The lack of humidity can also be uncomfortable. For example, in winter your eyes, nose, throat, and skin can dry out as a result of low humidity. Understanding what controls humidity and how it influences the weather and life on Earth is an important part of meteorology.

Clouds in the atmosphere are one of the most visible forms of change in the air. A cloud is formed by billions of tiny water droplets and ice particles that have condensed in the atmosphere. Clouds play an important role in maintaining the Earth's heat balance by reflecting solar radiation back into space and absorbing heat radiated from the planet's surface. They are also an important part of the water cycle, which transports water over long distances and returns it to the surface by forming precipitation. The different forms of clouds in the atmosphere often create beautiful displays across the background of the blue sky and help to signal the type of weather that may be approaching.

17.1 *Humidity*

Atmospheric Moisture

Humidity is a measure of how much water vapor is present in the atmosphere. **Water vapor** is the invisible gaseous form of water, also called atmospheric moisture. When water enters the air as vapor, it coexists with the other gases that make up the Earth's atmosphere. In tropical regions near the ocean, the atmospheric moisture content can be as high as 4%. This means that if you took a sample of air, 4% of it would be composed of water vapor. The ability of water vapor to exist in the atmosphere can depend on the temperature of the air. Generally an increase in air temperature increases the ability of water vapor to exist in the atmosphere, meaning that warmer air has a higher atmospheric moisture level than colder air. This is apparent during different times of the year.

In the colder months of winter, the air has a low atmospheric moisture level and is considered dry. This causes air in winter to dry out the mucous membranes in your eyes and throat. People develop colds during winter because they lose the protective barrier that mucus provides. Many people use humidifiers in their homes during the colder months to increase the atmospheric moisture content of the air they breathe. Cooler air with low atmospheric moisture creates the clear, dry weather that is associated with high atmospheric pressure.

In contrast, warmer air has a much higher atmospheric moisture content and is considered humid. During warm, humid summer days, you can almost feel the moisture in the air. The visibility of the air is also greatly reduced as a result of the high amount of water vapor present in the atmosphere. The water vapor scatters light in the atmosphere, making it difficult to see over long distances. Because warm air is often associated with air that has a high atmospheric moisture content, it also tends to be lower in barometric pressure.

Sources of Atmospheric Moisture

Water vapor in the atmosphere is an important part of the hydrological cycle, and there are three main ways by which water can enter into the air. The main source of atmospheric moisture is evaporation. Evaporation is the process by which liquid changes into a gas. Much of the water that enters the atmo-

EARTH SYSTEM SCIENTISTS *HENRY CAVENDISH*

Henry Cavendish was born in France in 1731 and spent most of his life in England. Born the son of rich parents, Cavendish was free to pursue his interests in mathematics and physics. In 1776 he published his first scientific paper, which reported his discovery of the element hydrogen. He also studied many other properties of gases and proved that water was not an element but a compound of hydrogen and oxygen. In 1783 Cavendish discovered that the composition of the atmosphere was the same all over the globe. He went on to describe the actual composition of the atmosphere as approximately 79% nitrogen, 21% oxygen, and trace amounts of other gases. Cavendish also experimented with electricity and gravity and derived the gravitational constant. Much of Cavendish's work went unpublished and was not discovered until more than 100 years after his death.

Figure 17–1 One of the main sources of atmospheric moisture is the evaporation of water from the ocean surface. *(Courtesy of PhotoDisc.)*

sphere as water vapor is evaporated from the surface of the ocean (Figure 17–1). This is because approximately 70% of the Earth's surface is covered by the oceans. Insolation strikes the ocean and heats the water molecules at its surface, which then causes the water to evaporate into the atmosphere. Another way that water enters the atmosphere is by **evapotranspiration** (Figure 17–2). Evapotranspiration is the movement of water from the soil into a plant's root system and up through the body of a plant; it then evaporates off the surface of the leaves.

Evapotranspiration is an often overlooked source of atmospheric moisture that contributes large amounts of water vapor to the air. Approximately 8800 gallons per day of water is transpired by 2.5 acres of corn plants. Extensive evapotranspiration over large forested areas helps to regulate the local climate. When the forest is removed by logging or for agriculture, a rapid reduction in atmospheric moisture takes place as a result of the loss of evapotranspiration. This can cause a shift in climate that often results in drier weather.

The third way in which water enters the atmosphere is by **sublimation.** Sublimation is the process of ice changing directly into water vapor; it occurs in areas that experience snowfall. When snow is exposed to bright sunlight, it absorbs the radiant energy and begins to sublimate (Figure 17–3). Large amounts of snow can sublimate on a cold, clear day, greatly reducing the depth of the snow.

Relative Humidity

The most common way to measure the amount of water vapor in the atmosphere is to determine the relative humidity. Relative humidity is the ratio of the amount of water vapor in the air compared with its saturation point at a specific temperature and pressure. Saturation is the total amount of water a specific parcel of air can hold. When air becomes

Figure 17–2 Evapotranspiration transports water from the soil, through a plant, and into the atmosphere where it evaporates off the leaf surface.

Figure 17–3 The dramatic effects of the sublimation of a field of snow on a sunny day. *(Courtesy of PhotoDisc.)*

saturated, it can no longer hold any more water vapor, much like a sponge that has absorbed all the water it can hold. Relative humidity is expressed in the form of a percentage. A relative humidity of 80% means that the air contains 80% of the total amount of water vapor it can hold at that specific temperature. When the relative humidity reaches 100%, water must then condense in the air and form precipitation.

Dew Point Temperature

Another term that is related to atmospheric moisture and relative humidity is the **dew point.** The dew point is the temperature to which the air must be cooled for saturation to occur (Figure 17–4). For example, if the dew point is 65° Fahrenheit, when the air temperature reaches 65° Fahrenheit, the air becomes saturated. Both the dew point and the relative humidity can be calculated by using an instrument called a psychrometer.

A psychrometer consists of two thermometers. One thermometer records the dry bulb temperature, which is equal to the current air temperature. The other thermometer records the wet bulb temperature. It is called the wet bulb temperature because the end of thermometer is covered with a damp cloth, which records the cooling temperature of the evaporation of the water from the cloth into the air. When water evaporates, it removes heat energy from the surrounding air, causing it to cool; therefore evaporation is a cooling process. Generally the drier the air, the lower the wet bulb temperature. This is because more water is evaporating from the cloth, therefore causing more evaporative cooling. Air containing more water vapor causes the wet bulb temperature to be higher because less evaporative cooling occurs. The difference between the dry bulb and wet bulb temperatures is called the dew point depression. The dew point depression can then be used to calculate the dew point temperature and the relative humidity. Generally the greater the dew point depression, the drier the air. When the dew point depression becomes smaller, it signifies that the air contains more water vapor. This is often used as an indicator to predict cloud formation and the formation of precipitation. By using the difference between the wet bulb and dry bulb temperatures and a dew point or relative humidity table, you can easily determine the dew point temperature and relative humidity of the surrounding air (Tables 17–1 and 17–2).

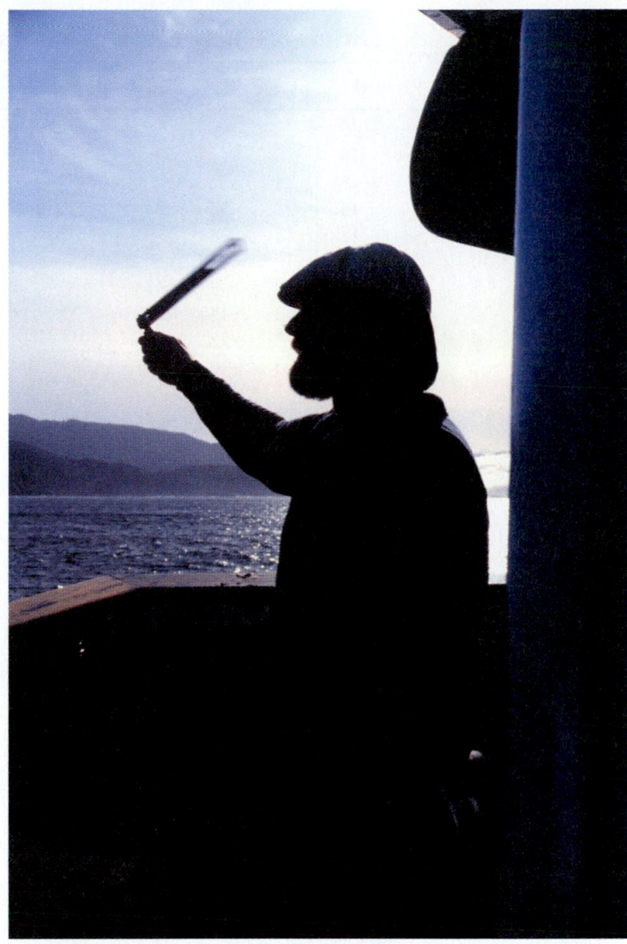

Figure 17–4 A psychrometer determines the wet bulb and dry bulb temperatures, which are used to establish the relative humidity and dew point temperature of the air. *(Courtesy of NOAA.)*

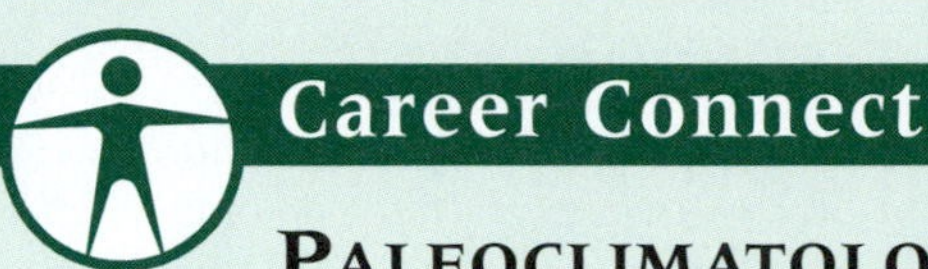

Career Connections

PALEOCLIMATOLOGIST

A paleoclimatologist studies the climate history of the Earth. This includes piecing together the ancient composition of gases in the atmosphere, along with estimating regional temperatures, wind direction, humidity, and precipitation patterns. The work of a paleoclimatologist involves the study of the fossil record, landscapes, and rock formations to search for clues to past climates. These scientists also explore glaciers to learn about past glaciations and ice ages. Paleoclimatologists travel the world in search of evidence that may be used to infer what the Earth's climate was like throughout its history. This research involves the study of ice cores, tree rings, coral reefs, and ocean sediments, which all help to reveal the Earth's past climates. A paleoclimatologist needs a college education in many different scientific disciplines, including paleontology, meteorology, geology, biology, and chemistry. These scientists are often employed by the petroleum and mineral industries, where they help to reveal places on the Earth that once had climates that might have led to oil formation or specific mineral deposits. Paleoclimatologists also conduct academic research and can work for government agencies.

TABLE 17-1 The chart used to determine dew point temperature

Dry Bulb Temperature (°C)	Difference Between Wet Bulb and Dry Bulb Temperatures (C°)															
	0	1	2	3	4	5	6	7	8	9	10	11	12	13	14	15
−20	−20	−33														
−18	−18	−28														
−16	−16	−24														
−14	−14	−21	−36													
−12	−12	−18	−28													
−10	−10	−14	−22													
−8	−8	−12	−18	−29												
−6	−6	−10	−14	−22												
−4	−4	−7	−12	−17	−29											
−2	−2	−5	−8	−12	−20											
0	0	−3	−6	−9	−15	−24										
2	2	−1	−3	−6	−11	−17										
4	4	1	−1	−4	−7	−11	−19									
6	6	4	1	−1	−4	−7	−13	−21								
8	8	6	3	1	−2	−5	−9	−14								
10	10	8	6	4	1	−2	−5	−9	−14	−28						
12	12	10	8	6	4	1	−2	−5	−9	−16						
14	14	12	11	9	6	4	1	−2	−5	−10	−17					
16	16	14	13	11	9	7	4	1	−1	−6	−10	−17				
18	18	16	15	13	11	9	7	4	2	−2	−5	−10	−19			
20	20	19	17	15	14	12	10	7	4	2	−2	−5	−10	−19		
22	22	21	19	17	16	14	12	10	8	5	3	−1	−5	−10	−19	
24	24	23	21	20	18	16	14	12	10	8	6	2	−1	−5	−10	−18
26	26	25	23	22	20	18	17	15	13	11	9	6	3	0	−4	−9
28	28	27	25	24	22	21	19	17	16	14	11	9	7	4	1	−3
30	30	29	27	26	24	23	21	19	18	16	14	12	10	8	5	1

TABLE 17-2 The chart used to determine relative humidity

Dry bulb Temperature (°C)	Difference Between Wet Bulb and Dry Bulb Temperatures (C°)															
	0	1	2	3	4	5	6	7	8	9	10	11	12	13	14	15
−20	100	28														
−18	100	40														
−16	100	48														
−14	100	55	11													
−12	100	61	23													
−10	100	66	33													
−8	100	71	41	13												
−6	100	73	48	20												
−4	100	77	54	32	11											
−2	100	79	58	37	20	1										
0	100	81	63	45	28	11										
2	100	83	67	51	36	20	6									
4	100	85	70	56	42	27	14									
6	100	86	72	59	46	35	22	10								
8	100	87	74	62	51	39	28	17	6							
10	100	88	76	65	54	43	33	24	13	4						
12	100	88	78	67	57	48	38	28	19	10	2					
14	100	89	79	69	70	50	41	33	25	16	81					
16	100	90	80	71	62	54	45	37	29	21	14	7	1			
18	100	91	81	72	64	56	48	40	33	26	19	12	6			
20	100	91	82	74	66	58	51	44	36	30	23	17	11	5		
22	100	92	83	75	68	60	53	46	40	33	27	21	15	10	4	
24	100	92	84	76	69	62	55	49	42	36	30	25	20	14	9	4
26	100	92	85	77	70	64	57	51	45	39	34	28	23	18	13	9
28	100	93	86	8	71	65	59	53	47	42	36	31	26	21	17	12
30	100	93	86	79	72	66	61	55	49	44	39	34	29	25	20	16

EARTH MATH

1) WHICH PARCEL OF AIR CONTAINS MORE WATER VAPOR: PARCEL A WITH A DRY BULB TEMPERATURE OF 73° FAHRENHEIT AND A WET BULB TEMPERATURE OF 57° FAHRENHEIT, OR PARCEL B WITH A DRY BULB TEMPERATURE OF 79° FAHRENHEIT AND A WET BULB TEMPERATURE OF 58° FAHRENHEIT?

SECTION REVIEW

1. Define the term *humidity.*
2. What is the highest percentage of water vapor in the Earth's atmosphere?
3. What type of humidity is associated with low air pressure and with high pressure?
4. Describe the three ways by which water enters the atmosphere.
5. What is relative humidity?
6. Describe how the dew point depression can be used to determine the amount of moisture in the air?
7. Who was Henry Cavendish?

For more information go to this Web link:
<http://ww2010.atmos.uiuc.edu/(Gh)/guides/mtr/cld/home.rxml>

17.2 *Clouds and Precipitation*

Cloud Formation

A **cloud** is a large mass of condensing water droplets and ice crystals that forms in the atmosphere when the air is cooled to its dew point. Remember that the dew point is the temperature of the air at which it becomes saturated. Once air becomes saturated, the water vapor it contains begins to condense. **Condensation** is the process of water vapor changing into a liquid. When water vapor condenses in the atmosphere, it forms tiny droplets of water or ice on small solid particles that float through the air. These microscopic particles are composed of sea salt, ash, dust, and other substances and are called **condensation nuclei.** The small water droplets or ice crystals that form on the condensation nuclei together make up a cloud. The process that cools air to its dew point begins when a parcel of air rises up into the atmosphere. The rising air begins to expand and then cools. This expansion and cooling of rising air is called adiabatic cooling. The rate at which air cools adiabatically in the atmosphere is called the **lapse rate** (Figure 17–5). The rate at which a moist parcel of air cools as it rises and expands is known as the moist adiabatic lapse rate, which is approximately 3.3° Fahrenheit for every 1000 feet of elevation. The moist adiabatic lapse rate helps to determine the altitude at which a cloud forms. For example, a parcel of air at the surface with a temperature of 70° Fahrenheit and a dew point of 60° forms a cloud when it is raised to a height of approximately 3000 feet.

Clouds form when air is forced to rise and cool; therefore it is important to understand the processes that cause air to rise (Figure 17–6). Rising air is also known as **uplift.** The warming of air from the heat radiated by the Earth's surface is one way that air is forced to

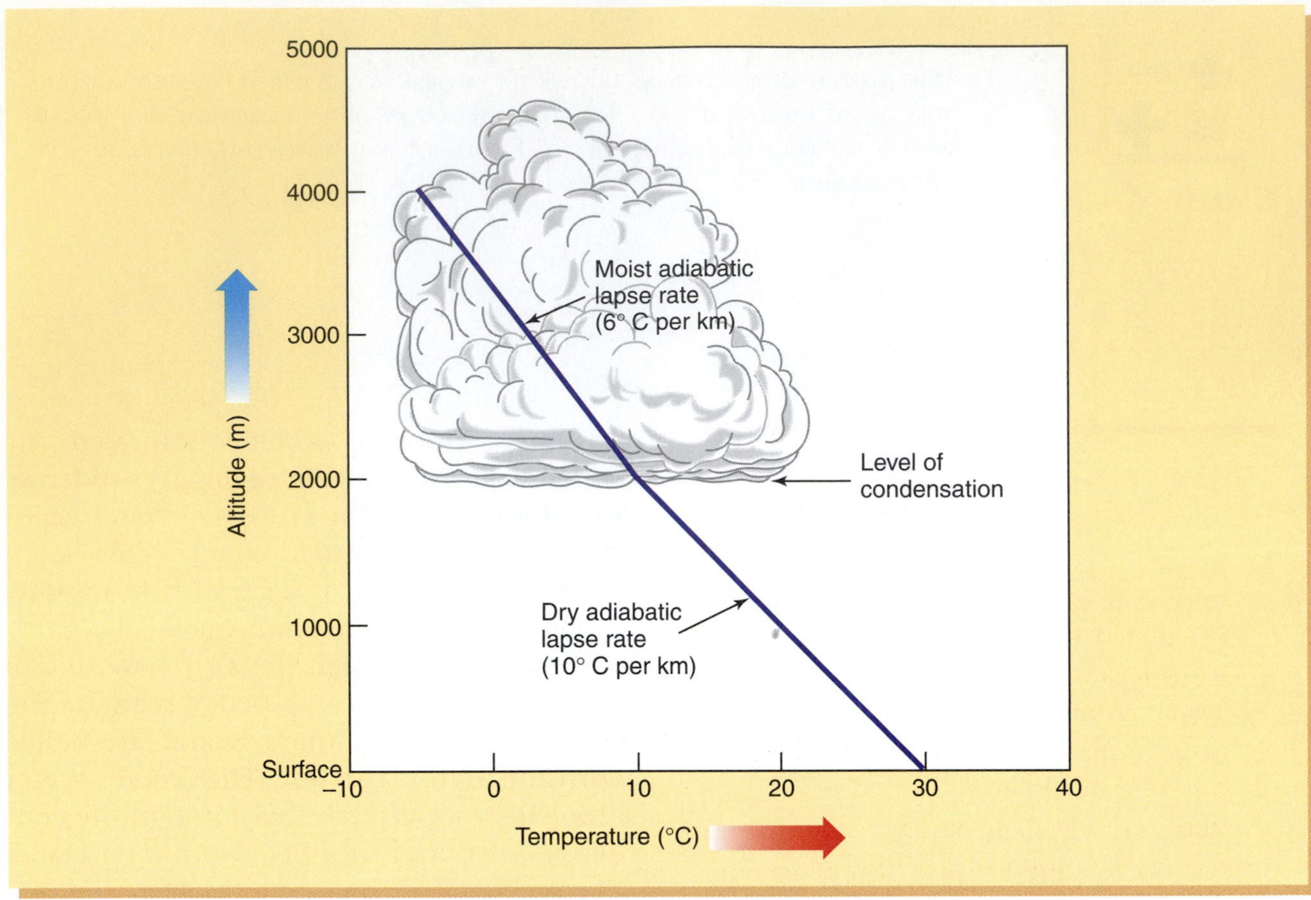

Figure 17–5 The lapse rate measures the decline in temperature with an increase in altitude of a parcel of air. The lapse rate can be used to determine at what height clouds are likely to form.

rise in the atmosphere. Warm air becomes less dense than the air around it and therefore begins to rise. This same process is used to raise hot air balloons into the air. The balloon traps air that is heated by a flame, which raises the temperature of the air inside the balloon. This causes the air to become less dense than the cooler air outside the balloon and forces it to rise upward. In the atmosphere, hot rising air can form massive clouds that produce strong thunderstorms.

Another way that air is forced to rise in the atmosphere to form clouds is by topography. Air that encounters a mountain is forced to rise as the elevation of the land increases (Figure 17–7). This type of lifting process is also called orographic lifting. Orographic lifting often causes the tops of mountains to be hidden in the clouds.

Another type of lifting mechanism that causes air to rise, cool to its dew point, and then form clouds, is the **convergence** of surface winds. Areas on the Earth's surface where winds converge, or come together, can force air upward into the atmosphere. Wind often converges along the shoreline of a large body of water, when fast-moving wind traveling over the relatively smooth water comes into contact with the rough surface of the land. The friction from the rough land surface causes the wind velocity to slow, resulting in a pile-up of air, or a convergence zone. This then can force air upward into the atmosphere.

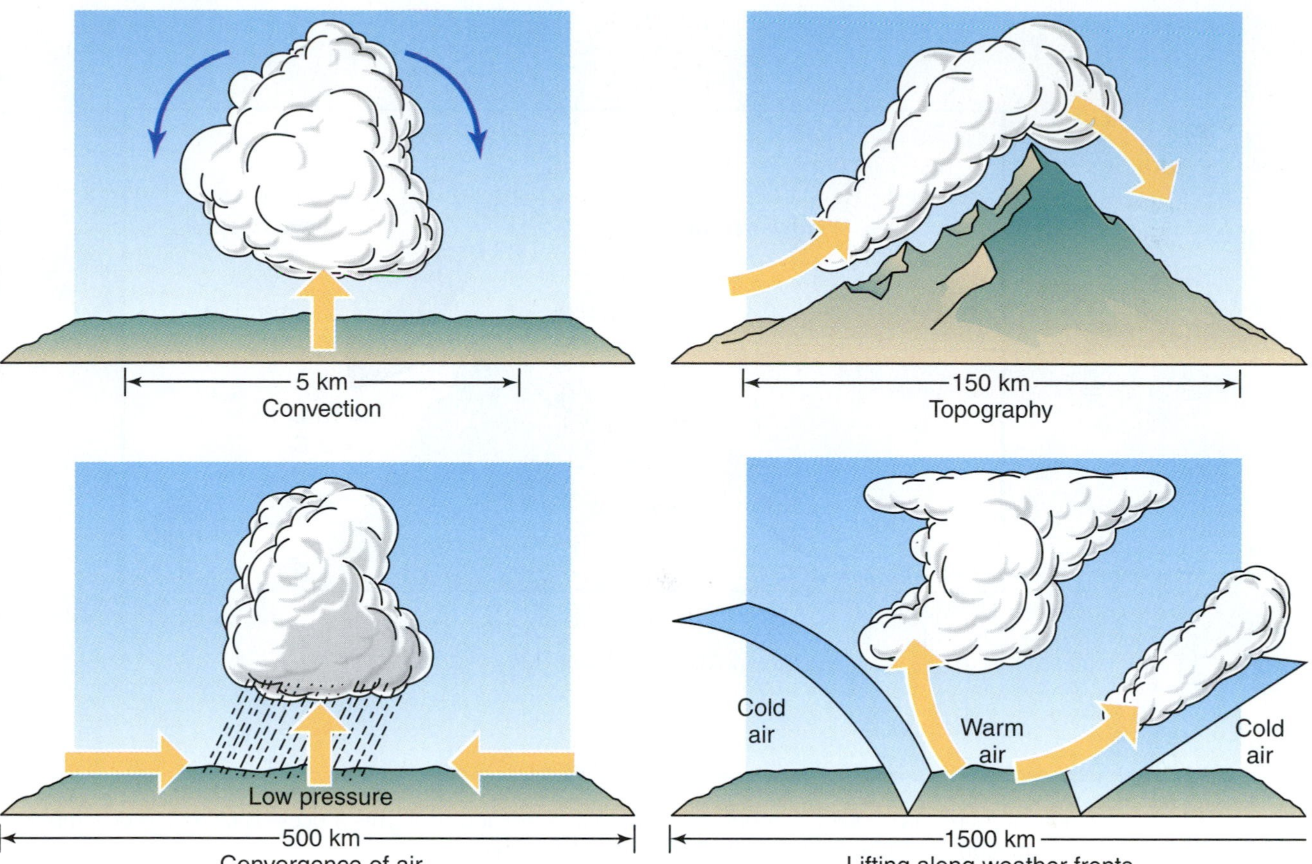

Figure 17–6 A cloud forms as a result of warm air rising and cooling to its dew point, causing water vapor to condense in the atmosphere; this occurs by four different processes.

The last way that uplift causes air to rise and cool is by the action of a cold air mass moving underneath a warm air mass. Because cold air is more dense than warm air, it can wedge itself underneath warmer air and force it to rise. It occurs as cold fronts collide with a warm air mass in front of them. This often result in the formation of clouds.

EARTH SYSTEM SCIENTISTS *JOHN TYNDALL*

John Tyndall was an Irish physicist who was born in 1820. Tyndall became a teacher of mathematics and eventually a professor of physics. In 1869 he discovered that when light passes through certain substances it can be scattered. This led to his conclusion that sunlight passing through the atmosphere was scattered with invisible particles, causing the sky to appear blue. Tyndall also worked on the interaction of specific gases with radiant energy. He proposed that gases in the atmosphere, such as water vapor and carbon dioxide, are capable of absorbing heat energy. This work led to the discovery of how the Earth's atmosphere is heated. Tyndall also experimented with the notion that the air held microscopic bacteria that could spoil food. He then derived ways of sterilization by using heat.

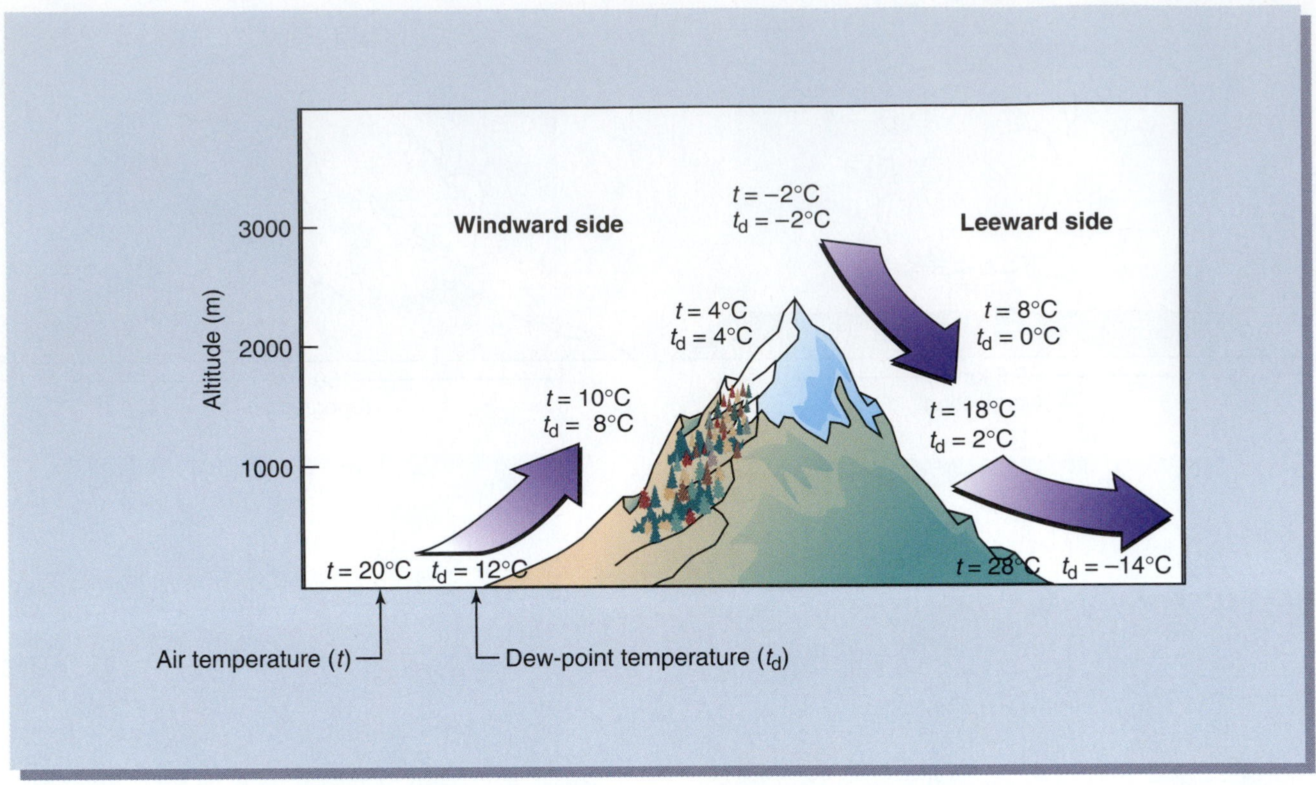

Figure 17–7 When an air mass encounters a mountain, it is forced to rise and cool. This process is known as orographic lifting and leads to the formation of clouds on the windward side of a mountain.

Types of Clouds

The condensation of water vapor into liquid droplets or ice in the atmosphere forms many unique types of clouds, which are classified both by general appearance and height of formation (Figure 17–8). Clouds that form at levels just above the surface and to a height of 6500 feet are called low clouds. These clouds form when the temperature of the atmosphere is above 23° Fahrenheit. Because of this they are often composed of tiny liquid water droplets. Low clouds include cumulus clouds, which appear like large cotton balls and are often separated by clear blue sky. Cumulus clouds form on clear days when convection begins to force warm air upward, causing the water vapor in the air to condense. As the day progresses, cumulus clouds grow bigger. Eventually, large cumulus clouds may produce light rain or snow showers. When the sun goes down and the atmosphere cools, cumulus clouds begin to dissipate as the water droplets they contain vaporize. Stratus, another type of low cloud, consists of one gray, uniform layer that covers the entire sky. Nimbostratus is a low cloud that is thicker than a stratus cloud and much darker; they produce light precipitation for extended periods. Cumulonimbus clouds begin as small cumulus clouds and evolve into large clouds with high vertical development. These clouds form thunderstorms and heavy precipitation. Some cumulonimbus clouds can develop vertical heights of more than 23,000 feet.

The next major cloud type is the middle clouds. Middle clouds form at elevations between 6600 and 23,000 feet where temperatures fall between 32° Fahrenheit and −13°

Fahrenheit. Middle clouds are composed of a mixture of supercooled water droplets and ice crystals. Common forms of middle clouds include the altostratus and altocumulus clouds (Figure 17–9). Altostratus clouds are composed of one uniform layer that appears white or gray. Altocumulus clouds are thick, white, puffy clouds that form long bands in the sky. Altostratus and altocumulus clouds rarely produce precipitation that reaches the ground (it usually evaporates before it reaches the surface).

The highest clouds to form in the atmosphere, also known as high clouds, are called cirrus clouds. These highest of clouds form at heights above 23,000 feet. Cirrus clouds are composed of ice crystals and appear like thin wisps that float through the sky (Figure 17–10). Cirrostratus and cirrocumulus clouds form thin veils high in the atmosphere through which the sun can easily penetrate.

Sometimes when the temperature of humid air above the ground is lowered to its dew point as a result of radiational cooling, a cloud forms. When a cloud forms close to the surface of the Earth, it is known as **fog.** The formation of fog usually occurs in the morning or at night when the ground is cool enough so that it lowers the temperature of air above it to its dew point. Fog then rapidly dissipates when sunlight begins to heat the ground once again.

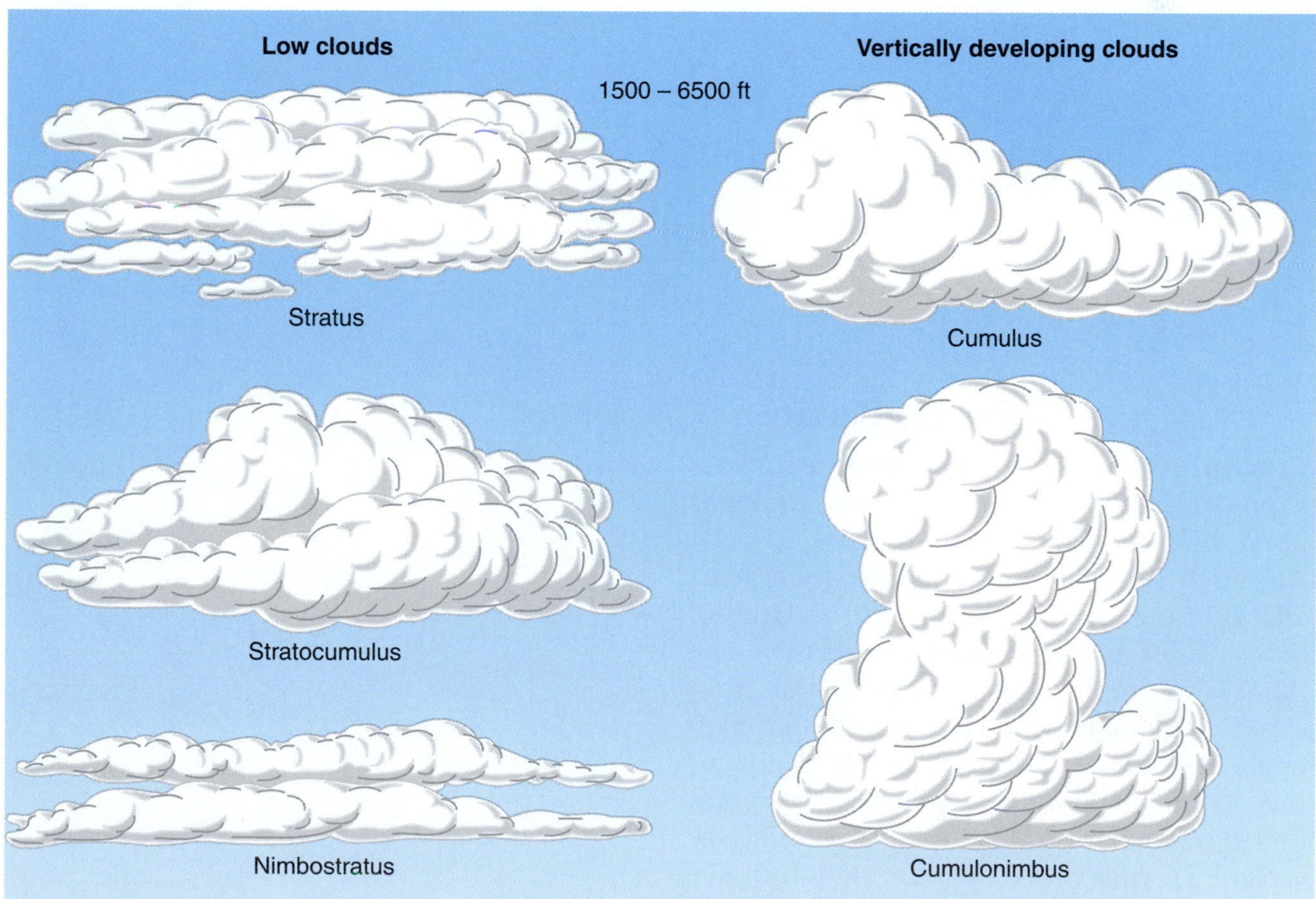

Figure 17–8 The different types of low clouds that form at elevations between 1500 and 6500 feet above the ground.

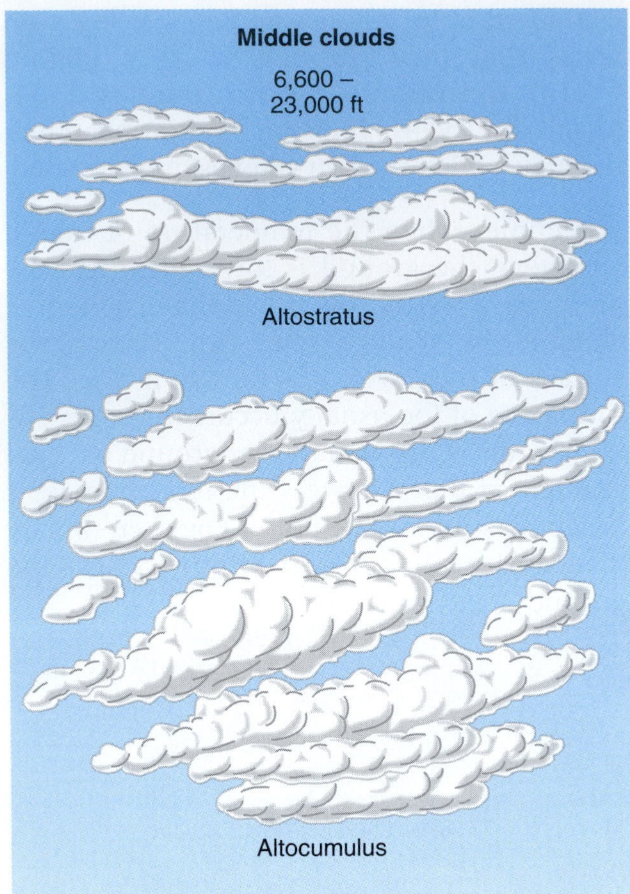

Figure 17–9 Altostratus and altocumulus are types of middle clouds that form between 6,600 and 23,000 feet above the ground.

Formation of Precipitation

Precipitation is the process by which water vapor in the atmosphere condenses and falls back to the Earth's surface (Figure 17–11). Although this process seems simple, it actually is quite complex and is the result of many different atmospheric processes. The tiny water droplets and ice crystals that form clouds are so light that they would float high up in the atmosphere indefinitely if not for the action of updrafts. **Updrafts** are winds that flow upward from the Earth's surface as a result of convection. The formation of cumulus clouds are usually associated with updrafts. As cumulus clouds begin to build, the water droplets and ice crystals that form from condensation begin to get knocked around by updrafts. This causes the water droplets and ice crystals to collide into one another and stick together, resulting in the formation of larger water droplets and ice crystals that continue to collide and grow in size. Eventually the growing water droplets or ice crystals become too heavy, and the updrafts can no longer keep them suspended in the cloud. They then fall toward the Earth's surface as precipitation.

Types of Precipitation

Once the water droplets or ice crystals begin to get heavy enough to become precipitation, they can fall to Earth in many different forms

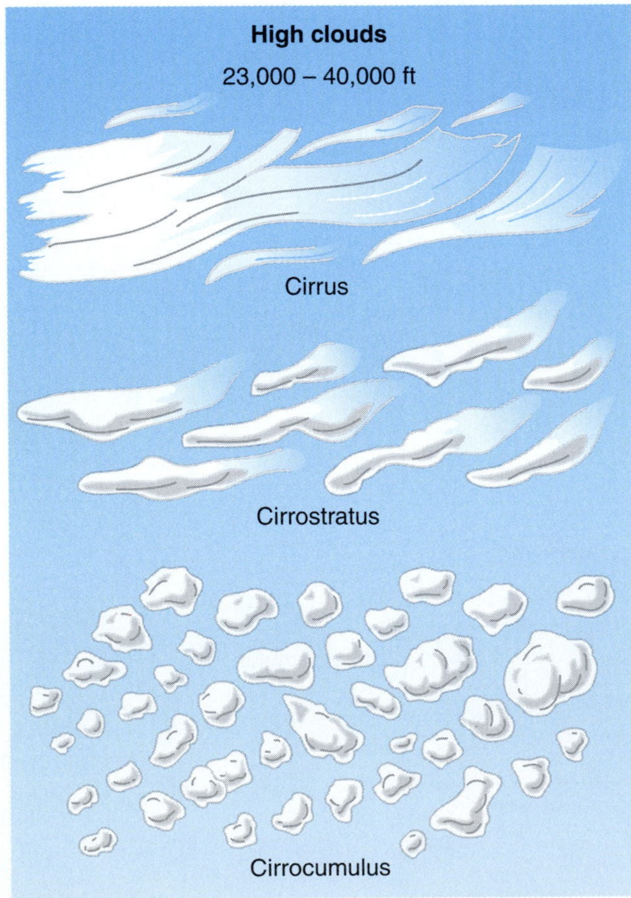

Figure 17–10 Cirrus clouds are classified as high clouds, forming at heights between 23,000 and 40,000 feet above the ground.

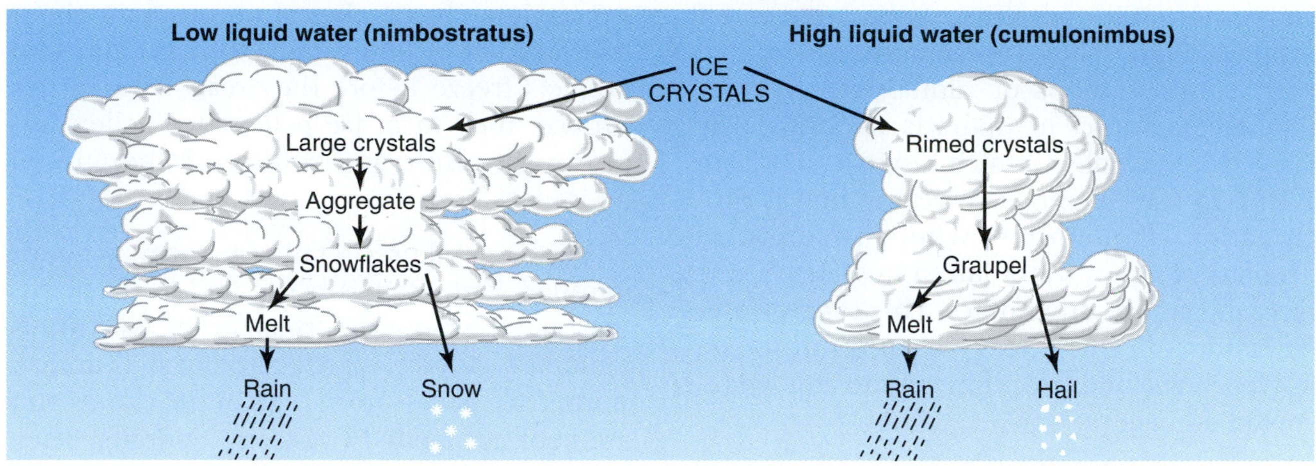

Figure 17–11 Precipitation forms in clouds when ice crystals and tiny water droplets clump together to the point that they are heavy enough to fall to the Earth's surface.

(Figure 17–12). Drizzle is precipitation that consists of very tiny water droplets no bigger than 0.5 mm in diameter. Drizzle develops in stratus clouds, which do not have sufficient

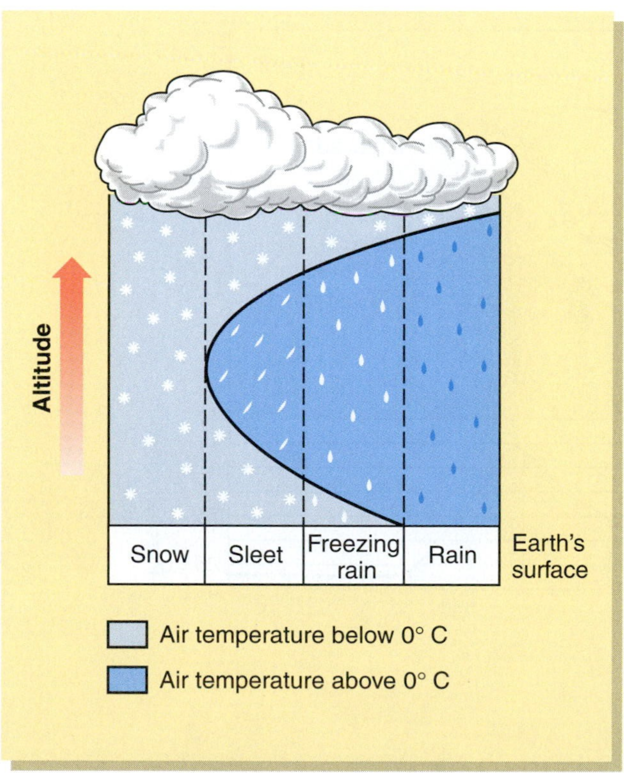

Figure 17–12 Different types of precipitation can form as a result of the air temperature above the surface.

Career Connections

Agricultural Meteorologist

Agricultural meteorologists are highly specialized atmospheric scientists who apply a knowledge of weather to the field of agriculture. These researchers study weather patterns and how they affect crop growth. Atmospheric phenomena such as droughts, floods, and storms cause millions of dollars of crop damage each year. These occurrences might be prevented, or at least the damage they cause reduced, with a better understanding of the interaction of weather with agricultural crops. Research in this field includes the possibilities of controlling rainfall by seeding clouds, along with the study of the adverse affects of atmospheric pollution on crop growth. Agricultural meteorologists are also interested in long-term climate changes that might affect agricultural regions throughout the world. These scientists require a college education in both meteorology and agricultural science. Agricultural meteorologists can work in private industry, in academic research, or for government agencies such as the U. S. Department of Agriculture or the National Weather Service.

convection and updrafts to form large water droplets. Rain is liquid precipitation associated with cumulonimbus or nimbostratus clouds that create sufficient updrafts to form fairly large water droplets, which range in size from 1 to 6 mm in diameter. Freezing rain is precipitation that forms when supercooled droplets of liquid water fall to the Earth's surface and freeze on contact. This type of precipitation can form heavy coatings of ice on power lines and tree limbs, causing widespread damage.

Water that condenses as only ice crystals that stick together and grow larger fall to the ground as snowflakes. Snowflakes usually become larger in moist air, when there is sufficient water to allow for large ice crystals to form. In drier air, very fine snow flakes form. Approximately 4 inches of snow equals about half an inch of liquid rain. When liquid rain droplets freeze before they reach the Earth's surface, they form ice pellets. Ice pellets appear to be white and bounce off the ground on impact.

The last type of precipitation to form in clouds is called hail. Hail is associated with clouds that have strong updrafts, such as cumulonimbus clouds, which also form thunderstorms. Ice pellets form high in the clouds and are kept suspended by strong updrafts. They periodically melt and then refreeze to form concentric rings much like an onion. The stronger the updraft, the larger the size of the hail that can form. Hail is commonly no larger than one half inch in diameter when it finally falls to the ground, but if the updrafts within

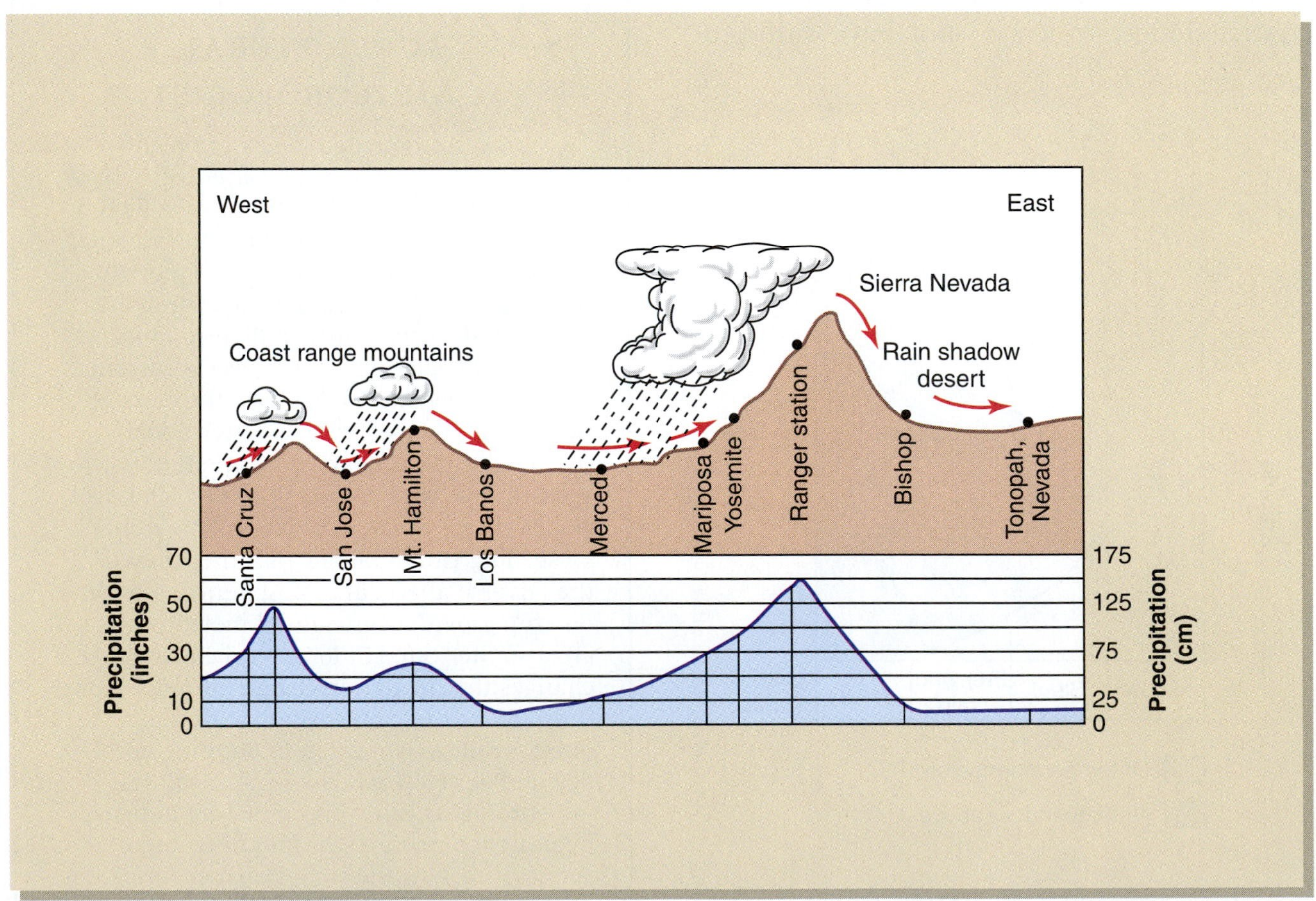

Figure 17–13 Orographic precipitation affects the amount of precipitation received by the windward and leeward sides of mountains.

the cloud are extremely powerful, large hailstones can form. Hailstones are very large hail that can be as big as a golf ball. The largest hailstones ever observed to fall to the Earth were equal in size to a small grapefruit.

Orographic Precipitation

Orographic precipitation is precipitation that forms as a result of the lifting of an air mass over a mountain. When moist air encounters the higher elevation of a mountain, it is forced to rise. This causes the air to cool and reach its dew point very quickly. This occurs on the side of the mountain that is facing the wind that is associated with the movement of the air mass, also known as the windward side of the mountain. The result is the rapid formation of clouds and precipitation. As the air continues to rise up the windward side of the mountain, more precipitation falls. Eventually, as the air moves over the top of the mountain, it begins to decrease in elevation. The descending air is compressed and therefore warms. This causes the temperature of the air mass to rise above its dew point. Precipitation then ceases, and clouds begin to dissipate. This whole process is known as orographic precipitation, which causes the windward sides of mountains to be cooler in temperature and receive high amounts of precipitation (Figure 17–13), whereas the leeward, or downwind, side of the mountain experiences warmer temperatures and drier air. Orographic precipitation controls the weather and climate of many different geographic areas around the world that are located either on the windward or leeward sides of mountain ranges.

SECTION REVIEW

1. Describe the process by which clouds form.
2. Explain the moist adiabatic lapse rate.
3. What are three things that cause air to rise in the atmosphere?
4. Explain the three basic types of clouds and the heights where they form.
5. Describe how precipitation forms in clouds.
6. What are the six forms of precipitation?
7. What is orographic precipitation?
8. Who was John Tyndall?

For more information go to these Web links:

<http://www.wrh.noaa.gov/Flagstaff/science/clouds.htm>

<http://ww2010.atmos.uiuc.edu/(Gh)/guides/mtr/cld/home.rxml>

<http://www.cloudman.com/>

<http://ww2010.atmos.uiuc.edu/(Gh)/guides/mtr/cld/prcp/home.rxml>

CHAPTER SUMMARY

Humidity is a measure of how much water vapor is present in the atmosphere. Water vapor is the gaseous form of water; it enters the

EARTH MATH

1) USING THE APPROXIMATE MOIST ADIABATIC LAPSE RATE OF 3.3° FAHRENHEIT FOR EVERY 1000 FEET IN ELEVATION, DETERMINE THE HEIGHT AT WHICH A CLOUD FORMS IN A PARCEL OF AIR AT SEA LEVEL THAT HAS A TEMPERATURE OF 52° FAHRENHEIT AND A DEW POINT OF 32° FAHRENHEIT.

atmosphere in three principal ways: evaporation of water from the oceans or from surface freshwater, evapotranspiration by plants, and sublimation of snow and ice. Generally the warmer the air, the more water vapor it can hold. This causes warm air to be more humid than cold air. Near the tropics, the water vapor content of the air can be as high as 4%. The amount of water vapor in the air is usually measured by determining the relative humidity. Relative humidity is the ratio of how much water vapor is in the air compared with its saturation point at a specific temperature and pressure. It is usually expressed in the form of a percentage.

The relative humidity of the atmosphere is related to the air's dew point temperature. The dew point is the temperature to which the air must be cooled for saturation to occur. The dew point temperature can be calculated by using a psychrometer, which records the wet bulb and dry bulb temperatures. The closer the temperature is to the dew point, the more water vapor that is in the air. The difference between the dew point and the air temperature can be used to predict the formation of clouds and precipitation.

A cloud is a large mass of condensing water droplets and ice crystals that forms in the atmosphere when the air is cooled to its dew point. For a cloud to form, moist air from the surface must rise into the atmosphere, where it cools to its dew point temperature and causes water vapor to condense. There are four ways by which air can be forced to rise to form clouds. Warm air at the surface becomes less dense than the surrounding air and rises. Topography with high elevations can also cause air to rise, therefore causing it to cool. The convergence of wind at the surface of the Earth can force air upward, and the movement of a cold air mass into a warm air mass can cause air to rise. This occurs when the cold air wedges itself underneath the warm air, forcing it upward. A cloud that forms at the Earth's surface is called fog.

Once clouds form in the atmosphere, they can produce precipitation. Precipitation forms when the tiny water droplets and ice crystals in clouds clump together and fall back to the Earth's surface. There are six different forms of precipitation: drizzle, rain, freezing rain, snow, ice pellets, and hail. Orographic precipitation occurs when moist air comes into contact with a mountain. The mountain forces the air to rise, cool, and condense. This causes the formation of clouds and precipitation on the windward side of the mountain. Once the air moves over the mountain, it descends, compresses, and heats. This causes the clouds and precipitation to dissipate on the leeward side of the mountain.

CHAPTER REVIEW

Multiple Choice

1. Most moisture enters the atmosphere by the processes of:
   a. Convection and conduction
   b. Condensation and radiation
   c. Reflection and absorption
   d. Evapotranspiration and evaporation

2. Which area on the Earth is most likely to have an atmospheric moisture content of 4%?
   a. The poles
   b. 45 degrees north and south
   c. The tropics
   d. Nowhere

3. As compared with warm air, cold air:
   a. Is lower in pressure
   b. Is drier
   c. Is less dense
   d. Holds more water

4. The movement of water from the soil up through a plant and off its leaf surface is called:
   a. Condensation
   b. Evapotranspiration
   c. Evaporation
   d. Sublimation

5. Snow changing phase into water vapor occurs by the process of:
   a. Condensation
   b. Evapotranspiration
   c. Evaporation
   d. Sublimation

6. The change in phase from water vapor to liquid water is called:
   a. Condensation
   b. Evapotranspiration
   c. Evaporation
   d. Sublimation

7. Which temperature and dew point indicate the lowest humidity?
   a. Temperature 17° Celsius, dew point 12° Celsius
   b. Temperature 13° Celsius, dew point 12° Celsius
   c. Temperature 17° Celsius, dew point 11° Celsius
   d. Temperature 21° Celsius, dew point 7° Celsius

8. What is the approximate relative humidity if the dry bulb temperature is 12° Celsius and the wet bulb temperature is 7° Celsius?
   a. 28%
   b. 35%
   c. 48%
   d. 65%

9. What is the dew point of the air if the wet bulb temperature is 10° Celsius and the dry bulb is 14° Celsius?
   a. −25° Celsius
   b. 6° Celsius
   c. 3° Celsius
   d. 4° Celsius

10. For clouds to form, which of the following must occur?
    a. Cool air must sink and warm.
    b. Cool air must rise and warm.
    c. Warm air must rise and cool.
    d. Warm air must sink and cool.

11. The formation of clouds requires air to be:
    a. Saturated and have condensation nuclei
    b. Saturated and have no condensation nuclei
    c. Unsaturated and have condensation nuclei
    d. Unsaturated and have no condensation nuclei

12. Why is it possible for no rain to be falling from a cloud?
    a. There are no condensation nuclei in the cloud.
    b. The cloud is water vapor.
    c. The dew point has not yet reached the cloud.
    d. The water droplets are too small to fall.

13. If the air temperature is 10° Celsius, which dew point results in the highest probability of precipitation?
    a. 8° Celsius
    b. 6° Celsius
    c. 0° Celsius
    d. −4° Celsius

14. Which of the following is a form of liquid precipitation that is less than 0.5 mm in diameter?
    a. Rain
    b. Drizzle
    c. Fog
    d. Sleet

15. Compared with the windward side of a mountain, the leeward side is:
    a. Cooler and drier
    b. Warmer and drier
    c. Moist and cool
    d. Moist and warm

continued

Matching *Match the terms with the correct definitions.*

a. humidity
b. water vapor
c. evapotranspiration
d. sublimation
e. dew point
f. cloud
g. condensation
h. condensation nuclei
i. lapse rate
j. uplift
k. convergence
l. fog
m. precipitation
n. updrafts
o. orographic precipitation

1. ____ The term to describe the phase change from a solid to a gas.
2. ____ A common term that refers to the amount of water vapor content in the air.
3. ____ A type of precipitation that occurs as a result of orographic lifting, which causes moist air to cool and condense when it is forced to rise up the windward side of a mountain.
4. ____ The temperature to which the air needs to be cooled to become saturated with water at a specific atmospheric pressure.
5. ____ Strong, vertical winds that move upward through a cloud.
6. ____ The gaseous form of water.
7. ____ Liquid or solid water formed in clouds that falls to the surface of the Earth.
8. ____ A large mass of condensing water droplets and ice crystals in the atmosphere.
9. ____ The important pathway by which water moves from the soil, through the body of a plant, and evaporates off the leaf surface back into the atmosphere.
10. ____ Microscopic particles floating in the atmosphere on which water condenses to form clouds.
11. ____ A cloud that forms near or at the Earth's surface and is composed of tiny water droplets.
12. ____ The change in phase from a gas to a liquid.
13. ____ The process of coming together.
14. ____ The rate at which temperature or moisture decreases with height.
15. ____ The process of lifting something up.

Critical Thinking

1. Explain why pilots should try to avoid flying through developing cumulonimbus clouds.
2. Why do think your skin feels hotter on a day that has 99% relative humidity, compared with a day that has 50% relative humidity?

CHAPTER 18 Wind, Air Masses, and Fronts

Section 18.1 – Wind Objectives

Pressure Gradient • Planetary Winds • Pressure Systems • Mesoscale Winds • Local Winds • The Jet Stream • Wind Measurement

After reading this section you should be able to:

- Define the term *wind.*
- Determine the pressure gradient between two points on the Earth's surface.
- Describe the formation of planetary winds on the Earth.
- Identify the locations of the low- and high-pressure centers on the Earth that are associated with prevailing winds.
- Describe the Coriolis effect and how it affects winds on the Earth.
- Identify the specific wind circulation around high- and low-pressure centers in the Northern Hemisphere.
- Explain the conditions that lead to the formation of land and sea breezes.
- Define the term *jet stream.*

Section 18.2 – Air Masses and Fronts Objectives

Air Mass Formation • Source Regions and Classification of Air Masses • Fronts • Mid-latitude Cyclones

After reading this section you should be able to:

- Define the term *air mass.*
- Identify the different regions that form unique air masses.
- Describe the characteristic temperature, moisture, and pressure associated with the five different air mass types.
- Identify the four different types of fronts and their characteristic changes in temperature, pressure, and precipitation.
- Describe the development of a mid-latitude cyclone.

TERMS TO KNOW

wind	**Coriolis effect**	**cold front**
pressure gradient	**pressure center**	**warm front**
vector force	**jet stream**	**occluded front**
planetary winds	**air mass**	**stationary front**
Hadley cell	**front**	**cyclone**

INTRODUCTION

For thousands of years, human civilizations have harnessed the wind to move ships and power machines. The prevailing winds that blow across the oceans opened up new lands for both trade and colonization and forever changed our relationship with the sea. Humans have long respected the winds on the planet, for they not only provide easily accessible power, but they can be some of the most damaging aspects of the environment. Events such as hurricanes, tornadoes, and thunderstorms can generate winds with deadly force. Even our modern technology cannot prevent the potential destruction that strong winds can cause. Wind is also an important aspect of the Earth's climate. It is the natural way by which our planet distributes energy and heat throughout the atmosphere. Most everyone on Earth can appreciate the cooling effects of a gentle breeze on a hot summer day or the power of the wind to raise a kite into the sky. The movement of air across the Earth is an important part of weather. This is especially noticeable when a hot, humid summer day suddenly cools off as a result of a cold front passing through the area. The passing of fronts is often the cause for change in weather, but what actually is a front, and what causes air to have unique characteristics of temperature and moisture? All these phenomena are part of the complex interaction of the atmosphere, oceans, and land surface that creates weather and climate on the Earth.

18.1 *Wind*

Pressure Gradient

The only time we can really feel the atmosphere around us is when the wind blows. **Wind** on Earth is the horizontal movement of air across the Earth's surface. Wind is also known as a pressure gradient force, because it is caused by a difference in air pressure. A **pressure gradient** is a change in air pressure over a specific distance (Figure 18–1). Air on Earth always moves from areas of high atmospheric pressure to areas of low atmospheric pressure. This movement of air from high pressure to low pressure is called wind.

Wind is a **vector force,** which means that it has both a direction and a velocity. The direction that wind moves is always recorded from where it originates. A wind that is blowing from the northwest is actually traveling from the northwest toward the southeast. Because wind is caused by changes in air pressure, atmospheric pressure centers on the Earth can reveal a lot about the force and direction of wind. Rapid changes in air pressure over short distances result in a strong pressure gradient, which causes strong winds. The more slowly the pressure changes over a distance, the weaker the pressure gradient, therefore creating weaker winds. Winds on the Earth are divided into three main categories: planetary winds, mesoscale winds, and local winds.

Planetary Winds

Planetary winds are the large-scale wind patterns that flow across the Earth as a result of the unequal distribution of insolation received by the Earth's surface. Common planetary winds, also called prevailing winds, include the trade winds, which blow from the northeast between 30 degrees north latitude and the equator (Figure 18–2). The westerlies are also prevailing planetary winds; they blow from the

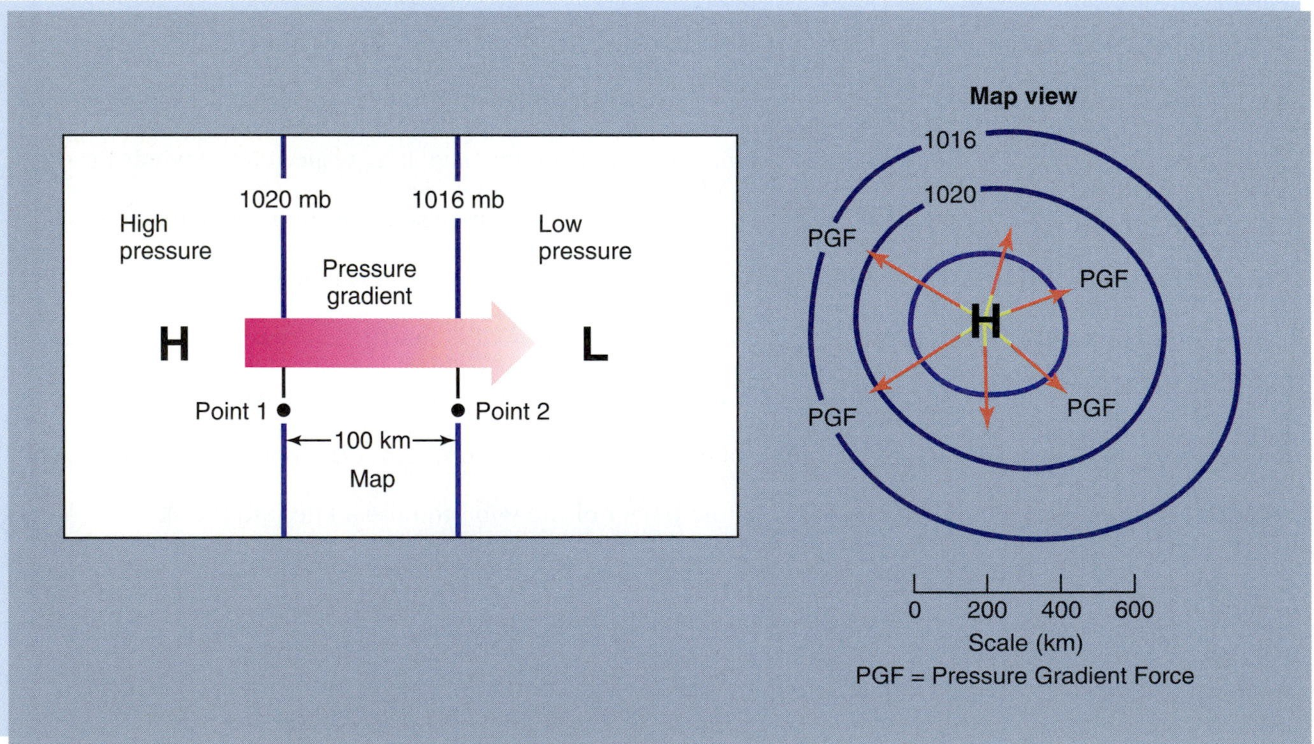

Figure 18–1 The formation of winds by the flow of air from areas of high pressure toward areas of low pressure over a specific distance on the Earth, also called the pressure gradient.

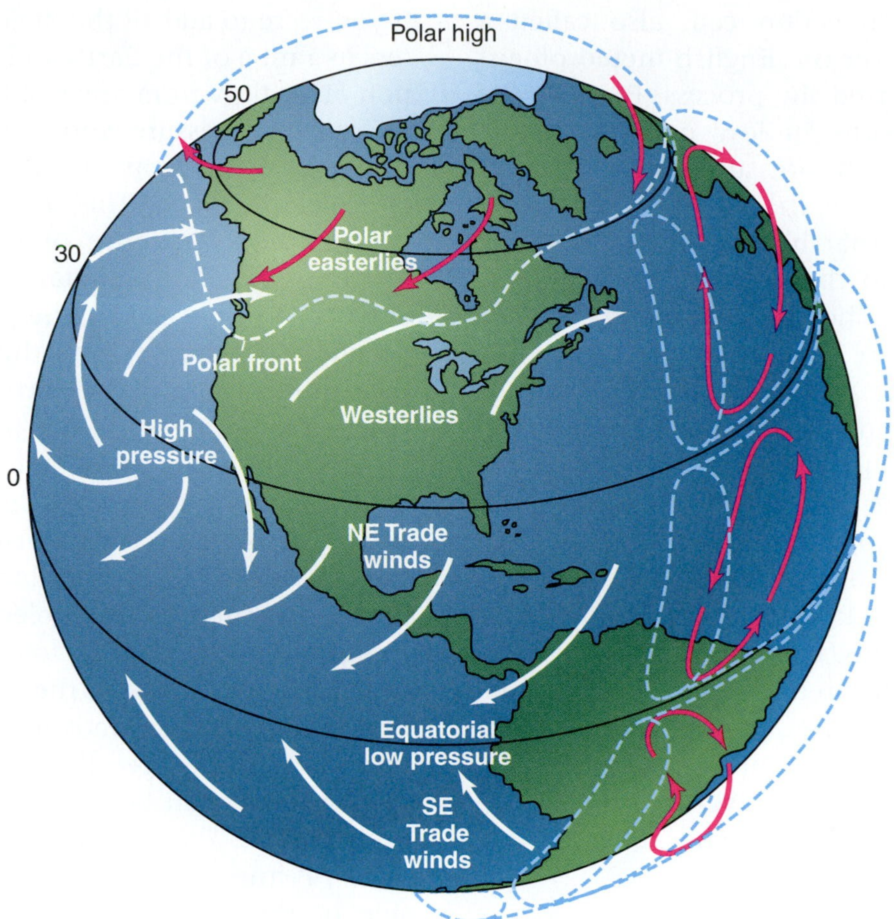

Figure 18–2 Planetary winds on the Earth's surface, also known as prevailing winds.

southwest across the Earth's surface between 30 degrees and 60 degrees north latitude. Many explorers who sailed the ocean in ships took advantage of these prevailing winds, which steered them to their destinations.

Planetary winds are caused by large pressure differences that occur near the Earth's surface as a result of the unequal distribution of heat on Earth. At the equator, the sun strikes the Earth at nearly a 90-degree angle all throughout the year, which causes rapid heating of the surface near the tropics. This results in the formation of warm air at the surface, which begins to rise in the atmosphere as a result of its lower density. The area of warm, rising air is also known as an area of convergence because air molecules are converging, or coming together, at the surface to replace the air that is rising into the atmosphere. This warm, rising air causes areas of low pressure to form near the equator. As this air continues to rise into the atmosphere, it begins to cool as it expands. The cooling air then becomes more dense and begins to sink back toward the surface of the Earth. This cool, dense, sinking air returns to the surface at about 30 degrees north and south latitude. As the cooler air sinks, it begins to form areas of high pressure at the surface. These areas are called areas of divergence, because the cool, dense air is pushing downward and causing the air at the surface to spread apart, or diverge.

The result of the warm air rising and cooler air sinking is the formation of a large-scale atmospheric convection cell, also called a **Hadley cell,** after the English meteorologist who first theorized the process. Because the Hadley cell results in low pressure at the equator and high pressure at 30 degrees north and south of the equator, large-scale winds develop that blow from the areas of high pressure to the areas of low pressure (Figure 18–3). These are planetary winds. The low-pressure area near the equator is also called the intertropical convergence zone (ITCZ), and the areas of high pressure that form near 30 degrees north and south of the equator are known as the subtropical highs. Farther north and south of these pressure areas, at approximately 60 degrees north and south latitude, another low pressure system forms, called the subpolar lows. The cooler air located at the poles forms areas of high pressure, known as the polar highs. This causes air to flow from the polar highs to the subpolar lows, creating planetary winds between the higher latitudes.

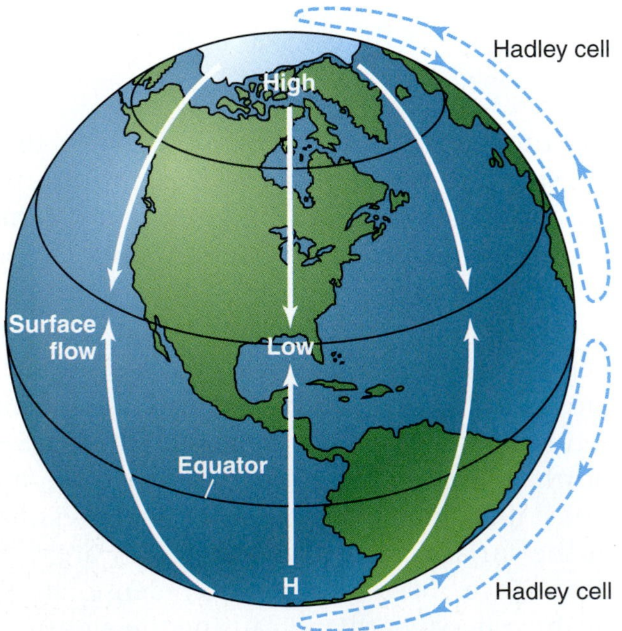

Figure 18–3 The formation of large-scale convection currents, called Hadley cells, creates distinct areas of low and high pressure at the Earth's surface. This results in the formation of planetary winds.

If you were to add all the major pressure centers to a map of the Earth and draw the direction of air flow from areas of high pressure to areas of low pressure, you would find that the planetary winds blow directly from the north or from the south. This, however, does not explain the occurrence of the easterly and westerly prevailing planetary winds that actually occur on the Earth. The prevailing winds do not blow directly from the north or from the south because of the Earth's rotation. The Earth is spinning on its axis at a speed of approximately 1000 miles per hour near the equator, which causes winds on the Earth's surface to experience the **Coriolis effect.** The Coriolis effect causes free-floating objects on the Earth, such as air molecules, water, or things in the atmosphere, to move to the right or left as they travel on the Earth. This is actually caused by the movement of the Earth underneath the free-floating objects. The Coriolis effect can be illustrated by two children playing on a merry-go-round (Figure 18–4). Picture one child standing on the outside of the merry-go-round and the other child standing at the center. If the child at the center of the merry-go-round tries to throw a ball straight toward the child spinning around on the outside, the ball will never reach the friend, because she would have moved away before it gets to her. If you observed this from above the merry-go-round looking down, it would appear that the ball was being deflected to the right or left, depending on which way the merry-go-round was spinning as it was thrown from the center. This is exactly what causes the Coriolis effect on the Earth. Because the Earth is spinning on its axis, its surface moves under objects in the atmosphere, which causes them to be deflected to the right or left of travel. The first evidence of the Coriolis effect was noticed by French artillery officers when they were practicing firing their cannons at targets located long distances away. No matter how well they aimed their cannons, the artillery shells always landed to the right of the target, even when they adjusted for the wind. After care-

EARTH SYSTEM SCIENTISTS *GEORGE HADLEY*

George Hadley was born in England in 1685. He was a meteorologist who took the ideas of Edmund Halley and applied them to his theory of atmospheric circulation. In 1735 Hadley proposed that the trade winds were formed when hot, rising air near the equator moved north and south and eventually cooled and sank back toward the Earth's surface. His theory explained how areas of high and low atmospheric pressure formed at the surface, leading to the formation of planetary winds. Hadley also theorized that the rotation of the Earth on its axis caused these planetary winds to be deflected to the right of travel in the Northern Hemisphere and to the left of travel in the Southern Hemisphere. These large-scale convection cells that Hadley discovered are known as Hadley cells.

ful investigation, the effect was finally explained by French physicist Gustave-Gaspard Coriolis in 1835: The Earth was actually rotating underneath the long-range artillery shells, causing them to miss their target every time.

The Coriolis effect also affects wind on the Earth and results in the path of the wind being deflected to the right of travel in the Northern Hemisphere and to the left of travel in the Southern Hemisphere (Figure 18–5). If you

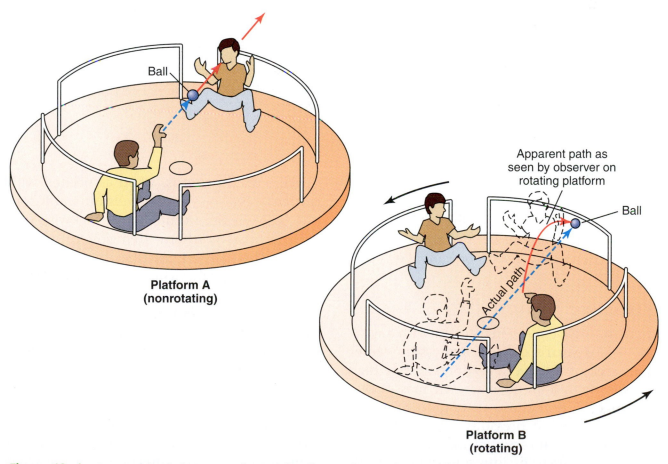

Figure 18–4 An example of what causes the Coriolis effect can be seen by two children playing catch on a merry-go-round. As the ball is thrown from the center, it appears to curve to the right of motion as the child on the outside moves away from it.

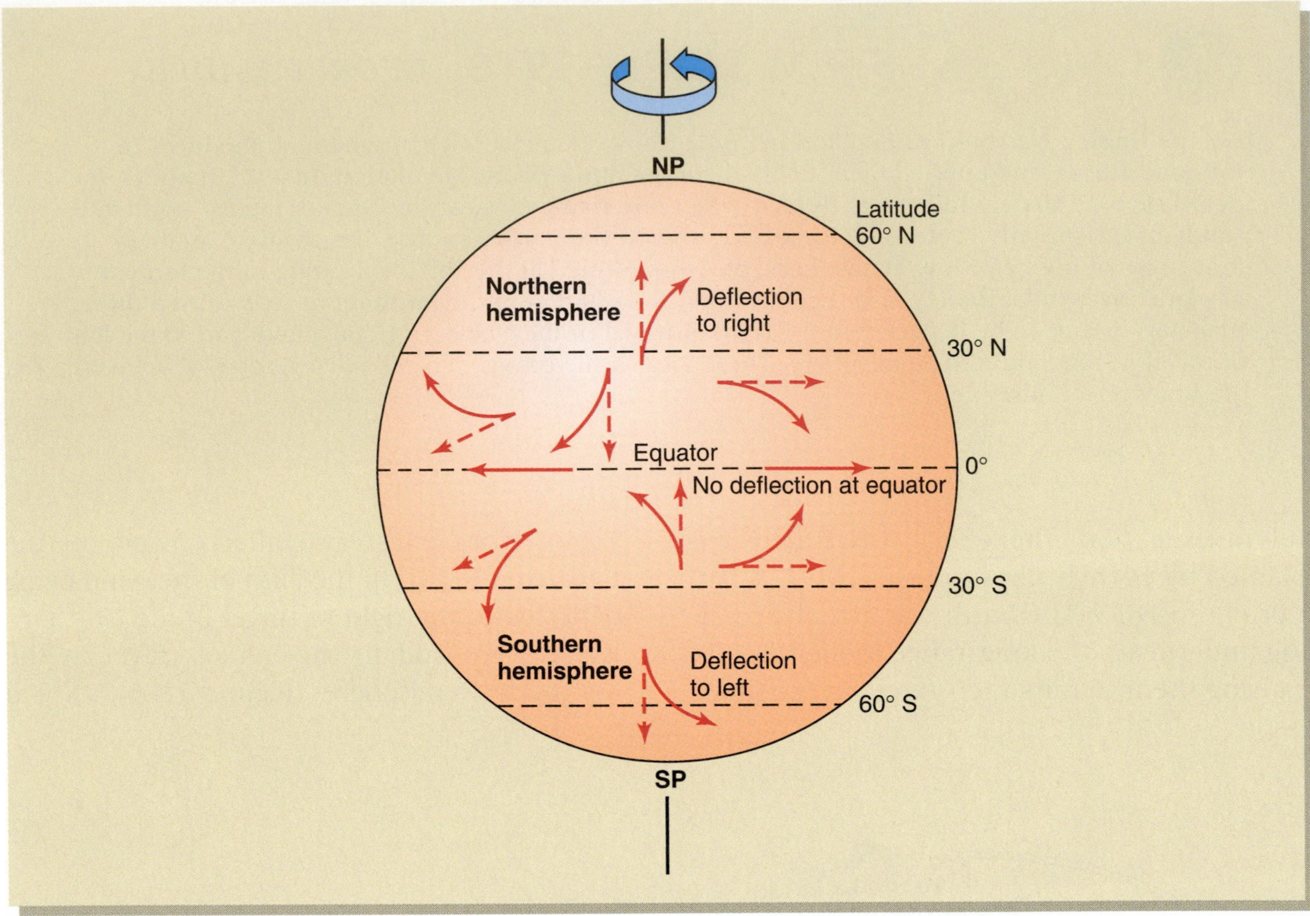

Figure 18–5 The Coriolis effect causes winds to be deflected to the right of travel in the Northern Hemisphere and to the left of travel in the Southern Hemisphere as a result of the Earth's rotation.

then adjust your map of planetary winds by curving them to the right of travel in the Northern Hemisphere and to the left of travel in the Southern Hemisphere, you can see the correct direction of the Earth's prevailing winds.

Pressure Systems

A pressure system is a mass of air with a well-defined **pressure center.** A pressure center is an area of relatively low pressure or high pressure within an air mass. Air masses that possess cold, dry, dense air tend to contain a high atmospheric pressure center. Because wind moves from areas of high atmospheric pressure to areas of low atmospheric pressure, pressure centers are also associated with specific wind patterns. Wind associated with a high-pressure system travels outward from the high-pressure center in a clockwise pattern in the Northern Hemisphere (Figure 18–6). This is also known as an anticyclone. The reason that the wind spirals outward in a clockwise direction around a high-pressure center has to do with the Coriolis effect that is caused by the Earth's rotation. Recall that winds are deflected to the right of travel in the Northern Hemisphere. The air that is moving outward from the high-pressure center is deflected to the right, causing a clockwise rotation. High-pressure centers are noted on a weather map by a blue capital *H* and can indicate the direction that the wind is blowing.

Air masses that contain warm, humid, less dense air tend to contain a low atmospheric pressure center. Wind patterns associated with low-pressure systems spiral inward and counterclockwise in the Northern Hemisphere (Figure 18–7). This is also known as a cyclone. Low-pressure centers are marked on weather maps by a red capital *L* and can also indicate the direction that the wind is moving. Warm, humid air masses that contain areas of extremely low pressure can form into tropical storms and hurricanes, which can generate high wind speeds that spiral inward around the low-pressure center.

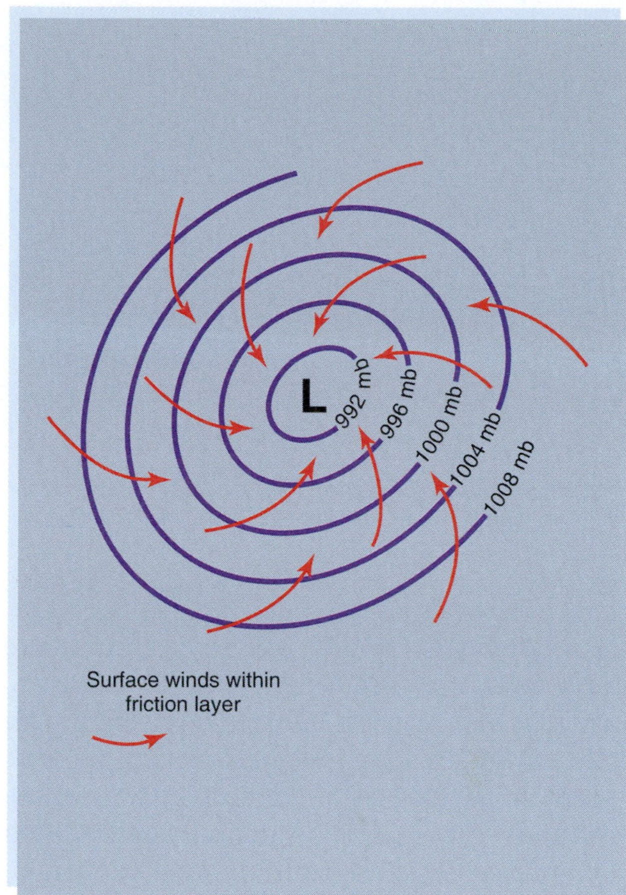

Figure 18–7 Winds spiral inward in a counterclockwise direction around a low-pressure center in the Northern Hemisphere.

Meteorologists are interested in identifying pressure centers because they can often indicate the type of weather that might form in a particular region. In general, areas of high atmospheric pressure bring clear and dry weather. Areas of low atmospheric pressure tend to be associated with clouds and precipitation. Utilizing the knowledge of how air moves around areas of high or low pressure can help meteorologists forecast the wind direction.

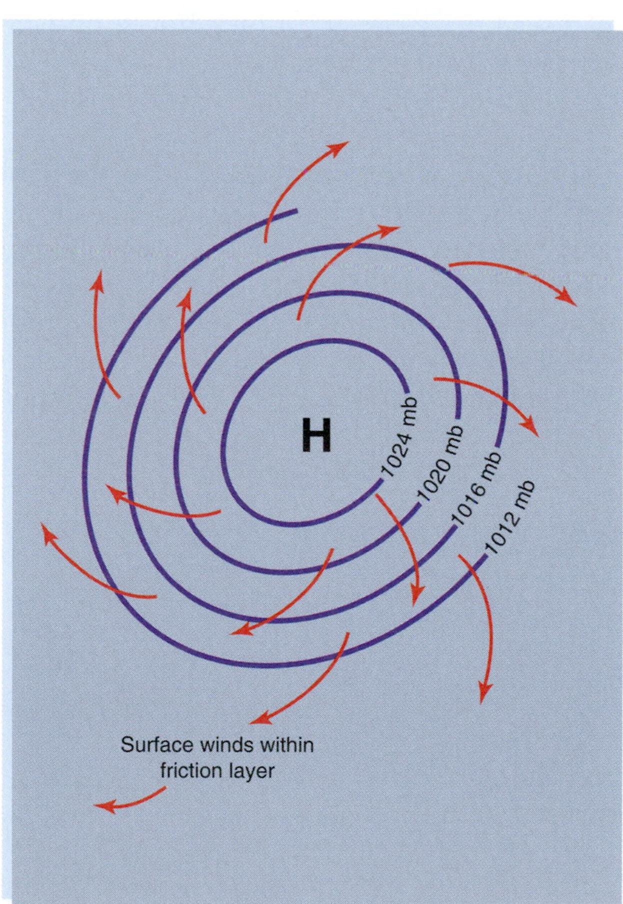

Figure 18–6 Winds spiral outward in a clockwise direction around a high-pressure center in the Northern Hemisphere.

Mesoscale Winds

Mesoscale winds, also known as regional winds, develop as a result of smaller scale interactions of changing pressure near the Earth's surface. A good example of the formation of mesoscale winds is the land and sea breeze. On a hot day, insolation received by the land surface located next to a large body

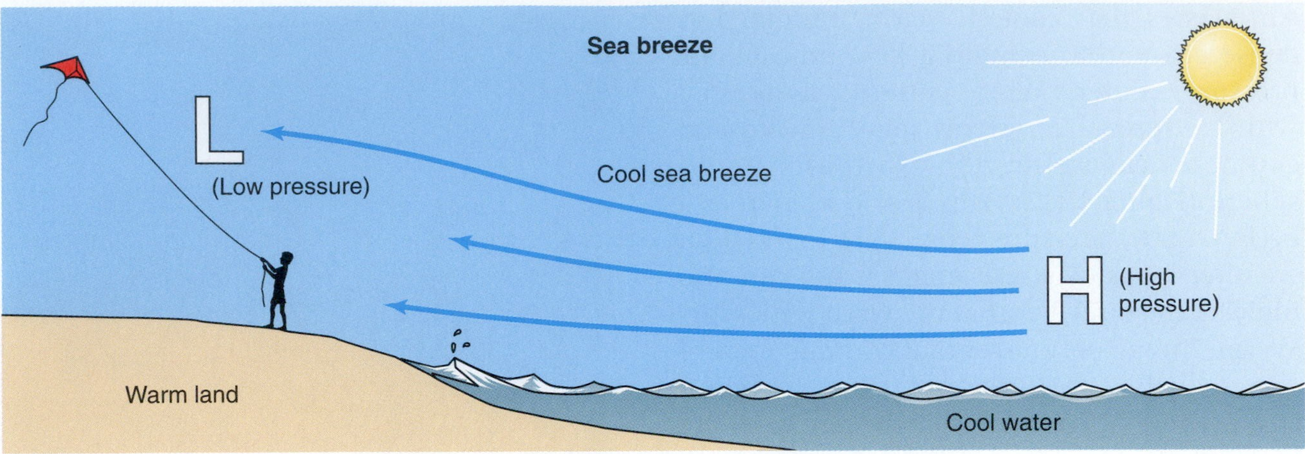

Figure 18–8 A sea breeze develops as a result of low pressure forming over the rapidly heating land surface during the day and high pressure forming over the cooler water.

of water heats rapidly. This causes hot air to form near the surface; the hot air begins to rise as it becomes less dense. The rising hot air creates an area of low barometric pressure near the surface, as air molecules are heated and move upward into the atmosphere. Next to land, cooler air is forming over the water, which is not heating as rapidly as the land surface. This is a result of the water's high heat capacity. This cooler, more dense air begins to form an area of high pressure over the water. The result of the formation of high pressure over the cool water and low pressure over the warm land is wind that travels from the high pressure toward low pressure. This is also called a sea breeze, which blows cool air toward the coast on a hot day. The formation of a sea breeze can occur on a hot, sunny day wherever land lies next to a large body of water (Figure 18–8).

Eventually, as the sun goes down, the land surface begins to cool quickly and cool air begins to form over the land surface (Figure 18–9). The temperature of the water stays relatively the same when the sun is down and results in warmer air forming over its surface as compared with the land. After the temper-

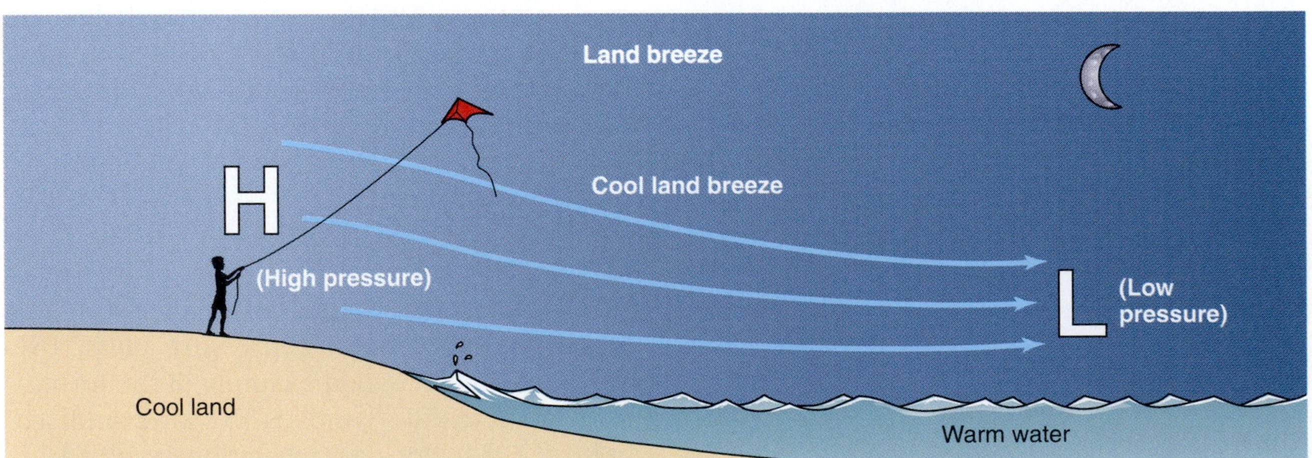

Figure 18–9 A land breeze develops as a result of high pressure forming over the cooler land surface at night and low pressure forming over the warmer water.

ature of the land cools to a temperature below that of the water, high pressure begins to form over the land surface. The warmer water then causes the air over it to form a low pressure area. The result is the development of wind that blows from the high pressure over the land to the low pressure over the water. This is called a land breeze, which can occur on cool summer nights.

These mesoscale winds can greatly influence the weather around the area in which they occur. Everyone on the beach during the summer appreciates the benefit of a cool sea breeze blowing in from the water, which is the pleasant result of changing atmospheric pressure.

Career Connections

Operational Meteorologist

Operational meteorologists gather information about the pressure, wind direction and speed, temperature, and humidity of specific points on the Earth. They then use this information to make short-term and long-term forecasts about regional weather. Operational meteorologists apply the principles of atmospheric science to predict weather conditions. By using sophisticated computer models, radar, satellite images, and ground-based observations, these scientists provide detailed predictions about the state of the atmosphere in a particular place on Earth. Many aspects of daily life depend on the work of operational meteorologists, especially in the transportation industry. Forecasting potentially dangerous events such as tornadoes, hurricanes, blizzards, and flooding helps to prevent the loss of life and property and is one of the most important aspects of this type of career. Operational meteorologists require a college degree and can find work in the armed forces, transportation industry, and academic research or in government agencies such as the National Weather Service.

Local Winds

Local winds are created by local geographic features. When air rides up and over a mountain range and picks up colder temperatures, it then descends into a warmer valley, creating a cool mountain breeze. The opposite effect can occur when warmer air developing in a low valley rises up the slopes of the surrounding cooler mountains, creating a warm valley breeze. Similar effects can also occur when cold air moving over a high plain spills down into adjacent lower regions, creating a cold katabatic wind. *Katabatic* means "to fall." Local katabatic winds generated in Greenland can rush down off the ice fields at speeds of more than 100 miles per hour.

Another type of local wind is the famous Santa Ana winds of California. This wind forms over the hot desert and picks up speed as it moves through dry canyons and valleys. Eventually it reaches the populated areas of Los Angeles as a hot, harsh, dry wind. Local topography or buildings can also create local winds. Chicago is called the Windy City because cold winds blowing across the great lakes slam into the buildings of the city and swiftly move through the streets, creating strong cold wind gusts. Whirlwinds or dust devils, another example of a local wind, are spiraling winds that form in certain areas as moving air rapidly circles around, bringing with it dust and debris. These local wind gusts usually dissipate as quickly as they form. The famous nor'easters that blow in from the Atlantic Ocean along the northeastern coast of the United States are composed of high winds that blow in from the North Atlantic. These large-scale local winds can cause damaging winter storms in New England.

The Jet Stream

Winds on Earth do not only occur near the surface; some winds form high up in the

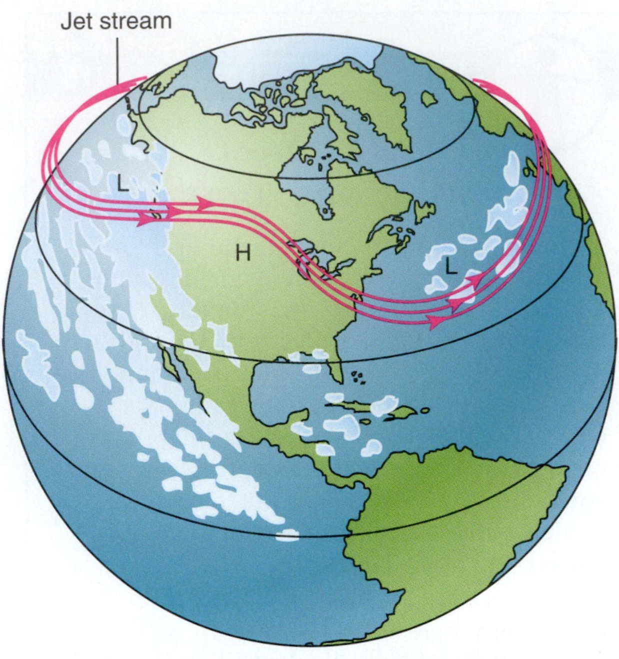

Figure 18–10 Rapidly moving winds high in the tropopause form the polar front jet stream.

atmosphere at the level of the tropopause. Areas high in the tropopause, where cool air begins to descend back toward the surface, form belts of high winds called the **jet stream** (Figure 18–10). Jet streams are bands of high-speed winds located high in the atmosphere. The rapid decline in pressure associated with the descending air of a Hadley cell can generate winds of more than 100 miles per hour. Jet streams are located 6 to 9 miles up in the atmosphere over the subtropical and polar highs. These fast-moving streams of air are approximately 200 miles wide and less than a mile thick. Over the United States the polar jet stream travels from west to east and helps to move air masses and their associated weather in a westward to eastward direction. The subtropical jet stream, which is located over the lower latitudes, plays an important role in moving warm tropical air, and many hurricanes, along the east coast of the United States.

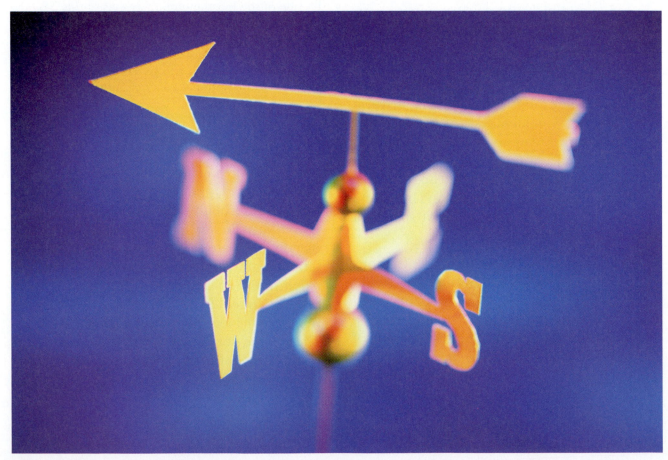

Figure 18–11 A weather vane is one of the oldest weather instruments; it is used to record wind direction. *(Courtesy of PhotoDisc.)*

Wind Measurement

The movement of wind across the Earth's surface is measured in both its direction and speed. The wind direction is recorded in degrees from which direction the wind is blowing. Wind blowing directly from the north is recorded as 0 degrees, and wind blowing directly from the south has a direction of 180 degrees. The direction of the wind is recorded by a wind vane. This instrument consists of a long metal arrow that freely swings on a vertical pole. The wind vane is then calibrated to the directions of the compass (Figure 18–11). Another way to measure wind direction is by using a wind sock, a cloth tube through which the wind blows and points in the direction the wind is blowing. The velocity of the wind is a measure of how fast the air is moving across the Earth's surface. Wind speed is usually measured in knots. One knot is equivalent to 1.15 miles per hour. Wind velocity, or the speed of the wind, is measured by an instrument called an anemometer, which is composed of three metal cups that are allowed to swing freely around a vertical pole. The speed with which the cups spin around the pole, caused by the force of the wind, is calibrated to record the wind speed. An aerovane is a modern device that uses a propeller mounted on a weather vane to record both the wind speed and direction.

The relative wind speed can also be recorded by using the Beaufort scale of wind force (Table 18–1). This uses common observations

TABLE 18–1 The Beaufort wind scale measures wind by its effects on the environment

Beaufort Number	Description	Wind Speed: ml/hr	Wind Speed: knots	Wind Speed: km/hr	Observations
0	Calm	0–1	0–1	0–2	Smoke rises vertically
1	Light air	1–3	1–3	2–6	Direction of wind shown by drifting smoke, but not by wind vanes
2	Slight breeze	4–7	4–6	7–11	Wind felt on face; leaves rustle; wind vanes moved by wind; flags stir
3	Gentle breeze	8–12	7–10	12–19	Leaves and small twigs move; wind will extend light flag
4	Moderate breeze	13–18	11–16	20–29	Wind raises dust and loose paper; small branches move; flags flap
5	Fresh breeze	19–24	17–21	30–39	Small trees with leaves begin to sway; flags ripple
6	Strong breeze	25–31	22–27	40–50	Large tree branches in motion; whistling heard in power lines; umbrellas used with difficulty
7	High wind	32–38	28–33	51–61	Whole trees in motion; inconvenience felt walking against wind; flags extend
8	Gale	39–46	34–40	62–74	Wind breaks twigs off trees; walking is difficult
9	Strong gale	47–54	41–47	75–87	Slight structural damage occurs (signs and antennas blown down)
10	Whole gale	55–63	48–55	88–101	Trees uprooted; considerable damage occurs
11	Storm	64–74	56–64	102–119	Winds produce widespread damage
12	Hurricane	≥75	≥65	≥120	Winds produce extensive damage

on a scale from 0 to 12 to estimate the speed of the wind. For example, a Beaufort wind scale score of 2 is considered a light breeze that can be felt on your face and may rustle leaves on the ground. In contrast, a Beaufort wind scale score of 9 is a strong gale force wind that can produce slight structural damage and cause waves to crest and roll over.

SECTION REVIEW

1. What is the definition of *wind?*
2. Describe the process that forms planetary scale winds.
3. Describe the process that causes sea breezes to form.
4. What are three examples of local winds?
5. What are jet streams, and where are they located?
6. Who was George Hadley?

For more information go to this Web link:
<http://www.2010.atmos.uiuc.edu(Gh)/guides/mtr/fw/home.rxml>

18.2 *Air Masses and Fronts*

Air Mass Formation

An **air mass** is a large body of moving air in the troposphere that has similar characteristics of temperature, pressure, and moisture. Air masses derive their characteristics from source regions, which are geographical areas that give an air mass its unique qualities. Generally if an air mass forms in an area located in the higher latitudes, it has a cooler air temperature. Conversely, air masses that form near the equator possess warmer temperatures. The atmospheric moisture of an air mass also is related to its source region. Air masses that form over the ocean have a higher atmospheric moisture content and are considered moist. Air masses that form over the continents tend to be lower in atmospheric moisture content and are considered dry.

Source Regions and Classification of Air Masses

The source region from which an air mass originates can reveal a lot about the weather it may create. Air masses are classified in five categories. Four of those categories are shown in Table 18–2. The first is a continental polar air mass, which develops over a land mass near the poles. This particular air mass contains cool, dry air. Continental polar air masses that affect the United States

Earth Math

1) Using the following information, determine the pressure gradient that helps to form the planetary winds: The pressure at the equator (ITCZ) is 1008 millibars, and the pressure at 30 degrees north latitude (subtropical high) is 1020 millibars. The distance between the two pressure zones is approximately 3000 miles. Remember: gradient = change in value/change in distance.

TABLE 18–2 Classification of air masses and their unique characteristics

Source Region	Polar (P)	Tropical (T)
Land continental (c)	cP cold, dry, stable	cT hot, dry, stable air aloft; unstable surface air
Water maritime (m)	mP cool, moist, unstable	mT warm, moist; usually unstable

often originate over northern Canada (Figure 18–12). A maritime polar air mass originates near the poles over the ocean. This type of air mass contains cool, moist air and is often associated with cold winter rains. Source areas for maritime polar air masses include the North Atlantic and North Pacific oceans. Continental tropical air masses develop over a land mass near the equator and contain warm, dry air; those that affect North America form over Mexico and Texas. Maritime tropical air masses form over the ocean near the equator and contain warm, humid air; they are often associated with the development of hurricanes. Source regions for these air masses include the South Atlantic Ocean and the Gulf of Mexico. The arctic air mass, which contains extremely cold and dry air, forms north of 60 degrees north latitude on the ice fields of Siberia,

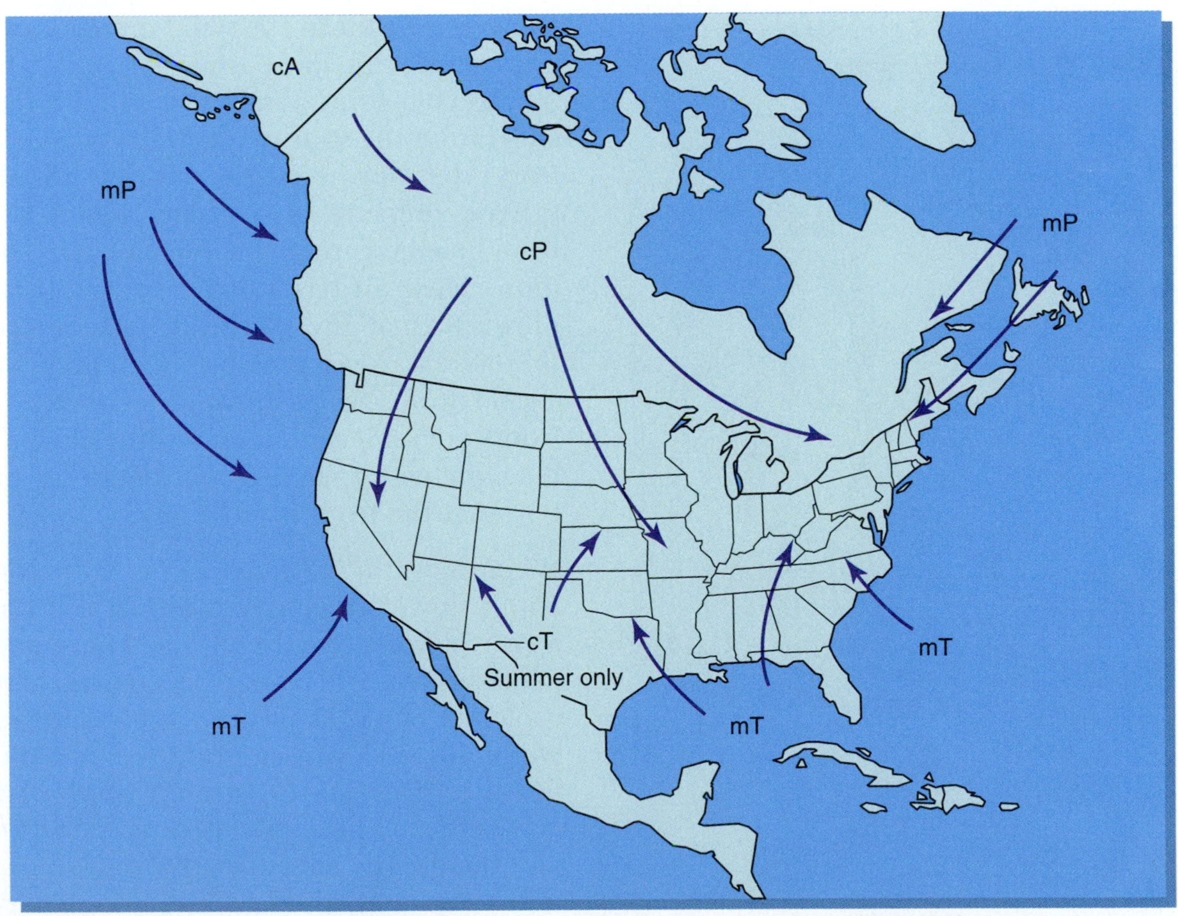

Figure 18–12 Air mass source regions that affect North America, and their general paths of movement.

Greenland, and the Arctic Ocean. Arctic air masses can bring extremely cold and dry weather to the United States during winter, known as the arctic express.

Fronts

The Earth's surface is covered with many different types of air masses moving relative to one another. Not all air masses move at the same speed; some are moving faster than others. When one air mass comes into contact with another air mass, the area where they meet is known as a **front.** The different characteristics of temperature, barometric pressure, and moisture that each air mass have come into contact with one another in the area of a front. This causes characteristic weather, unique to each type of front, to form. When an advancing air mass with cooler temperatures comes into contact with a slow-moving warmer air mass, a **cold front** forms (Figure 18–13, Table 18–3). Because the colder air is more dense than the warmer air, it wedges itself under the warm air and forces it to rise. The rising warm air, which is being pushed up into the atmosphere by the advancing cold front, cools adiabatically. This adiabatic cooling, or cooling of rising air by expansion, causes the air temperature to meet the dew point and form clouds. As a result of the rapid rising air, caused by the wedging action of cold air, cumulonimbus clouds quickly form, creating heavy precipitation and thunderstorms. Therefore cold fronts are often associated with weather that forms cumulus clouds with strong vertical development, producing heavy rains for a short period.

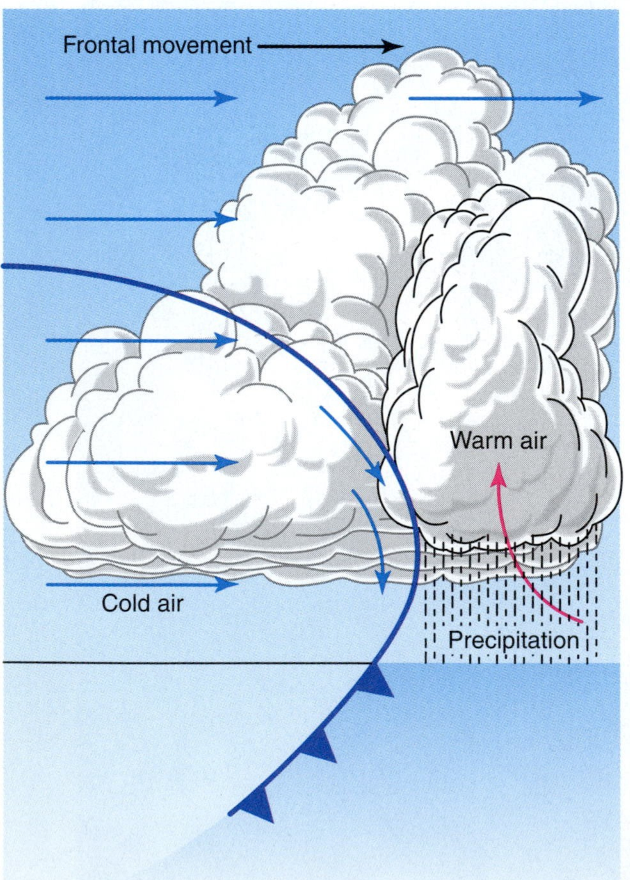

Figure 18–13 The unique cross section of a cold front.

Fast-moving cold fronts produce a band of intense thunderstorms called a squall line. Slower moving cold fronts cause the warm air ahead to rise slowly, which produces brief showers. A shift in the wind direction is also associated with the passage of a cold front. As the cold front passes, the wind direction shifts from the southwest and begins to blow from the northwest. A change in atmospheric pressure also occurs when a cold front moves through a region. The cooler, more dense air has a higher atmospheric air pressure than the warm air it is replacing; therefore a rise in pressure often occurs after a cold front passes through a region. Cold fronts are represented on weather maps by a blue line with triangles pointing in the direction that the front is moving.

When a warm air mass comes into contact with a slower moving cold air mass, it is called a **warm front** (Figure 18–14). Warm air that makes up a warm front is less dense than the cooler air of a cold air mass, which causes the warm air to slowly override the cold air beneath it. This causes the formation of high clouds first, then mid-level clouds, and finally low clouds. The advancing warm front is preceded by a gradually thickening layer of clouds that produce light precipitation for ex-

TABLE 18-3 Characteristic weather conditions associated with a cold front

Weather Element	Before Passing	While Passing	After Passing
Winds	South-southwest	Gusty, shifting	West-northwest
Temperature	Warm	Sudden drop	Steadily dropping
Pressure	Falling steadily	Minimum, then sharp rise	Rising steadily
Clouds	Increasing Ci, Cs, then either Tcu or Cb	Tcu or Cb*	Often Cu
Precipitation	Short period of showers	Heavy showers of rain or snow, sometimes with hail, thunder, and lightning	Decreasing intensity of showers, then clearing
Visibility	Fair to poor in haze	Poor, followed by improving	Good except in showers
Dew point	High; remains steady	Sharp drop	Lowering

**Tcu* stands for towering cumulus, such as cumulus congestus; *Cb* stands for cumulonimbus; *Ci* stands for cirrus; *Cs* stands for cirrostratus; and *Cu* stands for cumulus.

tended periods. This type of light rain is good for agriculture because it allows the precipitation to infiltrate the soil. If a warm front is producing precipitation in the form of snow, it can lead to heavy accumulations. As the warm air overrides the cold air around a warm front, fog may also form along the thin boundary that separates the warm air from the cold air just above the Earth's surface. This is called a prefrontal fog.

The shift in wind that is associated with the passage of a warm front changes from the southeast to the southwest (Table 18–4). The less dense warm air that is replacing the cooler, more dense air also brings a change in barometric pressure. The pressure usually drops after a warm front passes through an area. Warm fronts are represented on weather maps by a red line with half circles pointing in the direction the front is traveling.

Another type of front that forms between two different air masses is an **occluded front** (Figure 18–15). An occluded front forms when a rapidly moving cold front moves

EARTH SYSTEM SCIENTISTS *SVANTE ARRHENIUS*

Svante Arrhenius was born in Sweden in 1869. Most of his life's work was devoted to the study of chemistry. His research involved the study of the electrical conductivity of solutions. He received the Noble Prize for chemistry in 1903 as a result of this pursuit. Arrhenius' contribution to atmospheric science came from what he called a hobby. In 1895 he presented a scientific paper titled "On the Influence of Carbonic Acid in the Air Upon the Temperature on the Ground." This groundbreaking hypothesis revealed the effects of carbon dioxide gas and water vapor on the surface temperature of the Earth. In his paper, Arrhenius unlocked the mechanisms of the greenhouse effect. He also hypothesized that changes in the composition of carbon dioxide in the atmosphere might have led to ice ages in the past.

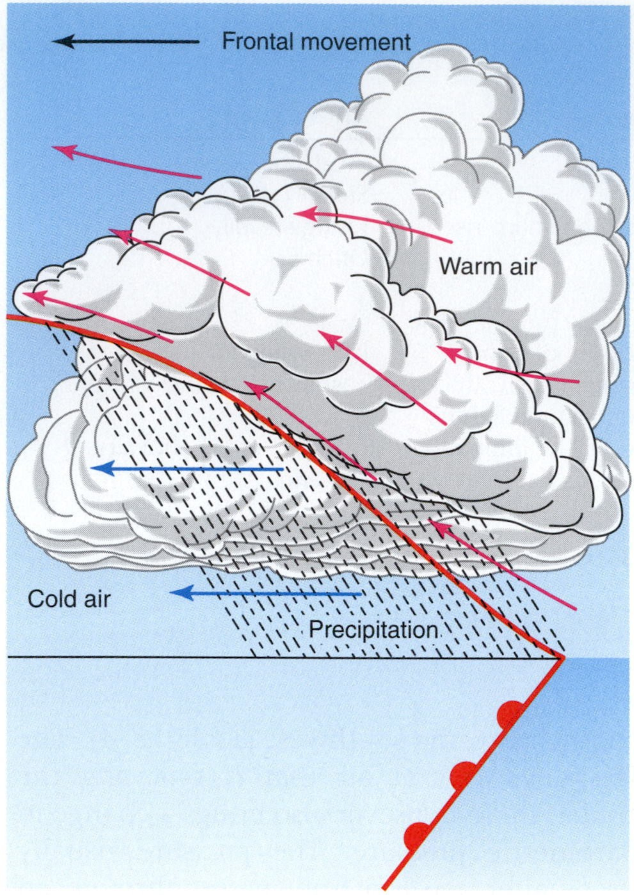

Figure 18–14 The unique cross section of a warm front.

Weather Observer

A weather observer is a trained individual who makes hourly observations about the present state of the atmosphere. Surface weather observations form the base for all types of meteorology, by providing current atmospheric conditions at many points around the world. These technicians record hourly temperature, humidity, cloud cover, precipitation, winds, pressure, and visibility at their weather stations. This data is then collected and used to create synoptic weather maps. These maps are then analyzed, and weather forecasts are made. Weather observers work in weather stations located in airports and colleges, onboard ships, and in other special locations around the globe. Many weather observers gain their training while serving in the military, or they can gain experience in college by majoring in meteorology.

under a slow-moving warm front and causes it to uplift. The rapid movement of the cold air completely raises the entire warm air mass off the surface.

This results in widespread precipitation, which can be intense or sustained, depending on the lifting action of the cold air (Table 18–5). Wind shifts associated with the passage

TABLE 18-4 Characteristic weather conditions associated with a warm front

Weather Element	Before Passing	While Passing	After Passing
Winds	South-southeast	Variable	South-southwest
Temperature	Cool, cold, slow warming	Steady rise	Warmer, then steady
Pressure	Usually falling	Leveling off	Slight rise, followed by fall
Clouds	In this order: Ci, Cs, As, Ns, St, and fog; occasionally Cb in summer	Stratus-type	Clearing with scattered Sc; occasionally Cb in summer
Precipitation	Light-to-moderate rain, snow, sleet, or drizzle	Drizzle or none	Usually none; sometimes light rain or showers
Visibility	Poor	Poor, but improving	Fair in haze
Dew point	Steady rise	Steady	Rise, then steady

Ci, cirrus; *Cs*, cirrostratus; *As*, altostratus; *Ns*, nimbostratus; *St*, Stratus; *Cb*, cumulonimbus.

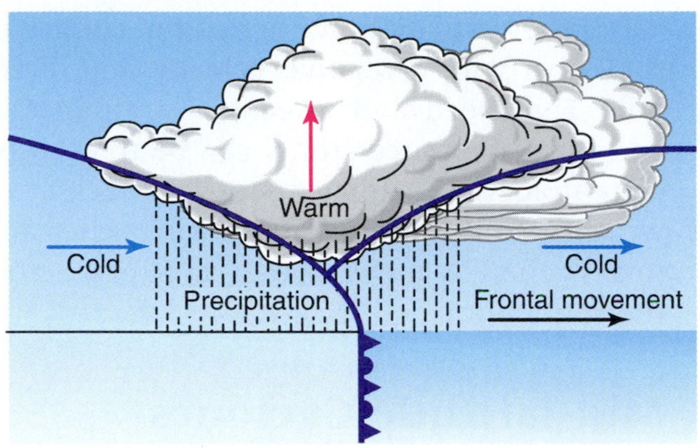

Figure 18–15 The unique cross section of an occluded front.

of an occluded front are usually from the southeast to the northwest, which is almost a 180-degree shift in winds. An occluded front is represented by a purple line with both triangles and half circles pointing in the direction of travel of the front.

Not all air masses run into one another to form cold, warm, or occluded fronts; sometimes air masses slide past one another. The type of front that forms when cold air and warm air sit next to each other is called a **stationary front** (Figure 18–16). This type of front usually creates clear weather with no appreciable cloud development. This is because the cold air stays on one side of the front and the warm air stays on the other, with no interactions occurring. Eventually one of the air masses begins to overtake the other, and the stationary front becomes either a cold front or a warm front. Stationary fronts are represented on weather maps by a blue and red line with blue triangles pointing toward the cold air mass and red half circles pointed at the warm air mass.

TABLE 18-5 Characteristic weather conditions associated with an occluded front

Weather Element	Before Passing	While Passing	After Passing
Winds	Southeast-south	Variable	West to norhtwest
Temperature			
Cold type	Cold, cool	Dropping	Colder
Warm type	Cold	Rising	Milder
Pressure	Usually falling	Low point	Usually rising
Clouds	In this order: Ci, Cs, As, Ns	Ns, sometimes Tcu and Cb	Ns, As, or scattered Cu
Precipitation	Light, moderate, or heavy precipitation	Light, moderate, or heavy continuous precipitation or showers	Light-to-moderate precipitation followed by general clearing
Visibility	Poor in precipitation	Poor in precipitation	Improving
Dew point	Steady	Usually slight drop, especially if cold/occluded	Slight drop, although may rise a bit if warm/occluded

Ci, cirrus; *Cs*, cirrostratus; *As*, altostratus; *Ns*, nimbostratus; *Cb*, cumulonimbus.

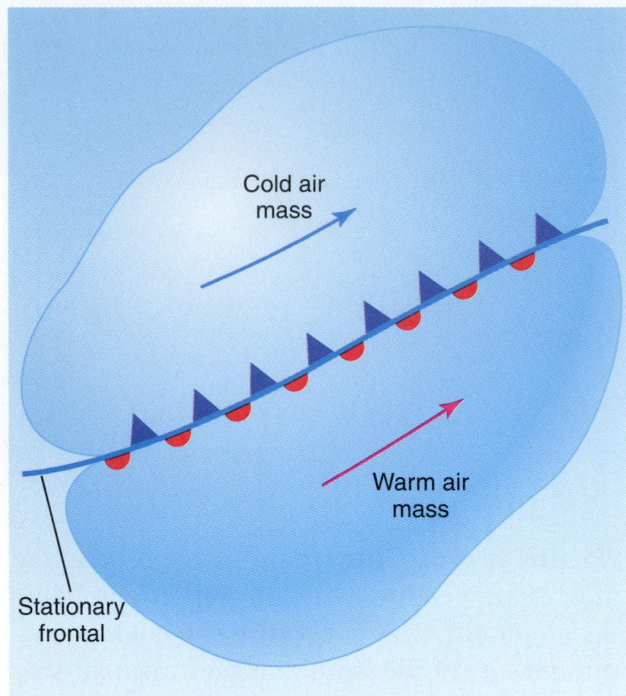

Figure 18–16 A stationary front forms when a cold air mass moves alongside a warm air mass.

Air masses usually move across the surface of the Earth with a relative velocity and direction. The speed and direction of air masses and their associated fronts can be tracked by meteorologists, helping them to forecast the weather. In the United States, air masses and fronts generally move from west to east across the continent.

Mid-latitude Cyclones

The formation of fronts along the boundaries of air masses constantly undergoes change as the temperature, humidity, and pressure of the colliding air masses also change. The type of changes associated with interacting fronts help to form a unique storm system called the mid-latitude cyclone. A mid-latitude cyclone forms in the Northern Hemisphere around a low-pressure system. A **cyclone** is a counterclockwise, inward rotation of air around a

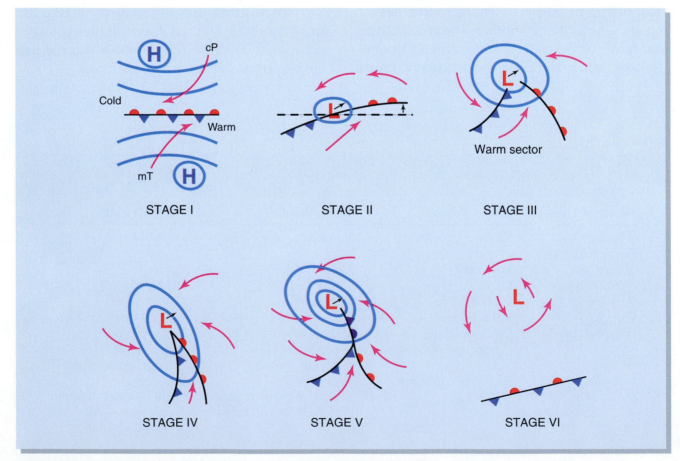

Figure 18–17 The different stages of development of a mid-latitude cyclone.

low-pressure center. Mid-latitude cyclones form when an advancing cold air mass moves up behind a slower moving warm air mass. Both frontal systems are centered around the spiraling low-pressure center. The clouds and precipitation associated with the two frontal boundaries cause a wide band of precipitation that precedes the fronts (Figure 18–17).

Eventually, as the low-pressure system intensifies, the precipitation begins to spiral inward toward the pressure center, forming a unique cloud formation called a comma cloud. The comma cloud gets it name from the shape of the clouds that result from the cold and warm fronts spiraling around the center of low pressure. As the air masses move across the land surface, the cold front begins to overtake the warm air mass and forms an occluded front. The pressure system then begins to weaken, and the cyclone breaks apart as it moves out over the Atlantic Ocean. Many of the strong storms that affect the northeastern United States, creating heavy precipitation, are caused by the formation of mid-latitude cyclones.

For more information go to this Web link:
<http://www.2010.atmos.uiuc.edu(Gh)/guides/mtr/af/home.rxml>

SECTION REVIEW

1. Define the term air mass.
2. What are the five source regions for air masses?
3. Define the term *front.*
4. What is a cold front? Describe some of it characteristics.
5. What is a warm front? Describe some of its characteristics.
6. What is an occluded front? Describe some of its characteristics.
7. What is a stationary front? Describe some of its characteristics.
8. Draw a rough sketch of a mid-latitude cyclone, using the correct fronts.
9. Who was Svante Arrhenius?

CHAPTER SUMMARY

Wind is the horizontal movement of air across the Earth's surface. It is caused by differences in atmospheric pressure as air moves from areas of high pressure to areas of low pressure. The speed of the wind is affected by the pressure gradient. Pressure gradient is the change in pressure that occurs over a specific distance. The greater the pressure gradient, the greater the wind speed over an area. Wind is known as a vector force because it has both direction and velocity. Large-scale winds that move air through the atmosphere are called planetary winds. Planetary winds form as a result of the unequal heating of the Earth. Warm, rising air at the equator forms areas of low atmospheric pressure. The air eventually cools and sinks back toward the surface at approximately 30 degrees north and south latitude, forming areas of high atmospheric

EARTH MATH

1) If a cold front is located in Chicago at noon and arrives in Cleveland (363 miles away) at 11:00 P.M., approximately how fast is the front moving?

pressure. Air then flows from the high pressure toward the low pressure, creating planetary winds. This also occurs at the higher latitudes when air flows from the polar highs toward the subpolar lows.

Wind on Earth is affected by the Coriolis effect, which is caused by the Earth's rotation. This results in winds being deflected to the right of travel in the Northern Hemisphere and to the left of travel in the Southern Hemisphere. The Coriolis effect also influences wind that is associated with pressure centers on the Earth. It causes winds to spiral inward and counterclockwise around low pressure and outwards and clockwise around high pressure in the Northern Hemisphere.

Mesoscale winds can form as a result of local pressure differences. A sea breeze occurs when air flows from the higher pressure forming over cooler water toward low pressure forming over the rapidly heated land surface. At night the opposite, called a land breeze, occurs. This forms when air flows from high pressure over the cooler land toward low pressure over the warmer water. Other smaller scale winds that can form at the Earth's surface include mountain and valley breezes, along with cold katabatic winds.

An air mass is a large mass of air that has uniform temperature, pressure, and moisture. Air masses are characterized by where they form, known as air mass source regions. These regions produce five basic types of air masses: continental tropical, continental polar, maritime tropical, maritime polar, and arctic. When two air masses come into contact with one another, they form a front.

Fronts are areas where unique weather develops as a result of the collision of two different air masses. There are four types of fronts: cold, warm, occluded, and stationary. All fronts have their own characteristic change in temperature, pressure, wind direction, cloud type, and precipitation. In North America a unique type of weather pattern forms as a result of the interaction between cold and warm air masses. This is known as a mid-latitude cyclone. A cyclone is a low-pressure system. The mid-latitude cyclone goes through unique stages of development that bring stormy weather to the United States.

CHAPTER REVIEW

Multiple Choice

1. Winds blow from regions of:
   a. high air temperature to regions of low air temperature
   b. high air pressure to regions of low air pressure
   c. high precipitation to regions of low precipitation
   d. convergence to regions of divergence

2. The primary cause of winds on the Earth's surface is the:
   a. unequal heating of the Earth's atmosphere
   b. uniform density of the atmosphere
   c. friction between the atmosphere and lithosphere
   d. rotation of the Earth

3. The wind speed between two nearby locations is affected most directly by the differences in the:
   a. latitude between the two locations
   b. longitude between the two locations
   c. air pressure between the two locations
   d. Coriolis effect between the two locations

4. The Coriolis effect is caused by the:
   a. movement of the Earth in relation to the Milky Way
   b. movement of the Earth in relation to the Moon
   c. revolution of the Earth around the Sun
   d. the rotation of the Earth on its axis

5. In the Northern Hemisphere the wind blowing from the north is deflected toward which direction as a result of the Coriolis effect?
   a. northwest
   b. northeast
   c. southwest
   d. southeast

6. What are the planetary winds that form between 30 degrees north latitude and the equator?
   a. the ITCZ winds
   b. the westerlies
   c. the trade winds
   d. the katabatic winds

7. In the Northern Hemisphere, winds move around a low-pressure system:
   a. spiraling outward and clockwise
   b. spiraling inward and clockwise
   c. spiraling outward and counterclockwise
   d. spiraling inward and counterclockwise

8. Which type of wind results from high pressure forming over the ocean and low pressure forming over the land?
   a. a land breeze
   b. a sea breeze
   c. a cyclone
   d. a katabatic wind

9. In the Northern Hemisphere a high-pressure center is also known as:
   a. a cyclone
   b. an anti-cyclone
   c. a whirlwind
   d. a mid-latitude cyclone

10. What type of air mass forms over central Canada?
    a. continental tropical
    b. continental polar
    c. maritime tropical
    d. maritime polar

11. What type of air mass forms over the South Atlantic Ocean?
    a. continental tropical
    b. continental polar
    c. maritime tropical
    d. maritime polar

12. When a warm air mass overtakes a cold air mass, what type of front forms?
    a. a cold front
    b. a occluded front
    c. a warm front
    d. a stationary front

13. When a cold air mass completely lifts a warm air mass off the surface, it forms:
    a. a cold front
    b. an occluded front
    c. a warm front
    d. a stationary front

14. Generally fronts move across the United States from:
    a. north to south
    b. south to north
    c. east to west
    d. west to east

15. Heavy, short-duration rain and lowering then rising air pressure signal the passing of:
    a. a cold front
    b. an occluded front
    c. a warm front
    d. a stationary front

continued

Matching *Match the terms with the correct definitions.*

a. wind
b. pressure gradient
c. vector force
d. planetary winds
e. Hadley cell
f. Coriolis effect
g. pressure center
h. jet stream
i. air mass
j. front
k. cold front
l. warm front
m. occluded front
n. stationary front
o. cyclone

1. ____ A front that develops between two air masses that are not moving.
2. ____ A large body of air that has similar temperature and moisture characteristics.
3. ____ The horizontal movement of air across the Earth's surface, from areas of high atmospheric pressure to areas of low atmospheric pressure.
4. ____ An area where two different air masses come together.
5. ____ A term for winds that spiral inward and around a low-pressure center.
6. ____ The change in atmospheric pressure that occurs over a specific distance.
7. ____ The zone where a cold air mass overtakes and replaces a warm air mass.
8. ____ A force that has both speed and direction.
9. ____ A frontal boundary that usually occurs when a rapidly moving cold air mass overtakes a warm air mass by wedging underneath it and lifting upward, creating very unstable weather.
10. ____ Large-scale winds that circulate air around the Earth, such as the trade winds or the westerlies.
11. ____ A front that develops when a warm air mass replaces a cold air mass.
12. ____ A region of high-velocity winds that exist high up in the atmosphere.
13. ____ The large-scale convection cell that forms low pressure at the equator and high pressure at 30 degrees north and south of the equator, producing planetary winds.
14. ____ A region on the Earth's surface around which air circulates, where the atmospheric pressure is relatively low or high compared with the surrounding air.
15. ____ The deflection of free-floating objects to the right of travel in the Northern Hemisphere and to the left of travel in the Southern Hemisphere as a result of the Earth's rotation.

Critical Thinking

1. Describe the rotation of the winds around the low-pressure center of a hurricane that forms in the Southern Hemisphere and slowly moves over the equator into the Northern Hemisphere.

CHAPTER 19 Storms and Weather Forecasting

Section 19.1 – Storms Objectives

Thunderstorms • Tornadoes • Hurricanes

After reading this section you should be able to:

- Describe specific aspects of all three stages of thunderstorm development.
- Identify the dangerous weather that accompanies a thunderstorm.
- Define the term *tornado* and describe how they form.
- Explain how the Fujita scale is used to classify a tornado.
- Describe the conditions that lead to the formation of a hurricane.
- Differentiate between a tropical disturbance, tropical storm, and a hurricane.
- Define the term *storm surge* and explain how it forms.
- Explain how the Saffir-Simpson scale is used to classify hurricanes.

Section 19.2 – Weather Forecasting Objectives

Weather Data Collection • Synoptic Weather Maps • Weather Forecasts • Weather Radar and Satellites

After reading this section you should be able to:

- Describe the three processes involved in making weather forecasts.
- Identify the specific weather data that is commonly observed and collected at weather stations around the world.
- Decode the information recorded in a station model.
- Define the terms *isotherm* and *isobar*.
- Describe the information displayed on weather forecast maps, and differentiate between long-term and short-term weather forecasts.
- Describe how radar is used to forecast the weather.
- Describe some simple observations of the environment that may be used to forecast the weather.

TERMS TO KNOW

thunderstorm	**tropical disturbance**	**station model**
lightning	**tropical depression**	**isotherms**
tornado	**tropical storm**	**isobars**
hurricane	**storm surge**	**radar**
cyclone	**typhoon**	**infrared satellite images**

INTRODUCTION

The most deadly aspect of the atmosphere is the formation of strong storms at the Earth's surface. The advances in meteorology over the past 50 years have been centered around understanding how and when storms form. Deadly storms such as thunderstorms, hurricanes, and tornadoes can cause severe damage to populated areas and lead to the loss of life. Hurricane Andrew struck southern Florida in 1992, causing more than 30 billion dollars in damage and taking the lives of 25 people. Predicting these dangerous storms involves the constant observation of changing weather conditions by meteorologists around the country. These observations are then used to make maps and computer models of the atmosphere, which are used to forecast the weather. Today's society is highly dependent on weather forecasts for many aspects of modern life.

19.1 Storms

Thunderstorms

Probably the most widespread type of storm to affect the United States is the **thunderstorm.** Thunderstorms form as a result of intense convection associated with the heating of the Earth's surface during spring, summer, and fall. Thunderstorms go through a typical life cycle that is divided into three stages of development (Figure 19–1). The beginning stages of a thunderstorm involve the formation of a cumulus cloud. The rapid heating of the land during a hot day causes warm air to form at the surface. The warm air begins to rise because it is less dense than the surrounding air. Eventually the rising air begins to cool, causing water vapor to condense and form a cloud. As the ground continues to be heated, the cloud grows bigger and small-scale convection cells form. The rising air from the surface creates an area of low pressure beneath the cloud, which draws in air from the surrounding area. This begins to form winds at the surface. During the beginning stages of growth, cumulus clouds can experience rapid vertical development of more than 26,000 feet in only 15 minutes; eventually its base can cover a distance of more than 6 miles. As the cloud grows, the convection intensifies, creating strong winds at the surface of the cloud as air rushes in to replace the warm air that is rising into the cloud. The rapidly rising air also creates strong updrafts within the cloud. These updrafts cause the water droplets and ice crystals that make up the cloud to clump together, forming precipitation. When the top of the cloud reaches a height of more than 7 miles, it begins to produce heavy precipitation and is known as a cumulonimbus cloud. The thunderstorm is now entering in the mature stage of formation.

During the mature phase, strong updrafts and downdrafts of wind continue to intensify in the cloud. Downdrafts are formed by cooler air sinking toward the Earth's surface. The updrafts and downdrafts cause the separation of lighter ice crystals near the top of the cloud and heavier hail at the lower part of the cloud. The ice crystals become positively charged with static electricity, and the hail becomes negatively charged. A large electrical potential is formed between the two different particles in the cloud. This potential is eventually released in the form of **lightning** (Figure 19–2).

Lightning is an electrical discharge of more than 100 million volts that heats the sur-

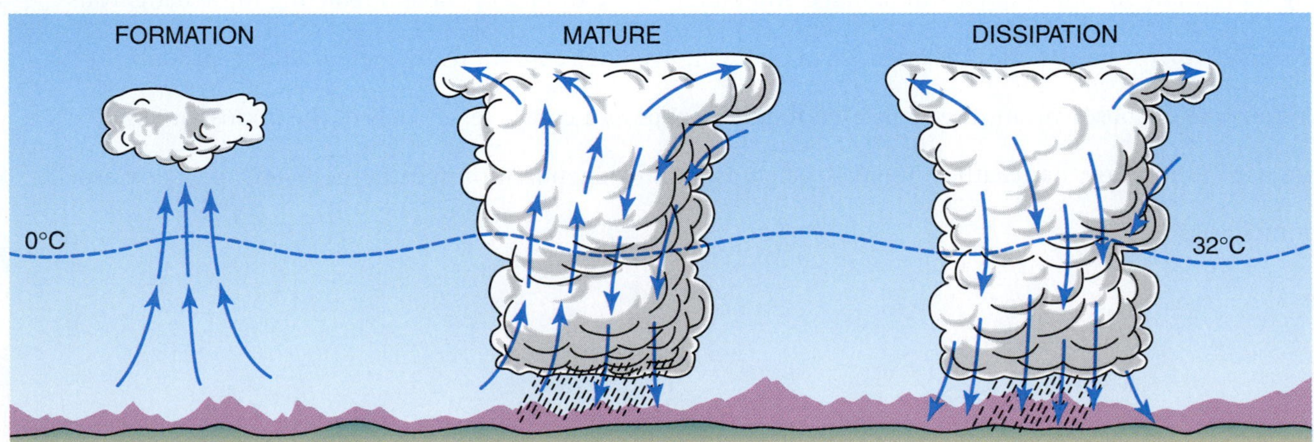

Figure 19–1 The three main stages in the life cycle of a thunderstorm.

Figure 19–2 The formation of lightning within a thunderstorm as a result of opposite electrical charges forming within the cloud.

Figure 19–3 When lightning strikes, it can generate more than 100 million volts of electricity and heat the surrounding air to more than 45,000° Fahrenheit. *(Courtesy of PhotoDisc.)*

rounding air to about 45,000° Fahrenheit (Figure 19–3). This causes the cool air in the cloud to rapidly expand and create a loud noise known as thunder. The lightning that is produced during a thunderstorm is classified by where it travels. Lightning can travel many different ways, including cloud to cloud, cloud to ground, and cloud to air. All forms of lighting can be potentially dangerous. Lightning can start fires, destroy electrical equipment, and kill or injure people. One of the most dangerous things a person can do during a thunderstorm is to talk on the telephone. Lightning can strike telephone lines, travel into homes, and electrocute a person using a phone that has a cord. Cordless phones and cell phones are safe to use during thunderstorms. It also dangerous to stand beneath trees during a thunderstorm, because lighting is often attracted to the tops of tall objects such as trees. The safest place to be during a thunderstorm is indoors.

EARTH SYSTEM SCIENTISTS *WILHELM BJERKNES*

Wilhelm Bjerknes was born in Norway in 1862. Today many meteorologists regard him as the father of weather forecasting and modern meteorology. Bjerknes's early work in physics and electricity had him working with Heinrich Hertz in Germany. Their research on the flow of electrical currents helped lead to the development of the radio. Upon his return to Norway, Bjerknes turned his attention to the study of atmospheric and oceanic circulation. In 1905 he presented his theory of weather prediction, which proposed that the weather could be predicted by applying mathematical models to the present state of the atmosphere. This led to the development of modern forecasting techniques. Bjerknes was one of the first meteorologists to set up a network of weather observation points to collect data about the state of the atmosphere at a particular time over a wide geographic region. Later in his life, while working with his son Jakob, he also discovered the phenomena known as air masses and fronts and helped to create a model that explains the stages of the formation of a mid-latitude cyclone.

Current Research

A joint study between scientists at the National Aeronautic and Space Administration (NASA) and the University of Georgia have revealed the effects of urban heat islands on thunderstorm formation. An urban heat island is an area where natural forests and vegetation have been replaced by buildings and blacktop roads. The roads and buildings absorb a greater amount of heat energy than vegetation, causing an increase in the surface temperature of the area. The study centers around the metropolis of Atlanta, Georgia, where more than 350,000 acres of land has been replaced by urban sprawl since 1973. The heat island effect caused by the urbanization of Atlanta has raised the surface temperatures there by as much as 10°. This excess heat has a number of adverse effects on the region. Because of the higher temperatures, people must use more electricity to cool their homes with air conditioners. The researchers also revealed that Atlanta is creating its own thunderstorms. The excess heat causes air to rise rapidly, developing into violent storms. Other urban areas around the world are also believed to be experiencing the heat island effect. The team of scientists suggest that the effect could be reduced if the roofs of buildings are painted white, if lighter colored construction materials are used, and if more trees are planted.

Another deadly aspect of thunderstorms can be the strong, damaging winds they can generate. Many lives are lost and property damaged by the strong winds that thunderstorms can bring to an area. Often thunderstorms travel so quickly there is little warning and little time for people to seek shelter. Lives are lost each summer in campgrounds where thunderstorms cause trees to blow over onto tents and campers who are seeking shelter from the quickly arriving storm. Heavy rains produced by thunderstorms can also cause flash flooding. The downpours that accompany thunderstorms quickly dump large amounts of water on the surface of the Earth, causing rapid flooding of low-lying areas. Eventually the strong convection that formed the thunderstorm begins to diminish, and the storm enters the dissipating stage. This results in a decline in winds and rain as the cloud begins to dissipate and the skies clear.

Tornadoes

A **tornado** is a rapidly spiraling column of air that comes into contact with the ground; it is one of the most deadly and unpredictable

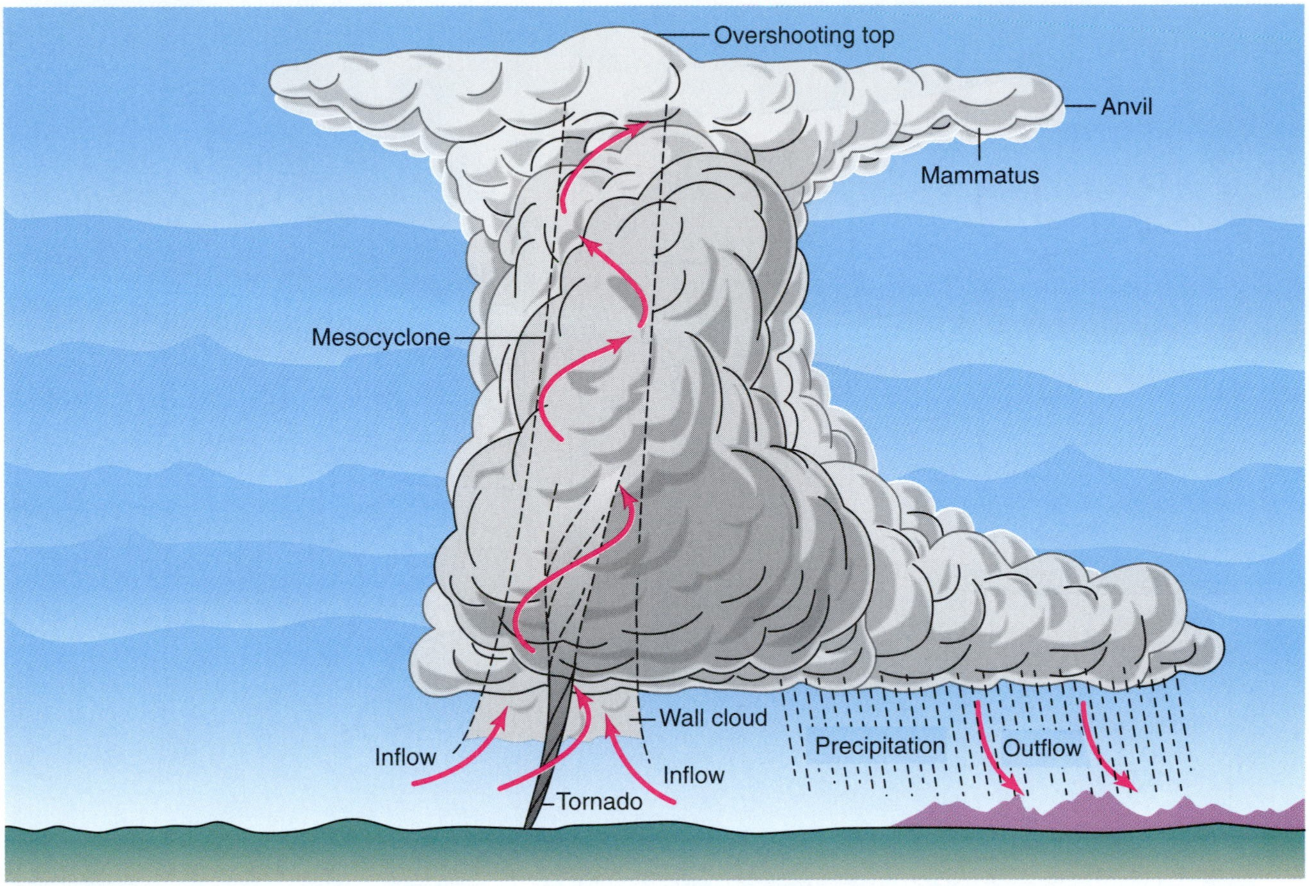

Figure 19–4 The formation of a tornado at the base of a thunderstorm results from strong updrafts.

forms of weather on the Earth. Tornadoes form at the base of thunderstorms when an extremely steep pressure gradient develops as a result of strong convection (Figure 19–4). Tornadoes produce visible columns of swirling air called funnel clouds, which are composed of rapidly moving dust and debris that is being picked up from the ground by the extremely powerful winds.

Tornadoes can produce wind speeds of up to 300 miles per hour (Figure 19–5). Although extremely violent and deadly, tornadoes usually last only a few minutes; however, some tornadoes have lasted up to 2 hours. Every year in the United States approximately 800 tornadoes are formed. Most tornadoes develop during the warmest part of the day, between 10:00 A.M. and 6:00 P.M., and occur between the months of March and July. Tornadoes have been reported in every state in the United States, but most occur in the south central part of the country, known as Tornado Alley (Figure 19–6).

Figure 19–5 A tornado is one of the most powerful natural forces on the Earth. *(Courtesy of NOAA.)*

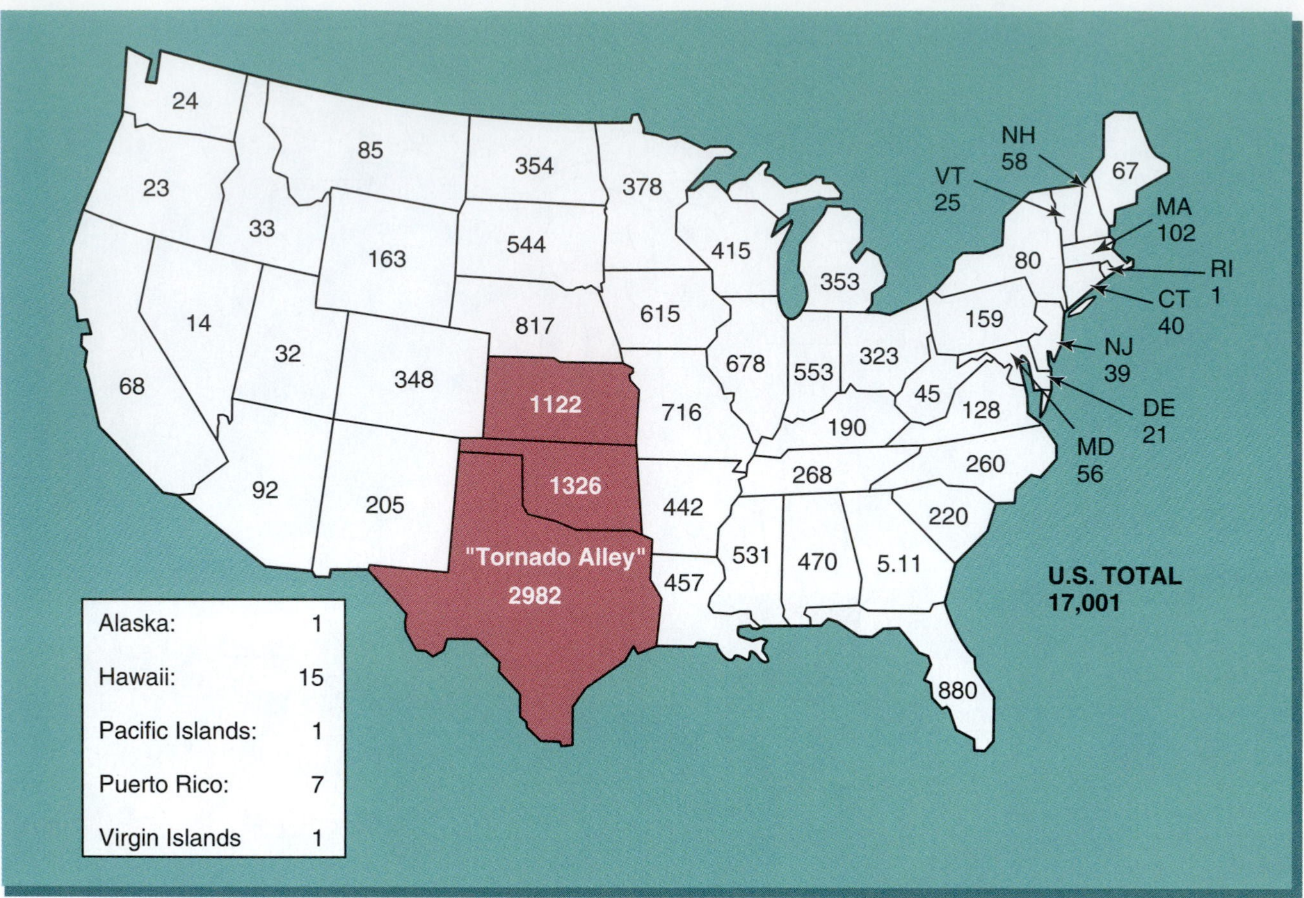

Figure 19–6 Tornado occurrences during a 25-year period for each state. Notice the high number of tornadoes in the Midwest, known as Tornado Alley.

Professor T. Theodore Fujita from the University of Chicago created what is called the Fujita scale of tornado intensity. This scale rates tornadoes by their wind speed on a scale from F0 to F5 (Table 19–1). A tornado classified as F0 on the Fujita scale produces winds of up to 73 miles per hour. An F5 tornado can sustain winds between 261 to 318 miles per hour. When tornadoes move over water, they produce what is known as a waterspout. This occurs as water is drawn up into the spiraling winds that make up the tornado. Both waterspouts and tornadoes can move across the landscape with speeds from 30 to 75 miles per hour. Because of the unpredictable nature of tornado formation and their rapid movement, they are the most difficult storms to predict, allowing only a few minutes of warning to those who might be affected by the tornado.

Hurricanes

A **hurricane** is a violent storm that is associated with a large low-pressure system, also called a **cyclone.** Cyclones rotate counterclockwise around a center of low pressure in the Northern Hemisphere, with winds spiraling inward at high speeds (Figure 19–7). Hurricanes begin to form over the Atlantic Ocean near the equator as a center of low pressure. As the pressure center moves over the ocean toward the United States, it picks up energy and moisture from the warm surface water. At this point the low-pressure sys-

Career Connections

Synoptic Meteorologist

A synoptic meteorologist helps to create new methods of studying and forecasting the weather. This involves the use of new technologies such as satellites, Doppler radar, and complex computer models. Synoptic meteorologists must have a solid background in weather forecasting and the principles of atmospheric science. They then use this knowledge to discover new and more effective methods of weather prediction. Synoptic meteorology is searching for ways to better understand the chaotic nature of weather, which might make it easier to predict. This research is especially important when it is applied to forecasting potentially dangerous storms. An interest in technology and computers is required for this profession because technical advances have greatly improved the forecasting abilities of synoptic meteorologists. Persons in this occupation can find employment in private industry and academic research or with government agencies. It also requires at least a 4-year college degree in meteorology.

tem is called a **tropical disturbance.** The winds speeds associated with a tropical disturbance are less than 20 miles per hour.

As the disturbance moves over the warm ocean, it gathers more heat and moisture and begins to intensify. At this point the storm is known as a **tropical depression,** with winds speeds between 21 and 39 miles per hour. As the storm moves across the ocean, its clouds begin to thicken and rotate around the center of low pressure. Wind speeds then increase to between 40 and 70 miles per hour and the system is known as a **tropical storm.** When the sustained winds of the cyclone reach more than 74 miles per hour, the tropical storm becomes a hurricane. The characteristic spiraling clouds of a hurricane swirl around the center of low pressure, which is also called the eye, and the storm strengthens. Winds generated by a strong hurricane can be in excess of 250 mile per hour.

The rotation of the winds and clouds in a hurricane is the same for any low-pressure system in the Northern Hemisphere: They move inward and counterclockwise. Because of this counterclockwise rotation, the strong winds build up high waves on the northwestern edge of the hurricane as it moves westward toward the coast. This pileup of water ahead of the hurricane is called a **storm surge,** which can raise tides along the coast by as much as 16 feet. Waves generated by the strong winds can also top 10 feet. When the hurricane strikes land, the storm surge it creates often causes disastrous flooding.

TABLE 19-1 The Fujita scale classifies the intensity of tornadoes

Scale	Category	Mi/Hr	Knots	Expected Damage
F0	Weak	40–72	35–62	Light: tree branches broken, sign boards damaged
F1		73–112	63–97	Moderate: trees snapped, windows broken
F2	Strong	113–157	98–136	Considerable: large trees uprooted, weak structures destroyed
F3		158–206	137–179	Severe: trees leveled, cars overturned, walls removed from buildings
F4	Violent	207–260	180–226	Devastating: frame houses destroyed
F5		261–318	227–276	Incredible: structures the size of autos moved more than 100 meters, steel-reinforced structures highly damaged

Figure 19–7 A satellite image shows the characteristic cyclonic movement of clouds around the eye of a hurricane. *(Courtesy of NOAA.)*

Approximately 90% of all deaths associated with hurricanes are caused by flooding as a result of the storm surge.

Because hurricanes form in the Atlantic Ocean and move toward the coast, it is possible to predict the speed and direction that the hurricane is moving (Figure 19–8). Hurricanes typically travel between 6 and 12 miles per hour. This information is extremely important to the people that live along the coast, who can then be warned and have time to evacuate the area. Once a hurricane strikes land, it usually begins to weaken as its energy and moisture supply is cut off. Besides strong winds and storm surges, hurricanes can also cause inland areas to flood as a result of heavy rains. Rainfall amounts associated with hurricanes can be as high as 10 to 20 inches in 24 hours. Hurricanes can also generate tornadoes from any of the thunderstorm systems associated with the mass of clouds that surround the eye.

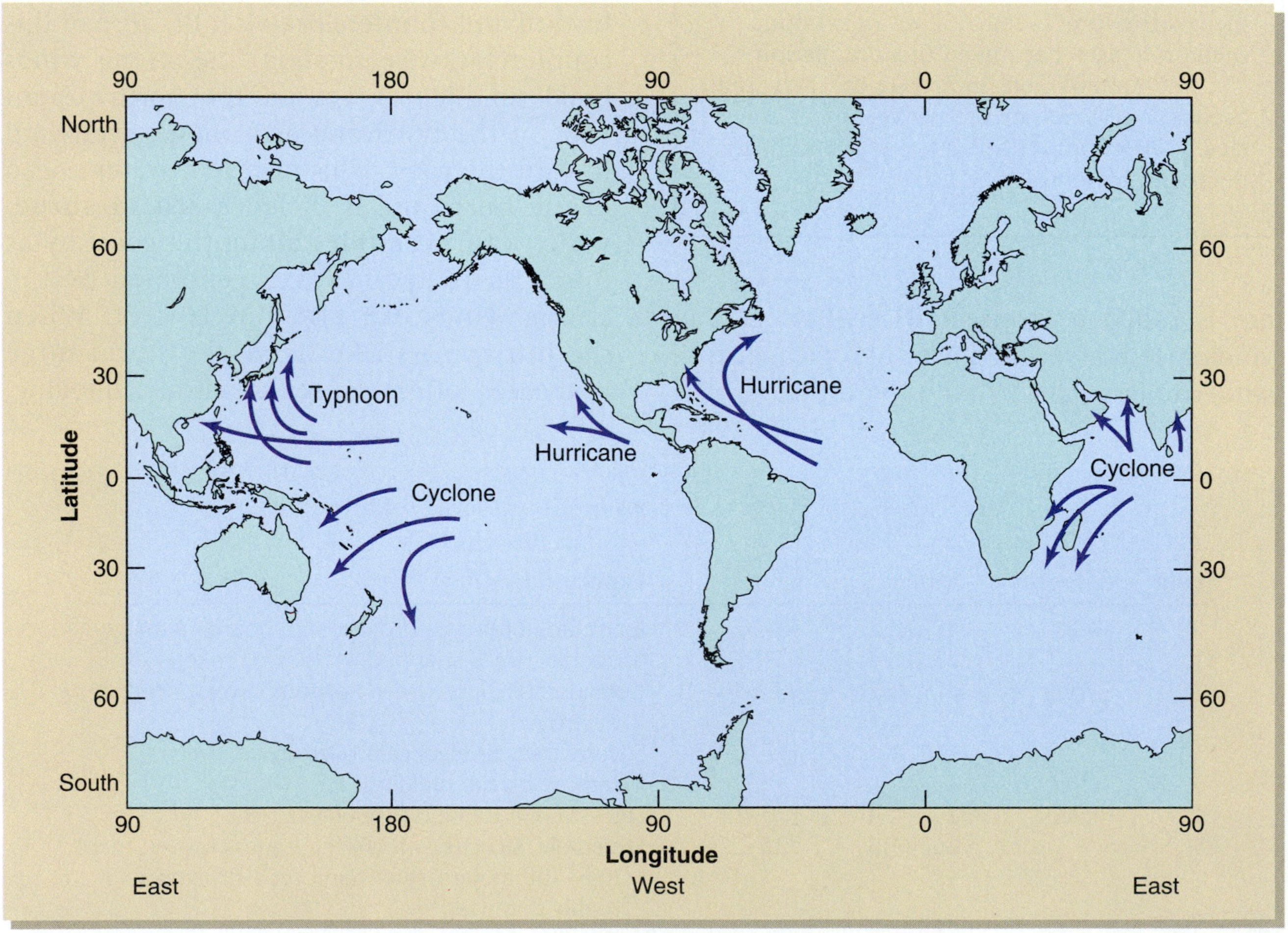

Figure 19–8 The average paths of movement for hurricanes around the world.

The intensity of hurricanes is often measured using the Saffir-Simpson scale, which measures the pressure, wind speed, and storm surge of hurricanes on a scale from 1 to 5 (Table 19–2). Hurricanes usually occur during specific times of the year in the United States. This is called the hurricane season, which begins in June and ends in late November. When a tropical depression that may develop into a hurricane forms in the Atlantic, it is given a name. The naming of hurricanes is done alphabetically and alternates between male and female names each year. Strong low-pressure centers that form hurricanes can

TABLE 19-2 The Saffir-Simpson scale used to classify hurricanes

Scale Number (Category)	Central Pressure mb	Central Pressure in.	Winds mi/hr	Winds knots	Storm Surge ft	Storm Surge m	Damage
1	≥980	≥28.94	74–95	64–82	4–5	~1.5	Damage mainly to trees, shrubbery, and unanchored mobile homes
2	965–979	28.50–28.91	96–110	83–95	6–8	~2.0–2.5	Some trees blow down; major damage to exposed mobile homes; some damage to roofs of buildings
3	945–964	27.91–28.47	111–130	96–113	9–12	~2.5–4.0	Foliage removed from trees; large trees blown down; mobile homes destroyed; some structural damage to small buildings
4	920–944	27.17–27.88	131–155	114–135	13–18	~4.0–5.5	All signs blow down; extensive damage to roofs, windows, and doors; complete destruction of mobile homes; flooding inland as far as 10 km (6 mi); major damage to lower floors of structures near shore
5	<920	<27.17	>155	>135	>18	>5.5	Severe damage to windows and doors; extensive damage to roofs of homes and industrial buildings; small buildings overturned and blown away; major damage to lower floors of all structures less than 4.5 m 915 ft) above sea level within 500 meters of shore

also form in the Pacific Ocean; however, they are referred to as **typhoons;** therefore typhoons strike Hawaii, and hurricanes strike the East Coast. The worst hurricane to hit the United States in the past 100 years struck the coast of Texas in September 1900. This occurred before hurricanes were named and produced winds in excess of 130 miles per hour. The storm surge created at the northwestern edge of the hurricane rose tides more than 18 feet, killing more than 6000 people.

SECTION REVIEW

1. Describe the process of thunderstorm formation.
2. What are the hazards associated with thunderstorms?
3. What are tornadoes, and where can they occur in the United States?
4. Describe the process that leads to the formation of a hurricane.
5. What are the hazards associated with hurricanes?
6. Who was Wilhelm Bjerknes?

For more information go to these Web links:

<http://ww2010.atmos.uiuc.edu/(Gh)/guides/mtr/svr/home.rxml>
<http://www.spc.noaa.gov/>
<http://www.noaa.gov/tornadoes.html>
<http://hurricanes.noaa.gov/>
<http://observe.arc.nasa.gov/nasa/earth/hurricane/splash.html>

19.2 *Weather Forecasting*

Weather Data Collection

The knowledge of all the processes that occur in the atmosphere is put to use by meteorologists who try to forecast the weather. Forecasting weather involves three main processes. The first is the gathering of current weather information. Second, this information is recorded on charts and maps or entered into computer models. Last, the charts, maps, and computer models are analyzed and the weather is predicted, or forecasted. The first step in the forecasting process involves the gathering of current weather data from around the region. In the United States, there are almost 1000 weather stations around the country that gather weather data (Figure 19–9).

These stations are mostly run by the National Weather Service, airports, the military, and private citizens. Many ships at sea and automated buoys in the ocean also record weather data. Information about the current state of the weather is usually recorded every hour, using Greenwich Mean Time. The weather data that is recorded each hour by these weather stations includes the surface air temperature, dew point, barometric pressure, precipitation, cloud cover, visibility, and wind speed and direction. Some weather stations use weather balloons to record data about weather conditions high in the atmosphere. These balloons carry radio transmitters that send back weather data at specific altitudes, including air temperature, dew point, wind direction and speed, and barometric pressure.

EARTH MATH

1) A WESTWARD-MOVING TROPICAL STORM, LOCATED 200 MILES OFF THE EAST COAST, BECOMES A HURRICANE AT 1:00 A.M. AND IS MOVING AT APPROXIMATELY 12 MILES PER HOUR. AT APPROXIMATELY WHAT TIME WILL IT STRIKE THE COAST?

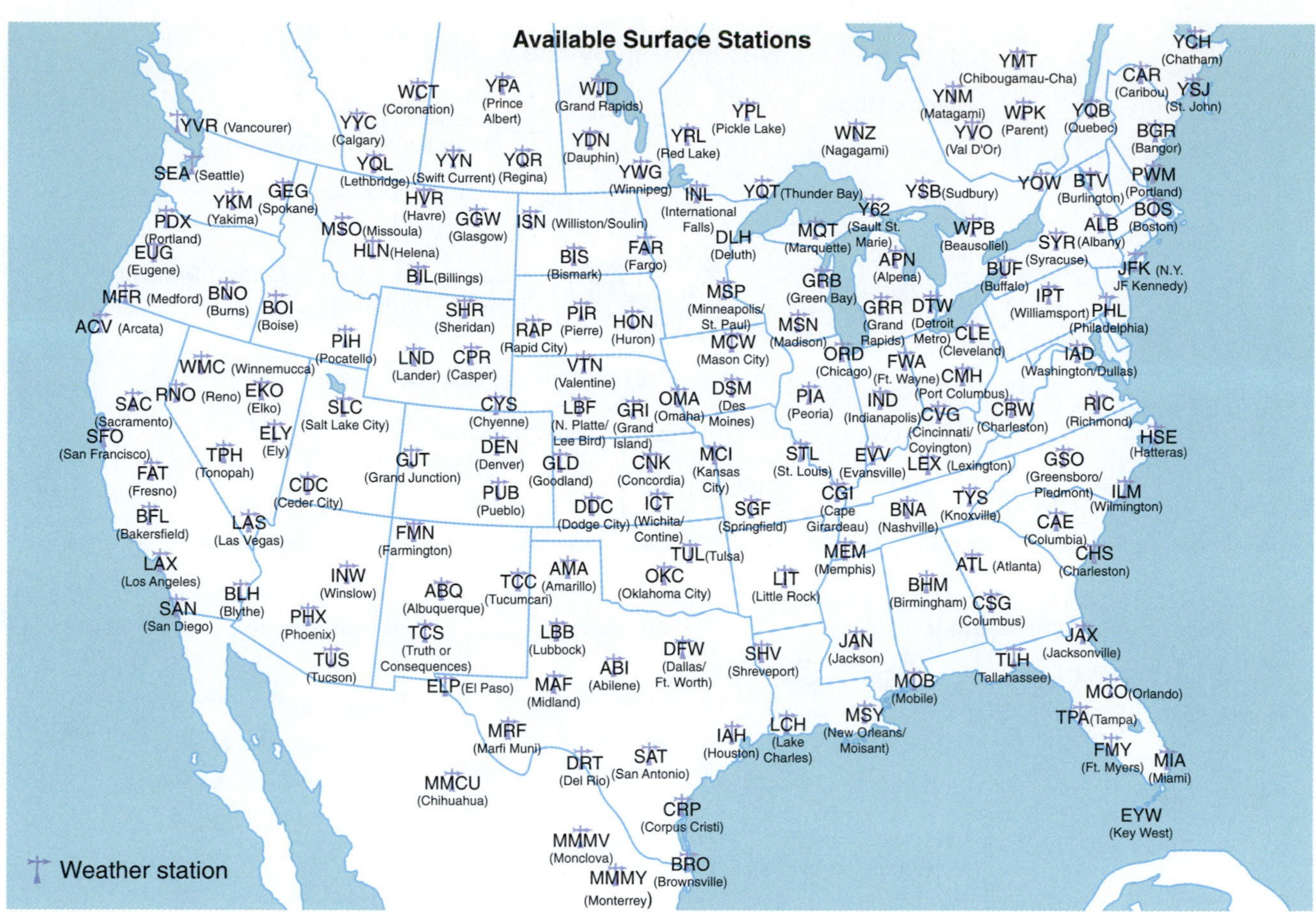

Figure 19–9 The network of weather stations around the United States that records hourly weather observations.

Synoptic Weather Maps

All the weather data gathered by each weather station is then plotted on weather maps every 3 hours. The data is displayed on these maps using the **station model,** which shows the current weather data for each station in a coded form (Figure 19–10). Meteorologists then use the information plotted on the maps to construct detailed weather maps. Surface weather maps usually show the locations of high- and low-pressure systems, temperatures, fronts, and precipitation (Figure 19–11). Often surface temperature maps are created using **isotherms,** which are lines that connect equal values of temperature along the surface. Pressure maps can also be created by using **isobars,** which are lines that connect values of equal pressure. Both temperature and pressure maps can also be generated for different levels in the atmosphere; these are called upper air maps. Today computers are used to create many weather maps that depict current weather every three hours.

Weather Forecasts

Meteorologists use the synoptic weather maps they have created from weather observations to try to predict the weather. Their predictions are often based on the previous direction and movement of weather across a specific region. Even with all the advanced technology that exists in gathering and analyzing weather data, it is often very difficult to predict the weather exactly. Currently weather forecasts are divided into two categories: long term and

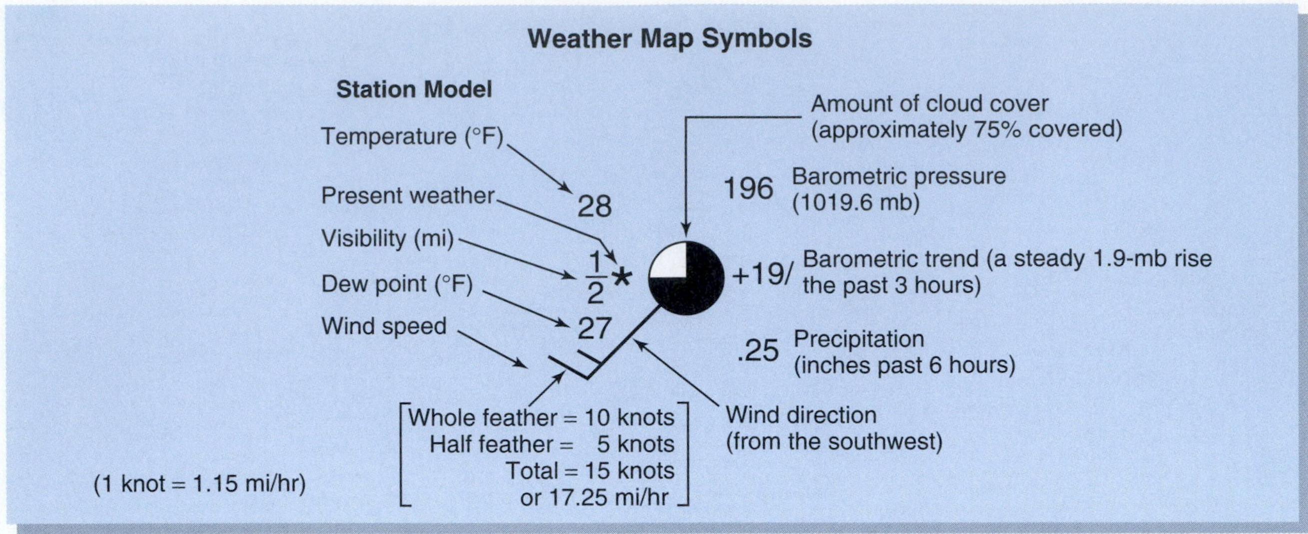

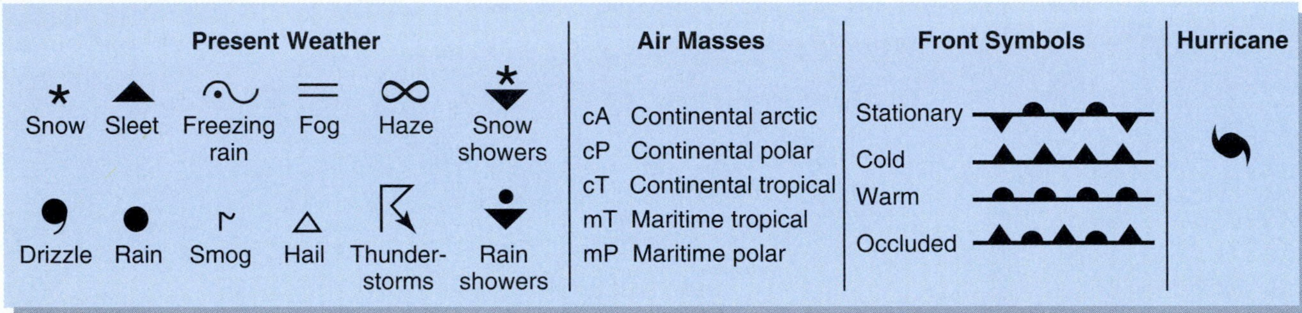

Figure 19–10 The weather station model is used on weather maps to encode weather observations.

short term. Short-term weather forecasts, which are usually quite reliable, involve predicting the weather up to 6 hours ahead.

Long-term forecasts are much more difficult to make, because they may be made for a few days, weeks, or even months. These types of forecasts are often very general. The use of computer models of the atmosphere is becoming an important aspect of weather forecasting. Many computer models exist that try to make both long-term and short-term forecasts. There are also computer models used to predict long-term climate changes on the Earth many years into the future.

Weather Radar and Satellites

The use of **radar** to predict weather has been extremely successful over the past 20 years. Weather radar uses the reflection of radio waves off clouds and precipitation to help locate areas of poor weather. The United States is nearly covered by regional radar stations that can give an accurate view of the current precipitation patterns across the country. These up-to-date radar images are easily accessible today on the internet and can help determine when and where precipitation will occur. New radar systems can also differentiate between different forms of precipitation, making it easier for meteorologists to predict where snow or freezing rain may form. The use of weather satellites is also an important advancement in technology that can aid in the forecasting of weather. **Infrared satellite images** can identify cloud formations and height across the country during both day and night.

Weather prediction does not always require sophisticated technology and a network of weather stations. Some simple local observations can also be used to predict the weather. Because low air pressure is often associated with poor weather, a drop in barometric pressure may signal the coming of poor weather and an increase in barometric pressure may indicate good weather. The type of clouds that are present in the sky can also indicate the

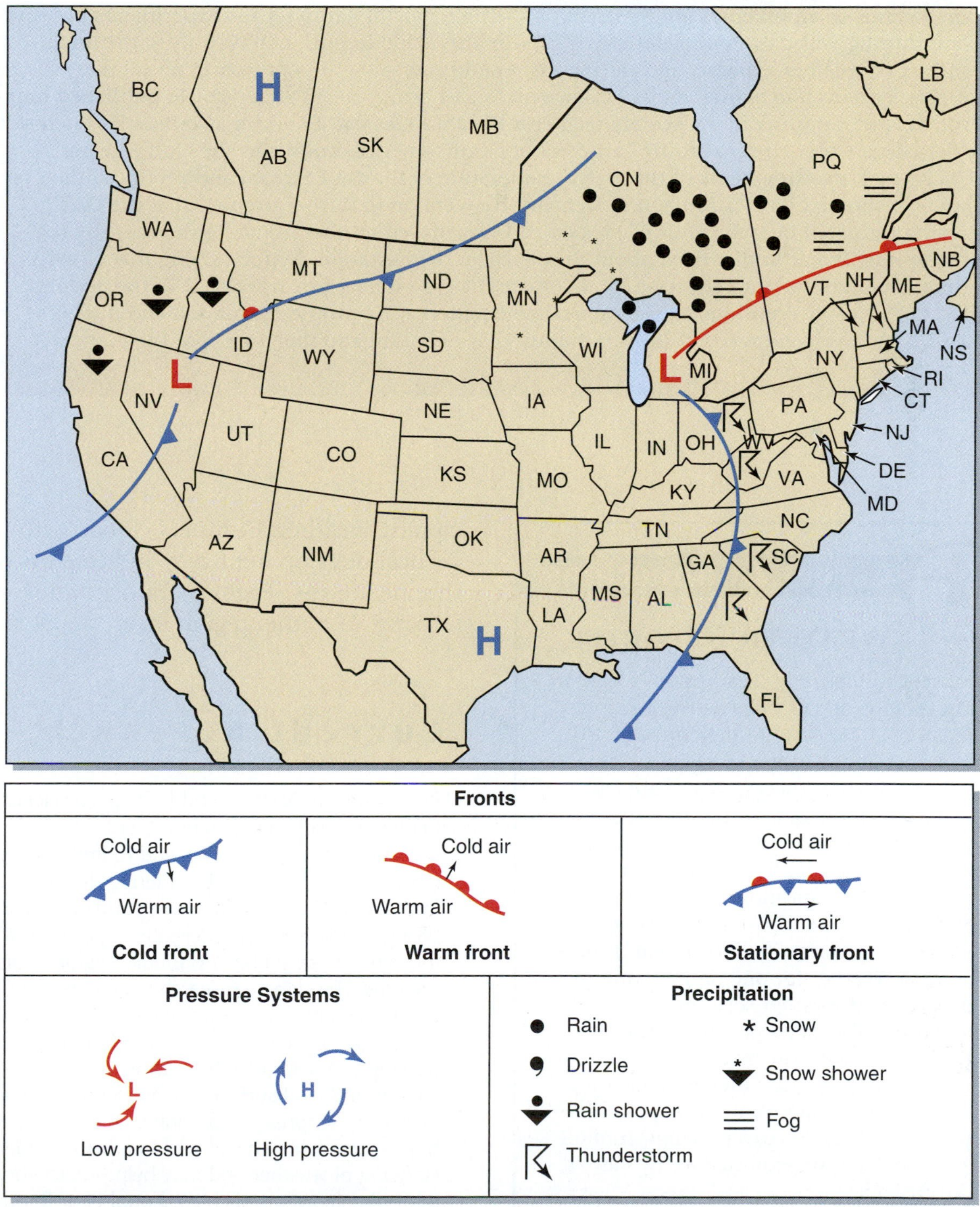

Figure 19–11 A typical surface weather map showing the locations of pressure centers, fronts, and precipitation.

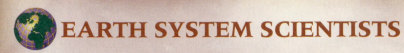

EARTH SYSTEM SCIENTISTS

BENJAMIN FRANKLIN

Benjamin Franklin was born in Boston, Massachusetts in 1706. Franklin was to become one of the most famous men of his time, highly respected in both America and Europe. Franklin was a self-educated man who had an interest in many things well beyond his role in the American Revolution. He spent much of the first part of his life as a printer and writer. Later he would become famous as an inventor, politician, and scientist, most notably for his experiments to prove that lightning was a form of electricity. Franklin also made important advances to the understanding of weather, climate, and the oceans. Franklin was the first person to reveal that weather systems and storms moved from west to east across North America. He published long-term weather forecasts in his widely read *Poor Richard's Almanac.* Franklin also became interested in the warm water current off the east coast of North America, known as the Gulf Stream. In 1775 he took measurements of the water temperature of the Gulf Stream and produced the first chart that showed its location and movement. He went on to further propose that the Gulf Stream originated near the Gulf of Mexico and was driven northward along the coast by the trade winds. Franklin theorized about the effects of deforestation, hypothesizing that cleared farmlands might have an effect on local climate. In 1784 he became interested in the possible effects of volcanic eruptions on regional climate patterns. He proposed that ash and dust produced from a volcano would block incoming solar radiation and therefore cool the planet.

coming weather. Cumulus clouds with strong vertical development may indicate oncoming thunderstorms. Higher stratocumulus clouds moving into the region may signal an ap-

Career Connections

Weather Reporter

A career in reporting weather on television or radio is one of the most well known occupations in the field of meteorology. No news report on television or radio is complete without a weather forecast. There are also specialty television stations that deal with only weather-related topics. A weather reporter requires a solid background in meteorology and communications and can find employment all over the country. Another fast-growing opportunity in this field is forecasting the weather for one of the many internet-related weather services. There are many World Wide Web sites that provide up-to-date weather forecasts 24 hours a day. These companies require meteorologists to make forecasts, create weather maps, and to keep their Web pages current. Opportunities in all these fields are expected to increase in the future.

Current Research

Researchers at NASA's Goddard Space Science Center are using a supercomputer to predict short-term climate changes and to improve long-term weather forecasts. By entering data on surface temperature and moisture conditions and applying mathematical models, the supercomputer then crunches numbers to predict regional rainfall patterns. The climate research team uses the Cray T3E supercomputer, which makes billions of calculations per second. The models that the team develop use data from the atmosphere, hydrosphere, and land surface. So far their predictions have been surprisingly accurate. One day the team hopes their research will lead to a better understanding of weather and may help predict short-term climate events such as droughts or floods.

proaching warm front bringing extended precipitation to the region. Clearing night skies during winter can also signal an extreme drop in temperature.

SECTION REVIEW

1. What are the three main processes used to forecast the weather?
2. What weather information is commonly recorded by weather stations?
3. What are the common weather maps that depict the current state of the weather?
4. Describe how radar and satellites are used to forecast the weather.
5. What are some common changes that occur in the local conditions that can be used to forecast the weather?
6. Draw a station model for the following weather observation: temperature 24° Fahrenheit, dew point 20° Fahrenheit, 50% cloud cover, winds from the southwest at 25 knots, atmospheric pressure 1011.1 millibars.
7. Who was Benjamin Franklin?

For more information go to these Web links:

<http://ww2010.atmos.uiuc.edu/(Gh)/guides/maps/home.rxml>
<http://ww2010.atmos.uiuc.edu/(Gh)/guides/mtr/fcst/home.rxml>
<http://www.nws.noaa.gov/>

CHAPTER SUMMARY

Storms in the atmosphere can form dangerous conditions, leading to the damage of property and loss of life. There are three main types of potentially dangerous storms on the Earth: thunderstorms, tornadoes, and hurricanes. Thunderstorms form when the Earth's surface is heated, causing warm air to rise rapidly. This causes the formation of a small-scale convection cell that leads to the creation of a cumulonimbus cloud. Eventually the cloud grows large enough to create heavy precipitation and generate strong winds. The vertical winds inside a thunderstorm are called updrafts and downdrafts. These winds cause the separation of lighter ice crystals near the top of the cloud and heavier hail near the cloud base. The two different forms of precipitation build up opposite electrical charges within the cloud, eventually forming lightning. Lightning can strike from cloud to cloud, cloud to air, or cloud to ground. When the convection that formed the thunderstorm ceases, the cloud begins to dissipate and the thunderstorm is over. The most damaging aspects of a thunderstorm involve heavy winds and rain, along with lightning.

Some very strong thunderstorms cause wind to rush in toward the base at such high velocities that they can form a tornado. A tornado is a rapidly spiraling column of air that comes into contact with the ground. Tornadoes can generate wind at speeds more than 300 miles per hour, making them one of the most

EARTH MATH

1) THE FOLLOWING ATMOSPHERIC PRESSURE READINGS IN MILLIBARS WERE RECORDED AT A WEATHER STATION OVER THE PAST 12 HOURS: 1005, 1006, 1012,1014, 1017, AND 1024. WHAT WAS THE RATE OF CHANGE IN PRESSURE OVER THIS TIME?

deadly forces on Earth. Tornadoes usually only last for a few minutes and are very difficult to predict.

Hurricanes are another type of storm system that can generate strong winds and heavy rain. These are the largest storms that form on Earth and can affect hundreds of square miles. Hurricanes form from developing low-pressure systems over warm ocean water near the tropics. They gain energy and moisture as they move across the ocean, leading to the formation of a massive cyclone around which clouds rotate. The winds of a hurricane spiral around the central eye of a hurricane in a counterclockwise motion in the Northern Hemisphere. This causes water to build up on the northwestern edge of the hurricane, forming a storm surge. This wall of water can raise tides by as much as 16 feet, causing massive flooding when the hurricane strikes land. Eventually, if the hurricane moves over land, it loses its source of energy and begins to dissipate. Hurricane season in the eastern United States is usually from June to November.

Meteorologists try to predict storms by studying weather observations taken from weather stations all around the country, recording the surface air temperature, dew point, barometric pressure, precipitation, cloud cover, visibility, and wind speed and direction at specific locations. They then enter this data into computers and plot it on weather maps, which are then analyzed to make forecasts about the weather. Data on surface weather maps is in a form called the station model, which encodes surface weather information for each specific weather station. Meteorologists also use radar and satellite images to make predictions about the weather. Some simple observations made of the surrounding environment can also be used to forecast the weather. These include observing changes in atmospheric pressure, approaching clouds, and sky conditions.

CHAPTER REVIEW

Multiple Choice

1. In the United States, during which month are thunderstorms most likely to occur?
   a. January
   b. March
   c. August
   d. November

2. Which conditions are favorable for the formation of thunderstorms?
   a. cool, sinking air
   b. an approaching warm front
   c. an approaching stationary front
   d. hot, rising air

3. The strong vertical winds inside a thunderstorm are called:
   a. updrafts
   b. a tornado
   c. prevailing winds
   d. cyclones

4. The separation of hail at the base of the cloud and tiny ice crystals near the top of the cloud results in:
   a. updrafts
   b. downdrafts
   c. lightning
   d. tornadoes

5. The most difficult type of weather event to predict a few hours in advance of its development is a:
   a. thunderstorm
   b. tornado
   c. hurricane
   d. cyclone

6. Which storm usually lasts the shortest time?
   a. a thunderstorm
   b. a tornado
   c. a hurricane
   d. a cyclone

7. What conditions are favorable for the formation of a hurricane?
   a. cool, moist air
   b. a continental tropical air mass
   c. warm ocean water
   d. rapidly heating land

8. The circulation of clouds and wind around the eye of a hurricane in the Northern Hemisphere is best described as:
   a. counterclockwise and inward
   b. counterclockwise and outward
   c. clockwise and inward
   d. clockwise and outward

9. Another term for the low-pressure system that forms a hurricane is:
   a. anticyclone
   b. typhoon disturbance
   c. maritime polar air mass
   d. cyclone

10. A tornado that can cause severe damage with 200 mile per hour winds has a Fujita intensity of:
    a. F1
    b. F2
    c. F3
    d. F4

11. Hurricanes that form in the Atlantic Ocean near the equator travel in what general direction?
    a. west
    b. southwest
    c. northeast
    d. southeast

12. During which month is a hurricane most likely to form near the equator in the Atlantic ocean?
    a. January
    b. March
    c. September
    d. April

13. A hurricane with an atmospheric pressure reading in the eye of 28.00 inches of mercury is classified on the Saffir-Simpson scale as what type:
    a. 1
    b. 2
    c. 3
    d. 4

14. Lines that connect equal temperatures on a weather map are called:
    a. isobars
    b. isotachs
    c. isotherms
    d. contour lines

15. Lines that connect equal pressure on a weather map are called:
    a. isobars
    b. isotachs
    c. isotherms
    d. contour lines

16. The atmospheric pressure on a station model is recorded as 164. This is equal to:
    a. 164 inches of mercury
    b. 916.4 millibars
    c. 1016.4 millibars
    d. 1016.4 inches of mercury

continued

17. Decreasing atmospheric pressure usually indicates what type of approaching weather?
    a. fair weather
    b. partly cloudy weather
    c. poor weather
    d. cold weather

Matching *Match the terms with the correct definitions.*

a.	thunderstorm	**f.**	tropical disturbance	**k.**	station model
b.	lightning	**g.**	tropical storm	**l.**	isotherms
c.	tornado	**h.**	storm surge	**m.**	isobar
d.	hurricane	**i.**	tropical depression	**n.**	radar
e.	cyclone	**j.**	typhoon	**o.**	infrared satellite images

1. ____ A rapidly rotating, funnel-shaped column of air located at the base of a cumulonimbus cloud, resulting in deadly high-velocity winds.
2. ____ Isolines that are used to connect points of equal temperature on a weather map.
3. ____ A large-scale tropical cyclone with winds in excess of 74 miles per hour.
4. ____ An organized group of thunderstorms associated with a strong low-pressure system with cyclonic winds between 40 and 70 miles per hour.
5. ____ A low atmospheric pressure system that forms in the tropics over the ocean, consisting of group of thunderstorms and cyclonic winds between 20 and 40 miles per hour.
6. ____ An organized group of thunderstorms with cyclonic winds less than 20 miles per hour.
7. ____ A hurricane that forms and is located over the western Pacific ocean.
8. ____ Satellite imagery created by sensing the infrared energy given off by a substance, which allows images to be recorded at night without the presence of light.
9. ____ A coded symbol used on weather maps to display specific atmospheric variables.
10. ____ A small-scale storm caused by the formation of a cumulonimbus cloud, producing strong winds, heavy rain or hail, lightning, and thunder.
11. ____ A term for winds that spiral inward and around a low-pressure center.
12. ____ A term that stands for "**r**adio **d**etecting **a**nd **r**anging," which describes a device that bounces radio waves off an object to track its location, speed, and movement.
13. ____ The rise in water level that is caused by the strong winds associated with an approaching hurricane.
14. ____ A naturally occurring electrical discharge in the atmosphere usually associated with thunderstorms.
15. ____ An isoline that connects points of equal atmospheric pressure on a weather map.

Critical Thinking

1. If predictions are correct about the Earth's increasing surface temperatures, explain three ways you think this will affect the weather in the United States.
2. What observations of the local weather might be used to predict the formation of a tornado?

CHAPTER

20 Global Climate Change

Objectives

Revealing the Earth's Past Climate • Ice Ages and Glaciations • The Milankovitch Cycle • Hot House Climates • Goldilocks Syndrome • Humans and Global Climate Change

After reading this chapter you should be able to:

- Define the term *climate.*
- Identify six ways that paleoclimatologists can infer what the Earth's past climate was like.
- Differentiate between a hot house climate and an ice age.
- Describe the effects that an ice age has on the planet.
- Explain the three aspects of the Milankovitch cycle and how they relate to global climate.
- Describe the effects that a hot house climate has on the Earth.
- Explain what scientists believe caused the Earth's temperature to increase in the past.
- Describe the Goldilocks syndrome.
- Explain the evidence used to suggest that humans are causing an increase in global temperatures.
- Describe the possible effects of increasing global temperatures on the Earth.

TERMS TO KNOW

- climate
- hot house climate
- ice age
- paleoclimatology
- tree ring
- pollen
- ice cores
- precession
- eccentricity
- Goldilocks syndrome
- fossil fuels
- rice paddies
- chlorofluorocarbons
- nitrous oxide

INTRODUCTION

The long-term weather conditions that occur in specific regions are known as the **climate.** The climate of the Earth is directly the result of the state of the atmosphere and its relationship to the Earth's surface. Research has revealed that the Earth's climate has gone through many changes in its long history. Many think that these changes are part of the natural cycle of interactions that occur among our planet's atmosphere, hydrosphere, and lithosphere. Changes in these aspects of the Earth's systems may be the result of variations in atmospheric gases, ocean currents, insolation, and tectonic plate movements. The concept of global climate change is not new. In fact, scientists have determined that the Earth has experienced changes in global climate many times in its 4.5 billion year history.

Revealing the Earth's Past Climate

Geologists refer to the Earth's climate in two general terms: **hot house climate** and **ice age.** A hot house climate occurs when the average surface temperature of the Earth is much warmer than it is today. An ice age occurs when the average surface temperature of the planet drops well below that of today, resulting in many portions of the Earth becoming covered with ice and snow. Researchers can reveal what the average surface temperature of the Earth was like before humans used instruments to record global temperatures and other climate data. This branch of science is called **paleoclimatology.** Paleoclimatology uses several different methods to hypothesize about what the climate was like during different times in the Earth's past. One of these techniques involves careful study of the fossil record. By looking at fossil evidence, paleoclimatologists can determine what the climate was like when that particular organism flourished. For example, fossilized palm trees and ferns found in Alaska suggest that the climate was much different in that region when these organisms were alive (Figure 20–1).

Figure 20–1 Fossils of warm climate tree leaves found in areas that are extremely cold today suggest that the climate was different when the trees were alive. *(Courtesy of PhotoDisc.)*

Studying coral reefs can also reveal much about the past temperatures of the ocean. Coral reefs are large colonies of individual organisms called corals. When a coral dies, it is covered by new generations of corals, leading to the formation of a large reef. By drilling into a coral reef and extracting a core sample, scientists can study its chemical composition to learn about the conditions that existed in the oceans when the ancient corals were alive. Studying **tree ring** data is another means by which climate change can be revealed (Figure 20–2). Trees grow thick rings during wet, warm summers and thinner rings during drier seasons. By analyzing the thickness of a tree's rings, scientists can infer what the climate was like during the time that it was growing.

In addition, lake and ocean sediments can reveal a lot about the climate that existed when they were deposited in the water. This is because different types of sediments are associated with changes in the landscape. Analyzing fossilized **pollen** samples is also another method by which scientists can reveal past climates. Certain species of plants grow only in specific climates. By identifying the pollen that existed in an area, scientists can piece together what plants grew there and therefore infer what the climate may have been like. **Ice cores** drilled into glaciers and ice sheets are another important way to determine past climate. When snow falls and compacts to form a glacier, it traps tiny air bubbles that represent the composition of the atmosphere during the time it snowed. Scientists can analyze these trapped air bubbles and determine what the global atmosphere and climate may have been like (Figure 20–3). Using all these methods has enabled scientists to piece together the climate history of the Earth with reasonable accuracy.

Figure 20–2 Tree rings can provide clues to the Earth's past climate by examining its growth rates in the past. *(Courtesy of PhotoDisc.)*

Ice Ages and Glaciations

Research has revealed that at several times in the past the Earth was much colder than it is today. This resulted in the widespread formation of ice sheets and glaciers on many of the continents. When this type of event lasts more than a million years, it is called an ice age. Over shorter periods, such as thousands of years, these events are known as glaciations. It is estimated that four major ice ages have occurred on the Earth. The earliest of these occurred about 800 to 600 million years ago in the late Proterozoic eon. Another major ice age took place during the late Ordovician and early Silurian periods between 430 to 460 million years ago. The late Carboniferous and early Permian periods also experienced an ice age between 250 to 350 million years ago. The last known ice age on Earth occurred approximately 4 million years ago, during the late Tertiary period. During this ice age, approximately 30% of the land surfaces on the Earth were covered in ice. The most extensive ice coverage of the continents during this time occurred approximately 20,000 years ago. Since then the glaciers have been retreating, leading to the formation of much of the landscapes that we recognize today. Evidence based on global temperature averages during the past 100,000 years suggests that the Earth is still in a cooler climate period. It is only in the past 1000 years or so that the Earth's climate has begun to warm up (Figure 20–4).

The Milankovitch Cycle

One question scientists are trying to answer is what causes an ice age. Scientists agree that

Figure 20–3 Ice cores contain tiny bubbles of gas that represent the composition of the Earth's atmosphere during the past. They can be analyzed to reveal the Earth's past climates. *(Courtesy of PhotoDisc.)*

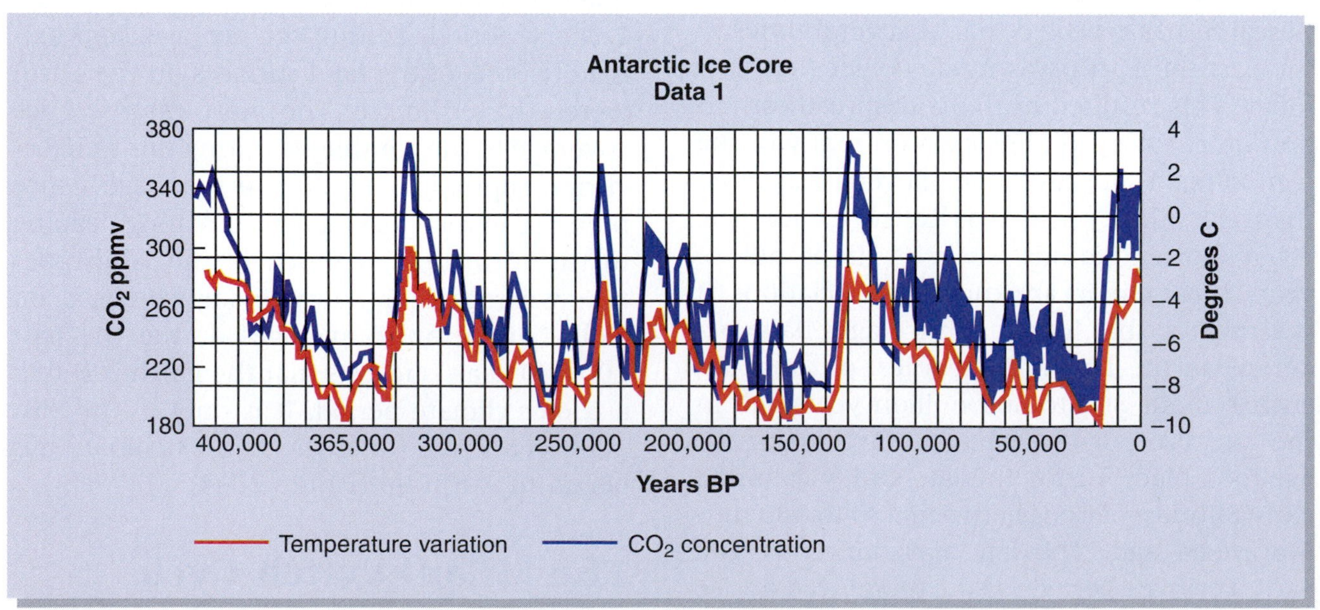

Figure 20–4 Ice core data show the relationship between carbon dioxide concentration in the Earth's atmosphere and global temperature. This record reveals the changes in the Earth's climate that have occurred for the past 420,000 years.

the interactions among the oceans, land, and atmosphere play an important role; however, changes in the Earth's position relative to the Sun may also influence the occurrence of ice ages. This hypothesis was first proposed by Milutin Milankovitch in the mid-nineteenth century. Milankovitch revealed that changes in the Earth's tilted axis, along with its orbit around the Sun, occur on periodic scale. The Earth's axis does not remain at a constant tilt of 23.5 degrees. It actually fluctuates between a maximum tilt of 24.5 degrees and a minimum tilt of 22 degrees every 41,000 years. This causes changes in the insolation received at higher latitudes, which could lead to glaciations (Figure 20–5).

Milankovitch also determined that the Earth wobbles on its axis, which is known as precession (Figure 20–6). Precession can be easily demonstrated by observing a spinning top. As the top spins on its axis rapidly, it also slowly wobbles back and forth. This is precession. The Earth's rapid rotation on its axis also causes it to slowly wobble back and forth on a periodic scale. This change in the orientation of the Earth to the Sun is a result of precession. The time it takes to for the Earth to wobble back and forth once is approximately 22,000 years.

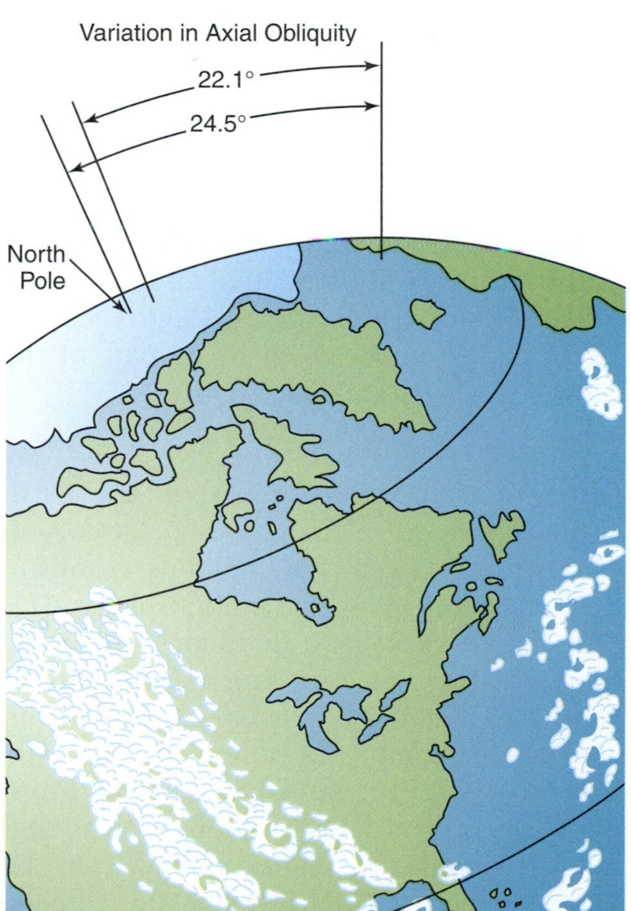

Figure 20–5 The periodic change that occurs in the Earth's tilt between 22 degrees and 24.5 degrees every 41,000 years is part of the Milankovitch cycle.

Figure 20–6 The wobble of the Earth as it spins on its axis, known as precession, occurs on a 22,000-year cycle. This is part of the Milankovitch's theory of climate change.

The **eccentricity** of the Earth's orbit around the Sun is a measure of how close to a perfect circle the orbit is (Figure 20–7). The Earth's orbital eccentricity also changes on a periodic scale approximately every 100,000 years. This results in a periodic change in the distance between the Earth and the Sun, which causes variations in the intensity of insolation

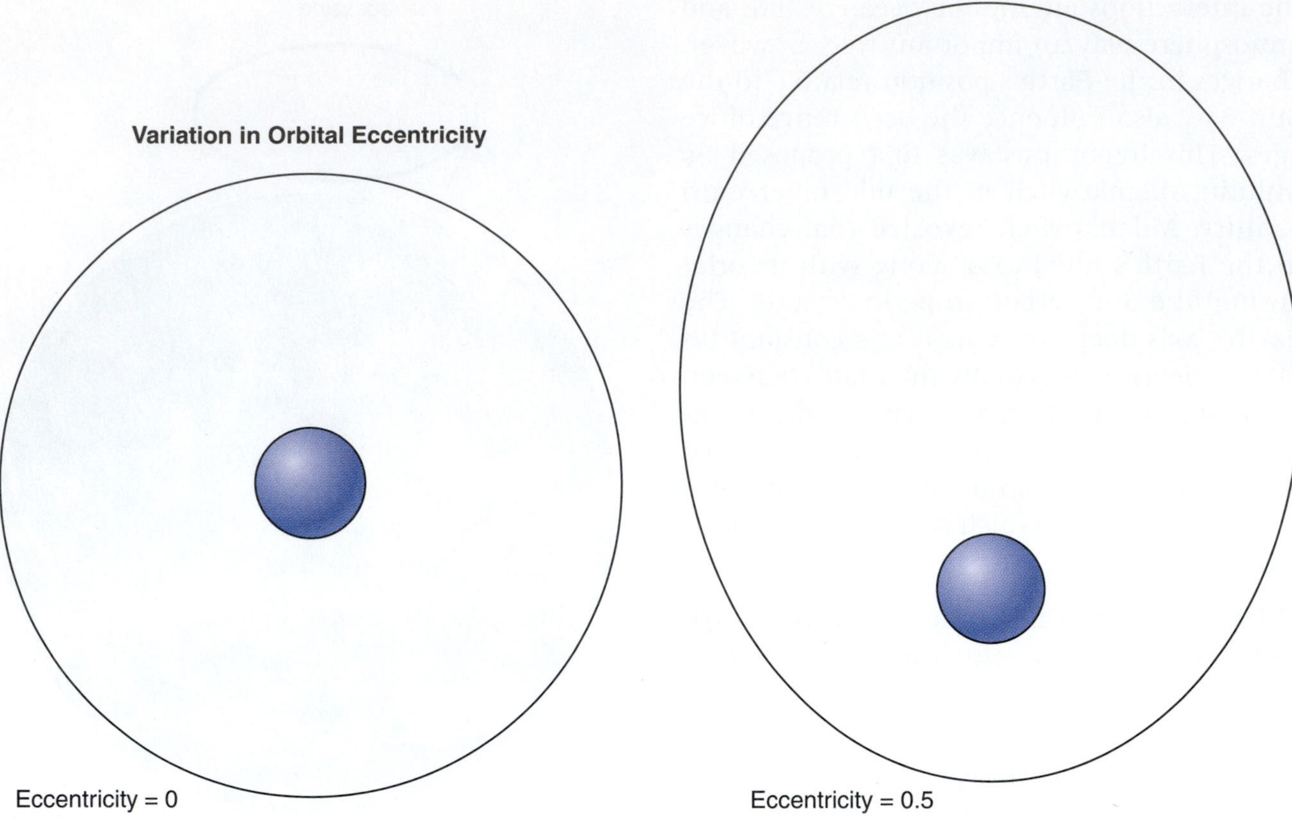

Figure 20–7 The changing eccentricity of Earth's orbit was also proposed by Milankovitch to influence the Earth's climate; these exaggerated views show how changes in eccentricity affect orbital paths.

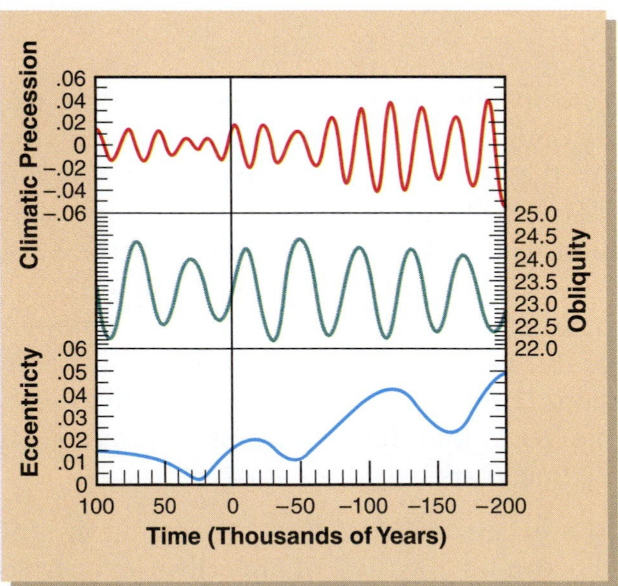

Figure 20–8 Milankovitch used the periodic changes in the Earth's tilted axis, precession, and orbital eccentricity, known as the Milankovitch cycle, to explain the changes in the Earth's climate for the past 200,000 years. Compare his data to the ice core data in Figure 20–4 and see if you can find a relationship between temperature and the Milankovitch cycles.

received by the Earth. Together these changes in the Earth's position relative to the Sun, are called Milankovitch cycles (Figure 20–8). These cycles are believed to be part of the cause of ice ages and glaciations.

Other, smaller scale changes on the Earth can also lead to global cooling. Large volcanic eruptions can spew millions of tons of dust and ash high into the atmosphere. This dust can act as a shield that blocks incoming solar radiation. The resulting reduction in solar energy received by the Earth's surface can cool the planet and alter the climate. Volcanoes can put large amounts of sulfur dioxide gas into the atmosphere. This gas helps to reflect sunlight away from the planet, which can help to cool the climate. These examples illustrate how the lithosphere and plate tectonics can also help to alter global climate.

EARTH SYSTEM SCIENTISTS *MILUTIN MILANKOVITCH*

Milutin Milankovitch was born in Serbia in 1879. He was educated at the Vienna Institute of Technology and later became a professor of applied mathematics. Milankovitch is best known for his theory of how the Earth's motions may be related to climate change. Known as the Milankovitch theory, he proposed that the periodic change in the motions of the Earth around the Sun and the Earth's tilted axis lead to periodic changes in the insolation received by the Earth. He then applied a mathematical model to these cyclical changes and developed an insolation and average temperature record for the Earth dating back 600,000 years. Milankovitch applied this record to try to explain past ice ages. Unfortunately, Milankovitch's theory of climate change and the Earth's orbit was ignored by the scientific community for more than 50 years. In 1976 new research revealed that Milankovitch's model correlated with past temperatures derived from studying deep sea sediments. The Milankovitch theory has since been widely accepted as one of the major factors affecting the Earth's climate.

Hot House Climates

The opposite of an ice age, a hot house period is when the Earth's climate is significantly warmer than today. This has occurred several times in the Earth's past. The most ancient of these hot house climates occurred during the mid-Cretaceous period approximately 90 to 120 million years ago. During this time, tropical plants and animals lived as far north as 55 degrees north latitude. Today this region is near the Arctic Circle, where only cold-hardy plants can exist. Another hot house period occurred approximately 125,000 years ago, during the Penultimate Interglacial period. During this time, global temperatures were approximately 4° to 6° Fahrenheit warmer than today. Since then, two more warming periods occurred on the Earth: the mid-Holocene warm period approximately 6000 years ago and the Medieval warm period between 600 and 1100 years ago.

All these hot house episodes are believed to have been caused by an increase in carbon dioxide gas in the Earth's atmosphere. During all the warm periods, levels of carbon dioxide have been found to be 2 to 4 times greater than today. Carbon dioxide gas is a greenhouse gas that helps traps the Earth's heat. What caused an increase in carbon dioxide concentration in the past is a question that scientists are still trying to answer. Volcanoes, along with the weathering of carbonate rocks, can put large amounts of carbon dioxide into the atmosphere. Changes in the populations of organisms that use carbon dioxide for photosynthesis or produce it by respiration can also impact global levels. The exact reason for the past increases in carbon dioxide may always remain a mystery.

Goldilocks Syndrome

The role that carbon dioxide gas plays in determining the surface temperature of our planet is often referred to as the greenhouse effect. The greenhouse effect occurs as a result of our atmosphere's ability to allow incoming short-wave radiation from the Sun to strike the Earth's surface. The gases in the atmosphere transmit short-wavelength visible light radiation. This radiation is absorbed by the Earth's surface and is then reradiated back into the atmosphere as long-wave, infrared radiation. Greenhouse gases such as carbon dioxide trap this energy, which regulates the

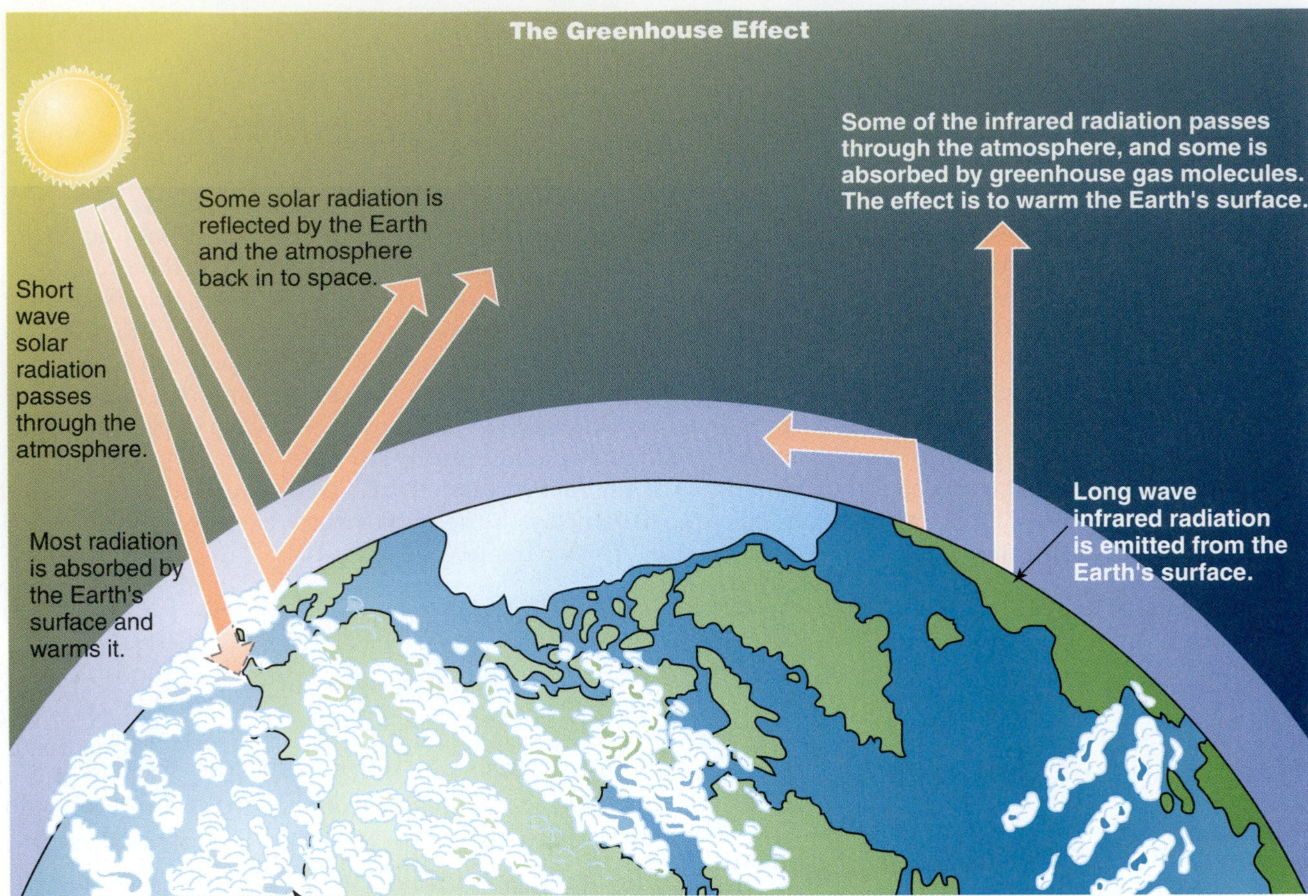
The Greenhouse Effect
Some of the infrared radiation passes through the atmosphere, and some is absorbed by greenhouse gas molecules. The effect is to warm the Earth's surface.
Some solar radiation is reflected by the Earth and the atmosphere back in to space.
Short wave solar radiation passes through the atmosphere.
Most radiation is absorbed by the Earth's surface and warms it.
Long wave infrared radiation is emitted from the Earth's surface.

Figure 20–9 The greenhouse effect regulates the surface temperature on the Earth and depends on the amount of greenhouse gases, such as carbon dioxide, that are present in the atmosphere.

temperature of the atmosphere and the Earth's surface (Figure 20–9).

The important link between carbon dioxide and surface temperatures does not exist only on Earth. Astronomers refer to the effect of carbon dioxide on a planet's surface temperature as **Goldilocks syndrome** (Figure 20–10). "Goldilocks and the Three Bears" is the fable of a girl who tasted three different bowls of porridge. The first was too hot, the second was too cold, and the third was just right. This analogy is also used to describe the surface temperatures of the planets Venus, Mars, and Earth. Venus has a thick atmosphere that is 95% carbon dioxide, which causes the surface temperature to reach 800° Fahrenheit. This planet has too much carbon dioxide and is therefore too hot. Mars has a very thin atmosphere of carbon dioxide, which causes it to have an average surface temperature of −72° Fahrenheit. This planet has too little carbon dioxide and is therefore too cold. Earth, however, maintains a careful balance of carbon dioxide in its atmosphere, which makes it just right. This simple example reveals the important link between atmospheric carbon dioxide concentration and surface temperature.

Humans and Global Climate Change

During the past 1000 years, global temperatures have been slowly increasing (Figure

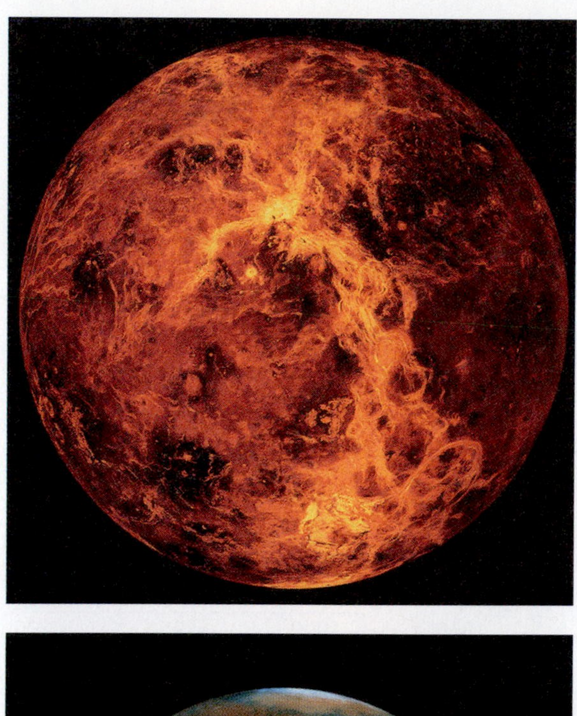

Figure 20–10 Goldilocks syndrome illustrates how the surface temperatures of Venus, Earth, and Mars are affected by carbon dioxide concentration in their atmospheres. *(Courtesy of PhotoDisc.)*

20–11). The greatest increases in temperatures have occurred during the past 100 years and may be the result of human activity.

The use of **fossil fuels** such as coal and oil have been causing an increase in atmospheric carbon dioxide levels since the end of the nineteenth century (Figure 20–12). Fossil fuels are hydrocarbon compounds, which, when burned, produce carbon dioxide gas. The increases in both industrial carbon dioxide production and atmospheric carbon dioxide concentration are most certainly linked (Figure 20–13).

Carbon dioxide is not the only greenhouse gas whose levels have been increasing in the Earth's atmosphere. The levels of methane gas

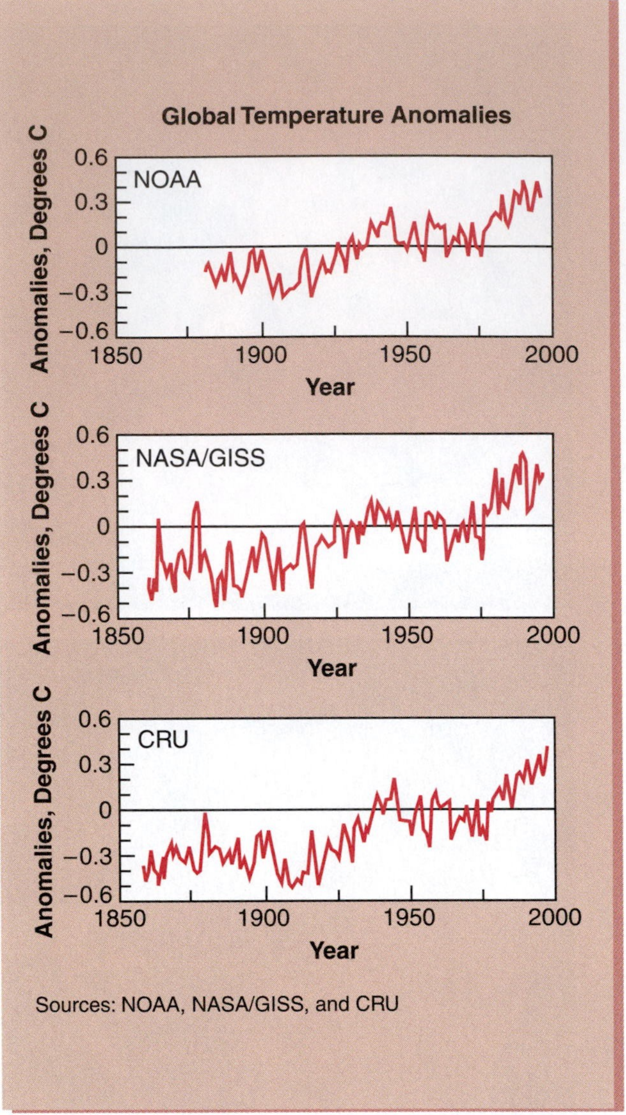

Figure 20–11 The average change in global temperature from three different sources going back to 1860 reveals that the Earth's average surface temperature is increasing.

have also increased in the past 100 years (Figure 20–14). On Earth, methane is produced mainly by livestock and **rice paddies.**

In addition, the levels of greenhouse gases such as **chlorofluorocarbons** (CFCs) and **nitrous oxide** are increasing. Evidence of rising global temperatures and increased concentrations of greenhouse gases suggests that human beings are indeed altering the climate of our planet. Increased temperatures around the globe may have worldwide effects. As the Earth begins to get warmer, seasonal storms such as thunderstorms, hurricanes, and tornadoes may become more powerful and occur more frequently. A shift in general climate types may also occur, resulting in widespread drought in some places and excess rainfall in others. A change in climate types may also impact regional vegetation. Warmer varieties of plants will be able to grow in higher latitudes than today. A changing climate also may cause ice sheets and glaciers to melt, which would lead to a gradual rise in the sea level around the globe as the melting ice adds more water to the world's oceans.

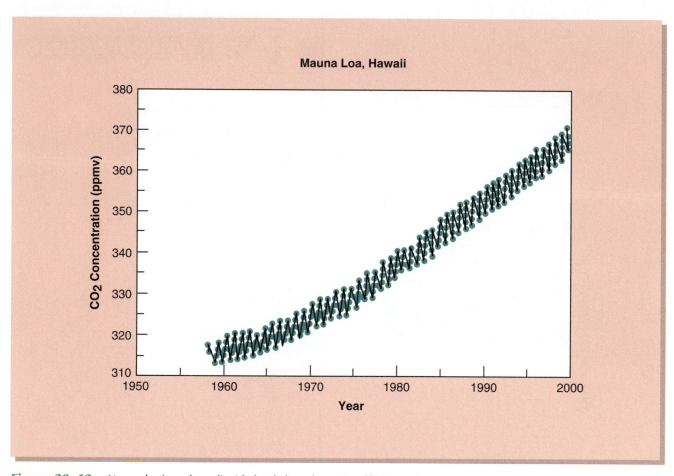

Figure 20–12 Atmospheric carbon dioxide levels have been steadily increasing since 1960, as these data show. The cyclical nature shown on the graph is the result of the periodic drop in carbon dioxide that occurs every spring when vegetation removes carbon dioxide from the atmosphere to grow new leaves.

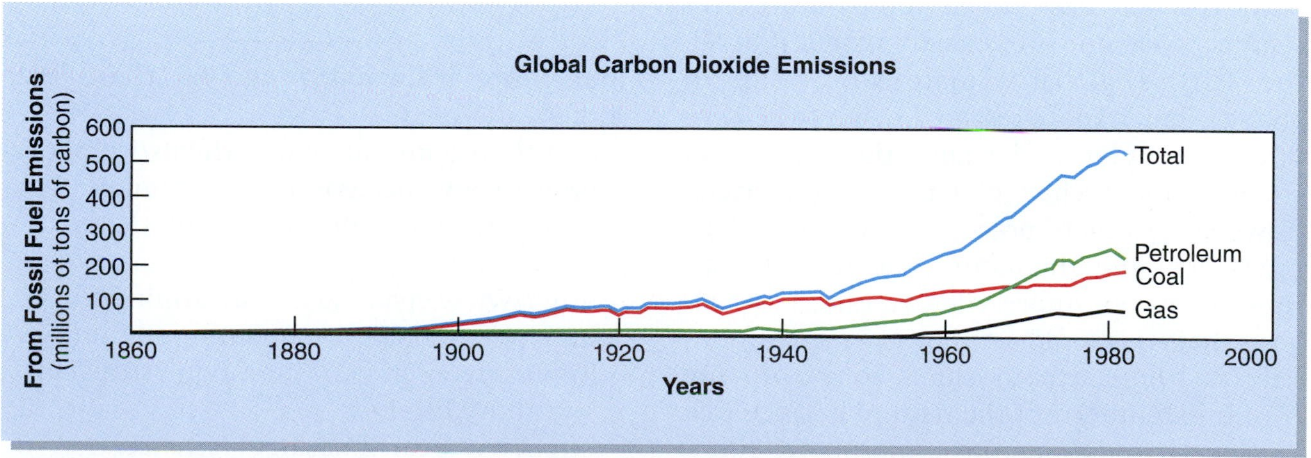

Figure 20–13 The rise in the production of carbon dioxide gas since 1860 as a result of the burning of fossil fuels.

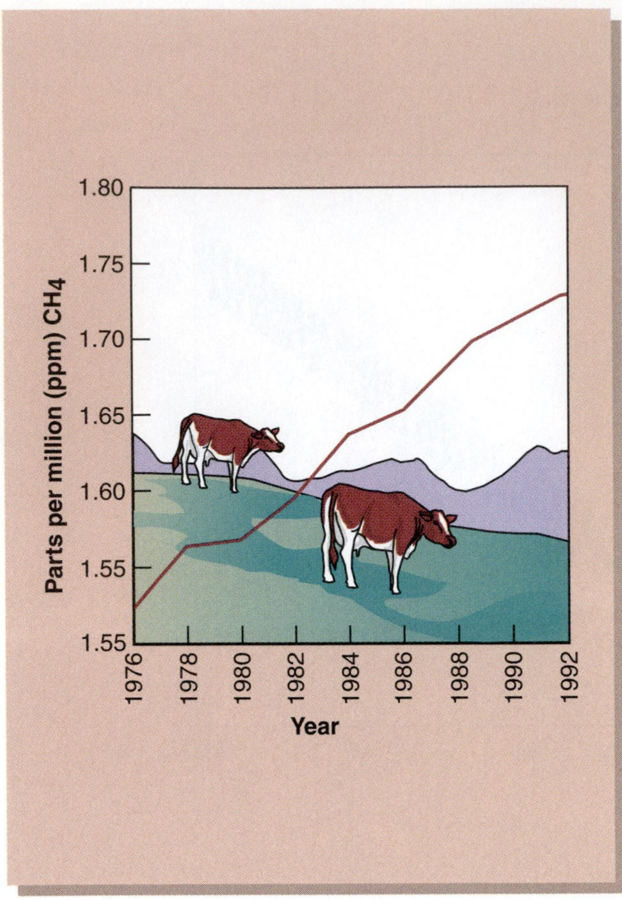

Figure 20–14 The atmospheric concentration of methane, a greenhouse gas, has steadily increased since 1976.

Although the levels of carbon dioxide and other greenhouse gases are increasing, along with the average global temperature, researchers are not sure exactly when, if at all, the Earth's global climate will begin to change. Much more research needs to be conducted to fully understand the impacts of global climate change. One thing is clear, however: Human practices and technology are interrupting the natural balance of the atmosphere at an unprecedented rate. This suggests that we should attempt to lessen our impact on the Earth's systems so we do not continue to interrupt the natural balance that has evolved among the atmosphere, lithosphere, and hydrosphere over billions of years.

Career Connections

Geographic Information Systems Specialist

A geographic information systems specialist uses computers and geographic information system (GIS) software to analyze all types of geographic data. The job includes the creation, retrieval, and viewing of specific geographic information. GIS-created maps display layers of different information to reveal any trends or relationships that are occurring in unique geographic areas. This includes data on population, vegetation, water use, disease, droughts, floods, utilities, sewer systems, agriculture, and any other information being researched. GIS systems specialists must be able to use GIS software to sort data and to create colorful layered maps, as well as be able to operate digital cameras, scanners, plotters, and printers. They must also have knowledge of coordinate systems, latitude and longitude, cartography, topography, and map scales. A college degree with an emphasis on geography and computer programming is required for this type of career. Jobs within government agencies or in private industry are becoming more widespread as the use of GIS software increases around the world. Many free versions of GIS software are available for download via the internet, which can be explored to see how this new technology is used and what type of work a GIS systems specialist performs.

For more information go to these Web links:

<http://museum.state.il.us/exhibits/ice_ages/>
<http://www.noaa.gov/climate.html>
<http://www.cotf.edu/ete/modules/climate/GCmain.html>
<http://www.epa.gov/globalwarming/>
<http://www.noaa.gov/greenhouse.html>
<http://www.cotf.edu/ete/modules/carbon/earthfire.html>
<http://www.earth.nasa.gov/science/Science_ecosystems.html>

EARTH MATH

1) IF THE FOUR MAJOR ICE AGES OCCURRED APPROXIMATELY 700, 445, AND 300 MILLION YEARS AGO, WHAT IS THE AVERAGE NUMBER OF YEARS BETWEEN ICE AGES?

2) THE CONCENTRATION OF CARBON DIOXIDE GAS IN THE EARTH'S ATMOSPHERE WAS APPROXIMATELY 320 PARTS PER MILLION (PPM) IN 1960 AND 365 PPM IN 2000. DETERMINE THE RATE OF CHANGE FOR CARBON DIOXIDE GAS CONCENTRATION FOR THIS PERIOD.

REVIEW

1. What is paleoclimatology?
2. What methods do scientists use to determine the Earth's past climates?
3. What is the difference between an ice age and a glaciation?
4. How many ice ages have occurred during the Earth's past, and when did they occur?
5. What may cause global cooling of the planet?
6. What are hot house periods, and when have they occurred in the past?
7. What may cause global warming?
8. Describe Goldilocks syndrome.
9. How have humans contributed to possible global warming?
10. Who was Milutin Milankovitch?

CHAPTER SUMMARY

The long-term weather on Earth is known as climate. Scientists believe that the Earth's climate has undergone drastic changes in the past. Scientists who study the Earth's past climate are known as paleoclimatologists. They use several techniques to infer what the past climate was like at different periods of geologic time. These techniques include examining fossils, tree rings, ice cores, sediment deposits, pollen, and coral reefs. The results of their investigations reveal several different times in the Earth's past when the climate was either much warmer than today or much colder. Ice ages occurred at different times in the past when temperatures were much colder than today. They resulted in much of the continents being covered in ice sheets or glaciers.

The Milankovitch cycle proposes that variations in the Earth's orbit and axis of rotation cause changes in the amount of insolation received by the Earth on a periodic basis. This is believed to cause ice ages. Hot house climates occur when the Earth's past temperature was much higher than today. This has also occurred at different times in the past. Scientists believe the main cause for the increase in temperature during hot house climates is an increase in carbon dioxide gas in the atmosphere. Carbon dioxide is a greenhouse gas that helps to trap heat on the planet. The relationship between atmospheric carbon dioxide gas concentration and the surface temperature of the planet can be illustrated by observing our two neighboring planets. Venus has too much carbon dioxide in its atmosphere and therefore has a very high surface temperature. Mars has too little carbon dioxide in its atmosphere and has a very cold surface temperature. Recent evidence suggests that the Earth has been experiencing a gradual rise in surface temperatures over the past 100 years. This is thought to be the result of human activity. The increasing use of fossil fuels by industry and for transportation has led to a gradual increase in atmospheric carbon dioxide levels on Earth. Other greenhouse gases, such as methane and CFCs, have also been increasing in their atmospheric concentration. An increase in global temperatures may result in a change in global climate. This may lead to an increase in the intensity of storms, widespread drought, and rising sea levels.

CHAPTER REVIEW

Multiple Choice

1. Long-term patterns of temperature and moisture in a particular region are called:
   a. weather
   b. paleoclimatology
   c. climate
   d. meteorology

2. How can studying sediment deposits reveal past climates?
   a. Different landscapes form different sediments.
   b. Sediments only form in warm climates.
   c. Sediments trap gas bubbles that can analyzed.
   d. Sediments only form in cold climates.

3. What effect do volcanic ash and sulfur dioxide have on climate?
   a. They both cool the climate.
   b. They both increase the surface temperature.
   c. They cause hot house climates and ice ages.
   d. They create a balance in global temperatures.

4. The wobbling of a spinning object on its axis is known as:
   a. eccentricity
   b. tilting
   c. precession
   d. rotation

5. What is the result of the Milankovitch cycle?
   a. varying orbital speed of the Earth
   b. varying insolation received by the Earth
   c. a change in the axis of the Sun
   d. the slowing down of the Earth

6. How long ago did the last ice age cover much of North America with glaciers?
   a. 5000 years
   b. 10,000 years
   c. 20,000 years
   d. 100,000 years

7. What is believed to be responsible for the creation of hot house climates in the past?
   a. the Milankovitch cycle
   b. volcanic ash
   c. carbon dioxide gas
   d. CFCs

8. For the past 100 years, scientific evidence suggests that:
   a. Earth is about to enter another ice age
   b. Earth's average temperature is increasing
   c. Earth's average temperature is decreasing
   d. Earth's climate is stabile

9. An increase in carbon dioxide in the atmosphere over the past 100 years is the result of:
   a. rice paddies
   b. melting glaciers
   c. volcanic eruptions
   d. burning of fossil fuels

10. Which of the following is not a greenhouse gas?
    a. carbon dioxide
    b. methane
    c. water vapor
    d. nitrogen gas

Matching *Match the terms with the correct definitions.*

a. climate
b. hot house climate
c. ice age
d. paleoclimatology
e. tree ring
f. pollen
g. ice cores
h. precession
i. eccentricity
j. Goldilocks syndrome
k. fossil fuels
l. rice paddies
m. chlorofluorocarbons
n. nitrous oxide

1. ____ Long cylindrical sections of ice that are removed from glaciers by drilling and can be used to study the Earth's past climate.
2. ____ A chemical compound that is the form of a gas containing two atoms of nitrogen and one atom of oxygen (N_2O).
3. ____ The long-term weather patterns of a specific region on the Earth, usually defined by the area's annual temperature and precipitation values.
4. ____ The wobbling motion of the axis of a rapidly rotating body, such as the Earth, which causes the tilt of the axis to change periodically.
5. ____ Tiny, dustlike male reproductive cells produced by a flower.
6. ____ The mathematical expression of how far an ellipse is from a perfect circle, which can be determined by dividing the distance between the foci by the length of the major axis.
7. ____ A term used to describe a period in the Earth's history when the average surface temperature was much warmer than today.
8. ____ A class of human-created molecules commonly used as refrigerants and in electrical manufacturing and foam production that is responsible for ozone destruction and is also greenhouse gas.
9. ____ A specific period in the Earth's history when the average surface temperature was much lower than today, causing the widespread formation of glaciers.
10. ____ A illustration of the relationship among the atmospheric carbon dioxide concentrations on Venus, Earth, and Mars and their surface temperatures.
11. ____ A type of agricultural field used to grow rice; it is periodically flooded with water.
12. ____ The scientific discipline that studies the history of the Earth's climates.
13. ____ A term used to describe hydrocarbon fuels such as coal and oil, which were formed from the remains of once living organisms.
14. ____ The ringlike growth of new wood in the trunk of tree that marks the occurrence of one growing season.

Critical Thinking

1. The Milankovitch cycle involves the changing tilt of the Earth's axis between 24.5 and 22 degrees. Explain how the climate would be affected in the United States if the Earth's axis was tilted 21.5 degrees instead of its present 23.5 degrees.

CHAPTER 21

Acid Precipitation and Deposition

Objectives

The pH of Precipitation • Formation of Precipitation • Anthropogenic Gases • Long Distance Transport of Acid-Causing Pollutants • Effects of Acid Deposition on Aquatic Ecosystems • Effects of Acid Precipitation on Terrestrial Ecosystems • Effects of Acid Deposition on Human Beings and Building Materials • Control of Acid Deposition

After reading this chapter you should be able to:

- Define the term *acid precipitation.*
- Explain why the pH of unpolluted rain is slightly acidic.
- Identify three natural sources that produce acid-forming compounds.
- Describe the process by which nitric acid and sulfuric acid are formed in the atmosphere.
- Identify the two main anthropogenic sources of sulfur dioxide and nitrogen oxides gases.
- Describe how acid precipitation is transported over long distances.
- Identify the main source area for sulfur dioxide gas, which creates acid precipitation that affects the northeastern United States.
- Explain two ways by which aquatic ecosystems are negatively affected by acid precipitation.
- Define the term *acid shock.*
- Explain two ways by which terrestrial ecosystems are negatively affected by acid precipitation.
- Identify five building materials that can be damaged by acid precipitation.

TERMS TO KNOW

- **acidic**
- **pH**
- **alkalinity**
- **acid precipitation**
- **sulfur dioxide**
- **sulfuric acid**
- **nitric acid**
- **fossil fuels**
- **hydrocarbon**
- **combustion**
- **anthropogenic**
- **acid deposition**
- **toxic heavy metals**
- **acid shock**
- **buffer**

INTRODUCTION

One of the most shocking examples of how human activities can alter the chemistry of the atmosphere, resulting in widespread damage to the environment, is the formation of acid precipitation. Acid precipitation is a form of global pollution that is the direct result of human activities altering the Earth's systems. Following the pathways through the environment that pollutants take to form acid precipitation reveals the complex interaction that occurs among human activity, the atmosphere, the hydrosphere, and the biosphere.

The pH of Precipitation

Under normal conditions, precipitation is naturally **acidic.** This is due to the effect of carbon dioxide gas in the atmosphere. Carbon dioxide dissolves in water high up in the atmosphere to form weak carbonic acid. Because of this, normal, unpolluted rain has a pH value between 5.2 and 5.6. The **pH** scale measures the acidity or **alkalinity** of a solution on a scale from 0 to 14. A pH of 7 is considered neutral. Any solution that is less than 7 is acidic, and any solution that is greater than 7 is considered basic or alkaline (Figure 21–1). **Acid precipitation** is formed when atmospheric pollutants create rain, snow, or fog with a pH level of 5 or less.

Formation of Acid Precipitation

Some natural events also create slightly acidic precipitation. Volcanic eruptions can emit **sulfur dioxide** gas into the atmosphere, which can also mix with atmospheric moisture. This combination forms **sulfuric acid.** Forest fires and lightning can naturally lower the pH of precipitation (Figure 21–2). These natural phenomena form nitrogen oxide compounds that can mix with moisture in the atmosphere to form **nitric acid.** Together, these natural forms of acid precipitation occur on such a small level that they pose no threat to the environment; however, human technology is altering this process.

Anthropogenic Gases

The combustion of **fossil fuels** such as gasoline, coal, or natural gas is adding large amounts of nitrogen oxides and sulfur dioxide to the Earth's atmosphere. Most automobiles and trucks get their power by mixing a **hydrocarbon** fuel, such as gasoline, with oxygen in the air. This mixture is then ignited to create a controlled explosion that powers

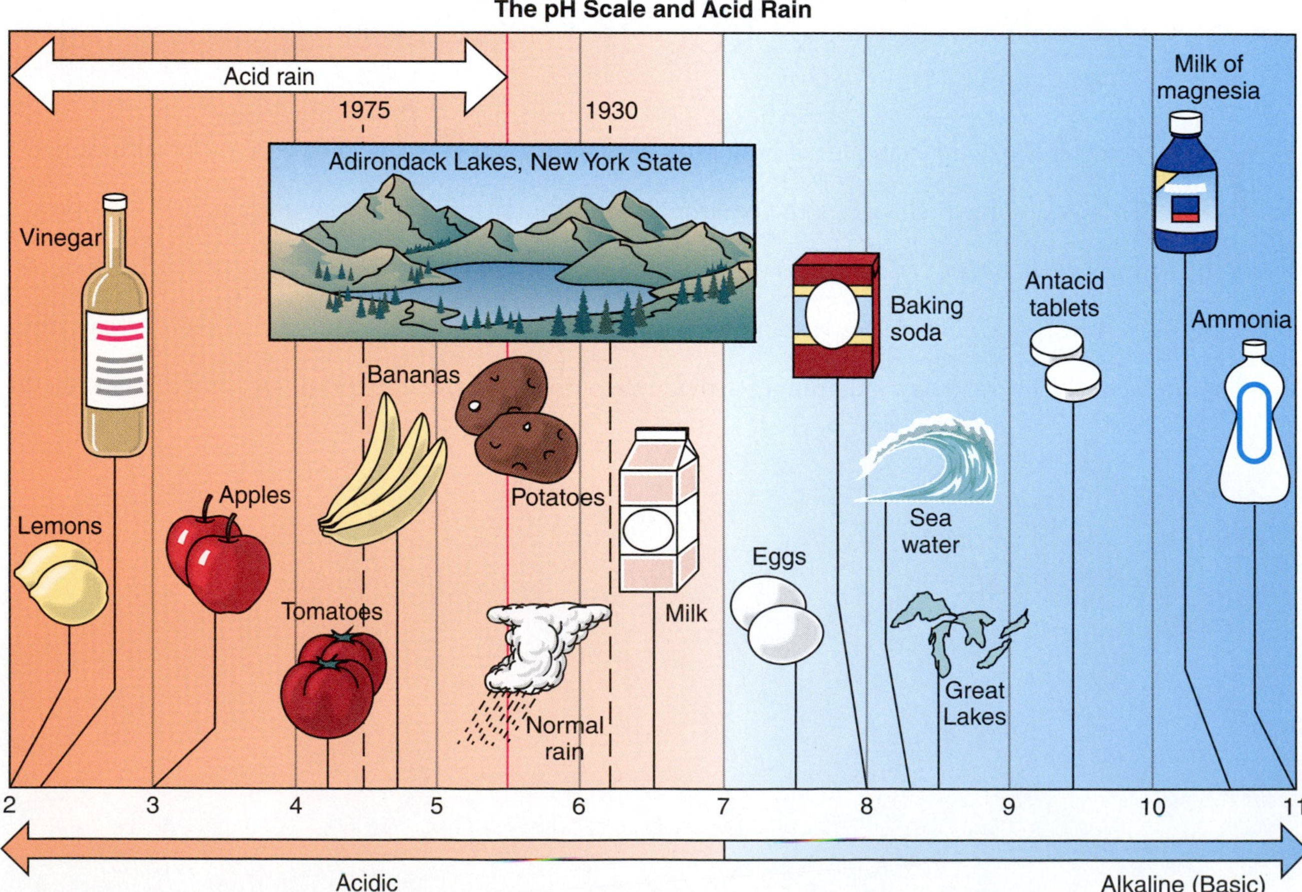

Figure 21–1 The pH scale and the acidity or alkalinity of some common substances.

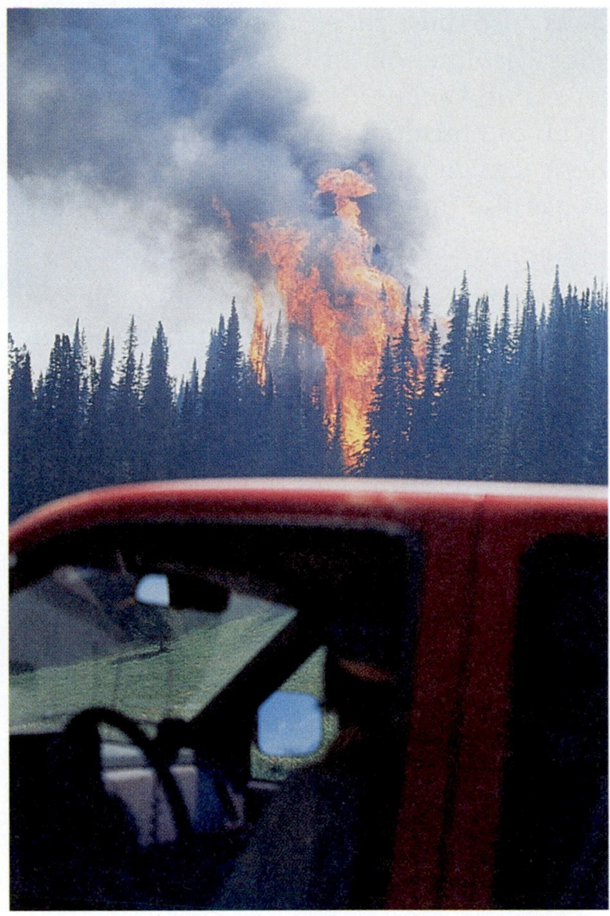

Figure 21–2 Volcanic eruptions and forest fires are natural sources of acid-forming compounds in the atmosphere. *(Left image courtesy of PhotoDisc. Right image courtesy of Boise National Forest.)*

the vehicle. The burning of a substance is also known as **combustion.** Because 78% of the atmosphere is nitrogen gas, the combustion inside of an engine combines this nitrogen with atmospheric oxygen to form nitrogen oxide compounds. These nitrogen oxides are the byproduct of combustion and end up in the Earth's atmosphere. In a different process,

EARTH SYSTEM SCIENTISTS

ROBERT ANGUS SMITH

Robert Angus Smith was born in Scotland in 1817. He was educated as a chemist and became interested in the pollution that was being created by the industrial revolution. Smith became one of the first scientists concerned about the negative effects of industry on the environment. He theorized that factory smoke and fumes could mix with precipitation and lower its acidity. In 1852 he presented his findings in a paper titled "On the Air and Rain of Manchester." Smith was the first scientist to call this phenomenon acid rain; it is now one of the most damaging forms of air pollution around the world. In 1872 Smith published *Air and Rain: The Beginnings of Chemical Climatology,* a book that presented his theories of how precipitation formed. Later in his life he also became interested in the effects of pollution on public health.

coal is also burned as a source of energy. Unlike gasoline, oil, or natural gas, coal contains sulfur. When this sulfur is heated as a result of combustion, it is joined with atmospheric oxygen to form sulfur dioxide gas. This gas also enters the Earth's atmosphere.

Both the nitrogen oxides and sulfur dioxide that are created as a result of technological combustion are known as **anthropogenic** gases. The term *anthropogenic* means "human created." Both these gases rise into the atmosphere and combine with atmospheric moisture to form nitric and sulfuric acids. It is these anthropogenic acids that create acid precipitation (Figure 21–3).

Acid deposition is a different process than acid precipitation. Acid deposition occurs when dry nitrogen and sulfur compounds are deposited directly onto the Earth's surface. This causes a reaction with the surrounding environment that leads to the formation of strong acids. Acid deposition is also caused by anthropogenic gases.

Long Distance Transport of Acid-Causing Pollutants

Since 1987 more than 50 million tons of sulfur dioxide gas has been emitted into the atmosphere by utility companies burning coal in the United States, and more than 210 million tons of sulfur compounds fall to the Earth's surface every year. The combustion of coal is used to create steam that generates electrical power. The majority of the sulfur dioxide emissions in the United States originate in the Midwest. About 90% of all the acid precipitation and deposition that occurs on the Earth is derived from sulfur dioxide. Nitrogen oxide emissions into the atmosphere from the United States since 1987 equal about 250 million tons. It is estimated that more than 56 million tons of nitrogen compounds are deposited on the Earth's surface annually. The nitrogen compounds are derived from both transportation and industrial combustion of fossil fuels. The amount of these gases that have entered the atmosphere is altering the chemistry of precipitation.

Because of the prevailing winds that move weather across North America from west to east, the northeastern portions of the continent receive the greatest amount of acid precipitation (Figure 21–4). Nitrogen oxides and sulfur dioxide gas that are produced west of the Appalachian Mountains rise into the atmosphere and mix with atmospheric moisture to form acid precipitation. By the time weather systems reach the northeast, the pH of the precipitation is between 4.3 and 4.5. This is called long-distance transport of acid precipitation. A similar process occurs in Europe, where the western European countries create the pollutants that cause acid precipitation to fall on eastern Europe.

Effects of Acid Deposition on Aquatic Ecosystems

Acid precipitation can adversely affect the Earth in a number of ways. The acids that fall from the atmosphere can damage aquatic ecosystems. The pH values for unpolluted, healthy ecosystems fall between 6.0 and 8.0. Lakes and ponds that are exposed to acid deposition and precipitation for extended periods experience a decreasing pH level. Almost all aquatic organisms die when the pH is 4.0 or lower (Figure 21–5).

More than 70% of the lakes in the Adirondack Mountains of New York are being adversely affected by acid precipitation. In many of these lakes the acid precipitation is lowering the pH of the lakes below 4.0. Acids that collect in aquatic ecosystems can also leach **toxic heavy metals** such as mercury, aluminum, and cadmium into the water. These metals are then absorbed into the tissues of the organisms that reside there, causing adverse health effects. If humans eat these organisms, the toxic metals can build up in human tissue, causing health problems. One of the worst effects of acid deposition on aquatic ecosystems occurs during

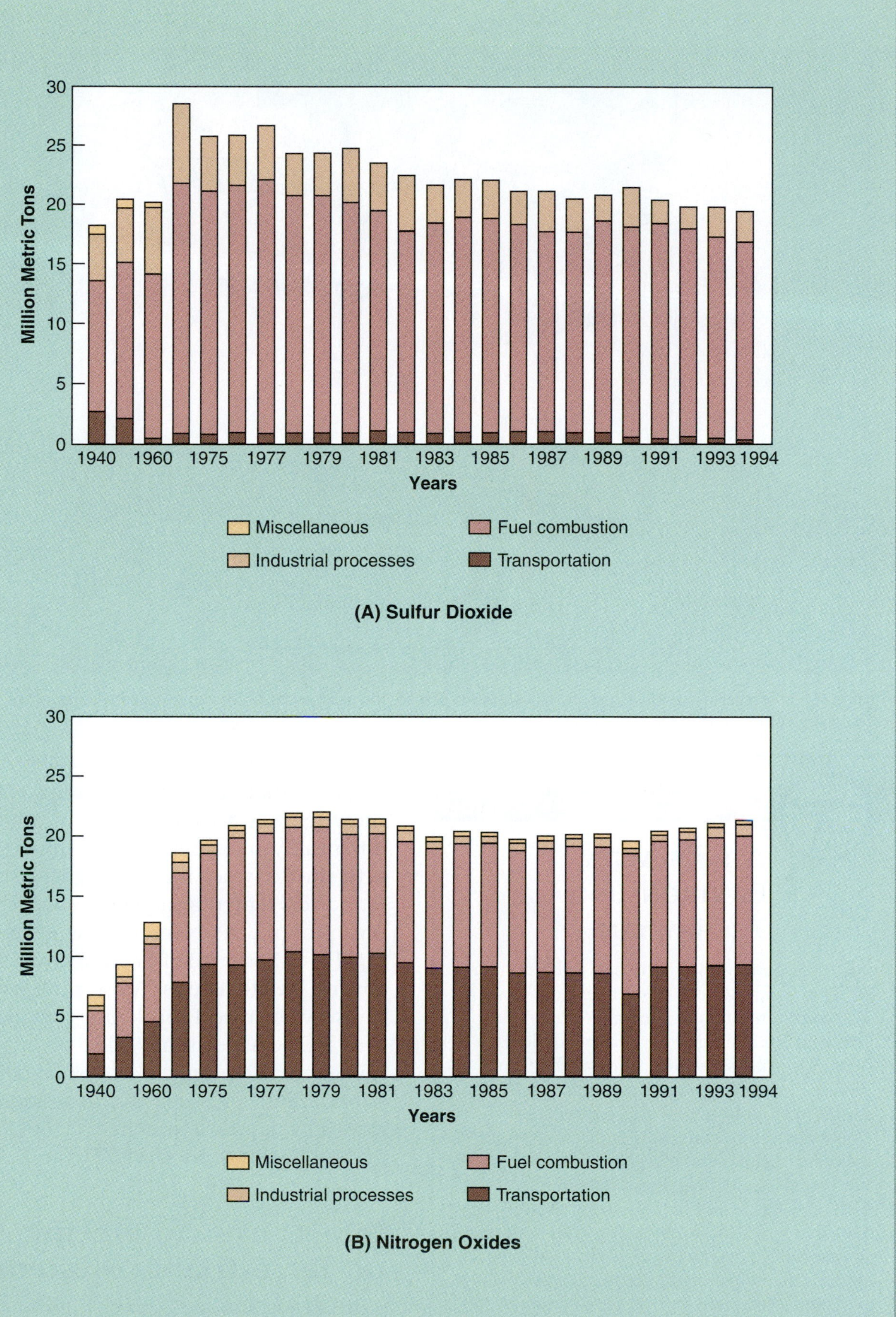

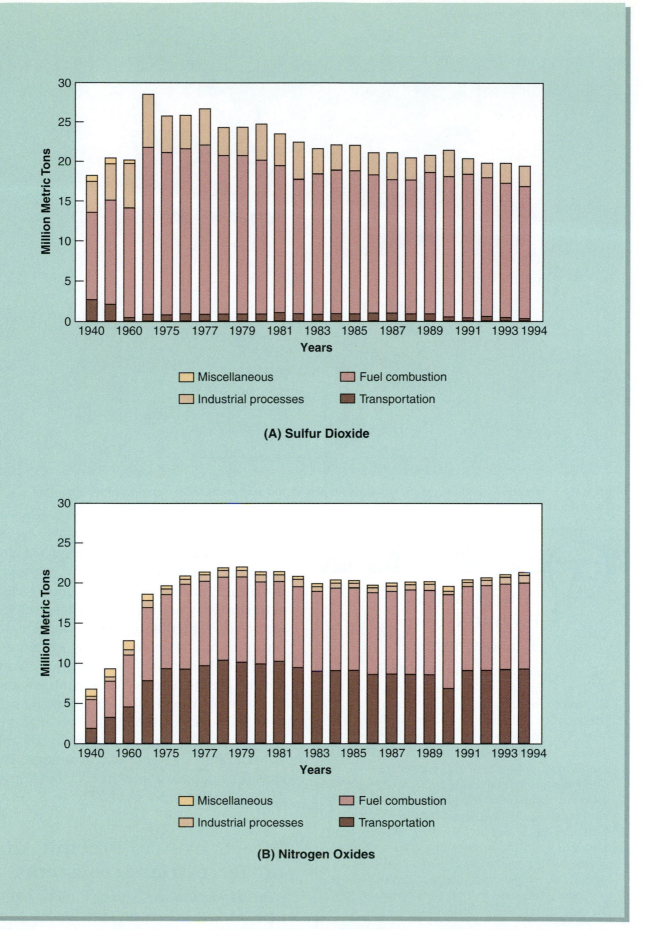

Figure 21–3 Anthropogenic sources and their annual emissions of sulfur dioxide and nitrogen oxides, which form acid precipitation.

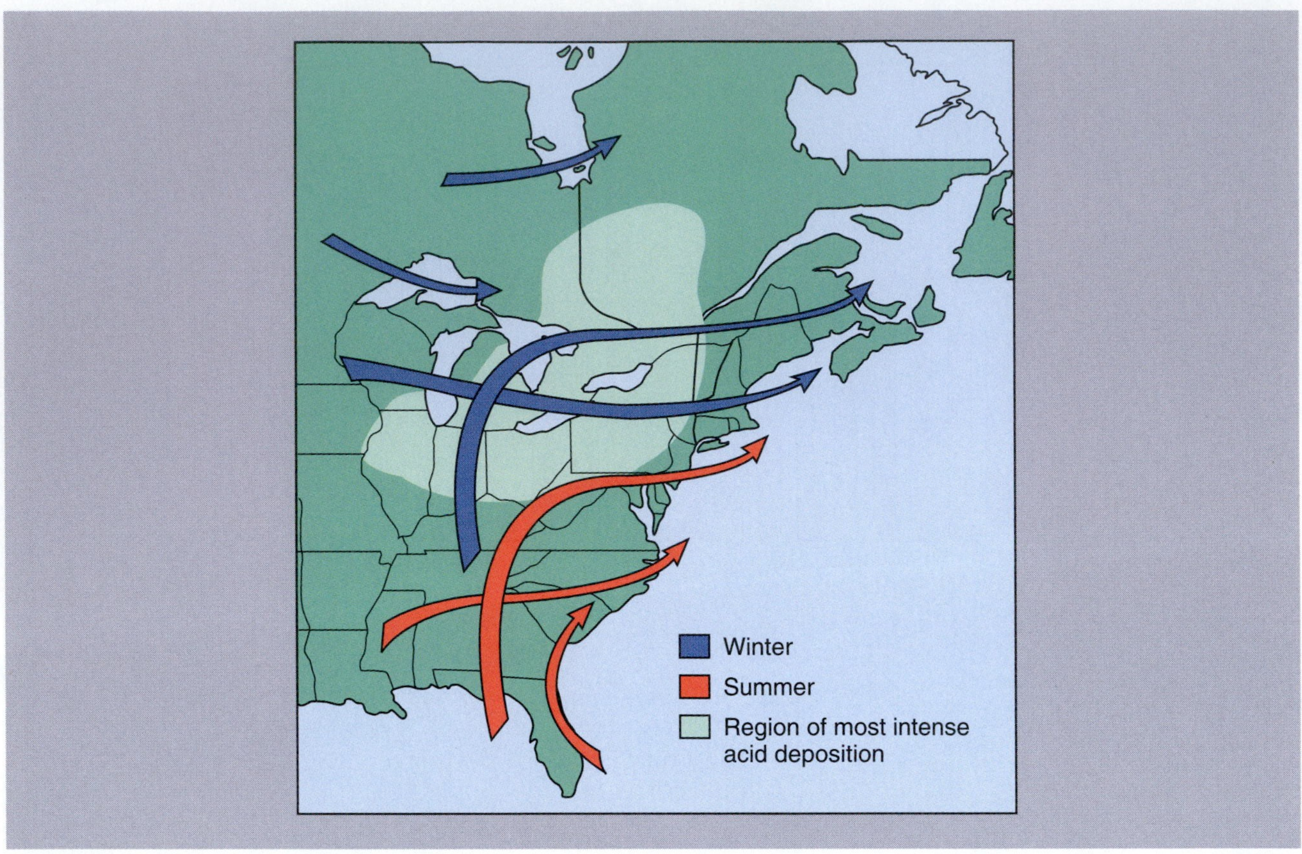

Figure 21–4 Prevailing winds across the United States provide the mechanism for the long-distance transport of acid-forming compounds in the atmosphere.

> **Career Connections**
>
> ### Air Quality Environmental Technician
>
> An air quality environmental technician applies a knowledge of science, technology, and communication to the monitoring of acid precipitation and other air quality problems. This job includes the collection, labeling, and storing of both air and precipitation samples for use in the monitoring of acid precipitation. These environmental technicians also operate, calibrate, maintain, and repair monitoring equipment in the field; input data into computer databases; and help to analyze samples in the laboratory. This profession requires a 2- or 4-year college degree in environmental science; persons meeting this requirement can seek employment with state or local government agencies and within academic institutions.

the winter. Over the winter, acidic snow accumulates; then during the spring thaw this snow releases a large amount of acidic water into nearby lakes and streams. This is called an **acid shock,** which is caused by acidic melt water rapidly entering aquatic ecosystems. The organisms that live there are then exposed to a rapid drop in the pH of the water, which causes adverse health effects. Most adult organisms can survive an acid shock, but younger organisms and eggs are usually killed off by this phenomenon. Some aquatic ecosystems are able to neutralize the effects of acid deposition. This is caused by natural limestone rock outcrops that act as a **buffer** to the acids (Figure 21–6).

Effects of Acid Precipitation on Terrestrial Ecosystems

Acid deposition and precipitation can also have adverse effects on land-based environments, also known as terrestrial ecosystems.

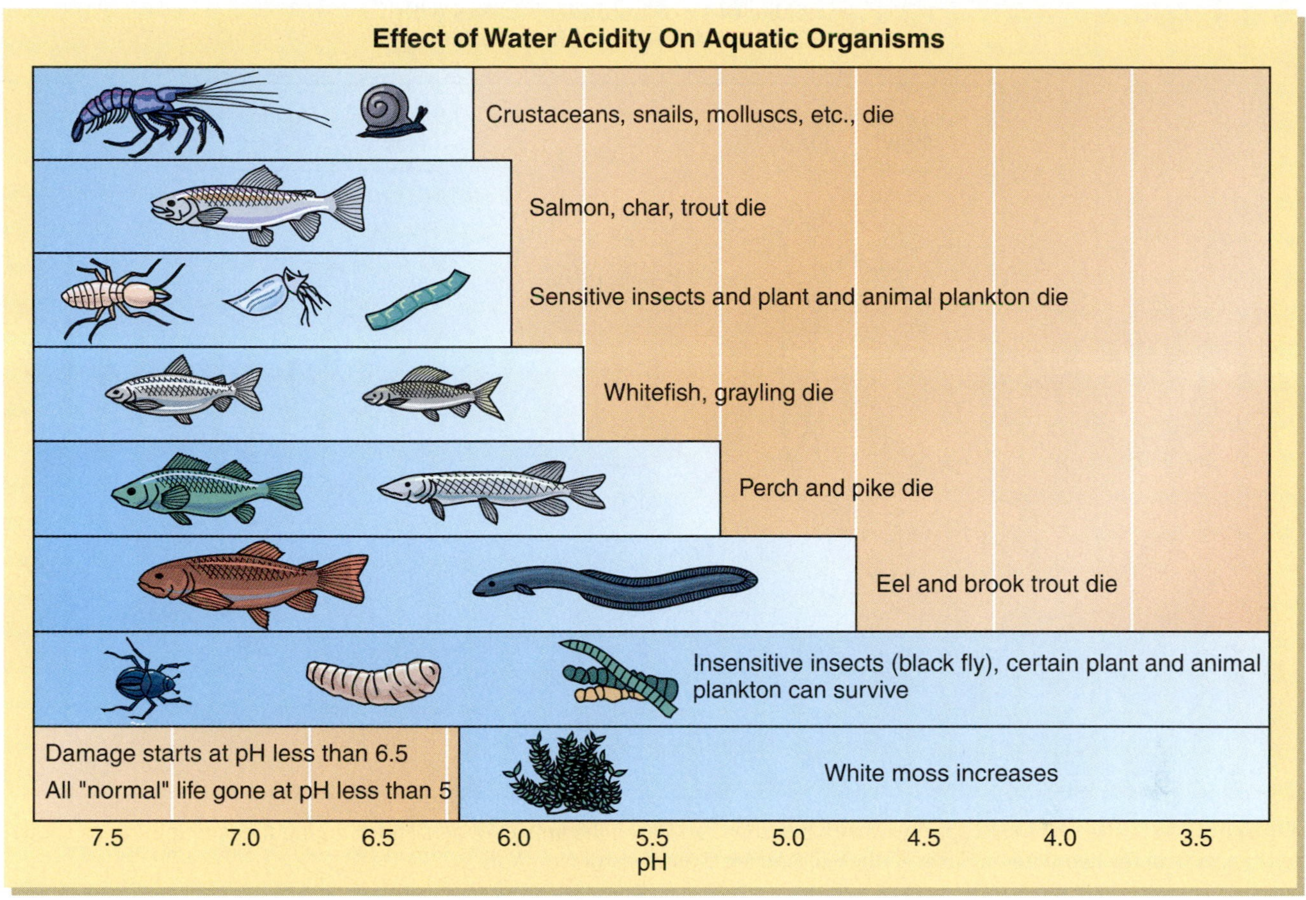

Figure 21–5 The effect of increased acidity on aquatic organisms.

Many plants and trees experience stress from acids burning their leaves and stems. The acids break down the protective waxy layer on many plants, which exposes their leaves and stems to disease and causes them to lose water. Trees that are located at higher elevations and are constantly bathed in fog and mist are especially susceptible to acid damage. Acids can also leach important plant nutrients from the soil, including potassium, calcium, and magnesium. Other elements in soil can be released by the increased acidity, which causes plants to become damaged by their exposure. Aluminum is responsible for the die off of many evergreen trees that are exposed to acid precipitation and deposition (Figure 21–7).

Acid precipitation that falls on soil can also kill off beneficial microorganisms, such as bacteria and fungi, that add to soil fertility. Extremely acidic soils can also inhibit the germination of plant seeds, directly affecting agricultural production.

Effects of Acid Deposition on Human Beings and Building Materials

Human beings can also be affected by acid precipitation and deposition. Humans who consume fish that live in acidic waters may be affected by the intake of toxic heavy metals. Also, because acid precipitation affects plants, it can affect food production and forest products, such as timber, pulp, and maple sugar. Trees and plants that are damaged by acid precipitation are not as productive. Humans can also develop respiratory illness and irritation of the eyes, nose, and throat when they are exposed to sulfur dioxide gas and other acid-forming compounds.

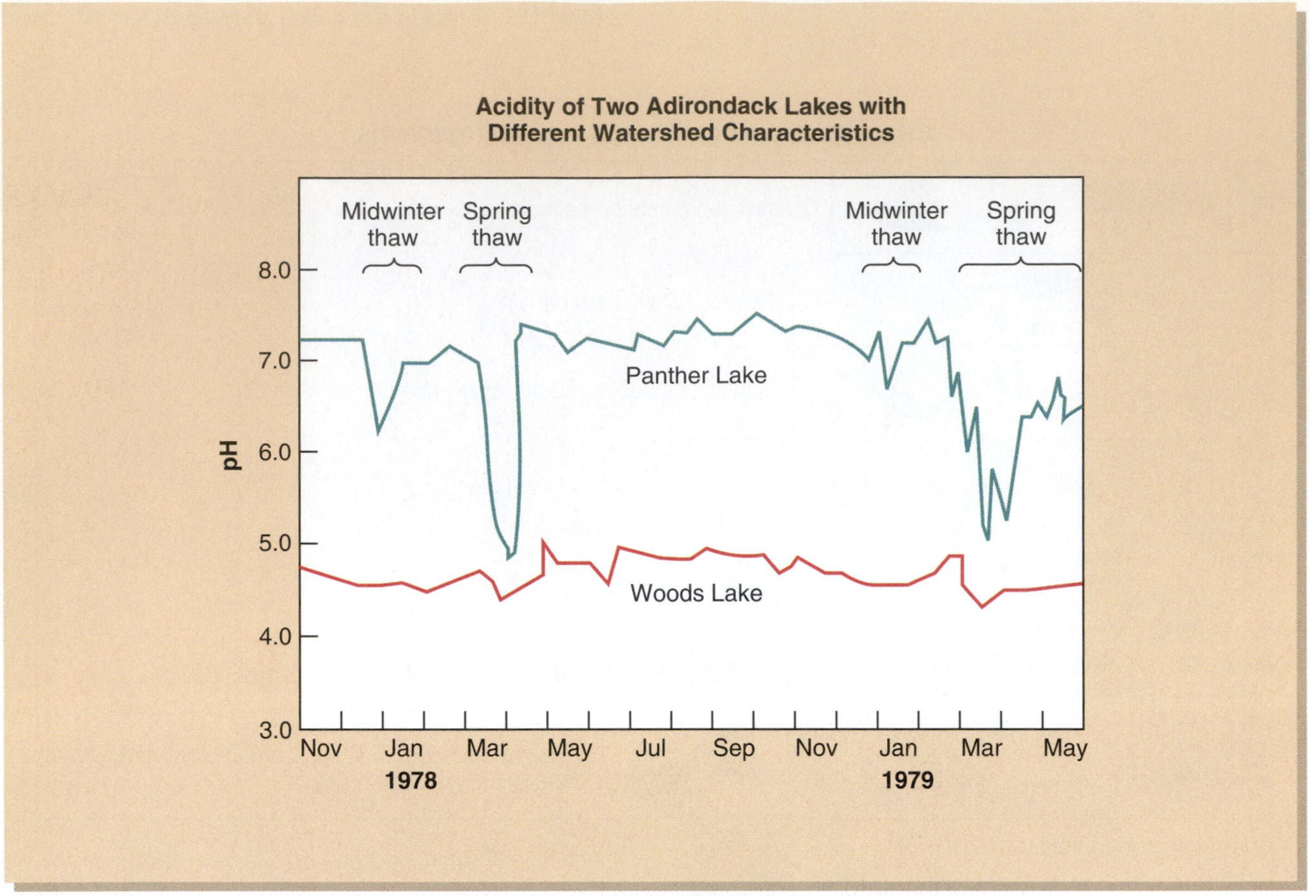

Acidity of Two Adirondack Lakes with Different Watershed Characteristics

Midwinter thaw
Spring thaw
Midwinter thaw
Spring thaw
pH
8.0
7.0
6.0
5.0
4.0
3.0
Panther Lake
Woods Lake
Nov Jan Mar May Jul Sep Nov Jan Mar May
1978
1979

Figure 21–6 The buffering capacity of rock outcrops can lessen the impact of acid shock during the spring thaw, as shown by the pH data for two different lakes in the Adirondack Mountains of New York.

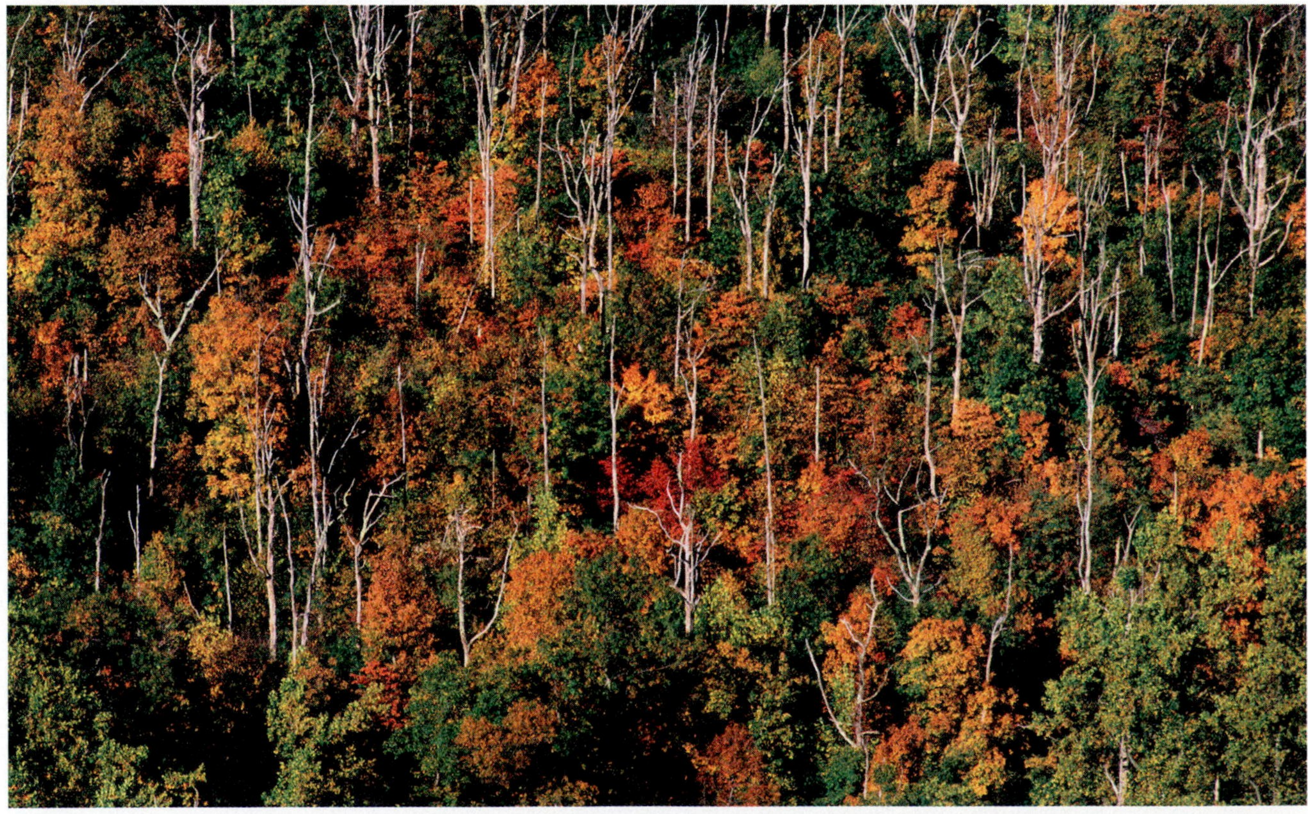

Figure 21–7 The damage caused to trees in the mountains of New York as a result of acid deposition. *(Courtesy of PhotoDisc.)*

Living organisms are not the only things that are affected by acid precipitation. Structures and building materials can also be damaged. Concrete and stone can experience enhanced weathering as a result of exposure to acids. Bridges, buildings, and other concrete structures are slowly dissolved and weaken with exposure to acids. Other materials, such as stone and brick, are broken down over time. Many old gravestones made from stone are almost unreadable today as a result of acid precipitation damage (Figure 21–8). Many ancient stone statues around the world are also slowly wearing away as result of acid damage.

Metals such as steel, iron, and copper are corroded by the acids that fall from the atmosphere. Paints, varnish, rubber, and ceramics are damaged by acid deposition. The costs associated with maintaining or replacing these structures or artifacts reach billions of dollars annually.

Another negative aspect of excess acids in the atmosphere is the reduction of visibility. Nitrogen oxides and sulfur oxides greatly reduce visibility. This is becoming quite a problem for many of the national parks. Millions of people travel to these parks to experience their amazing visual beauty; however, these pollutants are causing reduced visibility at many of these parks.

Control of Acid Deposition

The federal and state governments are attempting to regulate the amount of nitrogen oxides and sulfur dioxide gas that are emitted into the atmosphere. Northeastern states are especially concerned about controlling these atmospheric pollutants. Forests in eastern

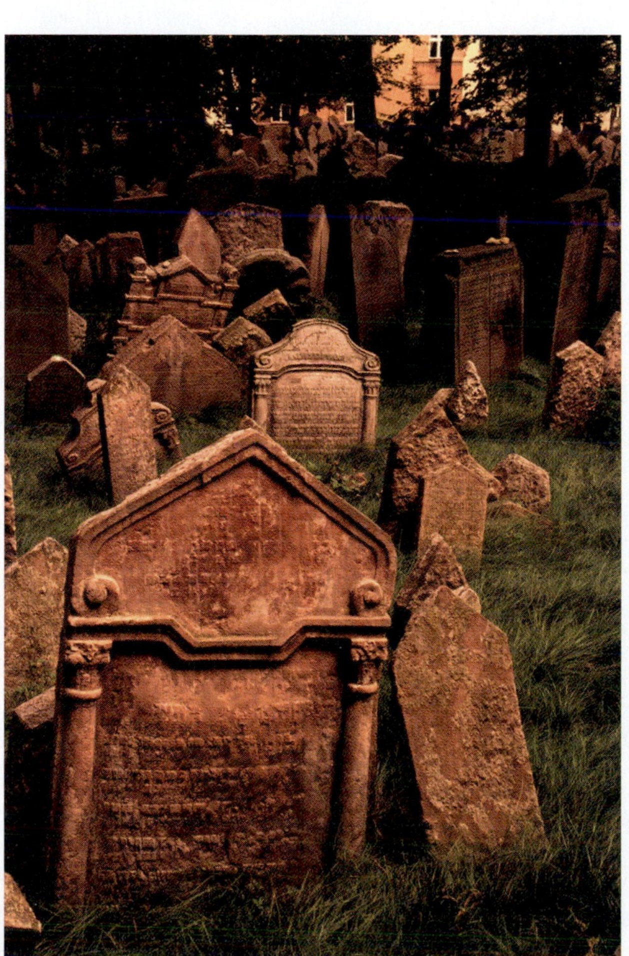

Figure 21–8 Many gravestones in cemeteries across the Northeastern United States have been severely degraded by the accelerated chemical weathering caused by acid precipitation. *(Courtesy of PhotoDisc.)*

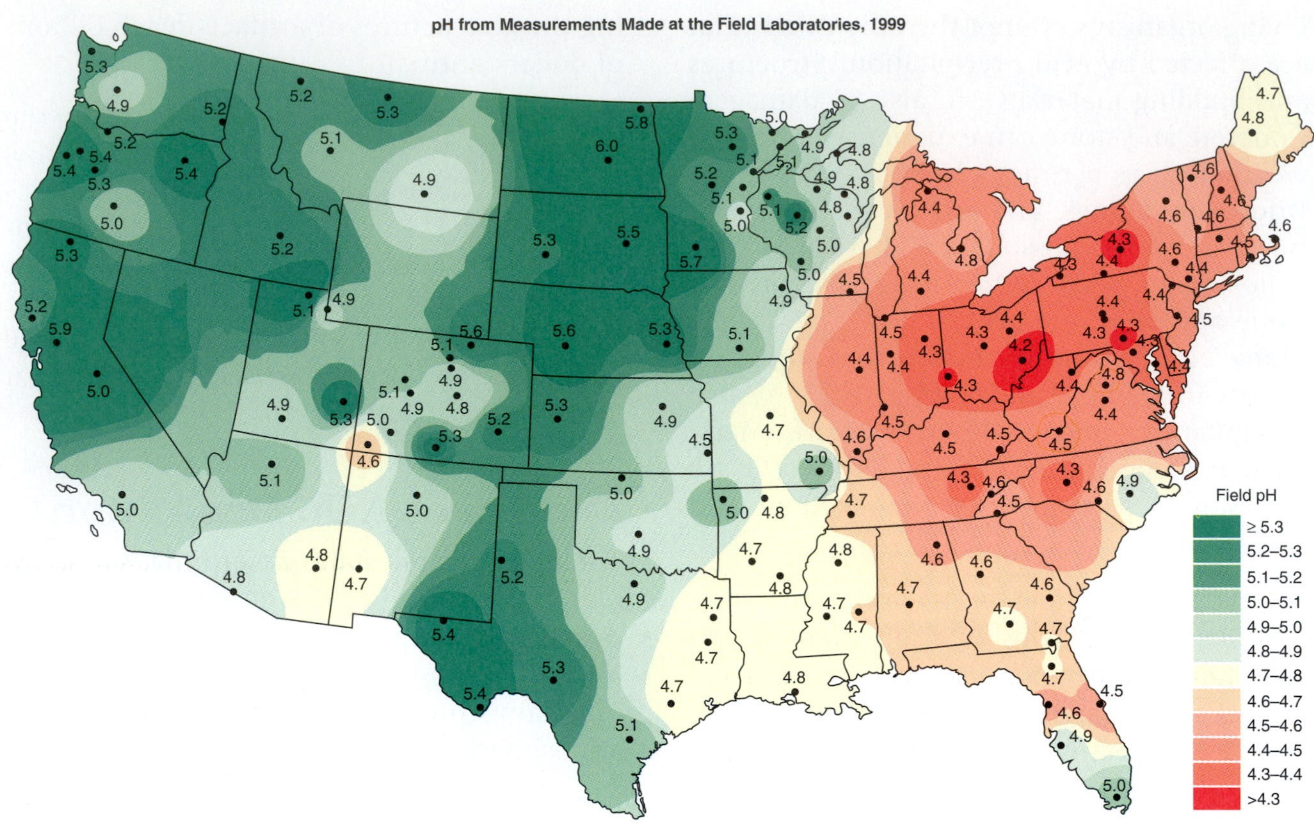

Figure 21–9 A map of the United States showing the average pH levels for different regions.

New York and western Vermont are experiencing the worst damage by acid deposition (Figure 21–9). The interactions that occur among human technology, the atmosphere, and the biosphere are responsible for the damaging affects of acid precipitation. Now that the process by which this pollution is formed is fully understood, efforts must be made to stop it before the damage becomes too great.

EARTH MATH

1) IF 250 MILLION TONS OF NITROGEN OXIDES WERE PUT INTO THE ATMOSPHERE BETWEEN 1987 AND 1997, APPROXIMATELY HOW MANY TONS OF NITROGEN OXIDES ENTER THE ATMOSPHERE EACH YEAR?

2) BY THE YEAR 2010 THE ENVIRONMENTAL PROTECTION AGENCY PLANS TO HAVE UTILITY COMPANIES REDUCE THEIR NITROGEN OXIDE EMISSIONS BY 2 MILLION TONS EACH YEAR. IF THEIR ANNUAL LEVELS OF NITROGEN OXIDES ARE APPROXIMATELY 22 MILLION TONS, BY WHAT PERCENTAGE WILL THEIR EMISSIONS BE REDUCED?

REVIEW

1. What is the normal, unpolluted pH value of precipitation?
2. What causes unpolluted precipitation to be slightly acidic?
3. What is the definition of acid precipitation?
4. How does acid precipitation form?
5. What processes produce the gases that form acid precipitation?
6. How does acid precipitation affect aquatic ecosystems?
7. How does acid precipitation affect terrestrial ecosystems?
8. How can human beings be affected by acid precipitation?
9. Explain how building materials are damaged by acid precipitation.
10. Who was Robert Angus Smith?

For more information go to this Web link:
<http://www.epa.gov/airmarkets/acidrain/>

CHAPTER SUMMARY

Acid precipitation occurs when the pH of rain is lower than 5.0. The pH of a solution measures its acidity or alkalinity on a scale from 0 to 14, with 7 being neutral. A pH of less than 7 is considered an acid, and a pH greater than 7 is considered alkaline or basic. Unpolluted rain is naturally acidic, with a pH between 5.4 and 5.6, because it can mix with carbon dioxide in the atmosphere to form weak carbonic acid. Other natural events, such as volcanic eruptions and forest fires, also naturally lower the pH of rain. Acid precipitation forms when chemicals such as sulfur dioxide and nitrogen oxides mix with water in the atmosphere to form sulfuric and nitric acids. These then return to the surface in the form of rain or snow with a greatly reduced pH. Human activities such as the burning of fossil fuels create these anthropogenic gases.

The two major sources of anthropogenic gases are the burning of coal and the combustion of gasoline used for transportation. Acid precipitation can cause many adverse effects on the environment. When acid precipitation falls into aquatic ecosystems, the pH of the water can be reduced. This can cause stress or even death to the organisms that reside in the water. When the pH of an aquatic ecosystem drops below 4.0, nothing will survive. Many lakes in the northeastern United States have been adversely affected by acid precipitation.

Terrestrial ecosystems can also be harmed by acid precipitation. The leaves of trees and plants can be damaged by the acids, as can organisms residing in the soil, such as bacteria. Many buildings and structures built from materials such as concrete, stone, brick, and iron are damaged by exposure to acid precipitation. These building materials are rapidly degraded by the acids. The structures damaged by this pollution cost millions of dollars to repair or replace. Human health also is affected by exposure to acid precipitation. The eyes, nose, throat, and lungs can be irritated by exposure to the acids in the atmosphere.

Much of the acid precipitation that falls on the United States is formed by power companies in the Midwestern part of the country. These utility companies burn coal to produce electricity, which emits millions of tons of acid-forming gases into the atmosphere. These gases are then transported over long distances toward the East Coast by the prevailing winds, which move weather across the country. The long-distance transport of acid precipitation has caused eastern New York and western Vermont to experience the worst damage by acid rain. New regulations and cleaner forms of transportation are required to reduce this threat to our environment.

CHAPTER REVIEW

Multiple Choice

1. Which of the following atmospheric gases causes precipitation to be naturally acidic?
   a. nitrogen
   b. oxygen
   c. carbon dioxide
   d. argon

2. Lightning, forest fires, and vehicle exhaust all produce this gas, which combines with water in the atmosphere and lowers its pH:
   a. oxygen
   b. nitrogen oxides
   c. sulfur dioxide
   d. argon

3. The burning of coal and volcanic eruptions produce this gas, which combines with water in the atmosphere and lowers its pH:
   a. oxygen
   b. nitrogen oxides
   c. sulfur dioxide
   d. argon

4. These two acids are primarily responsible for the formation of acid precipitation:
   a. acetic and carbonic
   b. citric and carbonic
   c. vinegar and soda
   d. nitric and sulfuric

5. Most acid rain that falls on Vermont and New York originates in:
   a. New York City
   b. the Midwest
   c. California
   d. Canada

6. Most all aquatic organisms die when the pH of the water drops below:
   a. 6.0
   b. 5.0
   c. 4.0
   d. 3.0

7. In the northeastern United States the annual spring thaw can cause:
   a. acid rain
   b. acid shock
   c. acid melt
   d. algal bloom

8. Which type of aquatic organism is most affected by lowering the pH of water?
   a. young
   b. old
   c. carp
   d. bacteria

9. What percentage of lakes in the Adirondack Mountains of New York are affected by acid precipitation?
   a. 10%
   b. 25%
   c. 50%
   d. 70%

10. Which element can be leached from rock and soil into water by acid precipitation, harming organisms?
    a. nitrogen
    b. sulfur
    c. aluminum
    d. calcium

Matching *Match the terms with the correct definitions.*

a. acid precipitation
b. pH
c. acidic
d. alkalinity
e. sulfur dioxide
f. sulfuric acid
g. nitric acid
h. fossil fuels
i. hydrocarbon
j. combustion
k. anthropogenic
l. acid deposition
m. toxic heavy metals
n. acid shock
o. buffer

1. ____ A gaseous chemical compound composed of one atom of sulfur and two atoms of oxygen (SO_2).
2. ____ The deposit of acidic substances on the Earth's surface.
3. ____ A term used to describe hydrocarbon fuels such as coal and oil, which were formed from the remains of once living organisms.
4. ____ A substance that is capable of stabilizing the acidity or alkalinity of a solution.
5. ____ Precipitation that contains high concentrations of sulfuric or nitric acids and has a pH of 5.0 or lower.
6. ____ Rapid introduction of acidic water into lakes and streams caused by melting snow.
7. ____ The unit of measurement used to measure the acidity or alkalinity of a solution.
8. ____ Naturally occurring, poisonous metal elements such as lead or mercury.
9. ____ A chemical reaction that results in light and heat, commonly called burning.
10. ____ A solution that possesses a pH lower than 7.0.
11. ____ A term that describes any substance that is created or introduced into the environment by human activity.
12. ____ A solution that has a pH greater than 7.0, also called a base.
13. ____ A type of chemical compound composed of hydrogen and carbon and is commonly associated with fuels.
14. ____ A strong acid (H2SO4) that forms in the atmosphere when sulfur dioxide gas reacts with atmospheric moisture.
15. ____ A strong acid (HNO3) that is formed in the atmosphere when rain mixes with nitrogen compounds to create acid precipitation.

Critical Thinking:

1. Some researchers believe that acid rain may actually be beneficial to agriculture. Why do you think they believe this true?

CHAPTER

22 Ozone Depletion

Objectives

Ozone Gas • The Ozone Layer • Measuring Stratospheric Ozone • Effects of Ozone Depletion • The Ozone Hole • Reducing Ozone Depletion

After reading this chapter you should be able to:

- Describe how ozone gas forms in the atmosphere.
- Identify the location of the ozone layer within the atmosphere.
- Explain the process of how ozone gas prevents ultraviolet radiation from striking the Earth's surface.
- Identify the gases that are responsible for the destruction of the ozone layer.
- Describe the process by which ozone is being destroyed.
- Explain the negative effect that ozone depletion has on the Earth's surface.
- Define the term *ozone hole.*
- Describe two ways by which humans can be affected by the destruction of the ozone layer.
- Explain three things that you can do to protect yourself from the loss of ozone.

TERMS TO KNOW

ozone
stratosphere
ultraviolet radiation
ozone layer
regeneration
dynamic equilibrium
depletion
chlorofluorocarbon
Freon
immune systems
mutated
ozone hole

INTRODUCTION

The discovery of a decrease in atmospheric ozone is one of the most serious threats that our planet has faced since the last mass extinction 65 million years ago. The layer of ozone gas that surrounds the planet acts as a protective barrier that prevents ultraviolet radiation from striking the Earth's surface at deadly levels. Today the ozone layer is being depleted around the world as a result of human activity, which is directly threatening life on the surface of the Earth. Understanding how ozone forms, what part it plays in the Earth's systems, and how we can prevent its loss is one of the most important scientific topics of the twenty-first century.

Ozone Gas

Ozone is a an unstable gas composed of three atoms of oxygen that naturally exists between 12 and 15 miles up in the **stratosphere.** Ozone gas is formed when atmospheric oxygen, which is composed of two atoms of oxygen (O_2), is bombarded by incoming **ultraviolet radiation** from the Sun. This high-energy radiation breaks apart the two oxygen atoms, which recombine with other individual atoms of oxygen to form a molecule of ozone gas (O_3). As a result of this process, ozone gas has built up in the stratosphere over time to form what is called the ozone layer (Figure 22–1). The **ozone layer** is an important part of the Earth's atmosphere because it acts as a shield that protects the surface from deadly ultraviolet radiation emitted by the Sun.

The Ozone Layer

When an ozone molecule is struck by high-energy ultraviolet radiation, it breaks apart and converts the energy into heat (Figure 22–2). The release of this heat is what causes the temperatures to increase at higher altitudes in the stratosphere. This process prevents the high-energy ultraviolet radiation from reaching the surface of the Earth. The individual atoms of oxygen that are split apart by the ultraviolet radiation then recombine to form more ozone. The process of the destruction and **regeneration** of ozone molecules in the stratosphere has reached a state of **dynamic equilibrium.** This means that over time, the amount of ozone gas that is destroyed by ultraviolet radiation equals the amount that is reformed. This natural bal-

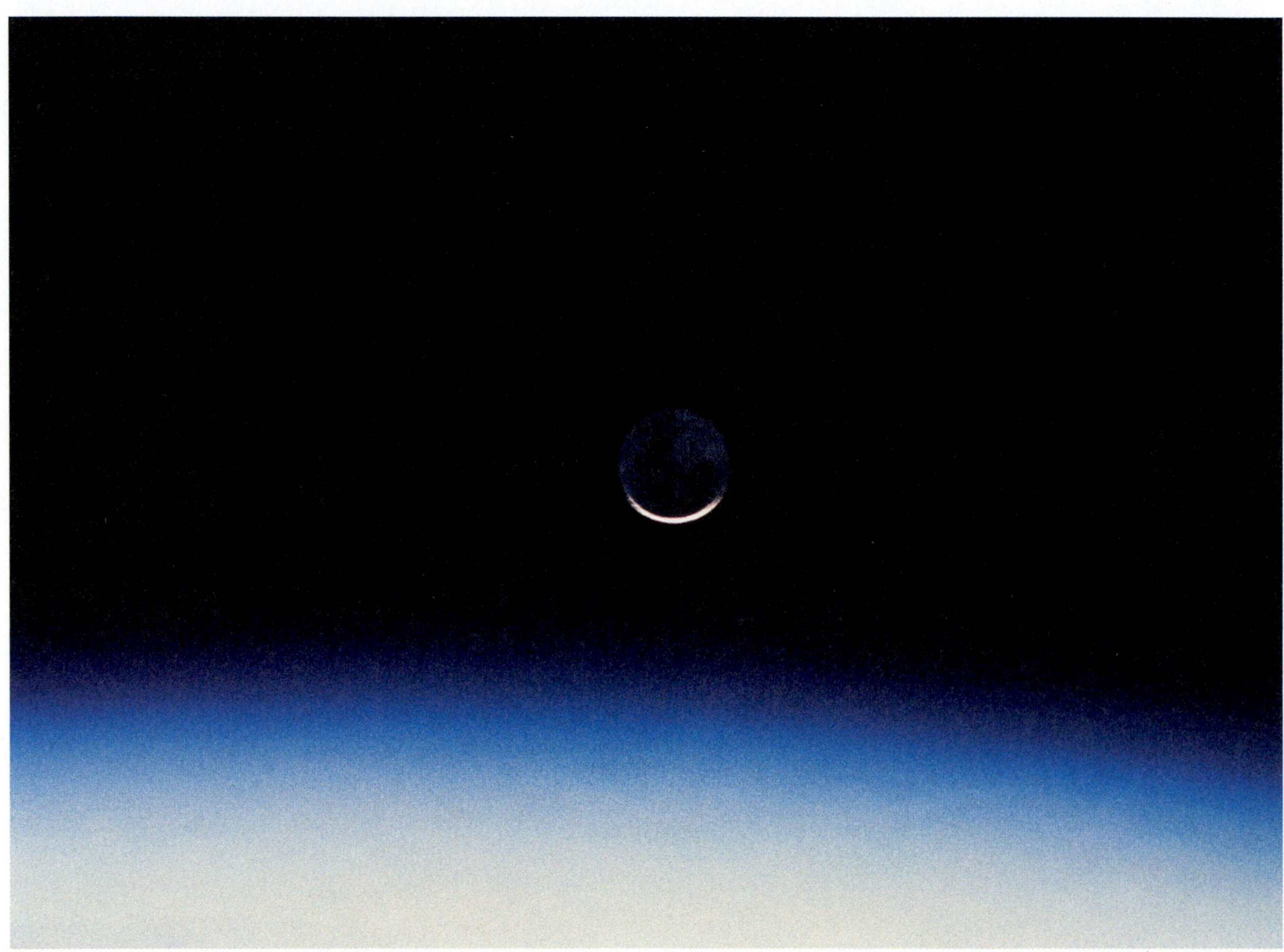

Figure 22–1 The location of the ozone layer within the Earth's atmosphere. *(Courtesy of PhotoDisc.)*

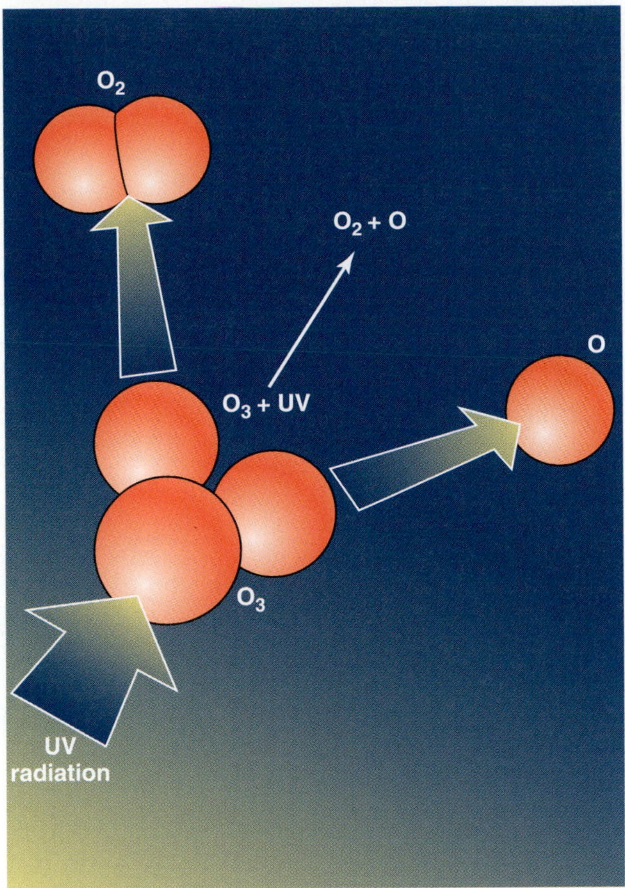

Figure 22–2 The destruction of a molecule of ozone as it absorbs ultraviolet radiation.

ance has existed in the atmosphere for billions of years.

Before the Earth's atmosphere contained oxygen, there was no ozone layer. This allowed deadly ultraviolet radiation to strike the Earth's surface, making it impossible for life to exist on land. Ultraviolet radiation does not penetrate through water, so during the first 2 billion years of Earth's history, life could only exist in the oceans. Eventually, as the atmosphere became filled with oxygen, the ozone layer began to form, enabling life to exist on land for the first time.

Measuring Stratospheric Ozone

To study how much ozone exists in the stratosphere, scientists set up a series of ground-based instruments that determined the amount of ozone over specific spots on the Earth. This was done by measuring the amount of ultraviolet radiation that was reaching the surface at each particular location. The more radiation received at the surface, the lesser the amount of ozone gas present in the ozone layer above that location. The lesser the amount of radiation at the surface, the greater the amount of ozone gas. This research revealed that ozone levels naturally change with the seasons and with the global circulation of the atmosphere. These natural changes were minimal and showed no threat

Aeronomer

An aeronomer is a specialized scientist who works to improve the ability to observe, understand, predict, and protect the quality of the atmosphere. The job of an aeronomer is to study the atmosphere in three distinct ways. These include theoretical research, field monitoring, and laboratory science. Theoretical research involves an improved knowledge of atmospheric movement and the transport of chemicals in the air. Field monitoring involves the measurement of the chemical and physical properties of the atmosphere around the world. Laboratory science investigates the chemical processes and reactions that occur within the atmosphere and how they are being altered by human activity. All three of the areas that make up the science of aeronomy work together to create an improved understanding of how the atmosphere operates and how it is being impacted by human activity. Major research topics include the El Niño, southern oscillation, stratospheric ozone depletion, ozone in the troposphere, the effects of aircraft on the atmosphere, and global warming. Scientists in the field of aeronomy must have a college education in atmospheric science and are employed in academic research or with a government agency such the National Oceanic and Atmospheric Administration or the National Aeronautics and Space Administration (NASA).

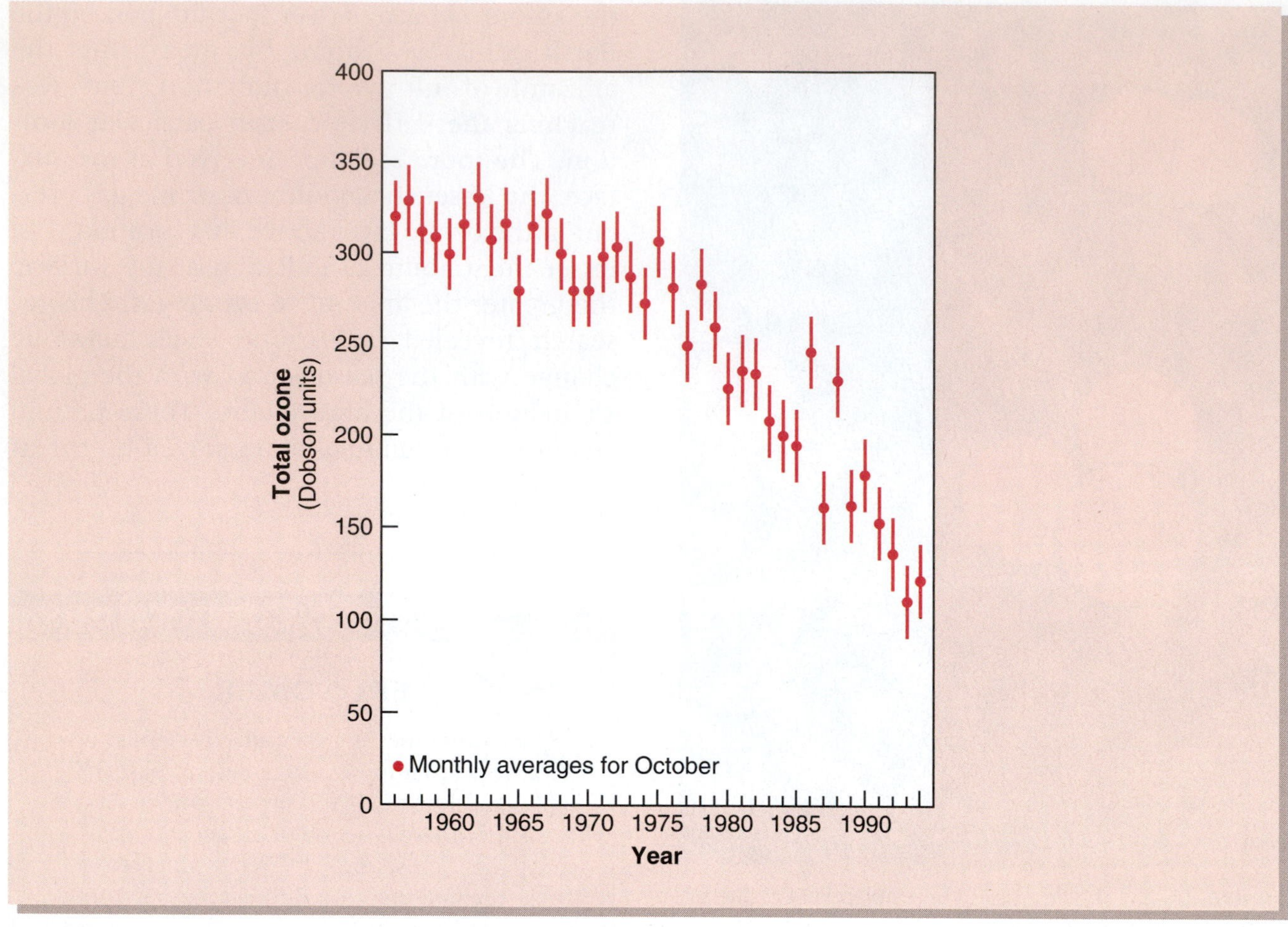

Figure 22–3 The decline in the level of ozone in the stratosphere between the years 1955 and 1995.

to the Earth's surface. However, in the 1970s it was noticed that ozone levels were beginning to decrease. This depletion was not understood at first, but as time went on, researchers realized that the ozone layer was indeed becoming thinner in some regions (Figure 22–3).

The **depletion** of the ozone layer was found to be the result of human-created gases that are interrupting the natural chemistry of the ozone layer in the stratosphere. During the early part of the twentieth century, scientists were experimenting with different gases that could be used for refrigeration. These gases needed to be able to efficiently remove heat from the air and enable it to cool to a low temperature. The first gas used for refrigeration was ammonia; however, this could be deadly to the surrounding environment if it leaked out. Chemists then synthesized a gas that could be used for refrigeration but was not deadly if it leaked. This gas was called a **chlorofluorocarbon,** or CFC. One popular CFC gas used in refrigerators and air conditioners was **Freon.** CFCs were also used for industrial processes, as propellants in aerosol spray cans, in foam production, and for cleaning electrical parts. Over time it began to increase in concentration in the atmosphere. When CFCs are released into the atmosphere, it takes approximately 2 years for them to rise up to the level of the stratosphere. At low atmospheric levels, CFC gas does not react with other molecules and is harmless to life; however, as it rises up into the stratosphere, it undergoes chemical changes that lead to ozone destruction.

EARTH SYSTEM SCIENTISTS *GORDON DOBSON*

Gordon Dobson was born in England in 1889. In 1920 he became a lecturer in meteorology at Oxford University. His research involved the properties of the stratosphere. In 1924 Dobson became interested in the levels of ozone gas within the stratosphere. He developed and built a device known as the Dobson ozone spectrometer, which he used to measure ozone levels around the world. In 1925 he discovered that ozone levels experienced seasonal changes around the globe. Later in his career he became interested in the effects of pollutants on the atmosphere. Dobson's work on ozone helped pave the way to an understanding of what role ozone gas plays in the Earth's systems. He also helped to identify changes in the levels of ozone that led to an understanding of ozone depletion. The Dobson unit is still used today to measure the amount of ozone within the stratosphere.

A CFC molecule is composed from atoms of fluorine, chlorine, and carbon. When a CFC molecule enters the stratosphere, it is exposed to high-level ultraviolet radiation (Figure 22–4). Like a molecule of ozone, high-energy radiation strikes the CFC molecule and breaks it apart. When this occurs, chlorine atoms are freed and begin to join with the oxygen atoms in the ozone layer. This creates chloride compounds that remove individual oxygen atoms from the stratosphere, causing a decrease in oxygen molecules that normally combine to create protective ozone gas. Therefore CFCs are primarily responsible for depleting the ozone layer. Other chemical compounds, such as bromide and halons, are also responsible for some ozone destruction.

Effects of Ozone Depletion

A reduction of ozone in the stratosphere results in a greater amount of ultraviolet radiation striking the Earth's surface. This can lead to potential damage to the world's ecosystems, as organisms become exposed to this high-energy radiation. Ultraviolet radiation can damage cells that are exposed to the atmosphere. The high-energy radiation strikes the cells much like a baseball striking a tower made of blocks. This results in damage to the structure of the cell. In human beings this can lead to skin cancer, degradation of the eyes, and weakened **immune systems.** Some researchers believe that many species of amphibians, such as frogs and salamanders, are especially susceptible to damage from depleted ozone. These organisms have sensitive skin, making them more likely to be damaged by ultraviolet radiation. Widespread incidence of **mutated** frogs may be the first signs of how a depleted ozone is affecting the biosphere.

The Ozone Hole

Today the National Aeronautics and Space Administration (NASA) and the National Oceanic and Atmospheric Administration use advanced satellites to monitor ozone levels around the world. The continued monitoring of stratospheric ozone revealed an alarming decrease in ozone gas over Antarctica (Figure 22–5). For the past 20 years, data has revealed that ozone over the South Pole has been depleted rapidly. The low concentration of ozone gas over Antarctica is known as the **ozone hole.** The formation of the ozone hole over Antarctica occurs during the Antarctic spring between the months of September and November. Since the 1950s the seasonal ozone over Antarctica has been reduced by 60%.

A similar ozone loss also occurs over the North Pole during the Northern Hemisphere's spring season. These reductions have amounted to

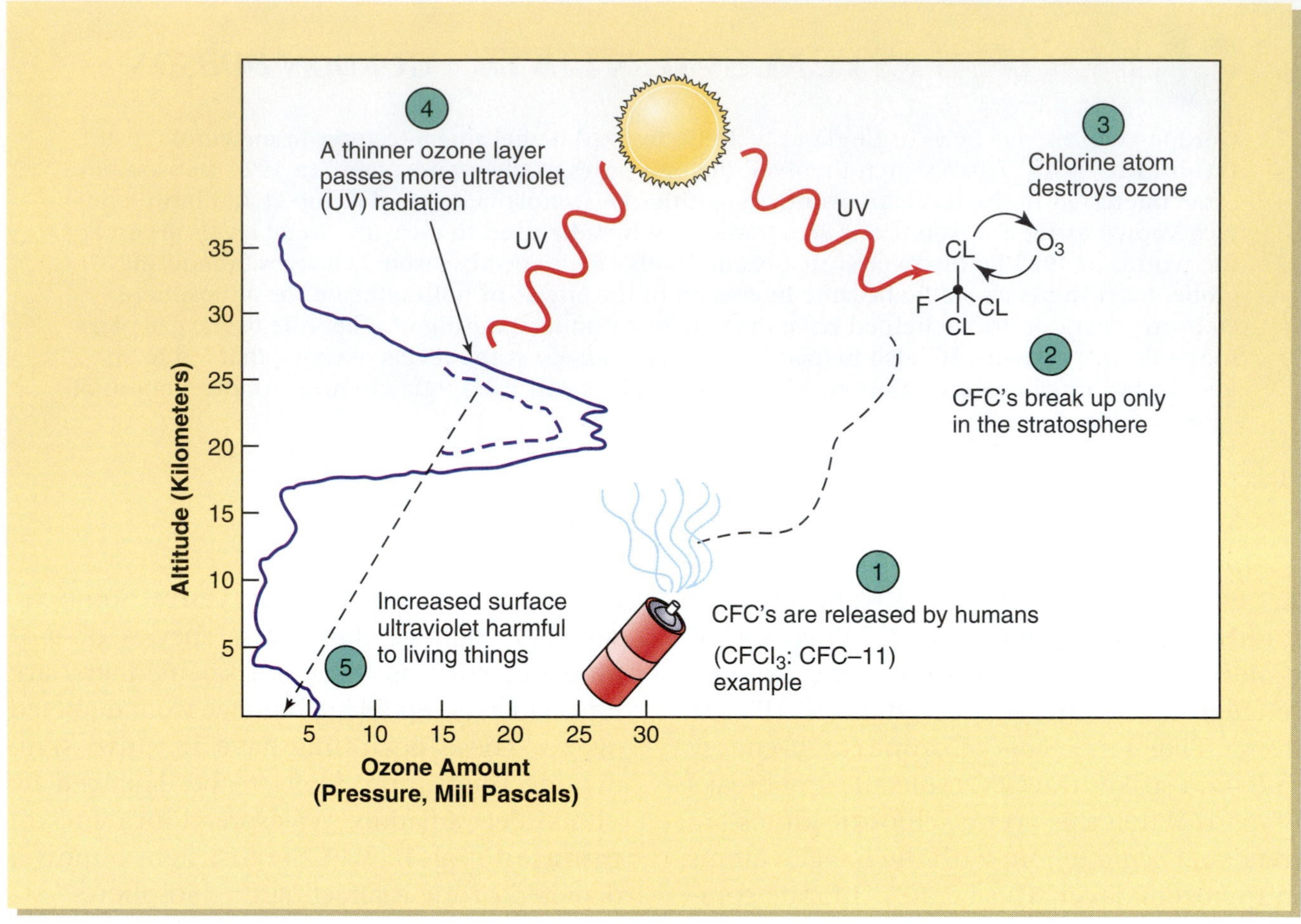

Figure 22–4 The pathway by which CFC gases travel through the atmosphere to destroy ozone in the stratosphere, thereby allowing more ultraviolet radiation to strike the Earth's surface.

approximately a 20% to 25% loss of ozone. Other parts of the globe are also experiencing ozone loss. The mid latitudes typically have experienced a reduction in ozone of approximately 4% to 6%. Areas between the equator and 20 degrees north and south latitude have experienced no loss of ozone.

Studies reveal that the loss of ozone is related to an increase in ultraviolet radiation reaching the surface. On average, a 10% reduction in ozone can lead to a 10% increase in ultraviolet radiation at the surface. A 30% loss of ozone can lead to a 50% increase in ultraviolet radiation striking the surface, and a 60% loss of ozone, such as in Antarctica, can result in an increase of ultraviolet radiation at the surface of almost 150%. Doctors have found that there is a direct relationship between increased exposure to ultraviolet light and skin cancer in humans. These statistics reveal the actual threat to the Earth's surface that ozone depletion can cause.

Reducing Ozone Depletion

As scientists discovered that the chemistry of ozone destruction was being caused by chemicals such as CFCs, governments around the world began to control the use of these gases. In 1996 CFCs were completely banned around the world; however, leftover CFCs that are still in the atmosphere continue to

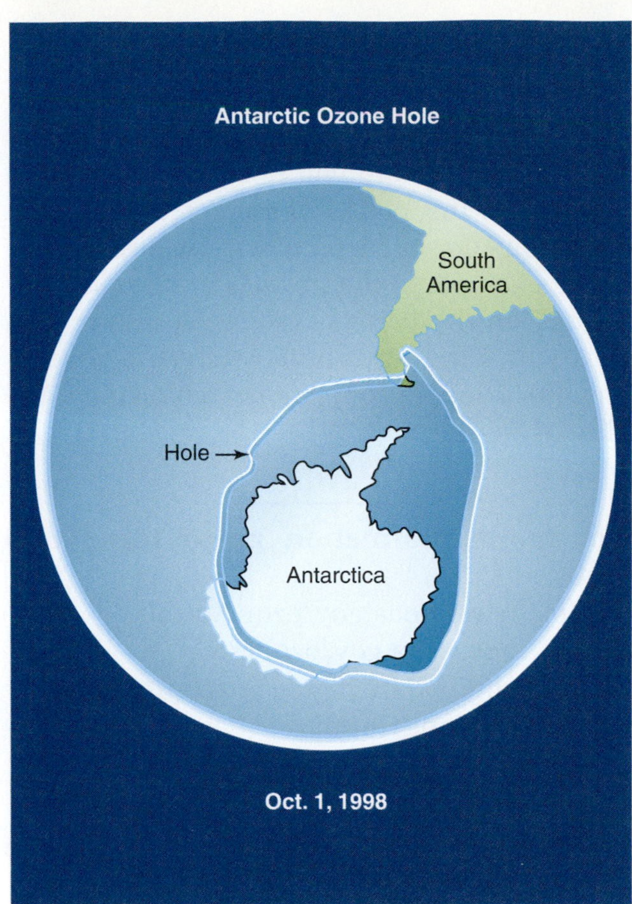

Figure 22–5 The location and size of the ozone hole over Antarctica.

Figure 22–6 Loss of stratospheric ozone results in an increase in harmful ultraviolet radiation reaching the Earth's surface. This has been linked to increasing incidence of skin cancer in humans. *(Courtesy of PhotoDisc.)*

destroy ozone. Old refrigerators and air conditioners in wrecked cars located in landfills and junkyards will continue to leak CFCs into the atmosphere for many years. These chemicals will take between 50 and 100 years to be removed from the atmosphere by natural processes. It is hoped that the ozone layer will then begin to rebuild itself and return to its natural state. Until then, careful monitoring of ozone depletion will be necessary to provide information about how much ultraviolet radiation is reaching the Earth's surface.

Protecting exposed skin from the Sun can help lessen the impact of these potentially harmful forms of radiation (Figure 22–6). This includes the use of sunscreen, wearing hats and sunglasses, and reducing the time that your skin is exposed to the Sun. Scientists hope the biosphere will not be too adversely affected by this period of increased ultraviolet radiation. Ozone depletion provides yet another example of the impacts of human technology on the Earth's systems. It also reveals that with careful scientific study and governmental cooperation, many environmental threats can be controlled.

For more information go to these Web links:

<http://www.noaa.gov/ozone.html>
<http://www.epa.gov/ozone/>
<http://observe.arc.nasa.gov/nasa/exhibits/ozone/Ozone1.html>

REVIEW

1. What is ozone gas, and where is it found naturally on the Earth?
2. What is the ozone layer, and why is it important?
3. What is causing the depletion of the ozone layer?
4. How is ozone gas destroyed?
5. How much of the ozone layer has been lost in different parts of the world?
6. What is the ozone hole?
7. What are some of the negative effects of increased ultraviolet radiation striking the Earth's surface?
8. Who was Gordon Dobson?

CHAPTER SUMMARY

Ozone is an unstable molecule that is composed of three atoms of oxygen (O_3). It exists naturally in the stratosphere, where it forms the ozone layer. The ozone layer acts as a protective barrier to deadly ultraviolet radiation. It prevents high levels of this radiation from striking the Earth's surface, which enables life to flourish here. In the late 1950s scientists detected reduced levels of ozone gas in the stratosphere. They discovered that human-created gases such as chlorofluorocarbons were responsible for the depletion of the ozone layer. These compounds were rising into the stratosphere and releasing atoms of chlorine, which combined with the oxygen there. As a result, the amount of free oxygen

EARTH MATH

1) IF THE OZONE CONCENTRATION OVER ANTARCTICA HAS DECREASED BY 60% FROM 1960 TO 2001, BY WHAT PERCENTAGE DID THE OZONE DECREASE OVER THE SOUTH POLE EACH YEAR?

that normally formed ozone was becoming depleted. The reduction of ozone caused an increase in the amount of ultraviolet radiation striking the Earth's surface. A large amount of ozone was lost over Antarctic; this is known as the ozone hole.

The increase in ultraviolet radiation striking the Earth's surface as a result of ozone destruction can damage an organism's cells. The eyes and skin are especially susceptible to damage by increased exposure to this high-energy radiation. This can lead to the development of skin cancer in humans. The discovery that gases such as CFCs were responsible for ozone destruction led governments around the world to ban their use. Although they are no longer used, CFCs still exist in the atmosphere and will continue to deplete ozone in the stratosphere. Until they are removed by natural processes, it will be necessary to take precautions to protect your skin and eyes from increased levels of ultraviolet radiation.

CHAPTER REVIEW

Multiple Choice

1. Ozone gas is formed from how many molecules of oxygen?
   a. one
   b. two
   c. three
   d. four

2. Ozone exists naturally in which part of the atmosphere?
   a. troposphere
   b. stratosphere
   c. mesosphere
   d. thermosphere

3. The ozone protects the Earth's surface from:
   a. insolation
   b. visible light radiation
   c. infrared radiation
   d. ultraviolet radiation

4. What happens to the oxygen molecules that compose ozone when it breaks apart?
   a. They recombine to form more ozone.
   b. They fly off into space.
   c. They turn into heat energy.
   d. They combine with water vapor.

5. As the amount of ozone gas in the ozone layer decreases, what occurs on the Earth?
   a. The stratosphere gets hotter.
   b. More ultraviolet light strikes the surface.
   c. Oxygen in the atmosphere decreases.
   d. The surface temperature increases.

6. What is responsible for the depletion of the ozone layer?
   a. climate change
   b. burning of fossil fuels
   c. CFCs
   d. acid precipitation

7. Approximately how much time does it take for ozone-destroying compounds to rise up to the ozone layer?
   a. 6 months
   b. 2 years
   c. 50 years
   d. 100 years

8. Which part of the Earth is not experiencing a depletion in the ozone layer?
   a. the area between the equator and 20 degrees north and south latitude
   b. the mid latitudes
   c. the North Pole
   d. the South Pole

9. Which part of the Earth is experiencing a reduction in the ozone layer between 4% and 6%?
   a. the area between the equator and 20 degrees north and south latitude
   b. the mid latitudes
   c. the North Pole
   d. the South Pole

10. Where is the ozone hole located?
    a. the area between the equator and 20 degrees north and south latitude
    b. the mid latitudes
    c. the North Pole
    d. the South Pole

11. A 30% reduction of ozone in the ozone layer can increase the level of ultraviolet radiation striking the Earth's surface by how much?
    a. 10%
    b. 30%
    c. 50%
    d. 150%

12. How does the depletion of the ozone layer affect humans:
    a. It is altering the climate.
    b. It is raising sea level.
    c. It is increasing the threat of skin cancer.
    d. It is polluting the air we breathe.

13. What is likely to occur if humans stop polluting the atmosphere with ozone-destroying compounds?
    a. The ozone layer will return to its normal levels.
    b. The ozone level will remain at its present levels.
    c. Ozone will continue to be destroyed.
    d. Scientists are not sure what will happen.

Matching *Match the terms with the correct definitions.*

a. ozone
b. stratosphere
c. ultraviolet radiation
d. ozone layer
e. regeneration
f. dynamic equilibrium
g. depletion
h. chlorofluorocarbon
i. freon
j. immune system
k. mutated
l. ozone hole

1. ____ The system in the body that is responsible for fighting off disease.

2. ____ The process of reconstructing something.

3. ____ A colorless, gaseous compound composed of three atoms of oxygen (O_3).

4. ____ A class of human-created molecules commonly used as refrigerants, in electrical manufacturing, and in foam production that is responsible for ozone destruction; it is also a greenhouse gas.

5. ____ The loss of something.

6. ____ A type of chlorofluorocarbon gas that was commonly used in air conditioners and refrigerators.

7. ____ An area over Antarctica that has a reduced level of ozone gas in the stratosphere.

8. ____ The specific area in the stratosphere, located at an altitude between 10 and 20 miles, that contains a high concentration of ozone gas.

9. ____ To be altered by a change in the structure of an organism's genes.

10. ____ A layer in the Earth's atmosphere where the ozone layer exists and temperature increases with an increase in altitude.

11. ____ A balance between two opposing processes that occur at the same rate in an energetic system such as a river or stream.

12. ____ A specific high-energy form of electromagnetic radiation emitted from the Sun that can be harmful to living things.

Critical Thinking

1. If astronomers are searching for life on other planets in the solar system, what should they look for that might indicate that life could exist there, and why?

UNIT 5

The Hydrosphere

Topics to be presented in this unit include:

- The Distribution of Water on the Earth
- Properties of the Water Molecule
- The Hydrological Cycle
- Circulation of Water in the World's Oceans
- Zones of Life in the Ocean
- The Classification of Lakes
- Properties of Flowing Surface Water
- Groundwater Flow and Distribution
- Properties of Glaciers
- Pollution of the Hydrosphere
- El Niño and the Southern Oscillation
- Coral Bleaching

OVERVIEW

In 1968 the mission of the Apollo 8 spacecraft was to allow human beings to orbit the Moon for the first time. Mission commander Jim Lovell knew he was making lunar history; the view of his home planet rising from the Moon's horizon forever changed the view human beings have of the Earth. Lovell stared through the small window of the Apollo 8 spacecraft and saw the Earth as a tiny blue marble hanging in the vast emptiness of black space. The pictures he took of the Earth during that mission more than 30 years ago still remind us today of how precious our planet is. The Earth appears to be a blue marble when viewed from outer space because most of its surface is covered in water, its most abundant resource.

CHAPTER 23 Earth, The Water Planet

Section 23.1 – The Blue Marble Objectives

The Earth's Oceans • Distribution of Freshwater on Earth

After reading this section you should be able to:

- Define the term *hydrosphere.*
- Describe the approximate area and depth of the water in the world's principal oceans.
- Describe the distribution of all the water on the Earth.

Section 23.2 – The Amazing Water Molecule Objectives

The Water Molecule • Adhesion and Cohesion • Heat Capacity of Water • Properties of Ice • Water as a Solvent

After reading this section you should be able to:

- Explain how the shape of the water molecule contributes to its unique properties.
- Define the terms *adhesion, cohesion,* and *capillary action.*
- Describe the heat capacity of water and why it is important to life on Earth.
- Differentiate between the density of liquid water and ice.
- Describe the importance of water's ability to dissolve many substances.

TERMS TO KNOW

hydrosphere
seawater
freshwater
glacier
groundwater
polar molecule
cohesion
adhesion
hydrophilic
capillary action
heat capacity
ice
water vapor
solvent
dissolve

INTRODUCTION

The term *Earth* is derived from the Old English word eorthe, which means "ground" or "soil." We use a derivation of this term as the name for our planet, which is ironic because most of it is covered in water. The Earth is indeed a water planet. Water surrounds us in a multitude of ways. If you were to remove all the water from a plant, you would find that it would lose almost 90% of its weight. It would be almost the same for any living organism on Earth. Water falls from the sky, collects in puddles, and soaks the ground. The sky is filled with water, both visible as clouds and invisible as water vapor. Every day we depend on water to be where we want it to be, so we can use its special properties. The water molecule is unique among other molecules on the Earth. It is water's unique properties that allow life to flourish on our planet and shape the world in which we live.

Figure 23–1 The planet Earth as viewed from space appears as like a blue marble because of the abundance of water that covers its surface. *(Courtesy of PhotoDisc.)*

23.1 The Blue Marble

The Earth's Oceans

Together, in all its forms, all water on Earth is known as the **hydrosphere.** More than 70% of the Earth's surface is covered in liquid water (Figure 23–1). All this water, which makes the planet appear blue from space, comprises the planet's three main oceans (Figure 23–2). The largest of Earth's oceans is the Pacific Ocean, which covers more than 64 million square miles, or 35% of the Earth's surface, with an average depth of 14,045 feet. The next largest ocean is the Atlantic, which stretches for more than 33 million square miles, or 21% of the Earth's surface, and has an average depth of 12,254 feet. The third largest of the Earth's oceans is the Indian Ocean. This covers more than 28 million square miles, or 15% of the planet's surface,

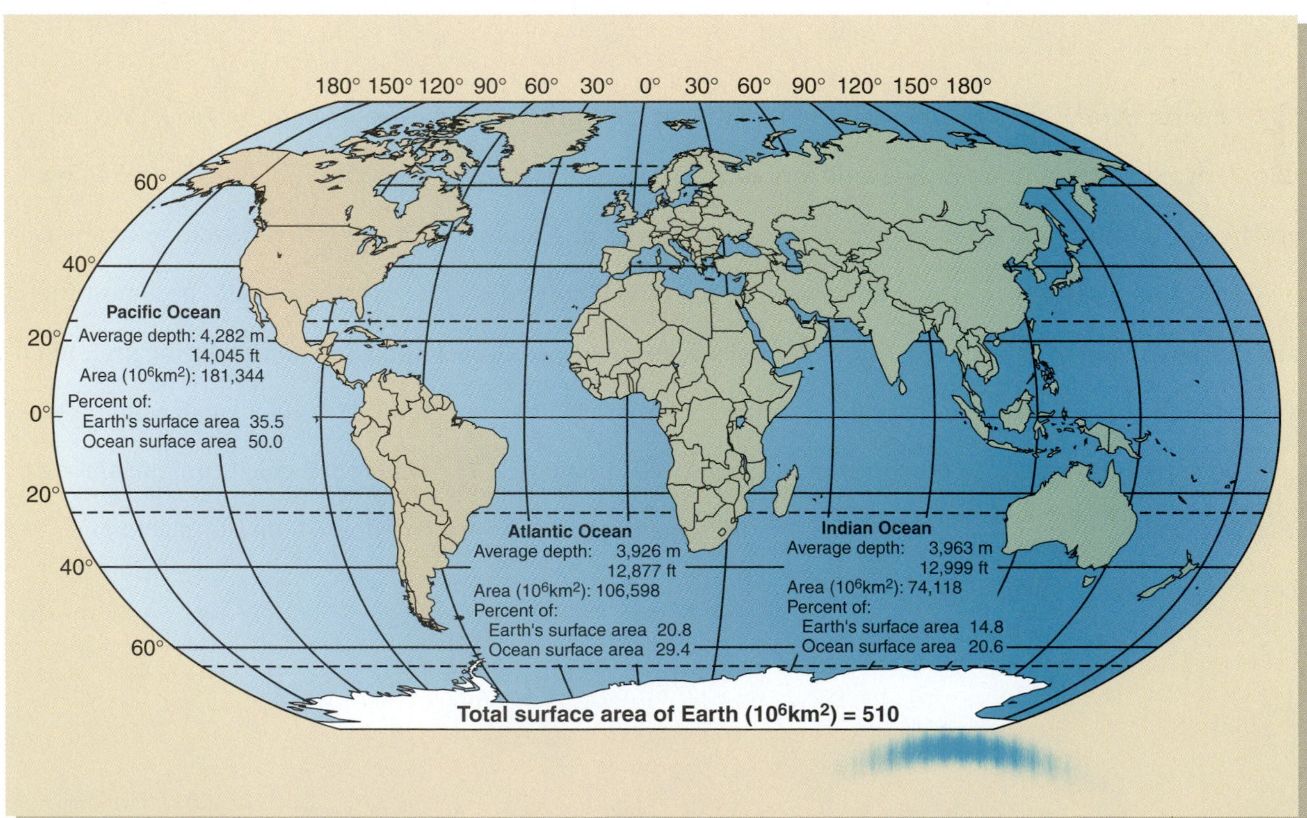

Figure 23–2 The average size and depth of the world's three principal oceans, which together cover approximately 70% of the Earth's surface.

EARTH SYSTEM SCIENTISTS *JACQUES-YVES COUSTEAU*

Cousteau was born in France in 1910 and began his lifelong attachment to the sea by joining the French Navy. During World War II, he began to experiment with a variety of diving techniques that would enable a person to remain underwater for extended lengths of time. After two failed attempts in trying out experimental diving equipment that nearly cost him his life, Cousteau finally achieved his dream in 1943. With Emile Gagnan, Cousteau invented the first self-contained underwater breathing apparatus, or SCUBA, which enabled him to breathe while underwater. In 1951 Cousteau set out on the first of many expeditions to explore the world's oceans. Using his now famous ship, the Calypso, he revealed the mysteries of the underwater world. His expeditions were filmed and eventually made into a series of popular television shows. No other person has provided the world with a better understanding of the oceans than Jacques Cousteau. As a result of his underwater expeditions, Cousteau turned his attentions to the health of the oceans and the influence of human pollutants on sea life. He created the Cousteau Foundation to both explore and protect the world's water resources and to educate the public about sustaining the life within them. Cousteau also experimented with the possibilities of living under the sea. He researched the long-term effects of living deep below the ocean in specially created submersible laboratories; this was known as the Conshelf program. In 1965 three of Cousteau's men lived at a depth of 330 feet below the surface for 3 weeks. Over the next three decades, Cousteau conducted countless expeditions, published books in more than 20 languages, and produced film documentaries about all aspects of the Earth's hydrosphere. Cousteau died in 1997. He will be forever remembered as an inventor, explorer, and environmentalist and the man who introduced humanity to the wonders of the world's oceans.

at an average depth of 12,704 feet. In total, all the oceans combined cover much of our planet with water that is more than 2 miles deep! So it is not hard to imagine why water is Earth's most abundant resource.

Distribution of Freshwater on Earth

The world's oceans do not contain pure water, but rather a mixture of dissolved minerals and gases that together make up what is called **seawater.** Seawater, also known as saltwater, tastes salty because it contains sodium chloride, or table salt. Although the oceans cover the planet with a vast amount of seawater, this is not the only form of water on the Earth. All the water on Earth can be divided into two categories, seawater and **freshwater.** Freshwater is what humans require to drink, and makes up only 3% of all the water found on Earth. The remaining 97% is seawater. The

Career Connections

CHEMICAL OCEANOGRAPHER

Chemical oceanographers study the chemistry of water within the ocean. Their research involves the understanding of chemical processes that occur all throughout the world's oceans. This includes the chemistry of both ocean sediments and the open water. Today much of the work of a chemical oceanographer involves the study of the effects of pollutants in the ocean. Research on the effects of oil spills, garbage, human waste, and toxic chemicals on ocean water is becoming increasingly important. A chemical oceanographer must spend time at sea and in the laboratory to successfully perform their research. This occupation requires a college education in oceanography, chemistry, and hydrology. Chemical oceanographers can work in private industry, in the academic fields, or for a government agency.

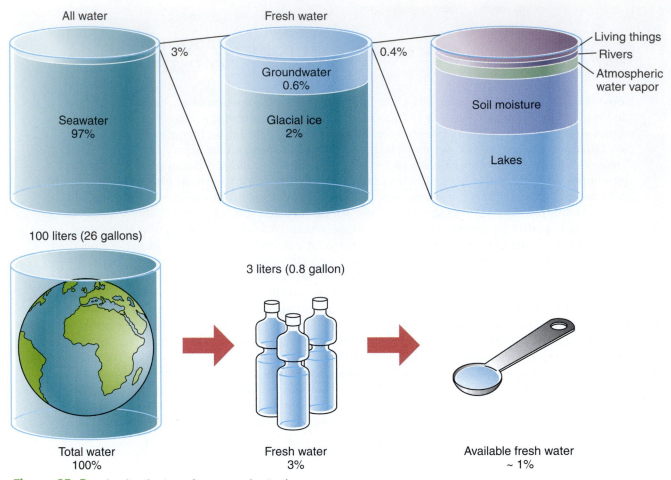

Figure 23–3 The distribution of water on the Earth.

Earth's freshwater supply is found in three major places: glaciers, surface water, and groundwater. **Glaciers** hold approximately 2% of all the world's water and are the largest storehouse of freshwater on the planet. Surface waters, which includes all lakes, rivers, ponds, and streams, only hold 0.4% of Earth's total water. And **groundwater** makes up 0.6% of our planet's water supply. Some of the Earth's water is present in the atmosphere in the form of water vapor, and all the organisms that live on the Earth contain water in their bodies. Even though most of our planet is covered in water, only a small amount of it is available as freshwater, which is required for growing food and drinking (Figure 23–3). This is why many people are so concerned about maintaining the quality of

EARTH MATH

1) HOW MANY SQUARE MILES DO ALL THREE OF THE EARTH'S LARGEST OCEANS COVER?
2) USE THE AVERAGE DEPTHS OF THE PACIFIC, ATLANTIC, AND INDIAN OCEANS, IN FEET, TO DETERMINE THE AVERAGE DEPTH OF THE EARTH'S OCEANS IN MILES.

our freshwater supply. The blue marble of the Earth is indeed like Samuel Coleridge's *The Rime of the Ancient Mariner:* "Water, water everywhere, nor any drop to drink."

SECTION REVIEW

1. How much of the Earth's surface is covered by water?
2. What is the average depth of the world's oceans?
3. How much of the Earth's water is freshwater?
4. Who was Jacques-Yves Cousteau?

For more information go to these Web sites:

<http://www.ec.gc.ca/water/en/nature/e_nature.htm>

<http://interactive2.usgs.gov/learningweb/students/homework_hydrology.asp>

<http://water.usgs.gov/>

23.2 *The Amazing Water Molecule*

The Water Molecule

Not only is water the most abundant resource on Earth, it is also one of the most interesting molecules, possessing many unique features. The water molecule, or H_2O, consists of two hydrogen atoms attached to one oxygen atom. The arrangement of these atoms creates a unique shape that resembles the head of the Mickey Mouse cartoon character. This unique shape also contributes to water's unique properties (Figure 23–4). The hydrogen atoms, or the "ears" of the water molecule, carry a positive electrical charge, whereas the oxygen atom, or "head" of the molecule, possesses a

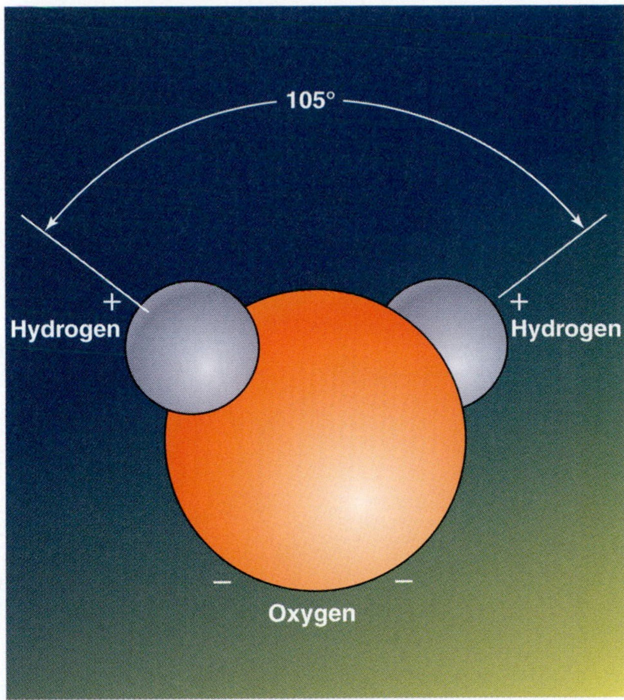

Figure 23–4 The characteristic shape of the water molecule gives it unique properties.

negative electrical charge. These opposite charges make water a **polar molecule.** A polar molecule is a molecule that has a positive charge on one end and a negative charge on the other end. Items such as batteries and magnets are also polar.

Adhesion and Cohesion

The result of water's polarity creates its first unique property, **cohesion.** Cohesion is the attraction of water molecules to one another (Figure 23–5). This occurs as a result of the attraction of the positively charged hydrogen molecules to a neighboring water molecule's negatively charged oxygen atom. This attraction occurs because opposite electrical charges are attracted to one another. Cohesion tends to clump water molecules together in chains, which is why drops of liquid clump together as they falls toward the ground, forming a characteristic raindrop shape.

Another unique property of the water molecule is **adhesion.** Adhesion is the attraction

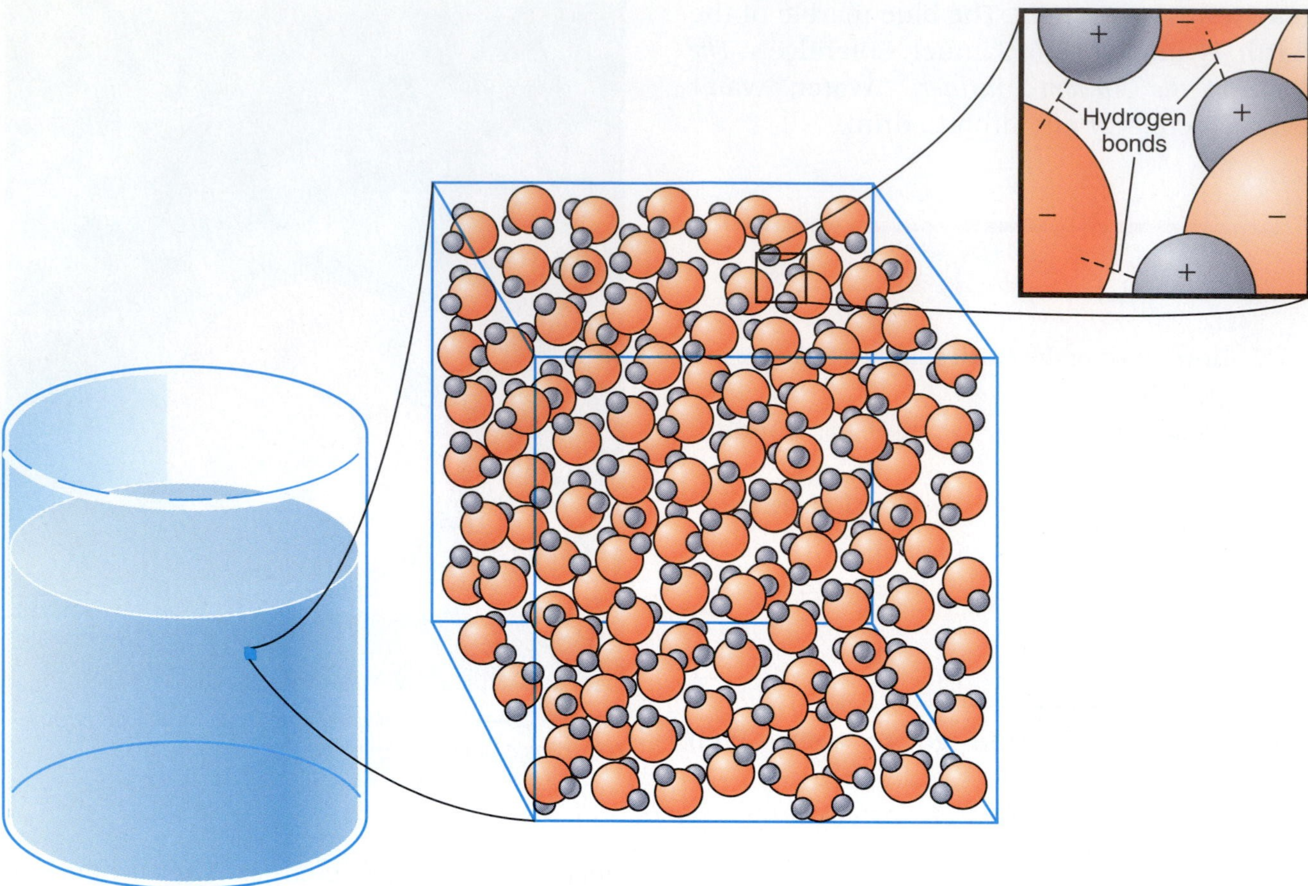

Figure 23–5 The attraction of water molecules to one another as a result of their polarity is called cohesion.

of water molecules to molecules of other substances (Figure 23–6). Adhesion is also caused by the polarity of the water molecule and results in water clinging to certain surfaces. These surfaces are **hydrophilic,** meaning "water loving." Adhesion can be seen when you look at the surface of water inside a glass. You will notice that the water is higher near the edges of the glass. This is the result of the water molecules "climbing" up the glass because of adhesion. You can also observe adhesion after you empty a glass of water and hold it upside down. Water droplets will still cling to the sides of the glass as a result of adhesion.

Together, adhesion and cohesion cause **capillary action.** Capillary action is the ability of water to move upward in small tubes or vessels (Figure 23–7). The smaller the diameter of the tube, the higher the water column rises. This is caused by the attraction of water to other substances (adhesion) and the attraction of water to itself (cohesion). The attraction of water molecules to a hydrophilic substance causes water to pull other water molecules along behind it in a long chain, which results in the whole water column rising upward. Capillary action is a fundamental life process that helps organisms distribute water throughout their bodies. You can observe the power of capillary action if you take a dry sponge and stand it upright in a small puddle of water. Without applying any force at all, water in the puddle will be drawn into the vessels of the sponge by capillary action.

Heat Capacity of Water

The water molecule also possesses a high **heat capacity.** Heat capacity is the ability of

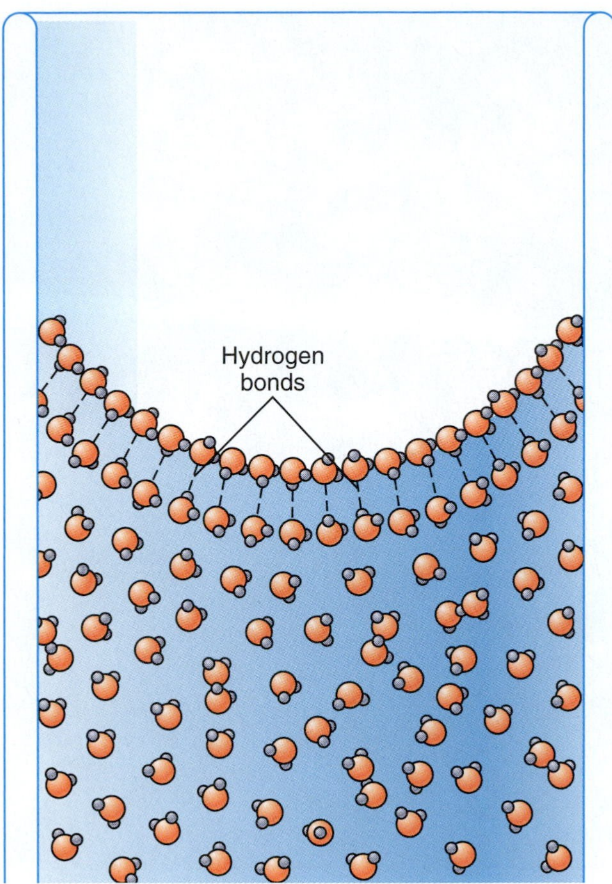

Figure 23–6 The attraction of water molecules to specific substances, such as glass, forming a meniscus is called adhesion.

Figure 23–7 The ability of water to rise upward in tiny tubes or vessels as a result of adhesion and cohesion is known as capillary action.

a substance to store energy without changing its temperature. Water has a high heat capacity; in contrast, metal has a low heat capacity. For example, if you filled a metal pot with water and set it on a stove, what would happen when heat was applied? The metal pot would become hot very fast, while the water would remain cool for a much longer time. This is the result of heat capacity. The metal pot has a low heat capacity; therefore its temperature is easy to change when heat energy is applied to it. The result is that the metal quickly becomes hot. The water, however, takes a much longer time to become hot because of its high heat capacity. Water's high heat capacity makes it a valuable reservoir for storing heat on Earth. The world's oceans are vast storehouses of heat that has been absorbed from the sun and is important for global climate. Storms such as hurricanes and typhoons are one way that the oceans give off this stored heat. The strength of these tropical storms provides a good example of how much energy is actually stored in the world's oceans.

Properties of Ice

Water has yet another unique feature: It is one of the only substances that expands when it freezes. Most substances on Earth contract, or reduce their size, when they become colder. Water, however, expands, or increases its size, when it freezes. For example, if you fill a plastic bottle with water and then place it in your freezer, what will happen? The water will freeze in the bottle, expand, and crack the plastic container. Water increases its volume by approximately 9% when it freezes.

Figure 23–8 Ice is the solid form of water, which is less dense than liquid water, causing it to float. *(Courtesy of PhotoDisc.)*

Other substances usually contract, or become smaller, when they get colder, such as steel bridges. Engineers use large expansion joints when they construct large bridges because bridges shrink during cold weather and expand in hot weather. If bridge engineers did not install these expansion joints, the bridges would eventually pull themselves apart! Water's ability to expand when it freezes also makes it less dense (Figure 23–8). Most substances become more dense when they freeze, unlike water. Frozen water is less dense than liquid water, which allows ice to float. The ability of ice to float is important for aquatic life on Earth, because it allows life to flourish beneath the ice in lakes and ponds during winter. If ice were more dense than liquid water, every winter the aquatic plants and animals that reside in ponds and lakes would be crushed by the sinking ice. Unlike most substances on Earth, which have their greatest density in the solid form, water's density is greatest when it is in liquid form at 4° Celsius.

Water is unique because it is one of the only substances found on Earth that exists in all three states of matter (Figure 23–9). Solid water forms **ice** and snow in colder parts of the planet. Liquid water fills the oceans and lakes and flows in rivers and streams. The gaseous form of water, called **water vapor,** exists in the atmosphere and creates weather and climate. The interactions between all of water's three states of matter help to drive many of the processes that shape our Earth and enable life to exist on our planet.

Water as a Solvent

The last unique property of water is its ability to be a **solvent.** Water is sometimes referred to as the universal solvent because it

Career Connections

Hydrologist

A hydrologist applies scientific knowledge to all aspects of water-related problems that affect society. This includes maintaining healthy surface waters and improving water quality. Hydrologists help with the management of municipal water supplies and all surface freshwater resources. These scientists monitor the water quality of reservoirs, rivers, and any type of surface water for human use or recreation. They are also concerned with irrigation of agricultural lands and the problems associated with flooding. Hydrologists monitor snow packs in the mountains, along with local precipitation, to determine the future availability of surface water. They are concerned with the chemical study of surface waters to maintain their ability to support healthy wildlife populations. Some surface water hydrologists also monitor water for pollutants and are charged with keeping water safe for recreational uses such as swimming, boating, and fishing. These scientists must have a college education in hydrology and environmental science and can find employment with state and local governments, as well as with private industry.

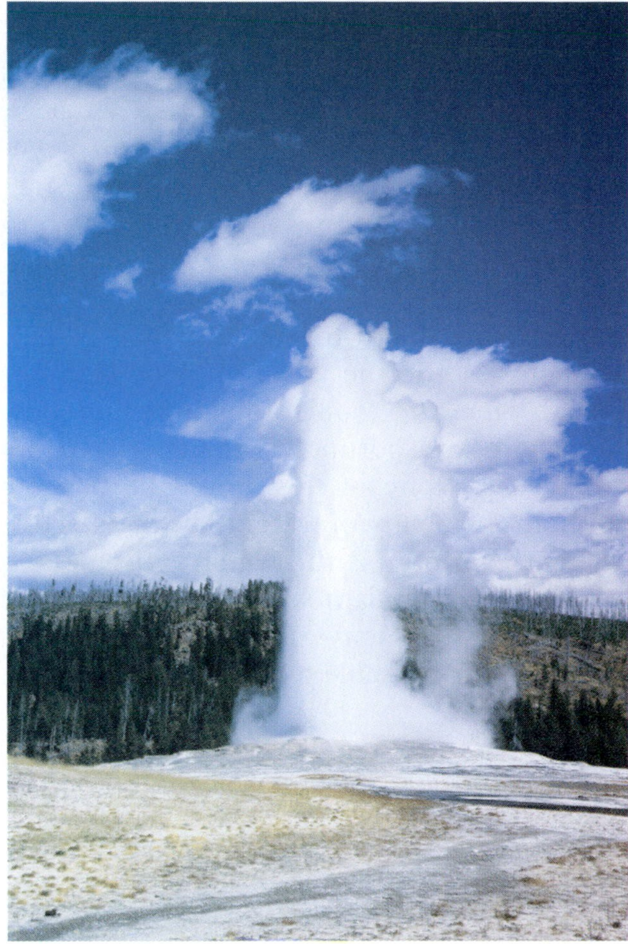

Figure 23–9 Water exists in all three states of matter on Earth. *(Courtesy of PhotoDisc.)*

can **dissolve** many substances. All water on the planet, even freshwater, contains dissolved minerals and gases. Fish and other aquatic organisms depend on dissolved gases in water, such as oxygen, to live. The ability of water to dissolve substances makes it extremely important for life on Earth. Plants depend on water to distribute dissolved nutrients and

EARTH SYSTEM SCIENTISTS *SIR JOHN ROSS*

John Ross was born in England in 1777. He was best known as an arctic explorer who was in search of a northwest passage from the Atlantic Ocean to the Pacific Ocean. While at sea, Ross took many bottom measurements and began to develop some of the first accurate charts that revealed the depth of the ocean. He also took samples from the bottom of the ocean to learn about the types of organisms that lived there. In 1817 he recovered living organisms from the ocean floor at a depth of more than 1 mile. Previous to this discovery, scientists thought that no life could survive in the deep ocean. Later, while exploring the waters off Antarctica, Ross recovered bottom-dwelling organisms from depths of more than 4 miles. His research proved that life could exist in the deep, dark, cold, and high-pressure environment of the ocean floor.

EARTH MATH

1) If the density of ice is 0.9170 grams per cubic centimeter and the density of liquid water at 20° Celsius is 0.9982 grams per cubic centimeter, what is the difference in density between the liquid and solid forms of water in grams per cubic centimeter?

2) If liquid water becomes water vapor at a temperature of 212° Fahrenheit, what is the difference in temperature between water vapor and ice?

manufactured food throughout their bodies. Humans rely on the water in blood to circulate oxygen and food throughout the body, while also transporting waste materials away from cells. None of this would be possible without the ability of water to be an effective solvent. Not only is the water molecule one of the most abundant substances on Earth, it also possesses many unique qualities that make it indispensable to life on Earth.

SECTION REVIEW

1. Draw the shape and label the polarity of the water molecule.
2. Describe the difference between adhesion and cohesion.
3. What are the three states of matter for water?
4. Explain why ice floats.
5. Who was John Ross?

For more information go to these Web links:

<http://interactive2.usgs.gov/learningweb/students/homework_hydrology.asp>

<http://www.epa.gov/students/water.htm>

CHAPTER SUMMARY

The Earth is a water planet. More than 70% of the Earth's surface is covered in water, including three major oceans that have an average depth of more than 2 miles. Water on the Earth is divided into two main categories: seawater and freshwater. Seawater contains many dissolved minerals, including sodium chloride, and makes up 97% of all the water on the Earth. Freshwater composes the remaining 3% of all the world's water and is what we use to drink and for agriculture. Freshwater occurs in three different forms on the Earth: glaciers, groundwater, and surface water.

Water is a unique substance because it possesses many special properties. The water molecule, H_2O, is made up of two molecules of hydrogen and one molecule of oxygen. This results in the unique shape of the water molecule that also makes it a polar molecule, meaning that one side of the water molecule has a positive charge and the other has a negative charge. The polarity of the water molecule causes it to be attracted to other water molecules. This is known as cohesion. Polarity also causes water to be attracted to certain hydrophilic, or water-loving, substances. This is known as adhesion. Together adhesion and cohesion create capillary action. Capillary action is the ability of water molecules to rise upward in tiny tubes or vessels and is an important mechanism by which living things distribute water throughout their bodies.

Another unique property of the water molecule is its high heat capacity. Heat capacity is the ability of substances to absorb heat without changing their temperature. Water's high heat capacity allows it to store large amounts of heat. This enables the oceans to store great amounts of heat energy, helping to moderate the planet's climate. Water is also unique because it becomes less dense when it freezes to form ice. Almost all substances on the Earth become more dense when they are in their solid forms. Because ice is less dense than liquid water, it floats. The most dense form of water is the liquid form at 4° Celsius.

Water exists on the Earth in all three states of matter, which is also unique. The interaction among ice, liquid water, and water vapor affects many processes on the Earth. Water is considered the universal solvent, because it can dissolve many substances. This property of water is essential to all living things on the planet.

CHAPTER REVIEW

Multiple Choice

1. What percentage of the Earth's surface is covered by water?
a. 15%
b. 21%
c. 35%
d. 70%

2. What percentage of the Earth's surface does the Pacific Ocean cover?
a. 15%
b. 21%
c. 35%
d. 70%

3. The average depth of the world's oceans is approximately:
a. 100 feet
b. 5000 feet
c. 13,000 feet
d. 21,000 feet

4. Where is most of the world's freshwater located?
a. glaciers
b. lakes and ponds
c. groundwater
d. plants and animals

5. The water molecule is a polar molecule, which means that it:
a. formed near the poles
b. has an electrical charge
c. sinks when frozen
d. has a low heat capacity

6. The ability of water molecules to be attracted to one another is called:
a. adhesion
b. cohesion
c. polarity
d. capillary action

7. Water rising into a sponge is an example of:
a. adhesion
b. cohesion
c. polarity
d. capillary action

8. Water sticking to the sides of a glass is an example of:
a. adhesion
b. cohesion
c. polarity
d. capillary action

9. Water's high heat capacity enables it to:
a. boil rapidly
b. quickly change temperature
c. store heat
d. form water vapor

10. Water's density is greatest when:
a. it is frozen solid, at 0° Celsius
b. it is liquid, at 4° Celsius
c. it is packed inside a glacier
d. it is in the form of hail

11. When it freezes, water's volume increases by approximately:
a. 1%
b. 3%
c. 9%
d. 50%

continued

Matching

Match the terms with the correct definitions.

a. hydrosphere
b. seawater
c. freshwater
d. glacier
e. groundwater
f. polar molecule
g. cohesion
h. adhesion
i. hydrophilic
j. capillary action
k. heat capacity
l. ice
m. water vapor
n. solvent
o. dissolve

1. ____ A term that describes a substance that attracts water molecules.

2. ____ The gaseous form of water.

3. ____ All the water on the Earth.

4. ____ To enter into a solution.

5. ____ Water that has a high concentration of minerals dissolved in it, also known as saltwater.

6. ____ A substance that is capable of dissolving another substance.

7. ____ The attraction of water molecules to one another.

8. ____ The solid form of water, which is less dense than liquid water, causing it to float.

9. ____ Water on or below the Earth's surface that contains a small amount of dissolved mineral salts and is good for drinking.

10. ____ The attraction of water molecules to a hydrophilic substance.

11. ____ A long-lasting, large mass of snow and ice that forms over land from the accumulation and compaction of snow that creeps down slope.

12. ____ A molecule that has a weak positive and negative electrical charge.

13. ____ The movement of water molecules upward in tiny tubes as a result of adhesion and cohesion.

14. ____ Naturally occurring freshwater that flows or is stored underground in rock or sediments.

15. ____ The ability of a substance to absorb, contain, and release heat energy.

Critical Thinking

1. If water were truly a universal solvent, how would this affect life on Earth?

CHAPTER

24 The Hydrologic Cycle

Objectives

Evaporation • Water Vapor and Condensation • Precipitation and Surface Water • Runoff • Evapotranspiration • Infiltration and Groundwater

After reading this chapter you should be able to:

- Define the term *hydrologic cycle.*
- Describe the process by which water enters the atmosphere.
- Explain what happens to water vapor when it is in the atmosphere.
- Describe the five pathways that water can take when it falls back to the surface.
- Define the term *evapotranspiration.*
- Identify the factors that affect the rate at which water infiltrates the ground.
- Identify the sources of energy that drive the hydrologic cycle.

TERMS TO KNOW

hydrologic cycle
evaporation
water vapor
condensation
precipitation
surface water
runoff
absorbed
evapotranspiration
infiltration
groundwater
irrigation
flowing springs

INTRODUCTION

One of the most important aspects of understanding the role that the hydrosphere plays on Earth is examining the function of the **hydrologic cycle.** The hydrologic cycle is the circular movement of water between the oceans, atmosphere, and land surface. The term *hydrologic* means "the study of water"; therefore the hydrologic cycle is the study of how water moves through environment and the role it plays in the Earth's systems.

Evaporation

The term *cycle* refers to the circular movement of something, so there really is no starting point for the hydrologic cycle on the Earth; however, because the world's oceans contain 97% of all the world's water, this is a good theoretical starting point for the hydrologic cycle (Figure 24–1). Radiant energy from the Sun strikes the surface of the oceans and begins to heat the liquid water molecules. Eventually there is enough heat energy absorbed by the water molecules to cause them to evaporate off the ocean surface. **Evaporation** is the phase change from a liquid to a gas. The gaseous form of water is known as **water vapor.** Millions of gallons of water evaporates from the surface of the oceans each day around the world. This is by far the largest contributor of water into the Earth's atmosphere.

Water Vapor and Condensation

Once water from the ocean evaporates into water vapor, it then becomes an important part of the atmosphere (Figure 24–2). Water vapor makes up anywhere from 0% to 4% of the atmosphere, depending on the location and time of the year. The warmer areas closer to the equator typically have the greatest amount of water vapor in the atmosphere. Areas close to the poles of the Earth have the least amount of water vapor in the atmosphere. This is the result of warmer air having the ability to absorb more water than cooler air. Although water vapor in the atmosphere composes only 0.001% of the Earth's total water supply, it is an important pathway by which water moves from the oceans to the land surface. All the surface freshwater on the

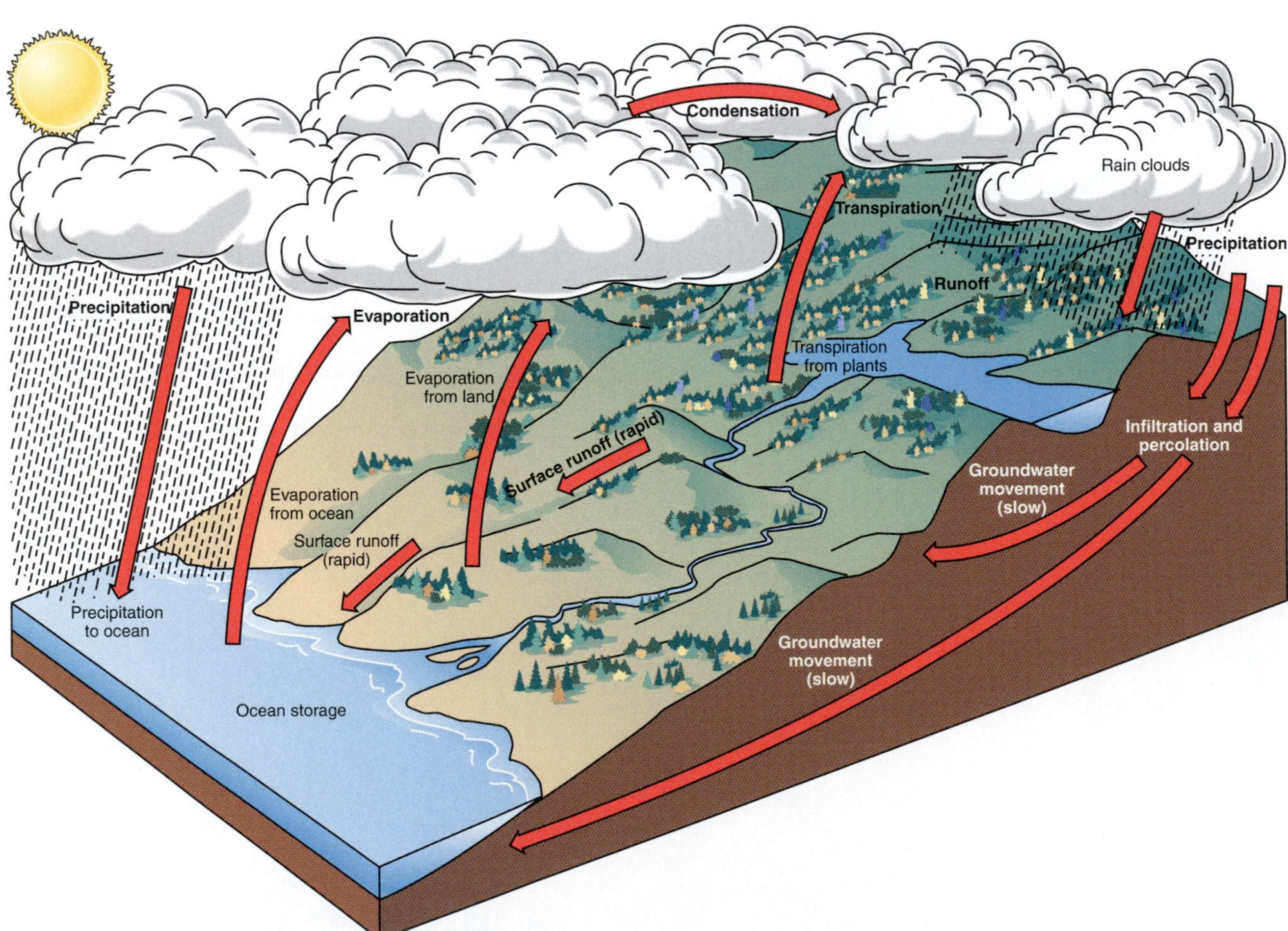

Figure 24–1 The hydrologic cycle is a model of the movement of water through the environment.

Figure 24–2 The evaporation of water from the ocean, forming water vapor that eventually condenses in the atmosphere to create clouds. This process transports water from the oceans to the land surface. *(Courtesy of PhotoDisc.)*

planet, which is located in glaciers, rivers, lakes, and groundwater, was once atmospheric water vapor. This illustrates the important link that the atmosphere provides between the water in the oceans and the water on land. Once water vapor enters the atmosphere, it rises and begins to cool. This is because moist air is less dense than dry air. The cooling action of the upper portions of the atmosphere causes water vapor in the rising air to condense.

The average time water spends in the Earth's atmosphere is 10 days, because eventually air will cool enough and cause the water vapor to condense. **Condensation** is the phase change from a gas to a liquid. Eventually enough water vapor condenses in the atmosphere and begins to collect as clouds. As the clouds begin to grow, so does the liquid water contained within them. The clouds are driven over the land by atmospheric winds, and soon liquid water begins to fall toward the land surface. This is known as **precipitation.**

Precipitation and Surface Water

Precipitation can be either in the form of a liquid (rain) or a solid (snow). Once the precipitation arrives on the land surface, it can take one of five pathways (Figure 24–3). The first path it can possibly take is to collect on the surface in lakes or ponds. These freshwater storage areas can be extremely large. Lake Superior in North America covers more than 31,000 square miles and has an average depth of 3264 feet. **Surface water** can also be very shallow, such as a pond, which can cover only 1000 square feet. The surface water of lakes and ponds can also evaporate and return the water to the atmosphere once again to repeat the cycle.

EARTH SYSTEM SCIENTISTS *ROBERT HORTON*

Robert Horton was born in Michigan in 1875. After receiving a college education, he began working with his uncle, a civil engineer who helped construct the Erie Canal. Their work together involved the measurement of stream flow in rivers and creeks in New York State. In 1900 Horton became the New York district engineer for the U. S. Geological Survey. His research of stream flow helped him to develop theories on the interaction of precipitation with the ground. He proposed that water can take one of four pathways when it reaches the Earth's surface. Horton discovered that precipitation could either run off, infiltrate the ground, transpire through plants, or evaporate. This important discovery helped form a better understanding of the hydrologic cycle. Horton's research in hydrology revealed that stream flow during the drier months was attributed to groundwater discharge. He also helped to establish flood stages for major flowing waterways in New York State to help prevent flood damage and the loss of life.

Figure 24–3 Precipitation returns water back to the Earth's surface. *(Courtesy of PhotoDisc.)*

The second pathway that water can take when it falls to the Earth is to collect as snow and ice to form glaciers. In this way freshwater can be stored on the Earth's surface for thousand of years. Approximately 2% of all water on the Earth is locked in the ice that composes the world's glaciers. Eventually, when the leading edge of a glacier meets the sea, the ice melts and the water once again returns to the oceans.

Runoff

The next pathway that precipitation can take when it reaches the land surface is called **runoff.** Rain or melting snow is driven by the force of gravity to collect in streams and rivers that eventually find their way back to the ocean (Figure 24–4). This is another way that the hydrologic cycle can complete itself. The largest of the world's rivers is the Nile River in Africa, which is more than 4100 miles long. All the minerals that are dissolved in the water of the oceans were transported there by the world's river systems.

Evapotranspiration

The fourth pathway that precipitation can take once it arrives at the surface is **absorption** by plants. Plants take water up through their root systems and distribute it throughout their bodies. Some of this water eventually reaches the leaves, where it evaporates back into the atmosphere. The process of water entering the root system of a plant, moving through the plant body, and then evaporating off the leaf surface is called **evapotranspiration,** also known simply as transpiration (Figure 24–5). Evapotranspiration can often play an important role

Figure 24–4 Precipitation that does not collect in surface water or infiltrate into the ground can run off the surface into streams and rivers. *(Courtesy of PhotoDisc.)*

in determining local climate by adding water vapor to the atmosphere. Large forests can add millions of gallons of water into the atmosphere by the process of evapotranspiration. One birch tree alone can add approximately 70 gallons of water, and one mature corn plant contributes more than 1 gallon of water each day to the atmosphere by evapotranspiration.

Infiltration and Groundwater

The final pathway that precipitation can take once it strikes the land surface is called **infiltration.** Infiltration is the movement of water into soil or rock. The rate of infiltration of water into the ground depends on a few factors. The size of the pore spaces within the ground greatly affects the rate at which water infiltrates. Generally, the larger the size of the pores, the greater the rate of infiltration. Also the amount of water within the pore spaces affects the rate of infiltration. The more water contained in the pore spaces within the ground, the slower the rate of infiltration. Pore spaces that are completely saturated with water will prevent any water from infiltrating into the soil, causing it to run off instead. The arrangement of particles in a soil also affects the rate of infiltration. Particles of mixed size tend to pack tightly together and reduce the amount of pores in the ground. This then reduces the infiltration rate. Sorted particles of similar size increase the pore spaces and therefore increase the infiltration rate.

Once the water infiltrates the ground, it is called **groundwater.** Groundwater holds approximately 0.6% of Earth's total water, which is a greater volume than all the

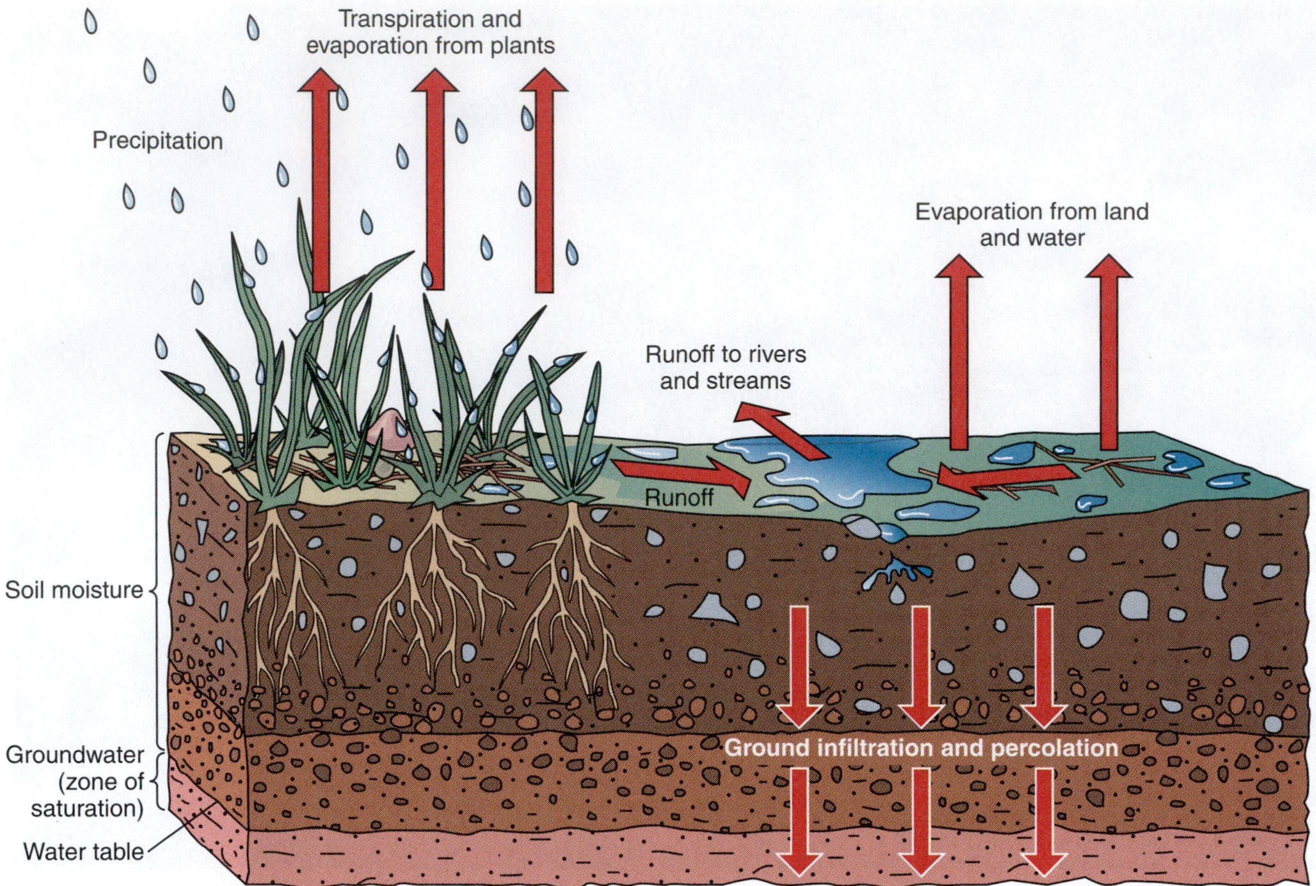

Figure 24–5 The pathways that water can take when it comes into contact with the soil include runoff, infiltration, and evapotranspiration.

Career Connections

Water Resource Engineer

A water resource engineer works in the planning, design, construction, and operation of all aspects of water use. This includes the management of all engineering projects that use water resources. Water resource engineers design systems for specific water resource projects such as flood control, dam construction, sewer systems, municipal water supplies, irrigation, and water transportation. These engineers work both in the field and in the laboratory, where they conduct surveys, perform tests, design projects, oversee construction, and monitor systems. They require a knowledge of engineering, hydrology, and construction and can find employment in private industry or with state and local governments.

world's lakes and rivers combined. Groundwater is an important source of freshwater for drinking and **irrigation.** Most groundwater eventually returns to the surface as **flowing springs** or is pumped out by mechanical wells. Approximately 50% of all Americans get their water from wells supplied by groundwater.

No matter which pathway precipitation takes when it reaches the land surface, eventually all water on Earth returns to the oceans, and the hydrologic cycle begins again. This may take days, years, or thousands of years depending on the specific pathway the water is traveling. The processes that help to move water around the planet involve the use of a great amount of energy. Mostly it is the power of the Sun that drives the hydrologic cycle, along with the force of gravity.

EARTH MATH

1) IF ONE PLANT TRANSPIRES 80 MILLILITERS OF WATER PER DAY, HOW MANY LITERS OF WATER DOES THE PLANT TRANSPIRE IN A WEEK?

2) IF IT TAKES A WATER MOLECULE 88 MINUTES TO TRAVEL 1 MILE DOWN THE NILE RIVER, HOW MANY DAYS WILL IT TAKE WATER TO TRAVEL THE ENTIRE LENGTH OF THE RIVER?

REVIEW

1. How much of the Earth's water is located in the oceans?
2. Define the term *evaporation* and explain its importance to the hydrologic cycle.
3. Describe the five pathways water can take when it reaches the land surface.
4. Who was Robert Horton?

For more information go to these Web links:

<http://observe.arc.nasa.gov/nasa/earth/hydrocycle/hydro1.html>

<http://danpatch.ecn.purdue.edu/~epados/ground/src/cycle.htm>

<http://www.earth.nasa.gov/science/Science_global.html>

CHAPTER SUMMARY

The hydrologic cycle is the circular movement of water through the environment. The theoretical starting point for the hydrologic cycle is in the oceans. Solar energy heats the ocean surface and causes the water to evaporate. The water enters the atmosphere in the gaseous state called water vapor. Eventually the water vapor in the atmosphere rises, cools, and condenses. This forms clouds, which then move through the atmosphere and form precipitation. Precipitation returns water to the Earth's surface in the form of rain or snow.

Once water strikes the land surface, it can take five possible pathways. These include collecting as surface water in lakes or ponds or as the snow and ice in glaciers. Precipitation can also run off the land surface, forming streams and rivers that flow toward the oceans. Some precipitation can infiltrate into the rock and soil, forming groundwater. The rate at which water infiltrates the ground is affected by the size of the pore spaces within the ground and by the amount of water that is already there. Once water enters the ground, it can be stored there for long periods or can be taken up into the root systems of plants. The movement of water through the roots of plants and up through the plant body, eventually evaporating off the leaf surface back into the atmosphere, is called evapotranspiration. All these pathways eventually return water either to the ocean or to the atmosphere, which starts the cycle over again. The driving force of the hydrologic cycle is energy from the Sun, along with gravity.

CHAPTER REVIEW

Multiple Choice

1. The movement of water from the oceans to the atmosphere is by the process of:
   a. evaporation
   b. condensation
   c. precipitation
   d. evapotranspiration

2. Generally, the amount of water vapor in the atmosphere is greatest near the:
   a. North Pole
   b. mid latitudes
   c. equator
   d. South Pole

3. The process by which water forms clouds in the atmosphere is called:
   a. evaporation
   b. condensation
   c. precipitation
   d. evapotranspiration

4. Ponds, lakes, and glaciers are all examples of:
   a. groundwater
   b. precipitation
   c. stored surface water
   d. runoff

5. Which of the following increases the amount of runoff?
   a. small pore spaces in an unsaturated soil
   b. small pore spaces in a saturated soil
   c. large pore spaces in an unsaturated soil
   d. large pore spaces in a saturated soil

6. Which type of soil particles have the greatest infiltration rate?
   a. mixed saturated particles
   b. mixed unsaturated particles
   c. sorted saturated particles
   d. sorted unsaturated particles

7. Plants move water from the ground by the process of:
   a. infiltration
   b. condensation
   c. precipitation
   d. evapotranspiration

8. Approximately how long does water vapor remain in the atmosphere?
   a. 1 day
   b. 10 days
   c. 1 month
   d. 1 year

9. What is the driving force for much of the hydrologic cycle?
   a. adhesion and cohesion
   b. the Sun
   c. capillary action
   d. prevailing winds

Matching *Match the terms with the correct definitions.*

a. hydrologic cycle
b. evaporation
c. water vapor
d. condensation
e. precipitation
f. surface water
g. runoff
h. absorbed
i. evapotranspiration
j. infiltration
k. groundwater
l. irrigation
m. flowing spring

1. ____ The phase change when a liquid changes into a gas.

2. ____ The rapid loss of soil, sediments, or other substances as a result of being washed away by rain or melting snow.

3. ____ The circular pathway of water molecules as they move through the environment, also called the water cycle.

4. ____ An area where groundwater is discharged at the surface and flows freely.

5. ____ The gaseous form of water.

6. ____ An artificial means of supplying water to plants.

7. ____ To take in or soak up matter or energy.

8. ____ Naturally occurring freshwater that flows or is stored underground in rock or sediments.

9. ____ The change in phase from a gas to a liquid.

10. ____ The important pathway by which water moves from the soil, through the body of a plant, and evaporates off the leaf surface back into the atmosphere.

11. ____ Liquid or solid water formed in clouds that falls to the surface of the Earth.

12. ____ The process of infiltrating, or entering into something.

13. ____ Water that is located on the surface of the Earth.

Critical Thinking

1. Explain how cutting down a forest might affect the hydrologic cycle.

CHAPTER

25 Oceanography

Objectives

Seawater • Ocean Currents • Deep Ocean Circulation • Life Zones in the Ocean • Continental Shelves • Intertidal Zone

After reading this chapter you should be able to:

- Describe the approximate salinity of the ocean.
- Identify some of minerals that compose seawater.
- Describe the mechanism that drives surface ocean currents.
- Identify at least four major surface ocean currents.
- Explain the process of thermohaline circulation.
- Define the terms *upwelling* and *thermocline*.
- Identify the four main life zones within the ocean.
- Describe some of the characteristics of the continental shelves.
- Define the term *intertidal zone*.

TERMS TO KNOW

seawater
salinity
wind-driven current
gulf stream
California current
upwelling
thermohaline circulation
euphotic zone
disphotic zone
aphotic zone
benthic zone
thermocline
hydrothermal vents
continental shelf
intertidal zone

INTRODUCTION

Of all water on Earth, 97% resides in the oceans, which cover more than 70% of the planet. These vast storehouses of water have an average depth of more than 2 miles and remain largely unexplored. Some oceanographers argue that we know more about other planets in our solar system than we do about our own oceans. It is believed that life on Earth began within the ocean. This vast resource has also helped to sustain life for more than 3 billion years with its abundant resources. Our dependence on the ocean continues to this day. The oceans of the world also store incredible amounts of heat energy, which help to regulate global climate. Understanding the physical and chemical properties of the ocean is important to understanding its role in the Earth's complex systems and how it influences life on Earth.

Seawater

The water that resides in the oceans is called **seawater.** Seawater is a mixture of water and more than 70 other chemical elements. The measure of the amount of chemical elements in seawater is called **salinity.** The salinity of the ocean is approximately 3.5%, which means that 3.5% of seawater is made up of dissolved minerals (Table 25–1). For example, if a little more than 100 gallons of seawater evaporates, approximately 3.5 pounds of mineral salts is left behind. The principal mineral salts in seawater are sodium, chloride, sulfur, and magnesium. The salinity of the ocean remains fairly constant all around the globe, although near the polar regions it is slightly higher. This is the result of the freezing of water that forms the ice caps, which helps to increase the concentration of mineral salts dissolved in the ocean around the poles.

Ocean Currents

Another important feature of Earth's oceans involves the circulation of seawater around the planet. The circulation of water throughout the oceans is known as currents, which are divided into two main types (Figure 25–1). The first type of ocean current is called a **wind-driven current.** Wind-driven currents move the upper parts of the ocean horizontally by wind action that strikes the ocean surface. The planetary-scale prevailing winds that circulate the atmosphere create these currents. Major wind-driven currents begin near the equator and then move along the edge of the continents, bringing warm equatorial water northward or southward toward the poles. Other wind-driven currents transport cold water from the poles back toward the equator, where it is heated once again. Wind-driven currents help to transport heat energy from the Earth's equatorial regions and bring it to the colder high latitudes. The **gulf stream** is a major wind-driven current that moves warm water from the Caribbean Sea up along the east coast of the United States, where it eventually reaches Greenland.

TABLE 25-1 The Major components of seawater

Constituent	Percent of Substances by Mass
Oxygen	85.4
Hydrogen	10.7
Chlorine	1.85
Sodium	1.03
Magnesium	0.127
Sulfur	0.087
Calcium	0.040
Potassium	0.038
Bromine	0.0065
Carbon	0.0027
Nitrogen	0.0016
Strontium	0.00079
Boron	0.00043
Silicon	0.00028
Fluorine	0.00013

The warm waters that the gulf stream current moves northward help to moderate the climate of England. Although the British Isles are located at high colder latitudes, their climate is mild because of the influence of the gulf stream. Other wind-driven currents, such as the **California current,** move colder arctic waters southward. The California current brings cold water down the western coast of North America, which can also influence local climate.

Another important aspect of wind-driven ocean currents involves the vertical distribution of heat throughout the ocean. This is called **upwelling.** Upwelling occurs when winds move warm surface waters away from the equator or the coasts of continents (Figure 25–2). The cooler deep water moves upward and then replaces the warm surface waters. Major upwelling zones occur along the western coast of South America and along the equator in the Pacific Ocean. Upwelling also helps bring cold, nutrient-rich water up to the surface, where it is used by aquatic organisms. Changes in the upwelling zones near the

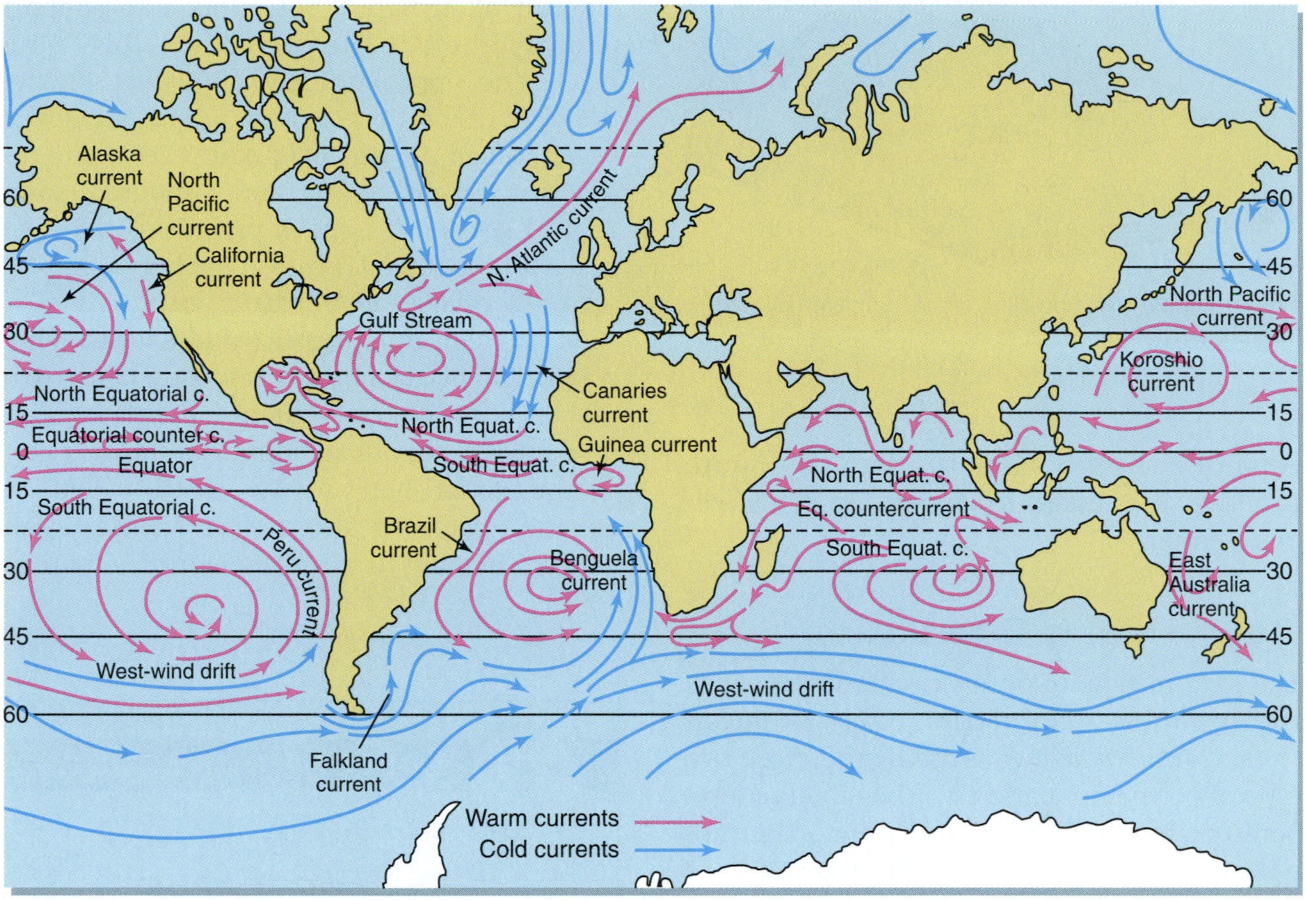

Figure 25–1 The world's major surface ocean currents.

 EARTH SYSTEM SCIENTISTS *MATTHEW MAURY*

Matthew Maury was born in 1803 in Virginia and joined the Navy as a young midshipman. His love for the sea was rivaled only by his love of knowledge. Early in his naval career, Maury was aboard the first U. S. ship to circumnavigate the globe, in 1830. During his life he wrote many scientific papers that dealt with navigation and the physical aspects of the oceans. In 1839 his leg was badly broken in a stagecoach accident, which ended his career at sea. He then became the superintendent of the U.S. Naval Observatory in 1844. While there he began to analyze the data that were contained in the thousands of ships logs stored at the observatory. Maury recognized that these logs held valuable information about winds, currents, and weather conditions at many points in the ocean. Over the next 10 years, Maury created the first detailed navigation charts of the world's oceans. By using the information contained in the ship's logs, Maury charted ocean currents, prevailing winds, and other important information that could be used by ship captains. In 1855 he published The Physical Geography of the Sea, which became the first book to present the science of oceanography. Maury was also responsible for developing standardized methods for recording weather and nautical information gathered at sea. Because of his lifelong work, Maury is often regarded as the father of modern oceanography.

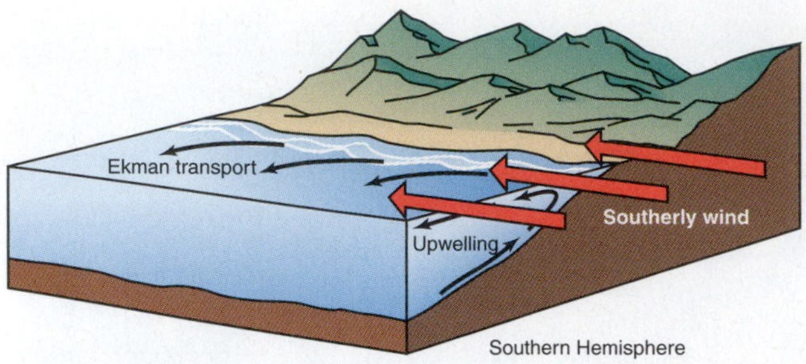

Figure 25–2 The upwelling of cold, nutrient-rich water from the deep ocean.

equator, off the western coast of South America, are linked to changes in global climate patterns.

Deep Ocean Circulation

The second type of ocean circulation is called **thermohaline circulation.** *Thermo* means "heat," and *haline* means "salt"; therefore this type of ocean circulation is driven by the temperature and salinity of the water. Thermohaline circulation begins in the cold waters of the North Atlantic Ocean, where the rapid freezing of water near the poles creates cold, saline-rich water that sinks to the bottom of the ocean. This cold, dense water is called North Atlantic deep water, which creeps along the bottom of the ocean until it reaches upwelling zones. In this way, nutrients and heat are circulated throughout the world's oceans.

Life Zones in the Ocean

Because the ocean is so deep, scientists also study vertical layers, or zones, that affect marine life (Figure 25–3). The top zone, or layer, of the ocean, is called the **euphotic zone.** This is where enough sunlight reaches into the ocean to support photosynthesis. The euphotic zone usually extends no deeper than 600 feet, although that varies depending on the regional clarity of ocean water. The next layer of the ocean is called the **disphotic zone.** The disphotic zone receives a small amount of light but not enough to support photosynthesis. The depth of this zone reaches to approximately 3000 feet. Below the disphotic zone lies the largest layer of the ocean, called the **aphotic zone.** This zone receives no light at all and supports many strange species of aquatic organisms that live in a world of total darkness. The bottom of the ocean is called the **benthic zone,** which is subject to extreme pressure and near freezing water. The ocean bottom is mostly composed of a thick mudlike sediments, which form as the bodies of tiny dead aquatic organisms sink to the ocean bottom. The fecal pellets of aquatic organisms also collect on the ocean bottom, helping to form the benthic environ-

Career Connections

Physical Oceanographer

A physical oceanographer studies all the physical aspects of the world's oceans. This includes ocean temperatures, currents, salinity, wave formation, density, and tidal forces. These marine scientists are interested in revealing the interactions that occur within the oceans. They collect data about the ocean by using remote sensing satellites, sonar, radar, and a wide array of different sensors. They also collect water samples from different parts of the world and at different depths from within the ocean. Physical oceanographers are interested in the relationship among the ocean, the atmosphere, and global climate. Specialized physical oceanographers study the topography and composition of the ocean floor. These scientists create accurate maps of the ocean floor. Some physical oceanographers search for mineral resources or oil that might be located in ocean sediments. Careers in this field require at least a 4-year college degree. Work can be found within academic research, government agencies, or private industry.

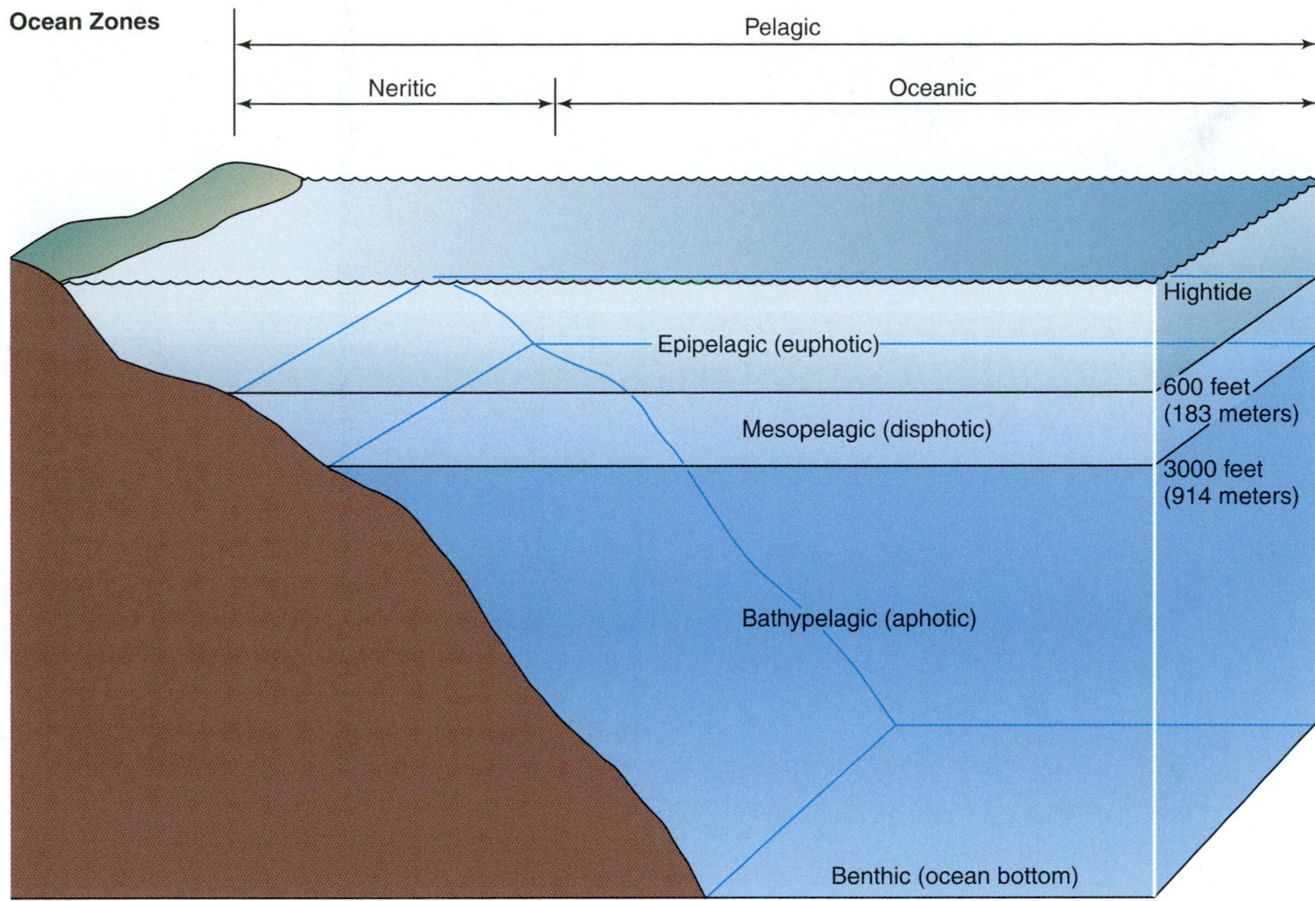

Figure 25–3 The aquatic life zones within the ocean.

ment. Some of the loose sediments that cover the ocean floor can be more than 1 mile thick.

Another important feature of the ocean is called the **thermocline.** The thermocline is the area where warm, nutrient-poor surface water mixes with cold, nutrient-rich bottom water. This is an area of extreme temperature change (Figure 25–4). The depth of the thermocline varies depending on the time of year and latitude. In tropical ocean waters near the equator, the thermocline can be located more than 600 feet from the surface. In the higher latitudes of temperate climates, the thermocline can be found only 50 feet from the surface. Many aquatic organisms live within the thermocline, seeking the nutrients that it provides. These organisms also migrate with the

Current Research

Researchers from Japan and Australia believe that fertilizing the Pacific Ocean off the coast of South America may help to curb global warming. Australian oceanographer Ian Jones and a Japanese engineering firm are proposing that nitrogen fertilizer pumped into the ocean off the coast of Chile would help to boost the plankton population residing in its coastal waters. Plankton removes carbon dioxide from the atmosphere as it builds its body, which could result in a reduction in global atmospheric carbon dioxide levels as the plankton population grows. The so-called ocean enrichment plan is reported to be less expensive and easier to carry out than planting trees or using industrial scrubbers to remove carbon dioxide. Skeptics of the plan are not sure what the impact would be if nitrogen was added to the ocean, therefore increasing the plankton population.

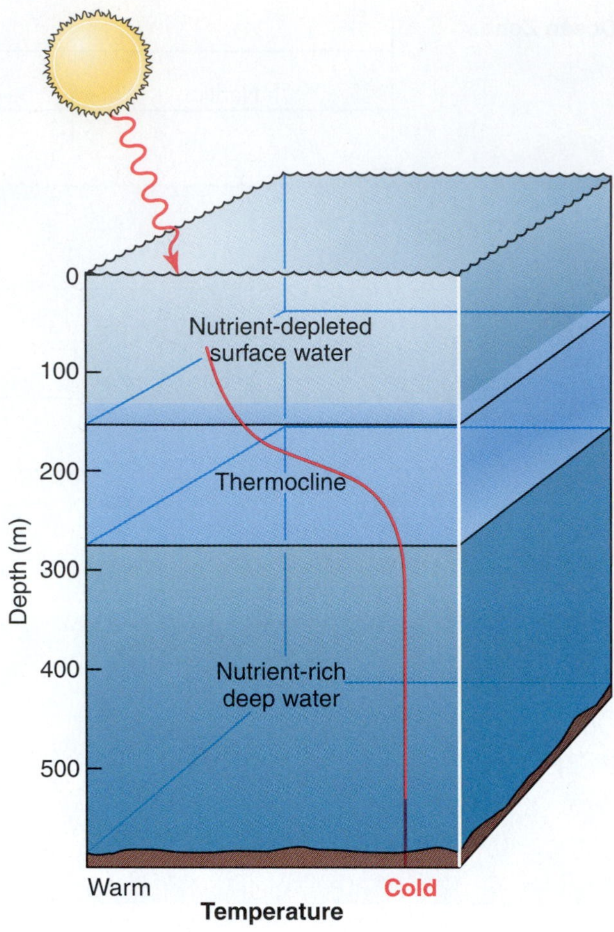

Figure 25–4 The location of the thermocline, which is the depth at which the temperature of the water declines rapidly.

Figure 25–5 A hydrothermal vent community that survives in the intense heat and crushing pressure of the ocean floor, near a volcanic vent. *(Courtesy of NOAA.)*

thermocline as its depth fluctuates throughout the year.

Although much of the benthic zone in the world's oceans is located almost 3 miles below the surface, it still is home to many varieties of living organisms. Recently oceanographers have been studying a unique feature of the deep ocean benthic zone called **hydrothermal vents.** Hydrothermal vents are areas where volcanic activity emits rich minerals and extreme heat into the cold bottom waters of some parts of the oceans (Figure 25–5). The pressure around these benthic zone vents can be 300 times that of the Earth's surface, and temperatures can be as high as 750° Fahrenheit. The vents, however, support a

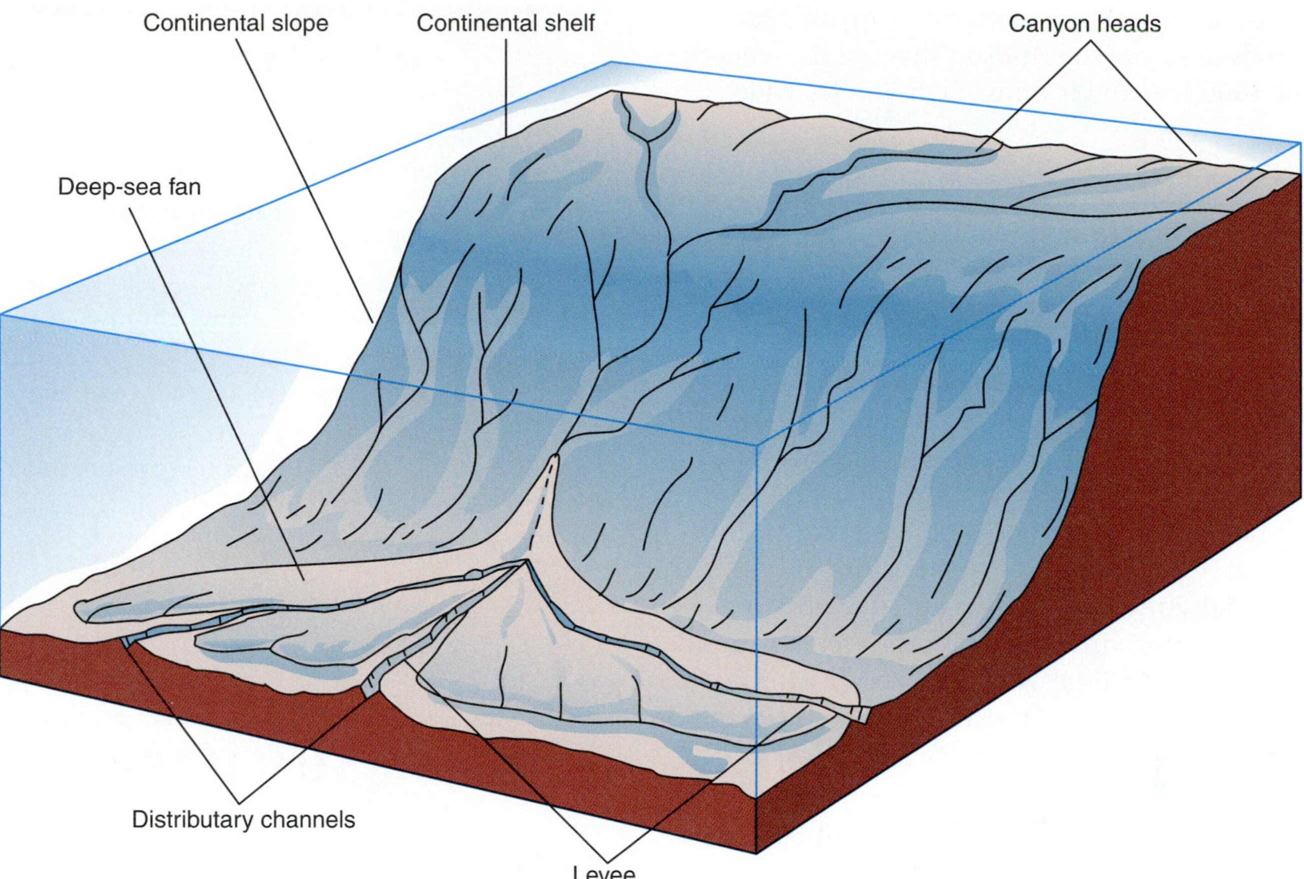

Figure 25–6 The location of a submarine canyon near the edge of a continental shelf.

variety of organisms, such as tube worms, clams, crabs, and bacteria, that thrive in these deep, harsh conditions.

Continental Shelves

Most organisms that reside in the world's oceans survive in the relatively shallow waters of the continental shelves. A **continental shelf** is a shallow, sloping area located around the margins of continents. The average depth of a continental shelf is approximately 400 feet. Most of the continental shelves around the world lie within the euphotic zone, which supports a variety of aquatic life. Approximately 90% of all the fish and shellfish harvested from the oceans comes from the continental shelves. On average, the world's continental shelves stretch out from the land about 45 miles; however, some continental shelves can be as wide as 900 miles. Along the edges of the continental shelves, where the depth of the ocean increases rapidly, there exist large, deep canyons called submarine canyons (Figure 25–6). Some of these canyons can have depths of more than 12,000 feet. Many submarine canyons are associated with the world's large rivers. Submarine canyons are believed to have formed as part of the large valleys that were carved by rivers when the sea level was lower during the last ice age. Today these canyons are still being shaped by the sediments and currents that these river systems bring into the oceans. The Hudson

Canyon is a large submarine canyon that extends out from the Hudson River; it has a depth of 3600 feet and is more than 5 miles wide.

Intertidal Zone

Another important feature of the oceans is the **intertidal zones.** The intertidal zones are areas along the shore that lie within the reaches of the low and high tides. These areas also harbor an abundance of life and consist of shore lines and tidal pools. The rise and fall of the water, wave action, strong currents, and periodic exposure to the air make this one of the harshest aquatic environments. The organisms that live in the intertidal zone are well adapted to this ever-changing environment. Their are two basic types of intertidal zones. A rocky intertidal, which is composed mostly of large rocks that line the shore, and a sandy intertidal, which is made up of mostly sand. Both types of intertidal zones have their own unique organism communities (Figure 25–7).

For more information go to these Web links:
<http://www.noaa.gov/ocean.html>
<http://www.noaa.gov/charts.html>
<http://enchantedlearning.com/subjects/ocean/>
<http://www.noaa.gov/coasts.html>

REVIEW

1. How much of the world's water resides in the oceans?
2. What percentage of seawater is composed of minerals?
3. What is the name of the wind-driven ocean current that moves along the eastern United States?
4. Describe the process of upwelling in the ocean.
5. List the four main life zones within the ocean.
6. Who was Matthew Maury?

CHAPTER SUMMARY

Of all the water on Earth, 97% is in the form of seawater. Seawater is a mixture of dissolved minerals and water. The amount of minerals dissolved in water determines its salinity. The salinity of seawater in the oceans is approximately 3.5%. The oceans of the world are composed of seawater and cover approximately 70% of the Earth's surface. The prevailing planetary winds that circulate the atmosphere drive major ocean currents. These are known as wind-driven currents, which help to distribute warm water away

EARTH MATH

1) How many pounds of minerals does a 100-pound sample of seawater contain?
2) If the depth of the euphotic zone of the ocean is 333 feet in summer and 568 feet in winter, what is the average depth of the euphotic zone?
3) If pressure increases by 14.7 pounds per square inch for every 33 feet of depth in the ocean, what is the pressure at a depth of 14,850 feet?

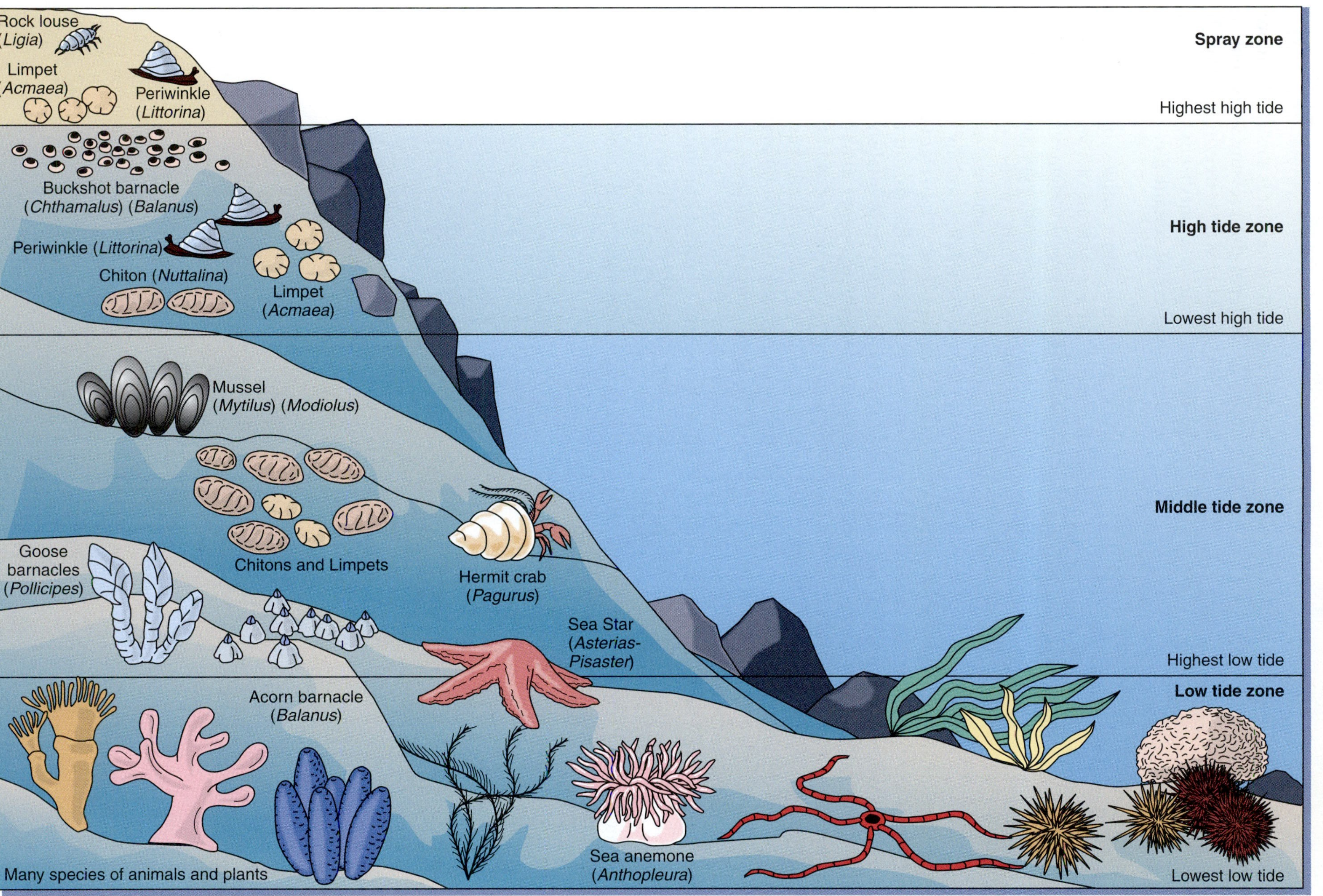

Figure 25–7 The community of organisms that reside in the harsh conditions of the rocky intertidal zone.

from the equator and cold water away from the poles.

Ocean currents also cause upwelling, which is the rising of nutrient-rich colder water from the bottom of the ocean, replacing warm surface waters that are moved away by prevailing winds. This is known as the vertical mixing of the ocean. Another type of ocean circulation, called thermohaline circulation, moves cold, saline-rich water from the poles down along the bottom of the ocean toward the equator.

The thermocline is the layer of the ocean where the temperature drops rapidly with depth. This is where cold, nutrient-rich water mixes with warmer surface water. The ocean is divided into four distinct vertical layers called life zones. The upper most layer of the ocean is called the euphotic zone. This is where enough light penetrates into the water to support photosynthesis. Below this layer is the disphotic zone, where a small amount of light reaches but cannot support photosynthesis. Most of the ocean exists in the aphotic zone, which is the deep zone where no light reaches. The bottom of the ocean is called the benthic zone. In some places this can be more than 5 miles deep; the pressure is extremely high, and the water is near freezing. Although the benthic zone is a harsh environment, it supports a variety of aquatic life.

There are some communities of organisms that reside on the bottom of the ocean near volcanic vents. These nutrient-rich areas are known as hydrothermal vent communities, where the temperature of the water can be more than 200° Fahrenheit. Although the water is extremely hot, specially adapted organisms thrive there. The relatively shallow regions of the ocean that surround the continents are called the continental shelves. These areas are usually no deeper than 400 feet and support a great amount of aquatic life. Near the edges of the continental shelves are natural deep canyons called submarine canyons, which are extensions of large river systems that drain into the oceans. The area closest to the shore line is called the intertidal zone. This area exists between the lowest and highest tides, where wave action, strong currents, and periodic exposure to air make it difficult to survive.

CHAPTER REVIEW

Multiple Choice

1. Approximately how much of the Earth's surface is covered by seawater?
a. 30%
b. 50%
c. 70%
d. 90%

2. If you were to weigh out 1000 pounds of seawater, approximately how much of it would be minerals?
a. 1 pound
b. 2.5 ponds
c. 3.5 pounds
d. 35 pounds

3. What drives the major surface ocean currents?
a. the Earth's rotation
b. temperature differences
c. convection
d. prevailing winds

4. A major ocean current that brings cool water south toward the equator is:
a. the California current
b. the gulf stream
c. the North Atlantic current
d. the Brazil current

5. A major ocean current that brings warm water away from the equator is called:
a. the Labrador current
b. the gulf stream
c. the Peru current
d. the Canaries current

6. Which area in the ocean has the highest salinity?
a. the equator
b. the poles
c. the mid latitudes
d. the central Pacific

7. Upwelling results in:
a. cooler ocean water sinking
b. warmer ocean water rising
c. cooler ocean water rising
d. warmer ocean water sinking

8. If you were to scuba dive through the thermocline, the temperature of the water would:
a. increase rapidly
b. increase gradually
c. decrease gradually
d. decrease rapidly

9. Which life zone in the ocean is in total darkness?
a. euphotic
b. disphotic
c. aphotic
d. benthic

10. Which life zone in the ocean supports photosynthesis?
a. euphotic
b. disphotic
c. aphotic
d. benthic

11. Hydrothermal vents are located in which life zone?
a. euphotic
b. disphotic
c. aphotic
d. benthic

12. Which of the following is associated with the continental shelves?
a. submarine canyons
b. hydrothermal vents
c. intertidal zones
d. the aphotic zone

13. What type of ocean environment is most affected by exposure to air and strong currents?
a. submarine canyons
b. hydrothermal vents
c. intertidal zones
d. the aphotic zone

continued

Matching *Match the terms with the correct definitions.*

a.	seawater	**f.**	upwelling	**k.**	benthic zone
b.	salinity	**g.**	thermohaline circulation	**l.**	thermocline
c.	wind-driven currents	**h.**	euphotic zone	**m.**	hydrothermal vents
d.	gulf stream	**i.**	disphotic zone	**n.**	continental shelf
e.	California current	**j.**	aphotic zone	**o.**	intertidal zone

1. ____ The relatively shallow region of the ocean surrounding a continent.

2. ____ The uplift of cold ocean water from the bottom to the surface.

3. ____ Water that has a high concentration of minerals dissolved in it.

4. ____ An aquatic life zone that exists near the shoreline between the area of the highest and lowest tides.

5. ____ A wind-driven surface ocean current that brings cold water from the North Pacific south toward the equator along the west coast of North America.

6. ____ Ocean water that seeps through cracks in the sea floor and is superheated by magma that is close to the surface.

7. ____ A measure of the mineral salt content of a solution.

8. ____ The layer of water located below the surface where the temperature drops rapidly.

9. ____ Surface ocean currents that are formed by planetary winds.

11. ____ The vertical distribution of water in the oceans that is caused by differences in the salinity and temperature of the water.

10. ____ An aquatic life zone located on the bottom of a body of water.

12. ____ A wind-driven, surface ocean current that brings warm water from the equator northward along the east coast of North America; it was first discovered by Benjamin Franklin.

13. ____ The uppermost life zone in an aquatic ecosystem that receives enough light to support photosynthesis.

14. ____ A particular zone in an aquatic ecosystem where there is no light present.

15. ____ The dimly lit portion of an aquatic ecosystem that cannot support photosynthesis.

Critical Thinking

1. If you were to plot a course to sail from New York to England and then back to Florida, what surface ocean currents would you use to help you on your journey?

CHAPTER 26

Fresh Surface Water and Groundwater

Section 26.1 – Fresh Surface Water Objectives

Lakes • Lake Productivity • Life Zones in Lakes • Watersheds and Rivers • Stream Features • Floodplains • Life Cycle of Rivers

After reading this section you should be able to:

- Explain the unique characteristics of the three lake classifications based on their productivity.
- Identify the four aquatic life zones that exist in lakes.
- Define the term *watershed.*
- Describe the relationship between the slope and velocity of a river.
- Define the term *meandering* and describe the four primary stream features.
- Define the terms *dynamic equilibrium* and *discharge rate.*
- Describe the three stages of the life cycle of a river.

Section 26.2 – Groundwater Objectives

Groundwater Recharge and the Water Table • Groundwater Flow • Aquifers and Groundwater Discharge • Groundwater Pollution

After reading this section you should be able to:

- Explain the process by which groundwater is recharged and define the term *water table.*
- Describe what causes groundwater to flow and what controls the rate at which it flows.
- Differentiate between an aquifer and a confined aquifer.
- Identify three sources of groundwater pollution.

TERMS TO KNOW

oligotrophic lakes
mesotrophic lake
eutrophic lake
watershed
meandering
dynamic equilibrium
discharge rate
floodplain
channel
water table
aquifer
impermeable
confined aquifer
artesian wells

INTRODUCTION

The distribution and movement of freshwater, both above and below the ground, is an important part of the Earth's systems. All the freshwater on Earth that is not locked up in glaciers makes up less than 1% of the water on the planet. This small percentage, however, plays a crucial role in the hydrological cycle, shaping the Earth's surface and supporting life. Today it is rare, if not impossible, to actually drink from any surface water supply because they have all been polluted to some degree by human activity. Almost every lake, stream, pond, and river in the United States could potentially make you sick if you drank from it without treating the water. The only water supply that humans can safely drink from without any type of filtration is groundwater. The world's groundwater supply is more than 90 times greater than all the world's fresh surface water supplies combined. Every day in the United States more than 78.5 billion gallons of freshwater are pumped from the ground to be used for drinking or irrigation. This water supplies more than 40% of our nation's drinking water. This important water resource is often overlooked because it is out of sight and has been abused by both pollution and overuse. Understanding the dynamics of surface freshwater and groundwater—how we can improve their quality and sustain their use—is vital to the future of our planet.

26.1 Fresh Surface Water

Lakes

The Earth's fresh surface water is either flowing or standing. The standing freshwater on the surface of the Earth is commonly known as lakes, ponds, and swamps. Lakes can be extremely large and deep, such as Lake Baikal in Russia, which covers more than 12,000 square miles and is more than 5000 feet deep. The largest of the Great Lakes in North America, Lake Superior, covers more than 22,000 square miles, with an average depth of 1300 feet (Figure 26–1). Ponds, on the other hand, can be extremely small and shallow and are located all over the Earth's surface. Both lakes and ponds form when flowing surface waters become trapped and begin to accumulate. Lake and pond formation is a result of the local topography, glacial action, volcanoes, landslides, earthquakes, meteorites, and shifting river patterns. Some of the world's deepest lakes were formed by retreating glaciers. These large masses of ice carved deep into the Earth's crust. Later, when the glaciers melted, the deep canyons they left behind became deep glacial lakes.

Lake Productivity

Lakes are often classified by their age and productivity (Figure 26–2). Productivity is the amount of solar energy that is converted to plant material by the process of photosynthesis. Young, crystal clear lakes with low productivity are called **oligotrophic lakes.** This type of lake has limited amounts of aquatic organisms because of the low availability of nutrients. Oligotrophic lakes are usually

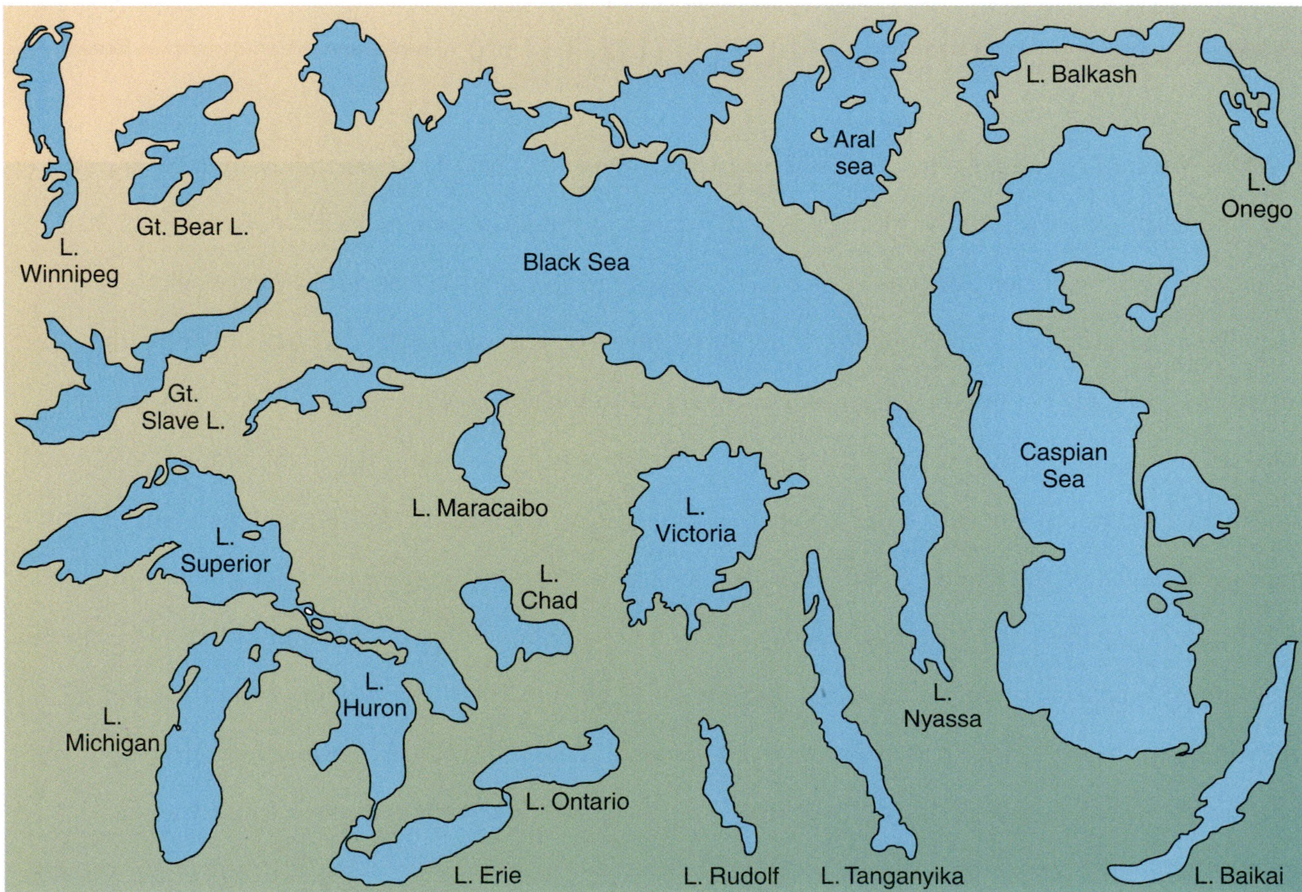

Figure 26–1 A comparison of the relative sizes of the major lakes of the world.

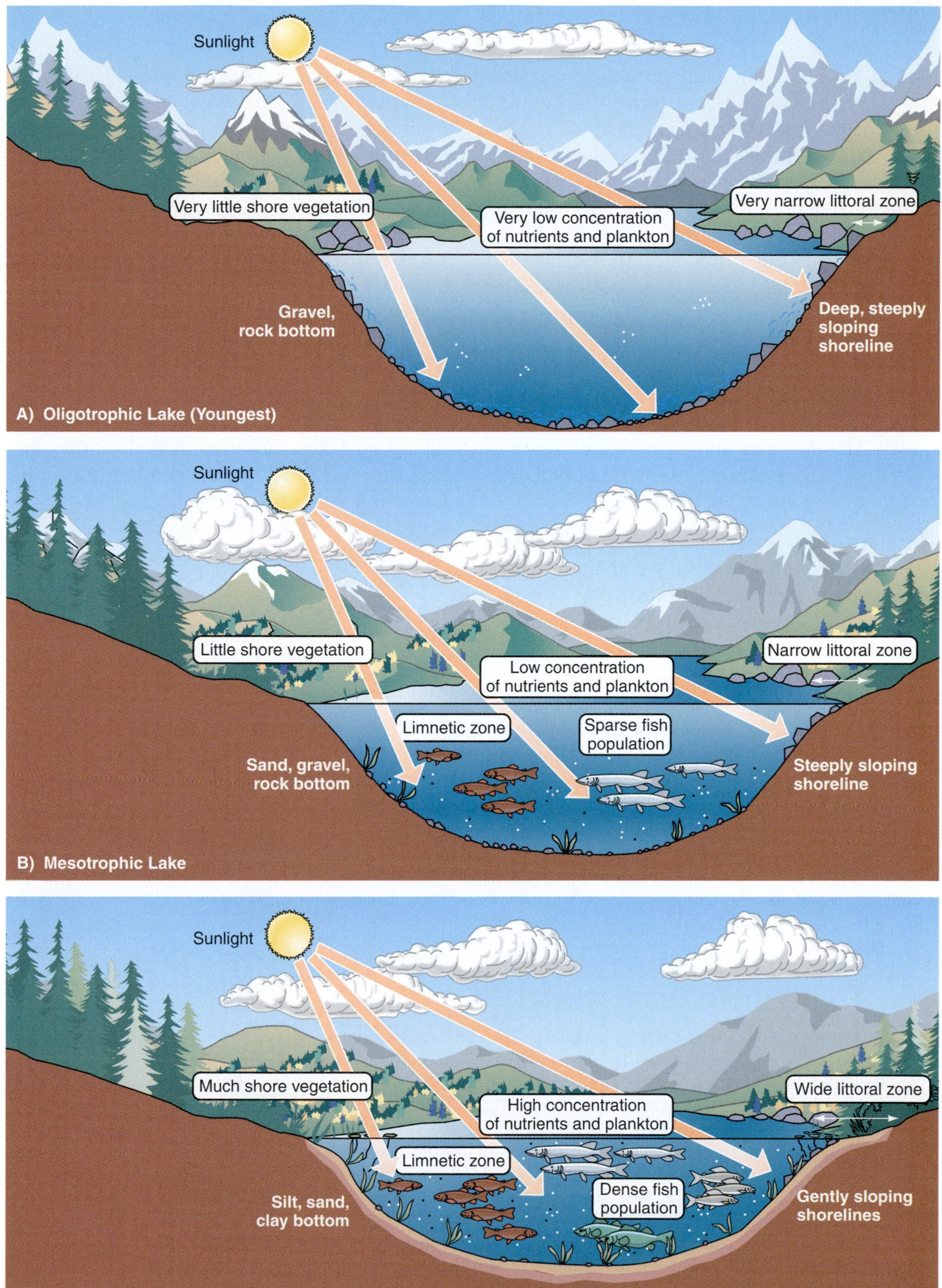

Figure 26–2 The unique characteristics of the three classifications of lakes.

located near the edges of retreating glaciers and therefore contain cold water throughout the year. The next lake classification is the **mesotrophic lake.** Mesotrophic lakes have medium productivity resulting from the increased availability of nutrients. The water is clear and usually cool, supporting an abundance of fish and other aquatic organisms. Many of the larger, clear lakes in the United States are mesotrophic lakes; these are used for water supplies and for recreation. The Finger Lakes in New York State are examples of mesotrophic lakes. These lakes formed after glaciers retreated back into Canada approximately 10,000 years ago. They are extremely deep and clear and support large populations of aquatic organisms.

The third classification for lakes is called eutrophic. **Eutrophic lakes** have an abundance of nutrients and high productivity (Figure 26–3). These lakes are usually very shallow with cloudy warm water, and support many forms of aquatic life. The high number of algae and aquatic plants usually identifies a lake as being in the eutrophic stage. Some eutrophic lakes have so many aquatic plants in them that their surface is completely covered in green mats of plants and algae. Lakes undergo a succession of development beginning with the oligotrophic stage. Over time, runoff from the surrounding area brings nutrients into the lake. Eventually the lake reaches a point at which it becomes mesotrophic. The time it takes to reach the mesotrophic stage is usually recorded in thousands of years. As a lake's nutrients continue to accumulate and it becomes filled with sediments, it eventually becomes eutrophic. This can take tens of thousand of years to occur, except if the lake is very shallow. Eventually all lakes become shallow swamps. Sediments washed into lakes from the surrounding land eventually fill up all lakes. Because of pollution, many mesotrophic lakes are becoming eutrophic lakes in a very short period. Lakes that were once clear are now being overgrown with aquatic plants and algae because of the increased productivity caused by the introduction of fertilizers and sewage leaking into the water as a result of human activity.

Life Zones in Lakes

Like the oceans, lakes also possess unique vertical layers called aquatic life zones (Figure 26–4). The shallow area that surrounds a lake

EARTH SYSTEM SCIENTISTS

ALEXANDER AGASSIZ

Alexander Agassiz was born in Switzerland in 1835 and eventually settled in the United States. After graduating from Harvard University in 1855, he took over the operation of a large copper mine. By 1875 Agassiz was a millionaire, and he turned his attention to the study of the world's oceans. He backed many scientific expeditions to research the oceans. In 1871 the British government launched an expedition that was designed to study the physical and biological aspects of the ocean. The ship, H. M. S. Challenger, sailed around the globe collecting important information about the oceans. Agassiz became involved with this epic voyage and helped to classify and organize much of the samples that were collected during the 4-year expedition. He also helped to finance many scientific studies conducted by researchers in the United States and aided in the acceptance of oceanography as a true science. Agassiz's talent as a scientist was in the design and construction of marine sampling devices and the ability to classify marine organisms.

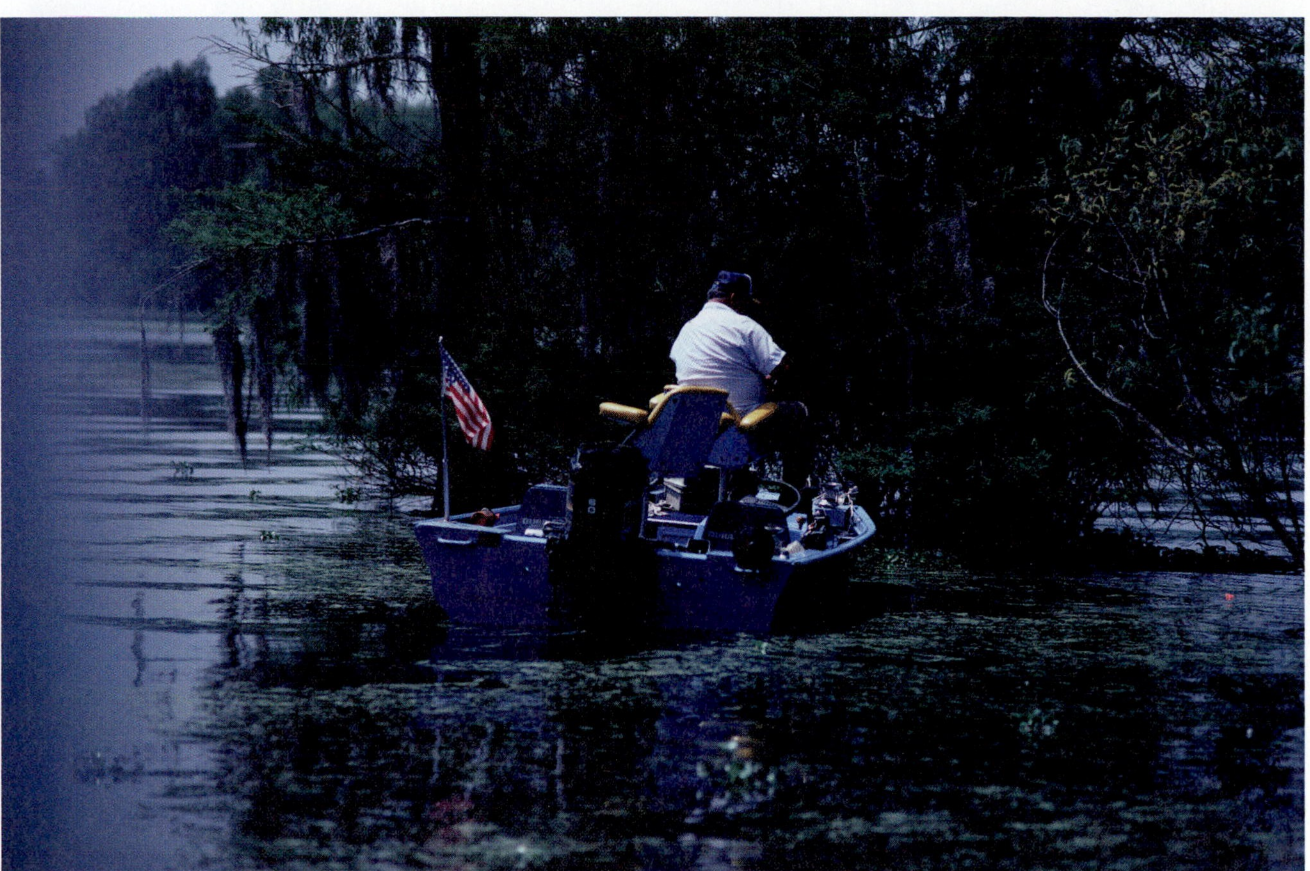

Figure 26–3 Large populations of aquatic plants and algae in a eutrophic lake as a result of the high amount of nutrients. *(Courtesy of PhotoDisc.)*

is the littoral zone, which supports many aquatic plants and animals. This zone is the most productive in the lake; it exists where sunlight can reach the lake bottom. The open waters of the lake, where light can penetrate and photosynthesis can occur, is called the limnetic zone. The depth of the limnetic zone can vary greatly depending on the lake. The clear waters of mesotrophic lakes can have deep limnetic zones reaching 100 feet or more. Cloudy eutrophic lakes can have a very shallow limnetic zone, where sunlight can only penetrate a few feet into the water. Very deep lakes may also possess a profundal zone sunlight cannot reach. In this zone the water is cold and is in perpetual darkness. Not all lakes have a profundal zone.

The bottom of the lake is called the benthic zone. The benthic zone can be home to many aquatic organisms, depending on the depth of the lake. Some glacial lakes have depths in excess of 600 feet. Many benthic zones of large lakes around the world have been virtually unexplored and may contain many unique forms of aquatic life. Lakes also posses a thermocline like the ocean. This is an area where cold deep water meets the warmer surface waters. Sometimes in early summer you can feel the thermocline when you jump into a lake, as your body plunges into the colder water below the surface. The thermocline is an important mixing zone where nutrients and oxygen are transferred from surface water to the deeper parts of the lake.

Watersheds and Rivers

The other classification of fresh surface water on Earth is known as flowing water. Flowing

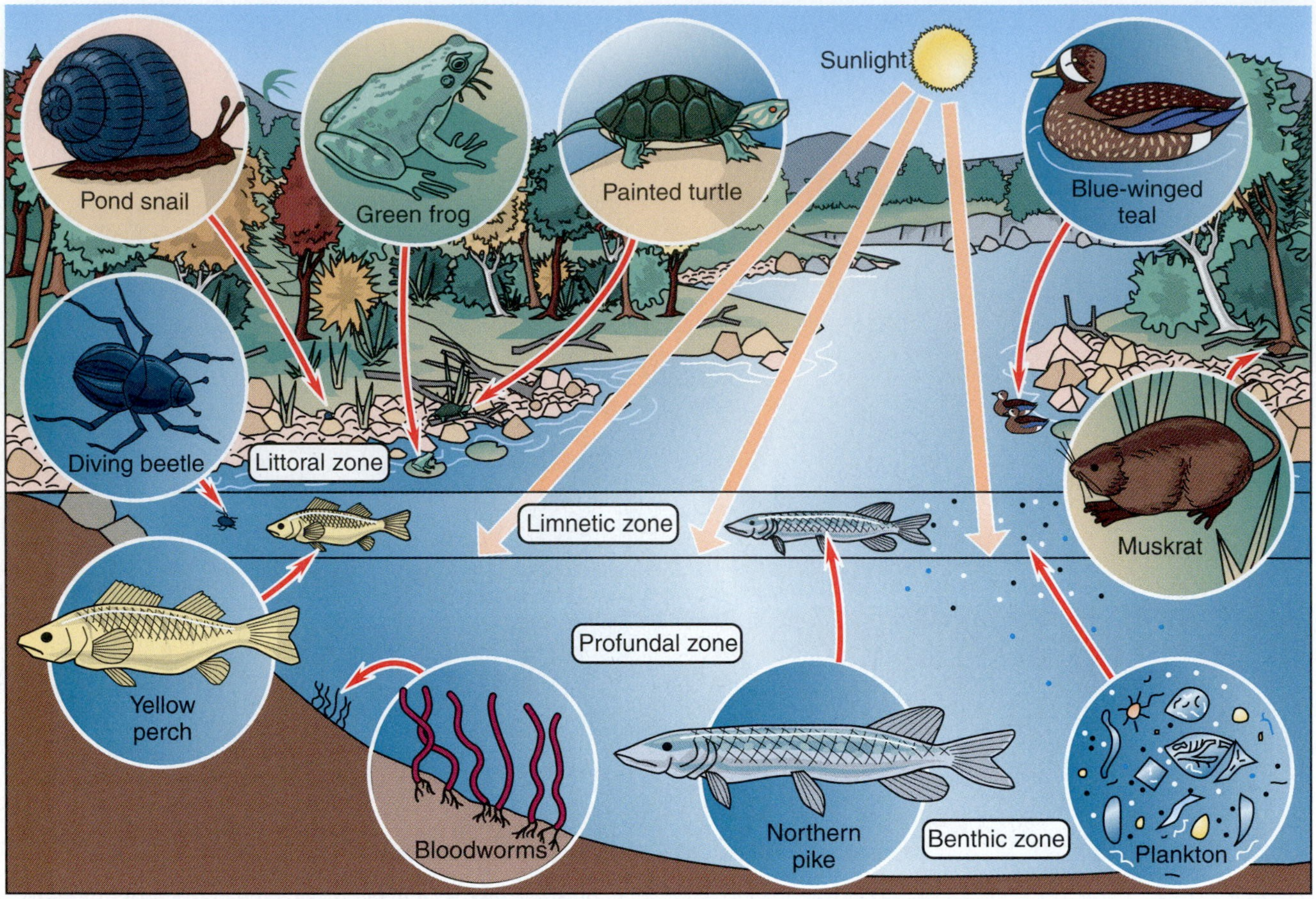

Figure 26–4 The aquatic life zones that exist in lakes.

Career Connections

Limnologist

Limnologists study the biological, chemical, geologic, and physical characteristics of inland freshwater systems. This includes all aspects of lakes, ponds, rivers, streams, and wetlands. Their research involves the circulation of water in a specific region, light transmission through water, bottom sediments, water temperature, seasonal changes, and the interaction of aquatic organisms with the physical aspects of freshwater systems. Limnologists also study the events that lead to the formation and evolution of inland bodies of freshwater. Current research also surrounds the effects of human society on freshwater systems, especially water use and pollution. Many of these aquatic scientists conduct research in the academic fields, but they can also find employment with state and local government agencies.

water is also known as rivers and streams. As precipitation falls onto the land surface, any water that does not infiltrate into the ground begins to flow toward lower elevations as a result of gravity. Eventually the flowing water, also called runoff, gathers in small creeks and streams. These streams begin to converge into larger flowing bodies of water called rivers, which are in constant motion flowing toward the sea. The total land area from which these flowing waters collect precipitation is called a drainage basin or **watershed** (Figure 26–5). Drainage basins can cover thousands of square miles for large river systems or a few square miles for small creeks. Small creeks and streams that contribute water to larger river systems are called tributaries. Most major river systems have thousands of smaller tributaries that feed water into them.

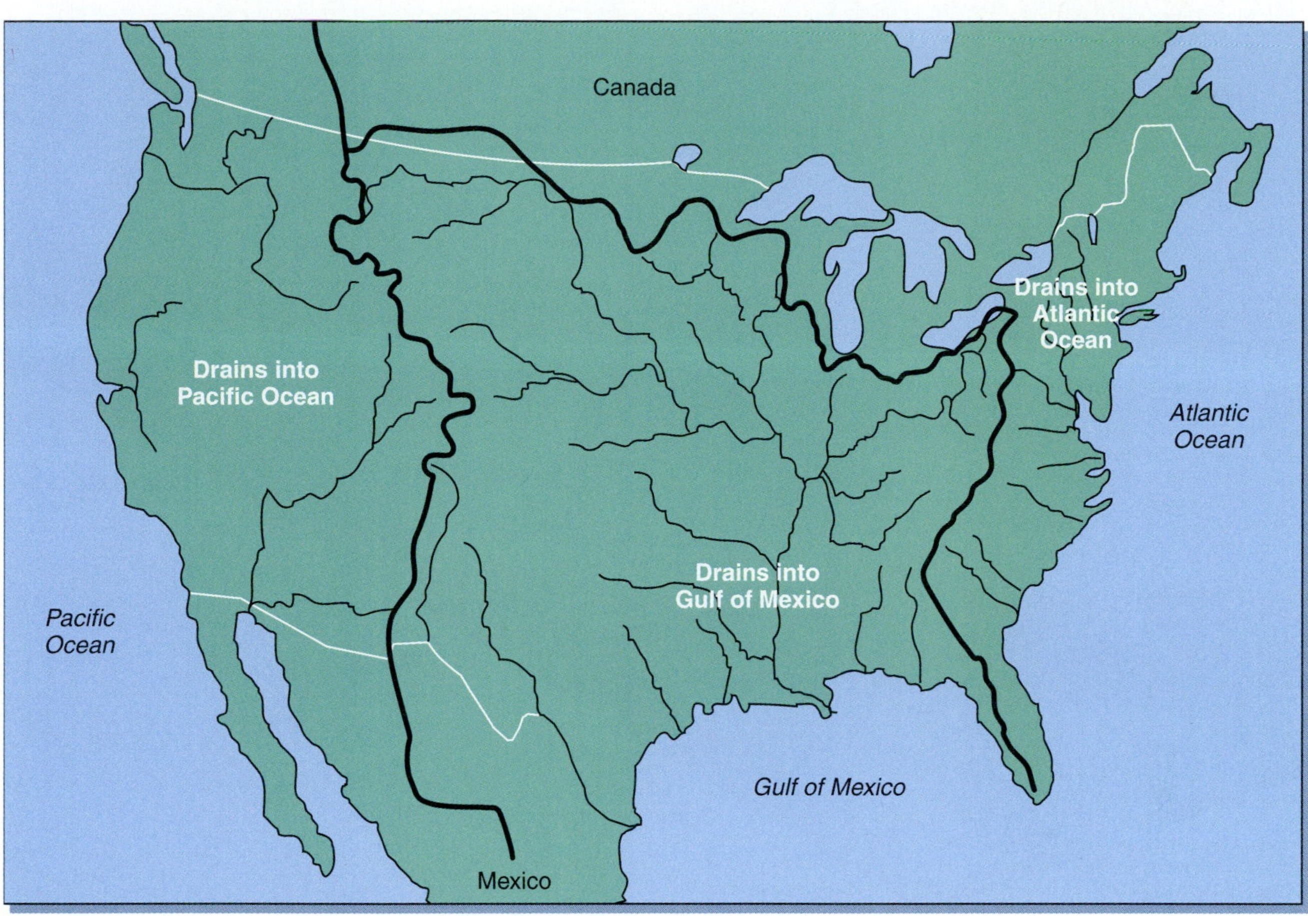

Figure 26–5 The drainage areas for the major river systems of the United States.

The largest river system in the United States is the Mississippi River, which stretches more than 3000 miles and drains water from more than one third of the United States. All river systems usually begin in areas of higher elevations, where creeks and streams collect melting snow or rain. This area is called the headwaters of a river system. Rivers are all powered by the force of gravity, which moves the water at a specific rate. This rate depends on a river's slope. The slope of a river is the elevation that the water drops over a specific distance (Figure 26–6).

Usually the headwaters of a river system have the greatest slope. The headwaters of the Arkansas River, in the Rocky Mountains of North America, drop approximately 4000 feet over a distance of 100 miles. The increased slope of a river's headwaters usually results in the formation of waterfalls and white-water rapids. As the river lowers in elevation, its velocity begins to decrease. This is because its slope is also decreasing. Eventually the river runs its course and empties into the ocean. All major river systems in the world terminate in an ocean. The location where a river meets the sea is called the mouth of a river and usually is marked by a wide delta. A delta is a large fan-shaped region of sediment that has been transported from the drainage basins of rivers (Figure 26–7). Deltas are usually getting larger as rivers continually transport sediments from inland areas. The principal action of river systems is to move water and sediment from

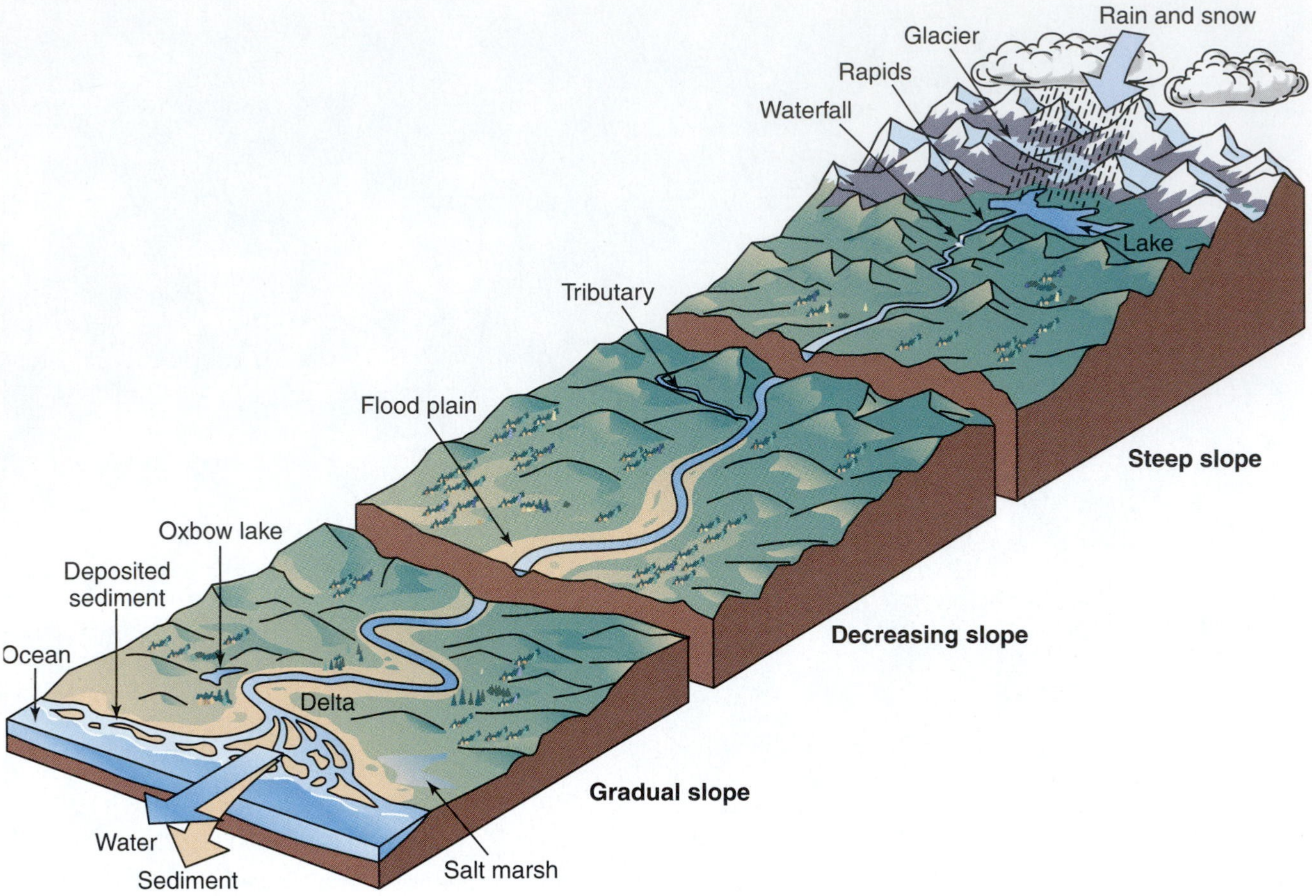

Figure 26–6 The slope of a river system gradually decreases from its source to its mouth.

Figure 26–7 Sediment is filling Lake Red Rock in central Iowa at rates much faster than anticipated. *(Courtesy of USDA/NRCS.)*

the inland regions of land masses and transport them to the sea.

Stream Features

Because river systems drop from areas of higher elevation to areas of lower elevation, they contain an abundance of energy. The power of flowing water has been used as a source of energy for hundreds of years. Many types of mills and factories were built next to rivers and streams to harness their immense power. The energy that river systems contain creates many unique stream features. Stream features are formed as a result of a river trying to dissipate its excess energy. The primary

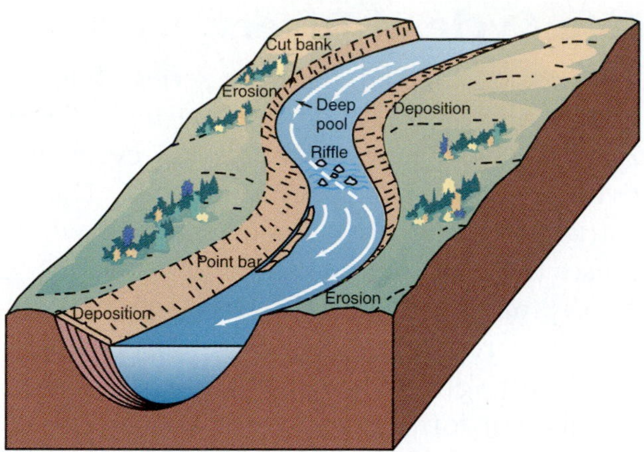

Figure 26–8 The meandering of a stream leads to the formation of deep pools, cut banks, point bars, and riffles as a result of the erosion and deposition of sediments by the flowing water.

way a river system dissipates energy is by **meandering.** Meandering is the series of S-shaped curves that make up a body of flowing water (Figure 26–8). Rivers rarely flow in a straight line, but tend to form a snakelike pattern. This meandering results in a continual pattern of erosion and deposition along the total length of a river.

Areas of erosion, or removal of rocks and soil along a river, are called cut banks. Cut banks are formed where the river cuts into the soil and rock along the bank and removes it. This area forms along the outside curve of a river. This is because the velocity of the water is greatest along the outside curve of a meander. This area of erosion also forms another unique stream feature called a deep pool. A deep pool is also formed on the outside curve of a meander where the river bottom is being eroded. Trout fisherman usually know the locations of these deep pools because this is where many fish like to reside on hot summer days. The force of the flowing water in the river moves the eroded material from the cut bank and deep pool downstream and deposits it as a point bar. A point bar is a shallow area of deposition located on the inside curve of a river. This is an area of deposition because the inside curve of a meander is where the velocity of the flowing water is at its least.

After the water moves through a meander, the river channel begins to straighten out. This part of a river is called a riffle. A riffle is a relatively straight shallow portion of a river located between meanders. Riffles are also areas of deposition where the material eroded from upstream is deposited. Riffles are shallow regions usually marked by white water and rapids. The riffle then enters into another meander and the whole process repeats itself. This series of erosions and depositions, which is caused by meandering, continues until the river reaches its mouth. The meandering process continually erodes and deposits rock and sediments all along a river and eventually transports the mountains to the sea.

A flowing river is in a state of **dynamic equilibrium,** because the erosion of material by the water is equal to the amount of material that it deposits. The amount of water that flows past a specific point in a river is called its **discharge rate.** Discharge rates are recorded in both cubic centimeters per second and gallons per second and are important indicators used for flood prediction (Figure 26–9).

Floodplains

Another important feature of a river system is called the floodplain. A **floodplain** is a gently sloping area that surrounds the **channel** of a river. The channel is the portion of the river that contains flowing water. Floodplains are formed by sediments that were deposited by a past flood event. Over time the sediments build up wide, flat areas alongside the river channel. Floodplains often contain a series of small steps that lead to the current river channel. These steps are called terraces and mark areas where the channel of the river was located during a past flood (Figure 26–10). Floodplains are usually wetland areas that contain low-growing vegetation. This is due to the periodic flooding that can occur along river systems. Because floodplains are

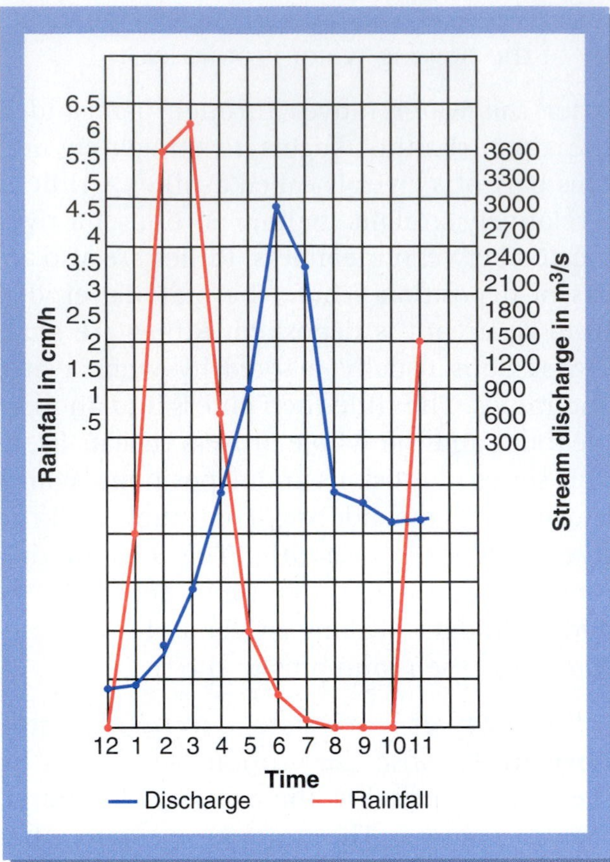

Figure 26–9 The relationship between rainfall and the discharge rate of a river shows the delay between the time of maximum rainfall and the time of maximum discharge.

naturally flat areas near flowing water, towns and farms often have been established near them. The periodic flooding that formed the floodplains can destroy lives and property located in these areas. It is important to identify floodplain areas to prevent the loss of life that occurs as a result of excess rainfall or snow melt.

Life Cycle of Rivers

All rivers go through a similar series of developmental stages called life cycles (Figure 26–11). The first stage of the life cycle of a river is called the youthful stage. This is marked by the river's steep slope, which causes the water to move at high velocity. The rapidly moving water cuts through the surrounding rock, forming deep canyons. The youthful river also has many rapids and waterfalls that form in a fairly straight channel, and they tend to cause more erosion than deposition. The rapidly moving headwaters of the Colorado River that cut through the mountains to form steep canyons are an example of a youthful river.

Eventually the slope of the river decreases, causing the velocity of the river to begin to slow. The river is now in the mature stage of development and begins to deposit sediments, forming a wide floodplain. The river also begins to form meanders as it cuts into the surrounding floodplain. The river is now moving through a wide river valley lined with gently sloping hills. The Hudson River is an example of a mature river that gently cuts through the wide, rolling hills of the Hudson Valley in New York.

The old age stage of river development occurs when the slope of the river is greatly reduced, causing it to flow slowly. Large meanders form in the flat surrounding floodplain. Some of these meanders get cut off from the main channel of the river to form what are called oxbow lakes. During the mature stage of development, deposition is occurring at a higher rate than

EARTH MATH

1) THE DISCHARGE RATE OF A RIVER IS DETERMINED BY MULTIPLYING THE VELOCITY OF THE WATER BY ITS CROSS-SECTIONAL AREA. IF A SECTION OF RIVER IS 75 FEET WIDE WITH AN AVERAGE DEPTH OF 5 FEET AND THE WATER IS FLOWING AT 0.7 FEET PER SECOND, WHAT IS THE DISCHARGE RATE FOR THIS RIVER?

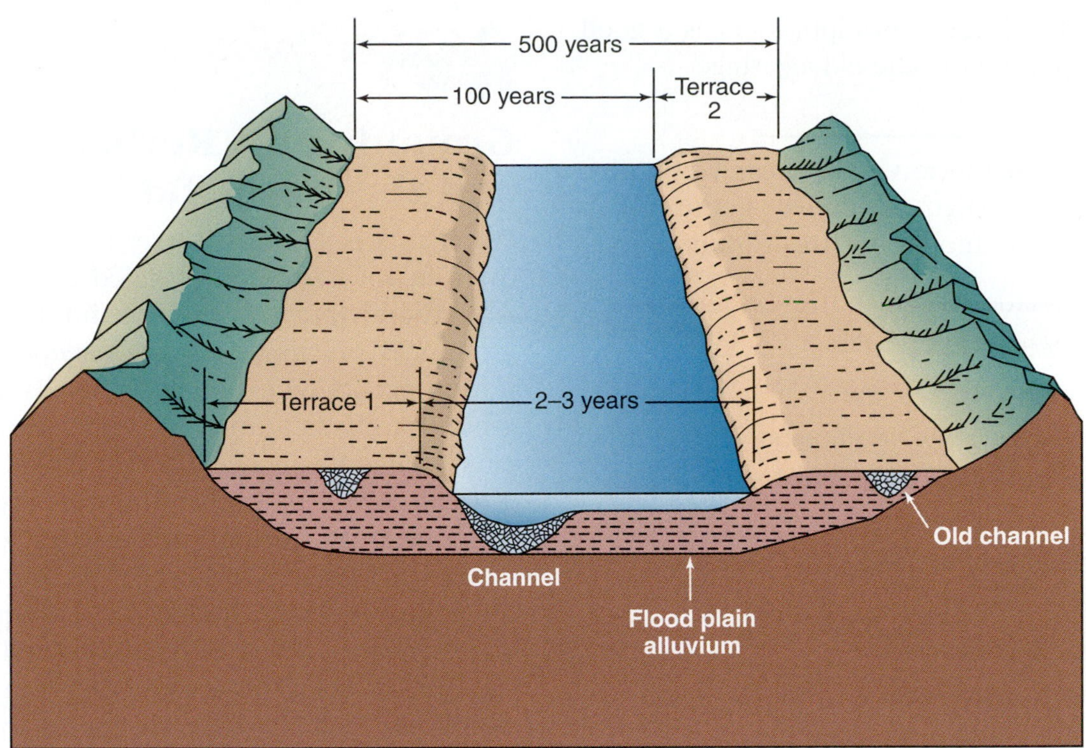

Figure 26–10 The cross section of a floodplain showing the location of the channel and terraces.

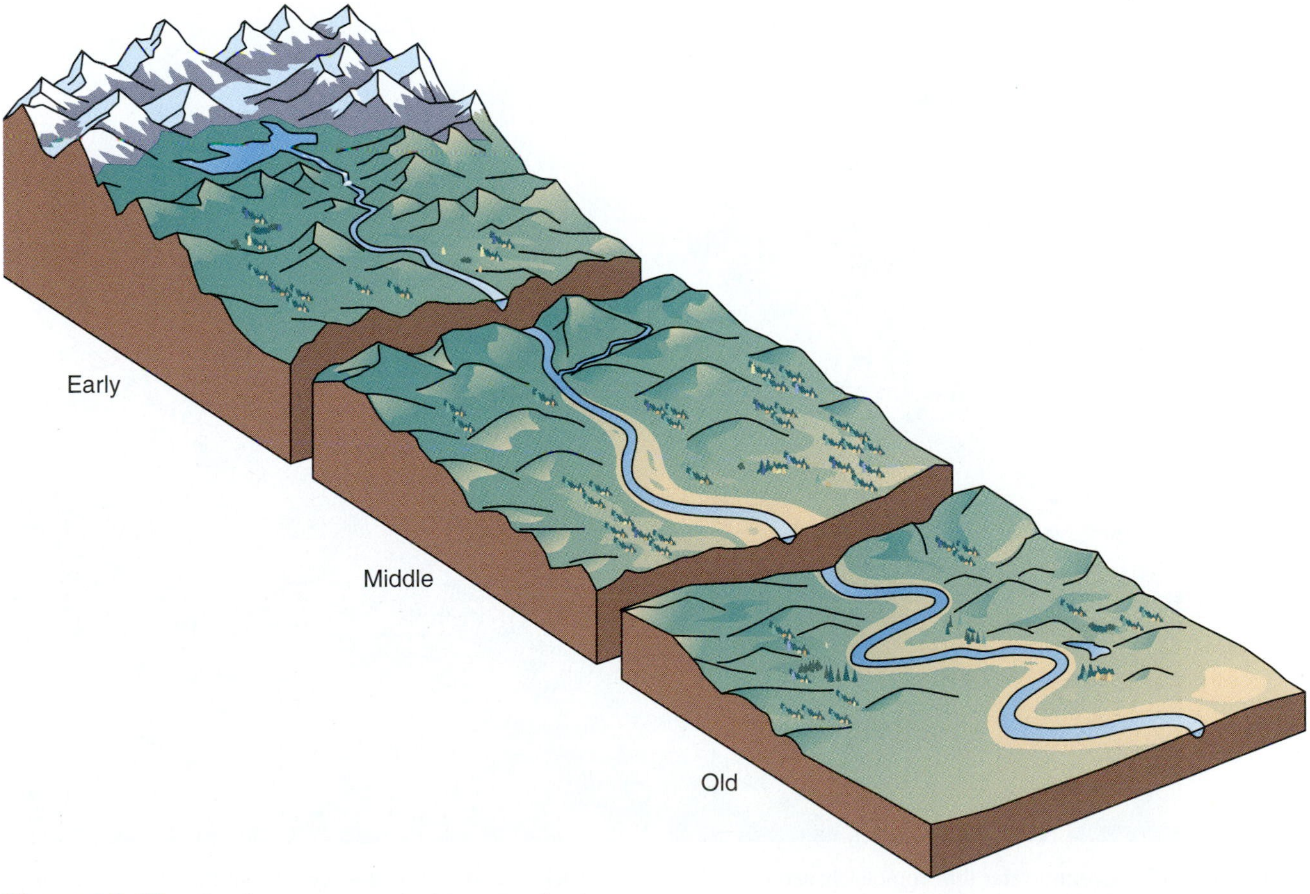

Figure 26–11 The three stages in the life cycle of a river.

erosion. The lower Mississippi River is a good example of a river in the old age stage.

For more information go to these Web links:
<http://www.epa.gov/students/wter.htm>
<http://mbgnet.mobot.org/fresh/index.htm>

SECTION REVIEW

1. List the three age classifications of lakes.
2. Describe the four zones found within a lake.
3. What causes rivers to flow?
4. Define the term *watershed.*
5. Describe the process of meandering.
6. What is a floodplain?
7. Who was Alexander Agassiz?

26.2 *Groundwater*

Groundwater Recharge and the Water Table

Water that is stored in pores and crevices located in rock and soil is called groundwater. All groundwater enters, or infiltrates, the ground as precipitation or as seepage. Infiltration occurs when precipitation falls to the ground and soaks into the soil. Seepage occurs as surface waters slowly leak into the ground. Areas of infiltration are also called recharge areas, because this is where groundwater supplies are recharged. As water infiltrates into the soil, the force of gravity drives it downward, where it begins to collect and saturate the soil. This area is called the zone of saturation and is where the pores in soil or rock underground are completely filled with

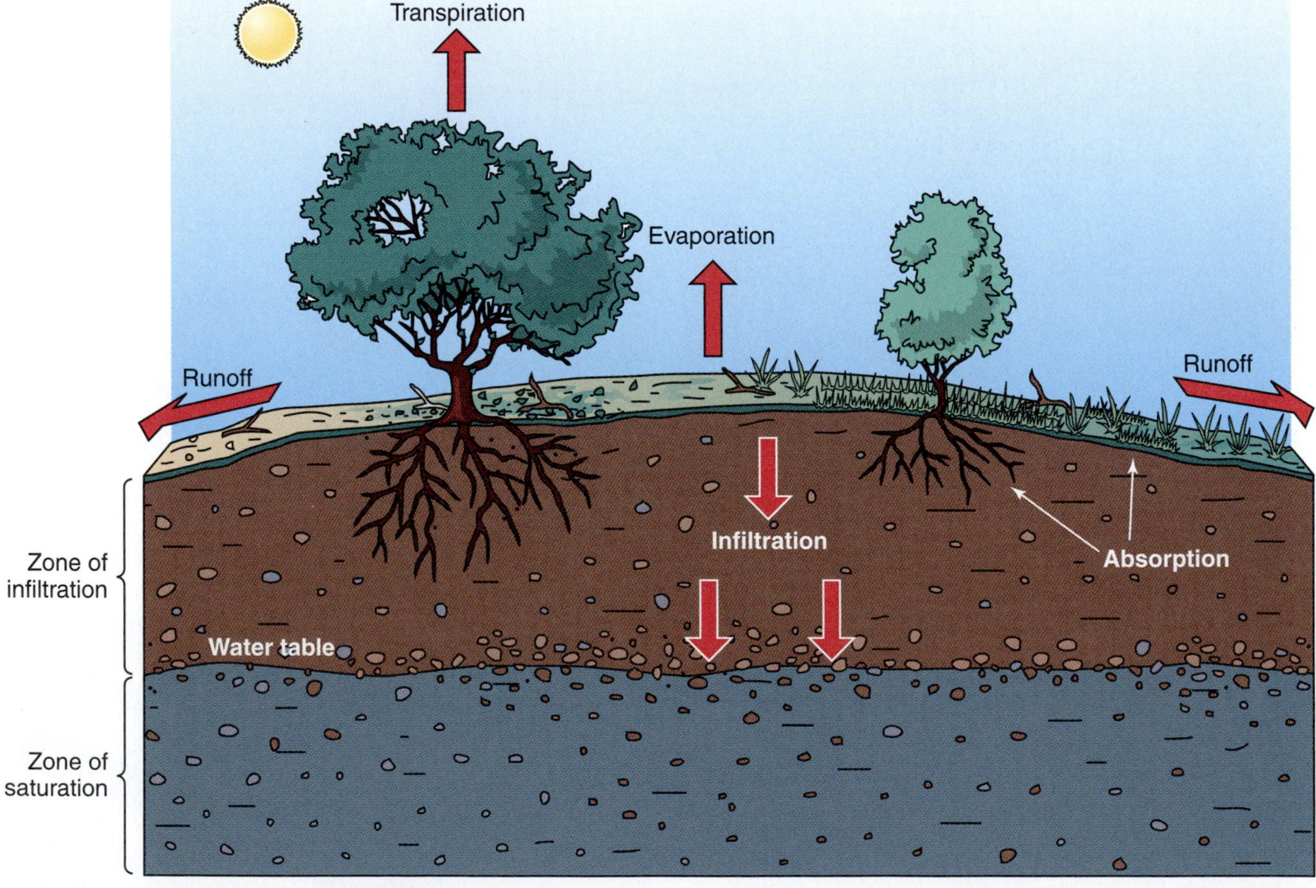

Figure 26–12 Groundwater that completely saturates the pores in rock and soil forms the zone of saturation. The top of the zone of saturation is known as the water table.

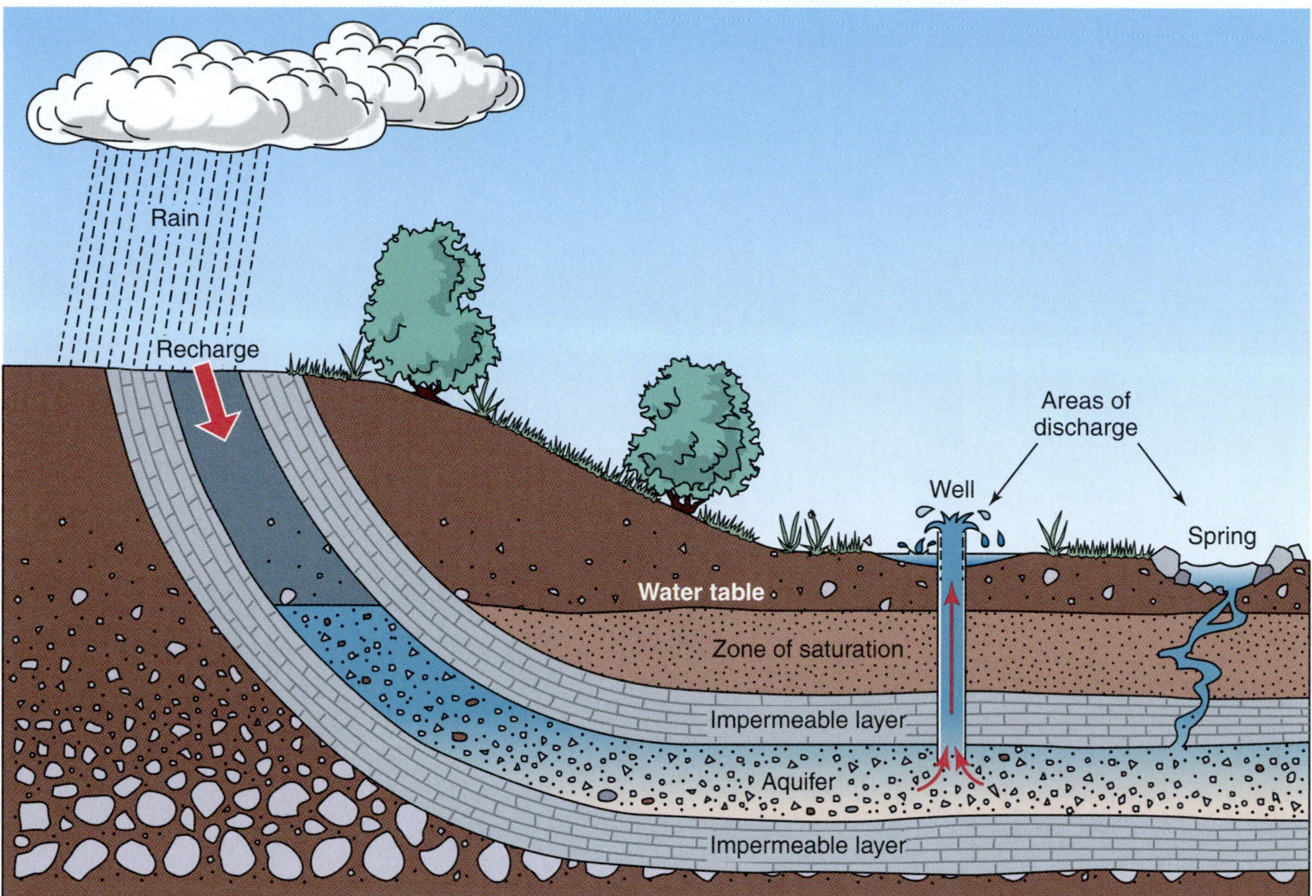

Figure 26–13 Groundwater flows from areas of recharge to areas of discharge.

water (Figure 26–12). The zone of saturation varies in its depth depending on the local climate. In some areas that receive high amounts of rainfall, the zone of saturation can be located only 8 to 10 feet below the surface. Other dry areas might have very deep zones of saturation. The top portion of the zone of saturation is called the **water table.** The water table can also vary in its depth seasonally. During rainy seasons the water table can move up toward the surface as infiltration is increased. Then during the drier months the water table may descend deeper as the infiltration rates decline.

Groundwater Flow

All groundwater, like flowing surface water, is in a constant state of movement, called groundwater flow. The flow of groundwater moves by the force of gravity; therefore it travels from areas of high elevation to low elevation. Eventually all groundwater exits the ground and returns to the surface. The area where groundwater reaches the surface is called the area of discharge (Figure 26–13). Common areas of discharge are flowing springs, where cool fresh groundwater flows out of the ground. During times of the year when there are low amounts of rainfall, the groundwater that is discharged from flowing springs continues to feed water into streams and rivers. This is what keeps these bodies of water flowing throughout the year. The time and distance that it takes for groundwater to travel from an area of recharge to an area of discharge can vary greatly and depends on the soil or rock in which it is flowing. Groundwater located in

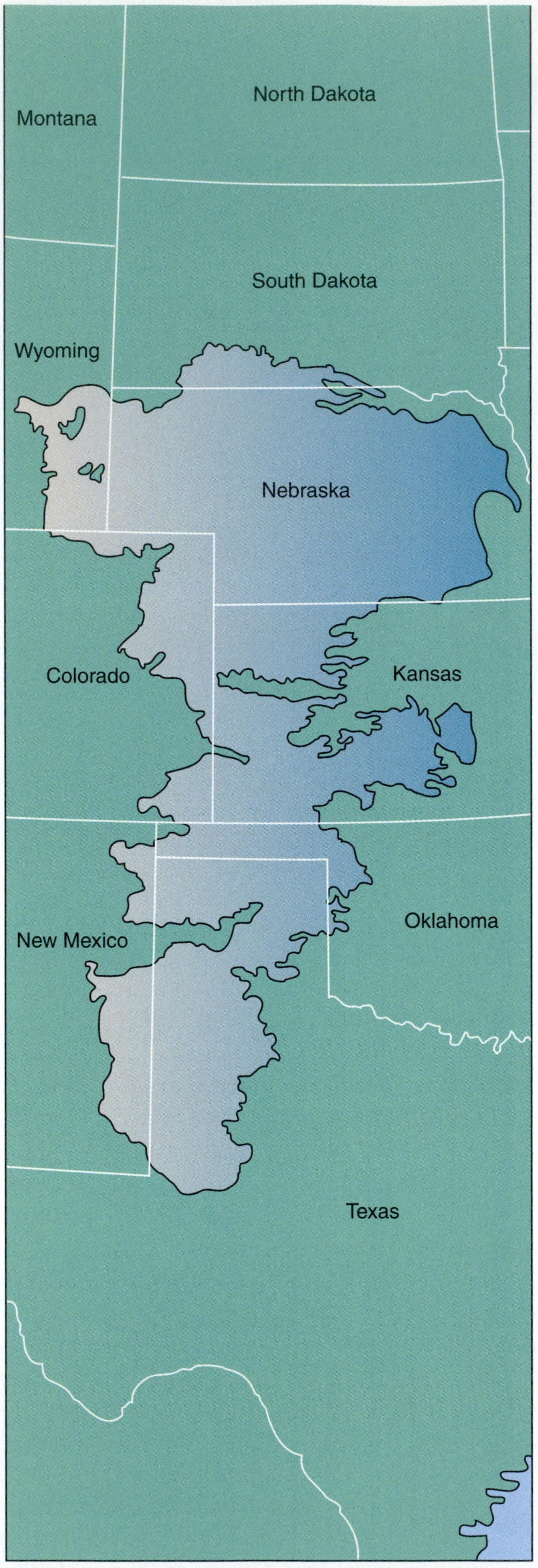

Figure 26–14 The Oglalla aquifer stores millions of gallons of water in the ground under large parts of the Midwest.

large porous rock formations can flow at a very high rate. If the groundwater is located in small, tightly packed pores of rock and soil, it flows at a much slower rate.

Aquifers and Groundwater Discharge

Another important aspect of groundwater is called the **aquifer.** Aquifers are areas of porous rock that store large amounts of groundwater. One of the largest aquifers in the United States, the Oglalla aquifer, underlies more than 174,000 square miles of Texas, New Mexico, Oklahoma, Kansas, Colorado, Nebraska, Wyoming, and South Dakota (Figure 26–14). This aquifer holds approximately the same amount of water as Lake Huron. The average thickness of the zone of saturation that makes up the Oglalla aquifer is approximately 200 feet. Most of the water that is stored in this huge aquifer infiltrated the ground as glaciers began to melt approximately 10,000 years ago.

Some aquifers are located between two layers of water-impermeable rock. **Impermeable** means that water cannot infiltrate the rocks. The trapped water between the two rock layers is called a **confined aquifer** (Figure 26–15). Wells drilled into confined aquifers are called flowing or **artesian wells,** because the trapped water becomes pressurized and flows forcefully from underground.

Groundwater Pollution

Unfortunately, many groundwater supplies can be easily contaminated or polluted. When

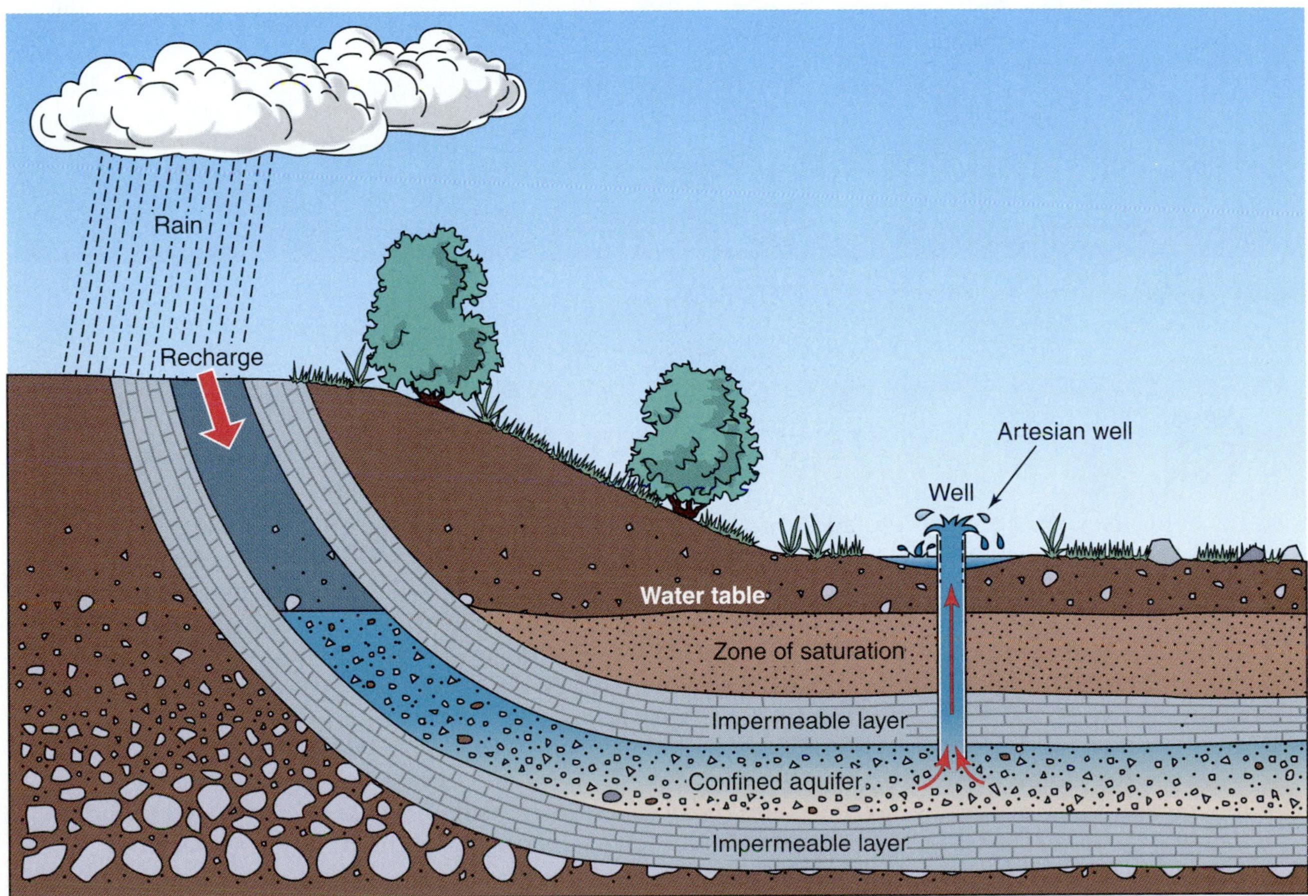

Figure 26–15 The location of a confined aquifer between two layers of impermeable rock.

water infiltrates the ground, it can carry with it many dissolved pollutants, such as chemicals, fertilizers, and heavy metals. Landfills, leaking underground chemical storage tanks, manure piles, and leaking septic systems are all sources of groundwater pollutants (Figure 26–16). Groundwater pollution can often be the most difficult form of water pollution to

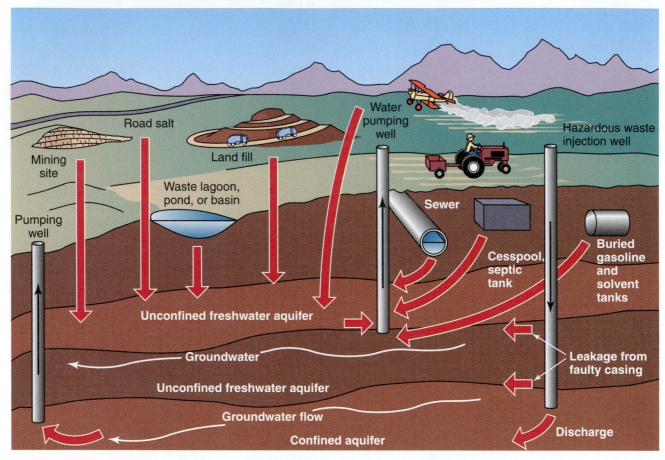

Figure 26–16 The main sources of groundwater pollution.

EARTH SYSTEM SCIENTISTS *RACHEL CARSON*

Rachel Carson was born in Philadelphia in 1907. She received her education in zoology and worked at the Woods Hole Oceanographic Institute as a teacher and researcher. She also became the head biologist for the U. S. Fish and Wildlife Service, where she wrote fisheries information for the government. Eventually she became a published writer of books that presented the biology of the sea. In 1962 her world-famous book *Silent Spring* was published. In this groundbreaking publication, Carson presented the ill effects of pesticides on the food chain. She wrote of the evils of DDT, a widely used pesticide that was moving through groundwater and surface waters and eventually entering the bodies of living organisms. Her study on the effects of pesticides on the environment led to the birth of the environmental movement in the United States. Because of Carson's research, the world was introduced to the dangers of chemicals being spread throughout the environment and the problems associated with industrialization and water resources.

Career Connections

HYDROGEOLOGIST

A hydrogeologist, also called a groundwater hydrologist, studies and monitors the world's groundwater resources. This includes the tracking of how much groundwater is stored in a particular region. These scientists determine the amount of groundwater that is available for use and the rate at which it is being used or replenished. They also help to locate underground water sources and determine where to drill wells. These researchers study the way that groundwater flows underground, from areas of recharge to areas of discharge. Hydrogeologists also research the pollution of groundwater resources. This aspect of their field involves tracking polluted groundwater, locating the source of pollution, and cleaning contaminated groundwater. Hydrogeologists study ways in which the pollution of groundwater can also be prevented. A career in this field requires a college degree with an emphasis on geology and hydrology. Hydrogeologists can find work in private industry or for state and local governments.

detect and clean up. This is because groundwater is not visible, which makes it difficult to locate the source and determine the extent of the groundwater pollution. The best way to protect the quality of groundwater is to prevent pollutants from entering the ground.

For more information go to this Web link:
http://www.epa.gov/students/water.htm>

SECTION REVIEW

1. What percentage of the United States' drinking water supply comes from groundwater?
2. Define the term *infiltration.*
3. What are the two areas called where groundwater enters and then leaves the ground?
4. Where is the water table located?
5. Define the term *confined aquifer.*
6. Who was Rachel Carson?

CHAPTER SUMMARY

The world's freshwater supply that is not locked up in glacial ice makes up approximately 1% of all the water on the planet. This water exists on the Earth in lakes and rivers and as groundwater. Lakes are large inland bodies of water that are often classified by their level of productivity. An oligotrophic lake is a cold and deep lake with crystal clear water that has very low productivity as a result of the lack of available nutrients. These lakes support a small amount of aquatic organisms and are often formed by melting glaciers. Mesotrophic lakes are cool, deep lakes with clear water and medium productivity. These lakes support a healthy population of both aquatic plants and animals as a result of the increased amount of nutrients in

EARTH MATH

1) HOW MANY GALLONS OF WATER ARE PUMPED FROM THE GROUND FOR DRINKING WATER AND IRRIGATION IN THE UNITED STATES EACH WEEK?
2) IF 1 CUBIC FOOT OF WATER EQUALS APPROXIMATELY 7 GALLONS OF WATER, HOW MANY GALLONS OF WATER DOES AN AQUIFER HOLD IF IT HAS AN AREA OF 12.5 MILLION CUBIC FEET?

the water. Eutrophic lakes are shallow, warm-water lakes that are very cloudy and have a muddy bottom. These lakes support a high number of aquatic plants and algae as a result of the increased amount of nutrients in the water.

Lakes also have unique aquatic life zones similar to those in the ocean. The area near the shore, where light can penetrate all the way to the lake bottom, is called the littoral zone. The limnetic zone is located out in the deeper water of the lake, from the surface to the depth where photosynthesis can occur. Some deep lakes have profundal zones where no light reaches, making it perpetually dark. The bottom of the lake is called the benthic zone. The depth of the lake at which a rapid drop in the water temperature occurs is called the thermocline.

Flowing bodies of water on the Earth's surface are called streams and rivers. They flow over the surface of the land by the power of gravity from areas of high elevation to areas of lower elevation. The total land area that a river system drains is called a watershed. All flowing bodies of water have a characteristic S-shaped curves called meanders. These form as a result of the flowing water trying to dissipate its energy. The meandering of a flowing body of water creates unique stream features. These include areas of erosion called cut banks and deep pools, which form on the outside curve of a meander. The point bar is a stream feature that forms on the inside curve of a meander where sediment is deposited. A riffle is the relatively straight, shallow area of deposition downstream from a meander.

A flowing body of water such as a stream or river is in a state of dynamic equilibrium, which means that its rate of erosion is equal to its rate of deposition. The amount of water that is flowing past a particular point in a river or stream is called its discharge rate.

All rivers go through a series of stages that together make up the life cycle of a river. The youthful stage is marked by the river's steep slope, which causes the velocity of the water to increase. This forms rapids and waterfalls and creates deep, narrow canyons through which the water flows. The mature stage of a river occurs when the slope is reduced and the river begins to form meanders that cut into its flat floodplain. Mature rivers also flow through wide river valleys. The old age stage of a river is marked by a very gradual slope, which causes the river to flow slowly. Large meanders develop as the river cuts through the wide floodplain. Some of these meanders get cut off from the main channel, forming oxbow lakes.

Water that seeps into the ground from surface water or from the infiltration of precipitation is called groundwater. Groundwater exists in the pore spaces of soil and rock. The top of the area where all the pore spaces are filled with water is called the water table. This important freshwater resource flows from areas of high elevation to areas of low elevation. The rate at which groundwater flows depends on the size and distribution of pore spaces within rock and soil.

A large amount of groundwater stored in rocks and soil is called an aquifer. Aquifers that are trapped between two layers of impermeable rock are known as confined aquifers. All groundwater eventually flows out of the ground from an area of discharge. These are commonly known as flowing springs. Groundwater supplies approximately 40% of drinking water in the United States; it can be polluted by leaking fuel tanks, fertilizers, sewage, and toxic chemicals.

CHAPTER REVIEW

Multiple Choice

1. Which of the following is a characteristic of a eutrophic lake?
   a. clear water and cold water
   b. little aquatic life
   c. low productivity
   d. high amounts of nutrients

2. A clear lake with medium productivity that supports a variety of aquatic life is called:
   a. oligotrophic
   b. mesotrophic
   c. eutrophic
   d. autotrophic

3. Which aquatic life zone exists in deep water where photosynthesis can occur?
   a. limnetic zone
   b. littoral zone
   c. profundal zone
   d. benthic zone

4. What aquatic life zone supports plants growing on the bottom of the lake?
   a. limnetic zone
   b. littoral zone
   c. profundal zone
   d. aphotic zone

5. Approximately one third of the United States is part of the Mississippi River's:
   a. watershed
   b. discharge area
   c. recharge area
   d. aquifer

6. An area of deposition located on the inside curve of a river's meander is called a:
   a. deep pool
   b. cut bank
   c. point bar
   d. riffle

7. This area of erosion located on the bottom of the outside curve of a meander is where a fisherman would most likely find trout on a hot day:
   a. deep pool
   b. cut bank
   c. point bar
   d. riffle

8. Rapidly moving water that cuts a deep, narrow canyon in the surrounding rock is in what stage of the life cycle of a river?
   a. youth
   b. maturity
   c. old age
   d. rejuvenation

9. The Hudson River in New York is an example of what type of river?
   a. youthful
   b. mature
   c. old age
   d. rejuvenated

10. When the amount of erosion equals the amount of deposition in river, it is:
    a. youthful
    b. in dynamic equilibrium
    c. mature
    d. meandering

11. What decreases in depth during wet seasons and increases its depth during dry seasons?
    a. soil pores
    b. the water table
    c. artesian wells
    d. flowing springs

12. Groundwater flows more rapidly when:
    a. pore spaces are large and the rock is permeable
    b. pore spaces are small and the rock is permeable
    c. pore spaces are large and the rock is impermeable
    d. pore spaces are small and the rock is impermeable

13. A large amount of water stored in the ground below impermeable rock is known as:
    a. soil pores
    b. the water table
    c. a confined aquifer
    d. flowing springs

14. Approximately what percentage of Americans get their drinking water from groundwater?
    a. 10%
    b. 20%
    c. 30%
    d. 40%

continued

Matching *Match the terms with the correct definitions.*

a. oligotrophic lake
b. mesotrophic lake
c. eutrophic lake
d. watershed
e. meandering
f. dynamic equilibrium
g. discharge rate
h. floodplain
i. channel
j. water table
k. aquifer
l. impermeable
m. confined aquifer
n. artesian wells

1. ____ The flat area of a river valley, located along both sides of a river channel, that is formed from the deposition of sediments during periodic floods.

2. ____ The total land area that is drained by a particular river system.

3. ____ A free-flowing well that discharges water from the ground that is recharged from a higher elevation.

4. ____ A classification of lake that has been recently formed from glacial melt waters; it has very clear, cold water and is very low in nutrients or aquatic life.

5. ____ The top of the zone of saturation.

6. ____ The portion of a moving body of water where water is currently flowing.

7. ____ Groundwater that is located below an impermeable rock layer.

8. ____ The classification for a middle-aged lake that is relatively clear, deep, and low in available nutrients.

9. ____ A term meaning *unable to pass through,* such as certain rocks that do not allow water to pass through them.

10. ____ A classification for a lake that is relatively cloudy, warm, and shallow and has an abundance of nutrients that support a large population of aquatic plants and animals.

11. ____ Large amounts of water stored in porous or fragmented rock underground.

12. ____ The reoccurring S-shaped curves of a river or stream.

13. ____ A balance between two opposing processes that occur at the same rate in an energetic system, such as a river or stream.

14. ____ The amount of water passing by a particular point in a flowing body of water.

Critical Thinking

1. If toxic chemicals were mixed in with river sediments, what do you think would happen to them over time?

CHAPTER

27

Glaciers

Objectives

Anatomy of a Glacier • Glacial Movement and Moraines • Types of Glaciers • Glaciers and Global Climate

After reading this chapter you should be able to:

- Define the term *glacier.*
- Differentiate among the zones of accumulation, flowage, and ablation on a glacier.
- Explain the process that causes a glacier to flow.
- Describe the location and formation of terminal, medial, and lateral moraines.
- Identify the five different types of glaciers and their unique characteristics.
- Describe the processes that lead to glacial advance and glacial retreat.
- Explain the relationship among global climate, glaciers, and sea level.
- Identify how much sea level has risen over the past 100 years and how much it is expected to rise in the future.

TERMS TO KNOW

glacier
zone of accumulation
glacial front
zone of flowage
glacial till
moraines
terminal moraines
zone of ablation
lateral moraines
medial moraines
icebergs
glacial advance
glacial retreat

INTRODUCTION

The largest storehouse of freshwater on Earth is located in the world's glaciers. Approximately one tenth of all the land surfaces on the Earth are covered in glaciers, and as recently as 11,000 years ago, more than 30% of the land surface was located under glacial ice. The entire continent of Antarctica is covered entirely by ice. The average thickness of the glacial ice there is approximately 7000 feet, with the thickest ice being more than 14,000 feet. Recent investigations of the glaciers that exist around the world suggest that they are all shrinking. Scientists fear that the melting of the world's glaciers might indicate a change in the global climate.

The amount of water that the world's glaciers hold is more than two times as much as all the world's groundwater and surface freshwater supplies combined. Scientists are not sure of what the impact would be on the environment if all this water were to melt. During the height of the last ice age approximately 18,000 years ago, global sea level was almost 400 feet lower than it is today. This reveals the important relationship between glaciers and sea level. Understanding the processes that form glaciers and how they interact with the environment is an important aspect of Earth system science.

Anatomy of a Glacier

Glaciers are large masses of ice formed by the compaction of snow over long periods. Glaciers form in the higher latitudes, where temperatures remain below freezing for much of the year and where more snow accumulates than melts. Eventually the snow builds up and becomes compressed by the accumulating weight of the snow that continues to collect on top of the glacier. The compressed snow begins to crystallize into glacial ice. As the ice begins to build up, crystallizing ice begins to change from the white color of snow to the bluish tint of glacial ice. The area where snow falls to form the glacier is called the **zone of accumulation.** This area is often located in higher elevations, where the temperature remains below freezing throughout the year, allowing snow to continually accumulate. When the ice reaches a thickness of more than 100 feet, the pressure from the zone of accumulation begins to force the ice below to flow outward. This causes glaciers to advance, or move away from the zone of accumulation. The flow of a glacier can be demonstrated by observing a piece of Play-Doh. If you took a ball of Play-Doh, laid it down on a table, and pressed your hand down on top of it, it would begin to spread apart all along its sides. This is the same way that a glacier flows. The weight from the zone of accumulation presses downward and causes the glacial ice to spread outward.

The advancing edge of glacier is called the **glacial front** (Figure 27–1). The lower portion of a glacier is the portion that undergoes the most movement or flow. This **zone of flowage** varies in its rate of movement by a few inches or up to 25 feet per day. Some areas of the Antarctic have zones of flowage that are rapidly transporting glacial ice toward the sea. Studies of these glaciers revealed that a layer of pressurized liquid was located below the glacier, which causes the ice to flow rapidly over the surface. These

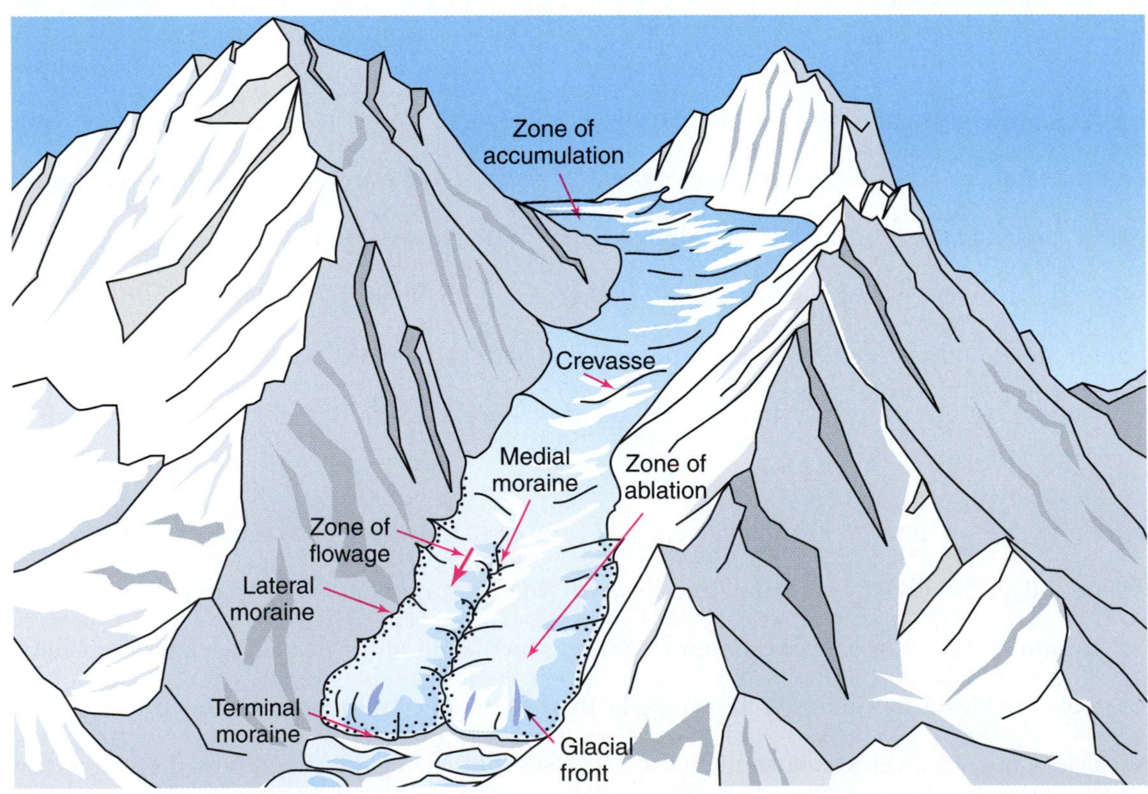

Figure 27–1 The main features of a glacier.

portions of Antarctica move like rivers of ice across the frozen landscape.

Glacial Movement and Moraines

As glaciers flow outward from their zone of accumulation, they usually move toward areas of lower elevation. Much like flowing rivers, glaciers tend to push their way through mountain valleys and eventually reach the oceans. As glaciers move through these valleys, their immense weight and forward movement grind up rocks and soil much like a large bulldozer. These rocks and debris become mixed up within the glacial ice to form **glacial till.** When the glacier begins to melt away, the glacial till is left behind as large mounds of crushed rock and soil deposits called **moraines.**

Terminal moraines are formed when the leading edge of the glacier begins to melt and retreat. This deposits large mounds of glacial sediments on the land surface that marked the position of the glacial front. The leading edge of a glacier is also known as the **zone of ablation.** Ablation is the loss of ice from the glacial front. This can occur on land or in water. **Lateral moraines** are formed along the sides of glaciers as they scrape the edges of mountain valleys. These result in long, dark lines of sediment that form along the side of glaciers as they remove rock from the valley walls (Figure 27–2). **Medial moraines** are

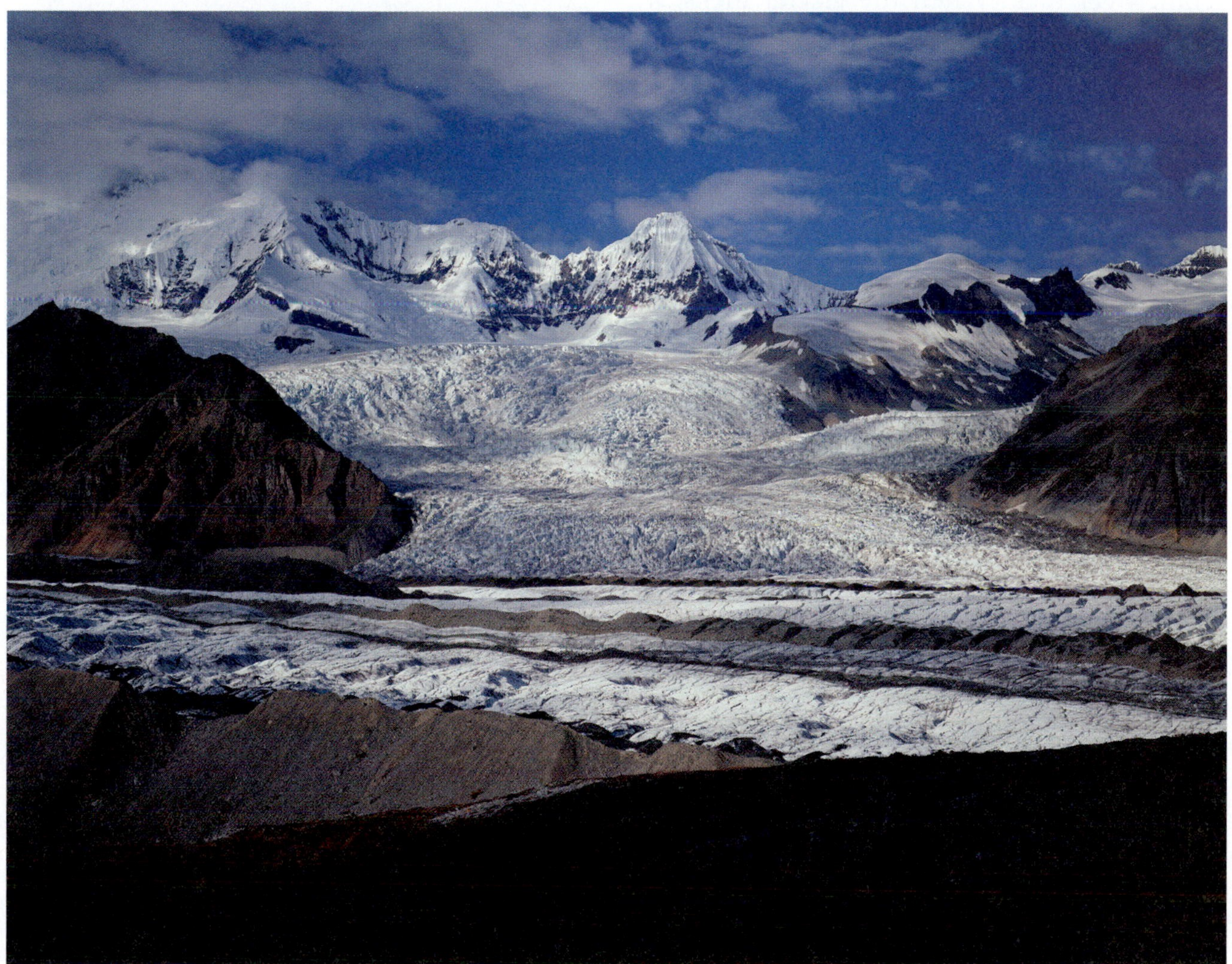

Figure 27–2 Lateral moraines form along the sides of a valley glacier as it scrapes along the valley walls. *(Courtesy of PhotoDisc.)*

formed in the center of glaciers where two separate valley glaciers join together to form one large glacier. Glacial moraines result in unique patterns in glacial ice that can often be used to identify its pattern of movement. In many parts of the United States, the remains of these moraines identify areas where glaciers once existed. Today many of these moraines are used as a source for gravel in construction. Glacial moraines can be distinguished because they consist of angular, unsorted sediments.

Some glaciers move large boulders in their ice that can be deposited when they melt. These large boulders are called glacial erratics, because they often exist in random locations. When rocks become trapped in glacial ice near the bottom of the glacier, they move and scrape over rock. When the glacier melts, it leaves behind characteristic scrapes on bedrock that can be used to identify the direction and movement of glaciers in the past. These marks still exist today in many rocks exposed at the Earth's surface, which reveal the movement of glacial ice during the last ice age.

> **Career Connections**
>
> **GLACIOLOGIST**
>
> A glaciologist studies the physical aspects of snow and ice. This includes research involving ice sheets, glaciers, and ice fields all over the world. These scientists monitor the changes that occur within glaciers, especially if they are advancing or retreating. Glaciologists also map glaciers to determine their size and elevation and study how glaciers move and flow. Their research is especially important in monitoring the effects of global climate change. Glaciologists must work in the often harsh, cold environments where glaciers are found. Much of their work involves collecting data on, around, and within a glacier. They take core samples of ice from glaciers to study their age and composition. A glaciologist requires a college background in geology and often works in research for colleges or government agencies.

Types of Glaciers

Glaciers can be divided into different categories, usually classified by their location and size. The largest glaciers, called continental glaciers, cover entire continents; they are also known as ice sheets. The two continental glaciers on Earth are located in Antarctica and Greenland. The Antarctic continental glacier covers more land area than the United States and Central America combined, at an average thickness of more than 7000 feet (Figure 27–3). The thickest portion of the Antarctic ice sheet is more than 2.5 miles!

The next classification of a glacier is called an ice field (Figure 27–4). Ice fields are large areas of connecting glaciers where only the tops of mountains extend from the surface of the ice. Ice fields are located in high-latitude mountain regions, such as the Juneau ice field in Alaska.

Ice caps are large glaciers that form over the top of a mountain peak and flow outward down the slopes of the mountain. Many ice caps exist in Iceland and usually appear like large, smooth domes. Valley glaciers form in mountain valleys as ice caps flow down into lower elevations (Figure 27–5). These are also called alpine glaciers and resemble rivers of ice flowing toward the sea. The last type of glacier is the piedmont glacier, which is formed from many valley glaciers joining together in a large, gently sloping plain. Piedmont glaciers can be found along the coast in Alaska.

The glacial front can occur in one of two locations: the land or the ocean. When the glacial front occurs in the ocean, large chunks of glacial ice break off the glacial front and float away as **icebergs.** The breaking off of large chunks of glacial ice to forms icebergs is called calving. When the glacial front occurs on

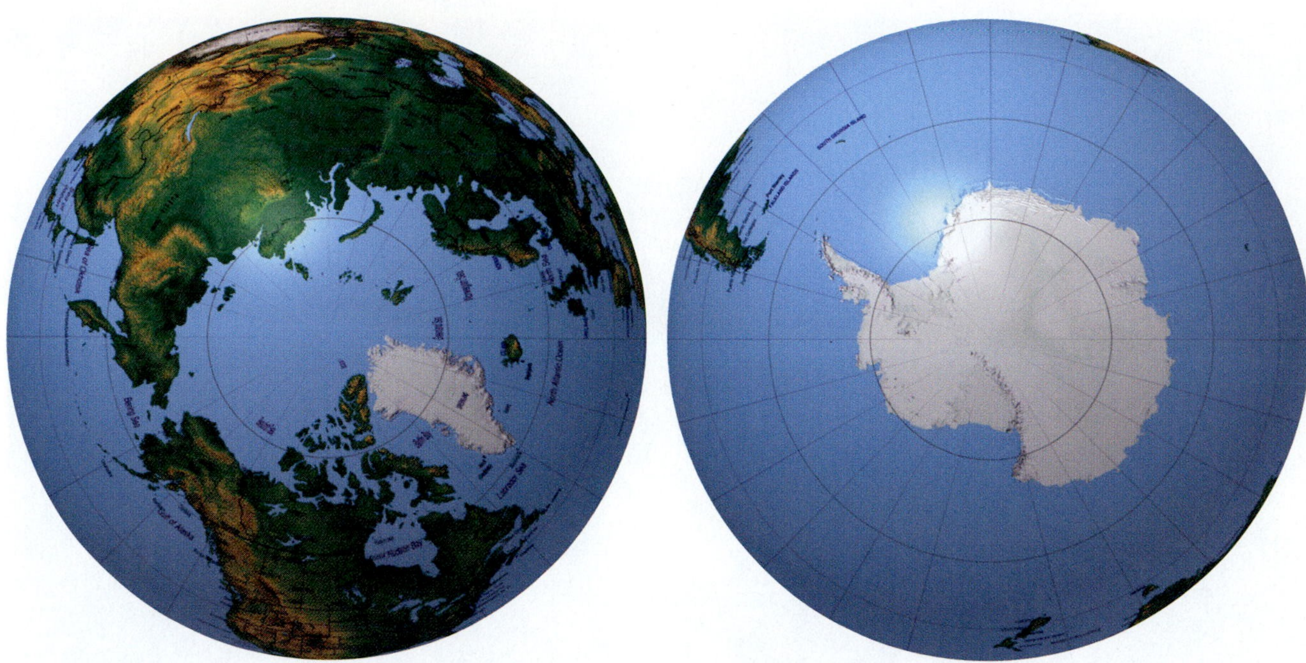

Figure 27–3 Ice sheets cover both Greenland and Antartica. *(Courtesy of EyeWire.)*

land, the melting glacial ice forms huge glacial lakes that flow into cold subglacial rivers. Some lakes formed from melting glaciers have characteristic bowl-like shapes that result from the deposition of glacial sediments around a large melting chunk of ice. These are called kettle lakes.

Glaciers can be also be classified according to their state of movement. A glacier is in a state

EARTH SYSTEM SCIENTISTS *JEAN AGASSIZ*

Jean Agassiz was born in Switzerland in 1807. He was educated in both comparative anatomy and paleontology, although his fame would later come from his interest in glacial geology. Early in his career, Agassiz helped to classify more than 1700 fossil fish species collected from around the world. His most famous work, however, involved the study of glaciers. In 1836 he proposed that glaciers were in a constant state of motion. He examined rocks that appeared to have been scarred by glacial movement and theorized that glaciers in Europe were once more widespread. This led him to develop the concept of the ice age. In 1840 he published *Studies on Glaciers,* which presented his theories on how much of Europe was once covered by ice and his theory of ice ages. He further proposed that these ice ages may have caused mass extinctions of living organisms. In 1847 he moved to the United States, where he became a professor at Harvard University. While there, he concluded that North America had also experienced ice ages in the past.

Figure 27–4 Many of the world's ice fields and glaciers terminate in the ocean. *(Courtesy of PhotoDisc.)*

Current Research

Research conducted by Lonnie Thompson, a professor of geology from Ohio State University, is revealing the disappearance of ice caps from Africa and South America. Thompson's research on Mount Kilimanjaro, in Africa, has shown a reduction in the ice caps by as much as 82% since 1912. His research has also shown a reduction of ice caps in the Andes Mountains in South America. If the present melting rate continues, the ice caps are expected to totally disappear in the next 10 to 15 years. Similar research has also shown a reduction of ice caps in both China and Tibet. Thompson compares the loss of these ice caps to the "canary in the coal mine" syndrome. Canaries were once used to detect deadly gases in coal mines; if the canary died, it signaled to miners the presence of the odorless gases. Thompson theorizes that the disappearance of ice caps from Africa and South America might be a signal of the changes that are occurring in the global climate.

of advance when the rate of melting of the glacial front is exceeded by the forward movement of the ice. **Glacial advance** occurs when accumulation of ice is greater than the ablation at the glacial front. Many glaciers in Greenland and Antarctica are in a constant state of advance. A glacier can be in a stationary state when the rate of melting at the glacial front equals the rate of advancement, or accumulation equals ablation. **Glacial retreat** occurs when melting at the front is faster than the forward advancement. This is the result of ablation occurring at a higher rate than accumulation. Glaciers that feed into Glacier Bay, Alaska, have been in a rapid state of retreat over the past few hundred years (Figure 27–6). It is estimated that these glaciers have retreated more than 60 miles during this period; some researchers believe this has been caused by increasing global temperatures.

Figure 27–5 A glacier flows from its zone of accumulation in the mountains of Alaska. *(Courtesy of PhotoDisc.)*

Glaciers and Global Climate

Recent research by scientists around the world suggests that many glaciers are retreating. This means that the water that they hold is melting and being added to the oceans. If all the water trapped in the world's glaciers melted into the ocean, it is estimated that sea level would rise more than 100 feet. Scientists have revealed that sea level has risen approximately 4 inches over the past 100 years. They predict that at the present rate of increase, the level of the oceans may rise by as much as 12 inches in the next 70 years. Most researchers believe this increase in sea level is the result of the melting of the world's glaciers. What they do not agree on, however, is what is causing the glaciers to melt. Some believe it is the result of global warming caused by human activity. Others suggest that it is the continued warming of the planet that led to the end of the last ice age 11,000 years ago. Whatever the cause, the impact that melting glacial ice will have on modern society may threaten many of the world's major cities. Most of the world's major cities and more than 50% of the planet's population lives near a coast. The predicted rise in sea level as a result of melting glaciers would be a disaster for these population centers. Many major coastal cities are located only a few feet above sea level; therefore this change could be catastrophic.

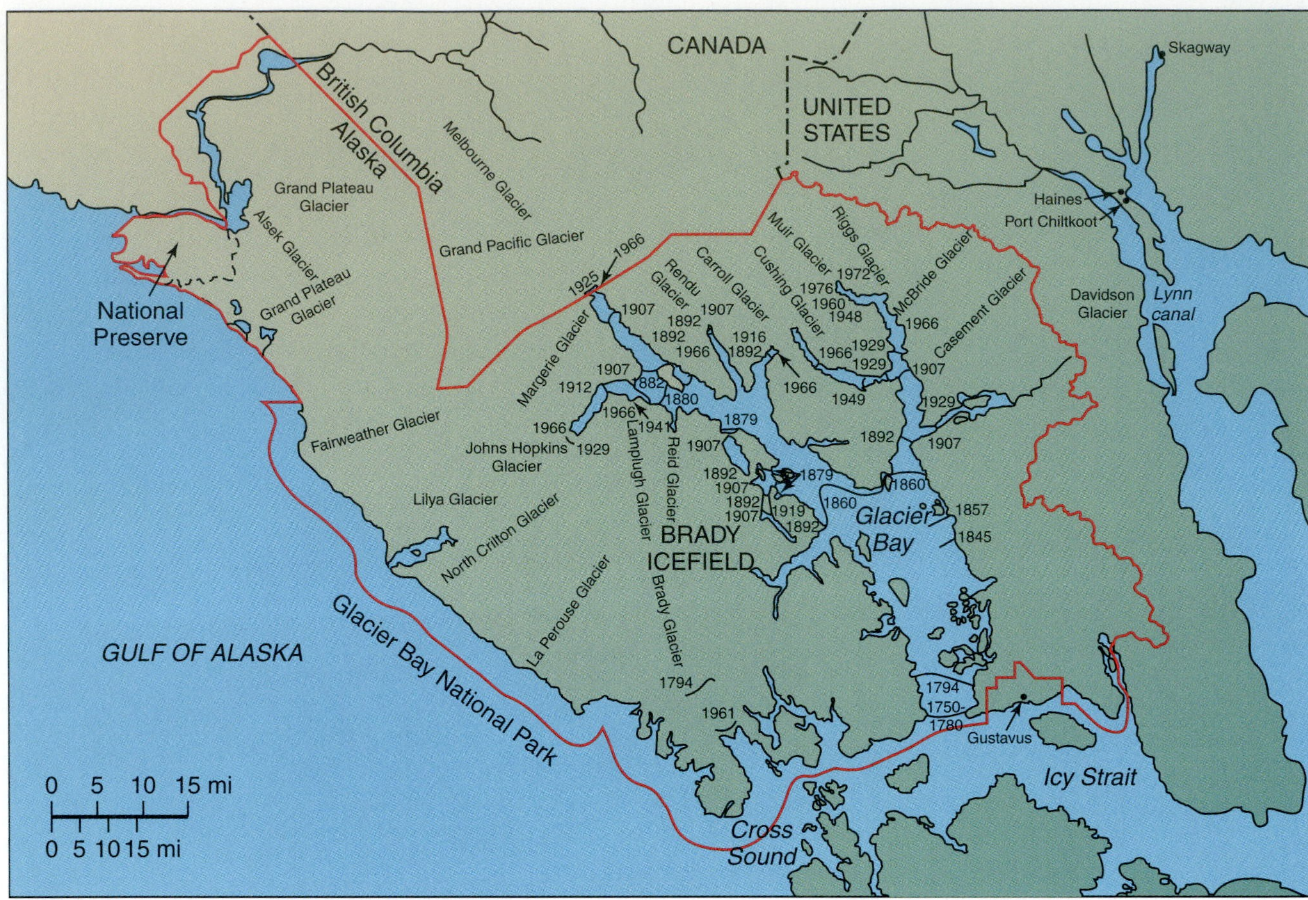

Figure 27–6 The specific years and locations of the glacial fronts of the retreating glaciers in Glacier Bay National Park, Alaska.

Although glaciers make up a cold, harsh environment on Earth, their role in the cycling and storage of water makes them an important aspect of the hydrosphere. Glaciers in the past have also helped to transform much of the world's landscape and have contributed to the formation of many fertile soils that support agriculture. More importantly, today glaciers may be indicators of a changing global climate.

For more information go to these Web links:

<http://www.glacier.rice.edu/>
<http://nsidc.org/glaciers/>
<http://ak.water.usgs.gov/glaciology>

REVIEW

1. Explain how glaciers form.
2. What are the three type of moraines associated with glaciers?
3. List the five types of glaciers that occur on Earth.
4. Explain the difference between a glacial advance and a glacial retreat.
5. Who was Jean Agassiz?

EARTH MATH

1. If a glacier is moving at a rate of 9 feet per day, how many days will it take to travel 1 mile?
2. How many gallons of water does a 1-square-foot, 7130-foot-long core section of the Antarctic ice sheet contain if 1 cubic foot of water equals approximately 7 gallons of water?

CHAPTER SUMMARY

Glaciers are large masses of ice that form by the accumulation of snow over long periods. The glaciers around the world contain approximately 2% of all the planet's freshwater supply. Glacial ice flows outward from the zone of accumulation. This is where snow accumulates and presses down on the glacier, causing it to flow outward. The area of the glacier where the ice is flowing is called the zone of flowage. The zone of flowage eventually reaches the zone of ablation, where the ice is lost from a glacier by melting. As glaciers flow, they scrape along rock to form a mixture of ice and sediment called glacial till. These sediments form unique glacial deposits called moraines.

Moraines that form along the sides of glaciers are known as lateral moraines. Terminal moraines form at the front of a glacier where sediments are deposited by the melting ice. Medial moraines form along the center of a glacier as a result of two valley glaciers joining together.

Glaciers are classified by their size and location on the Earth. Continental glaciers, also known as ice sheets, cover complete continents, such as in Antarctica. Ice fields are large glaciers that cover wide areas, leaving only the tops of mountains exposed. Ice caps are domelike glaciers that cover the top of a mountain and flow downward into surrounding valleys. Valley glaciers exist where ice flows down through mountain valleys. The last type of glacier, the piedmont, forms on flat plains where valley glaciers come together.

Glaciers that are growing are in a state known as glacial advance. This occurs when the accumulation of snow exceeds the ablation of ice, causing the glacial front to advance. Glacial retreat occurs when ablation occurs at a higher rate than accumulation, causing the glacial front to retreat. Scientists have linked the retreat of many of the world's glaciers to a change in global climate. As glaciers melt, they add water into the oceans, which causes a change in sea level. As a result of this, the level of the oceans has risen by 4 inches over the past 100 years and is expected to continue to rise as a result of glacial melting.

CHAPTER REVIEW

Multiple Choice

1. Glaciers during the last ice age lowered sea level by approximately how much?
   a. 4 inches
   b. 12 inches
   c. 100 feet
   d. 400 feet

2. The area of a glacier where snow collects and compacts to form glacial ice is called the:
   a. zone of accumulation
   b. zone of flowage
   c. zone of ablation
   d. zone of formation

3. In which area of a glacier would you most likely find icebergs?
   a. zone of accumulation
   b. zone of flowage
   c. zone of ablation
   d. zone of formation

4. Which part of a glacier is most likely to form a lateral moraine?
   a. zone of accumulation
   b. zone of flowage
   c. zone of ablation
   d. zone of formation

5. A pile of unsorted glacial sediment deposited by the glacial front is known as:
   a. lateral moraine
   b. terminal moraine
   c. medial moraine
   d. glacial erratic

6. What is the result when ablation occurs at a higher rate than accumulation?
   a. glacial advance
   b. glacial flow
   c. glacial retreat
   d. glacial equilibrium

7. The entire continent of Antarctic is covered by which type of glacier?
   a. ice sheet
   b. ice field
   c. ice cap
   d. ice dome

8. The type of glacier that covers much of Iceland is called:
   a. ice sheet
   b. ice field
   c. ice cap
   d. ice dome

9. When a glacier's rate of accumulation is higher than it is rate of ablation, what occurs?
   a. glacial advance
   b. glacial flow
   c. glacial retreat
   d. glacial equilibrium

10. Approximately how much is sea level expected to rise in the next 70 years as glaciers continue to retreat?
    a. 4 inches
    b. 12 inches
    c. 100 feet
    d. 400 feet

Matching *Match the terms with the correct definitions.*

a. glacier
b. zone of accumulation
c. glacial front
d. zone of flowage
e. glacial till
f. moraines
g. terminal moraines
h. zone of ablation
i. lateral moraines
j. medial moraines
l. icebergs
m. glacial advance
n. glacial retreat

1. ____ The face of a glacier where the ice breaks off and melts.
2. ____ Large chunks of glacial ice that break off the leading edge of a glacier and float into the sea.
3. ____ Unsorted glacial sediments that are located at the front or deposited near the front of a glacier.
4. ____ Glacial sediments that are located near the middle of a glacier.
5. ____ A large, long-lasting mass of snow and ice that forms over land from the accumulation and compaction of snow that creeps down slope.
6. ____ The zone of a glacier where ice is breaking off and melting.
7. ____ The forward movement of a glacier caused when snow accumulation is greater than melting.
8. ____ Glacial sediments that form along the sides of a glacier as it scrapes along rock.
9. ____ The area on a glacier where snowfall is building up, forming new glacial ice.
10. ____ The shrinking of a glacier caused when melting exceeds snow accumulation.
11. ____ Unsorted glacial sediments.
12. ____ The area on a glacier where glacial ice is currently flowing.
13. ____ An accumulation of glacial sediments.

Critical Thinking

1. Describe the effects on society and the environment if glaciers began to increase in size instead of shrink.

CHAPTER
28 Pollution of the Hydrosphere

Objectives

Sediment Pollution • Nutrition Pollution and Eutrophication • Toxic Organic Compounds • Toxic Inorganic Compounds • Disease-causing Agents • Thermal Pollution

After reading this chapter you should be able to:

- Define the term *sediment pollution* and describe two of its sources.
- Explain the negative effects of sediment pollution.
- Describe what causes an algal bloom and how this affects an aquatic ecosystem.
- Identify three sources of toxic organic water pollutants.
- Describe three toxic inorganic water pollutants and their sources.
- Define the terms *disease-causing agent* and *waterborne illness.*
- Describe two waterborne illnesses and their sources.
- Differentiate between the two different types of thermal pollution and describe their negative effects on aquatic ecosystems.

TERMS TO KNOW

pollution
sediment pollution
nutrient pollution
eutrophication
algal bloom
aerobic bacteria
polychlorinated biphenyls
landfills
disease-causing agents
waterborne illnesses
wastewater
thermal pollution
dissolved oxygen

INTRODUCTION

Pollution of the hydrosphere is one of the most life-threatening impacts that human beings have had on the Earth. Because of water's abundance and usefulness, it has been contaminated in a multitude of ways by a variety of human practices. The importance of water to living things cannot be overstated, because every organism alive on the planet requires clean water to survive. The availability of clean freshwater is especially important because it makes up such a small portion of the Earth's total water supply. Freshwater resources on the planet are unequally distributed around the globe, making it an even more precious resource. Today on our planet the freshwater that is used for drinking, agriculture, and industry often contains some form of pollutant. As the human population continues to grow, more and more people will depend on the small amount of freshwater that is available on the Earth. Understanding how water becomes polluted, preventing further pollution of this important resource, and conserving it for sustained use are vital to the future of our planet.

Sediment Pollution

The number one form of water pollution on the Earth is sediment pollution. **Sediment pollution** of water occurs when eroding soil particles are washed into water by the process of runoff (Figure 28–1). Runoff occurs when soil is exposed to precipitation that washes it into nearby lakes or streams. The main sources of sediment pollution are exposed agricultural fields, clear-cut logging operations, and construction sites. These practices clear land of the vegetation that naturally protects the soil from erosion. When the vegetation is removed, the soil is exposed and can be easily washed away. The sediment that is washed into nearby streams or lakes clouds the water and reduces the amount of sunlight that can penetrate the water. This then reduces the amount of photosynthesis that can occur in the water, which reduces the amount of food available to the aquatic organisms that live there. Introduction of sediments into water can also clog the gills of fish and other aquatic organisms. This reduces their ability to absorb oxygen from the water. Some fish species, such as trout, require clear water to live because they use their vision to hunt for insects. When the water becomes cloudy, they are unable to find food. Increased sediment in water can also negatively affect the spawning of fish, lowering their ability to reproduce. Sediment washed into water slowly builds up on the bottom of lakes and streams, making the water shallow over time. This can cause an eventual overflow of water to the surrounding land. Many rivers and lakes that are used for transportation by the shipping industry must be continually dredged to remove the accumulating sediments each year. It is estimated that more than 75 billion tons of sediments are washed into the water annually. Much of this sediment was once produc-

Figure 28–1 Clear-cut logging can contribute large amounts of sediments into surrounding bodies of water. *(Courtesy of PhotoDisc.)*

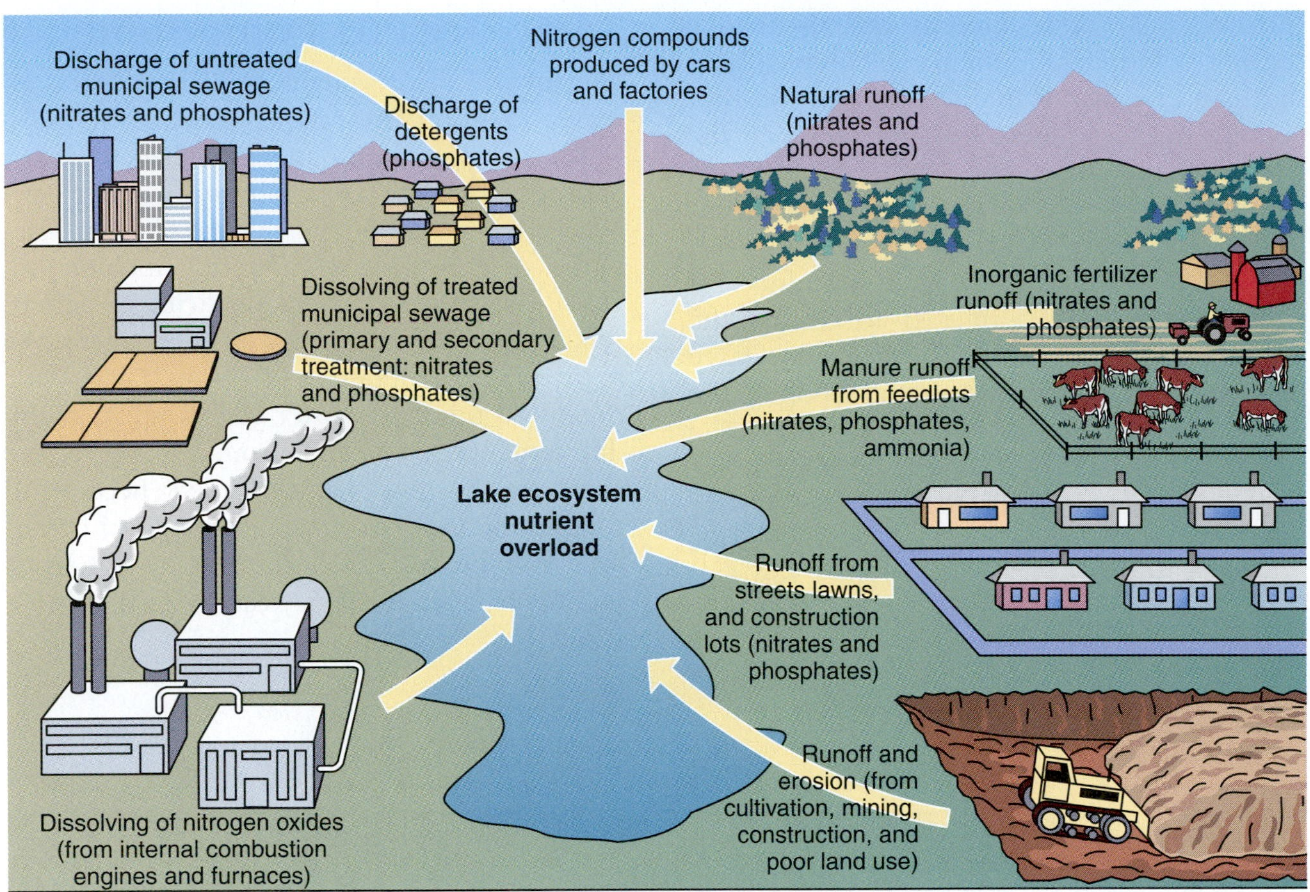

Figure 28–2 The major sources of nutrient pollutants in water.

tive soil that supported plant growth but has now been lost to the water.

Nutrient Pollution and Eutrophication

Another form of water contamination is called **nutrient pollution.** Nutrient pollution is caused by the increase of water soluble chemicals such as nitrates and phosphates in water (Figure 28–2). These chemicals act as nutrient fertilizers to aquatic plants and algae. The increasing amount of nutrients in a body of water is called **eutrophication.** The plants that live in water that are exposed to these nutrients begin to grow in abundance, which creates an imbalance in the aquatic ecosystem. This imbalance is known as an **algal bloom,** which is the rapid increase in the population of algae and aquatic plants caused by the increasing amount of fertilizer in the water.

Algal blooms also cause an increase in dying algae, which sink to the bottom of the body of water being affected. The decaying algae is then decomposed by **aerobic bacteria.** These bacteria require oxygen to live. The mass of dead algae causes a rise in the population of decomposing bacteria, which begins to remove oxygen from the water at a high rate. Sometimes the bacteria use so much oxygen to break down the dead algae that eventually the water contains no oxygen at all. At this point the body of water is classified as dead, because there is no oxygen for other aquatic organisms to breathe.

Algal blooms caused by nutrient pollution can cause severe problems for aquatic life in both standing and flowing water ecosystems

(Figure 28–3). Algal blooms can also choke the water with an abundance of aquatic plants that can clog water intake pipes and disrupt boating traffic. Many lakes and ponds in the United States have been affected by algal blooms. Communities near these affected bodies of water try to combat the increased amount of plants and algae in a number of ways. Some apply herbicides to control the growth of plants, whereas others use large harvesting machines that cut and scoop up the excess aquatic plants from the water. The source of nutrients that can cause algal blooms originates as plant fertilizers applied on lawns, gardens, and farm fields. Nutrient pollutants can also come from leaking septic systems, wastewater treatment plants, and detergents.

Toxic Organic Compounds

One of the most widely publicized forms of water pollution around the world are oil spills. This type of contamination is called toxic organic compound water pollution. The term *organic* refers to any molecule that contains carbon. Toxic organic compounds include oil, gasoline, solvents, and pesticides, which, when mixed with water, can be deadly to many different organisms (Figure 28–4). Toxic organic compounds enter the water in many different ways. Oil spills from leaking tanker ships are the most widely known because of their immense size, but smaller spills of chemicals can be just as harmful. The Environmental Protection Agency estimates that more

Figure 28–3 The rapid increase in the population of aquatic plants and algae, known as algal bloom, occurs as a result of nutrient pollution. *(Courtesy of PhotoDisc.)*

Figure 28–4 The pollution of water by toxic organic compounds such as oil can cause widespread contamination of the environment. *(Courtesy of PhotoDisc.)*

than 1 million underground fuel storage tanks are leaking into groundwater in the United States. This is resulting in widespread contamination of drinking water. These types of spills are difficult to trace because they often come from old abandoned fuel tanks. Because most of this pollution occurs underground, it is often hard to determine how much groundwater has been polluted and where it is traveling. Another source of toxic organic compounds is pesticides that are applied to plants to protect them from insects, weeds, and disease. If not applied at the correct time, pesticides can be washed into water by precipitation and runoff, where they can contaminate the water.

Toxic Inorganic Compounds

Toxic inorganic chemicals are another class of water pollutants. These include synthetic industrial chemicals and heavy metals such as lead, mercury, and arsenic. The presence of these chemicals in water can cause long-term health problems and even death. Many of these deadly chemicals have been introduced into water by poor waste disposal practices. Toxic chemicals stored in leaking metal drums enter both groundwater and surface water, causing widespread contamination. Unlike toxic organic chemicals, which usually break down in the environment after a few days or weeks, inorganic compounds and heavy metals can remain deadly for many years. The use of **polychlorinated biphenyls,** or PCBs, to manufacture electrical components led to one the worst forms of toxic inorganic water pollution. PCBs were washed into the Hudson River in New York State for many years (Figure 28–5). As a result, traces of PCBs are found in the tissue of many birds and fish that live in or along the Hudson River. Long stretches of the Hudson River are contaminated by PCBs, which resulted in a ban on fishing in the area for many years. Today the river's sediments still contain high concentrations of PCBs. Other sources of toxic inorganic chemicals include leaking **landfills** and old abandoned industrial sites.

Disease-causing Agents

The introduction of **disease-causing agents** into water supplies is a very serious water pollutant (Table 28–1). Viruses, bacteria, amoeba, protozoa, and parasitic worms are all agents of disease that can enter water supplies and affect human health. Most of these disease-causing agents enter water through human or animal waste. *Escherichia coli,* or *E. coli,* bacteria are present in the waste of all animals; however, when they are introduced into a water supply, they can be a serious health threat. **Waterborne illnesses,** such as typhoid, malaria, hepatitis, cholera, and dysentery, are all caused by these disease agents and result in sickness and death.

Figure 28–5 Contamination of water by industrial processes has decreased water quality all over the world. *(Courtesy of PhotoDisc.)*

TABLE 28-1 Common disease-causing agents and waterborne illnesses

Type of Organism	Disease	Effects
Bacteria	Typhoid fever	Diarrhea, severe vomiting, enlarged spleen, inflamed intestine; often fatal if untreated
	Cholera	Diarrhea, severe vomiting, dehydration; often fatal if untreated
	Bacterial dysentery	Diarrhea; rarely fatal except in infants without proper treatment
	Enteritis	Severe stomach pain, nausea, vomiting; rarely fatal
Viruses	Infectious hepatitis	Fever, severe headache, loss of appetite, abdominal pain, jaundice, enlarged liver; rarely fatal but may cause permanent liver damage
Parasitic protozoa	Amoebic dysentery	Severe diarrhea, headache, abdominal pain, chills, fever; if not treated can cause liver abscess, bowel perforation, and death
	Giardiasis	Diarrhea, abdominal cramps, flatulence, belching
Parasitic worms	Schistosomiasis	Abdominal pain, skin rash, anemia, chronic fatigue, and general ill health

Wastewater that enters into surface waters can carry these disease-causing agents into the environment, which can lead to potential health threats. Oysters growing in water contaminated by human waste often contain the hepatitis virus. This disease can then spread when the oyster is harvested and sold as food. People that often swim in water that contains any of these organisms can accidentally ingest it and become sick. It is extremely important to control these agents of disease and make sure they do not enter our water supplies, because the potential for widespread health problems is very real.

EARTH SYSTEM SCIENTISTS *LOUIS PASTEUR*

Louis Pasteur was born in France in 1822. He began his work as a scientist by studying chemistry. His attention turned to microbiology when he was asked to investigate the process that led to the formation of beer and wine. His resulting research unlocked the secrets of fermentation. His most famous work involved his germ theory of disease. By studying microorganisms, Pasteur concluded that bacteria present in food and water were the cause of disease and infection in animals. He later developed a method for sterilizing food and water by heating them to a temperature high enough to kill off microorganisms. Today this technique is known as pasteurization. His research into the cause of disease also led him to develop vaccines and the process of inoculation to prevent disease in humans. Pasteur's contributions to science and medicine make him one of the greatest scientists who ever lived.

Career Connections

Wastewater Treatment Operator

Wastewater treatment operators are responsible for the removal of harmful pollutants from industrial and domestic wastewater. They do this by managing a network of pipes, pumps, valves, tanks, filters, and chemical processes to cleanse wastewater. Their job also involves the periodic testing of water as it passes through the treatment process to maintain quality control. They must have a knowledge of the systems that together make up the wastewater treatment plant and should be able to perform routine maintenance or repairs when necessary. Wastewater treatment operators are usually employed by cities or towns that have wastewater treatment plants. Many of these jobs require a high school diploma and technicians receive on-the-job training. There are also 2-year college degree programs available in wastewater treatment technology. Large wastewater treatment facilities may employ a number of engineers, chemists, and technicians to operate these facilities. Because of the increasing importance of this type of service, job outlooks in this profession are expected to be excellent in the future.

Thermal Pollution

The last method of water pollution involves the introduction of excess heat into water. This form of water contamination is called **thermal pollution.** Many factories and electrical power plants are built near water because they use this resource to cool many of their industrial processes. Colder water taken from the environment is heated in these factories and then released back into the environment at a higher temperature. This can cause a drastic change in the aquatic ecosystems that exist near these areas. The amount of oxygen that is dissolved in water greatly depends on the temperature of the water (Figure 28–6). Colder water holds more **dissolved oxygen** than warmer water, and as the water is heated, this oxygen is driven out into the atmosphere.

Aquatic species such as fish require high amounts of dissolved oxygen in water to survive. When water is heated by factories and power plants and then reintroduced into the environment, it can lead to the reduction of oxygen in the water. As a result, water environments around these factories will no longer be able to support certain aquatic organisms. Nuclear power plants can cause extreme changes in the water temperature surrounding the plants (Figure 28–7). The change in water temperature caused by nuclear power plants located in Florida is caus-

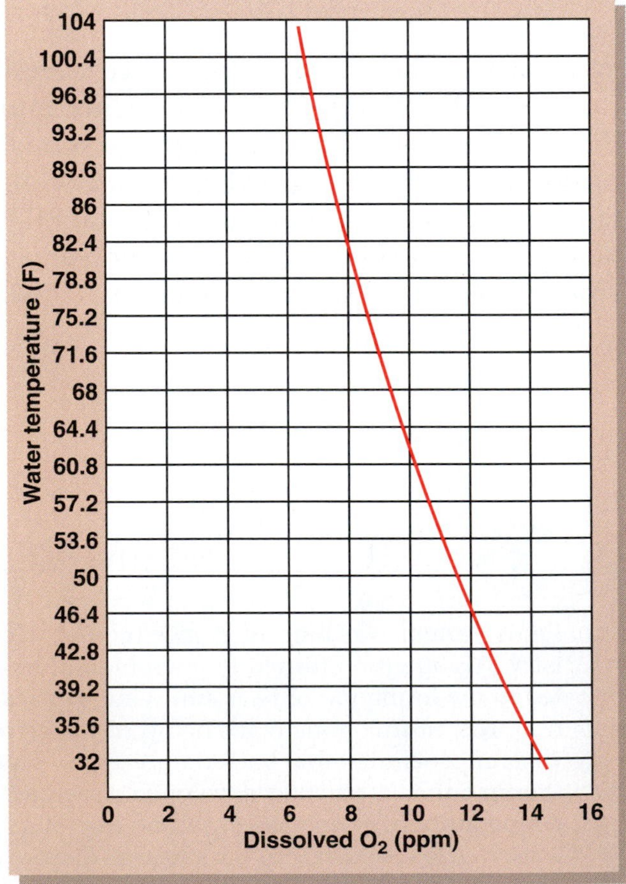

Figure 28–6 An increase in the temperature of the water caused by thermal pollution causes a decrease in the amount of dissolved oxygen available to aquatic organisms.

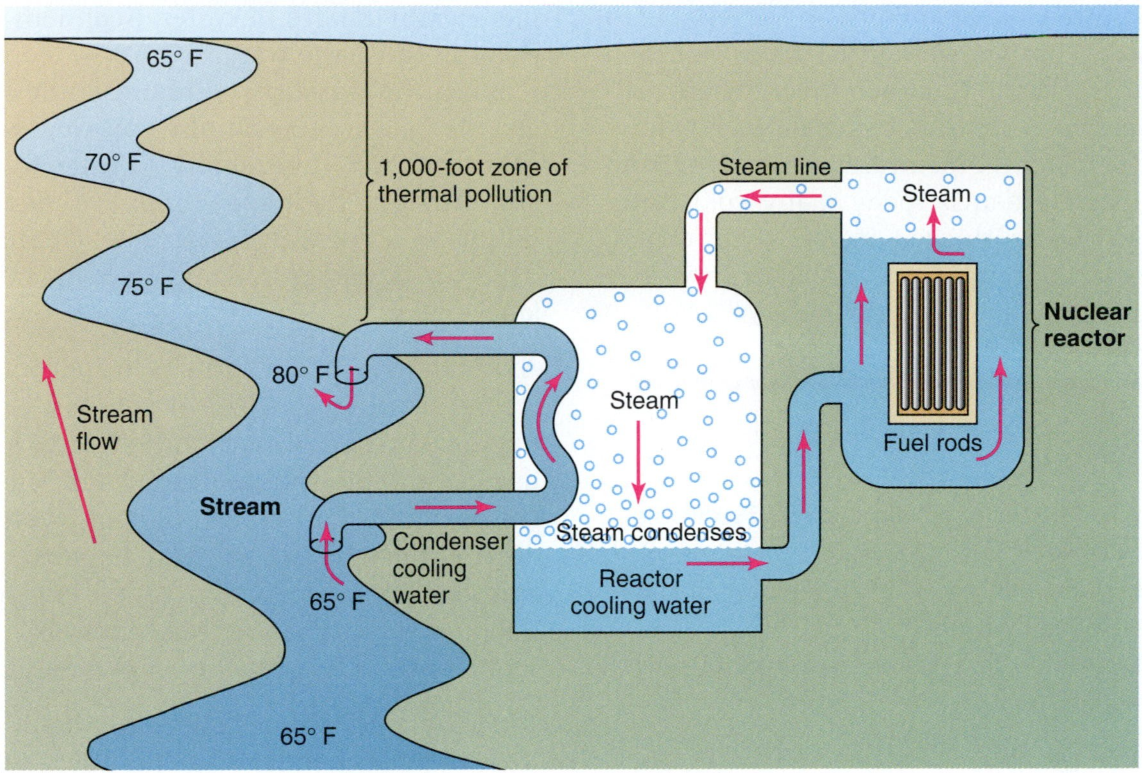

Figure 28–7 The effects of a nuclear power plant on the water temperature of a river illustrates the concept of thermal water pollution.

ing a disruption of the migration of manatees. Manatees are large aquatic mammals that reside in coastal rivers and bays; they normally migrate to areas with warmer water during the winter. As a result of thermal pollution, manatees are staying in warmer waters near the nuclear power plants. This interrupts their normal migration, which is causing overpopulation and interbreeding of the manatees.

Another type of thermal pollution can be caused by the construction of large dams on rivers. This type of thermal pollution is the result of the lowering of water temperature rather than the heating of the water. Large dams create deep artificial reservoirs of water behind them. The deep water at the bottom of the reservoir becomes much colder than the normal temperature of the water in the river. When the cooler water from the base of the dam is released, it can lower the temperature of the river water by 10° or more. The native fish species of the river cannot tolerate this

Earth Math

1) How many tons of sediment are washed into the world's oceans each month?

temperature change and are no longer able to live there. This type of thermal water pollution is occurring in the Colorado River, where artificial reservoirs are responsible for greatly lowering the water temperatures down river from these massive dams. Today many cold-water trout are being stocked in the river, where they would normally not be able to survive.

REVIEW

1. What are the two main sources of sediment pollution?
2. Describe the process of eutrophication and how it affects aquatic ecosystems.
3. What are three examples of toxic organic compounds?
4. What are three examples of toxic inorganic compounds?
5. How do disease-causing organisms contaminate water?
6. Explain how thermal pollution affects an aquatic ecosystem.
7. Who was Louis Pasteur?

For more information go to this Web link:
http://www.epa.gov/students/water.htm>

CHAPTER SUMMARY

Pollution of the hydrosphere has resulted directly from human activity. The world's worst form of water pollution is sediment pollution. This is occurs when sediment is washed into surrounding bodies of water by runoff. Human practices such as construction and agriculture often remove protective vegetation from the soil and expose it to the power of wind and rain, which washes rocks and soil into nearby bodies of water. Sediment pollution can cause the water to be clouded, which reduces the amount of photosynthesis that can take place there. It also can clog the gills of fish, reduce their ability to hunt for food, and interrupt their spawning. Sediments can fill up bodies of water over time, creating the need for periodic dredging to keep shipping lanes open.

Another widespread form of water pollution is nutrient pollution. This occurs as a result of the introduction of nutrient fertilizers such as nitrates and phosphates into water. These excess fertilizers cause algal blooms, which can disrupt the balance of aquatic ecosystems. Nutrient pollution is also called eutrophication and is caused by fertilizer runoff, leaking septic systems, wastewater, and detergents.

Toxic organic water pollution occurs when toxic organic chemicals such as fuels and pesticides leak into water. This form of pollutant can occur in both groundwater and surface water. Toxic inorganic water pollution is caused by the introduction of heavy metals such as mercury or lead into water, along with human-created compounds such as PCBs. These often can contaminate water supplies for long periods because they do not break down rapidly by natural processes. These chemicals can leak into water from old waste dumps, landfills, and industrial processes.

The introduction of disease-causing agents such as bacteria, parasitic worms, and viruses into water can pose a serious threat to the environment. Diseases associated with these organisms include dysentery, cholera, malaria, and hepatitis. Wastewater from human sewage is usually the source of these waterborne illnesses.

The last type of water pollutant is called thermal pollution. This occurs when the water temperature of a body of water is changed drastically as a result of human activity. Often, thermal pollution results in an increase in the temperature of the water, which leads to a decrease in the amount of dissolved oxygen available for aquatic organisms. This is caused by the water being heated when it is used for an industrial process or for generating electricity. Another form of thermal pollution is the result of the lowering of the temperature of the water, which is associated with the construction of large dams on rivers.

CHAPTER REVIEW

Multiple Choice

1. The most widespread form of water pollution is:
   - **a.** nutrient pollution
   - **b.** toxic inorganic compounds
   - **c.** sediment pollution
   - **d.** thermal pollution
2. The enrichment of fertilizers in aquatic ecosystems is called:
   - **a.** nutrient pollution
   - **b.** toxic inorganic compounds
   - **c.** sediment pollution
   - **d.** thermal pollution
3. Runoff from a construction site can result in:
   - **a.** increased water temperature
   - **b.** decreased photosynthesis by aquatic plants and algae
   - **c.** decreased water temperature
   - **d.** waterborne illnesses
4. Leaking septic systems located around lakes and ponds can cause:
   - **a.** nutrient pollution
   - **b.** toxic inorganic compounds
   - **c.** sediment pollution
   - **d.** thermal pollution
5. A rapid increase in the population of aquatic plants is known as:
   - **a.** eutrophication
   - **b.** an algal bloom
   - **c.** thermal pollution
   - **d.** a waterborne illness
6. Increasing amounts of nitrates and phosphate fertilizers in water causes:
   - **a.** eutrophication
   - **b.** an algal bloom
   - **c.** thermal pollution
   - **d.** a waterborne illness
7. As the temperature of the water increases, what happens to the level of dissolved oxygen within the water?
   - **a.** decreases
   - **b.** increases
   - **c.** remains the same
   - **d.** disappears
8. The main source of waterborne illness is:
   - **a.** run off
   - **b.** oil spills
   - **c.** leaking fuel tanks
   - **d.** wastewater
9. Polychlorinated biphenyls (PCBs) are an example of:
   - **a.** toxic organic compounds
   - **b.** toxic inorganic compounds
   - **c.** nutrient pollutants
   - **d.** thermal pollutants
10. An increase or decrease in water temperature caused by human activity is known as:
    - **a.** nutrient pollution
    - **b.** toxic inorganic compounds
    - **c.** sediment pollution
    - **d.** thermal pollution

continued

Matching *Match the terms with the correct definitions.*

a. pollution
b. sediment pollution
c. nutrient pollution
d. eutrophication
e. algal bloom
f. aerobic bacteria
g. polychlorinated biphenyls
h. landfills
i. disease-causing agents
j. waterborne illnesses
k. wastewater
l. thermal pollution
m. dissolved oxygen

1. ____ A rapid increase in the population of algae caused by the introduction of nutrient fertilizers into an aquatic ecosystem.
2. ____ Specific organisms that cause or spread disease.
3. ____ The amount of oxygen gas that is dissolved in water.
4. ____ An undesirable change in the quality of the environment that negatively affects the health of organisms living there.
5. ____ Bacteria that require oxygen to live.
6. ____ A form of pollution associated with a change in temperature.
7. ____ A form of water pollution caused by the rapid introduction of sediments into an aquatic ecosystem as a result of runoff.
8. ____ A class of human-created chemical compounds used in the production of electrical parts.
9. ____ A form of water pollution caused by an increase in nutrients in an aquatic ecosystem, leading to a rapid increase in aquatic plants and algae.
10. ____ A type of disease or sickness caused by an organism that lives in water.
11. ____ A term used to describe a place where large amounts of garbage are deposited.
12. ____ Polluted water that is unfit for drinking or introduction into the environment because it has been used for some purpose and made unclean.
13. ____ The rapid introduction of nutrient fertilizers into an aquatic ecosystem, leading to an increase in aquatic plants and algae.

Critical Thinking

1. Some people believe that thermal pollution is not a true form of water pollution because it does not affect human beings. Explain why you agree or disagree with this point of view.

CHAPTER

29 El Niño and the Southern Oscillation

Objectives

The South Equatorial Current and Upwelling • The Southern Oscillation • The Effects of an El Niño Event • La Niña Events • Monitoring the Pacific Ocean

After reading this chapter you should be able to:

- Describe the normal conditions that exist between the atmosphere and the ocean along the equator off the west coast of South America.
- Define the term *upwelling* and explain how it affects fish populations off the coast of South America.
- Explain the changes in the atmosphere and ocean that cause an El Niño event.
- Describe the relationship between atmospheric pressure and climate that is known as the southern oscillation.
- Identify three effects that an El Niño southern oscillation (ENSO) event has on the Earth's climate.
- Explain the changes in the atmosphere and ocean that cause a La Niña event.

TERMS TO KNOW

climate

trade winds

south equatorial current

upwelling

southern oscillation

El Niño southern oscillation (ENSO)

La Niña

buoys

INTRODUCTION

Humans beings have had a long relationship with the sea. The world's oceans have supplied us with a bountiful food supply for thousands of years. Today's ocean fishing industry is important not only for the food it supplies, but because it supports the economy of many communities that depend on the sea for their livelihood. Fishermen's observations of the natural rhythms of the sea have helped scientists understand the nature of the ocean and its role in the Earth's systems. In 1972 a massive die-off of fish populations off the west coast of South America nearly destroyed the communities that have depended on these fish for their survival. Similar events have occurred in the same region since then, and they have become known as El Niño events. After decades of study, it has been revealed that the changes that periodically occur in the ocean not only affect fish populations but impact the world's climate. As a result, El Niño has become a household phrase; it illustrates the interactions that occur between the hydrosphere and atmosphere and how they affect life on our planet.

The South Equatorial Current and Upwelling

The event known as El Niño is a **climate** phenomena that is linked to changes in the temperature of ocean currents off the western coast of South America. The name El Niño, which is Spanish for "the child," originated from local fishermen who first noticed the warmer waters that appeared around Christmas time off the South American coast. During normal conditions, planetary scale winds that blow from east to west along the equator, also known as the **trade winds,** drive the westerly moving ocean current called the **south equatorial current.** This ocean current moves warm equatorial waters from the coast of South America toward the western Pacific Ocean (Figure 29–1). As this current moves warm water westward, cold water moves up the continental slope off South America from the deep ocean to replace it.

The upward movement of deep ocean water toward the surface is known as **upwelling.** This process brings nutrient-rich, cooler ocean water up to the ocean surface. This nutrient-rich cold water sustains the food chain in the ocean waters off the Pacific coast of South America. Local fisherman rely on this cold water to provide nutrients for the fish that they catch for food. During an El Niño event, the easterly trade winds slow down, which also slows down the south equatorial current. The result is warmer water lying off the South American coast, which reduces the upwelling of colder, nutrient-rich water. The reduction of nutrients causes a decline in the fish populations off the coast of South America, which impacts the local fisherman (Figure 29–2).

The Southern Oscillation

The El Niño phenomena is linked to another climate variation called the **southern oscillation.** During the 1920s, a British meteorologist by the name of Sir Gilbert Walker was studying the formations of monsoons off the coast of southeast Asia. Monsoons are strong seasonal winds that often bring heavy rains

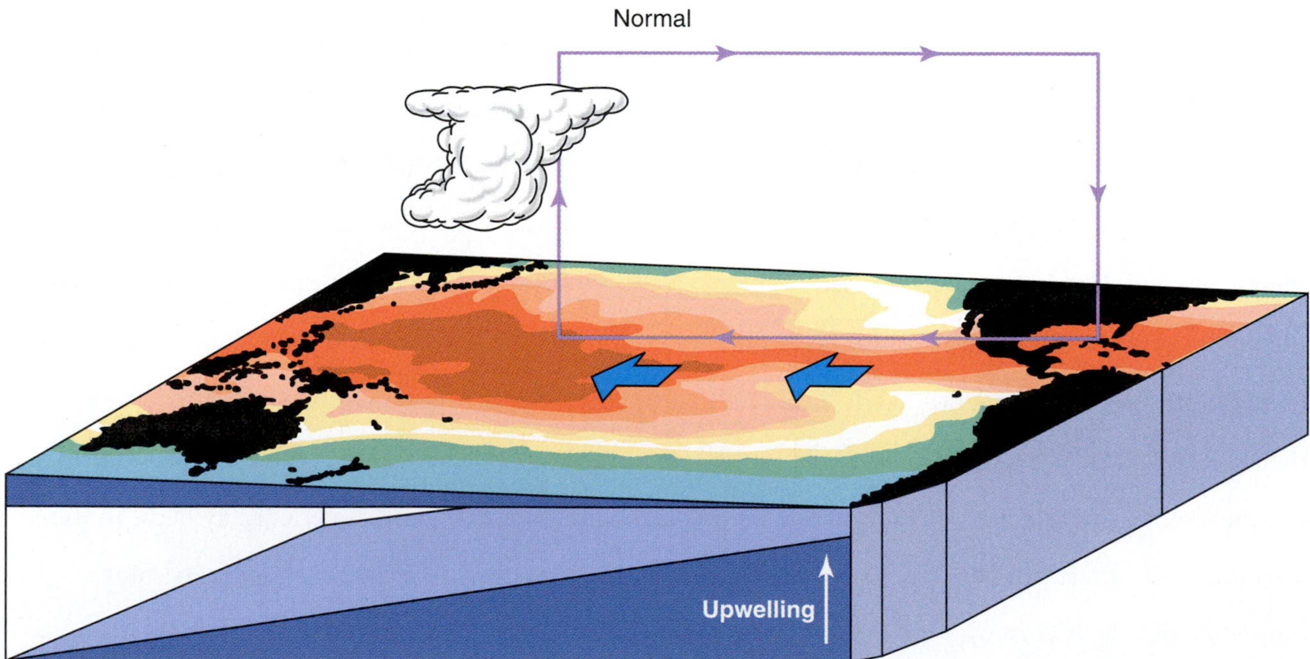

Figure 29–1 The normal circulation of surface ocean currents off the west coast of South America moves warm ocean water toward the west, where it is replaced by the upwelling of cold bottom water. This causes clouds and wet weather to form near the South Pacific.

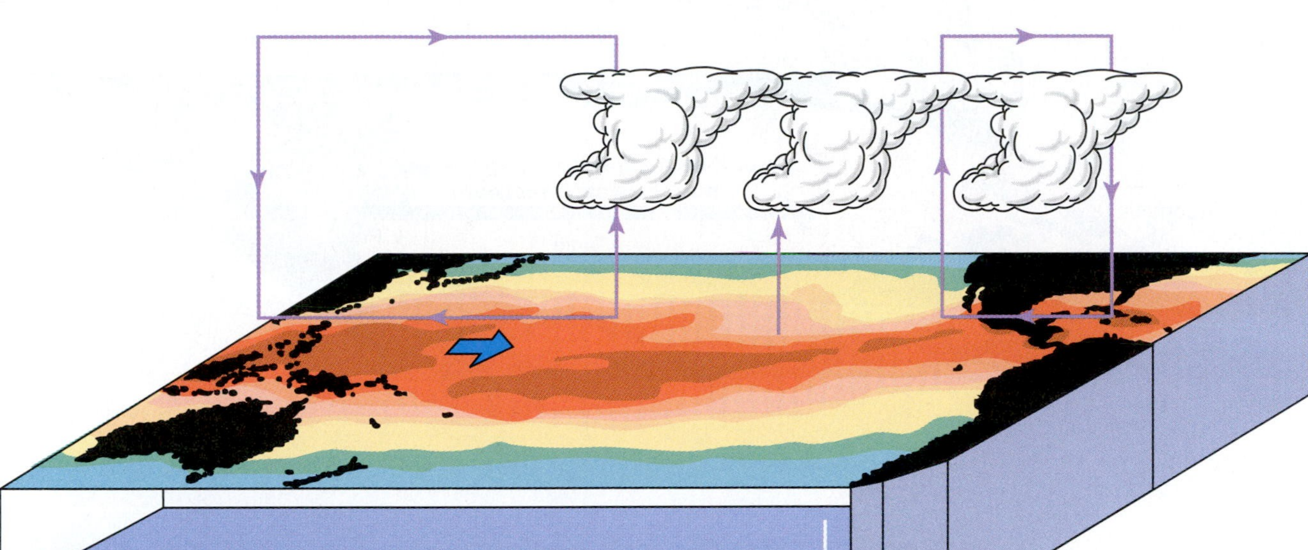

Figure 29–2 The stalled circulation of surface ocean currents during an El Niño event along the equator causes warm water to spread eastward, preventing upwelling along the west coast of South America. This causes clouds and wet weather to form farther east than normal.

to this region (Figure 29–3). He determined that they were associated with the development of low and high surface air pressure over the Pacific Ocean. Under normal conditions, a strong low atmospheric pressure system develops over Indonesia, bringing wet weather to the southwestern Pacific Ocean. On the other side of the Pacific, a high atmospheric pressure system develops over the eastern south Pacific Ocean, bringing cool and dry weather to the region.

Walker also noticed that these pressure systems tend to move in relation to ocean currents. During normal years, strong easterly trade winds develop between these two pressure systems. This causes the strong southern equatorial current to move warm water toward southeast Asia, forming the seasonal monsoons. For reasons unknown, this pattern shifts, which results in increased atmospheric pressure over the western Pacific and decreased pressure over the eastern Pacific. This reduces the strength of the trade winds. The movement of these pressure systems occurs on a 2- to 7-year cycle, which Walker called the southern oscillation. Sir Walker also hypothesized that this oscillation was responsible for the periodic droughts that occur in Australia, Africa, and India. Skeptics at the time of Walker's hypothesis laughed at the notion that weather could be influenced on a global scale.

During the 1950s another meteorologist, Jacob Bjerknes, determined that the change in the temperature of the ocean currents off the South American coast were related to Walker's southern oscillation (Figure 29–4). Bjerknes found that the warm water that normally flowed along the equator toward the western Pacific helped to form the moist low-pressure systems over Indonesia. He further hypothesized that when these currents were reduced,

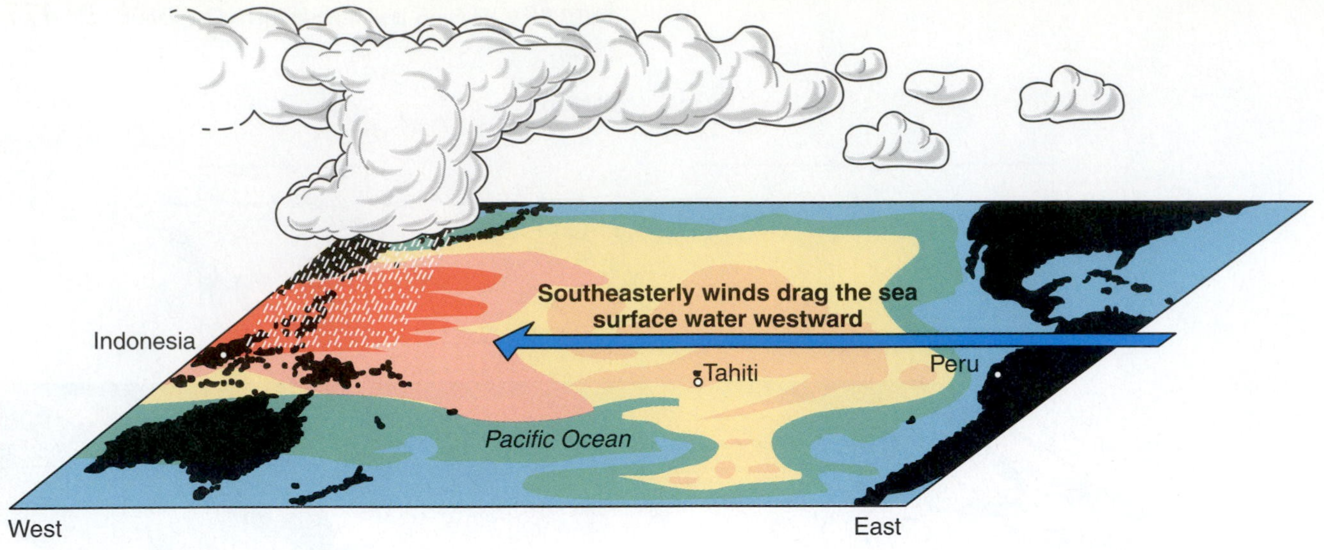

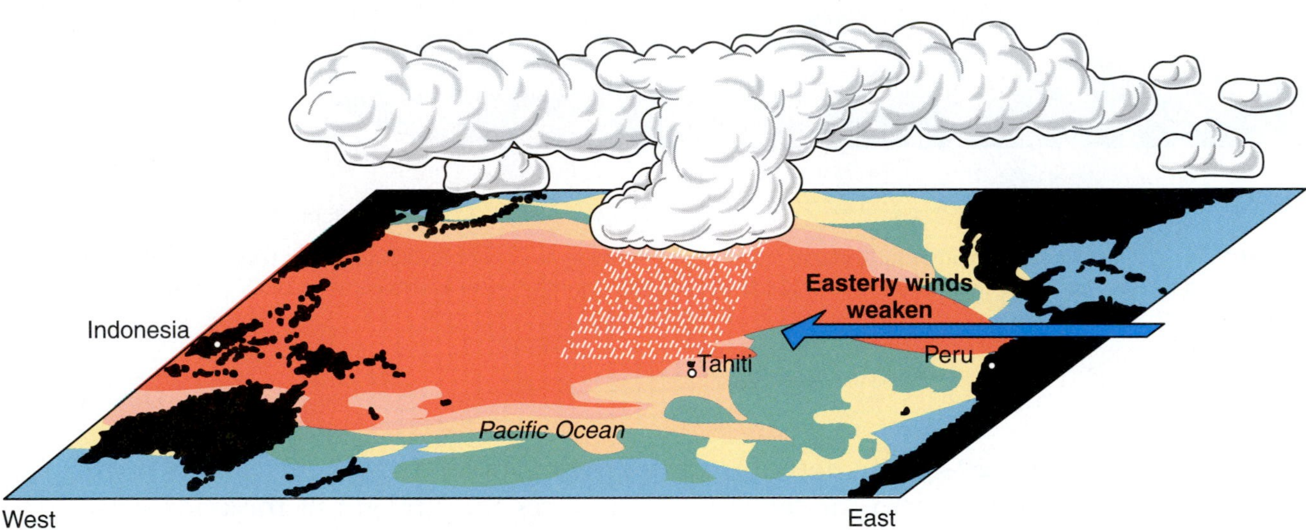

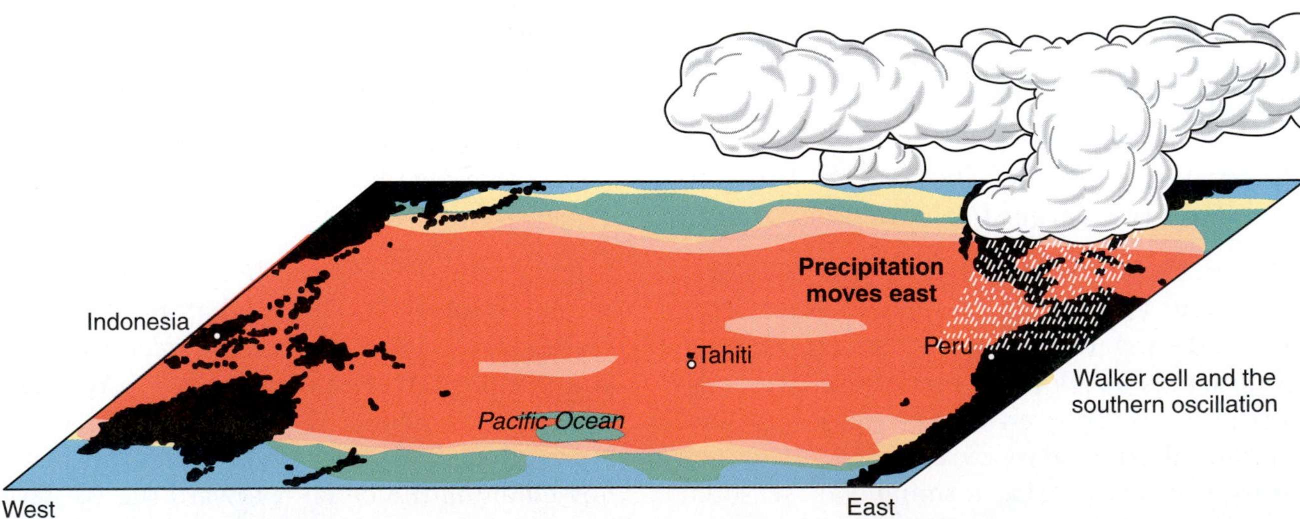

Figure 29–3 The movement of low-pressure systems and the precipitation they produce, which Gilbert Walker termed the southern oscillation. During normal years, the trade winds drive warm ocean water and low pressure westward. The reduction of the trade winds and the stalled movement of warm ocean water periodically causes the low pressure to form farther east. This phenomenon was later linked to the occurrence of an El Niño event.

EARTH SYSTEM SCIENTISTS *SIR GILBERT WALKER*

Gilbert Walker was a British meteorologist who was the head of the Indian Meteorological Service. In 1904 the British government asked Walker to investigate the seasonal monsoons that affected the Indian subcontinent. They were concerned with the periodic droughts that occurred in the region. Walker began to sift through the meteorological records of the area and discovered a relationship between the droughts and a change in atmospheric pressure, which he termed the southern oscillation. Walker noticed that when there was high atmospheric pressure in the western Pacific Ocean, there was low atmospheric pressure in the eastern Pacific Ocean. Walker then charted a pattern of precipitation that was associated with this phenomena, which is now called the Walker circulation. He also became the first scientist to propose that regional changes in weather can lead to changes in global climate. Later, scientists linked Walker's southern oscillation to the phenomenon known as El Niño.

the lows would form farther east than normal. The result was a change in global climate patterns. This was the first time that the El Niño event and the southern oscillation were linked together. The two phenomena have been linked ever since and are now known as the **El Niño southern oscillation,** or **ENSO.**

The Effects of an El Niño Event

The occurrence of an ENSO event can be traced back to the year 1567, when Peruvian fisherman were first known to experience the warm waters of an El Niño and the reduced fish populations that are associated with it. Since that time, it is believed that more than 25 ENSO events have occurred. Not all ENSO events are the same; they vary in their duration, water temperature, and effect on atmospheric pressure.

A ENSO event affects global climate by altering the moisture patterns that occur around the world (Figure 29–5). During normal conditions, low-pressure systems develop over the southwestern Pacific Ocean and bring rains to the region. During an ENSO event, the low-pressure system moves eastward, bringing the moisture with it. This causes drought conditions in parts of Australia, the South Pacific, and India. This also causes wet conditions and more storms to form over western North America and Central America. In northeastern North America, warmer winters are created, and the southern United States experiences more rainfall.

One of the most recent and most severe ENSO events occurred from 1982 to 1983 and resulted in terrible drought conditions in parts of Australia, South Africa, and Indonesia. It

Figure 29–4 The abnormally high sea surface temperatures along the equator during an El Niño event. *(Courtesy of Corbis.)*

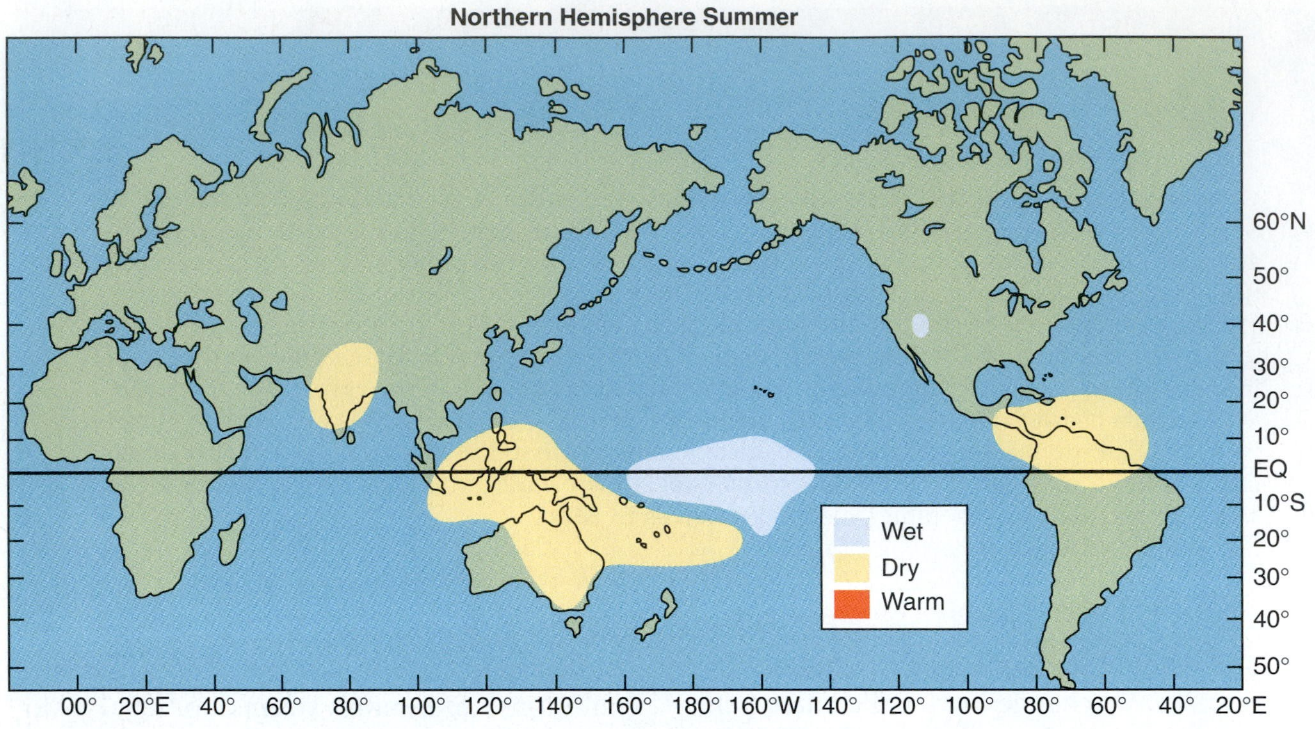

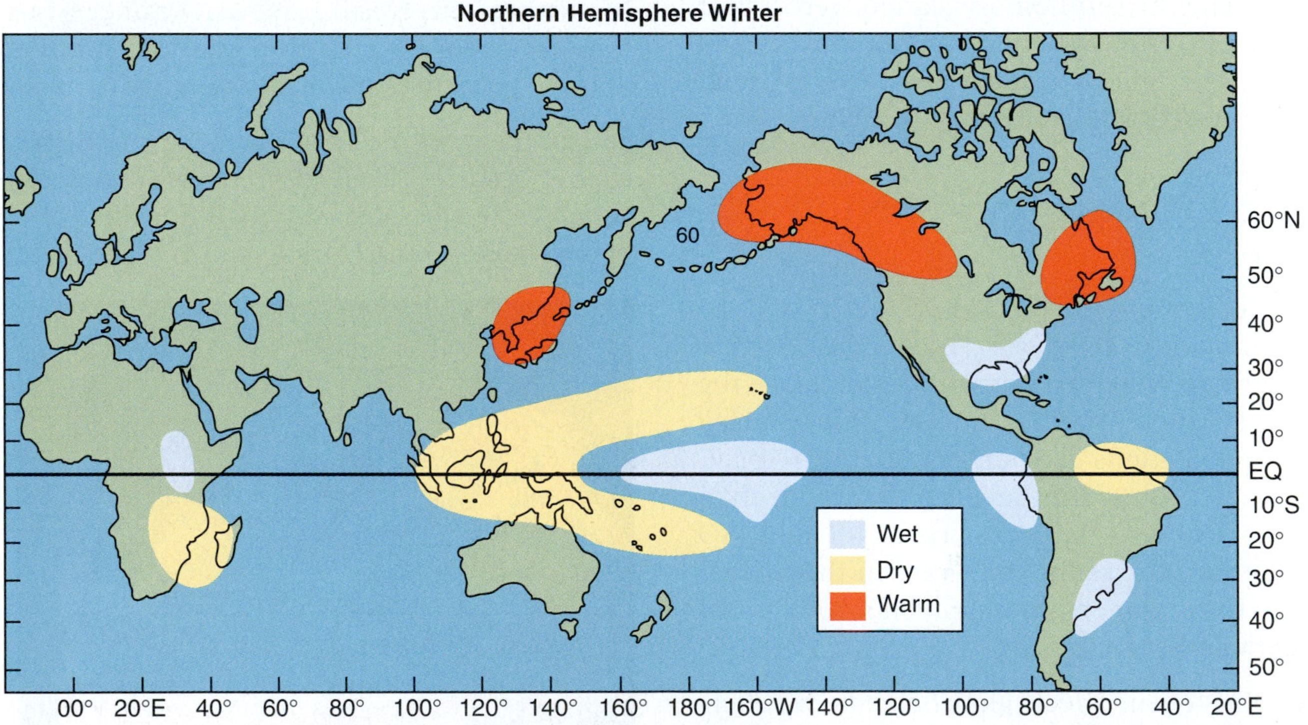

Figure 29–5 The effects on global climate during an El Niño event.

caused grain crop production in Australia to be cut in half. The abnormally warm waters off the coast of South America during this ENSO event caused a massive die-off of fish and sea birds, which was the result of the loss of cold, nutrient-rich water that affected their food supply. The commercial fish catch for that year was reduced by 50% compared with the previous year's catch. This event also resulted in excessive storms along the California coast, creating floods and widespread shoreline damage. The last El Niño event occurred in 1998, and it was the strongest ENSO event on record. It caused severe flooding in Chile and South America and heavy rainfall in parts of North America.

La Niña Events

The opposite of an ENSO event is known as **La Niña.** A La Niña event occurs when colder ocean currents spread farther east across the Pacific ocean.

A La Niña event causes a shift in atmospheric pressure and moisture, which has the opposite effects on climate as El Niño (Figure 29–6). In 1999 the last La Niña caused a

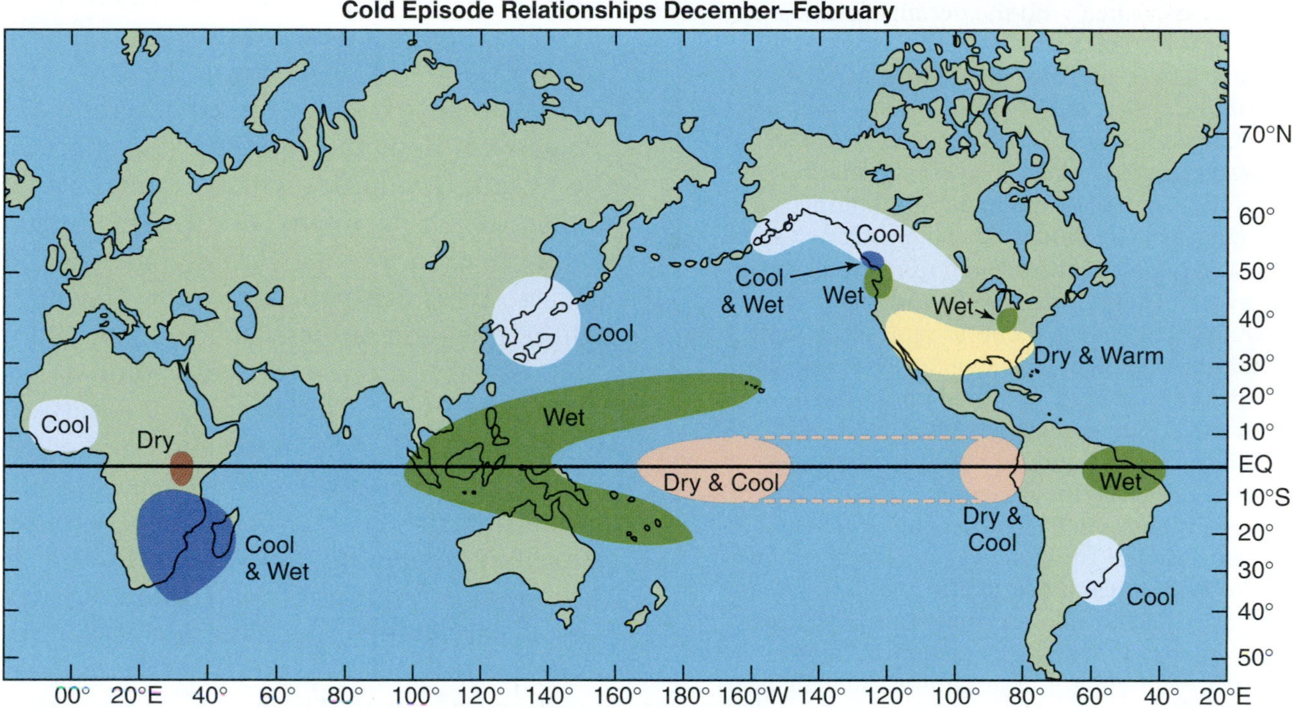

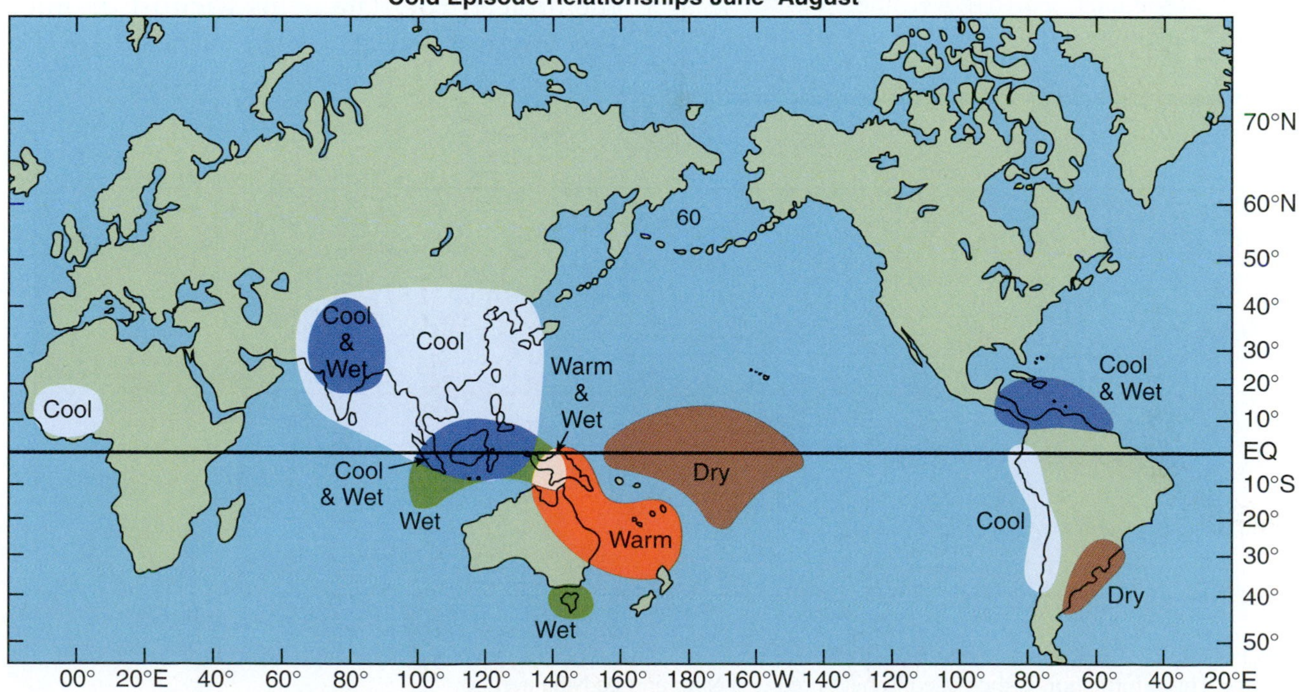

Figure 29–6 The effects on global climate during a La Niña event.

Career Connections

Marine Engineer

Marine engineers are specialized engineers that design equipment for studying all aspects of the ocean. They also design structures that must withstand the rough conditions associated with the ocean and along coastlines. This includes breakwaters, jetties, oil drilling platforms, docks, piers, and other marine structures. Their knowledge must include an understanding of the effects of saltwater, currents, tides, waves, and marine organisms on structural materials. Marine engineers also design and help construct scientific equipment that is used to research the oceans. These devices include specialized buoys that collect scientific data, water and marine organism sampling devices, remotely operated vehicles, and small research submarines. These engineers often work with other scientists and specialists to best meet their needs for the design and construction of specific equipment. Marine engineers often accompany the equipment they design out into the field to test it and to ensure it is operating properly. They also go along on scientific expeditions to make adjustments and repairs to their equipment if necessary. Marine engineers must have a college education specializing in ocean engineering. They can work for the military, academic research, government agencies, or private industry.

heavy monsoon season in Indonesia and dry conditions in the southern United States. There appears to be no connection between El Niño and La Niña years; however, these events are still not fully understood.

Monitoring the Pacific Ocean

After the strong ENSO event of 1983 to 1984, the National Oceanic and Atmospheric Administration (NOAA) began the Equatorial Pacific Ocean Climate Studies program. The project was designed to study the effects of the El Niño southern oscillation and possibly predict when it might occur. This would greatly help to forecast climate changes that are associated with these events. The highlight of this program was the development of the Tropical Atmosphere Ocean, or TAO, array (Figure 29–7). This consists of a network of 400 **buoys** that lie along the equator between 10 degrees north and 10 degrees south latitude in the Pacific Ocean. These buoys record air temperature, wind speed and direction, relative humidity, barometric pressure, ocean temperature at depths to approximately 1500 feet, and ocean current direction and speed. The buoys transmit their data via satellite to researchers working on the project. This data helps scientists monitor the physical aspects of the atmosphere and

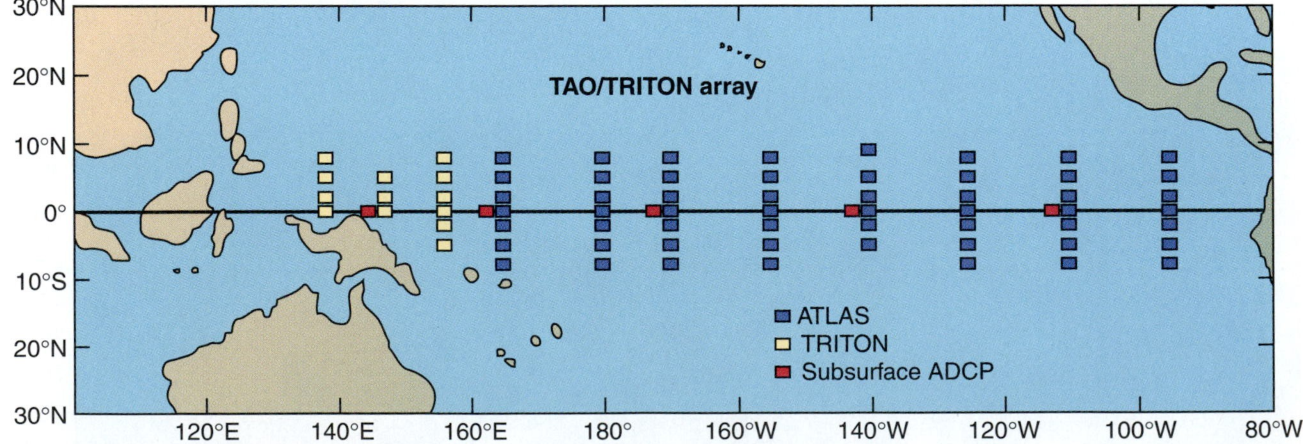

Figure 29–7 The location of the TAO array, which records data about the state of the atmosphere and the ocean along the equator. This information is then used to help predict El Niño and La Niña events.

ocean in that region and will be used to help forecast future ENSO events.

The ENSO and La Niña events are a perfect example of how the systems of the Earth interact to form unique patterns of global climate. The collaboration of meteorologists and oceanographers has made it possible to begin to understand this complex relationship that exists between the hydrosphere and the atmosphere.

REVIEW

1. Who were the first people to experience the effects of El Niño, and why?
2. What was the importance of Sir Gilbert Walker's work in the 1920s?
3. What did Jacob Bjerknes reveal in the 1950s about the southern oscillation?
4. Describe the conditions of the ocean and the atmosphere that lead to an ENSO event.
5. What effects does an ENSO event have on the Earth?
6. What is a La Niña event, and how does it affect the Earth?
7. What is the TAO array, and what does it monitor?

For more information go to these Web links:

<http://www.elnino.noaa.gov/>

<http://observe.arc.nasa.gov/nasa/earth/el_nino/elnino.html>

<http://www.cotf.edu/ete/modules/elnino/elnino.html>

CHAPTER SUMMARY

An El Niño event is linked to changes in the surface ocean temperatures near the equator off the western coast of South America. The term *El Niño* is Spanish for "the child," which was the term used by Peruvian fisherman to describe the occurrence of warm ocean water that occurred periodically during Christmas. During normal conditions, the trade winds move warm water from east to west along the equator. This produces the ocean current called the southern equatorial current. This movement causes cold, nutrient-rich ocean water to rise to the surface off the west coast of South America. This process is known as upwelling, which brings nutrients to fish and other marine organisms that live in this region.

An El Niño event occurs when the trade winds begin to slow down and cause the south equatorial current to also slow. This prevents the movement of warm ocean water off the coast of South America and reduces upwelling. The loss of nutrients as a result of reduced upwelling causes a decline in fish populations.

During the 1920s, British meteorologist Gilbert Walker began to study the periodic weather patterns that form seasonal monsoons in the South Pacific Ocean. He discovered that low atmospheric pressure was associated with warm water in the Pacific Ocean and high atmospheric pressure with cooler ocean water. He hypothesized that these pressure systems formed the seasonal monsoons and were linked to ocean currents along the equator. He called this phenomenon the

EARTH MATH

1) IF ENSO EVENTS OCCURRED IN THE YEARS 1900, 1902, 1915, 1920, 1941, 1973, 1983, 1988, 1995, AND 1998, WHAT IS THE AVERAGE PERIOD OF THEIR OCCURRENCE?
2) USING THE PERIOD YOU DETERMINED IN THE PREVIOUS QUESTION, WHEN DO YOU PREDICT THAT THE NEXT ENSO EVENT MIGHT OCCUR?

southern oscillation. Later it was discovered that the southern oscillation was linked to the changes in ocean temperatures that caused El Niño events. The change in surface ocean water temperatures was then linked to changes in regional climate and became known as the El Niño southern oscillation, or ENSO. An ENSO event can cause widespread changes in climate conditions, resulting in droughts, floods, changes in precipitation patterns, and reduced fish populations. The opposite of an ENSO event is called a La Niña event. This occurs when cooler ocean currents extend farther west in the Pacific ocean. This also affects worldwide climate patterns.

To better predict the occurrence of these events, scientists have deployed a series of buoys in the Pacific Ocean along the equator. These buoys record the physical aspects of both the atmosphere and ocean in this region to help researchers better understand these phenomena and better prepare the world for the resulting changes in climate they may cause.

CHAPTER REVIEW

Multiple Choice

1. Which planetary scale wind is associated with an El Niño event?
   a. the westerlies
   b. the trade winds
   c. the jet stream
   d. a sea breeze
2. What process moves cold ocean water toward the surface?
   a. the Coriolis effect
   b. monsoons
   c. upwelling
   d. tides
3. Which surface ocean current is associated with an El Niño event?
   a. the gulf stream
   b. the North Atlantic current
   c. the equatorial countercurrent
   d. the southern equatorial current
4. Warm ocean water in the Pacific Ocean is responsible for forming:
   a. high atmospheric pressure and dry weather
   b. low atmospheric pressure and wet weather
   c. upwelling
   d. La Niña
5. The periodic shifting of low- and high-pressure systems over the Pacific Ocean near the equator is called:
   a. La Niña
   b. the southern oscillation
   c. upwelling
   d. monsoons
6. An El Niño event is associated with:
   a. colder ocean water, low atmospheric pressure moving east, and increased upwelling.
   b. warmer ocean water, high atmospheric pressure moving west, and decreased upwelling
   c. warmer ocean water, low atmospheric pressure moving west, and decreased upwelling
   d. colder ocean water, high atmospheric pressure moving east, and increased upwelling
7. An ENSO event causes what changes in global climate?
   a. drought in southeast Asia and Australia
   b. heavy rain in southeast Asia and Australia
   c. drought in North America
   d. drought in Central America
8. During non–El Niño years, what normal climate conditions exist?
   a. drought in southeast Asia and Australia
   b. heavy rain in southeast Asia and Australia
   c. drought in North America
   d. drought in Central America
9. A La Niña event is associated with:
   a. colder ocean water, low atmospheric pressure moving east, and increased upwelling.
   b. warmer ocean water, high atmospheric pressure moving west, and decreased upwelling
   c. warmer ocean water, low atmospheric pressure moving west, and decreased upwelling
   d. colder ocean water, high atmospheric pressure moving east, and increased upwelling
10. During a La Niña event, what climate conditions exist?
    a. drought in Southeast Asia and Australia
    b. heavy rain in Southeast Asia and Australia
    c. heavy rain in North America
    d. increased hurricanes in Central America

Matching *Match the terms with the correct definitions.*

a. climate
b. trade winds
c. south equatorial current
d. upwelling
e. southern oscillation
f. El Niño
g. La Niña
h. buoy

1. ____ The uplift of cold ocean water from the bottom to the surface.

2. ____ When sea surface temperatures along the equator off the western coast of South America are cooler than normal, resulting in widespread change in climate.

3. ____ A wind-driven, surface ocean current that moves water westward along the equator away from the west coast of South America.

4. ____ The long-term weather patterns of a specific region on the Earth, usually defined by the area's annual temperature and precipitation values.

5. ____ An increase in the sea surface temperature off the western coast of South America near the equator that leads to changes in climate around the world.

6. ____ The periodic change in the locations of low and high atmospheric pressure systems over the Pacific Ocean near the equator.

7. ____ Planetary scale winds in the Earth's atmosphere that form when areas of high atmospheric pressure, located near 30 degrees north and south of the equator, move air toward low pressure at the equator.

8. ____ A floating marker that is anchored to a specific spot in a body of water.

Critical Thinking

1. Describe two ways in which the Sun is related to an ENSO event.

CHAPTER

30 Coral Bleaching

Objectives

Coral and Coral Reef Systems • The Bleaching of Corals • Occurrence of Coral Bleaching • Causes of Coral Bleaching

After reading this chapter you should be able to:

- ❖ Describe the environment in which coral reefs are located.
- ❖ Explain the symbiotic relationship between algae and coral.
- ❖ Describe the process by which coral reefs are formed.
- ❖ Define the term *coral bleaching*.
- ❖ Identify five possible causes of coral bleaching.

TERMS TO KNOW

coral reefs	**algae**	**herbicides**
corals	**coral bleaching**	**wastewater**
symbiotic relationship	**El Niño southern oscillation**	**infectious agents**

INTRODUCTION

Coral reefs are one of the most amazing and beautiful living structures on the Earth. Almost everyone has seen pictures or moving images of the variety of colors and diversity of the marine organisms that live there. Because of the incredible diversity of marine life that coral reefs support, they are often known as the rainforests of the sea. These beautiful natural treasures, however, are now being threatened around the world by a phenomenon known as coral bleaching. Studying the causes and effects of coral bleaching reveals the careful balance that exists between the living and nonliving aspects of the Earth and how human activity may be harming these delicate structures.

Coral and Coral Reef Systems

Coral reefs are located in shallow, warm ocean waters and are composed of colonies of small organisms known a **corals.** Corals are marine organisms that build protective shells from calcium carbonate and attach themselves to one another with a gluelike substance. They filter feed on tiny plankton and debris that float freely through the water. Corals maintain a **symbiotic relationship** with single-celled **algae** that reside in the protective shell of the coral. This algae help provide the coral with nutrients, while the coral provides carbon dioxide and a home for the algae. Over time, millions of these organisms form huge underwater structures called coral reefs (Figure 30–1).

The Bleaching of Corals

In recent years a phenomena known as **coral bleaching** has been occurring at high rate in reefs around the world. Coral bleaching is the whitening of corals caused by the death of the algae that reside inside. These algae give coral their unique colors, and when they die, the coral appears white, or bleached. The death of the algae within the coral can eventually lead to the death of the coral itself (Figure 30–2). If the environmental stress that causes the death of the algae ends, it may be possible for the coral to recover. However, the longer the coral survives without the symbiotic algae, the greater the chance that it will not be able to recover.

Figure 30–1 Coral reefs, the rainforests of the sea, are one of nature's most beautiful structures and are home to a variety of colorful marine life. *(Courtesy of PhotoDisc.)*

Figure 30–2 This set of photographs, taken from the same vantage point 10 years apart, illustrates the rapid degradation of coral reefs that has occurrred throughout the Florida Keys and Caribbean Sea. Carysfort Reef, the largest and most luxuriant reef in the Keys, has lost more than 92% of it living coral cover from pollution, disease, and physical damage. Ocean warming, which can cause bleaching, adds additional stress to already threatened reefs. *(Photos © P. Dustan.)*

Occurrence of Coral Bleaching

Scientists have been interested in coral bleaching because it is becoming more common around the world and may be caused by environmental stress (Figure 30–3). Since 1980 the occurrence of coral bleaching has dramatically increased, causing concern for the fate of coral reef systems around the world. More than 57% of coral deaths that occurred between 1979 and 1990 were attributed to coral bleaching. This is a substantial increase compared with a 4% coral death rate from bleaching during the previous 100 years.

Causes of Coral Bleaching

Scientists have identified several factors that may cause coral bleaching. Changes in the temperature of the water in which the corals reside can cause the death of algae. A sudden drop in temperature by 5° or 10° for a period of 10 days or more may cause the death of the algae. A sudden increase in water temperature by 2° to 4° for a period of months can also cause the die-off. Two factors could be leading to the temperature fluctuations in the oceans that are affecting the coral. The **El Niño southern oscillation,** which affects water temperatures near the equatorial Pacific Ocean, may impact some regional coral reefs. Increasing global temperatures may be leading to an increase of ocean temperatures in some regions. Satellite analysis of global sea surface temperatures shows that some regions of the world are experiencing a rise in temperature (Figure 30–4).

Increased exposure to ultraviolet radiation in shallow-water corals is believed to cause coral bleaching. The algae cannot survive the deadly effects of ultraviolet radiation on its sensitive tissue. The increase in ultraviolet radiation may be the result of ozone depletion in the stratosphere. Increased sediments entering the ocean as a result of erosion, runoff, and heavy rains may be leading to the death

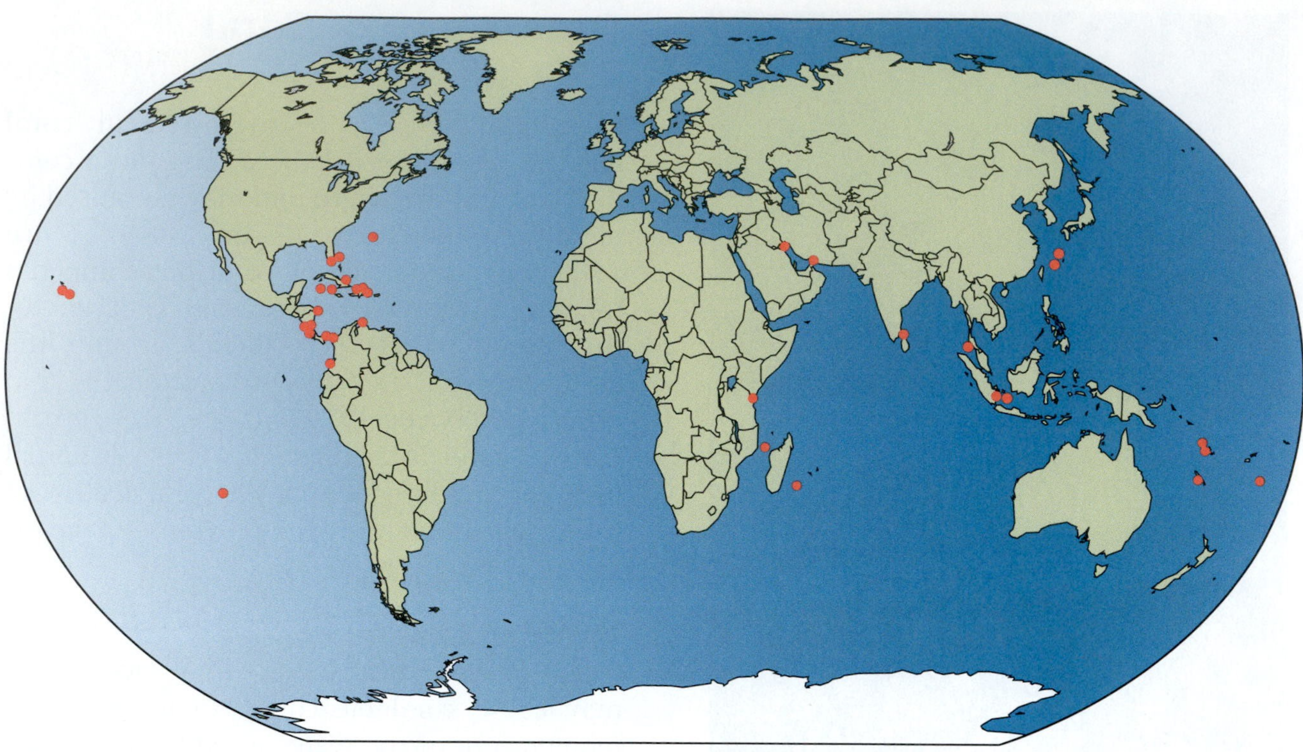

Figure 30–3 Areas around the world that are affected by coral bleaching.

of corals near the shoreline. The sediments cloud the water and prevent sunlight from reaching the algae within the corals. This reduces their ability to use photosynthesis to produce food. Also, toxic chemicals such as **herbicides,** oils, and other organic compounds that are washed into the ocean may cause coral bleaching. These compounds can kill off the algae and harm the coral. The input of **wastewater** from human sewage or animal manure can also negatively affect corals. Wastewater often contains **infectious agents** such as bacteria and viruses that impact the reef system.

EARTH SYSTEM SCIENTISTS *DR. SYLVIA EARLE*

Sylvia Earle was born in New Jersey in 1935 and received her education in marine biology. She has spent most of her life exploring the world's oceans. She was one of the first women scientists to use scuba gear to study life below the surface. Since 1961 Earle has authored more than 100 publications about life within the sea. In 1970 she participated in the Hydrolab project, which was an underwater laboratory experiment sponsored by the U. S. government. She spent 2 weeks living underwater and helped pave the way for the National Aeronautics and Space Administration's Skylab space station. Earle used the Hydrolab to study coral reefs off the coast of Florida. In 1979 she made the world's deepest untethered dive while strapped to the front of a submarine that took her to a depth of 1250 feet. She then walked around on the bottom of the ocean in a special dive suit and planted an American flag. In 1981 Earle co-founded Deep Ocean Engineering, a company that designs and builds underwater research equipment. She also became the first woman to serve as the chief scientist for the National Oceanographic and Atmospheric Administration (NOAA). Today Earle continues her study of the marine environment and her efforts to prevent its destruction.

Career Connections

CORAL REEF SCIENTIST

A coral reef scientist studies all aspects of a coral reef ecosystem. This includes an understanding of the life cycle and other biological properties of corals, as well as coral habitats and their distribution around the world and the interactions of other marine organisms with coral reefs and the reefs' relationship to other marine ecosystems. Today much of the work of a coral reef scientist involves the conservation of coral reef systems and the understanding of how human activity is affecting these important marine resources. Coral reef scientists must be certified scuba divers, because much of their work involves the underwater observation of coral reefs and the collection of both living and nonliving samples from the reef system. A love for working on and in the ocean, a willingness to travel to exotic locations, and a college education in marine biology and oceanography are musts for this type of occupation. Career opportunities are available in academic research or with government agencies.

Current Research

Research conducted by the National Center for Atmospheric Research has revealed that increased levels of carbon dioxide in the Earth's atmosphere may be causing the deaths of coral reefs. A report prepared by John Klepas has shown that in the laboratory, increased levels of carbon dioxide have caused the decline of coral. Carbon dioxide gas in the atmosphere is absorbed into the oceans, where it disrupts the natural chemistry of the water. The team of researchers has shown that the higher levels of carbon dioxide in the ocean interrupt the uptake of calcium carbonate that corals use to build their protective shells. This research has revealed the first possible negative effects of increasing levels of carbon dioxide on a marine ecosystem. The scientists hope that their research will help to reduce the global emissions of carbon dioxide and save the world's fragile coral reefs.

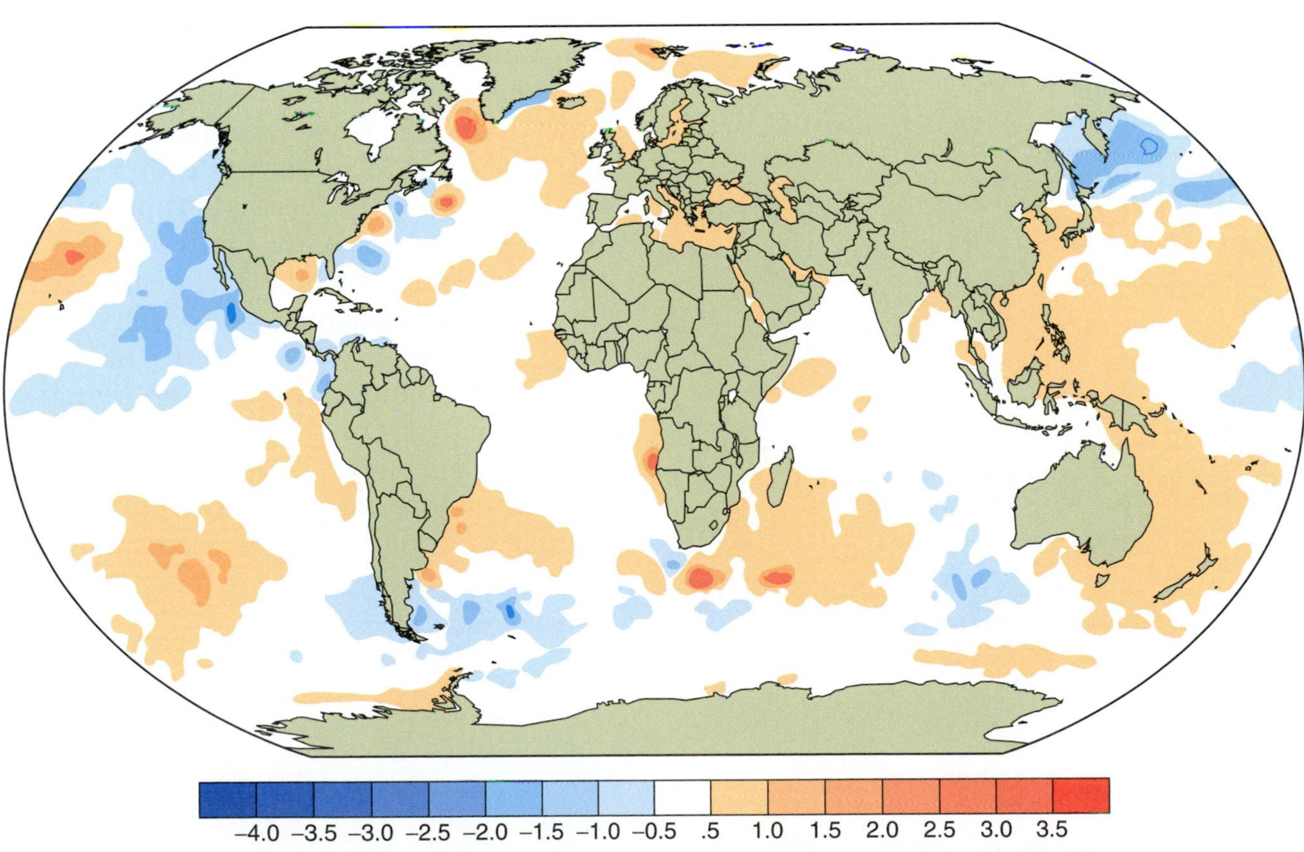

Figure 30–4 Change in sea surface temperatures around the world is one of the causes of coral bleaching.

EARTH MATH

1) A TYPICAL BRAIN CORAL GROWS ABOUT 0.14 INCHES PER YEAR. IF IT IS 36 INCHES TALL TODAY, APPROXIMATELY HOW OLD IS THIS CORAL?

2) CORAL REEFS OFF THE ISLAND OF JAMAICA RISE 40 FEET OFF THE OCEAN FLOOR AND TOOK APPROXIMATELY 5000 YEARS TO FORM. USING THIS INFORMATION, APPROXIMATELY HOW MUCH DOES THIS REEF GROW IN INCHES EACH YEAR?

The many environmental stresses that lead to coral bleaching may be the result of human activity and the disruption of the natural cycles within Earth's system. Not all the possible causes may be prevented by human action, but some human practices could be changed, which would lessen the impact of coral bleaching on the world's reef systems.

For more information go to this Web link:
<http://www.coralreef.noaa.gov/>

REVIEW

1. What is coral?
2. Why are coral reefs important, and where are they found?
3. What is coral bleaching?
4. What may cause coral bleaching?
5. Who is Dr. Sylvia Earle?

CHAPTER SUMMARY

Coral reefs support a great amount of diversity within the ocean and are one of nature's most beautiful structures. Coral reefs are composed of colonies of tiny marine organisms called corals. These creatures filter out plankton and organic debris from the water as a source of food. They also build protective shells out of calcium carbonate, which builds up over time to form large structures called coral reefs. Corals have a symbiotic relationship with algae that reside within the body of the coral. The algae produce food for the coral, while the coral provides carbon dioxide for the algae. Coral reefs are located in warm, shallow ocean water.

A phenomenon known as coral bleaching is killing off many corals around the world. Coral bleaching is caused by the death of the symbiotic algae that live within the coral. This causes the corals to appear white, or bleached. More than 50% of the deaths of corals around the world since 1980 are the result of coral bleaching. Coral bleaching is caused by many different environmental stresses, including changes in water temperature, increased amounts of sediments in the ocean, runoff of pesticides, toxic chemicals spilled in the ocean, and animal waste. Most of these causes are the result of human activity, which can be controlled to prevent further damage to the world's coral reef systems.

CHAPTER REVIEW

Multiple Choice

1. Coral reefs are located in what type of environment?
   a. deep, cold saltwater
   b. deep, cold freshwater
   c. shallow, warm saltwater
   d. shallow, warm freshwater

2. Which organism has a symbiotic relationship with coral?
   a. zooplankton
   b. tiny fish
   c. humans
   d. algae

3. Coral reefs are mostly composed of:
   a. calcium carbonate
   b. algae
   c. carbon dioxide
   d. sand

4. The white appearance of bleached coral is the result of:
   a. excess chlorine in the water
   b. excess carbon dioxide in the water
   c. the death of the symbiotic algae
   d. the death of the coral

5. Approximately what percentage of coral deaths since 1980 were caused by coral bleaching?
   a. 10%
   b. 27%
   c. 57%
   d. 100%

6. Which of the following does not cause coral bleaching?
   a. increased salinity
   b. change in water temperature
   c. sediment pollution
   d. ultraviolet radiation

continued

Matching *Match the terms with the correct definitions.*

a. coral reefs
b. corals
c. symbiotic
d. algae
e. coral bleaching
f. El Niño southern oscillation
g. herbicides
h. wastewater
i. infectious agents

1. ____ Single-celled or multicelled aquatic organisms that derive their energy from photosynthesis.

2. ____ A group of chemicals that are used to kill or control the growth of undesired plants.

3. ____ Large underwater structures located in warm, shallow saltwater that are built by colonies of coral and are composed primarily of calcium carbonate and sand.

4. ____ A relationship between two different species of organisms in which one or both organisms benefit from the action of the other.

5. ____ Polluted water that is unfit for drinking or introduction into the environment because it has been used for some purpose, and made unclean.

6. ____ The whitening of living coral as a result of the die-off of the coral's symbiotic algae, possibly leading to the death of the coral itself.

7. ____ An increase in the sea surface temperature off the western coast of South America near the Equator, which leads to changes in climate around the world.

8. ____ Any substance or organism that transmits disease.

9. ____ A group of benthic aquatic organisms that live in warm, shallow saltwater and may build shells from calcium carbonate.

Critical Thinking

1. Describe the specific human activities that have led to coral bleaching by increased runoff, increased global temperatures, and increased ultraviolet radiation reaching the Earth's surface.

UNIT 6

The Biosphere

Topics to be presented in this unit include:

- Ecological Systems
- World Biomes and Marine Ecosystems
- The Flow of Energy and Matter through Ecosystems
- Biological Succession
- Classification of the Living World

OVERVIEW

The interactions that occur between the living and nonliving world are an important part of the Earth system as a whole. Many of the physical aspects of our planet have been shaped by living organisms. The gases that compose the atmosphere are the result of biological activity. The soils that support all the world's plants are a mixture of both living and nonliving components. And the effects of the human race on the Earth are altering the planet at unprecedented rates. It is with this in mind that the biosphere becomes another important part of the Earth's systems.

CHAPTER 31
Ecological Systems

Objectives

Ecology • Habitats • Populations • Communities • Ecosystems

After reading this chapter you should be able to:

- Define the terms *ecology, biotic,* and *abiotic.*
- Identify the four aspects that make an organism's habitat.
- Define the terms *population* and *species* and provide two examples of a population.
- Differentiate between specialist and generalist populations.
- Define the term *community* and describe three examples of symbiotic relationships.
- Define the term *ecosystem* and identify four abiotic factors that exist within ecosystems.
- Describe the ecosphere.

TERMS TO KNOW

ecology
biotic
abiotic
habitat
population
species
community
symbiotic relationship
mutualism
lichen
commensalism
parasitism
ecosystem
ecosphere

INTRODUCTION

The study of living things and the way that they interact with the physical environment is called ecology. The word **ecology** is derived from the ancient Greek words *oikos,* meaning "house" or "place to live," and *logic,* meaning "to learn." Indeed, ecology is the study of where organisms live. Learning about how living organisms interact with the environment is important for determining what role they play in the Earth's systems and how best to maintain them for future generations.

Ecology

Two words that are important to the study of ecology are *biotic* and *abiotic.* The term **biotic** refers to any living organism, including plants, animals, bacteria, and fungi. **Abiotic** refers to the nonliving factors of the environment that interact with the biotic world. These include air, water, rocks, minerals, temperature, altitude, light, and soil. Scientists who study ecology are called ecologists. Ecologists study the interaction of the biotic and abiotic world in different levels of relationships on the Earth.

Habitats

The smallest level of relationship that ecologists study is an organism's **habitat.** *Habitat* refers to an organism's specific food, water, shelter, and space requirements. All organisms require these four fundamental things to survive (Figure 31–1). Food requirements for living organisms can vary greatly in the natural world; however, they all form the basis for how an organism derives energy. Some organisms, such as green plants, gain their energy from sunlight. Other organisms must consume other living things to gain energy.

Another important habitat requirement is a source of water. All living things on the Earth require water to survive. Some organisms spend their entire lives in water; these are called aquatic organisms. Organisms that live on the land, called terrestrial organisms, must absorb water from the environment.

The third habitat requirement for all living things is shelter. Shelter is the particular way that an organism protects itself from the environment. This can include protective body parts, such as shells, or the ability to find or build a shelter.

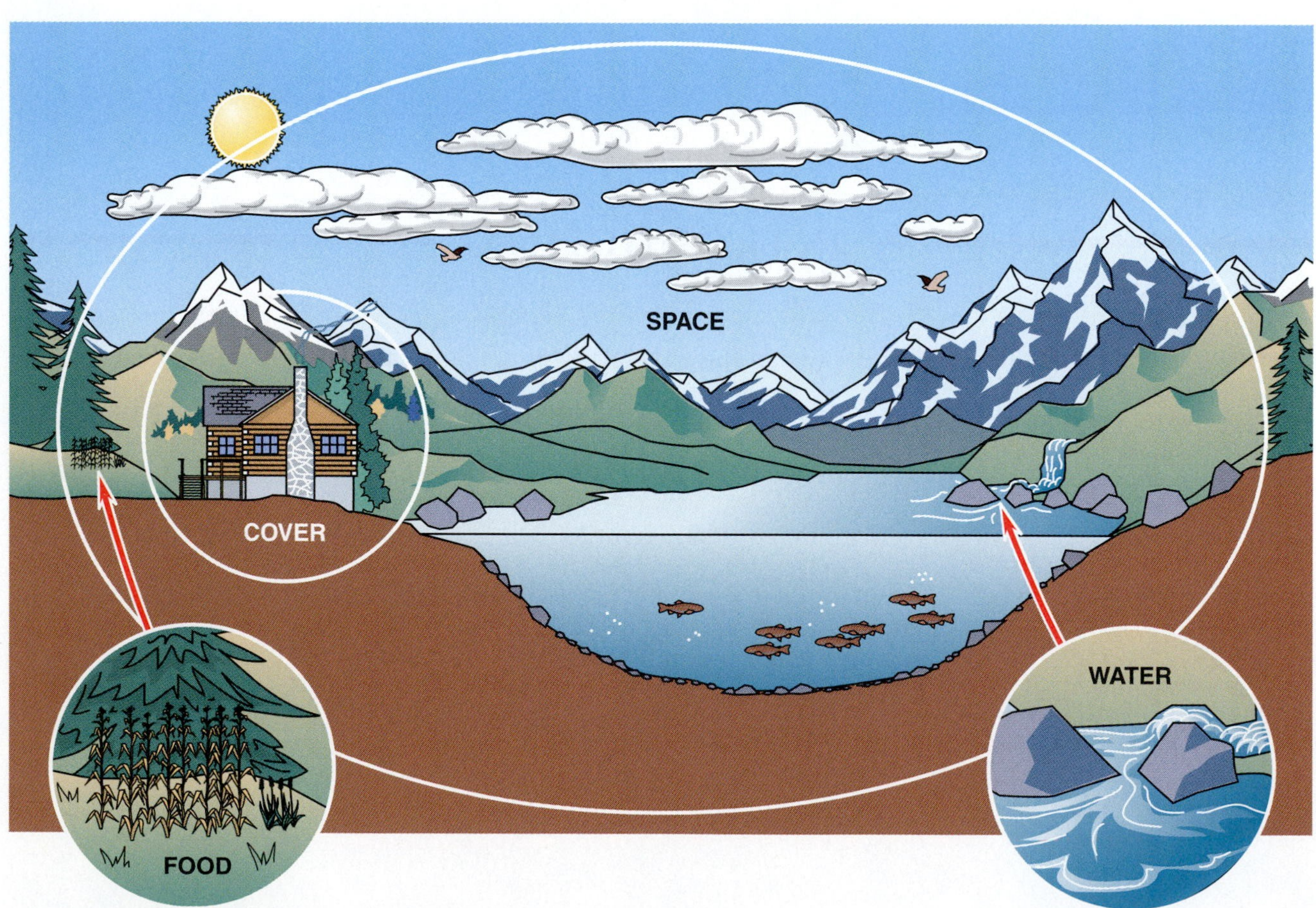

Figure 31–1 The four elements that together make up an organism's habitat.

Career Connections

ECOLOGIST

An ecologist studies the ecological issues associated with a wide range of environments. These scientists use this knowledge to solve problems with supplying food and shelter and also improving the health of human beings. Their job also includes monitoring and managing natural resources to sustain ecosystems and wildlife populations. Ecologists help to design sustainable land use plans and help develop environmental impact statements. Much of their work today involves using their knowledge to solve many environmental problems around the world, especially those associated with habitat destruction. These scientists often work in the outdoors gathering data about specific interactions that occur within ecosystems. The field of ecology requires a 4-year college degree and can find employment in research or with government agencies.

The final habitat requirement for all living things is space. All living things require space in which to live their lives. Some organisms, such as bacteria and algae, have extremely small space requirements. Other organisms, such as bears and migratory birds, require large amounts of space to live. The arctic tern is a bird that migrates from the North Pole arctic regions all the way to Antarctica every year. This is the largest space requirement for a living thing on the Earth.

Populations

The next level of relationship that ecologists study is the **population.** A population is a group of the same organisms living in a specific area (Figure 31–2). Groups of the same organism are also known as a **species.** A species is a group of organisms that resemble each other genetically and are able to reproduce with one another. Groups of the same species that live in a particular area are collectively known as a population. Populations can occupy an extremely small space, such as a population of bacteria on a particle of soil, or can be as large as all the human beings that live in the United States.

Many ecologists study specific populations to learn about the way that they interact with the environment. Some populations require special habitat requirements that make them very specialized. If there is an interruption or change in their habitat, these specialist populations can be threatened. Populations that include wolves, elephants, tigers, and many others are being threatened as a result of the loss of their habitats. Other populations, called generalists, can survive in a wide range of habitats. Examples of these types of populations include

Figure 31–2 A population of organisms consists of a group of the same species residing in a particular area.

Figure 31–3 A community consists of all the living things in a specific region.

human beings, squirrels, mice, rats, and many insects. Generalists can tolerate extreme change in their habitats and can easily adapt to their altered surroundings.

Communities

The third level of relationship that ecologists study is the **community.** A community is the interaction of different populations that reside in a particular area (Figure 31–3). Communities are composed of many different species of organisms that share similar habitats. Communities can exist in a very small area, such as the organisms found in one drop of pond water. Communities can also occupy extremely large areas, such as entire forests, lakes, or geographical regions. Often the removal of one species or population from a community can have a negative impact on the other members of the community.

Many populations that together form unique communities depend on each other for survival. The relationships that exist between populations in a community can sometimes be called symbiotic. A **symbiotic relationship** occurs when one or more organisms benefit from the actions of another organism (Figure 31–4). Another type of relationship that occurs in communities is **mutualism,** which is the interaction of two species that both benefit from actions of one another. An example of mutualism is the **lichen** community. A lichen is actually a community of two organisms, an alga and a fungus, that reside together on the surface of rocks and trees. The alga produces food from sunlight, in the form of sugar and starches, through the process of photosynthesis. The fungus consumes some of this food and in return secretes weak acids that help to break down the organic material and minerals from the surfaces on which the lichen reside. This is an important source of nutrients for the alga.

Another type of symbiotic relationship is called **commensalism.** This is a relationship between two organisms in which one organism benefits from another organism without any expense to its host. An example of commensalism is the relationship between the remora fish and their shark hosts. The remora harmlessly attaches itself to the body of a

Figure 31–4 Examples of the symbiotic relationships include mutualism between the algae and fungus of a lichen on rock (left) and parasitism between a tick (right) and a dog. *(Left image courtesy of PhotoDisc.)*

EARTH SYSTEM SCIENTISTS *CHARLES DARWIN*

Charles Darwin was born in Britain in 1809. At the age of 22, he signed on board the H. M. S. Beagle as a naturalist. The mission of the Beagle was to explore the west coast of South America and the islands of the south Pacific Ocean. Over the next 5 years, Darwin collected and recorded the natural history of many parts of the south Pacific, including the Galapagos Islands. Darwin became famous as a result of his journey. Over the next few years he compiled his records from his expedition and published a book about his discoveries. What Darwin saw on his expedition changed his life forever. His observations led him to develop his theory of natural selection, which proposed that species undergo a change in their form as a result of adaptations to their environment. This theory began the modern concept of evolution. Darwin published his theory in 1859 in his book *On the Origin of Species by Means of Natural Selection or the Preservation of Favoured Races in the Struggle of Life.* This controversial book forever changed the viewpoint on how life evolved on the Earth. In 1871 he applied his theory of natural selection and evolution to human beings. Darwin proposed that humans also underwent changes in their form over the years and that our early ancestors would be classified as primates. This sent shockwaves through the religious and scientific worlds but today has become a fundamental aspect of biology.

shark and benefits from the leftovers of the shark's meals.

A third type of symbiotic relationship between species is **parasitism.** This occurs when one organism harms a host organism by taking its nutrition from it. An example of this is a tick on a dog. The tick attaches itself to the dog and feeds on the dog's blood.

Ecosystems

The next level of relationship that ecologists study is called the **ecosystem.** *Ecosystem* is the short word for ecological system. An ecosystem is the interaction of a community of organisms with the abiotic factors in a specific area. Some examples of abiotic factors include the amount of sunlight a region receives and the amount of moisture in an area. The temperature of the environment where the organism resides is also an important abiotic factor, along with the surrounding landscape.

The particular size of an ecosystem varies depending on the region an ecologist is studying. Generally ecosystems cover large geographical areas such as entire forests or even parts of entire continents. The largest ecosystem that scientists study is called the **ecosphere.** This is the relationship between the biotic and abiotic components of the entire planet, which includes every living and nonliving thing on the Earth, and how they interact. Two broad categories of ecosystems on the planet include aquatic ecosystems and terrestrial ecosystems. Aquatic ecosystems exist in water, and terrestrial ecosystems are found on land (Figures 31–5 and 31–6).

EARTH MATH

1) SOME ECOLOGISTS BELIEVE THERE ARE 5 MILLION DIFFERENT SPECIES OF ORGANISMS ON THE EARTH; STILL OTHERS BELIEVE THERE ARE 30 MILLION SPECIES ON THE PLANET. WHAT IS THE AVERAGE ESTIMATED NUMBER OF LIVING SPECIES ON THE EARTH?

Figure 31–5 A terrestrial ecosystem consists of all the biotic and abiotic factors in a specific location on land.

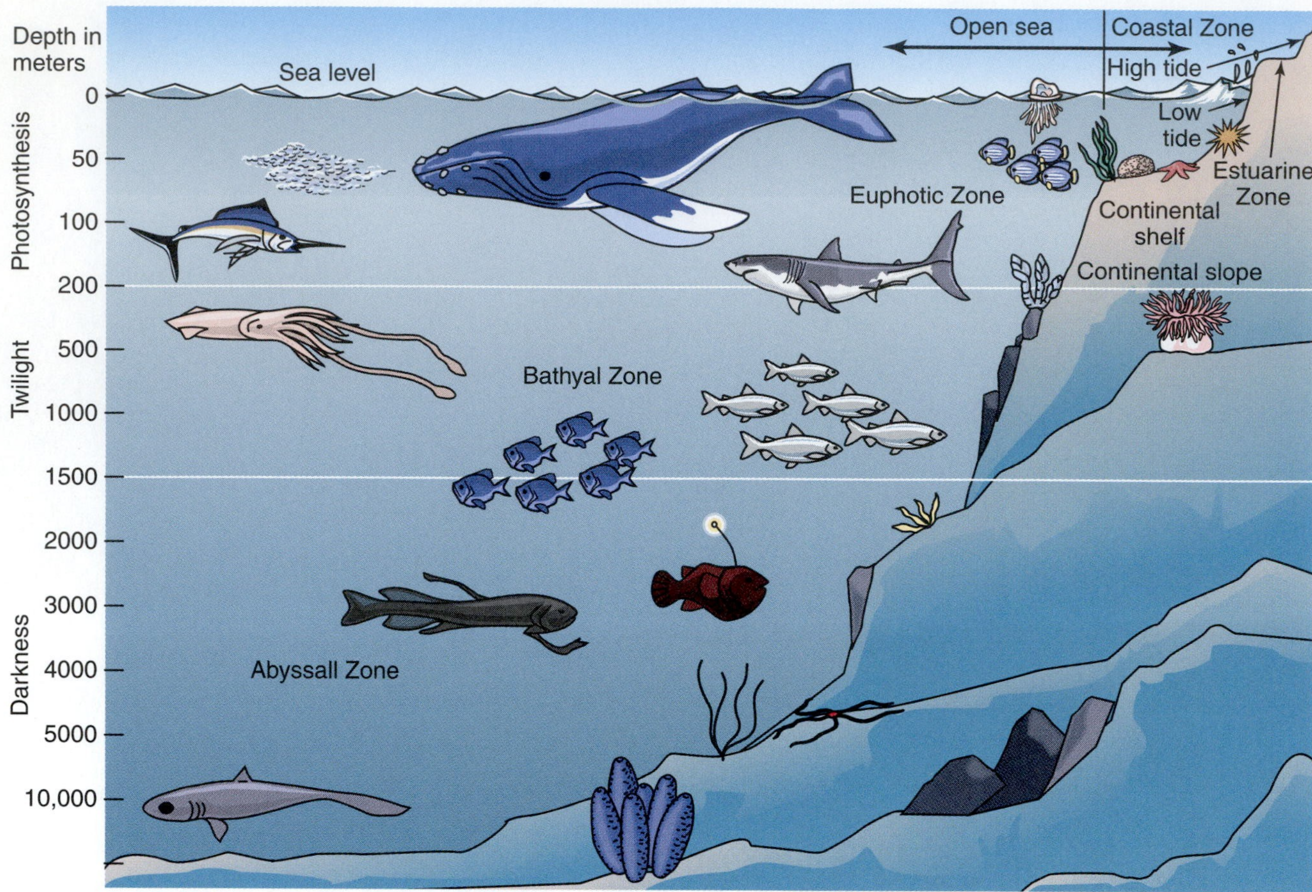

Figure 31–6 An aquatic ecosystem consists of all the biotic and abiotic factors in a specific location within a water environment.

REVIEW

1. Define the term *ecology.*
2. What are the four aspects of an organism's habitat?
3. What is a population of organisms?
4. Define the term *community.*
5. What are three examples of symbiotic relationships?
6. What is an ecosystem?
7. Who was Charles Darwin?

For more information go to this Web link:
<http://www.ucmp.berkeley.edu/glossary/gloss5/biome/>

CHAPTER REVIEW

Ecology is the study of the interactions between organisms and the environment in which they live. An ecologist is a specialized scientist who studies organisms and their environment. This includes a detailed knowledge of the biotic and abiotic factors in an organism's environment. The term *biotic* means "living"; *abiotic* means "nonliving."

Ecologists study the relationship of organisms with their environment on four main levels. The first level of study is called the habitat. A habitat is the specific water, food, shelter, and space requirements for a organism. Every living thing requires these four fundamental items to survive. The next level of relationship studied in ecology is the population.

A population is a group of the same species of organism that live in the same area. A species is a group of organisms that are genetically the same and can breed with one another. For example, all the brook trout that live in the Battenkill River in New York make up a population. Populations can be extremely small, such as bacteria in the soil, or very large, such as the human population.

The next level of ecology is called the community. A community is a group of different populations that live in the same geographical area and interact with each other. Communities can extend over a wide area, such as an entire forest, or exist in a very small space, such as a drop of pond water. The interactions between organisms in a community are important for the survival of many different species. Often if one species is removed from a community, it can harm the remaining populations.

Symbiotic relationships, in which two organisms benefit from one another, can exist within a community. Some specific types of symbiotic relationships include commensalism, mutualism, and parasitism.

The highest level of ecological relationship is called the ecosystem. An ecosystem is the interaction between a community and the abiotic factors in the environment. This relationship includes the interactions between all the living and nonliving things in a specific area. Example of abiotic factors in an ecosystem include the amount of sunlight, temperature, moisture, and nutrient elements. There are two general types of ecosystems recognized on the Earth: aquatic and terrestrial. Aquatic ecosystems exist in water, and terrestrial ecosystems are found on land. The largest ecosystem studied on the Earth is called the ecosphere. The ecosphere is the interaction between all the biotic and abiotic factors on the entire planet.

CHAPTER REVIEW

Multiple Choice

1. An example of an abiotic element is:
   a. lichen
   b. sunlight
   c. trees
   d. bacteria

2. Which of the following is not part of an organism's habitat?
   a. air
   b. space
   c. cover
   d. water

3. An example of a biotic element in an organisms habitat is:
   a. minerals
   b. sunlight
   c. plants
   d. water

4. Organisms that can breed with one another are known as:
   a. generalists
   b. species
   c. communities
   d. specialists

5. All the brook trout in a river are an example of:
   a. a habitat
   b. a population
   c. a community
   d. an ecosystem

6. Human beings are an example of:
   a. a generalist population
   b. a specialist population
   c. a community
   d. abiotic factors in an ecosystem

7. The interaction between the alga and fungus of a lichen is an example of:
   a. commensalism
   b. parasitism
   c. abiosis
   d. mutualism

8. A tick attached to the skin of a dog is an example of:
   a. commensalism
   b. parasitism
   c. abiosis
   d. mutualism

9. The interaction between the living and nonliving aspects of a particular area is called:
   a. a habitat
   b. a population
   c. a community
   d. an ecosystem

10. All the trees in a particular forest are an example of:
    a. a habitat
    b. a population
    c. a community
    d. an ecosystem

11. Which is the highest level of relationship studied by ecologists?
    a. habitat
    b. population
    c. community
    d. ecosystem

Matching *Match the terms with the correct definitions.*

a. ecology
b. biotic
c. abiotic
d. habitat
e. population
f. species
g. community
h. symbiotic relationship
i. mutualism
j. lichen
k. commensalism
l. parasitism
m. ecosystem
n. ecosphere

1. ____ A relationship between two different species of organisms in which one or both organisms benefit from the action of the other.

2. ____ The interaction between the biotic and abiotic factors in a specific area.

3. ____ All the species of organisms that reside in a particular area.

4. ____ The scientific discipline that studies how organisms interact with their environment.

5. ____ The specific food, water, cover, and space requirements for a particular organism.

6. ____ A taxonomical classification used to describe a fundamental group of organisms that can interbreed.

7. ____ A symbiotic relationship between two organisms in which the host organism is harmed by a parasite.

8. ____ A term that describes all the life zones that exist on Earth and is made up of the atmosphere, hydrosphere, lithosphere, and biosphere.

9. ____ A type of symbiotic relationship in which both organisms benefit from each other.

10. ____ A symbiotic organism consisting of an alga and fungus that lives on rocks and trees.

11. ____ A nonliving organism.

12. ____ A living organism.

13. ____ A specific type of symbiotic relationship in which one organism benefits from a host, and the host is neither helped nor harmed.

14. ____ A group of one species of organism that resides in a particular area.

Critical Thinking

1. Explain why the biosphere is an important aspect of the Earth's systems.

CHAPTER 32

World Biomes and Marine Ecosystems

Section 32.1 – World Biomes Objectives

Biomes • Tundra • Coniferous Forests • Temperate Forests • Grasslands • Savannas • Deserts • Tropical Rain Forests • Chaparral • Mountains

After reading this section you should be able to:

- Define the term *biome.*
- Identify the two main factors that create unique biomes on the Earth.
- Describe the unique features of the tundra biome.
- Differentiate between the coniferous forest and deciduous forest biomes.
- Describe the unique features of the grassland biome.
- Differentiate between the grassland and savanna biomes.
- Describe the unique features of the desert biome.
- Describe the unique features of the tropical rain forest biome.
- Describe the unique features of the chaparral and mountain biomes.

Section 32.2 – Marine Ecosystems Objectives

Coastal Wetlands • The Neritic and Intertidal Zones • The Oceanic Zone • The Benthic Zone • Hydrothermal Vent Communities

After reading this section you should be able to:

- Identify the four types of coastal wetlands.
- Differentiate between the neritic zone and the oceanic zones.
- Describe the harsh conditions that exist in the intertidal zone.
- Differentiate between plankton and nekton.
- Describe the conditions that exist in a hydrothermal vent community.

TERMS TO KNOW

terrestrial
biomes
tundra
permafrost
droughts
biological diversity
tropics
aquatic
marine
intertidal zone
phytoplankton
zooplankton
hydrothermal vent

INTRODUCTION

The large-scale interactions between the biotic and abiotic elements of specific regions are known as ecosystems. Much of the world that we live in has been shaped by the relationships that occur in the unique ecosystems that exist around the world. Often the nonliving aspects of a particular region determine what type of organisms reside in a specific ecosystem. Two types of ecosystems cover most of the world's surface: terrestrial ecosystems and aquatic ecosystems. Terrestrial ecosystems occur on the land surface, and aquatic ecosystems exist in water. The terrestrial ecosystems we live in today have been altered by human activity, which has interrupted the natural balance these systems have maintained for thousands of years. Understanding the natural interactions that occur in terrestrial ecosystems is vital to maintaining a healthy environment for humans and the organisms who share the ecosystems in which we live.

Although the oceans cover approximately 70% of the Earth's surface, we are just now beginning to unlock the secrets of the marine ecosystems that exist there. Recent research has revealed the complex interactions that occur within marine ecosystems and how they influence the entire planet. The existence of marine communities that thrive in the harsh environment of the deep ocean are revealing the ability of life to exist in many environments. Studying the interactions between the living and nonliving elements in the world's ecosystems is important for revealing the role that all living things play in the Earth's systems.

32.1 World Biomes

Biomes

A **terrestrial** ecosystem is the land-based relationship between organisms and the nonliving components in a specific area. Terrestrial ecosystems are also called **biomes.** Biomes are large terrestrial ecosystems that cover wide geographical areas. Ecologists recognize nine major biomes around the world, which are classified by two main abiotic factors. These factors include the average yearly temperature and precipitation each region receives.

Tundra

The **tundra** biome is one of the harshest environments on the planet; it is located in the higher latitudes. The tundra supports the northernmost limits of plant growth, with vegetation consisting of moss, lichen, hardy grasses, and small shrubs. Tundra biomes cover approximately 19% of the Earth's land surface and can be found in the higher latitudes in Northern Canada, Greenland, and Northern Asia. The average annual temperature for the tundra biome is 10° Fahrenheit. The soil of the tundra is classified as **permafrost,** which means that it is frozen throughout the year (Figure 32–1).

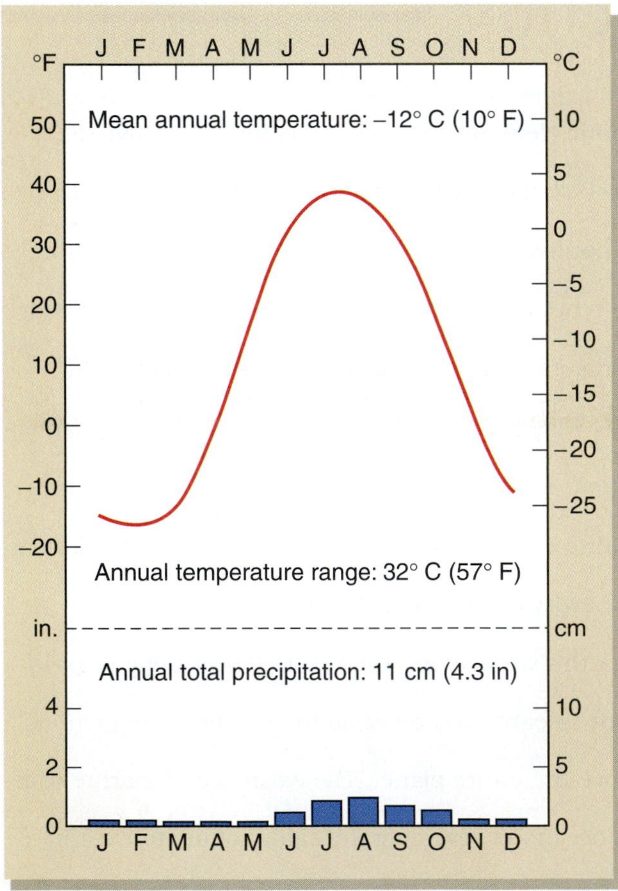

Figure 32–1 The monthly temperature and precipitation values over the course of a year for a tundra biome.

The top portion of the permafrost soil melts to form small pools of liquid water during the short summer months. This is the principal source of liquid water for the organisms that reside there. The tundra biome experiences long, cold winters with barely any sunlight, creating temperatures well below freezing. When the short summers do arrive in the tundra regions, the average high temperatures are around 50° Fahrenheit. The tundra is also considered arid, with less than 10 inches of precipitation each year on average (Figure 32–2). Large organisms that reside in the Tundra biome include musk ox, caribou, arctic fox, polar bear, arctic hare, and human beings.

Coniferous Forests

The next type of biome lies in the latitudes next to the tundra and is called the coniferous forest biome. This biome, also known as the taiga or boreal forest, is located at the higher latitudes just below the tundra and covers approximately 11% of the Earth's land surface. Coniferous forest biomes experience long, cold winters and short, hot summers. As the name states, this biome is home to many cone-bearing tree species, including pine, spruce, and fir. The coniferous forest receives higher amounts of precipitation than the tundra, with averages around 20 inches annually. Temperatures during the summer months can exceed 60° Fahrenheit and then can drop below freezing for most of the winter (Figure 32–3).

Figure 32–2 The low-growing, matlike vegetation of a tundra biome. *(Courtesy of EyeWire.)*

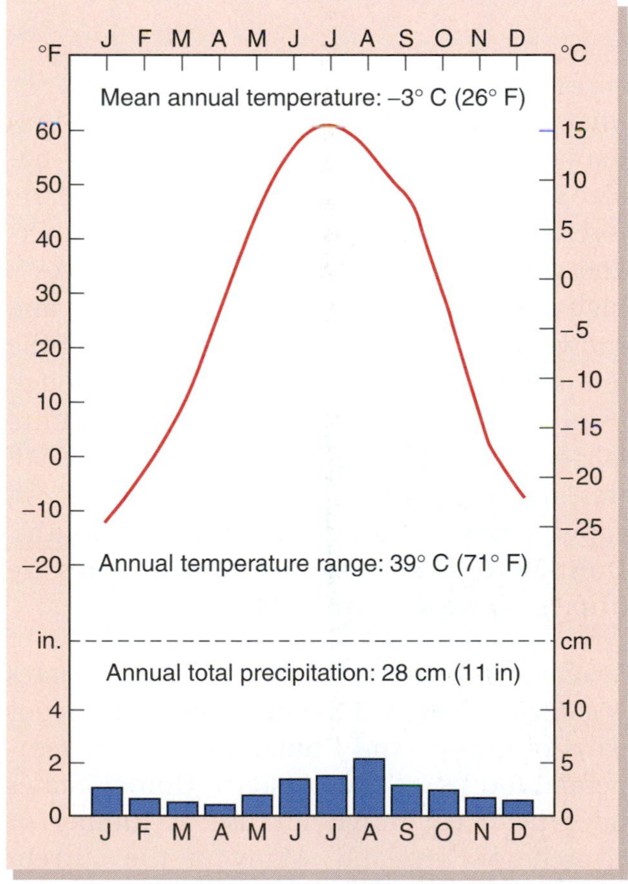

Figure 32–3 The monthly temperature and precipitation values over the course of a year for a coniferous forest biome.

Figure 32–4 Evergreen trees are the principal vegetation in the coniferous forest biome. *(Courtesy of PhotoDisc.)*

Coniferous forests support a wide variety of animal species, including wolves, bears, squirrels, rabbits, hawks, deer, moose, and of course humans (Figure 32–4). Large parts of North America, including parts of the United States and Canada, contain coniferous forests. Much of Northern Europe and Asia also have coniferous forest biomes.

Temperate Forests

The biome that lies adjacent to the coniferous forests, near the middle latitudes, is called the temperate forest biome. This type of climate is also known as a deciduous forest, because the primary form of vegetation growing there is the deciduous tree. A deciduous tree is a type of broad-leafed tree that drops its leaves every fall (Figure 32–5). Common deciduous trees found in the temperate forests include maple, birch, ash, hickory, beech, and oak. The annual change in the color of the leaves in a temperate forest each autumn is one of nature's most beautiful displays. The temperate forest biome experiences long, hot summers and cold winters. This biome also receives a high amount of precipitation in both rain and snow. The average annual temperature of the temperate forest is 49° Fahrenheit, although through most of the winter the temperature falls below freezing (Figure 32–6). Because of this, many parts of the temperate forests are snow covered during the winter months. During the summer months temperatures are usually above 70° Fahrenheit.

Much of the eastern coast of the United States is a deciduous forest biome, along with major parts of Europe and China. Most of the temperate forests of the eastern United States have been cut down for use as lumber or cleared for farmland. Many of the current

Figure 32–5 Deciduous trees lose their leaves each fall in the deciduous forest, also known as the temperate forest biome. *(Courtesy of PhotoDisc.)*

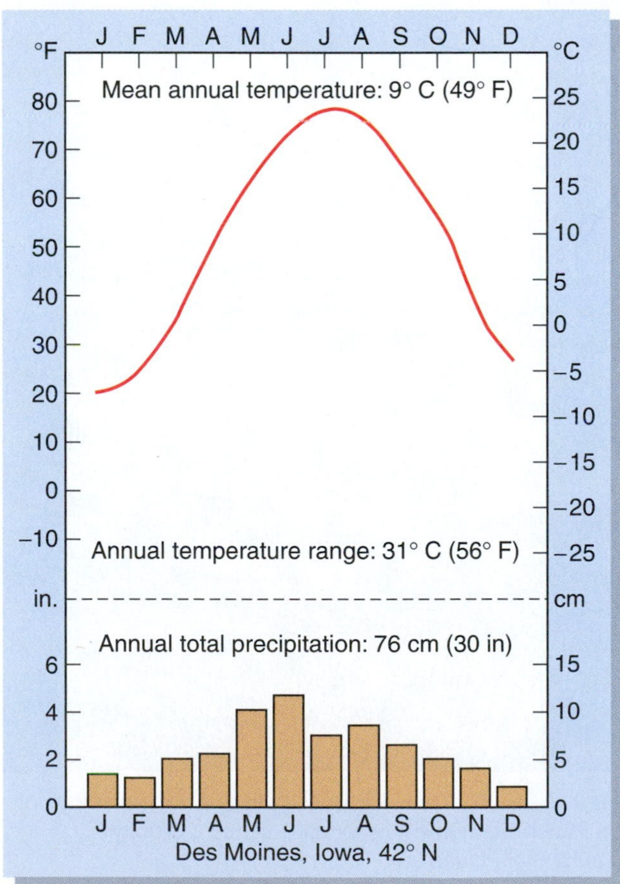

Figure 32–6 The monthly temperature and precipitation values over the course of a year for a deciduous forest biome.

Career Connections

BIOGEOGRAPHER

Biogeographers study the geographical distribution of living things on the Earth. Their work researches the way that regional geography and climate influence the types of plants and animals that live in particular region. Biogeographers also study the past relationships that have existed between the land and living things all through the Earth's history. This work involves the search for fossils and other clues to past climates. Biogeographers are also interested in plate tectonics and how the movement of land masses around the world has influenced the distribution of living things on Earth. Most of the work of a biogeographer is in academic research and requires a knowledge of geography, biology, and geology.

deciduous forests of the eastern United States are second-generation forests, which means they have grown back after they were originally cut down by Europeans when they settled North America. Less than 1% of all deciduous forests are old-growth forests that have not been altered by humans. The temperate forest biome also supports a variety of animals and birds.

Grasslands

The next type of terrestrial ecosystem, or biome, is called the grassland. The grassland biome also is located near the middle latitudes but is mostly found near the interior of continents. Grasslands experience long, hot summers that are accompanied by periodic **droughts** (Figure 32–7). During these drought periods, wildfires may be started by lightning, which quickly burns the dried vegetation. These periodic wildfires prevent larger plants such as trees from growing. This limits the type of vegetation that grows in these regions to grasses and shrubs. The grassland receives less than 15 inches of rain each year, with the summer months receiving as little as 1 inch per month (Figure 32–8).

Grassland biomes are located in the midwestern United States, parts of South America, Africa, and Asia. The grassland biome supports large herds of animals that graze on the wealth of grass found in the grasslands. This type of biome also supports much of the world's agriculture. The "bread basket" grassland in the midwestern United States is the most productive farmland in the world.

Savannas

Another type of biome that also supports a large amount of grassy vegetation is called the savanna. Savannas are also known as a tropi-

Figure 32–7 The wide-open plains of the grasslands of the American midwest experience periodic droughts. *(Courtesy of PhotoDisc.)*

EARTH SYSTEM SCIENTISTS ALDO LEOPOLD

Aldo Leopold was born in Iowa in 1887. He became interested in the outdoors and gained an appreciation for wildlife at a young age. Leopold graduated from Yale University with a degree in forestry. After college he worked for the U.S. Forest Service and began to develop an understanding of the interaction of different organisms in a specific area. His observations and research led to the concept known in ecology as a community. In 1933 he published his book titled *Game Management,* which quickly became one of the most influential books on wildlife management. Leopold began the modern science of restoring wildlife populations by using a variety of management techniques. As a result of his work, he has become known as the father of wildlife ecology. His most famous book, *A Sand County Almanac,* is a collection of essays devoted to restoring our nation's landscapes. Leopold believed that the same type of science and technology that led to the destruction of wildlife habitat should be applied to its restoration.

cal grassland, because they are found near the equator. Savannas are also located in the interior of continents, but they differ from the grassland biome in that they experience a long rainy season. Average rainfall is approximately 40 inches, but less than 2 inches falls over a 2-month period (Figure 32–9). This annual drought period reduces the vegetation to

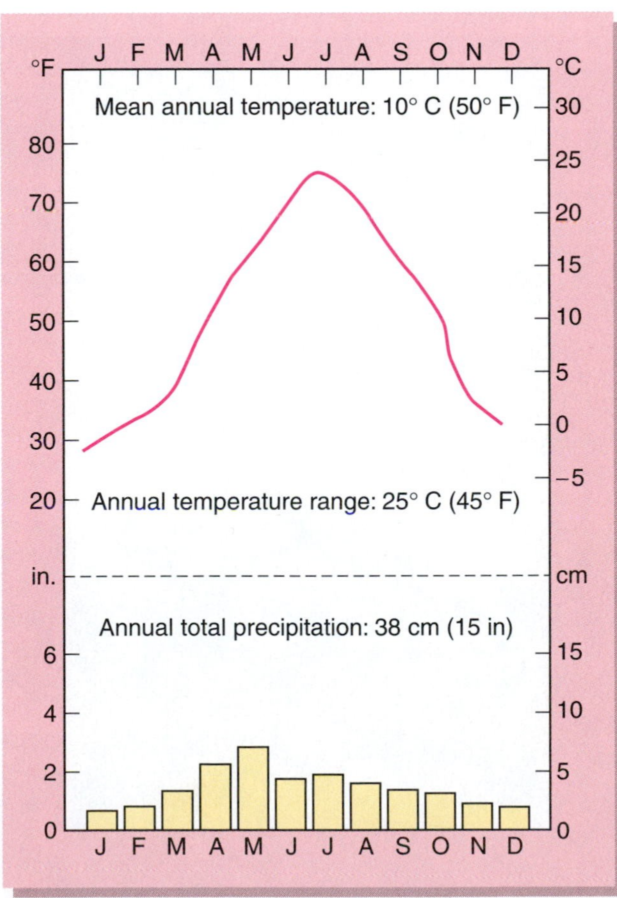

Figure 32–8 The monthly temperature and precipitation values over the course of a year for a grassland biome.

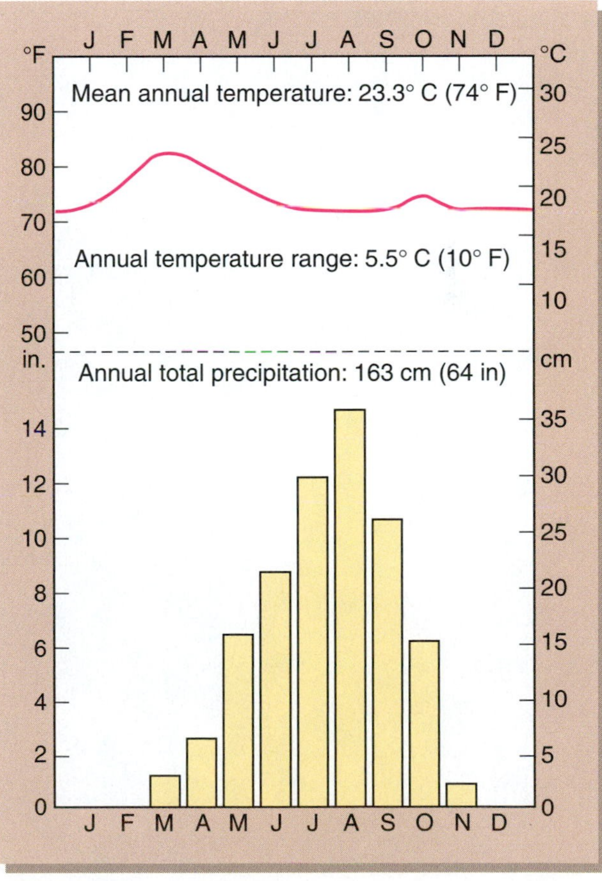

Figure 32–9 The monthly temperature and precipitation values over the course of a year for a savanna biome.

mostly tall grass, shrubs, and widely spread drought-resistant trees. Because the savanna is near the tropical regions of the world, the annual temperatures remain fairly constant, with an average of approximately 77° Fahrenheit throughout the year.

The famous Serengeti Plains of central Africa are a savanna that supports some of the world's most exotic animals (Figure 32–10). These include lions, giraffes, cheetahs, zebras, and elephants. Savannas also exist in South America and Southeast Asia.

Deserts

The desert biome is one of the Earth's harshest environments, because it receives very little rainfall throughout the year. Deserts are usually located near the interior of continents and are found at many different latitudes. The desert biome receives the least amount of precipitation of any biome on Earth. Annual rainfalls in most deserts of the world are less than 2 inches. This extremely dry climate is also known as an arid climate. Although most people believe that deserts are also hot, this is not always the case. Cold deserts also exist in the higher and middle latitudes. China's Gobi Desert is a cold desert where temperatures often fall below freezing. The entire continent of Antarctica is often classified as a polar desert, where temperatures are extremely cold throughout the year and very little precipitation is received.

Deserts can also be located at very high altitudes. The driest place on Earth is believed to be a cold desert located high in the Andes Mountains of South America. This desert is so dry that it has not rained there in nearly 100 years! The deserts of the United States are primarily located in the southwestern part of

Figure 32–10 The Serengeti Plains of Africa are an example of a savanna biome, also called a tropical grassland. *(Courtesy of PhotoDisc.)*

North America. These include the Sonoran and Mojave Deserts.

Desert biomes cover approximately 30% of all the land surfaces on the Earth. The world's largest hot desert is the Sahara, which is located in North Africa (Figure 32–11). Some deserts experience very wide temperature fluctuations throughout the day. During the daytime, temperatures can be as high as 100° Fahrenheit, and at night the temperature can drop to almost 40° Fahrenheit. This is due to the general lack of clouds that can act as a protective blanket helping to regulate heat on the Earth's surface. This extreme temperature change during a 24-hour period adds to the harsh environment of the desert biome.

Even though a desert is exposed to extremes in both temperature and lack of precipitation, it still supports a variety of living organisms. All of the plants and animals that reside in the deserts of the world have adapted to the extreme temperatures and lack of water (Figure 32–12). Desert plants such as cactus and animals such as the kangaroo rat can survive on very little water throughout the year.

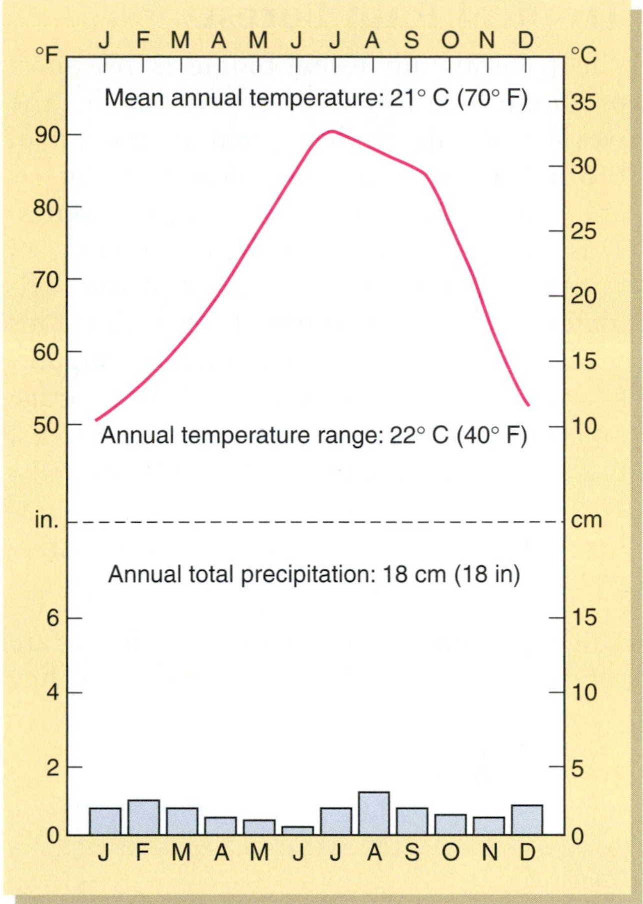

Figure 32–11 The monthly temperature and precipitation values over the course of a year for a hot desert biome.

Figure 32–12 The rugged landscape of the deserts of the American southwest show the drought-resistant plants that survive in a desert biome. *(Courtesy of PhotoDisc.)*

Tropical Rain Forests

The tropical rain forest biome is the most productive land-based ecosystem in the world and supports the greatest amount of **biological diversity.** Tropical rain forests are located near the equator in the **tropics.** This ecosystem receives high amounts of rainfall throughout the year and approximately 12 hours of sunlight every day. This results in rapid growth of plants that support a variety of animal species. The average amount of rainfall in the tropical rain forest biome is approximately 110 inches annually. This biome also maintains a stable temperature throughout the year of approximately 77° Fahrenheit (Figure 32–13).

The trees that make up the rain forest are called broad-leafed evergreens, because they have large leaves and keep them throughout the year, unlike deciduous trees. Tropical rain forests cover approximately 2% of the Earth's land surface and are found in Central America, South America, Africa, and Southeast Asia. Although the tropical rain forests cover a small amount of the land on the planet, they are home to more than 70% of all the species on the Earth. This makes them a valuable biological resource that must be preserved (Figure 32–14).

Chaparral

Another type of biome found on the Earth is called the chaparral. A chaparral is a warm coastal climate that experiences cool, rainy winters and hot, dry summers. The primary vegetation in a chaparral includes

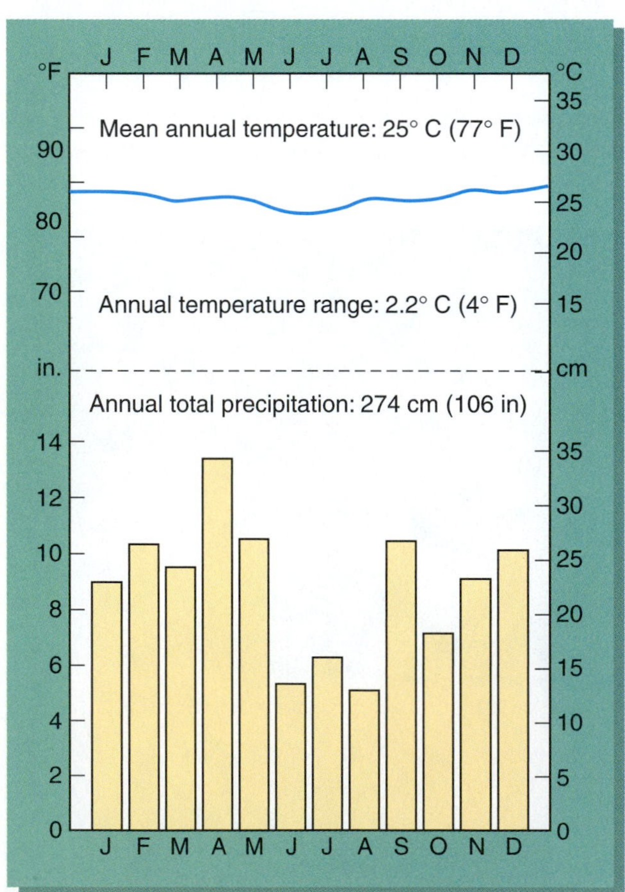

Figure 32–13 The monthly temperature and precipitation values over the course of a year for a tropical rain forest biome.

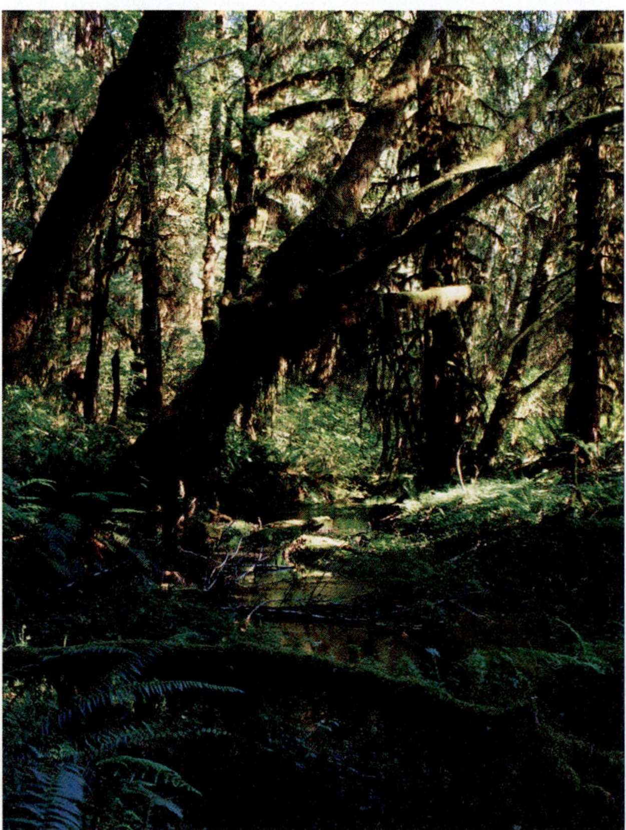

Figure 32–14 The lush vegetation of a tropical rain forest biome. This is the most productive biome on the Earth and supports a great amount of biological diversity. *(Courtesy of PhotoDisc.)*

Figure 32–15 The shrubs and small trees that grow in the chaparral biome. *(Courtesy of PhotoDisc.)*

shrubs, grasses, and drought-resistant trees (Figure 32–15).

Chaparral biomes can be found around the coast of the Mediterranean Sea and in southern California. Because the chaparral has a dry season, it also experiences periodic wildfires that can burn the dried vegetation. People who live in the chaparral biomes of California are susceptible to these wildfires, which can destroy life and property.

Mountains

The final biome type found on the Earth is called the mountain biome. A mountain biome is unique because it shares the same attributes as a coniferous forest biome and the tundra; however, the extremes in temperatures are caused by altitude, not latitude (Figure 32–16). Mountain biomes can be located anywhere on the planet where there are extremely high mountains. Mountain biomes contain coniferous trees at lower altitudes, which eventually give way to tundra vegetation if the altitude is high enough. Mount Washington in New Hampshire is one of the most extreme mountain biomes in North America. At the top of Mount Washington tundra vegetation exists where some of the world's coldest temperatures and strongest winds have been recorded. Mountain biomes can also be located near the equator, where mountain ranges reach high elevations (Figure 32–17).

Figure 32–16 A mountain biome exists at high elevations, where cooler temperatures create conditions similar to the coniferous forest and tundra biomes as a result of increasing altitude. *(Courtesy of PhotoDisc.)*

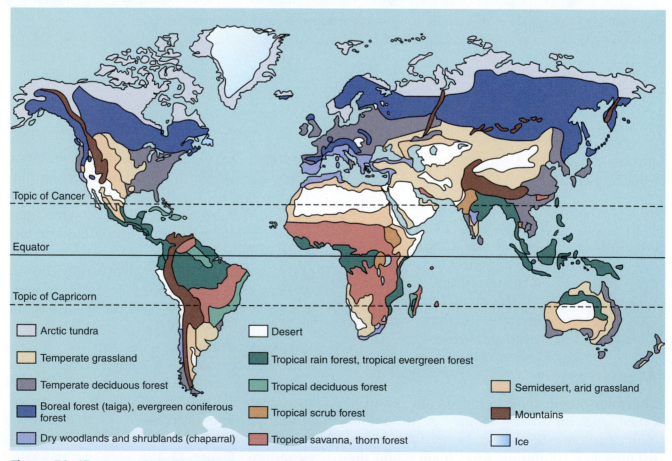

Topic of Cancer
Equator
Topic of Capricorn

Arctic tundra
Temperate grassland
Temperate deciduous forest
Boreal forest (taiga), evergreen coniferous forest
Dry woodlands and shrublands (chaparral)
Desert
Tropical rain forest, tropical evergreen forest
Tropical deciduous forest
Tropical scrub forest
Tropical savanna, thorn forest
Semidesert, arid grassland
Mountains
Ice

Figure 32–17 A map showing the locations of major biomes around the world.

EARTH MATH

1) IF THE TOTAL SURFACE AREA OF ALL THE LAND ON EARTH IS APPROXIMATELY 60 MILLION SQUARE MILES, DETERMINE THE APPROXIMATE SURFACE AREA OF THE FOLLOWING BIOMES: TROPICAL RAIN FORESTS (2%), TUNDRA (11%), AND DESERTS (30%).

SECTION REVIEW

1. What abiotic factors are used to classify the Earth's biomes?
2. Which two biomes experience harsh conditions throughout the year?
3. What type of trees make up most of the vegetation in a temperate forest biome?
4. Which biome supports the greatest amount of biological diversity on the Earth?
5. In which biome is most of the world's food crops grown?
6. Who was Aldo Leopold?

For more information go to these Web links:

<http://www.blueplanetbiomes.org/table_of_contents.htm>

<http://www.snowcrest.net/geography/slides/biomes/>

<http://mbgnet.mobot.org/sets/index.htm>

<http://www.cotf.edu/ete/modules/msese/earthsysflr/biomes.html>

32.2 *Marine Ecosystems*

Coastal Wetlands

Oceans cover 71% of the Earth's surface, and other parts of the land surface are covered by freshwater lakes, rivers, and wetlands. Together these **aquatic** ecosystems support a great variety of living organisms. The ecosystems that exist within the ocean are known as **marine** ecosystems. The marine ecosystems on Earth are classified by their locations within the ocean. The marine environment that lies closest to the shoreline, where the land meets the sea, is called the coastal wetlands (Figure 32–18). Coastal wetlands are important marine ecosystems where many marine organisms reproduce. These coastal marine breeding grounds are found in bays, salt marshes, lagoons, and mud flats and are partially or totally covered by salt water throughout the year. Approximately 3% of all wetlands in the United States are coastal wetlands. Although these are considered aquatic ecosystems, many of these marine environments support a large number of salt-tolerant grasses, which provide shelter for the variety of marine organisms that reside there.

The Neritic and Intertidal Zones

Although the coastal wetlands are an important breeding ground for many marine organisms, the bulk of the marine environment is located in the open oceans. The ecosystems of the open ocean are divided into two categories: the neritic zone and the oceanic zone. The neritic zone is the ecosystem that lies along the coasts. It begins at the shoreline, which experiences high and low tides each day (Figure 32–19).

The neritic zone is also called the **intertidal zone;** it is a marine ecosystem that is home to many well-known marine organisms such as mussels, hermit crabs, starfish, seaweeds, sea

Figure 32–18 The coastal wetlands located along the coast of the Atlantic Ocean off of Maryland. *(Courtesy of PhotoDisc.)*

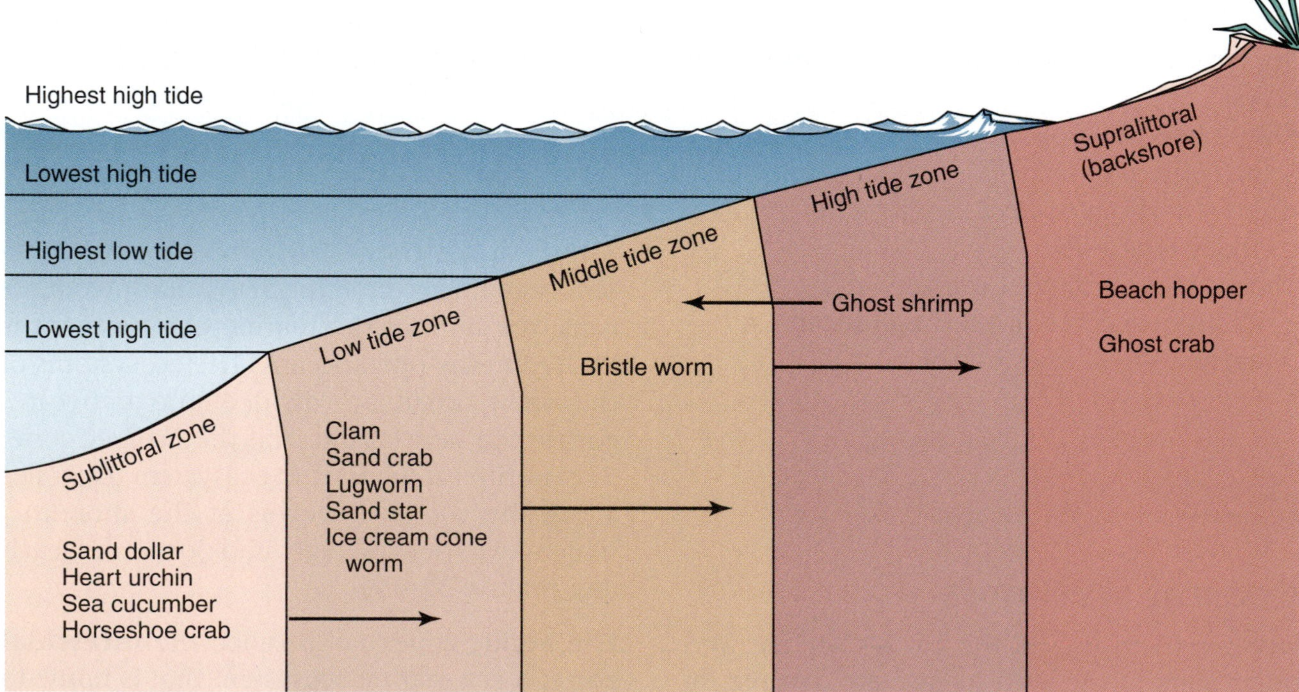

Figure 32–19 The different zones and the organisms that live along the shoreline in the intertidal ecosystem, which is located between the area of low and high tides.

anemones, algae, and barnacles (Figure 32–20). This ecosystem is one of the harshest marine environments, because the organisms that live there are exposed to dry periods when the tide is low and submerged conditions when the tide is high. This environment is also exposed to the constant impact of waves and tidal forces. Many of the organisms that reside in the intertidal ecosystem are extremely flexible, such as algae and grasses, or extremely hard, such as mussels and barnacles. These adaptations help them to survive in this harsh environment.

The Oceanic Zone

The neritic zone stretches out from the shallow waters near the shore into the ocean to a depth of about 600 feet. This marks the entry into the oceanic zone marine ecosystem. This is the open-ocean ecosystem that supports two main life forms, called plankton and nekton. Plankton are free-floating organisms that drift with the ocean currents. They include microscopic algae, also called **phytoplankton,** and single-celled animals called **zooplankton** (Figure 32–21). Larger plankton include the many species of jellyfish.

Most plankton float freely throughout the ocean to a depth of about 1000 feet. Although plankton can exist anywhere in the oceanic zone, the highest populations occur close to the continental shelves (Figure 32–22). This is due to the greater amount of nutrients available to the plankton there.

The other type of organism that resides in the oceanic zone is the nekton. Nekton are marine organisms that are capable of moving under their own power. These include all fish, squid, octopus, and marine mammals such as whales and dolphins (Figure 32–23). Nekton can live at

Figure 32–20 Marine organisms that live in the neritic zone of the ocean.

Figure 32–21 Examples of zooplankton and phytoplankton that live in the oceanic zone.

0.01 0.02 0.03 0.04 0.05 0.1 0.2 0.3 0.4 0.5 1.0 2.0 3.0 4.0 5.0 10 20 30 40 50 60

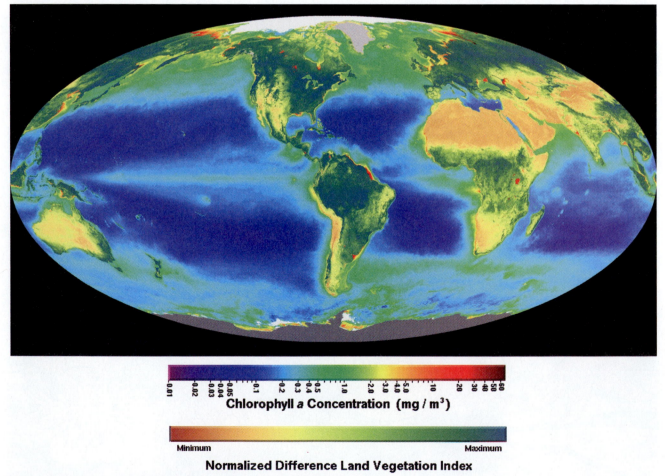

Chlorophyll *a* Concentration (mg / m^3)

Minimum Maximum

Normalized Difference Land Vegetation Index

Figure 32–22 A map showing the distribution of phytoplankton in the oceans around the world. *(Provided by the SeaWiFS Project, NASA/Goddard Space Flight Center and ORBIMAGE.)*

EARTH SYSTEM SCIENTISTS *VICTOR HENSEN*

Victor Hensen was born in Germany in 1835. Trained in anatomy and physiology, Henson became interested in life within the oceans. In 1887 he coined the term plankton to describe the tiny organisms that float freely in the sea and form the base of the aquatic food chain. Hensen led many marine expeditions to survey the extent of plankton within the oceans. In 1889 he revealed that the cooler, nutrient-rich waters of the Arctic and Antarctic supported higher populations than the warm, nutrient-poor waters of the tropics.

all depths in the open ocean; some prefer to stay near the surface to feed and live out their lives, whereas others survive in the total darkness of the deep ocean. Some marine nekton will live part of their lives in the ocean and the remaining part in freshwater ecosystems on land. The Atlantic eel is a marine fish that is born in the oceanic zone and eventually migrates into freshwater rivers to live. These types of marine organisms are classified as catadromous, which means they are born in saltwater and live in freshwater. The Atlantic and Pacific salmon do the opposite; they are born in freshwater streams and rivers and migrate out into the open ocean, where they live out their adult lives. These are called anadromous fish.

Figure 32–23 Examples of nekton that live in the oceanic zone.

The Benthic Zone

The benthic zone, another major marine ecosystem that exists within the world's oceans, includes all organisms that live at the bottom of the ocean. In some areas close to the shore and in shallow water, the benthic environment receives a great amount of sunlight. This creates highly productive marine communities known as coral reefs. Coral reefs are the largest living structures on the Earth. They are composed of tiny organisms called coral that secrete calcium carbonate to make a protective shell. Over time the colonies of coral grow and produce large structures composed of calcium carbonate and sand. When the older colonies die, new colonies grow on top of the remaining shells of the dead coral. This leads to the buildup of what is called the coral reef.

Coral maintain a symbiotic relationship with algae that together forms the base of the coral reef ecosystem. The single-celled algae reside within the protective shell of the coral. The algae produce nutrients for the coral while the coral offers protection and carbon dioxide to the algae. Coral can only exist in shallow water that receives ample amounts of sunlight and that maintains a temperature above 64° Fahrenheit. Because of this, they are only found in the tropical regions (Figure 32–24). Coral reefs form one of the most beautiful underwater environments on the planet, which supports a wide array of life.

Career Connections

Marine Biologist

A marine biologist, also known as a biological oceanographer, studies all the aspects of life within the world's oceans. This includes the behavior of marine organisms, their distribution in the ocean, the cycling of nutrients within marine ecosystems, and marine food chains. Marine biologists also study the communication of marine organisms, especially that of marine mammals such as whales and dolphins. Current research by biological oceanographers includes the impact of human activity on marine organisms. This work includes the study of pollution and the effects of boating traffic on specific ocean species. Many marine biologists are also interested in maintaining natural populations of fish species and preventing overfishing. Biological oceanographers need a college education with a sound base in oceanography and marine biology. Career opportunities exist in academic research and within government agencies.

Hydrothermal Vent Communities

Another interesting community that forms a benthic marine ecosystem resides in the extremely deep and dark waters of the ocean.

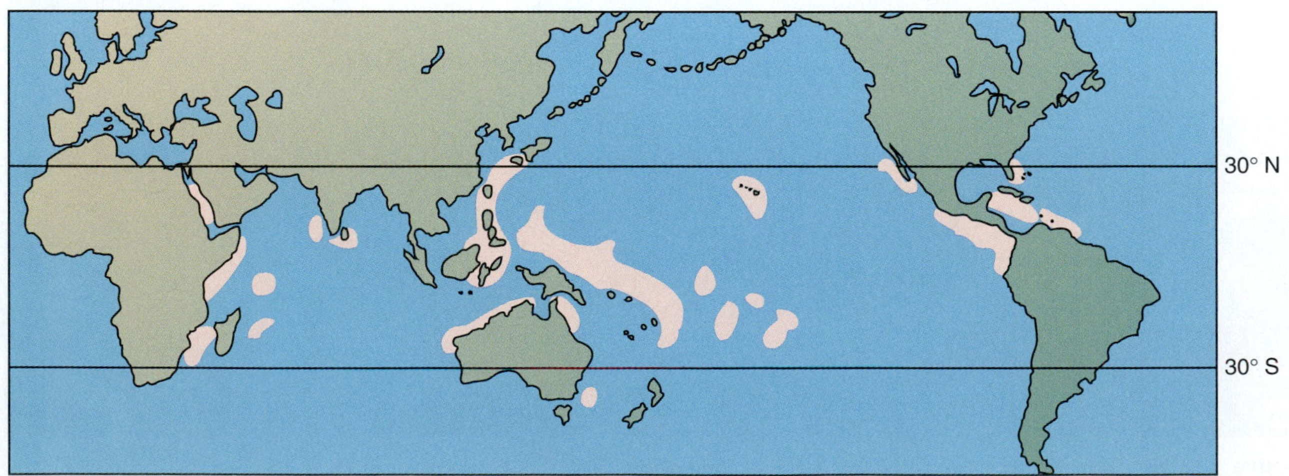

Figure 32–24 The locations of the world's major coral reef systems.

These are called deep sea hydrothermal vent communities. A hydrothermal vent is a chimneylike structure that spews out extremely hot water that is rich in minerals. The source of the water, which can reach temperatures of more than 600° Fahrenheit, is volcanic activity. Ocean water seeps into cracks within the Earth's crust and is then superheated by the Earth's hot interior. In 1977 the first hydrothermal vent community was discovered near the equator in the Pacific Ocean. Located at a depth of more than 8000 feet, and in total darkness, this aquatic community was unlike anything ever seen before. It consisted of large tube worms that were approximately 3 feet in length, large mussels, and white crabs. The whole community existed in total darkness, in water exceeding 200° Fahrenheit, and at pressures exceeding 5000 pounds per square inch.

As more of the ocean bottom began to be explored, many other hydrothermal vent communities were discovered, revealing that life can exist in the harshest of environments and without sunlight. At one hydrothermal vent community a video was taken that showed a wormlike creature wrapping itself around a temperature probe that a remotely operated submarine was deploying to record the water temperature. Unbelievably, the temperature of the water was 221° Fahrenheit!

For more information go to these Web links:
<http://mbgnet.mobot.org/salt/>
<http://oceanexplorer.noaa.gov/>
<http://www.nmfs.noaa.gov/>
<http://www.pmel.noaa.gov/vents/>
<http://seawifs.gsfc.nasa.gov/OCEAN_PLANET/HTML/oceanography_recently_revealed1.html>
<http://www.botos.com/marine/vents01.html>
<http://geosun1.sjsu.edu/~dreed/105/vents.html>

SECTION REVIEW

1. What are some of the ecosystems that are classified as coastal wetlands?
2. Describe two conditions that make the intertidal ecosystem extremely harsh.
3. How have organisms adapted to surviving in the intertidal ecosystem?
4. What are the two categories of organisms found in the oceanic zone? Give two examples for each.
5. Where is the benthic environment located within the ocean?
6. Under what conditions can coral reefs form?
7. Describe the relationship between algae and coral in a reef ecosystem.
8. What is a hydrothermal vent community, and under what conditions do they it exist within the ocean?
9. Who was Victor Hensen?

CHAPTER SUMMARY

The interaction between the living and nonliving elements of a specific region on the Earth is called an ecosystem. Two main types

EARTH MATH

1) IF A HYDROTHERMAL VENT COMMUNITY EXISTS AT A DEPTH OF 10,890 FEET BELOW THE OCEAN SURFACE AND PRESSURE INCREASES 14.7 POUNDS PER SQUARE INCH FOR EVERY 33 FEET OF DEPTH, WHAT IS THE PRESSURE IN POUNDS PER SQUARE INCH WHERE THIS COMMUNITY IS LOCATED?

of ecosystems are studied on the planet: terrestrial ecosystems and aquatic ecosystems. Terrestrial ecosystems exist on the land, and aquatic ecosystems are found in the water. Terrestrial ecosystems are also called biomes. Ecologists recognize nine major biomes on the Earth, which are determined by the temperature and precipitation they experience each year. The tundra biome is located in the higher latitudes where the temperature remains below freezing for most of the year. The tundra exists on permafrost, which is soil that is frozen throughout the year.

Located below the tundra biome is the coniferous forest biome. This ecosystem experiences long, cold winters and short, hot summers. The principal vegetation growing there is the coniferous tree, also known as evergreen trees. The deciduous forest biome, also called the temperate forest, is located near the middle latitudes and experiences cold winters and hot summers. The principal vegetation in the temperate forests are deciduous trees. These are trees that lose their leaves each fall.

Another type of biome is grassland. This ecosystem is located in the middle latitudes near the interior of continents. Grasslands experience periodic droughts during the year, which causes periodic wildfires to burn through the region. This allows grasses and shrubs to be the principal vegetation in this biome. Grasslands that grow near the equator are called savannas or tropical grasslands. These ecosystems receive greater amounts of precipitation and support grasses, shrubs, and drought-resistant trees.

The desert biome is one of the harshest ecosystems on the planet, receiving very little precipitation each year. These extremely dry environments can be hot or cold. The Sahara desert in North Africa is the world's largest hot desert. Cold deserts, such as the Gobi Desert in China, experience temperatures that can fall below freezing at certain times during the year. Even though deserts are very harsh, they still support life.

The chaparral biome is a coastal ecosystem that experiences hot, dry summers and cool, rainy winters. These biomes exist around the Mediterranean Sea and along the coast of California. The last major biome is called the mountain biome. These ecosystems resemble coniferous forests and tundra biomes but exist in the higher altitudes of mountain ranges. Marine ecosystems exist within the world's oceans. The coastal wetlands are areas that lie close to the shore and are covered with saltwater for most of the year. These marine ecosystems include mudflats, salt marshes, bays, and lagoons are important breeding grounds for many marine organisms. The open ocean is divided into two distinct types of marine ecosystems called the neritic zone and the oceanic zone. The neritic zone extends from the shoreline out to a depth of approximately 600 feet. It includes the area known as the intertidal zone, which lies between the zones of high and low tides. The organisms that live in this area must withstand a harsh environment exposed to the air, wave action, and tidal currents.

The oceanic zone includes all of the deep ocean. The organisms that live there are divided into two main categories, called plankton and nekton. Plankton are free-floating marine organisms such as algae or jellyfish. Nekton are capable of swimming, such as fish or whales.

The benthic zone is another type of marine ecosystem that exists on the ocean bottom. This can include the shallow water, where coral reefs exist, or the deep, dark ocean. Recently a unique benthic marine ecosystem was discovered that exists in total darkness near hydrothermal vents. These volcanic vents spew out superheated water that is rich in minerals and supports a community of marine organisms. The temperature of the water around hydrothermal vents can be more than 200° Fahrenheit. The organisms that thrive in this unique ecosystem illustrate the ability for life to exist in extreme environments.

CHAPTER REVIEW

Multiple Choice

1. Which biome exists in the higher latitudes in the zone of permafrost?
   a. tropical rain forest
   b. grassland
   c. coniferous forest
   d. tundra

2. Which biome is exposed to periodic wildfires during the summer?
   a. tropical rain forest
   b. grassland
   c. coniferous forest
   d. tundra

3. Trees that lose their leaves each fall are the principal vegetation in what biome?
   a. deciduous forest
   b. savanna
   c. chaparral
   d. desert

4. These type of biomes are also called tropical grasslands:
   a. deciduous forest
   b. savanna
   c. chaparral
   d. desert

5. The two most important abiotic factors that determine a biome are:
   a. altitude and latitude
   b. temperature and altitude
   c. temperature and precipitation
   d. precipitation and latitude

6. Which type of biome exists along the coast of California?
   a. deciduous forest
   b. savanna
   c. chaparral
   d. desert

7. The marine ecosystem that is periodically exposed to the air and wave action is called:
   a. the intertidal zone
   b. coastal wetland
   c. the oceanic zone
   d. the benthic zone

8. The marine ecosystem that exists from the shoreline to a depth of approximately 600 feet is called:
   a. the coastal wetland
   b. the neritic zone
   c. the oceanic zone
   d. the benthic zone

9. The free-floating organisms that live in the open ocean are known as:
   a. plankton
   b. nekton
   c. catadromous
   d. anadromous

10. Coral reefs and hydrothermal vent communities exist in which type of marine ecosystem?
    a. the coastal wetland
    b. the neritic zone
    c. the oceanic zone
    d. the benthic zone

continued

Matching

Match the terms with the correct definitions.

a. terrestrial
b. biomes
c. tundra
d. permafrost
e. droughts
f. biological diversity
g. tropics
h. aquatic
i. marine
j. intertidal zone
k. phytoplankton
l. zooplankton
m. hydrothermal vent

1. ____ A term that refers to anything associated with the oceans.
2. ____ Tiny animals that float freely in water.
3. ____ Another term for a land-based ecosystem, also called a terrestrial ecosystem.
4. ____ A term that refers to land.
5. ____ A type of plankton that uses photosynthesis to gain energy.
6. ____ A term that refers to the area located near the equator.
7. ____ An extended period when little or no precipitation is received by a specific region.
8. ____ Ocean water that seeps through cracks in the sea floor and is super heated by magma that is close to the surface.
9. ____ An aquatic life zone that exists near the shoreline between the area of the highest and lowest tides.
10. ____ Permanently frozen soil.
11. ____ The variety of living species on Earth and their unique genes.
12. ____ A term that describes anything that lives in water.
13. ____ A biome that is found in the higher latitudes, where the temperature is below freezing for most of the year, and supports matlike vegetation.

Critical Thinking

1. Identify the biome in which you live, and describe some of the specific biotic and abiotic factors in your particular biome.

CHAPTER 33

The Flow of Energy and Matter Through Ecosystems

Section 33.1 – Energy Flow Within Living Systems Objectives

Photosynthesis and Chemosynthesis • Autotrophs and Heterotrophs • Primary Production • Primary and Secondary Consumers • Food Chains and Webs • The Energy Pyramid

After reading this chapter you should be able to:

- Identify where most energy on the Earth comes from.
- Describe the photosynthesis and chemosynthesis chemical reactions.
- Differentiate between autotrophic and heterotrophic organisms.
- Describe the process of primary production.
- Explain the relationship among producers, primary consumers, secondary consumers, and decomposers in an ecosystem.
- Define the terms *herbivore, carnivore, omnivore,* and *detritivore.*
- Draw a diagram of a simple food chain.
- Describe the transfer of energy from one trophic level to another in a food pyramid.

Section 33.2 – Biogeochemical Cycling Objectives

Biogeochemical Cycling • Carbon Cycling • Oxygen Cycling • Nitrogen Cycling • Phosphorus Cycling

After reading this chapter you should be able to:

- Define the term *biogeochemical cycling.*
- Identify the four main biogeochemical cycles on the Earth.
- Describe three examples of how carbon moves through the carbon cycle.
- Explain the relationship between photosynthesis and respiration in the oxygen cycle.
- Describe the pathway by which nitrogen moves from the atmosphere into plants.
- Identify the main sources of phosphorus for ecosystems on the Earth.

TERMS TO KNOW

autotrophs
chemosynthesis
heterotrophs
primary production
biomass
herbivores
carnivores
omnivores
detritivores
decomposers
food chain
biogeochemical cycling
respiration
reservoir
limiting factor

INTRODUCTION

The interactions that occur in the world's ecosystems involve the exchange and movement of energy and matter between the living and nonliving worlds. The source of all the energy utilized in most of the world's ecosystems is the Sun. The pathways through which solar energy moves once it strikes the planet link all living things together. Every living thing on the Earth depends on the flow of energy through ecosystems for survival. The movement of matter through ecosystems is equally important for life to flourish on the Earth. The chemical elements that make up the nonliving world are the building blocks for life. These important molecules and nutrients are continually recycled between the nonliving and living world. Understanding the interactions between matter, energy, and life on the planet reveals the importance of the biosphere as part of the Earth's systems and how life on the Earth has flourished for more than 3 billion years.

33.1 Energy Flow Within Living Systems

Photosynthesis and Chemosynthesis

All life on Earth requires energy to live, and the way that an organism gains energy is vital for its survival. Organisms that can derive energy from sunlight or from chemical reactions are called **autotrophs,** which means "self-feeder." Autotrophs use two different processes to gain energy. The first is the process of photosynthesis. Photosynthesis is the chemical reaction by which organisms transform light energy from the sun into stored chemical energy (Figure 33–1). This is accomplished by creating sugars and starches. Photosynthesis is probably the most important chemical reaction on Earth. The photosynthesis reaction takes light energy from sunlight and combines it with carbon dioxide and water to form the sugar glucose, as well as oxygen as a byproduct. The oxygen is then released into the atmosphere.

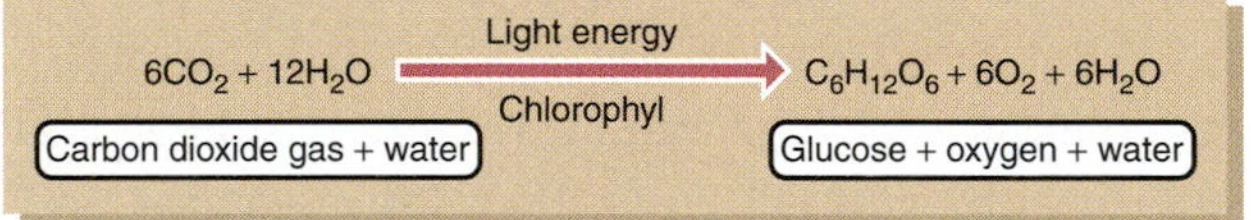

Figure 33–1 The photosynthesis reaction combines carbon dioxide and water in the presence of sunlight to form glucose sugar, water, and oxygen.

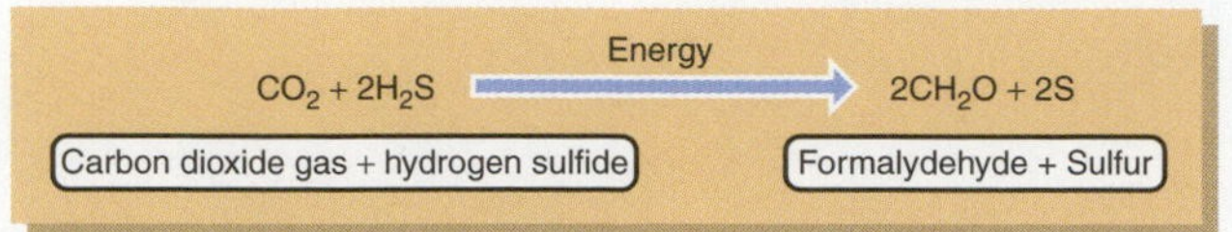

Figure 33–2 A chemosynthesis reaction that is used by bacteria near hydrothermal vents deep within the ocean utilizes hydrogen sulfide and carbon dioxide gas to gain energy while producing pure sulfur.

All the oxygen in the atmosphere is the result of photosynthesis. Glucose molecules produced by photosynthesis are then joined together to form long, chainlike molecules called starches. Green plants and phytoplankton are the two main types of organisms that utilize photosynthesis. **Chemosynthesis** is the other process by which autotrophs gain energy (Figure 33–2). Organisms that utilize chemosynthesis gain energy released from chemical reactions. Many organisms that utilize chemosynthesis derive their energy from sulfur-containing molecules. This is how bacteria that reside in deep sea hydrothermal vent communities gain their energy. These communities are unique on the Earth because they do not derive their energy from sunlight like most ecosystems.

EARTH SYSTEM SCIENTISTS *LINUS PAULING*

Linus Pauling was born in Oregon in 1901 and received an education in chemical engineering. Pauling's early work involved the understanding of chemical bonds. In 1939 he published this research in his book *The Nature of the Chemical Bond,* which opened the door to a better understanding of molecular structure. In 1954 his work earned him the Nobel Prize for chemistry. His later research involved the chemistry of living things and their molecular structure. His research helped to reveal the structure of the DNA molecule, proteins, and human blood. Pauling also became a proponent of the benefits of vitamin C to the health of the human body. In 1963 Pauling gained another Nobel Prize, this time for his work in helping to ban nuclear testing. As a result, he is the only person to receive two Nobel Prizes.

Autotrophs and Heterotrophs

Another method by which organisms gain their energy on Earth is the consumption of other organisms. These living things are called **heterotrophs,** which means "other feeder." Heterotrophs use the chemical energy stored in the bodies of other organisms, such as plants and animals, to gain energy. Human beings are heterotrophs; they must consume food to gain energy.

Primary Production

Although it is important to understand how individual organisms gain their energy, it is the flow of energy through entire ecosystems that enables life to exist as it does on our planet. Autotrophic organisms such as plants and algae form the base for energy in all ecosystems. With the exception of the deep sea hydrothermal vent communities, all energy flow in both aquatic and terrestrial ecosystems begins with green plants and algae. These autotrophs convert radiant energy from the Sun into stored chemical energy (Figure 33–3). This forms a vital link between the Sun and the rest of the organisms on Earth. Photosynthesizing autotrophs are also called producers because they produce the energy for the entire ecosystem. The amount of chemical energy that an autotroph converts from solar energy by the process of photosynthesis is called **primary production.** Primary production is measured by determining the amount of **biomass** a plant has.

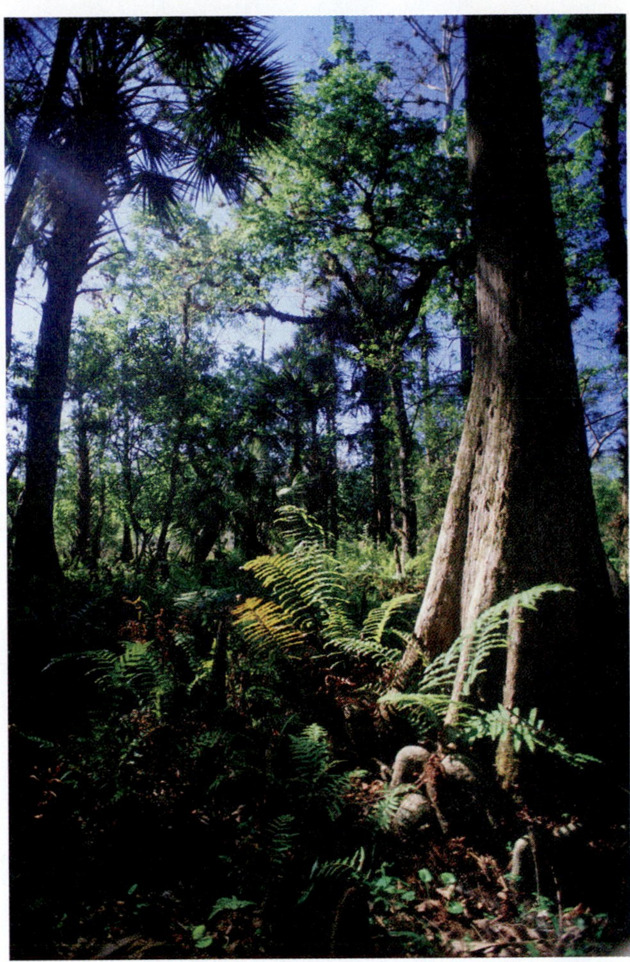

Figure 33–3 Example of green plants that are the producers for an ecosystem; they convert radiant energy from the Sun into stored chemical energy. *(Courtesy of PhotoDisc.)*

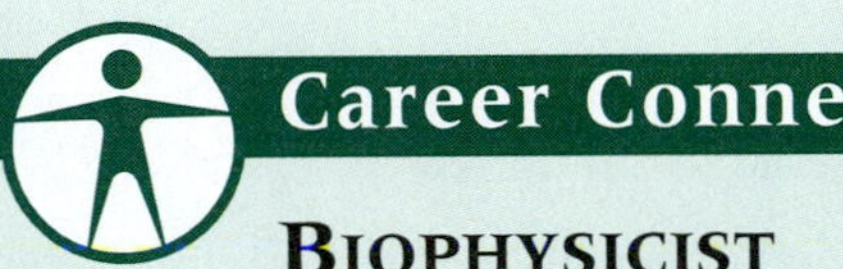

Biophysicist

A biophysicist studies the properties of molecules within organisms. This includes an understanding of their structure and their unique physical and electrical properties. These scientists also study the chemical reactions that are vital to life on Earth, such as photosynthesis and respiration. A biophysicist uses sophisticated laboratory equipment to unlock the secrets of molecules and their interactions, which are important to life. Biophysicists perform experiments and analysis on specific chromosomes and DNA molecules to further understand heredity and genetics. The work of a biophysicist may someday lead to cures for disease and improved health. Biophysicists also are researching ways to safely degrade toxic chemicals and oils that have polluted the environment. Biophysicists must have a college education with an emphasis on chemistry and biology. They can work in academic research, in medicine, or in the environmental sciences.

Biomass is a short term for biological mass, which is the total dry weight of an organism.

Primary and Secondary Consumers

The heterotrophic organisms that consume the producers are called primary consumers. These include all organisms that eat plants or algae as a source of energy. Known as **herbivores,** or "plant eaters," primary consumers come in all shapes and sizes in a specific ecosystem. Many insects are primary consumers in ecosystems around the world. Fish and larger animals such as deer, moose, and cows are also primary consumers, or herbivores that only consume plants (Figure 33–4).

Organisms that consume the herbivores are known as secondary consumers. These organisms are sometimes called **carnivores,** or "meat eaters." Tigers, killer whales, and hawks are all secondary consumers.

Sometimes an organism can act as both a primary consumer and secondary consumer by eating both producers and herbivores. These organisms are known as **omnivores,** or "all eaters." Humans are omnivores, as are bears and dogs (Figure 33–5). Some ecosystems contain higher level consumers who eat secondary consumers. These type of organisms are called tertiary consumers. The producers form the vital link between the Sun and the primary consumers in an ecosystem; however, there exists another important category of organisms that helps to recycle nutrients in a community. These organisms are called **detritivores,** or **decomposers,** which eat dead organisms (Figure 33–6). Although considered

Figure 33–4 Cattle are examples of primary consumers, also called herbivores, who eat the producers in an ecosystem. *(Courtesy of PhotoDisc.)*

Figure 33–5 A brown bear is an upper level consumer in an ecosystem, also known as an omnivore, who eats both plants and animals. *(Courtesy of PhotoDisc.)*

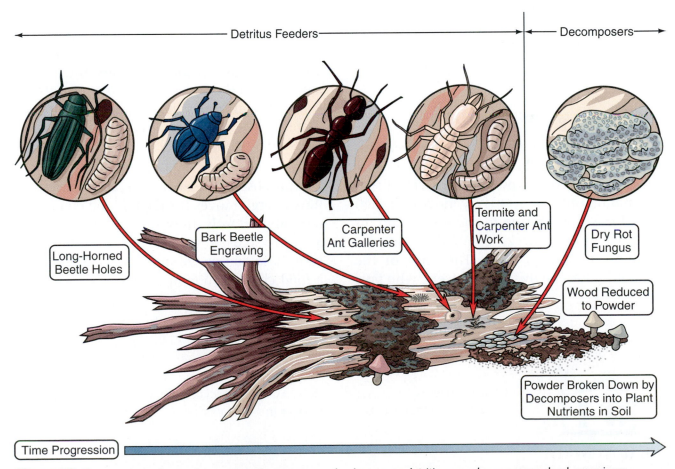

Detritus Feeders
Decomposers
Long-Horned Beetle Holes
Bark Beetle Engraving
Carpenter Ant Galleries
Termite and Carpenter Ant Work
Dry Rot Fungus
Wood Reduced to Powder
Powder Broken Down by Decomposers into Plant Nutrients in Soil
Time Progression

Figure 33–6 A variety of organisms act as decomposers, also known as detritivores, who consume dead organisms.

gruesome, detritivores perform an invaluable service to the ecosystem by breaking down, or consuming, the remains of dead organisms. This allows valuable nutrients to be recycled in an ecosystem. If not for the detritivores, things would be quite messy here on Earth. Decomposers are like nature's garbage collectors. Common detritivores include vultures, many types of insects, fungi, and bacteria.

Food Chains and Webs

The specific way that energy flows through a community of organisms in a ecosystem is called a **food chain** (Figure 33–7). The food chain is a series of eating processes by which energy and nutrients flow from one organism to another. A food web is complex interaction of food chains within a specific ecosystem.

All ecosystems have their own unique food chains or webs that sustain all the living things in an ecosystem's community (Figure 33–8). All food chains or webs always contain their own unique producers, primary consumers, secondary consumers, and decomposers.

The Energy Pyramid

The movement of energy through an ecosystem can also be illustrated by the energy pyramid. The energy pyramid is a visual representation of the way energy moves through a food chain. Each step in the food chain, illustrated by the energy pyramid, is called a trophic, or feeding, level. The base of the pyramid is formed by the producers, who take in radiant energy from the Sun and transform it into chemical energy in the form of sugar and starches.

The next level in the energy pyramid is occupied by the primary consumers, who eat the producers. This is the second trophic level. When a primary consumer eats a producer, approximately 10% of the total energy contained by the producer is gained by the primary consumer. Almost 90% of the total chemical energy contained in the producer is given off as heat when the organism digests, utilizes, and stores the chemical energy derived from the first trophic level. This "loss" of energy is an important part of the movement of energy through a food chain (Figure 33–9). The energy is really not lost, but it is no longer available for use by organisms in the food chain. Typically when an organism eats something, approximately 90% of the energy is given off to the atmosphere as heat.

Above the primary consumer in an energy pyramid is the third trophic level, which is occupied by the secondary consumer. Once again, when the secondary consumer eats the primary consumer, it only gains approximately 10% of the total chemical energy of the second trophic level.

The purpose of the energy pyramid is to show how energy moves through a food chain and, more importantly, how much of the energy in an ecosystem is lost to the atmosphere as heat. Each successive step of a trophic level in a food chain results in a large loss of energy.

Earth Math

1) If the first trophic level in a food chain contains 5000 units of energy, how many units of energy are transferred to the third trophic level?

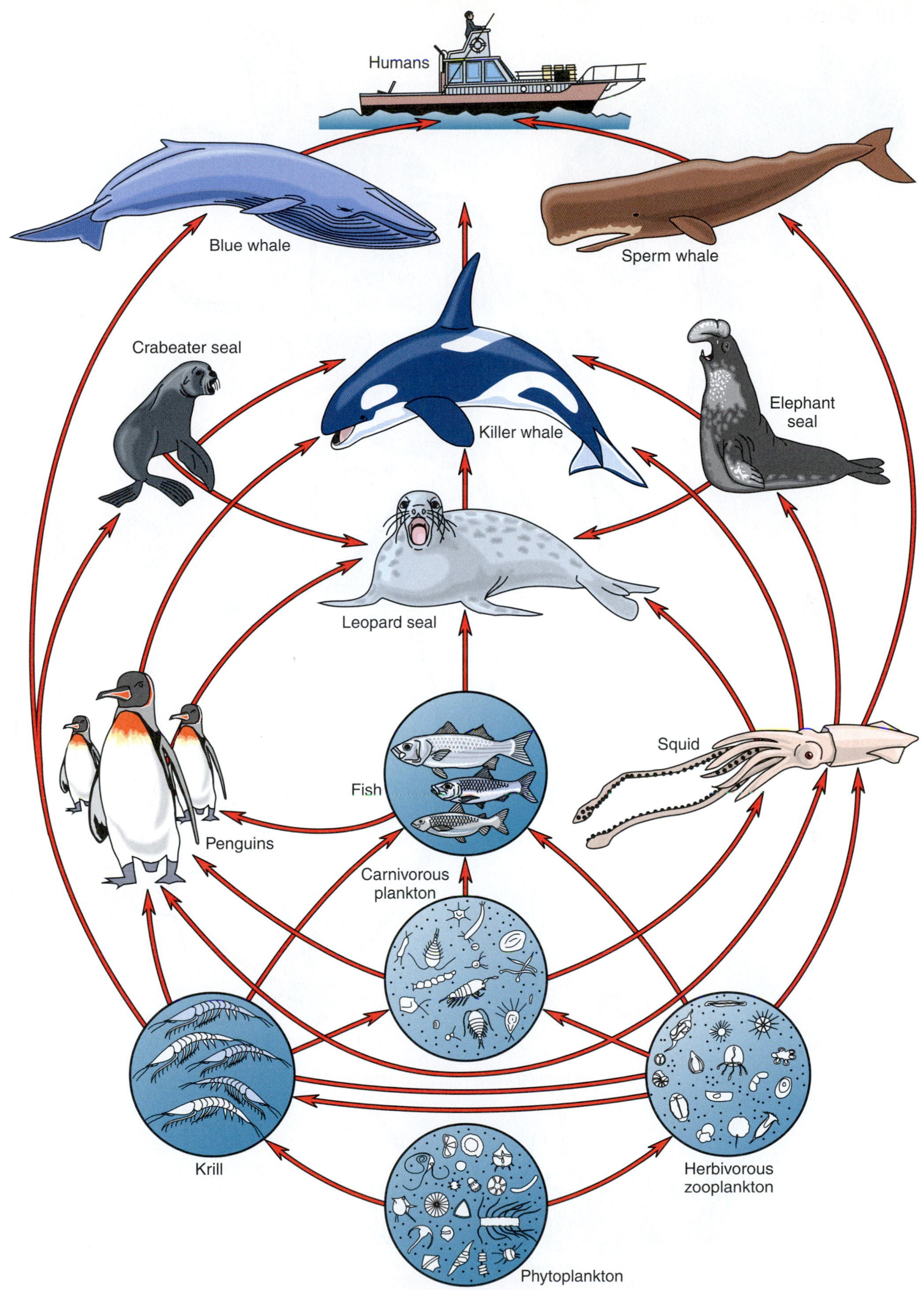

Figure 33–7 Interrelated food chains in the ocean form an aquatic food web.

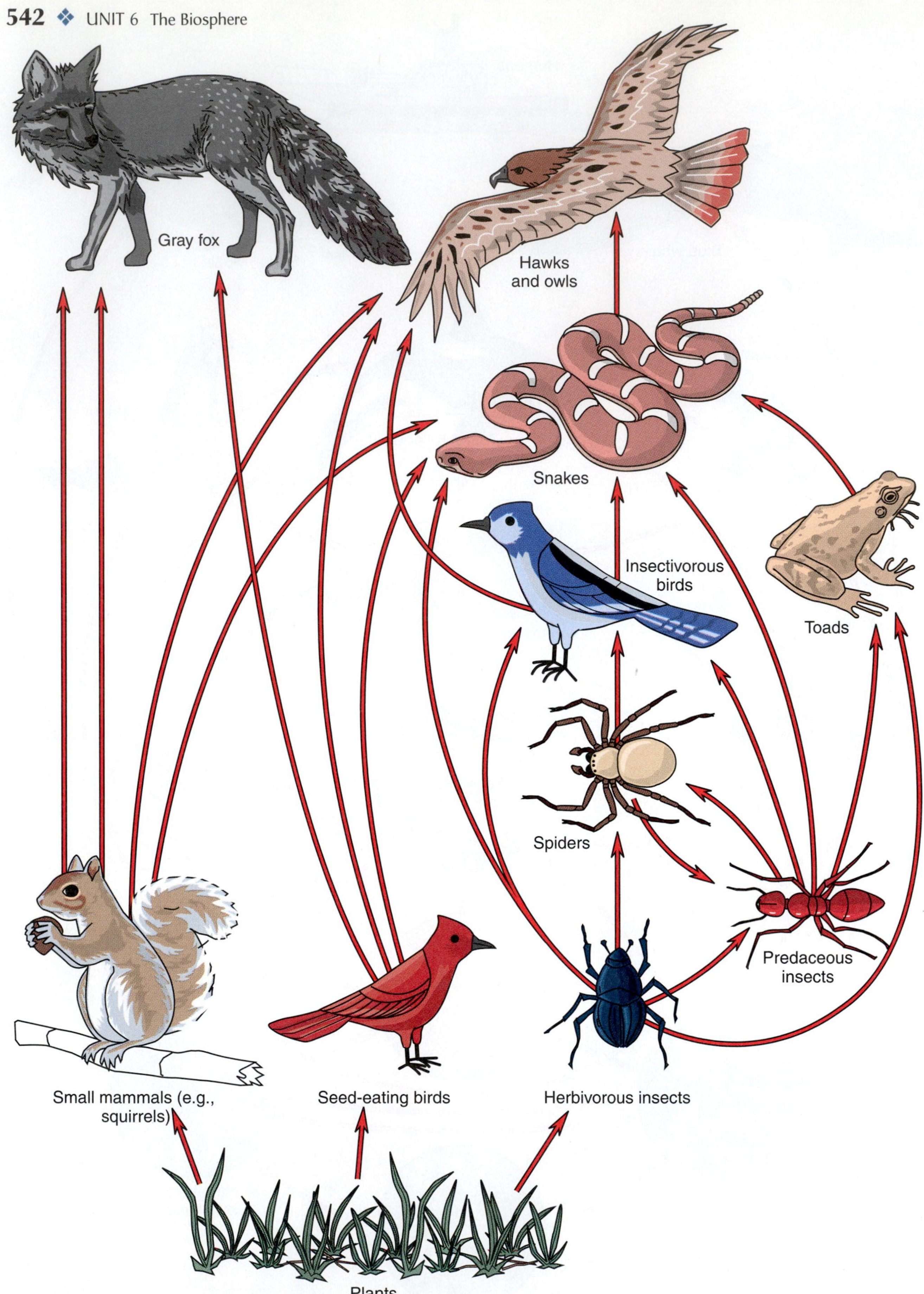

Figure 33–8 An example of a terrestrial food web.

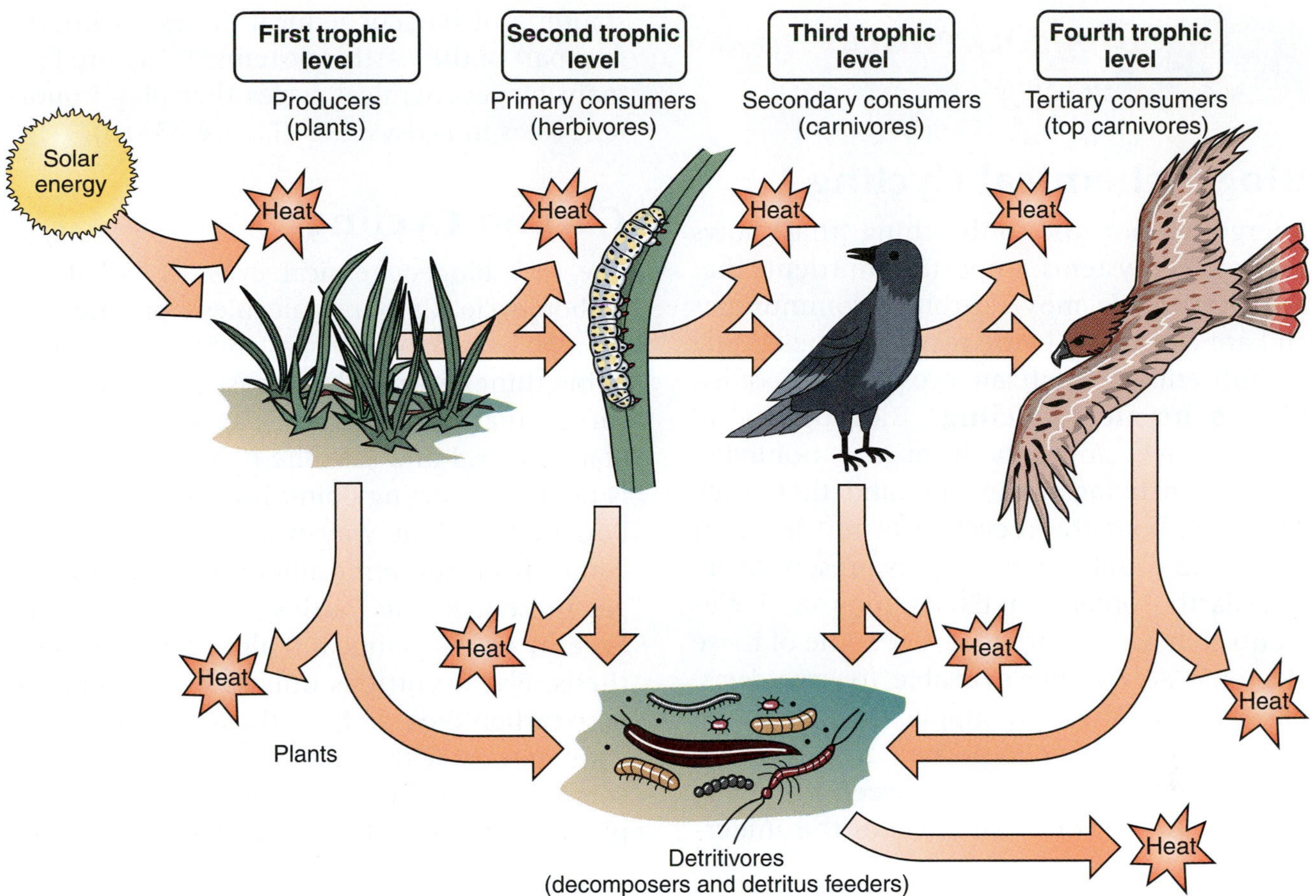

Figure 33–9 The movement of energy through a food chain reveals the loss of energy as heat to the atmosphere with each transfer between different trophic levels.

This is why upper level consumers must eat large quantities of food to survive. The other important aspect of the energy pyramid is how it illustrates that the Sun forms the base for all energy in an ecosystem. It also shows how the producers act as the important link between the Sun and all other organisms on the Earth.

For more information go to these Web links:

<http://www.flyingturtle.org/photosyn/photosynth.html>

<http://www.flyingturtle.org/me/pyramid.html>

<http://www.mesa.edu.au/friends/seashores/energy_pyramid.html>

SECTION REVIEW

1. Define the term *autotroph* and provide one example.
2. Describe the difference between chemosynthesis and photosynthesis.
3. Define the term *heterotroph* and provide one example.
4. Define the terms *producer, primary consumer,* and *secondary consumer* and explain how energy moves up through a food chain.
5. Provide an example for each of the following: herbivore, carnivore, omnivore, and detritivore.
6. Draw a simple food chain.
7. Approximately how much energy is lost between each trophic level in an energy pyramid?
8. Who was Linus Pauling?

33.2 *Biogeochemical Cycling*

Biogeochemical Cycling

Energy is not the only thing that flows through ecosystems. Important nutrients that are vital to life move through communities and are constantly being recycled. The cycling of nutrients through an ecosystem is called **biogeochemical cycling.** Biogeochemical cycles move nutrients from the nonliving world into living organisms, and then back again. Without the recycling of nutrients on Earth, life could not exist in its present form. The Earth contains a fixed amount of elements that are required for life. Some of these elements are readily available to organisms, but others fall in short supply. Over time, the availability of these important nutrients has reached a careful balance between the living and nonliving world. This makes the understanding of biogeochemical cycles an important part of the Earth's systems. There are four main biogeochemical cycles that play important roles in ecosystems (Figure 33–10).

Carbon Cycling

The first biogeochemical cycle is called the carbon cycle. Carbon molecules form the base for all life, making the availability of carbon to living things essential. Because a *cycle* refers to the circular movement of elements from the abiotic world and into the biotic world, there is really no starting point; however, most carbon that finds it way into living organisms comes from the atmosphere. Carbon dioxide gas present in the Earth's atmosphere is utilized by green plants and algae for photosynthesis. Photosynthesis utilizes solar energy to take carbon dioxide from the atmosphere and combines it with water to form sugars. A common form of sugar that is a product of photosynthesis is glucose. Glucose is classified

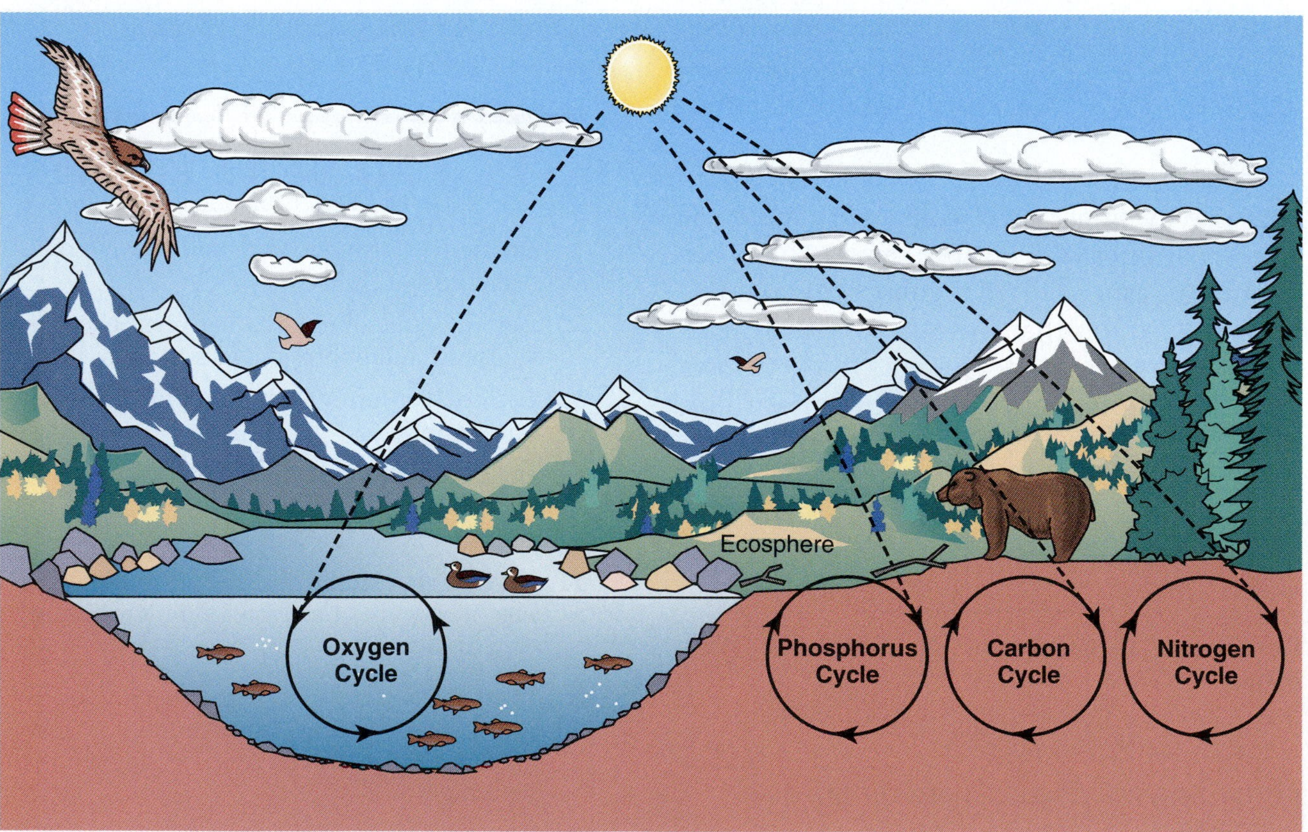

Figure 33–10 The four main biogeochemical cycles on the Earth.

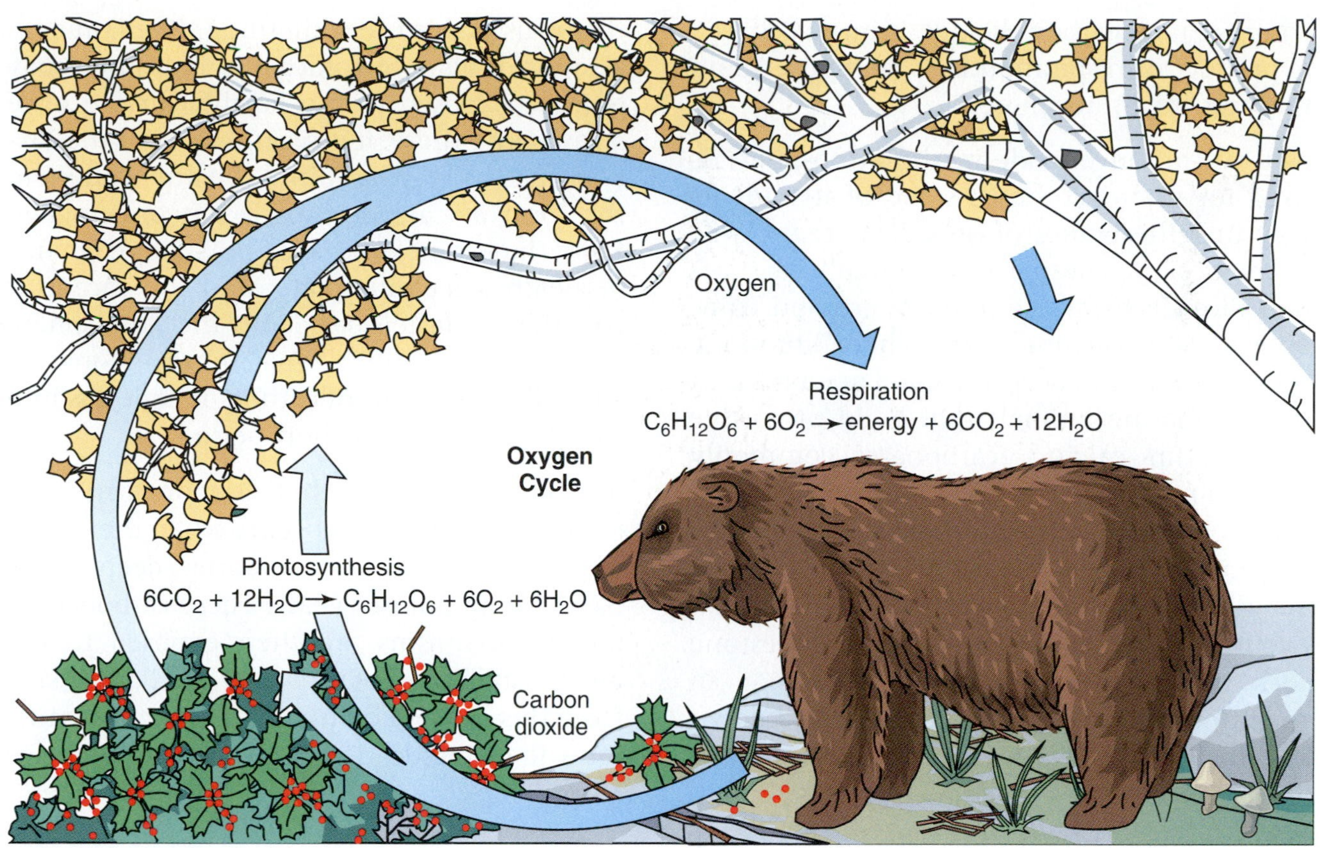

Figure 33–11 The exchange of carbon dioxide between the processes of photosynthesis and respiration sustain life on the Earth.

as an organic molecule, which means it contains carbon. The term *organic,* when applied to chemistry, literally means "carbon containing." The sugars that a plant creates are then used to build more complex molecules, such as starches. Starches are also known as carbohydrates, because they contain carbon, hydrogen, and oxygen.

Most of a plant's body is composed of carbohydrates. The carbon cycle continues when an animal then consumes a plant. The carbohydrates are then transferred into the body of the herbivore. Approximately 19% of the human body is composed of the element carbon. All this carbon was derived from eating food. Once the carbon moves its way into the food chain, it can take many pathways.

Heterotrophic organisms gain energy from organic molecules by the process of respiration. **Respiration** breaks down sugars and starches in the presence of oxygen and produces carbon dioxide as a byproduct. The respiration chemical reaction is the exact opposite of photosynthesis. It utilizes glucose, oxygen, and water to gain energy while also producing carbon dioxide gas. Every time human beings exhales air out of their lungs, they are removing this carbon dioxide from their bodies (Figure 33–11). Once carbon dioxide returns to the atmosphere, the pathway of the carbon cycle is complete.

Another path carbon can take in the food chain is to be eliminated from an organism in its waste. The waste then becomes part of the nonliving soil environment. Once there, the organic molecules that are in waste can be utilized by the decomposers such as fungi or bacteria. Eventually this carbon is also returned to the atmosphere as carbon dioxide. When organic molecules decay, such as in a compost pile, carbon dioxide is produced as a byproduct of respiration by decomposers and reenters the atmosphere.

Not all carbon dioxide in the atmosphere is used for photosynthesis; some of it finds its way into the bodies of organisms that reside in the oceans. These marine animals take carbon dioxide from the atmosphere and combine it with calcium to form hard limestone shells. Clams, mussels, and coral are all examples of marine animals that build protective shells from calcium carbonate, commonly called limestone (Figure 33–12). Over time these organisms die and their shells settle to the bottom of the world's oceans. This forms deep sediments of calcium carbonate that eventually form large layers of limestone rock. Limestone sediments store carbon in rocks for millions of years. This is also known as a **carbon reservoir.** A reservoir is anything that is used to store something. Over time, as a result of plate tectonics, the limestone may become exposed at the Earth's surface once again. Once there, it is weathered and the carbon finds its way back into the atmosphere in the form of carbon dioxide gas once again. Some limestone rock can be subducted underneath a tectonic plate, and the carbon it contains is blasted back into the atmosphere in a volcanic eruption. Volcanic eruptions can add millions of tons of carbon dioxide gas into the atmosphere in a short period.

An alternate path that carbon can take through an ecosystem occurs when the bodies of plants and animals are buried deep in the Earth for millions of years. These carbon-containing organisms are then exposed to the great heat and pressure below the Earth's surface. This causes chemical changes that transform the long-dead organisms into what we

Figure 33–12 Marine organisms that build shells from calcium carbonate are an important part of the carbon cycle. *(Courtesy of PhotoDisc.)*

call fossil fuels. Common fossil fuels include oil and coal. Oil is the remains of once living plankton that collected at the bottom of the ocean and became buried by sediments. Coal is the dead remains of plants that were buried in swamps millions of years ago.

All fossil fuels are also known as hydrocarbons, because they are mostly composed of hydrogen and carbon. When fossil fuels are removed from deep within the Earth by human activity and are then burned, the carbon that was locked in the Earth for millions of years is released back into the atmosphere as carbon dioxide gas. When something is burned it also called combustion. Fossil fuels act as a reservoir for carbon, which in the past 100 years has been utilized as an energy source and is causing an imbalance of carbon dioxide in the Earth's atmosphere. Many scientists believe this is leading to increased temperatures around the planet, which may be causing global warming.

Part of the carbon cycle also involves the creation of methane gas. Methane is a hydrocarbon molecule that is produced by anaerobic bacteria. *Anaerobic* means without oxygen. Anaerobic bacteria can only survive without the presence of oxygen and produce methane gas as a byproduct of their feeding process. Many herbivores, such as cows, contain anaerobic bacteria in their digestive tracts. These bacteria help to digest the tough grasses, while also producing methane. Another source of methane is the mud of rice paddies in Asia. Anaerobic bacteria reside there and release methane into the atmosphere. No matter which pathway carbon takes through the environment, the carbon cycle is an important aspect of many of the Earth's systems (Figure 33–13).

Oxygen Cycling

The movement of oxygen through ecosystems is an important biogeochemical cycle (Figure 33–14). Oxygen is another vital element for life on Earth. All heterotrophs on Earth require oxygen for respiration. Oxygen is the most abundant element in the Earth's crust, composing approximately 45% of the lithosphere. Oxygen is also the second most abundant gas in the atmosphere, with a concentration of approximately 21%. This important element also makes up about 63% of the human body. Some oxygen is combined with other elements such as silicon, iron, or aluminum to form many of the Earth's minerals and rocks. The water molecule, which is one of the most abundant molecules on Earth, is composed of one oxygen atom and two hydrogen atoms.

One of the most important interactions of the oxygen cycle occurs between the photosynthesis and respiration reactions. Photosynthesis takes in carbon dioxide gas, water, and sunlight to yield simple sugars, water, and oxygen. Respiration then utilizes the oxygen produced by plants and combines it with simple sugars and water to yield energy, water, and carbon dioxide. This relationship forms a vital link between plants and animals on Earth. At one time in the Earth's past, there was no oxygen in the atmosphere. It took the photosynthetic action of ancient algae to fill the atmosphere with oxygen to its current level. Oxygen can also be dissolved in water, where it is utilized by marine organisms for respiration.

The ozone layer high in the stratosphere is made up of oxygen. Ozone gas is composed of three atoms of oxygen and acts as a protective barrier that blocks harmful ultraviolet radiation from striking the Earth's surface. The cycling of oxygen from the biotic world into the abiotic world and back again is another in important aspect of the biosphere.

Nitrogen Cycling

The third biogeochemical cycle is called the nitrogen cycle. Nitrogen is an important element that is part of the DNA molecule, which is responsible for the replication of living cells. Nitrogen is also an important component of

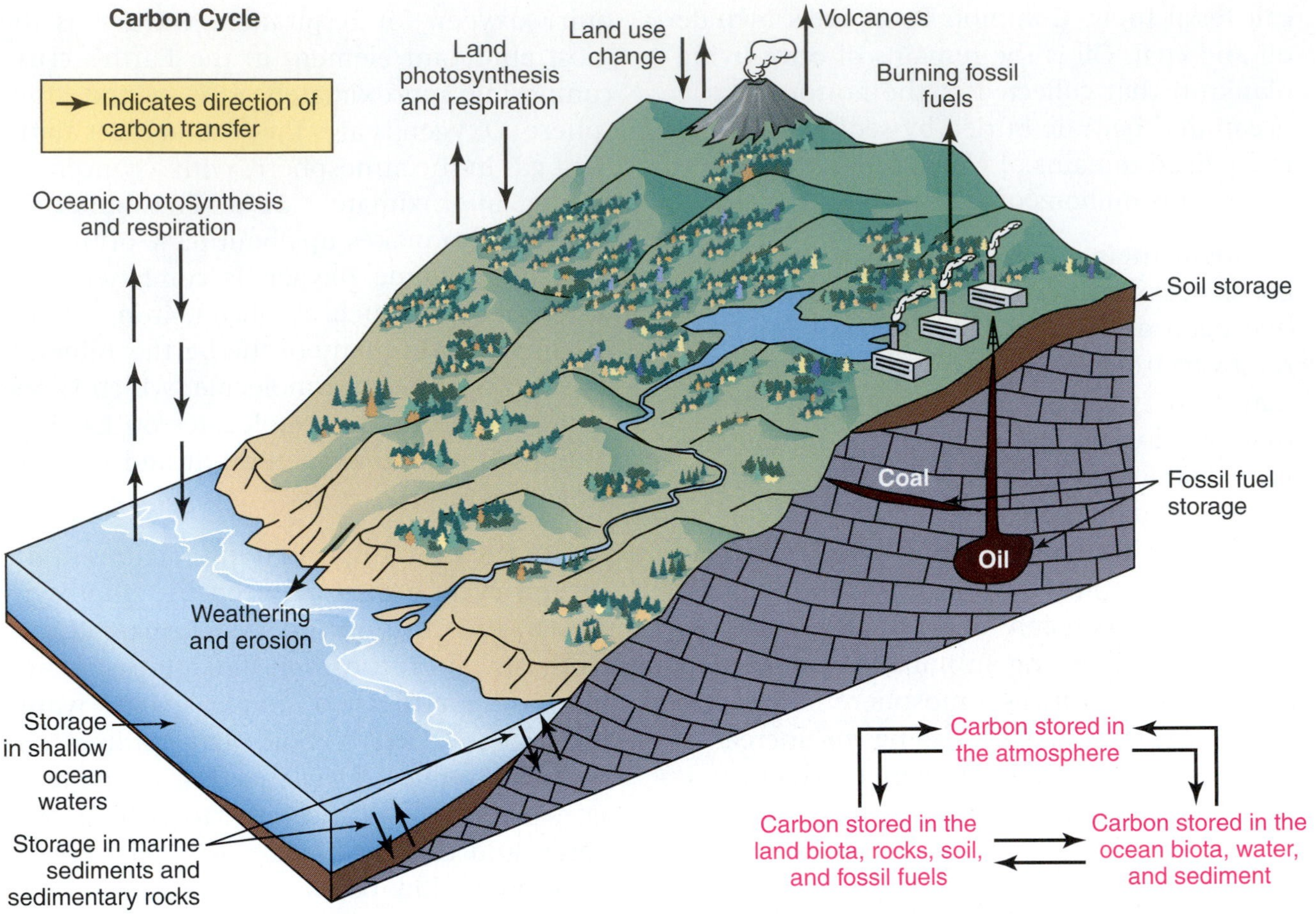

Carbon Cycle

→ Indicates direction of carbon transfer

Oceanic photosynthesis and respiration

Land photosynthesis and respiration

Land use change

Volcanoes

Burning fossil fuels

Soil storage

Coal

Oil

Fossil fuel storage

Weathering and erosion

Storage in shallow ocean waters

Storage in marine sediments and sedimentary rocks

Carbon stored in the atmosphere

Carbon stored in the land biota, rocks, soil, and fossil fuels

Carbon stored in the ocean biota, water, and sediment

Figure 33–13 The storage and movement of carbon through the environment is known as the carbon cycle.

EARTH SYSTEM SCIENTISTS *ROGER REVELLE*

Roger Revelle was born in Seattle, Washington, in 1901. He earned his college degree in geology and later studied oceanography. He received his doctorate by studying the bottom sediments of the Pacific Ocean. Revelle next became interested in studying the global carbon cycle, especially the role of the oceans in this important biogeochemical cycle. As a result of his research on the carbon cycle, he began to notice that global levels of carbon dioxide were rising steadily. His research lead him to write a groundbreaking scientific paper that proposed that the increasing levels of carbon dioxide gas were the result of burning fossil fuels. This broke open the concept of global warming and the effects of human technology on global climate. As a result of his research, Revelle's career turned to working with world governments and helping them to accept the fact that global warming was indeed a worldwide problem. Revelle concluded that global warming would lead to a melting of the polar ice caps and therefore cause sea levels to rise. He also worked on the problems of deforestation and its effect on global carbon dioxide concentration. The latter part of his career involved the applications of science and technology to food production.

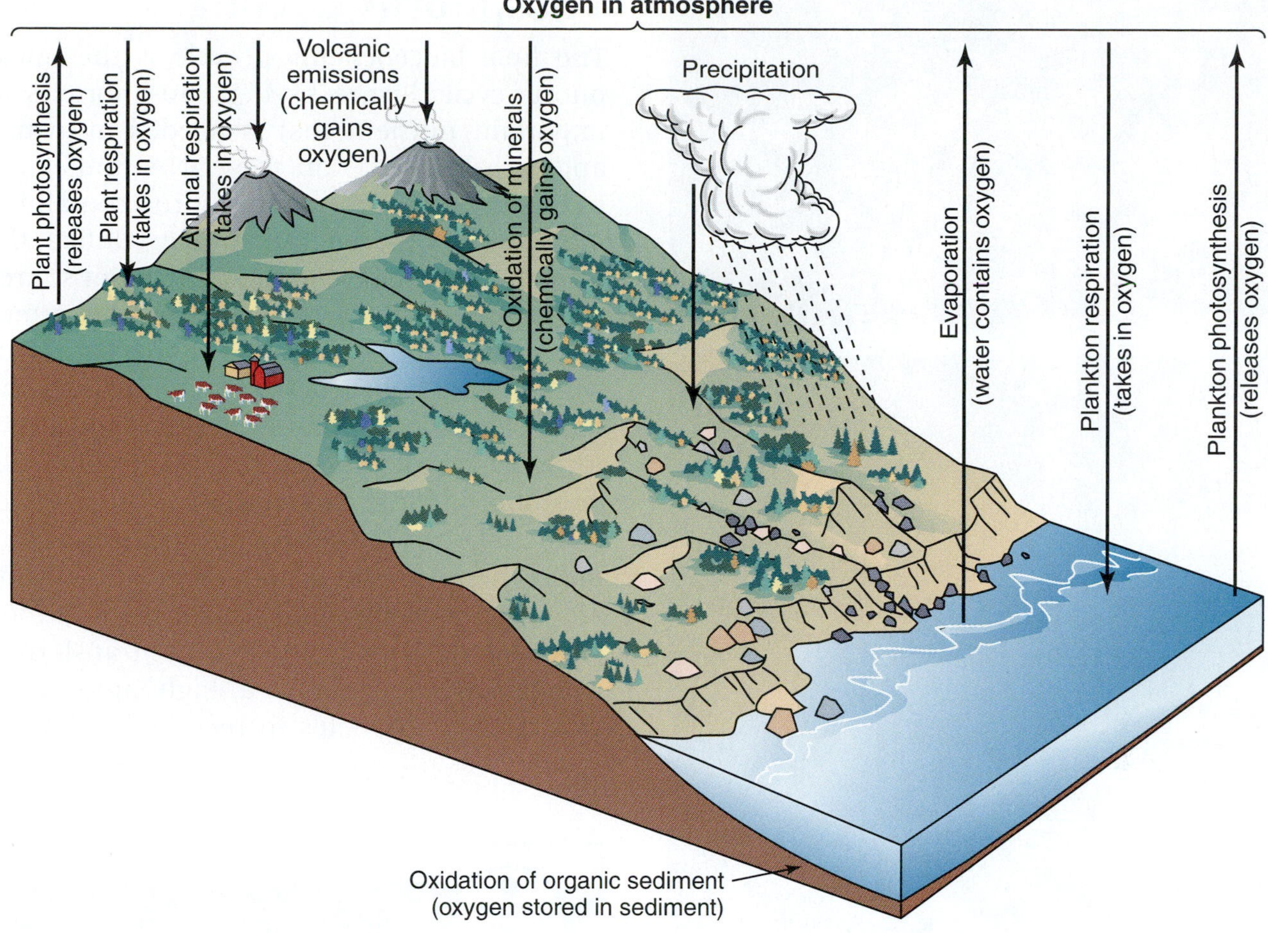

Figure 33–14 The oxygen cycle.

protein molecules. Approximately 5% of the human body is composed of nitrogen. The source of nitrogen for the world's ecosystems is the atmosphere. Earth's atmosphere is composed of 78% nitrogen. The nitrogen gas that fills the atmosphere is in a diatomic form. The term diatomic means that it is composed of two atoms of nitrogen.

The nitrogen cycle begins when microscopic bacteria located in the soil and in the root systems of specialized plants combines atmospheric nitrogen with other atoms to form nitrogen compounds. This process is called nitrification, and it produces nitrogen-containing molecules such as nitrate and ammonia. These nitrogen compounds can then be taken up by the root systems of plants to be used to make proteins. The plants that harbor this specialized bacteria in their root systems are called legumes (Figure 33–15). Small nodules located on the roots of legumes contain the nitrifying bacteria that convert atmospheric nitrogen into plant-available nitrogen. These can then be used to form proteins and other compounds within plants. Common legumes include clover, beans, peanuts, and alfalfa. Legumes are often used in agriculture to add nitrogen to the soil, and they act as natural fertilizers.

Once plants take up the nitrogen that the bacteria produce, it is then available for other organisms in the food chain (Figure 33–16). All the proteins and other nitrogen-containing molecules in our bodies were once part of

Figure 33–15 Legume nodules on the roots of a soybean plant contain bacteria that convert atmospheric nitrogen into nitrogen compounds that can be used by plants. *(Courtesy of USDA-ARS.)*

plants. Animals produce waste that contains nitrogen that is expelled from the body. This returns nitrogen compounds back into the environment, where they are decomposed by the detritivores. Some bacteria convert these nitrogen compounds back into diatomic nitrogen that reenters the atmosphere. These bacteria are called denitrifying bacteria and are also found in the soil. Some human activities, such as the burning of fossil fuels, causes atmospheric nitrogen to combine with oxygen to produce compounds called nitrogen oxides. These compounds can mix with water in the atmosphere to form nitric acid. This is the also known as acid precipitation, which has become one of the worst forms of air pollution.

Phosphorus Cycling

The final biogeochemical cycle is the phosphorus cycle (Figure 33–17). Phosphorus is an important nutrient that is needed by plants and is also part of the DNA molecule. Individual cells utilize phosphorus-containing molecules as an energy source. Unlike the other biogeochemical cycles, the major source of phosphorus is not found in the atmosphere, but rather in rocks. Phosphate-containing rocks that become exposed at the Earth's surface become weathered and begin to break down. This is the main source of phosphorus for the ecosystems of the world. Plants take up the phosphorus minerals and utilize them for their body functions. Plants are then eaten by other organisms in a food chain, and the phosphorus is transferred. Animal waste contains a high amount of phosphorus and helps to recycle this impor-

Career Connections

BIOCHEMIST

Biochemists study the chemistry of all living organisms. Their work includes an understanding of the chemical processes that take place in the body that are responsible for growth, reproduction, metabolism, and heredity. Biochemists also help to understand the cause and spread of disease and illness in the human body. Current research in biochemistry involves the use of living organisms to perform specific tasks. This is called biotechnology, and it is one of the most exciting aspects of modern science. Biochemists are also searching for ways to produce new drugs or to improve food sources. These scientists work in three major areas: medicine, agriculture, and nutrition. The main goal of all biochemists is to further understand the complex chemistry of living organisms. Biochemists need a college education in both chemistry and biology. Career opportunities are available in private industry or in academic research.

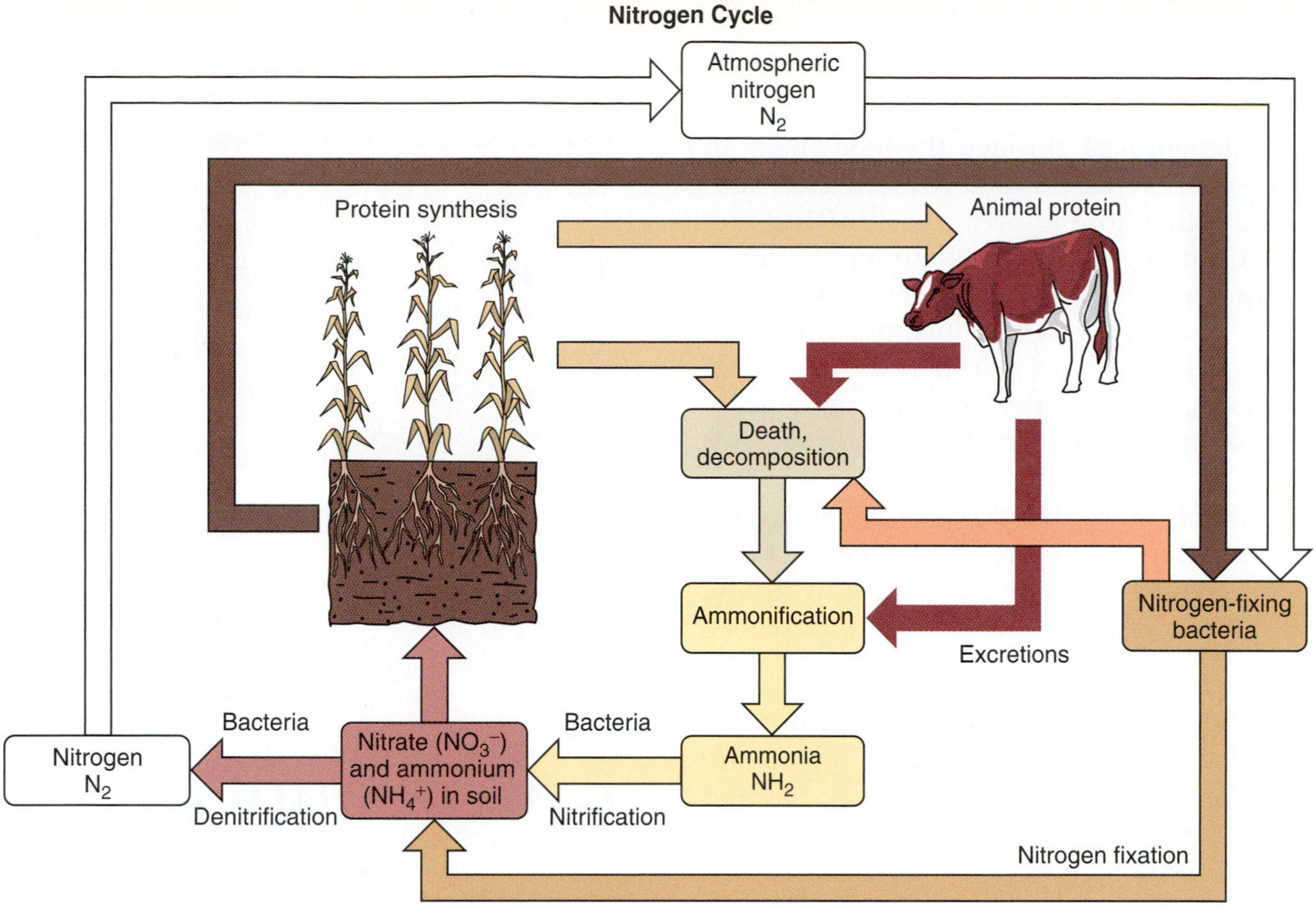

Figure 33–16 The nitrogen cycle. *(Courtesy of USDA-ARS.)*

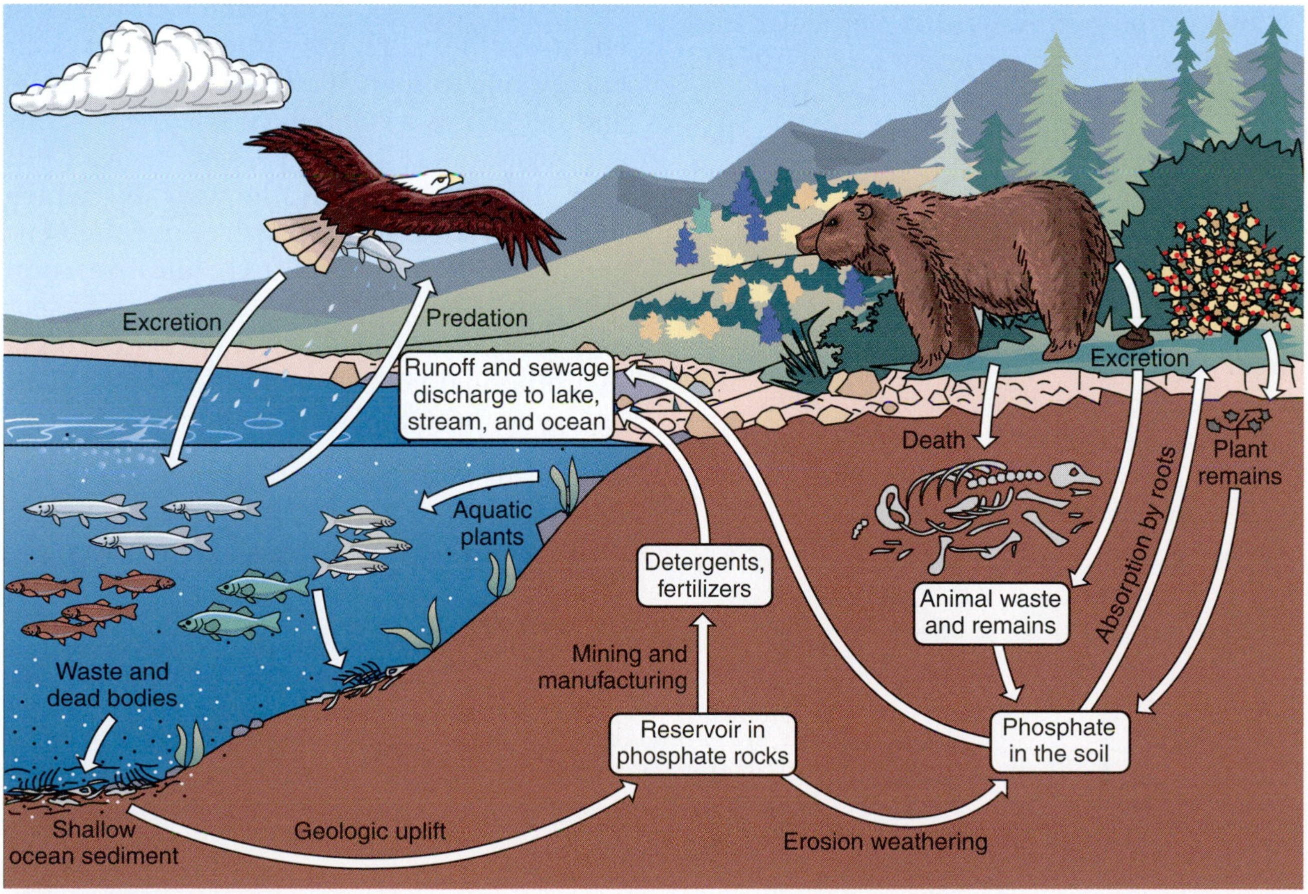

Figure 33–17 The phosphorus cycle.

tant element back through the food chain. In many ecosystems, phosphorus is in short supply; it is therefore known as a **limiting factor.** A limiting factor is something that limits the productivity of an ecosystem. As a result of human activity, which includes waste disposal and the use of phosphorus-containing fertilizers, phosphorus is entering aquatic ecosystems in higher levels than normal. This causes a rapid growth of aquatic plants and algae, which utilize the phosphorus as an important nutrient. The result is called an algal bloom, which chokes waterways with the overgrowth of algae and other aquatic plants.

Although the carbon, oxygen, nitrogen, and phosphorus cycles make up the four main biogeochemical cycles on Earth, there are many other elements and nutrients that also travel through ecosystems. Without the recycling of nutrients through the food chains of the world, and their movement from the abiotic environment to the biotic environment, life could not exist on Earth as we know it. Following the pathways that particular elements take through an ecosystem illustrates the complexities that exist in the Earth's systems and reveals the careful balance that has evolved over time between living things and their nonliving environment.

For more information go to these Web links:

<http://observe.arc.nasa.gov/nasa/earth/hydrocycle/hydro1.html>

<http://library.thinkquest.org/11226/>

<http://www.earth.nasa.gov/science/Science_ecosystems.html>

SECTION REVIEW

1. Define the term *biogeochemical cycle.*
2. Draw a simple diagram of the carbon cycle.
3. Draw a simple diagram of the oxygen cycle.
4. Draw a simple diagram of the nitrogen cycle.
5. Draw a simple diagram of the phosphorus cycle.
6. Who was Roger Revelle?

CHAPTER SUMMARY

The movement of energy through the ecosystems of the world is an important part of the Earth's systems. Organisms derive energy in two principal ways, from photosynthesis and chemosynthesis. Photosynthesis utilizes solar energy and carbon dioxide to create sugars and oxygen as a byproduct. All green plants and algae utilize photosynthesis to gain energy. Photosynthesis is the way by which most ecosystems on the Earth derive their energy. Chemosynthesis is a chemical reaction that gains energy from the breakdown of specific molecules. This is the way that deep sea hydrothermal vent communities derive their energy without access to sunlight.

Organisms that produce their own energy by these two processes are called autotrophs, or "self-feeders." Another way that organisms

EARTH MATH

1) IF APPROXIMATELY 19% OF THE HUMAN BODY IS COMPOSED OF CARBON, HOW MUCH CARBON DOES AN AVERAGE ADULT HUMAN WEIGHING 150 POUNDS CONTAIN?

derive energy on the Earth is by consuming other organisms. These type of living things are called heterotrophs, or "other feeders." The amount of chemical energy that is created by autotrophs is called primary production. It is measured in the amount of biomass produced by an organism. Autotrophs are also known as producers in an ecosystem because they produce food.

Organisms that consume the producers are called primary consumers. Because they consume plants, the primary consumers are also called herbivores. Other organisms that eat the primary consumers are called secondary consumers. These animals are known as carnivores because they eat meat. Organisms that consume both plants and animals, such as humans, are called omnivores. Some ecosystems have tertiary consumers who eat the secondary consumers; these are also known as upper level consumers. Another important aspect of an ecosystem is the decomposers. These organisms consume waste and dead organisms and are called detritivores.

The specific movement of energy through an ecosystem by a series of eating processes is called a food chain. Many interrelated food chains are known as food webs. An energy pyramid is used to illustrate the way that energy is used in a specific food chain. It is composed of a series of feeding levels called trophic levels. The movement of energy from one trophic level to another results in the loss of energy from a food chain as heat to the atmosphere. Energy pyramids also reveal the important link between the Sun and all living things within an ecosystem.

The recycling of matter within ecosystems is another important aspect of the Earth's systems. The Earth has a fixed amount of chemical elements that can be used by organisms. These elements form molecules, act as nutrients for organisms in ecosystems, and are continually recycled between the living and nonliving world. This is called biogeochemical cycling. There are four main biogeochemical cycles on the Earth: carbon cycle, oxygen cycle, nitrogen cycle, and phosphorus cycle. The way that these nutrients move through the environment and their constant reuse allow for life to flourish on the Earth.

CHAPTER REVIEW

Multiple Choice

1. Most energy on the Earth is derived from:
   a. ecosystems
   b. producers
   c. the Sun
   d. autotrophs

2. An organism that makes its own food is called:
   a. a secondary consumer
   b. an autotroph
   c. a heterotroph
   d. a decomposer

3. Bacteria in hydrothermal vent communities deep in the ocean utilize which process to gain energy?
   a. chemosynthesis
   b. primary production
   c. photosynthesis
   d. nitrification

4. Combining carbon dioxide with water and sunlight to produce glucose sugar is an example of:
   a. chemosynthesis
   b. respiration
   c. photosynthesis
   d. nitrification

5. The amount of biomass that an ecosystem produces is called:
   a. chemosynthesis
   b. primary production
   c. photosynthesis
   d. nitrification

6. An organism that consumes only plants is called:
   a. an omnivore
   b. an herbivore
   c. a carnivore
   d. a decomposer

7. The bacteria, fungi, and insects that eat waste and dead organisms are known as:
   a. omnivores
   b. herbivores
   c. carnivores
   d. decomposers

8. Approximately how much energy is given off to the atmosphere as heat between trophic levels in a food pyramid?
   a. 10%
   b. 30%
   c. 50%
   d. 90%

9. The movement of matter between the living and nonliving world is called:
   a. the food chain
   b. biogeochemical cycling
   c. the energy pyramid
   d. a food web

10. The source of most carbon in the carbon cycle is:
    a. the hydrosphere
    b. rocks in the Earth's crust
    c. the atmosphere
    d. fossil fuels

11. The process of gaining energy by breaking down sugars in the presence of oxygen is called:
    a. photosynthesis
    b. respiration
    c. chemosynthesis
    d. nitrification

12. Which geochemical cycle uses limestone rock and fossil fuels as a reservoir?
    a. the carbon cycle
    b. the nitrogen cycle
    c. the oxygen cycle
    d. the phosphorus cycle

13. The main source of phosphorus for the phosphorus cycle is:
    a. the hydrosphere
    b. rocks in the Earth's crust
    c. the atmosphere
    d. fossil fuels

14. The process by which bacteria convert atmospheric nitrogen into plant-available forms is called:
    a. photosynthesis
    b. respiration
    c. chemosynthesis
    d. nitrification

15. Which biogeochemical cycle is linked to global warming?
    a. the carbon cycle
    b. the nitrogen cycle
    c. the oxygen cycle
    d. the phosphorus cycle

16. The limiting factor in most ecosystems is associated with which biogeochemical cycle?
    a. the carbon cycle
    b. the nitrogen cycle
    c. the oxygen cycle
    d. the phosphorus cycle

17. Which of the following plants is not a legume that helps to fix nitrogen in the soil?
   a. grass
   b. clover
   c. alfalfa
   d. peanuts

Matching *Match the terms with the correct definitions.*

a. autotrophs
b. chemosynthesis
c. heterotrophs
d. primary production
e. biomass
f. herbivores
g. carnivores
h. omnivores
i. detritivores
j. decomposers
k. food chain
l. biogeochemical cycling
m. respiration
n. reservoir
o. limiting factor

1. ____ A classification of organisms that must consume other organisms to gain energy, such as humans.
2. ____ A classification of organisms that eat both plants and animals.
3. ____ A type of organism that produces its own food by photosynthesis, such as a plant or algae, or by chemosynthesis, such as certain bacteria.
4. ____ A specific nutrient that is lacking in an ecosystem and limits the growth of organisms.
5. ____ A method of deriving energy from the breakdown or formation of organic compounds.
6. ____ The model pathway that energy and matter takes through an ecosystem by a series of eating processes.
7. ____ A storage place for something.
8. ____ A type of animal that only eats plants.
9. ____ An organism that breaks down and decays dead organisms or waste.
10. ____ The chemical process by which carbohydrates are broken down in the presence of oxygen to derive energy and produce carbon dioxide.
11. ____ Meat-eating organisms.
12. ____ The process by which plants utilize photosynthesis to convert solar energy into chemical energy that is stored in plant material.
13. ____ A type of organism that consumes dead and decayed organisms or waste.
14. ____ The natural recycling of elements between the nonliving world and the living world.
15. ____ The total dry weight of an organism.

Critical Thinking

1. Explain the differences and similarities between a naturally occurring food web and a human-created agricultural food chain.

CHAPTER 34 Biological Succession

Objectives

Primary Succession • Primary Succession and Pioneer Communities • Secondary Succession

After reading this chapter you should be able to:

- Define the term *biological succession.*
- Differentiate between primary and secondary succession.
- Define the term *pioneer community.*
- Identify areas on the Earth where primary succession occurs.
- Provide an example of primary succession.
- Provide an example of secondary succession.

TERMS TO KNOW

dynamic
biological succession
primary succession
pioneer community
lichen
secondary succession
catastrophic
germinate
disturbed areas

INTRODUCTION

The interactions between a community of organisms and the environment in which they live is a **dynamic** system. Over time, ecosystems experience gradual changes in both their abiotic and biotic factors. The constant change that ecosystems experience is barely perceptible to human beings, but on a geologic time scale these changes are quite apparent. The geologic forces of weathering, erosion, and plate tectonics, along with the dynamic nature of life, cause gradual changes to occur in all aspects of the Earth's systems. These natural changes are continually reshaping the landscape and providing the regeneration necessary to sustain life all over the planet.

Biological Succession

One of the key scientific principles that helps to explain the changes ecosystems go through over time is called **biological succession.** Biological succession is the gradual replacement of one community of organisms by another in a slow, orderly, and predictable manner. Scientists have observed the process of biological succession in many parts of the world and now understand it as a natural force of change on the Earth. Biological succession is divided into two categories: primary succession and secondary succession.

Primary Succession and Pioneer Communities

Primary succession occurs when living organisms first began to inhabit a part of the Earth. As time goes by, communities of organisms begin to flourish and are slowly replaced by other communities of organisms. An example of primary succession can be seen in the wake of a retreating glacier (Figure 34–1). The land that was once covered by a thick ice sheet becomes exposed to the elements when the ice melts away as the glacier retreats. Over time, the barren rocks that were exposed by the melting glacier are colonized by a **pioneer community.** A pioneer community consists of a rapidly growing group of organisms that can reside in a place where no life presently exists. Examples of pioneer communities include microorganisms such as bacteria, mosses, and lichen.

Lichen are actually composed of two organisms, algae and a fungus, that maintain a symbiotic relationship (Figure 34–2). The algae utilizes the process of photosynthesis to create food for itself and its companion fungus. In re-

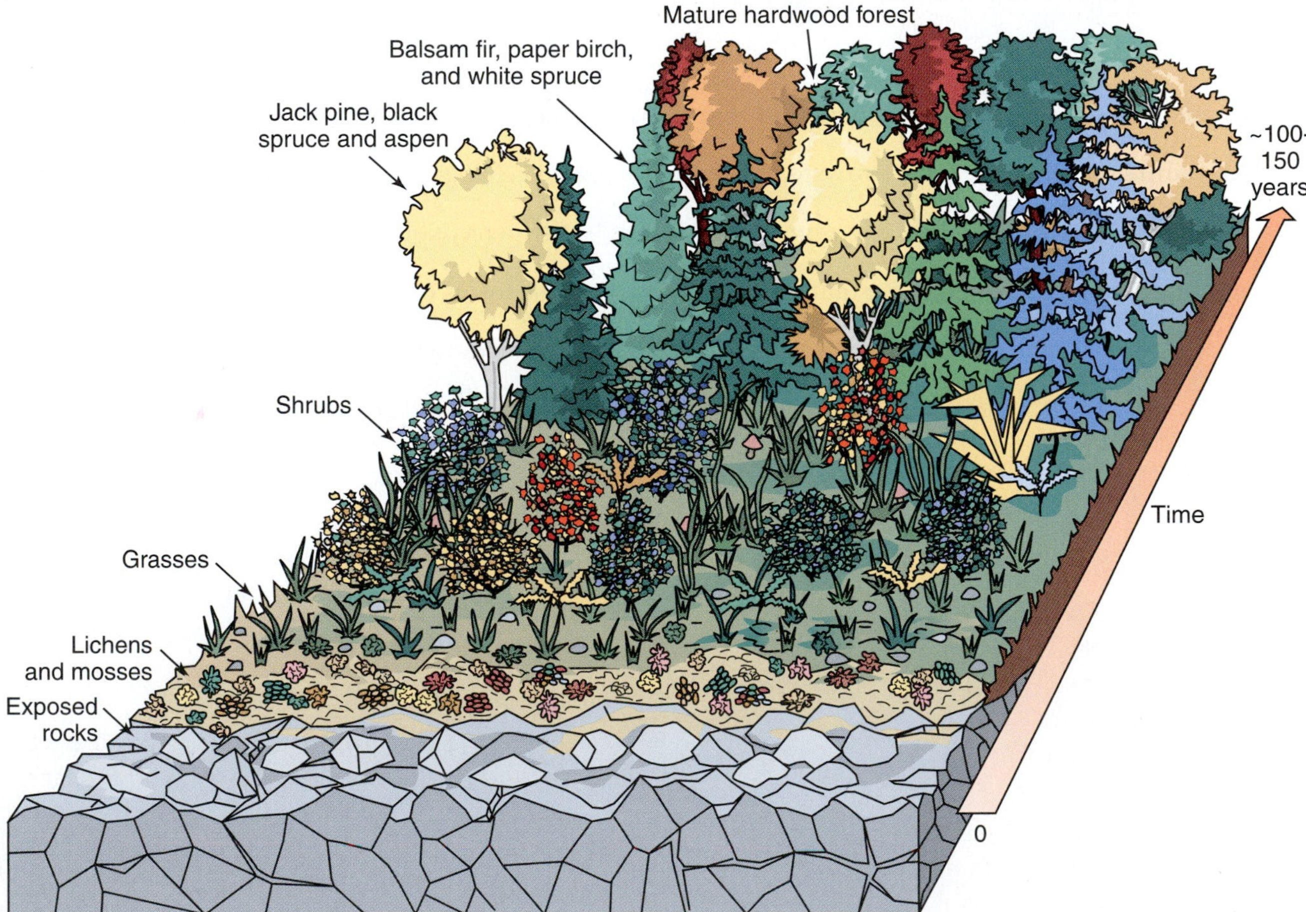

Figure 34–1 An example of the stages of primary succession on bare rock exposed by a retreating glacier.

Figure 34–2 Lichen growing on bare rocks are an example of a pioneer community that can live in harsh environments where no other life exists.

turn, the fungus secretes a weak acid that breaks down rock into usable minerals that the algae requires. When lichen establish themselves on barren rock, they slowly create an environment that becomes suitable for other organisms. Eventually, a thin soil layer forms over the rocks as the lichen colony breaks it down and mixes it with the decaying organic material of dead lichen. When the soil becomes thick enough, seeds and spores brought to the area by birds or on the wind can become established. Eventually the area can support hardy grasses, ferns, and small shrubs. This makes way for other organisms to inhabit the area, such as insects, rodents, birds, and migrating grazing animals. As more time passes, the community of organisms creates a rich environment that may begin to support trees. Over a period of a few thousand years, a mature forest ecosystem may become established where once there was no life at all. This is how the process of primary succession works.

Much of North America was once covered in thick ice sheets that were more than 1 mile thick. Then as the glaciers retreated approximately 11,000 years ago, the process of primary succession transformed the landscape into what it looks like today. Although the abiotic environment plays an important role in primary succession, it is the action of the living organisms that form communities that do much of the work. The interactions that occur between organisms and their environment are the driving force for change that causes succession.

Retreating glaciers are not the only force that can lead to primary succession. Shifting

EARTH SYSTEM SCIENTISTS *GIFFORD PINCHOT*

Gifford Pinchot was born in Connecticut in 1865. He graduated from Yale University and then attended the French Forestry School. While in France, he learned about the selective cutting of trees and responsible forest management. Upon his return to the United States, Pinchot was made the head of the Division of Forestry, which would later become the U. S. Forest Service. His job involved the management of all of the U. S. forest land. Under Pinchot's leadership, the Forest Service began the process of managing forests as a valuable natural resource. He realized that forests were vital to the U. S. economy and that they should be scientifically managed. Pinchot created management schemes that would ensure the continued availability of forest resources and helped to create millions of acres of national forest. His techniques prevented the overharvesting of trees and regulated the use of forest land. Pinchot, along with President Theodore Roosevelt, helped form the conservation movement of the early twentieth century. Later in his life Pinchot became a two-term governor of Pennsylvania.

rivers or shorelines can expose sandbars out of the water, which quickly become inhabited by pioneer communities. Volcanic islands in the oceans are also creating new landforms that can be shaped by primary succession. A new island is being formed off the Hawaiian Island chain in the south Pacific. Some day this new island will break the ocean's surface and be transformed into a tropical paradise, much like its sister islands, by the process of primary succession.

Secondary Succession

The other type of biological succession that occurs on Earth is known as **secondary succession.** Secondary succession is the gradual replacement of one community of organisms by another in a slow, predictable manner in an area where life has already flourished (Figure 34–3). The common cause of secondary succession is usually some type of **catastrophic** event such as a volcanic eruption, forest fire, or flood. Human activities have also interrupted the landscape to the point at which secondary succession can occur.

When a forest fire completely burns away a forest ecosystem and leaves behind nothing but burnt debris, secondary succession begins almost immediately. Seeds and spores carried on the wind **germinate** in the charred soil and begin to sprout. This leads to the establishment of grasses and ferns that quickly overtake the area. Eventually the grasses give way to sprouting trees, which shade out the grass. Small shrubs may also become established in the burnt remains of the forest. Soon the area that was once completely devastated harbors life again. As the trees grow taller, birds, insects, and mammals begin to inhabit the area. Approximately 70 to 100 years after the fire, the forest ecosystem completely reestablishes itself.

Secondary succession is an important way that the Earth heals itself after devastating events and truly shows how strong the force of life is on our planet. Scientists received a firsthand look at the process of secondary succession after the catastrophic volcanic eruption of Mount St. Helens in Washington State. The eruption of Mount St. Helens in 1980 destroyed thousands of acres of forest land and killed millions of aquatic and terrestrial organisms (Figure 34–4). After the eruption, the landscape was completely void of life. The blast of the volcano incinerated everything in its path and created devastating mudflows that choked lakes and streams.

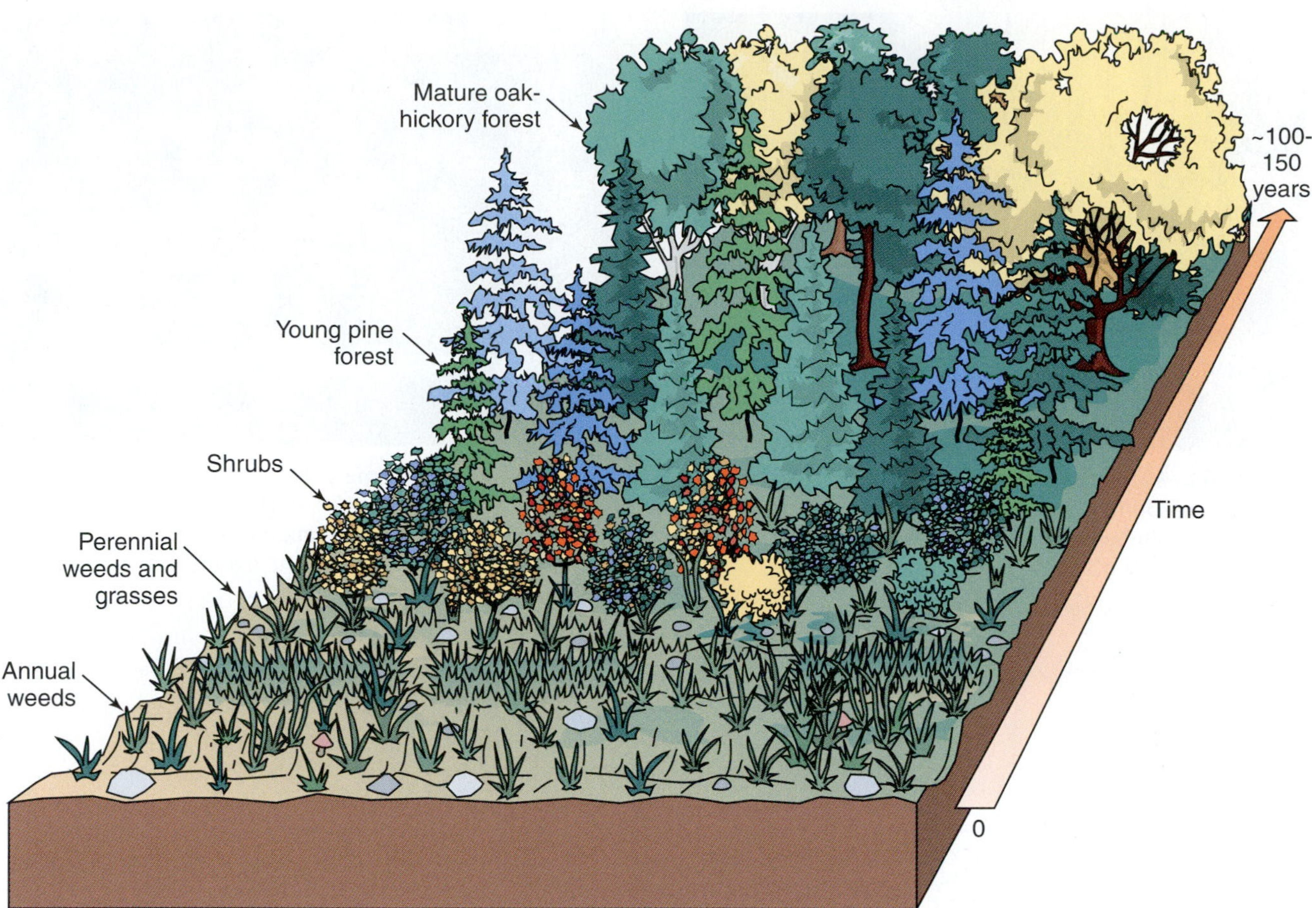

Figure 34–3 An example of the stages of secondary succession that occurred after a farm field was abandoned.

Not long after the disaster occurred, primary succession began. Seeds and insects blown in on the wind from surrounding areas began to inhabit the region once again. As time went by, the once black and gray charred remains of the area were transformed into the lush green colors of plant life. Today the landscape around Mount St. Helens has developed into a healthy ecosystem once again as a result of the powers of secondary succession (Figure 34–5).

Many parts of the northeastern United States have also experienced secondary succession over the past 100 years. Shortly after the American Civil War, the western frontier was opened up to settlers, which created a mass migration of people to the Great Plains. Many of these settlers abandoned their farms in the north for the much flatter, more forgiving soil of the midwest. Over time the abandoned farms of the northeast gave way to secondary succession. The fields that lay fallow eventually returned to the mixed hardwood deciduous forests that make up much of the landscape of New York and New England (Figure 34–6). As you walk through these forests today, you can still find the remains of old stone fences that once divided the farm fields. Most of the trees that inhabit these forests are the result of secondary succession.

Biological succession is the way that the Earth allows life to reclaim **disturbed areas.** It is a force that is driven by living things, that paves the way for other life forms to flourish in an area where they once could not. Understanding the process of biological succession has become an important management tool for reclaiming many disturbed areas. Whether it be old abandoned lots sprouting new growth

Figure 34–4 The catastrophic destruction caused by the eruption of Mount St. Helens in Washington State is an example of a disturbed landscape. *(Courtesy of P. Frenzen, USDA Forest Service.)*

Figure 34–5 The same disturbed landscape near Mount St. Helens shown in Figure 34–4. Here, 11 years after the volcanic eruption, the regeneration of the forest as a result of biological succession is evident. *(Courtesy of P. Frenzen, USDA Forest Service.)*

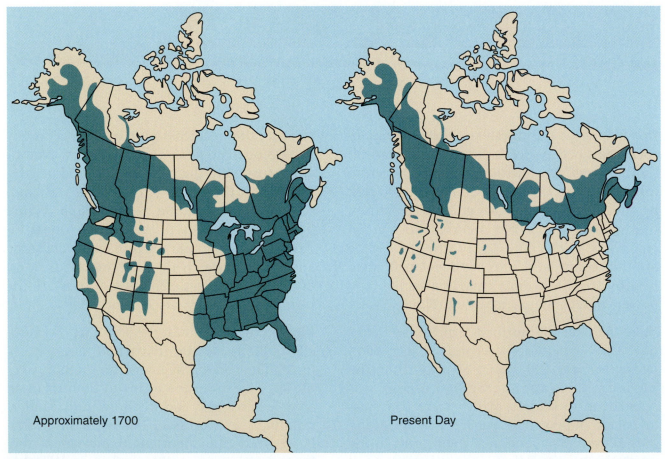

Figure 34–6 A map showing the comparison between the distribution of old growth forests of North America in 1700 and the present. This reveals that most of the forests that grow in the United States today are the products of secondary succession.

Career Connections

FORESTER

Foresters manage forests for both public and private land. They develop and carry out specific management techniques to sustain forest resources. Both private and public forest lands are managed for use in outdoor recreation, timber use, watersheds, wildlife habitat, and other forest resources such as maple sugar production. Forest management includes conducting forest inventories, creating maps, planting trees, studying wildlife, fire prevention, improving trails and roads, monitoring for disease and insects, and overseeing timber cutting. Foresters can work for private industry or for the state and federal government. Their work involves a love for the outdoors in all types of weather. There are managed forest lands in all parts of North America and throughout the world that employ foresters. Foresters need a college degree in forestry and natural resource management. Often forest technicians work with foresters to help carry out specific management techniques. Forest technicians can gain on-the-job experience with a professional forester or in a high school vocational conservation program. Many 2-year technical colleges offer instruction for careers as a forest technician.

in the middle of large cities or the sounds of birds in the once silent valley near Mount St. Helens, biological succession gives life the ability to flourish in the harshest of environments and paves the way for future communities.

REVIEW

1. Define the term *primary succession.*
2. What is a pioneer community?
3. Provide an example of primary succession.
4. Define the term *secondary succession.*
5. Provide an example of secondary succession.
6. Who was Gifford Pinchot?

For more information go to this Web link:
<http://jimswan.com/111/succession/succession.htm>

CHAPTER SUMMARY

Biological succession is the gradual replacement of one community of organisms by another in a slow, predictable manner. It is the natural way that Earth continues to renew itself and allows for life to flourish in disturbed areas. There are two main types of biological succession: primary and secondary. Primary succession takes place in a location where no life has existed previously. This is often associated with retreating glaciers, volcanic eruptions, and shifting bodies of water. The first organisms to inhabit an area during primary succession are known as a pioneer community. These are often hardy, rapidly growing

EARTH MATH

1) IF APPROXIMATELY 20,000 HECTARES OF FOREST WERE DESTROYED BY THE MOUNT ST. HELENS ERUPTION, WHAT WERE THE APPROXIMATE AMOUNT OF ACRES DESTROYED (1 HECTARE = 2.47 ACRES)?

species that can survive in harsh conditions. Eventually the pioneer community alters the area to the point at which other species, such as grass, trees, and shrubs, can grow. Over time the area that was once void of life supports a healthy ecosystem.

The other type of succession, called secondary succession, occurs in an area where life already exists. These areas are usually disturbed in some way by a natural or human-induced event. Forest fires commonly destroy areas and allow secondary succession to take over. The area that was affected by the fire eventually supports rapidly growing grasses and small plants. Over time, small trees and shrubs take over the area, and eventually the forest returns to its former state. This process can occur in as little as 70 to 100 years. Many of the forests of the northeastern United States are the result of secondary succession, which took place when farm fields were abandoned after the Civil War. Understanding and utilizing the processes of biological succession is often useful as a management technique to restore damaged areas.

CHAPTER REVIEW

Multiple Choice

1. The area inhabited by organisms shortly after a glacier has retreated is an example of:
   a. primary succession
   b. secondary succession
   c. pioneer communities
   d. forestry management

2. A rapidly growing, hardy population of organisms that colonize an area for the first time is known as:
   a. primary succession
   b. secondary succession
   c. a pioneer community
   d. forestry management

3. The regeneration of farm fields into a forest is called:
   a. primary succession
   b. secondary succession
   c. a pioneer community
   d. forestry management

4. Which of the following areas is not usually associated with primary succession?
   a. volcanoes in the ocean
   b. retreating glaciers
   c. a sand bar
   d. abandoned farm fields

5. Which organism would most likely begin the process of biological succession in a disturbed area?
   a. trees
   b. shrubs
   c. birds
   d. lichen

Matching *Match the terms with the correct definitions.*

a. dynamic
b. biological succession
c. primary succession
d. pioneer community
e. lichen
f. secondary succession
g. catastrophic
h. germinate
i. disturbed areas

1. ____ The gradual replacement of one community of species by another in a slow, predictable manner.
2. ____ An symbiotic organism that consists of an algae and fungus that live on rocks and trees.
3. ____ Energetic, always changing or moving.
4. ____ Ecosystems or parts of ecosystems that have been destroyed or disrupted.
5. ____ The introduction of a community of organisms into an area where life has not existed before.
6. ____ A great and sudden disaster.
7. ____ The process that a plant seed undergoes when it begins to grow.
8. ____ The slow, gradual, and predictable replacement of one community of organisms by another in a specific area where life has already existed.
9. ____ A group of organisms that populate an area on the Earth where no organisms have lived before.

Critical Thinking

1. Explain how you think the Earth would be affected if biological succession did not occur.

CHAPTER 35 Classification of the Living World

Objectives

Taxonomy • The Kingdom Monera • The Kingdom Protista • The Kingdom Fungi • The Kingdom Plantae • The Kingdom Animalia • Invertebrate Animals • Vertebrate Animals

After reading this chapter you should be able to:

- Describe the taxonomic system used to classify organisms on the Earth.
- Explain how organisms are divided into specific taxonomic groups.
- Identify the five main taxonomic kingdoms used to classify organisms on the Earth.
- Describe some of the characteristics of organisms within the kingdom Monera and provide an example of one of these organisms.
- Describe some of the characteristics of organisms within the kingdom Protista and provide an example of one of these organisms.
- Describe some of the characteristics of organisms within the kingdom Fungi and provide an example of one of these organisms.
- Differentiate between vascular and nonvascular plants that make up the kingdom Plantae and provide an example of each.
- Differentiate between invertebrate and vertebrate organisms within the kingdom Animalia.
- Identify three invertebrate organisms within the animal kingdom.
- Identify five vertebrate organisms within the animal kingdom.

TERMS TO KNOW

biological characteristics
taxonomy
siliceous ooze
neurotoxins
fermentation
mycorrhizae
gymnosperms
angiosperms
spores
invertebrates
bivalves
zebra mussel
gizzard
nocturnal
Homo sapiens

INTRODUCTION

The biosphere on Earth is home to more than 5 million different species of living things. The different shapes, sizes, and life cycles of all the organisms alive today illustrate the great diversity of life on our planet. Studying the fossil record has revealed an amazing variety of organisms that once lived on the Earth. As humans explored more of the planet, different forms of life were continually being discovered. In order to better classify organisms that compose the biosphere, a universal classification system was required. By formally grouping together all living things, scientists have discovered that all the creatures on the Earth share related characteristics. This has lead to the remarkable story of the evolution of life on our planet. Classifying organisms on the Earth is also important for identifying the role they play in ecosystems. Ecologists recognize that the greater the diversity of life in an ecosystem, the healthier it is. Appreciating the variety of life on the Earth and understanding the role each organism plays in the biosphere is an important aspect of the science of Earth's systems.

Taxonomy

Organisms on Earth are categorized into groups that possess similar **biological characteristics.** Today all the living things on the Earth are grouped into five broad categories also known as kingdoms. The classification of different organisms is called **taxonomy** and was first formally conceived by Carolus Linnaeus in the eighteenth century. He created a system for classifying animals that is known as the binomial method. *Binomial* means "two names," and this system uses two Latin words that represent the genus and species of a unique organism.

The taxonomic system that Linnaeus created is based on seven levels of classification that groups together similar organisms. This taxonomic system includes classification within a kingdom, phylum, class, order, family, genus, and species. When Linnaeus created his taxonomic key, he only had two kingdoms: the plants and the animals. Today five kingdoms are used to classify organisms: Monera, Protista, Fungi, Plantae, and Animalia (Figure 35–1).

The Kingdom Monera

The kingdom Monera includes all single-celled organisms, such as bacteria and blue-green algae, that do not have a cell nucleus. A nucleus is a membrane that surrounds a cell's DNA molecules. The organisms that are classified as Monera lack a nucleus, which allows their DNA to freely move throughout the cell. Bacteria are one of the most abundant life forms on the Earth; they are found living virtually everywhere. The human skin alone is home to more than 500 million bacteria.

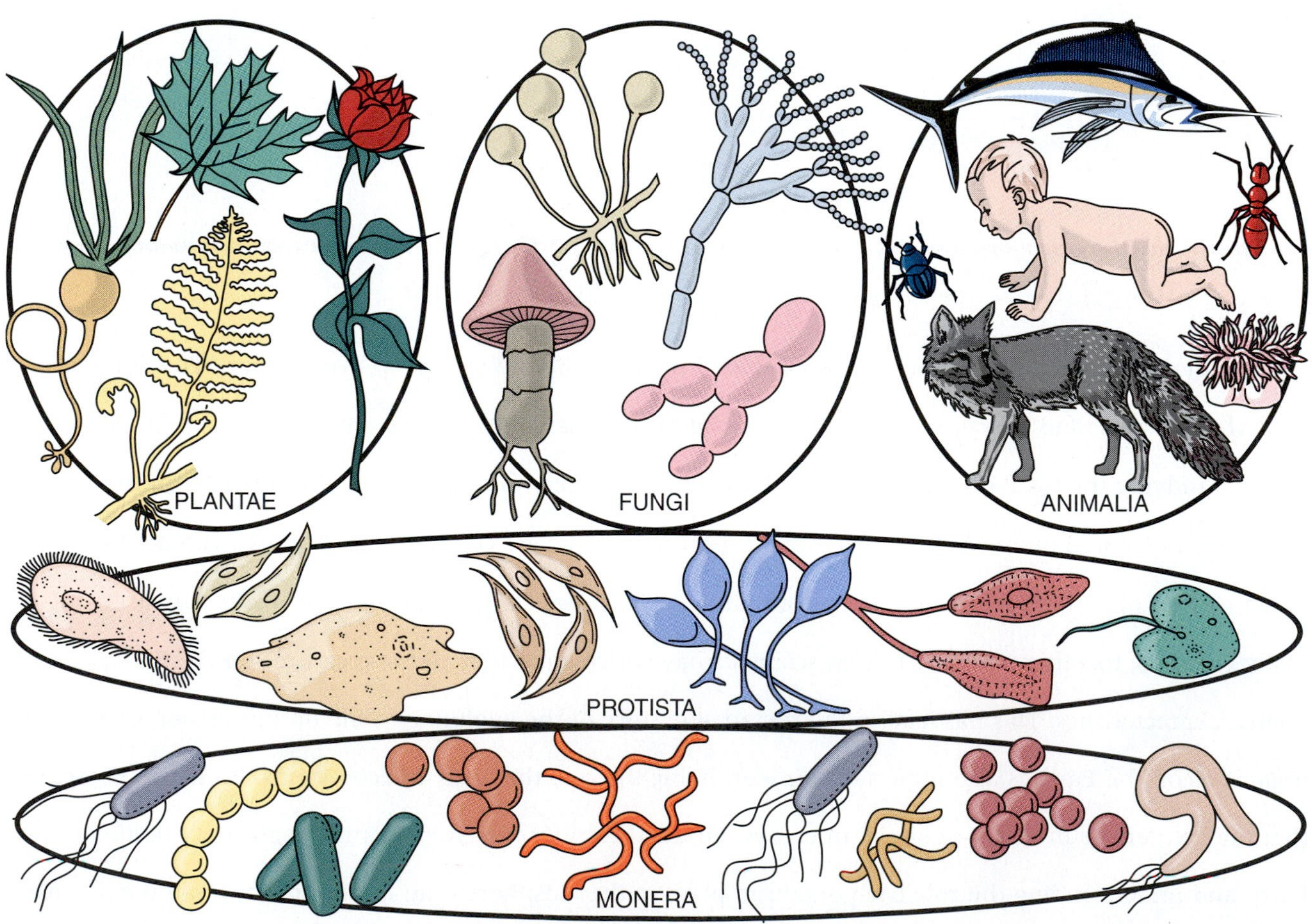

Figure 35–1 The five kingdoms used to classify organisms on the Earth.

EARTH SYSTEM SCIENTISTS *CAROLUS LINNAEUS*

Carolus Linnaeus was born in Sweden in 1707. Linnaeus became interested in the study of plants at a young age. He then went on to study and eventually teach botany. In 1735 he published a method for the formal classification of plants based on the anatomy of their flowers. In 1753 he developed his binomial system of nomenclature, which identified plants by using a specific genus and species name. His system of classification became known as the binomial system, which he applied to animals in 1758. The system that Linnaeus devised to classify plants and animals is still in use today. During his life, Linnaeus also wrote many books on the classification of the natural world, botany, and his travels in Europe.

Bacteria colonize deep in the soil and at the bottom of the ocean. Many bacteria act as important decomposers in the world's ecosystems and are responsible for the recycling of nutrients in a food chain. Bacteria are also an important part of the nitrogen cycle. They convert atmospheric nitrogen into forms that plants can use. Many bacteria live inside the intestines of animals and help to break down food. Some bacteria, however, can be deadly. Many kinds of bacteria can cause infections and sickness in both plants and animals.

Blue-green algae, also known as cyanobacteria, are also part of the kingdom Monera. These single-celled organisms produce their own food by the process of photosynthesis, although they are biologically different from plants and algae. Cyanobacteria can live in some of the harshest environments found on the Earth. Cyanobacteria grow in hot springs that reach 170° Fahrenheit and are also found alive in the ice of Antarctica (Figure 35–2).

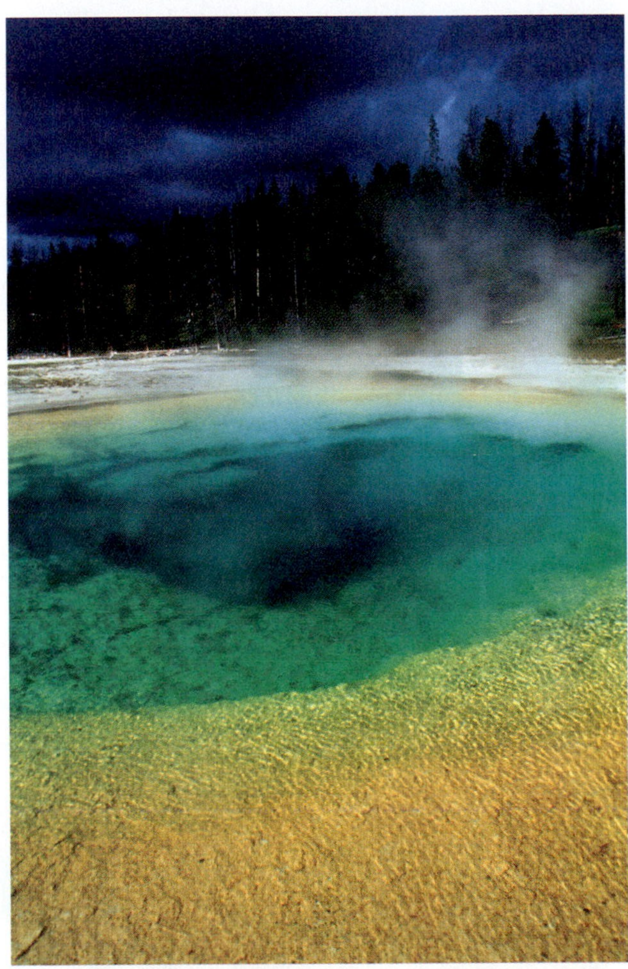

Figure 35–2 Brightly colored mats of bacteria growing in scalding water around a hot spring in Iceland are examples of organisms in the kingdom Monera. *(Courtesy of PhotoDisc.)*

Scientists believe that cyanobacteria are the oldest living things found alive today. Fossils of cyanobacteria colonies called stromatolites are the remains of the oldest living things that have ever been identified. Stromatolites are believed to have been the primary form of life on Earth more than 3.5 billion years ago. Today stromatolites can be found alive off the coast of Australia, making them the oldest living things on the planet. The cyanobacteria that lived more than 3 billion years ago may have been the source of oxygen in the atmosphere. Some researchers believe that cyanobacteria may one day be discovered on other planets or moons in our solar system. The planet Mars or Europa, one of Jupiter's moons, may be home to cyanobacteria. Scientists at the National Aeronautics and Space Administration (NASA) are designing remotely operated space vehicles to search these places for signs of life beyond the Earth.

The Kingdom Protista

The kingdom Protista includes all single-celled organisms that have a cell nucleus (Figure 35–3). These organisms can be both photosynthetic and heterotrophic. Most plankton that live in water are protists. Amoebae are also protists, which float through the water like small blobs and consume their food by completely surrounding it. Diatoms are also protists, which live in the world's oceans. These free-floating, single-celled organisms use photosynthesis to produce food from sunlight and construct protective shells made from silica.

Diatoms are an important part of the food chain within the open oceans. When diatoms die, their microscopic shells settle to the bottom of the ocean and collect over time to form sediments that are rich in silica. This is known as a **siliceous ooze,** which is found mainly near the equator and surrounding Antarctica. Both these locations support high populations of diatoms.

Another important protist is the single-celled algae, which form the base for many aquatic food chains around the world. These microscopic organisms convert sunlight into chemical energy stored in sugars and starches. They are also responsible for the production of atmospheric oxygen. Some biologists prefer to classify algae as part of the Plant kingdom because of their similar methods of photosynthesis; however, in this text they are considered protists.

Brown algae, commonly known as seaweed, include some of the most complex colonies of protists. Brown algae grow along the coast of the world's oceans and are used as a food by human beings. Other algae, such as red algae, also reside in the ocean and can produce **neurotoxins.** Neurotoxins are deadly poisons that affect the nervous systems of animals. Large populations of red algae, also called a red tide, can kill thousands of fish and be harmful to humans.

Protists also include the slime molds. These brightly colored, slimy colonies of protozoans grow in the damp debris on the forest floor.

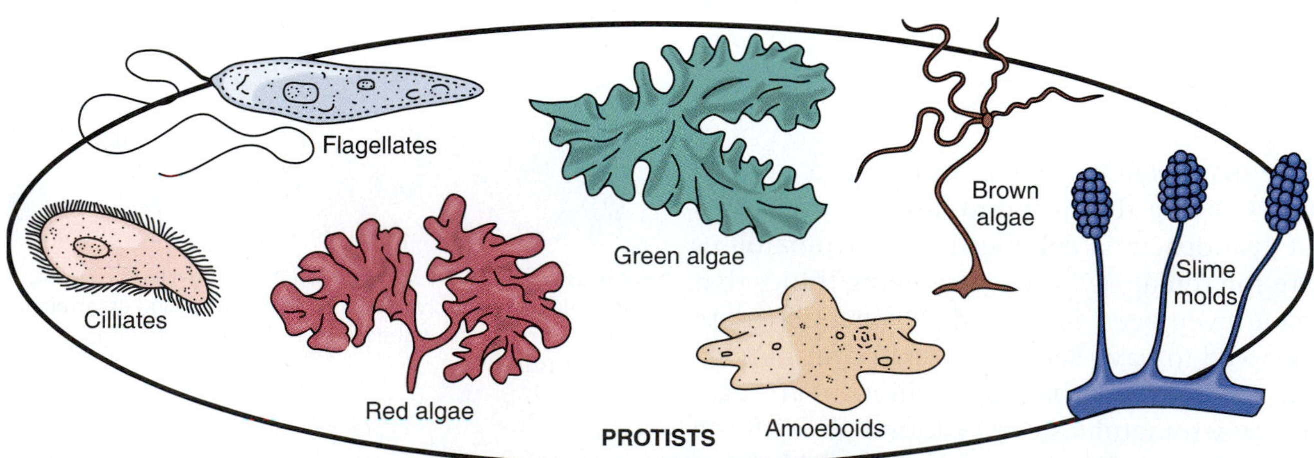

Figure 35–3 Examples of organisms in the kingdom Protista.

You can often find these protists when you look under a rotting log. Often mistaken for a fungus, slime molds actually move along the ground at extremely slow rates. Some protists can cause disease and sickness in animals. When these protists enter an organism's bloodstream, they can be deadly. Malaria is caused by a protozoan that is transferred into an organism's blood by a mosquito bite. The protozoan called Giardia is found in many freshwater lakes and streams in the United States and causes intestinal discomfort in humans. This is why it is wise to use some method of water purification when you are camping to kill protozoans that could be present in your drinking water.

The Kingdom Fungi

Organisms in the Fungi kingdom include yeasts, mildew, molds, and mushrooms (Figure 35–4). Many organisms in the fungi kingdom are important decomposers in the world's ecosystems. Yeasts are single-celled fungi that consume sugars and starches as food and produce both carbon dioxide and alcohol as byproducts. Humans use the power of yeasts for baking; cheese making; **fermentation** of beer, wine, and alcohol; and industrial processes. The use of a living organism to produce a usable product is called biotechnology. Yeasts are one of the most widely used organisms in biotechnology.

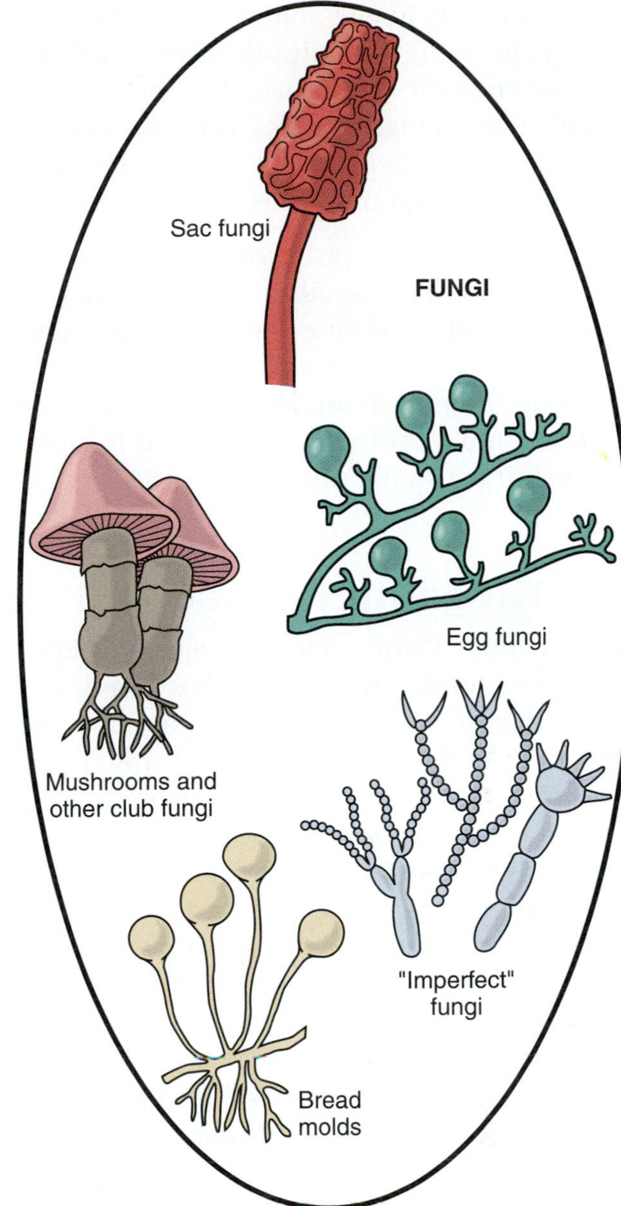

Figure 35–4 Examples of organisms in the kingdom Fungi include mushrooms and molds.

Other fungi include mildew, which are multicolored colonies of fungus that grow in damp areas. Almost everyone has encountered mildew trying to grow in a shower or in a damp corner of a basement. Mildew, like all fungi, reproduce by producing spores. Spores float on air and transport mildews all over the planet. Some people can become sick or have allergic reactions when they breathe in theses spores.

Mushrooms and molds are also fungi. These organisms are decomposers that consume the decaying organic material of both plants and animals. Most mushrooms are poisonous to humans when they are eaten, but some, such as white button mushrooms, are a popular food.

Many fungi secrete weak acids from their cells that help to break down the organic material on which they grow. This results in a solution of nutrients that the fungus uses for nutrition. Many fungi form mutualistic relationships with plants that they are growing near. These fungi are called **mycorrhizae,** which means "fungus roots." The fungus breaks down minerals in the soil by secreting weak acids. Plant roots then absorb the minerals for use in their bodies. In return, the plants supply organic material for the fungus to use as food. Scientists have determined that more than 90% of plants and trees have mycorrhizae.

Many fungi can be a problem in the storage of food. This is called spoilage, which is caused by molds that decompose food. Bread, fruits, and vegetables can be spoiled when molds begin to decompose them. We have all seen the effects of mold on food. When mold is found on food, it is usually uneatable.

The Kingdom Plantae

The kingdom Plantae includes all multicelled organisms that are autotrophs, known as green plants (Figure 35–5). In the Plantae kingdom organisms make their own food through the process of photosynthesis. Most plants are also terrestrial, which means they live on land. Some plants do reside in water, but by far most organisms in the Plantae kingdom live on land. Seaweeds and algae are similar to green plants but are not part of the Plantae kingdom. Many green plants also have a rigid cell wall around their cells, which gives them strength and stability. Green plants are divided into two main categories: vascular and nonvascular.

Vascular means that the organism has a network of tiny tubes that transports water and nutrients throughout the body (Figure 35–6). Vascular green plants include trees. Trees are some of the largest and oldest living organisms on the Earth.

The bristlecone pine that grows in the mountains of California and Nevada has been dated at 4600 years old. These trees are believed to live as long as 5500 years. The tallest trees include the giant redwoods also found in California. These trees reach almost 400 feet in height. Vascular green plants are classified into two broad categories: seed plants and seedless plants. Seed plants are green plants that reproduce by producing seeds. These in-

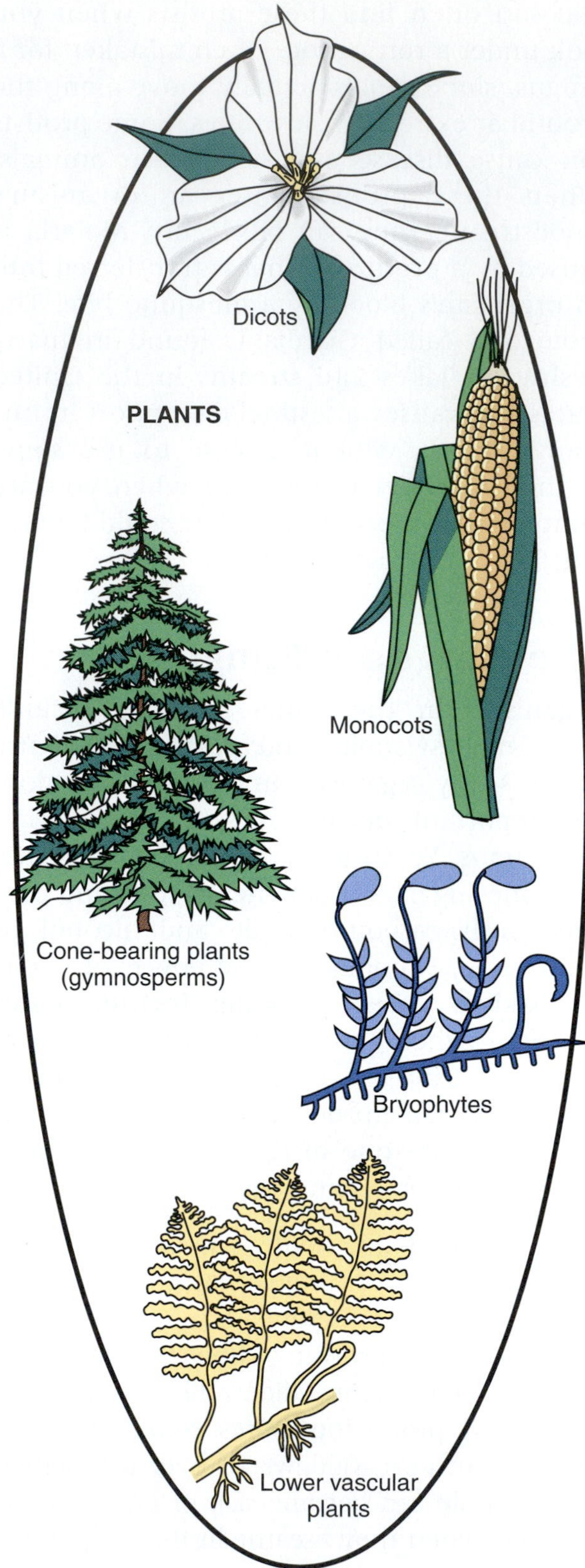

Figure 35–5 Examples of organisms in the kingdom Plantae.

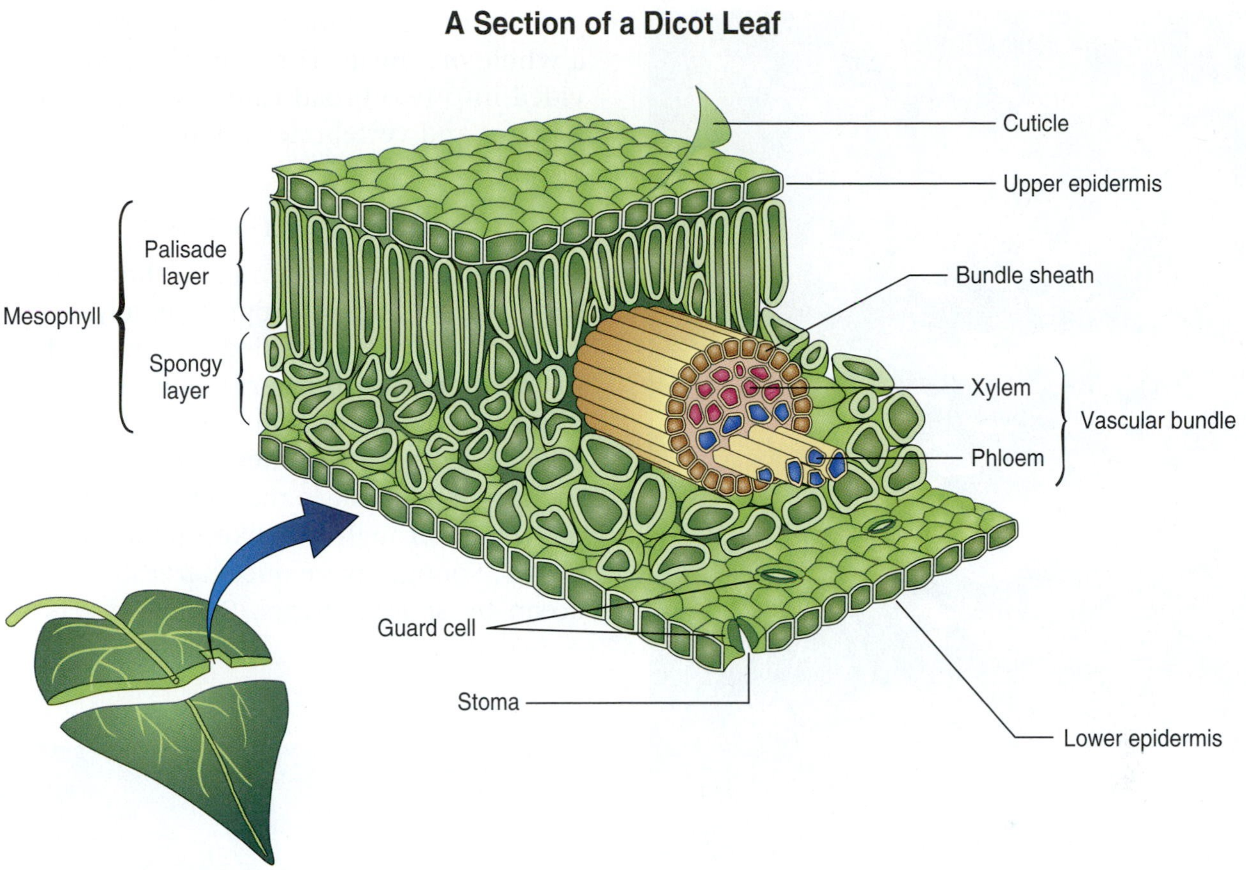

A Section of a Dicot Leaf

Cuticle
Upper epidermis
Mesophyll
Palisade layer
Spongy layer
Bundle sheath
Xylem
Vascular bundle
Phloem
Guard cell
Stoma
Lower epidermis

Figure 35–6 Cross section of the leaf of a vascular plant showing the tiny tubes that transport water and nutrients throughout the body of the plant.

clude categories of plants called angiosperms and gymnosperms.

Gymnosperms, which means "naked seed," are green plants that reproduce by producing seeds without seed coats. A seed coat is a hard cover that protects the seed. These include cone bearing trees, also known as conifers (Figure 35–7). Common conifers include pine and spruce trees.

Angiosperms, which means "enclosed seed," are also called flowering plants. These green plants reproduce by producing seeds with a protective shell. Angiosperms are also known as flowering plants because these organisms produce flowers that bear the seeds (Figure 35–8). Flowering plants must be pollinated before they can form a seed. Insects and wind are important for the pollination of flowering plants.

Once a flower is pollinated, it produces a fruit. A fruit is a fleshy organ that protects seeds and helps them to be transported. Many plant fruits are eaten by animals and are then dispersed. Humans use many fruits for their nutrition. Technically, a fruit is anything that contains a seed. Therefore common vegetables such as cucumbers, squash, and tomatoes are truly fruits. Seedless vascular plants include ferns, which reproduce by producing **spores.** Spores are tiny cells that float on the wind and eventually land on the ground, where they grow into a mature plant.

Nonvascular green plants include mosses and liverworts (Figure 35–9). These organisms form matlike green carpets along the soil. Mosses also use spores for reproduction.

The plant kingdom forms the base for primary productivity in all terrestrial ecosystems.

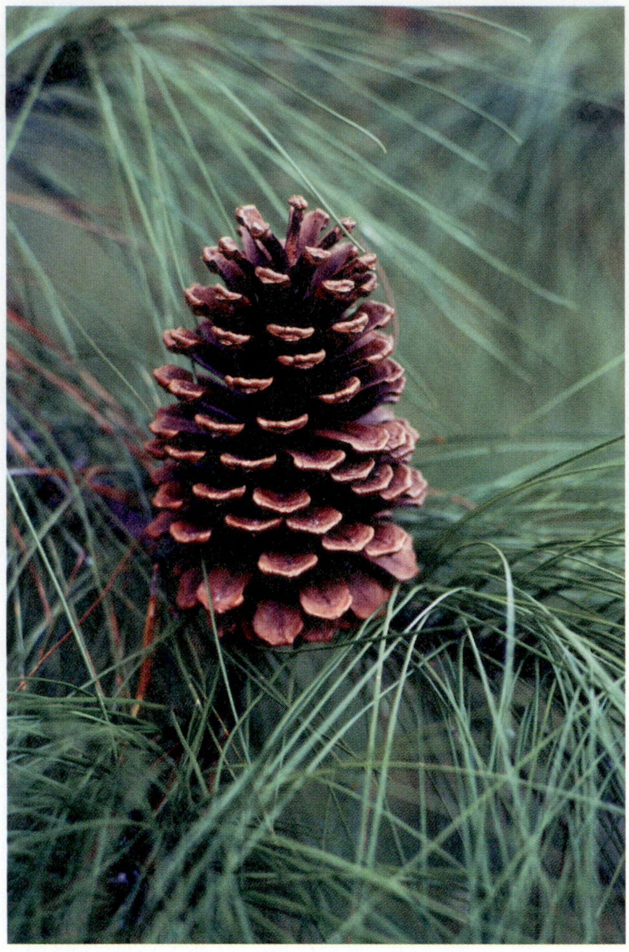

Figure 35–7 Pine cones protect the seeds of a gymnosperm such as a pine tree. *(Courtesy of PhotoDisc.)*

Many organisms rely on plants for food and shelter. Human beings cultivate specific plants for use as food and for industry. Trees are an important source of wood for construction and for fiber used in making paper. Other green plants are used for making cloths, such as cotton and flax, which is used to make linen. Green plants are also an important source of chemicals for use in medicine. It is estimated that green plants make up 90% of all the living material on land, making them an important part of the biosphere.

The Kingdom Animalia

The kingdom Animalia includes all multicellular organisms that gain their nutrition by ingestion. Ingestion means to take in parts of or a whole organism. The animal kingdom is divided into two broad categories called invertebrates and vertebrates (Figure 35–10).

Invertebrate Animals

Invertebrates are animals that have no backbone or spine. Approximately 95% of all animals are classified as invertebrates. Invertebrates are further divided into specific phyla based on similar characteristics. The phylum Porifera includes organisms such as sponges. Sponges are marine animals that filter water through porelike openings in their bodies. Sponges were once harvested from the ocean by sponge divers. Sponge divers were

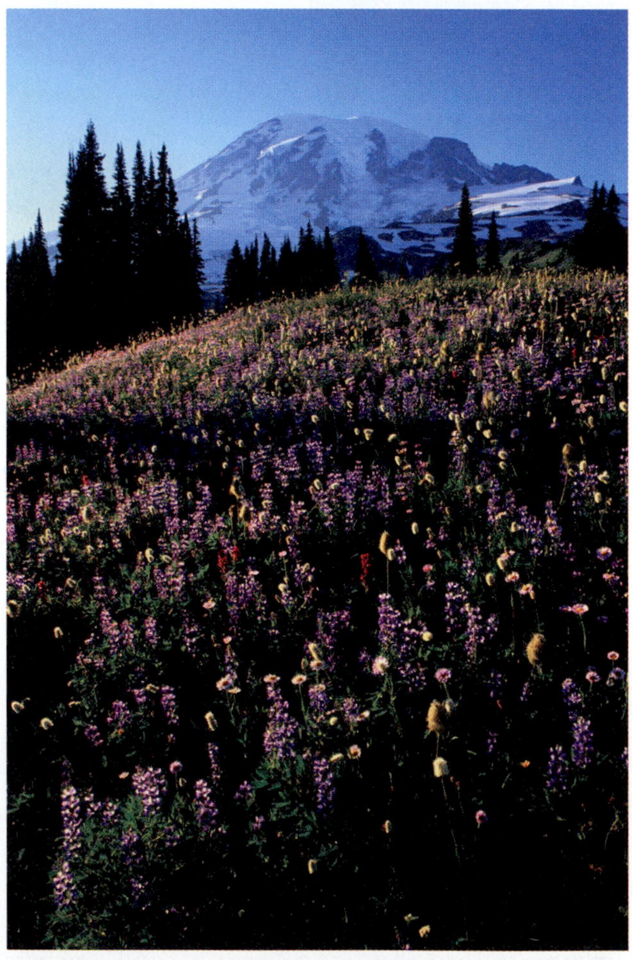

Figure 35–8 Angiosperms, also known as flowering plants, growing over a rolling hillside. *(Courtesy of PhotoDisc.)*

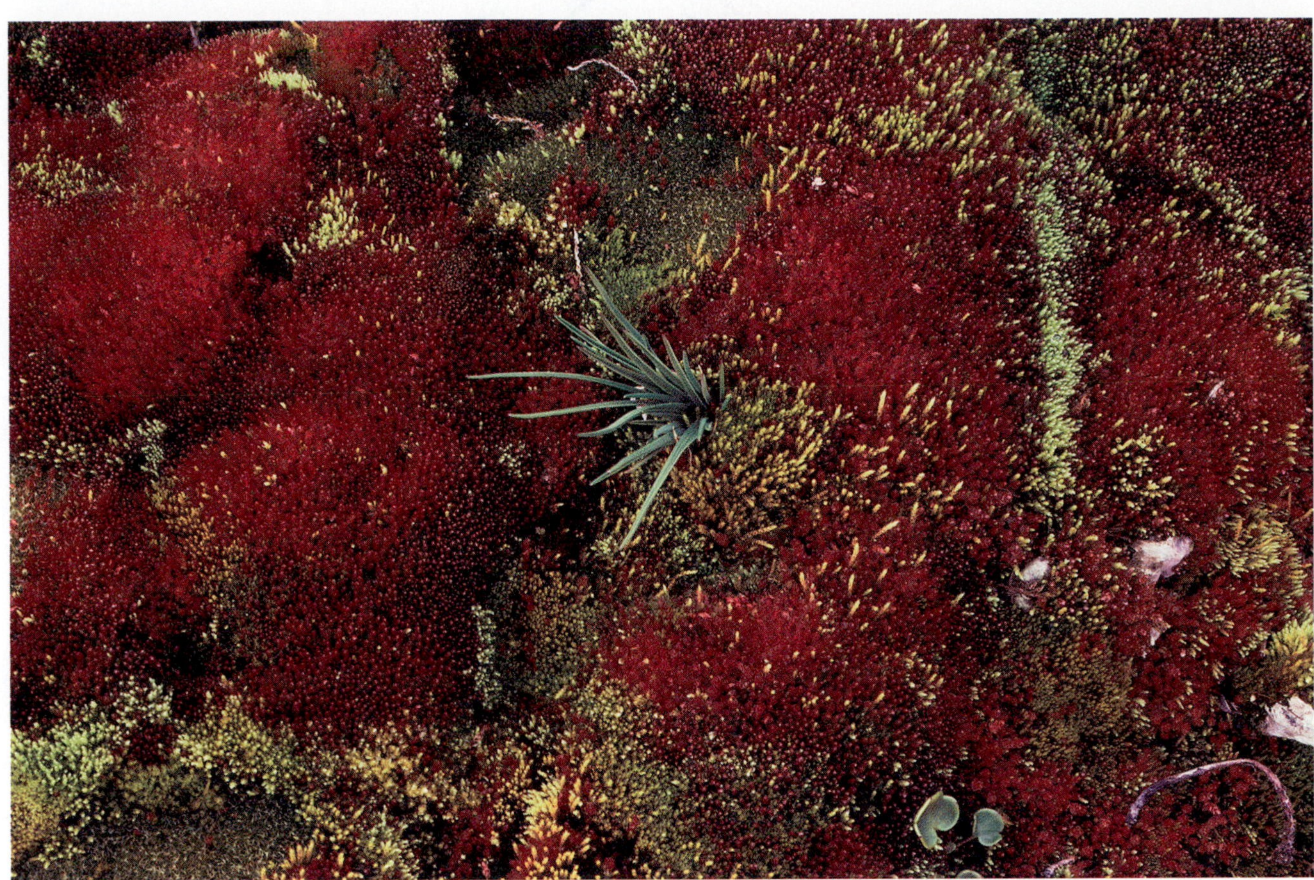

Figure 35–9 A close-up view of a moss, a nonvascular plant, producing spores. *(Courtesy of PhotoDisc.)*

people who could hold their breath for a long time and swim under water to retrieve sponges. The bodies of dead sponges were useful in absorbing great amounts of spilled liquids. The fibrous body material was also useful for cleaning. Today most sponges we use in homes are synthetic.

The phylum Cnidaria includes marine organisms such as jellyfish, coral, and sea anemones. Animals in this phylum are similar in that they have a body that surrounds a large digestive sac. Jellyfish float through the water and are often washed up on the beach. Some jellyfish can cause a painful sting if you accidentally touch them while swimming (Figure 35–11). One of the largest cnidarians is called the Portuguese man-of-war, which floats on the surface of the ocean and has tentacles that can reach 50 feet in length.

The phylum Platyhelminthes includes flat, wormlike animals such as flukes, tapeworms, and flatworms (Figure 35–12). These tiny organisms are often parasites that attach themselves to the wall of the intestines of larger animals or burrow into the skin and travel in the bloodstream. Some flatworms that live in the ocean are not parasites, can grow fairly large, and are quite colorful. These organisms are often found living on coral reefs.

The phylum Nematoda includes microscopic worms such as roundworms and nematodes. Nematodes reside in the soil and can infect the root systems of plants, causing damage to food crops (Figure 35-13). Roundworms can also infect humans by either being ingested or by burrowing into the skin. Trichinosis is a disease caused by eating meat that is infected by parasitic roundworms and is usually associated

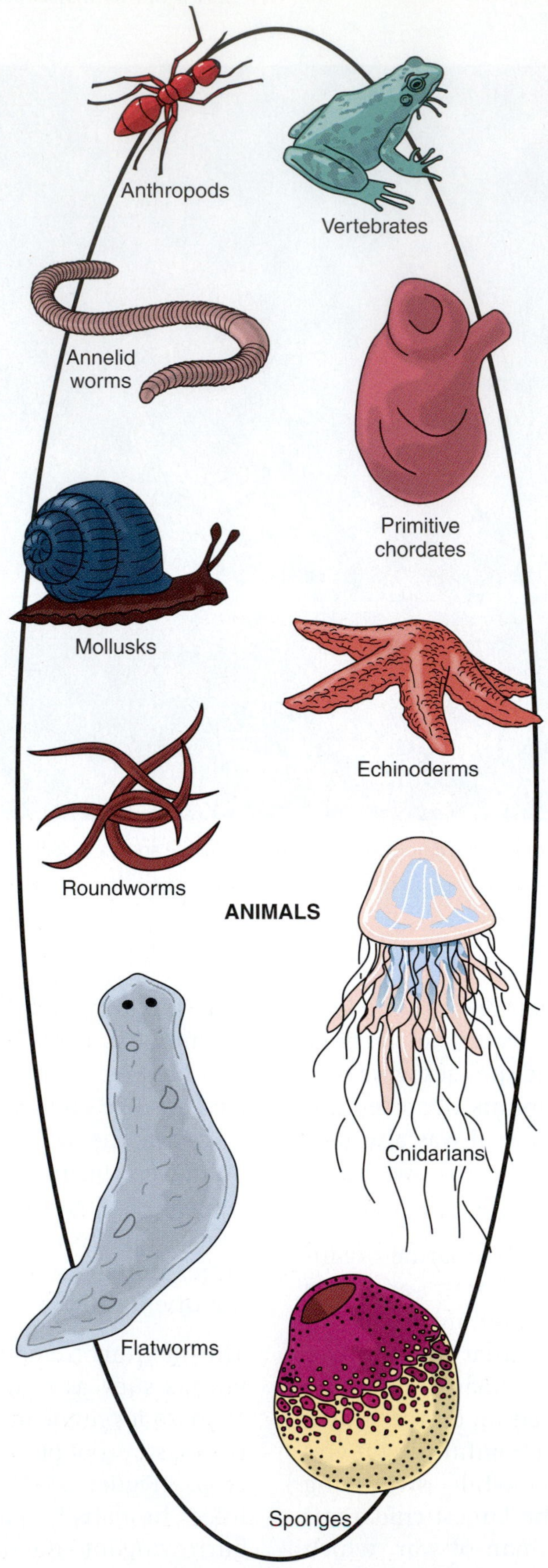

Figure 35–10 Examples of organisms within the kingdom Animalia.

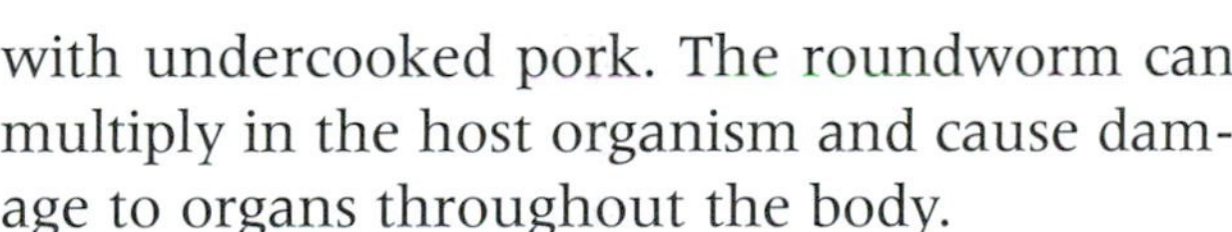

Figure 35–11 A jellyfish freely floating through the water is an invertebrate animal in the phylum Cnidaria. *(Courtesy of NOAA.)*

Career Connections

Zoologist

A zoologist is a scientist who studies all types of living organisms and their life processes, behavior, diseases, and origins. These scientists research animals in many different settings, including their natural habitat, in captivity, and in the laboratory. They also study the bodies of dead animals to learn more about their anatomy. Zoologists are often specialized and study particular types of organisms. A protozoologist studies single-celled organisms such as protozoans and bacteria. An entomologist works solely on insects. An ichthyologist studies all types of freshwater and marine fish. A herpetologist studies amphibians and reptiles. An ornithologist works with birds of all kinds, and a mammalogist studies mammals. All zoologists must have a college degree in biology and animal behavior, along with a special emphasis on a unique type of organism. Zoologists often work in research within an academic institution. Zoologists spend much of their time outdoors in the natural habitat, where their organisms live, and conduct observations and collect specimens. Employment opportunities are available in zoos, aquariums, and other animal-related parks and sanctuaries.

with undercooked pork. The roundworm can multiply in the host organism and cause damage to organs throughout the body.

The phylum Annelida includes animals such as earthworms or sea worms, also known as segmented worms (Figure 35–14). Earthworms are an important part of the soil environment, where they break down organic material. Thousands of earthworms can be found in just one acre of soil, where they help to turn over the soil and create channels through which air can pass into the ground. This helps to aerate the soil and create pore spaces for water to infiltrate more easily into the soil. Earthworms are usually forced to the surface of the soil after a heavy rain. This occurs when the small channels through which they are burrowing become saturated with water. The worms must surface so they can have access to atmospheric oxygen. These annelids are also raised in captivity on worm farms where they are sold as fish bait.

The phylum Mollusca includes a variety of snails, squid, octopus, and **bivalves.** Bivalves are mollusks that have shells with two halves. These include scallops, clams, mussels, and oysters. Scallops flap their hinged shells like wings, which pushes them through the water, making them one of the few bivalves that actually swims. Mussels attach themselves to wood, rocks, and other submerged substances by secreting a strong gluelike substance.

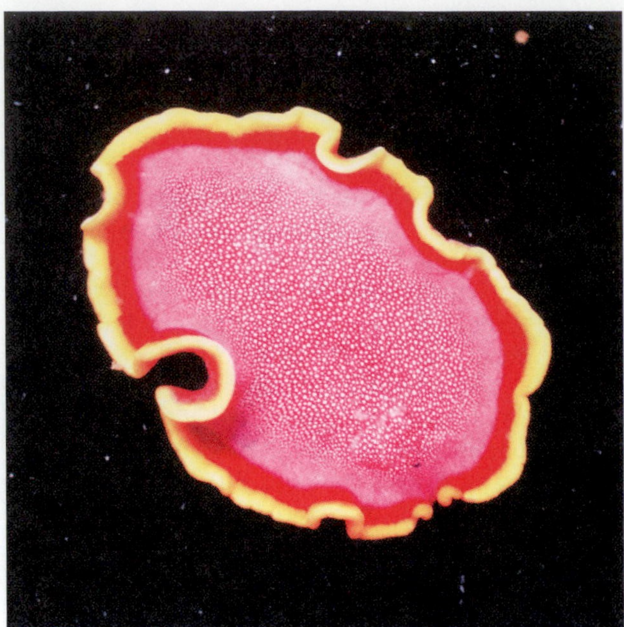

Figure 35–12 A flatworm grazing on a coral reef is an example of an animal in the phylum Platyhelminthes. *(Courtesy of NOAA.)*

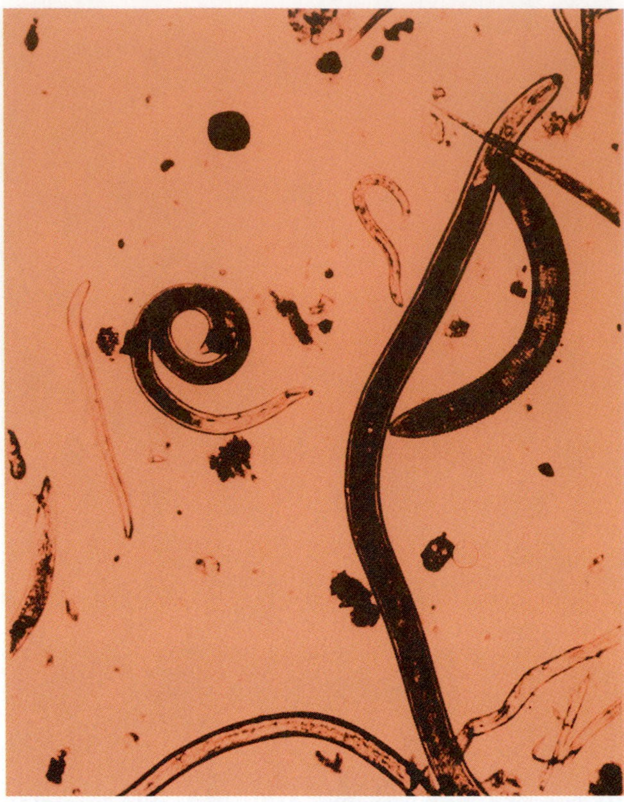

Figure 35–13 Microscopic nematodes reside in the soil and are part of the phylum Nematoda in the animal kingdom. *(Courtesy of Patricia Sanders and the Pennsylvania Turfgrass Council.)*

All bivalves are filter feeders. They pump water into their shells and filter out tiny plankton and organic material that they use for food. Oysters sometimes filter out tiny pieces of sand that become trapped in their bodies. As a defense, the oyster coats the sand grain with a hard substance that eventually forms a pearl. Although most pearls are white, some oysters make colored pearls such as the rare black pearl.

The clam, another bivalve, can grow very large. The giant clam can grow to almost 3 feet in length. The clam shell also contains growth rings similar to those found in trees. You can count these rings on the clam's shell to estimate its age.

The zebra mussel, a bivalve species, is a freshwater mollusk that lives in Europe (Figure 35–15). The tiny larvae of the zebra mussel were transported across the ocean in the ballast water of ships coming to the United States. Ballast water is pumped into a ship to add weight and stabilize it. The zebra mussel has spread quickly through freshwater lakes and is interrupting the balance of aquatic ecosystems in the United States. Because the zebra mussel has no natural predators in North America, its populations are increasing and the organism is taking over

Figure 35–14 Organisms in the phylum Annelida include earthworms. *(Courtesy of PhotoDisc.)*

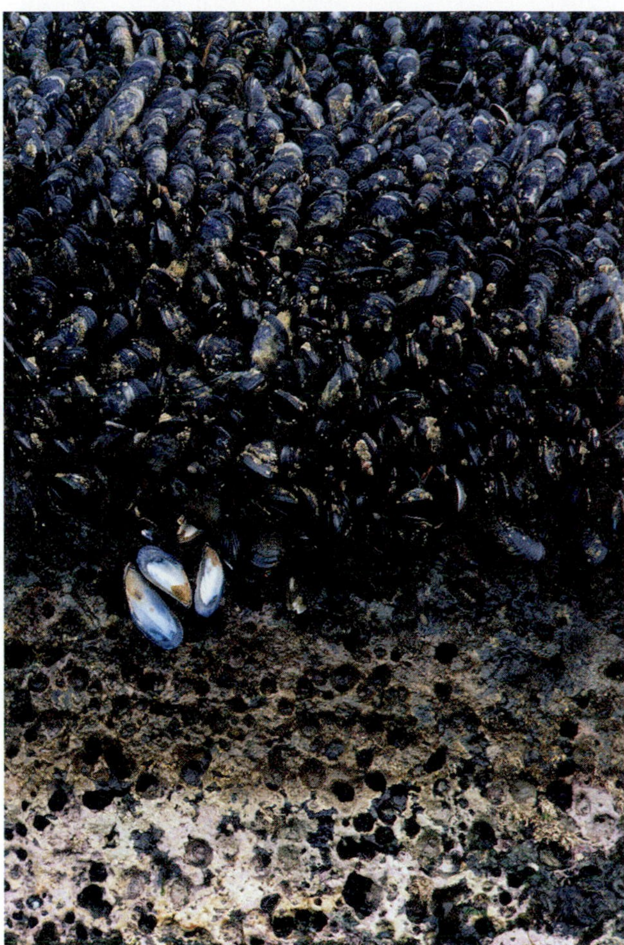

Figure 35–15 The zebra mussel, a bivalve mollusk, is an exotic species that is overpopulating lakes in North America. *(Courtesy of PhotoDisc.)*

whole lakes. Organisms that populate ecosystems where they do not naturally grow are called exotic species.

The Cephalopods are a class of mollusks that include octopus and squid. These organisms have long arms called tentacles. Octopus use these long tentacles to trap and hold their prey. These annelids also have a sharp beak, much like a bird, that is used to tear apart their food. Squid are also annelids and are the fastest swimming invertebrates in the world. Squids use jetlike propulsion that forces water quickly through their bodies, propelling them to speeds of more than 20 miles per hour. The largest squid species is called the giant squid. These cephalopods live deep in the ocean and can be more than 40 feet long. The giant squid also has the largest eyeball of any animal living on the planet. The eyeball of a giant squid is about the size of a volleyball. Many organisms in the annelid phylum are eaten by humans. Humans commonly ingest squid, clams, mussels, scallops, oysters, and occassionally even snails (Figure 35–16).

The phylum Arthropoda is one of the largest phyla on the planet, with more than 1 million species identified on Earth. Arthropods include the arachnids, which are land-dwelling arthropods such as scorpions, spiders, ticks, and mites (Figure 35–17). Crustaceans are another class of arthropods, which include aquatic organisms such as shrimp, crayfish, lobsters, and barnacles. Humans also dine on millions of pounds of crustaceans each year.

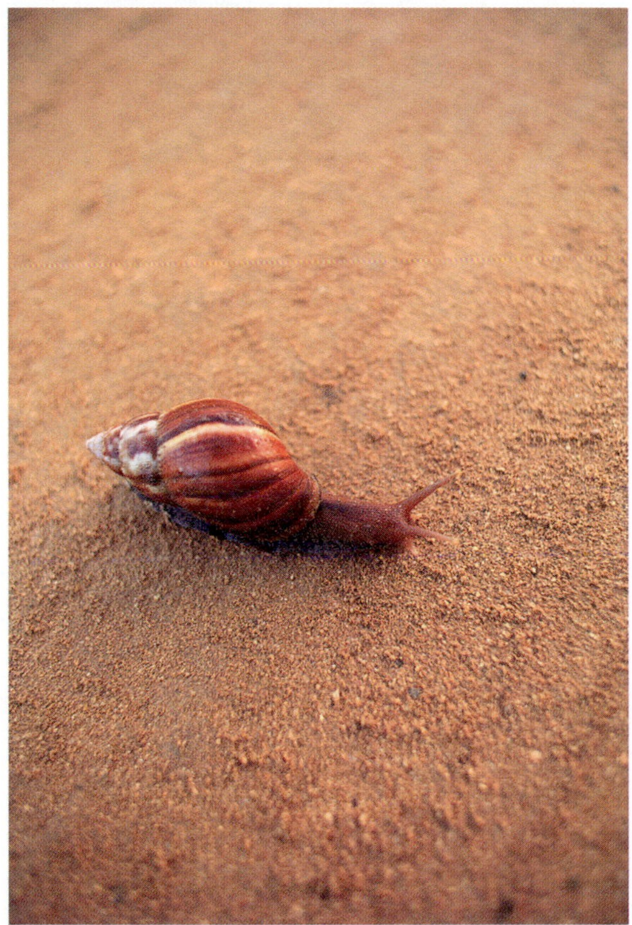

Figure 35–16 A common garden snail is part of the Gastropod class of mollusks. *(Courtesy of PhotoDisc.)*

Figure 35–17 A spider is an arachnid within the Arthropod phylum. *(Courtesy of PhotoDisc.)*

Millipedes and centipedes also fall under the Arthropoda phylum, as do aquatic organisms such as sea stars, sea cucumbers, sea urchins, and starfish. Although they appear lifeless, starfish are fearsome predators that attack mollusks such as oysters and mussels.

The largest class of arthropods is the Insecta class. This includes all insects of the world, which by far outnumber any other class of organisms on the planet. More than 900,000 insect species that have been identified on the Earth. Insects are some of the most adaptable organisms on the planet. They come in an amazing assortment of shapes and sizes. Insects play an important role in many ecosystems, especially as decomposers. They are also important for the pollination of flowering plants.

Vertebrate Animals

The other major category of organisms in the animal kingdom are called vertebrates. These are organisms that have a backbone and fall under the phylum Chordata. The chordates include many different classes of animals that all have a notocord. A notocord is a flexible rod that lies between the spinal chord and an animal's gut. The Agnatha class includes jawless organisms such as lampreys. These parasites attach themselves to the sides of fish with hooklike devices located in their mouths. They then pierce the skin of their host with a sharp tongue and suck out the animal's blood. Lampreys begin life as microscopic larvae that float in the water. Many freshwater lakes in the eastern United States have been invaded by freshwater lamprey that were brought over in the ballast water of ships arriving from Europe. In their new environment, the lamprey have no natural predators and are quickly overpopulating lakes and killing many native species of fish (Figure 35–18).

The class Chondrichthyes, also called the cartilaginous fish, include organisms that have skeletons made of cartilage. Cartilage is a soft, flexible, bonelike material. Sharks are probably the most well known of the cartilaginous fish (Figure 35–19). There are approxi-

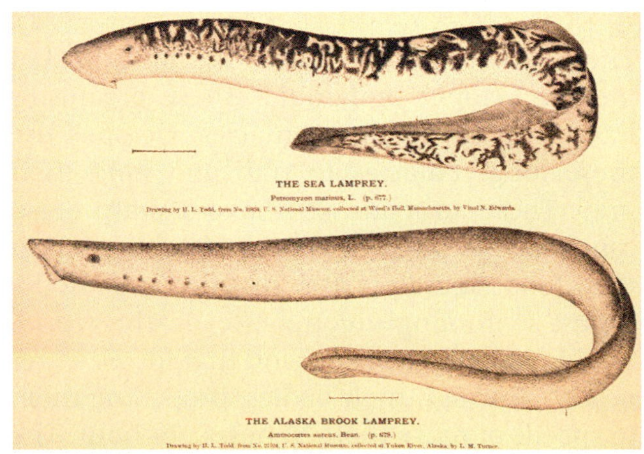

Figure 35–18 Lampreys are classified as a vertebrate in the Agnatha class. *(Courtesy of NOAA.)*

Figure 35–19 The vertebrate class Chondrichthyes, also called the cartilaginous fish, include organisms such as sharks. *(Courtesy of Sea Images, Inc.)*

mately 370 different species of sharks that live in the world's oceans. The largest shark is the whale shark, which grows to a length of more than 50 feet. The great white shark is also large, often growing more than 30 feet long. Sharks can be deadly when they attack humans. Their razorlike teeth can be more than 2 inches long. Approximately 100 people are attacked by sharks each year around the world. Scientists are interested in studying sharks because they can easily ward of diseases, especially cancer. Other cartilaginous fish include skates and rays, which "fly" through the water using winglike fins. The stingray has a sharp, swordlike tail that can pierce the skin, causing a stinging sensation. Manta rays can grow very large, with wingspans of more than 25 feet. Although very large, these creatures are harmless to humans, because that they only eat plankton and small fish that they filter out of the water in their mouths.

The chordate class Osteichthyes, also called the bony fish, include all fish that have skeletons made of bone. This is the largest class of vertebrate animals and includes more than 300,000 species. Because the ocean is difficult to explore, scientists believe there are even more species of bony fish that have not yet been identified. Bony fish come in many shapes and sizes and live in both freshwater and saltwater environments (Figure 35–20). All bony fish, however, have similar streamlined body shapes that are designed to propel them through the water. Almost all bony fish have a large fin near the tail, called a caudal fin, that pushes them forward through the water. Some fish use their powerful caudal fins to move at speeds near 60 miles per hour.

Bony fish are also an important food source for humans all around the world. The ocean fishing industry today is being regulated to prevent the overharvesting of fish. Once thriving fish populations in the Atlantic Ocean have been declining in recent years, making regulation necessary. Aquaculture, also called fish farming, is a method of raising fish species in captivity. Using aquaculture to provide fish for use as a food source is becoming more widespread in the United States. The three main bony fish that are raised on fish farms in the United States are catfish, trout, and salmon.

The vertebrate class Amphibia includes all amphibians, such as frogs and salamanders (Figure 35–21). Amphibians are unique creatures that begin their lives in water; when they mature, they can survive on land. Amphibians are believed to be the first vertebrates on the Earth to have emerged from the water to live on dry land. Amphibians have moist skin that must not be allowed to dry out. Although many amphibians can survive on land, they usually stay near some type of water source to prevent their sensitive skin from drying out. Amphibians transport oxygen through their skin for use in respiration. The largest amphibian in the world is the giant salamander, which lives in freshwater streams in Japan. This amphibian can grow as long as 6 feet. Scientists are discovering that many amphibians are becoming deformed and their populations are declining. It is believed that pollution or increased ultraviolet radiation reaching the Earth's surface resulting from ozone depletion may be causing this phenomenon, although much

Figure 35–20 Both freshwater and saltwater fish are classified in the chordate class Osteichthyes, also called the bony fish. *(Courtesy of PhotoDisc.)*

Figure 35–21 A common frog is an example of an amphibian. *(Courtesy of PhotoDisc.)*

research still needs to be done to prove or disprove this theory.

The Reptilia class of vertebrates includes all reptiles, such as snakes, lizards, turtles, and alligators. Reptiles are cold-blooded invertebrates, meaning that their body temperatures depend on the temperature of the environment. This causes reptiles to live in warmer climates or to go into dormancy during cold seasons. Many reptiles warm their bodies by lying on hot rocks in the sunlight (Figure 35–22). The largest reptiles on the Earth today include the American alligator and Komodo dragon. Both these organisms are carnivores, or meat eaters. Turtles are unique reptiles that protect themselves with hard shells. Some species of turtles can live extremely long lives, up to 100 years or more.

It was once thought that the long-extinct dinosaurs were reptiles, but recent debates have changed this viewpoint. Dinosaurs did have much in common with present-day reptiles. They both lay eggs to reproduce, and both have similar skin. The debate on whether dinosaurs were reptiles continues to this day. It is a common misconception that reptiles have slimy skin. All reptiles are covered with a dry, scaly skin that is rather soft and smooth to the touch. Some reptiles can be extremely deadly. Alligators have been known to attack humans and can be very dangerous. Poisonous snakes, such as rattlesnakes and cobras, can also kill a human being with their deadly venom.

The vertebrate class Aves includes all the birds in the world (Figure 35–23). Birds are unique organisms that possess many interesting

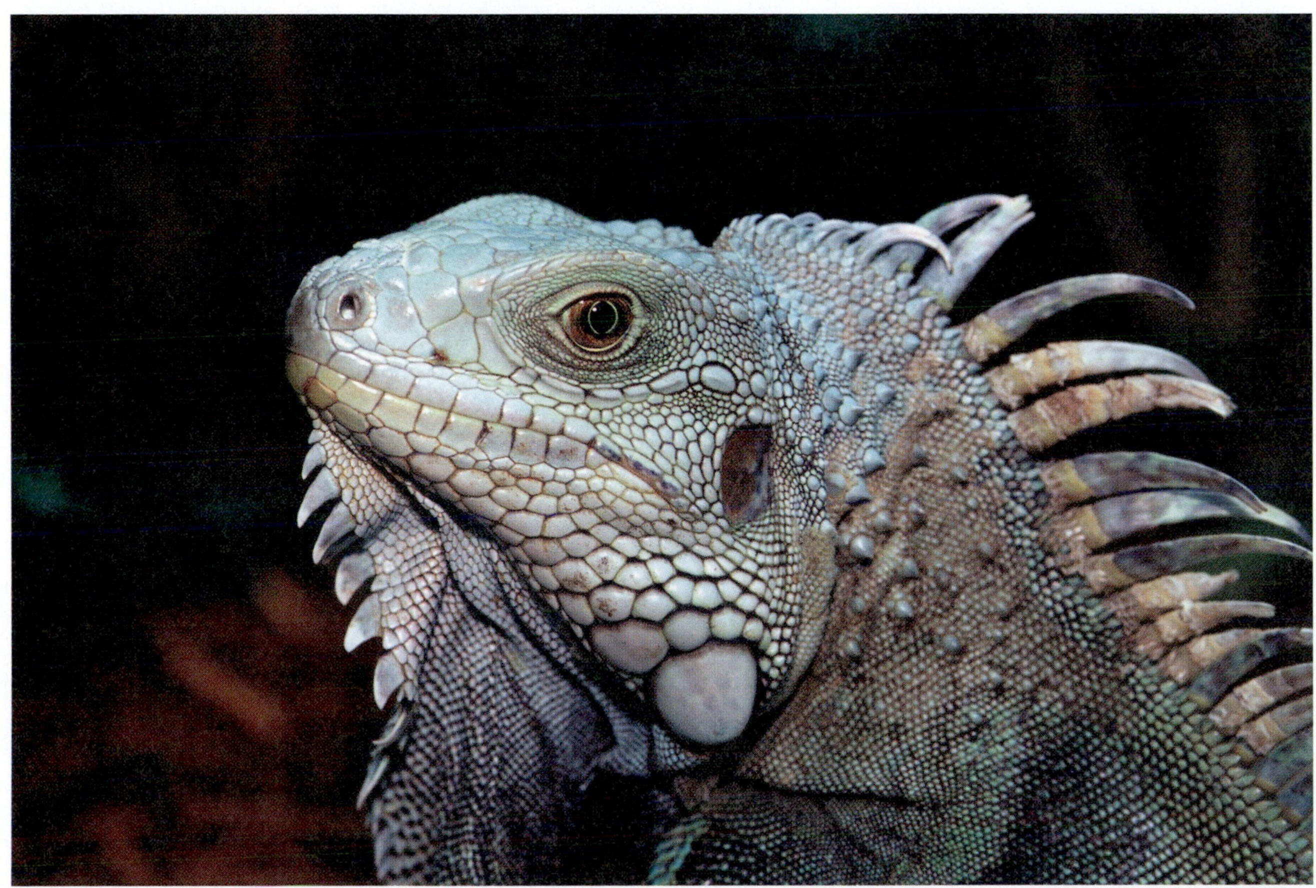

Figure 35–22 An iguana, a cold-blooded reptile, needs to warm its body in the Sun to raise its body temperature. *(Courtesy of EyeWire.)*

features. Almost all birds have wings that enable them to fly. Birds have extremely light, hollow skeletons that makes them lightweight. This is an advantage that helps them to fly. Birds are warm-blooded creatures that regulate their own body temperatures. Birds come in many varieties and sizes; some are flightless. Ostriches, emus, and penguins are examples of birds that do not fly. Penguins are also unique birds because they hunt for their food beneath the ocean surface. They use their wings much like fins.

Birds have well-developed brains that allow them to live fairly complex lives. Many birds communicate with a variety of songs and are able to mimic human speech. Birds also construct complex nests from which to raise their young. The smartest bird is believed to be the common crow, which has the ability to solve simple problems such as how to untie a knot to get to a food source.

Birds are also unique in that they are able to navigate over far distances. Many bird species migrate thousands of miles each year. The longest migration of any animal on Earth is performed by the arctic tern. The arctic tern spends its summer in the Arctic near the North Pole. Every year it travels all the way to Antarctica near the South Pole.

All birds have a unique covering on their bodies called feathers. Feathers are made from keratin, a protein that is found in fingernails. Feathers help to insulate a bird's body while also providing the extra surface area needed for flight. Another unique feature of birds is the **gizzard.** Birds do not have teeth with which to chew their food; instead they use a gizzard. The gizzard is a special saclike organ that crushes and grinds food before it reaches a bird's stomach. Birds must fill their gizzards with small stones, called grit, to help grind down their food.

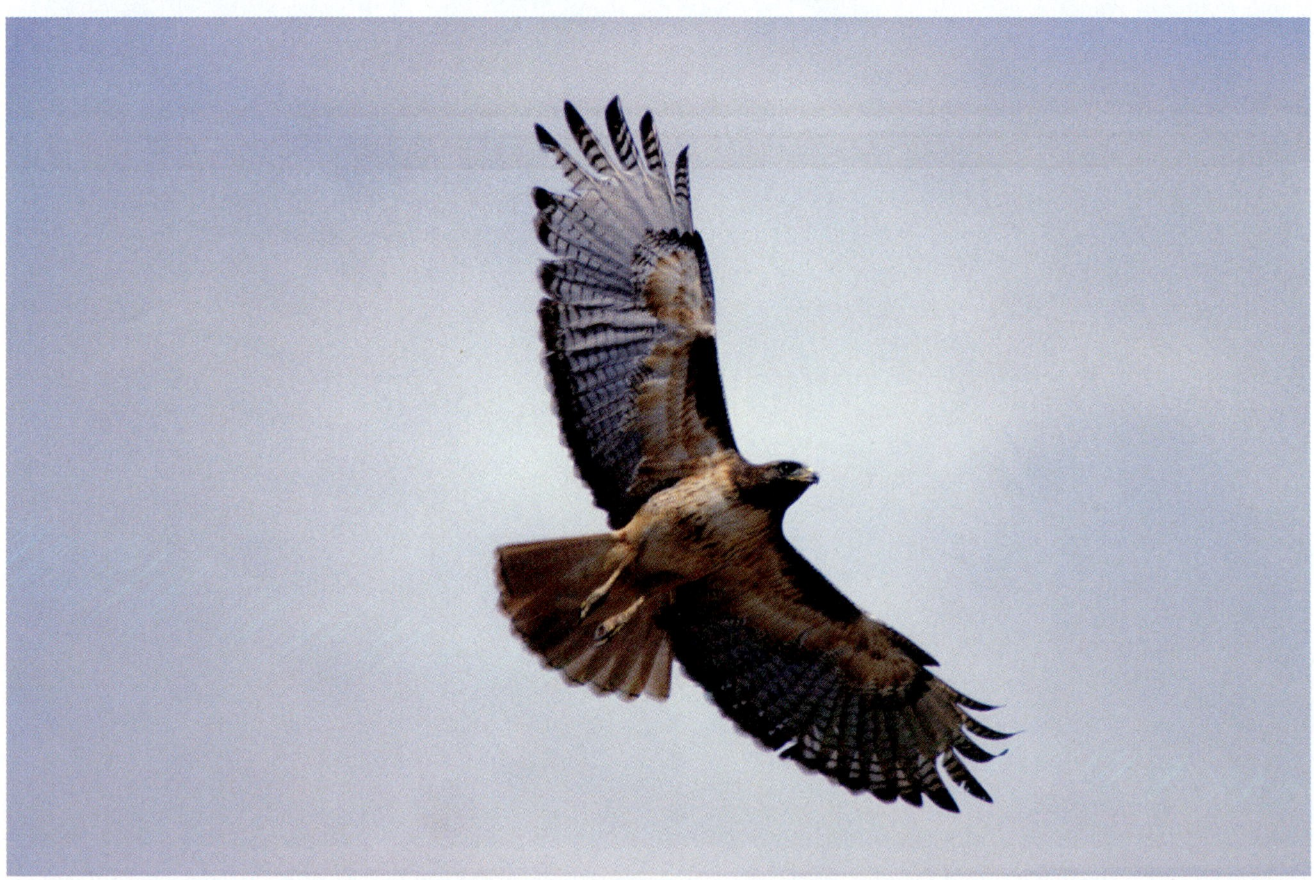

Figure 35–23 Birds belong to the vertebrate class Aves. *(Courtesy of PhotoDisc.)*

Figure 35–24 The platypus is an egg-laying mammal classified in the order Monotremata. *(Courtesy of Wernher Krutein/ photovault.com.)*

Eventually the grit within the gizzard becomes rounded and polished by abrasion.

Many scientists believe that dinosaurs are the ancestors of modern birds. Unlike reptiles, some dinosaurs were believed to be warm blooded like birds. The fossil remains of some dinosaurs have also been found with piles of polished gizzard stones, much like birds. The debate over what dinosaurs were related to, birds or reptiles, goes on to this day.

The final class of vertebrate animals is Mammalia, also called the mammals. Mammals are animals that have bodies that are mostly covered in hair. They also nourish their offspring with milk produced from mammary glands. This is where the Mammalia class derives its name. Human beings are part of the mammal class, which totals approximately 4000 different species. Mammals are further divided into specific orders that group together animals with similar characteristics. The order Monotremata include unique mammals such as the platypus and the spiny anteater. These organisms are unlike any other mammal because they lay eggs. The platypus has a bill, much like a duck, and resides near water (Figure 35–24). Both these strange mammals are only found on the island of New Guinea and in Australia.

Another order of mammals are the Marsupials. Marsupials carry their offspring in pouches on their bodies and include the koala bear, opossum, and kangaroo (Figure 35–25). When these organisms are born, they crawl along their mother's fur and into her pouch. They stay in the pouch until they are completely

Figure 35–25 The koala bear is an example of a Marsupial mammal. *(Courtesy of PhotoDisc.)*

developed. The continent of Australia has the largest number of marsupial species on the planet. The only other species of marsupial that lives in other parts of the world is the opossum.

The next order of mammals is the Artiodactyla. These include mammals that have an even number of toes or hooves on their legs. Organisms such as deer, moose, elk, cows, sheep, and goats are all examples of Artiodactyla. All the organisms in this order are also herbivores, meaning that they eat only plants. Another similar order of mammals is the Perissodactyla. This order includes organisms that have an odd number of toes or hooves on their legs (Figure 35–26). Horses, rhinoceroses, zebras, and donkeys fall under this order of animals. These organisms are also herbivores.

The Carnivora order of mammals includes the meat eaters, or carnivores. This order of mammals is further subdivided into two infraorders. One of these has retractable claws. This means that the claw can move in and out of the animal's foot. This infraorder includes all cats, such as lions, cheetahs, and house cats, along with the mongoose and hyenas (Figure 35–27). The other infraorder of carnivore includes all animals that have nonretractable claws. These are claws that cannot move in and out of the animal's foot. This includes wolves, dogs, bears, raccoons, foxes, skunks, weasels, seals, otters, walruses, and sea lions. Many carnivores have well-developed brains and show a complex social structure. For example, wolves depend on the coordinated group actions of their pack for survival. Domestic dogs are also intelligent carnivores that can complete many complex tasks.

Figure 35–26 Examples of an even-toed deer (top) in the order Artiodactyla and odd-toed Zebra (bottom) in the order Perissodactyla. *(Photos courtesy of PhotoDisc.)*

Figure 35–27 A male lion is an example of a mammal with retractable claws in the Carnivora order. *(Courtesy of PhotoDisc.)*

The next order of mammals is the Chiroptera, which includes bats. Bats are the only type of mammal adapted for flying. Bats are **nocturnal** animals, which means they are mainly active at night. Bats use a type of sonar, which bounces sound waves off objects to detect them (Figure 35–28).

Other mammals that use sonar include dolphins, which are in a different order of chordates. Whales, dolphins, and porpoises make up the Cetacea order of mammals (Figure 35–29). These aquatic mammals are unique in that they live their lives entirely in the water. Although they resemble fish, with their streamlined bodies and fins, the cetaceans all must breathe air to survive. These sea mammals all nourish their young with milk, unlike fish. Fossil evidence has revealed that the ancient ancestors of the dolphin were wolflike creatures who lived near the coastline on land.

Another order of aquatic mammals are the Sirenia. These include the manatee and dugongs, also called sea cows. These marine mammals are herbivores that dine on aquatic vegetation. Many manatees that live in the warm waters around Florida have become endangered as a result of their slow movement. This causes them to be run over by motorboats. These gentle sea mammals are believed to be the source of the mermaid myth. Sailors who were at sea for long periods may have mistakenly seen these creatures as half human, half fish.

The Edentata order of mammals are organisms that have few or no teeth. These include mammals such as armadillos, sloths, and anteaters (Figure 35–30). The tree sloth is an extremely docile animal. *Docile* means slow and gentle. The tree sloth moves so slowly through the tree tops that moss often grows in the hair on its back.

Figure 35–28 Flying mammals in the Chiroptera class, such as bats, are nocturnal and use a type of sonar to help them hunt and fly at night. *(Courtesy of PhotoDisc.)*

Figure 35–29 Dolphins and whales are aquatic mammals that belong to the Cetacea order of animals. *(Courtesy of PhotoDisc.)*

The Rodentia order of mammals groups together all rodents. Rodents are mammals that have front teeth that are continually growing. This causes them to be active chewers, which reduces the size of their teeth. Rodents include rats, mice, gerbils, gophers, hamsters, squirrels, and beavers (Figure 35–31).

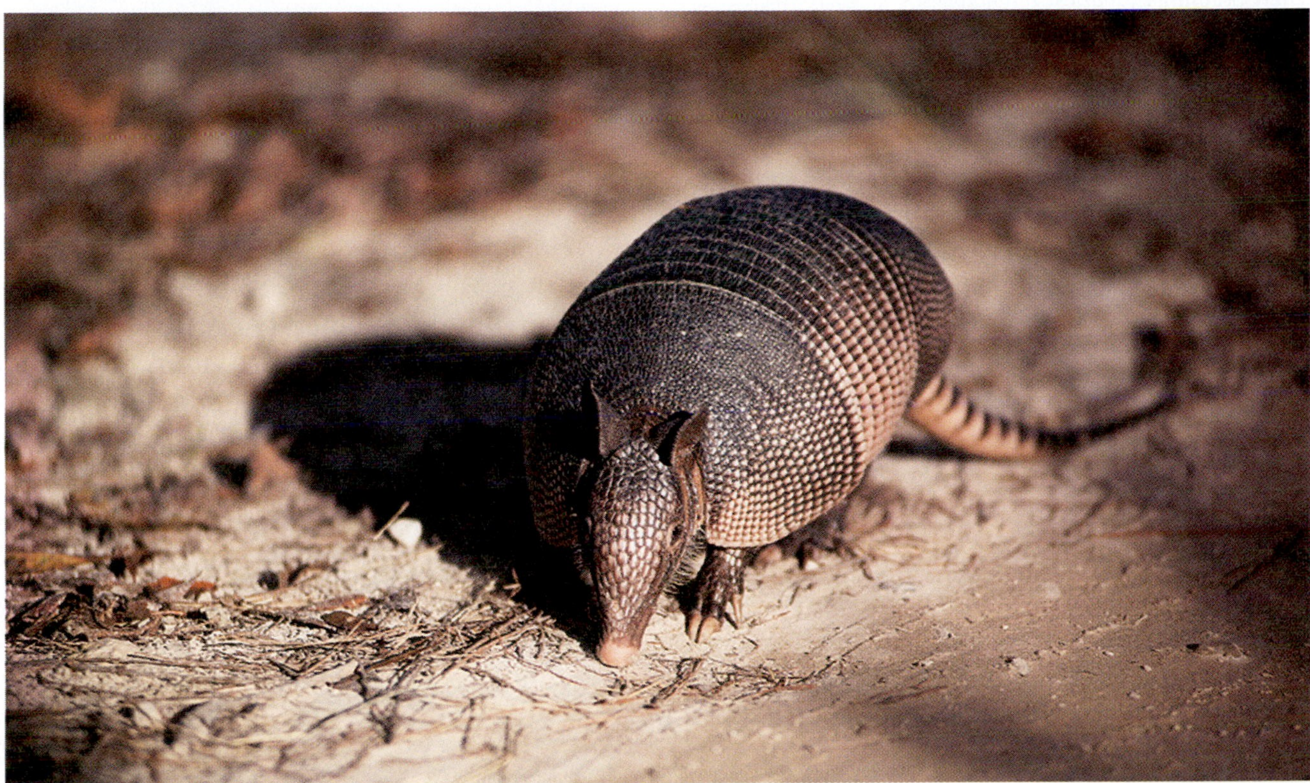

Figure 35–30 An armadillo is an animal with very little teeth that is grouped in the Edentata order of mammals. *(Courtesy of EyeWire.)*

Figure 35–31 Rodents, such as squirrels, are mammals that have constantly growing teeth, requiring them to continually chew on things. *(Courtesy of PhotoDisc.)*

Another similar class of mammals is the Lagomorpha order, which also has well-developed teeth. These include rabbits and hares. Animals in this order also have long legs, which helps them to move by jumping or hoping.

The Insectivora order of mammals include animals that eat only insects. Animals in this order are the moles, hedgehogs, and shrews.

The final order of mammals are the Primates. This order includes animals that have opposable thumbs, forward-facing eyes, and well-developed brains (Figure 35–32). An opposable thumb is one that can be used to grasp and manipulate objects. Animals in the Primate class include lemurs, monkeys, chimpanzees, gorillas, orangutans, and humans. All these mammals are highly intelligent and live in organized social groups. Human beings are

EARTH MATH

1) IF THEIR ARE 5 MILLION SPECIES OF ORGANISMS ON THE EARTH AND THERE ARE 4000 MAMMAL SPECIES, WHAT PERCENTAGE OF SPECIES ARE MAMMALS?

1) IF THERE ARE 5 MILLION SPECIES OF ORGANISMS ON THE EARTH AND 18% ARE CLASSIFIED AS INSECTS, HOW MANY INSECT SPECIES ARE THERE ON THE PLANET?

Figure 35–32 Primates are an order of mammals with highly developed brains that includes lemurs, monkeys, chimpanzees, gorillas, orangutans, and humans.

in the genus *Homo* and the *species sapiens*. ***Homo sapiens*** is Latin for "intelligent humans."

The taxonomic classification of living things that make up the Earth's biosphere is a very important part of understanding the Earth's systems. Taxonomy helps to organize living things by their similar characteristics It also illustrates the amazing variety of organisms on the Earth and the roles that they fill within their ecosystems. But more importantly, taxonomy helps us to see how all living things on the Earth are related.

For more information go to this Web link:
http://gened.emc.maricopa.edu/bio/BIO181/BIOBK/BioBookDivers_class.html>

REVIEW

1. What is taxonomy?
2. What is the binomial method of taxonomy?
3. Name the five kingdoms used to classify organisms on Earth.
4. What are stromatolites, in which kingdom do they belong, and why were they important?
5. What is a diatom and in which kingdom does it belong?
6. What are mycorrhizae and in which kingdom do they belong?
7. Describe the difference between vascular and nonvascular plants and provide one example for each.
8. List 10 organisms that are classified as invertebrates.
9. What are two examples of cartilaginous fish?
10. Give three examples of amphibians.
11. What are three examples of reptiles?
12. How do birds grind up food before it enters the stomach?
13. What are mammals?
14. To which phylum do humans belong?
15. What is the genus and species of a human being?
16. Who was Carolus Linnaeus?

CHAPTER SUMMARY

The biosphere consists of all the living things on the Earth. These organisms have helped to shape the world in which we live and perform important roles in the ecosystems in which they reside. All the living things on Earth are classified into specific groups based on their similar biological characteristics. The classification of things is called taxonomy. Botanist Carolus Linnaeus devised a system for classifying the world's organisms during the eighteenth century. He invented the binomial system of classification that identifies an organisms by its unique genus and species names.

Today organisms on the Earth are divided into five general categories called kingdoms: Monera, Protista, Fungi, Plantae, and Animalia. Organisms in the kingdom Monera include all single-celled organisms that have no cell nucleus. Examples of this type of organism are bacteria and blue-green algae; they are believed to be the first living things to inhabit the Earth. The kingdom Protista contains all the single-celled organisms that have a cell nucleus. Protists are free-floating organisms such as diatoms and algae, which play an important part in all aquatic ecosystems. The kingdom Fungi includes all the molds, yeasts, mildew, and mushrooms. These organisms act as decomposers in many of the world's ecosystems, helping to break down organic material.

The kingdom Plantae includes all the green plants. Plants are divided into two main categories: vascular and nonvascular. Vascular plants have a network of tiny tubes that transport water and nutrients throughout their bodies. Common vascular plants include trees, ferns, grass, and many of the plants we use for food. Nonvascular plants are matlike vegetation, including mosses and liverworts.

The kingdom Animalia consists of all the animals on the Earth that consume other organisms as food. The animal kingdom is subdivided into two main categories called vertebrates and invertebrates. Invertebrate animals have no backbone and include sponges, jellyfish, worms, mollusks, squid, crustaceans, spiders, and insects. The vertebrate animals all have a backbone and include organisms such as sharks, fish, amphibians, reptiles, birds, and mammals.

Human beings are classified in the genus and species *Homo sapiens,* within the class of mammals in the primate order. The taxonomic classification of living things that make up the Earth's biosphere is a very important part of understanding the Earth's systems. Taxonomy helps to organize living things by their similar characteristics It also illustrates the amazing variety of organisms on the Earth, the role that they fill within their ecosystems, and how all living things on the Earth are related.

CHAPTER REVIEW

Multiple Choice

1. Organisms on the Earth are classified according to:
   a. their size and shape
   b. their similar biological characteristics
   c. whether they live on land or in water
   d. their specific geographical location

2. The binomial classification of organisms identifies a specific living thing by its unique:
   a. kingdom
   b. adaptations
   c. genus and species
   d. phylum and order

3. The taxonomic classification of living things on the Earth is organized in which way?
   a. kingdom, phylum, class, order, family, genus, species
   b. kingdom, order, class, species, phylum, genus, family
   c. species, genus, family, order, class, phylum, kingdom
   c. phylum, order, species, genus, family, kingdom, class

4. Bacteria belong to which taxonomic kingdom?
   a. animalia
   b. protista
   c. monera
   d. fungi

5. The kingdom Protista includes which of the following organisms?
   a. cyanobacteria
   b. liverworts
   c. mildew
   d. algae

6. Which taxonomic kingdom includes the first forms of life believed to have existed on the Earth?
   a. animalia
   b. protista
   c. monera
   d. fungi

7. Flowering plants are classified under which category in the Plant kingdom?
   a. nonvascular
   b. angiosperm
   c. gymnosperm
   d. conifer

8. Which taxonomic kingdom composes approximately 90% of all living things on land within the biosphere?
   a. animalia
   b. protista
   c. monera
   d. plantae

9. Sponges, jellyfish, worms, squid, and insects are classified as which type of animals?
   a. vertebrates
   b. annelids
   c. invertebrates
   d. mammals

10. Which kingdom contains the class of organisms with the greatest number of species identified on the planet?
    a. animalia
    b. protista
    c. monera
    d. fungi

11. Sharks belong to which class of animals?
    a. bony fish
    b. cartilaginous fish
    c. cephalopods
    d. reptiles

12. The animals grouped in the vertebrate class Aves are commonly called:
    a. mammals
    b. amphibians
    c. birds
    d. reptiles

13. Which class of animals is believed to be the first to live on land?
    a. mammals
    b. amphibians
    c. birds
    d. reptiles

14. What mammal is classified as a marsupial?
    a. a bat
    b. an opossum
    c. a platypus
    d. a manatee

15. Humans are grouped in which order of mammals?
    a. primates
    b. carnivores
    c. *Homo sapiens*
    d. sirenia

continued

Matching *Match the terms with the correct definitions.*

a. biological characteristics
b. taxonomy
c. siliceous ooze
d. neurotoxins
e. fermentation
f. mycorrhizae
g. gymnosperm
h. angiosperm
i. spores
j. invertebrates
k. bivalves
l. zebra mussel
m. gizzard
n. nocturnal
o. Homo sapiens

1. ____ A type of symbiotic associated with the root systems of plants.
2. ____ The taxonomic classification of a specific type of plant that reproduces by flowers that produce seeds with a protective coat.
3. ____ A discipline of science that deals with the classification of organisms.
4. ____ The genus and species classification of human beings.
5. ____ An organism's unique physical or chemical features.
6. ____ Mudlike sediments composed of silicates that form on the ocean floor.
7. ____ The classification of a green plant that produces a seed without a seed coat in cone, such as a pine tree.
8. ____ A term used to describe an organism that is active during the night.
9. ____ A type of toxic chemical compound that affects the nervous system of animals.
10. ____ The reproductive cells of fungi.
11. ____ An anaerobic chemical process by which complex compounds, such as sugars and starches, are reduced into simple compounds, such as alcohol and carbon dioxide, by microorganisms.
12. ____ The organ in bird's digestive tract that helps to break down food by grinding it between stones.
13. ____ A classification of aquatic organisms that have a hinged shell that can open and close.
14. ____ A species of mollusk, named for its black and white shell, that is not native to North America and is currently overpopulating many fresh bodies of water in the northeastern United States.
15. ____ A taxonomic term used to describe all organisms within the animal kingdom that have no backbone.

Critical Thinking

1. Explain why you think biologists use the taxonomic classification of organisms to help support the theory of evolution.

GLOSSARY

A

A-horizon Uppermost layer of a soil, commonly called topsoil.

abiotic Nonliving.

abiotic factors Nonliving factors in the environment, such as wind, temperature, humidity, or minerals.

abrasion Physical weathering of rocks by the action of particles rubbing against one another.

absorbed Taken in or soaked up matter or energy.

absorption The process of absorbing, or when a solid, liquid, or gas takes in matter or energy and holds it.

accumulation To collect or build up a substance or substances.

acid deposition The deposit of acidic substances on the Earth's surface.

acid precipitation Precipitation that contains high concentrations of sulfuric or nitric acids and has a pH of 5.0 or lower.

acid shock Rapid introduction of acidic water into lakes and streams caused by melting snow.

acidic A solution with a pH lower than 7.0.

adaptations Adjusting to new conditions by the process of natural selection and evolution.

adhesion The attraction of water molecules to a hydrophilic substance.

adiabatic cooling The process by which rising air is cooled by expansion.

aerobic bacteria Bacteria that require oxygen to live.

aerovane A wind instrument that records both air speed and direction.

Agnatha The taxonomical classification of a jawless class of invertebrates that includes the lampreys.

Agricultural Revolution An historic period that included the development and improvement of agriculture that is believed to have begun approximately 10,000 years ago.

air The gases that together make up Earth's atmosphere; they include nitrogen, oxygen, argon, water vapor, carbon dioxide, and other trace elements.

air mass A large body of air that has similar temperature and moisture characteristics.

albedo The reflective ability of an object or surface.

alfisols A particular order of fertile soils that contain high amounts of aluminum and iron, with a well-developed topsoil; associated with deciduous forests.

algae Single-celled or multicelled aquatic organisms that derive their energy from photosynthesis.

algal bloom A rapid increase in the population of algae resulting from the introduction of nutrient fertilizers into an aquatic ecosystem.

alkalinity Having a pH greater than 7.0; also called a base.

alluvial soils Rich, fertile soils that are formed by the deposition of minerals by liquid water.

alpine glaciers Glaciers that are located within mountain ranges; also known as mountain glaciers.

altitude The particular elevation or height of something above sea level.

amber A yellow or brownish yellow petrified tree sap.

ammonia A colorless compound consisting of nitrogen and hydrogen (NH3); used by plants as a source of nitrogen and commonly used as fertilizer.

amoebae A type of free-moving single-celled Protoctista.

Amphibia The taxonomical classification of a class of vertebrates, commonly called amphibians, that include frogs and salamanders.

anadromous A type of fish that reproduces in freshwater and lives its adult life in saltwater, such as salmon.

anaerobic bacteria A type of bacteria that lives in an environment without oxygen.

anemometer An instrument that measures wind speed.

angiosperm The taxonomical classification of a specific type of plant that reproduces by flowers that produce seeds with a protective coat.

angle of insolation The specific angle at which incoming solar radiation strikes the Earth's surface.

Animalia A taxonomical classification of the kingdom of organisms that are commonly known as animals.

anions A negatively charged ion such as chloride (Cl-) or nitrate (NO3-).

Annelida A taxonomical classification for the phylum that includes all segmented worms such as the earthworm.

anthropogenic A term that describes any substance that is created or introduced into the environment by human activity.

anthropologists A scientist who specializes in the study of the origin, development, and culture of human beings.

anticline A type of fold in rock strata that forms an arched shape.

aphelion The point in the orbit of a planet when it is farthest from the Sun.

aphotic zone A particular zone in an aquatic ecosystem where there is no light.

Apollo asteroids A classification of asteroids with orbits that cross the Earth's orbit.

Appalachian orgeny The geologic event that formed the Appalachian Mountains approximately 290 million years ago by the collision of Africa with North America.

apparent motion The perceived movement of celestial objects caused by the Earth's rotation as they arc across the sky.

aquatic A term that describes anything that lives in water

aquatic organisms Living things whose habitat is water.

aquifer Large amounts of water stored in porous or fragmented rock under ground.

arachnids A taxonomical classification of arthropods that includes spiders and ticks.

archaeopteryx The fossilized remains of a 150 million year old birdlike reptile.

Archean eon A division of the geologic time scale marking the beginning of Earth's history from 4.6 to 2.5 billion years ago; also part of the Precambrian period.

archaeology A specialized division of anthropology that searches for and studies the remains of ancient human culture.

arctic air mass A large body of air that forms over the high latitudes and is usually very cold and dry.

armored fish A type of jawed fish that was covered with armor plates; it became extinct more than 362 million years ago.

artesian wells Free-flowing wells that discharge water from the ground and are recharged from a higher elevation.

Arthropoda A taxonomical classification for the phylum that includes insects, spiders, and crustaceans.

Artiodactyla A taxonomical classification for an order of mammals with even numbers of toes on each foot. These include sheep, deer, cattle, and pigs.

asteroid belt A region in the solar system located between the orbits of Mars and Jupiter where there are a high number of asteroids.

asteroid An object orbiting the Sun that is smaller than a planet and has no atmosphere.

asthenosphere An area of flowing plasticlike molten rock located directly below the Earth's crust to a depth of approximately 700 kilometers.

astronomical unit A unit of measurement used to mark the distance of a planet from the Sun. One astronomical unit is equivalent to 93 million miles, which is also the approximate distance of the Earth from the Sun.

astronomy The scientific discipline that studies all aspects of outer space.

Aten asteroids Asteroids with orbits that lie between the Earth and the Sun.

atmosphere The outer layer of gas that surrounds a planet.

atmospheric moisture The amount of water vapor that is present in the air.

atmospheric pressure The weight of a column of air at a specific point in the atmosphere, usually measured in millibars or inches of mercury.

atomic mass The exact mass of the protons and neutrons in the nucleus of an atom, measured in atomic mass units.

atomic number The number of protons in the nucleus of an atom.

atoms These are the smallest particles, also called elements, that can be combined to form compounds. Atoms consists of a nucleus that is made up of subatomic particles known as protons and neutrons, which are orbited by electrons.

australis borealis The Southern Hemisphere's equivalent of the northern lights, or aurora borealis, which is caused by streams of particles interacting with the Earth's magnetic field.

autotrophs Organisms that produce their own food by photosynthesis, such as plants or algae, or by chemosynthesis, such as certain bacteria.

autumnal equinox The time of the year, on or around September 21, when the Earth's axis of rotation is aligned perpendicular to the plane of the Sun and the length of daylight (12 hours) equals the length of darkness (12 hours).

avalanche A large mass of snow, ice, or rock rapidly sliding downhill.

Aves A taxonomical classification of a class of vertebrates commonly called birds.

axis A straight line around which an object rotates.

axis of rotation An imaginary line drawn through an object, around which it seems to rotate.

azimuth The measured angle of an object above the horizon .

B

B-horizon The second layer of soil, also known as the subsoil, that lies below the topsoil or A-horizon.

backbone The common term for the hard skeleton, or vertebrae, that makes up an organism's spinal column.

bacteria Microscopic, single-celled organisms that do not have a cell nucleus and are part of the kingdom Monera.

barometer An instrument that is used to measure atmospheric pressure, also called barometric pressure.

barometric pressure The measure of the weight of a column of air at a particular location on a planet; also known as atmospheric pressure.

barren An area that is lifeless or lacking vegetation.

basalt A common fine-grained volcanic rock that is dark, mafic, and dense.

bedrock A large mass of unbroken or undisturbed rock usually buried deep within the ground.

benchmarks A series of human-created markers or monuments that designate the precise latitude, longitude, and elevation of a specific point on the Earth's surface.

benthic zone An aquatic life zone located on the bottom of a body of water.

Big Bang A popular theory of the origin of the universe, which is believed to have begun after a violent explosion that propelled matter and energy away from one central point.

binomial An ancient Greek term meaning "two names," which is commonly used for the classification of organisms.

biogeochemical cycling The natural recycling of elements between the nonliving world and the living world.

biological characteristics An organism's unique physical or chemical features.

biological diversity The variety of living species on Earth and their unique genes.

biological succession The gradual replacement of one community of species by another in a slow, predictable manner.

biomass A shortened term for biological mass, which is the total dry weight of an organism.

biomes Another term for a land-based ecosystem, also called a terrestrial ecosystem.

biotechnology The scientific discipline that studies the use or manipulation of living organisms to perform specific tasks.

biotic Living.

bipedalism A term used to describe an organism that walks on two feet.

bivalve mollusks A classification of aquatic organisms that have a hinged shell that can open and close.

bivalves *See* bivalve mollusks.

black hole A theoretical celestial object with a strong gravitational attraction that prevents light from escaping its surface.

blue supergiant A classification of massive stars that are blue, extremely hot, and bright.

blue-green algae An aquatic organism that is similar to bacteria but gains its energy from photosynthesis; *see* cyanobacteria.

boiling point The specific temperature at which a substance begins to change its phase from a liquid to a gas; also known as vaporization.

boreal forest Another term for the coniferous forest biome.

broad-leafed trees A type of tree that photosynthesizes using large flat leaves.

Bronze Age A period beginning approximately 5000 years ago and ending 3000 years ago when bronze, a metal alloy composed of copper and tin, was widely used.

brown algae A brownish multicelled aquatic organism commonly called seaweed.

buffer A substance that is capable of stabilizing the acidity or alkalinity of a solution.

buoys Floating markers that are anchored to a specific spot in a body of water.

burial To cover something with soil or rock.

C

C-horizon The third layer of a soil, composed of unconsolidated rock or transported parent material.

calcium carbonate The chemical compound that makes up the mineral calcite, which forms limestone.

caldera A large crater (more than 1 mile in diameter) caused by a violent volcanic eruption.

California current A wind-driven surface ocean current that brings cold water from the North Pacific south toward the equator along the coast of California.

Cambrian period A period of the geologic time scale marking the appearance of the earliest shelled marine organisms that began approximately 544 million years ago and ended 490 million years ago.

capillary action The movement of water molecules upward in tiny tubes as a result of adhesion and cohesion.

carbohydrates Chemical compounds that are made up of carbon, hydrogen, and oxygen, many of which are produced by the process of photosynthesis, such as sugar and starch.

carbon dioxide A chemical compound, usually in the form of a gas, that is made up of one atom of carbon and two atoms of oxygen (CO2) and is the byproduct of respiration and combustion.

carbonation The addition of carbon dioxide gas to something.

Carboniferous period A period in the geologic time scale beginning approximately 362 million years ago and ending 290 million years ago, when an abundance of coal-forming trees and ferns lived.

Carnivora A taxonomical classification for an order of mammals that includes carnivores, which have sharp canine teeth, such as bears and lions.

carnivores Meat-eating organisms.

catadromous A type of fish that reproduces in saltwater and lives its life in freshwater, such as the Atlantic eel.

catastrophic A great and sudden disaster.

cations Positively charged elements such as sodium (Na+) or calcium (Ca2+).

celestial A term that pertains to something outside the Earth's atmosphere.

celestial object An object that is located outside the Earth's atmosphere.

cement A substance that binds things together.

cementation The process of binding something together

Cenozoic era A division of the geologic time scale that began 65 million years ago with the extinction of the dinosaurs and stretches to the present day.

centripetal force The force that keeps and object moving in a circular path.

Cephalopods A taxonomical classification for a class of mollusks, such as octopuses and squids.

Cetacea A taxonomical classification for an order of aquatic mammals, such as dolphins and whales.

CFCs *See* chlorofluorocarbons.

channel The portion of a moving body of water where water is currently flowing.

chaparral A type of biome that has hot dry summers and cool rainy winters.

chemical energy A form of energy that is stored in the chemical bonds between atoms and molecules.

chemical weathering The breakdown of rocks into smaller rock particles by a chemical process.

chemosynthesis A method of deriving energy from the breakdown or formation of organic compounds.

chlorofluorocarbons Human-created molecules, commonly used as refrigerants, in electrical manufacturing, and in foam production, that are responsible for ozone destruction; also a greenhouse gas.

Chondrichthyes A taxonomical classification for a class of chordates commonly known as cartilaginous fish, such as sharks.

Chordata A taxonomical classification for the phylum of animals that have a spinal chord.

chromosphere The specific layer of the Sun that is located above the photosphere.

cinder cone A small, straight-sided volcanic cone made up of pyroclastic material.

class A taxonomical classification for things that have at least one similar feature.

clay The smallest size classification of a rock particle that is between 0.00001 and 0.0004 centimeters in diameter.

cleavage The breaking of minerals along specific planes of weakness caused by the internal arrangement of their atoms.

climate The long-term weather patterns of a specific region on the Earth, usually defined by the area's annual temperature and precipitation values.

cloud A large mass of condensing water droplets and ice crystals in the atmosphere.

Cnidaria A taxonomical classification for the phylum of animals that includes jellyfish, corals, and sea anemones.

coal The rocklike fossilized remains of ancient plants, which is used as fuel.

coastal wetlands A type of wetland that is located near the coastline of a land mass.

cohesion The attraction of water molecules to one another.

cold front The zone where a cold air mass overtakes and replaces a warm air mass.

colonies A large grouping of a particular species of organism in a specific area.

color The specific appearance of an object that is caused by its ability to absorb or reflect visible light energy in specific wavelengths.

coma The gas and dust that is released from and surrounds the nucleus of a comet.

combustion A chemical reaction that results in light and heat, commonly called burning.

comet A mixture of frozen compounds and rock that orbits the Sun that has a distinct tail composed of vaporized gas and dust which always points away from the Sun.

comet's nucleus The frozen compounds and rock that form the core of a comet.

comet's tail The vaporized gas and dust that trail away from the nucleus of a comet.

commensalism A specific type of symbiotic relationship in which one organism benefits from a host and the host is neither helped nor harmed.

community All the species of organisms that reside in a particular area.

compaction The reduction in volume of a substance as a result of extreme pressure or weight.

composite cone volcano A volcano that is composed of both solidified lava and pyroclastic material.

composting The breakdown of organic material by decomposing bacteria, insects, and fungi into humus.

compounds Substances that are formed from joining two or more elements in specific proportions.

compression To press something together into a smaller space.

compressional primary wave (P wave) A seismic wave produced at the focus of an earthquake that is the fastest of all seismic waves and can travel through all states of matter.

conclusion The final summation of the scientific method that states what has been learned.

condensation The change in phase from a gas to a liquid.

condensation nuclei Microscopic particles floating in the atmosphere on which water condenses to form clouds.

conduction The transfer of heat energy by direct molecular contact.

confined aquifer Groundwater that is located below an impermeable rock layer.

coniferous forest A specific type of biome that experiences long, cold winters and short, hot summers, where evergreen trees (conifers) are the main vegetation.

conifers A cone-bearing tree, also called an evergreen.

constellations Specific groupings of stars that resemble mythological figures, objects, or animals.

contact metamorphism Metamorphism of rock by the heat of contact associated with an igneous intrusion.

contaminated To pollute or make impure.

continental crust The specific portion of the Earth's outer crust that is relatively thick, lower in density, and composed primarily of granitic rock that makes up the continents.

continental glaciers Large masses of accumulating snow and ice that cover an entire continent.

continental polar air mass A large mass of cold air that forms over land near the poles and has low moisture.

continental shelf The relatively shallow region of the ocean surrounding a continent.

continental tropical air mass A large mass of warm air that forms over land near the equator and has low moisture.

contour interval The specific change in elevation associated with each contour line on a topographical map.

contour lines Isolines that mark areas of equal elevation on a topographical map.

contraction The reduction in the volume of a substance that is usually associated with lowering its temperature.

convection The transfer of heat energy in a fluid as a result of a change in density associated with a change in temperature.

convection cell The circular movement of a fluid caused by a change in temperature and density that is associated with the transfer of heat.

convection currents The specific pattern of movement of a fluid caused by convection.

convective zone An inner layer of the Sun where heat from the core is convected outward.

convergence The process of coming together.

convergent plate boundaries An area on the Earth's surface where two tectonic plates come together.

coordinate system A system of precise location using a grid pattern of intersecting lines.

coral bleaching The whitening of living coral as a result of the die-off of the coral's symbiotic algae, which may eventually lead to the death of the coral itself. It is believed to be caused by exposure to environmental stress.

coral reefs Large underwater structures located in warm, shallow saltwater that are built by colonies of coral; composed primarily of calcium carbonate and sand.

corals Benthic aquatic organisms that live in warm, shallow saltwater and may build shells from calcium carbonate.

Coriolis effect The deflection of free-floating objects to the right of travel in the Northern Hemisphere and to the left of travel in the Southern Hemisphere as a result of the Earth's rotation.

corona The outermost layer of the Sun, which stretches millions of miles out into space, where temperatures are in the millions of degrees.

creep The slow, almost imperceptible movement of soil and rock down a slope.

Cretaceous period A period of the geologic time scale that began approximately 142 million years and ended 65 million years ago and marks the height of the age of dinosaurs and the appearance of flowering plants..

crinoids Animals, commonly called sea lilies, that use branching arms to capture tiny floating prey in the water; can be free floating or attached to the bottom by a stalk.

crust The solid outer layer of the Earth, composed of rock.

Crustaceans A taxonomical classification of arthropods that included lobster, shrimp, and crabs.

crystalline A substance or structure that is made up of crystals.

crystallization Rock changing phase from a liquid to a solid; also called solidification.

crystal A term that describes the distinct recurring pattern of the atoms or molecules in a specific mineral that form a unique recognizable structure.

currents A specific pattern of movement within a fluid.

cut bank A specific physical feature of a flowing body of water caused by the erosion of material away from the outside curve of a meander, resulting in a steep drop off.

cyanobacteria A photosynthetic, oxygen-producing single-celled organism; also known as blue-green algae.

cycads Ancient shrublike trees that are related to the conifers.

cyclic A periodic or reoccurring event.

cyclone A term for winds that spiral inward and around a low-pressure center.

D

D-horizon The fourth layer of soil; also known as the consolidated bedrock.

Data Specific information gathered for use or analysis.

daughter element The element that forms after the decay of a radioactive isotope.

deciduous forest A specific type of biome that is marked by long, hot summers and cold winters, where deciduous trees are the primary vegetation.

deciduous tree A type of tree that loses its leaves each autumn and goes into dormancy during winter.

declination The angular distance north or south of the celestial equator, used to mark the position of a celestial object.

decomposers Organisms that break down and decay dead organisms or waste.

decomposing organic material The breakdown of the remains of a living thing.

deep pools A physical feature of a flowing body of water that results when the moving water erodes and deepens the stream bottom on the outside curve of a meander.

deficient Lacking something essential to life.

deformed rock layers Rock strata exposed to outside forces that have caused them to be tilted, folded, faulted, or overturned.

delta The formation that results at the mouth of a river where sediments carried by the river have been deposited.

denitrifying bacteria A specific type of bacteria that converts nitrogen compounds found in soil into atmospheric nitrogen.

density The mass per unit volume of a substance, usually expressed in grams per cubic centimeter.

depletion The loss of something.

deposition To put down or place something.

derived units The combination of two or more fundamental units of measurement, such as miles per hour.

desert A specific type of biome that receives little or no moisture throughout the year and generally lacks an abundance of vegetation.

destruction The act of tearing down or ruining something.

detritivores Organisms that consume dead and decayed organisms or waste; *see* decomposers.

Devonian period A period of the geologic time scale that began approximately 418 million years ago and ended 362 million years ago, marking the appearance of the first amphibians and the rise of fish.

dew point The temperature to which the air would need to be cooled to become saturated with water at a specific atmospheric pressure.

dew point depression The difference between the wet bulb and dry bulb temperatures; also called the wet bulb depression.

diatomic A term that refers to a molecule that consists of two atoms.

diatoms Microscopic algae that build a protective silicate shell.

dinosaurs A class of reptile and birdlike organisms that were dominant on the Earth for approximately 200 million years and became extinct 65 million years ago.

disassociate To pull apart or separate.

discharge To unload or empty.

discharge rate The amount of water passing by a particular point in a flowing body of water.

disease-causing agents Specific organisms that cause or spread disease.

dislodged To be removed or forced out.

disphotic zone The dimly lit portion of an aquatic ecosystem that cannot support photosynthesis.

displaced rocks Rocks that have been moved or transported to a new location.

dissolve To enter into a solution.

dissolved oxygen The amount of oxygen gas that is dissolved in water.

disturbed areas Ecosystems or parts of ecosystems on the Earth that have been destroyed or disrupted.

diurnal tides High and low tides that occur only once every day.

divergent plate boundary A point in the earth's crust where two tectonic plates are moving away from one another.

diverging Moving apart or away from one another.

DNA The short form of deoxyribonucleic acid, which is the molecule that holds all of an organism's genes and is responsible for the replication of a cell.

domesticate To adapt for life with human beings.

downdrafts Strong vertical winds that blow downward from the base of a cloud.

drainage basin The total land area that contributes water to a river system, pond, or lake.

dredging A method of mining that involves the removal of sediments from the bottom of body of water.

drizzle A form of precipitation that consists of tiny liquid water droplets

droughts Extended periods when little or no precipitation is received by a specific region.

dry ice The solid form of carbon dioxide.

duration The length of time that it takes for an event to occur.

dust devils A term to describe very small cyclonic winds that kick up dust and whirl it across the landscape.

dynamic Energetic, always changing or moving.

dynamic equilibrium A balance between two opposing processes that occur at the same rate in an energetic system, such as a river or stream.

E

Earth's core The extremely hot and dense center of the Earth, which is believed to be composed of iron and nickel.

Earth's magnetic field The magnetic field that surrounds and protects the Earth from harmful radiation, which is believed to be generated by the Earth's core.

earthquake The violent, rapid shaking of the Earth caused by a rupture in the crust.

eccentricity The mathematical expression of how far an ellipse is from a perfect circle, which can be determined by dividing the distance between the foci by the length of the major axis.

eclipse The cutting off of all or part of the light of one celestial body by another.

ecliptic The imaginary line drawn from the center of the Sun to the Earth, also described as the apparent path of the Sun across the sky.

ecology The scientific discipline that studies how organisms interact with their environment.

ecosphere A term that describes all the life zones that exist on Earth, which is made up of the atmosphere, hydrosphere, lithosphere, and biosphere.

ecosystem The interaction between the biotic and abiotic factors in a specific area.

Edentata A taxonomical classification for an order of mammals that have reduced or no teeth, such as anteaters and sloths.

El Niño *See* El Niño southern oscillation.

El Niño southern oscillation (ENSO) An increase in the sea surface temperature off the western coast of South America near the equator, which leads to changes in climate around the world.

electrical energy A form of energy that flows by exciting electrons in specific elements called conductors, which produces an electric current.

electrical potential An element or ion's electrical charge.

electromagnetic radiation A type of energy that travels in the form of a wave and needs no medium for transfer.

electromagnetic spectrum The range of specific wavelengths and frequencies that identify the specific forms of electromagnetic energy.

electrons Negatively charged subatomic particles that surround the nucleus of an atom.

elements The 109 identified atoms that have a definite number of protons, neutrons, and electrons.

ellipse The flattened circular path of the orbits of most celestial objects around two foci, one of which is the Sun.

elliptical galaxy A type of galaxy whose stars form an elliptical shape.

energy The ability to cause change or perform work.

energy pyramid The mathematical model that describes the transfer of energy through a food chain, which illustrates the loss of energy at each trophic level.

engineer A skilled person who utilizes scientific principles to solve practical problems and design efficient systems.

eon The largest division of geologic time, which is measured in billions or hundreds of millions of years.

epicenter The point on the Earth's surface directly above the focus of an earthquake.

epoch The smallest division of geologic time, which is measured in millions or thousands of years.

equator The imaginary line, also known as zero degrees latitude, that divides the Earth in half into the Northern and Southern Hemispheres.

equatorial A term referring to a location near the equator.

equilibrium A state of balance.

era The second largest division of geologic time, which is measured in the hundreds of millions or millions of years and marks the mass extinction of species on Earth.

erosion The movement of rock particles or soil by wind, water, and the force of gravity.

eruption The sudden release of lava or pyroclastic material from a volcano.

euphotic zone The uppermost life zone in an aquatic ecosystem that receives enough light to support photosynthesis.

Eurasia The name for the large continent where Europe and Asia are located.

eurypterid An extinct marine organism that resembled a large scorpion; it flourished during the Permian and Devonian periods; also called the New York State fossil.

eutrophic lake A classification for a lake that is relatively cloudy, warm, and shallow and has an abundance of nutrients that support a large population of aquatic plants and animals.

eutrophication The rapid introduction of nutrient fertilizers into an aquatic ecosystem, which leads to an increase in aquatic plants and algae.

evaporation The phase change of a liquid changing into a gas.

evaporative cooling A cooling process caused by the evaporation of a liquid from a surface, which is the result of the evaporating liquid absorbing energy from the surrounding environment; therefore cooling it.

evapotranspiration The important pathway by which water moves from the soil, through the body of a plant, and evaporates off the leaf surface back into the atmosphere.

evidence A set of facts, proof, or demonstration used to support a theory.

exfoliation The process by which layers of rock peel off the parent rock.

exosphere The zone outside of Earth's atmosphere that is commonly known as outer space.

expand To increase the volume of a substance.

expansion The process of increasing the volume of a substance.

expansion joints Flexible material constructed into bridges, buildings, and other large structures to allow for the building material to expand and contract when exposed to changes in temperature.

experiment A controlled test to prove or disprove a hypothesis.

explosive eruption The violent release of pyroclastic material from a volcano.

extrusive rock A type of igneous rock that is formed at the Earth's surface from the solidification of lava.

F

family A taxonomical classification for a group of organisms that have similar characteristics.

fault A large break or crack in a rock mass formed by tectonic stress that results in the displacement of rock strata.

faulted rock Rock strata that contains a fault.

felsic rocks A classification of igneous rocks that are light colored and low density and contain silicates and aluminum.

fermentation An anaerobic chemical process by which complex compounds, such as sugars and starches, are reduced into simple compounds, such as alcohol and carbon dioxide, by microorganisms.

fertile soil A common term used to describe a soil that supports healthy crop growth.

fertilizers A term used to describe chemical compounds that are required by plants for healthy growth.

field capacity The moisture content of a soil when its pore spaces contain 50% air and 50% water, which is perfect for the growth of plants.

First Quarter Moon A phase of the Moon when it has completed the first quarter of its orbit around the Earth and half of its surface, as viewed from the Earth, is illuminated by the Sun; commonly called a Half Moon.

fissure eruption A specific type of volcanic eruption that occurs when lava flows out of a large crack in the Earth's surface.

floodplain The flat area of a river valley located along both sides of a river channel that is formed from the deposition of sediments during periodic floods.

flowering plants A classification of plants that reproduce by producing a flower that forms seeds with a protective coat.

flowing spring An area where groundwater is discharged at the surface and flows freely.

focus The point in the Earth's crust where a rock mass is broken or moved, causing an earthquake.

fog A cloud formed near or at the Earth's surface that is composed of tiny water droplets.

folded deformed rock Rock strata that has been folded by tectonic forces.

foliated The layered or wavy structure that forms in some metamorphic rocks.

food chain The model pathway that energy and matter takes through an ecosystem by a series of eating processes.

food web Interrelated food chains that exist in an ecosystem.

fossil The remnant or trace of a once living organism preserved in sedimentary rock.

fossil fuels A term used to describe hydrocarbon fuels, such as coal and oil, that were formed from the remains of once living organisms.

fracture The process of breaking.

freezing The phase change of a liquid to a solid.

freezing point The specific temperature at which a liquid changes phase into a solid.

freezing rain A form of precipitation that occurs when rain freezes after it comes into contact with the Earth's surface.

Freon A type of chlorofluorocarbon gas that was commonly used in air conditioners and refrigerators.

fresh surface water Water on the Earth's surface that contains a small amount of dissolved mineral salts and is good for drinking.

freshwater Water on or below the Earth's surface that contains a small amount of dissolved mineral salts and is good for drinking.

friction The rubbing of one surface or object against another.

front An area where two different air masses come together.

Fujita scale of tornado intensity A scale that rates the power of a tornado based on its wind speed and potential damage.

Full Moon A phase of the Moon at which it has reached the halfway point of its orbit around the Earth, and its surface is completely illuminated by the Sun as viewed from Earth.

fundamental measurements Any basic form of measurement such as mass, length, or time.

Fungi A taxonomical classification for the kingdom of organisms commonly known as fungus.

funnel cloud A rapidly moving vortex of wind and debris, also called a tornado, that has not yet reached the ground and is formed at the base of a cloud.

fusion reaction A specific type of nuclear reaction that results when two atoms are joined together to produce a great amount of energy.

G

galaxy A large grouping of stars.

gas A state of matter that has a very low density, where molecules and atoms are allowed to move freely.

gas giants A classification of planets that are extremely large and are composed mainly of gases.

gaseous planets *See* gas giants.

generalist population A group of organisms that can eat a variety of foods.

genus A taxonomical classification of organisms that includes several species with very similar characteristics.

geologic forces The main forces in geology that help to form rocks and transform the landscape, which include gravity, erosion, and tectonic activity.

geologic time scale The scale of time that divides the Earth's history into distinct periods based on geologic events and the appearance or disappearance of specific life forms.

geosyncline A bowl-shaped depression in the Earth's crust formed from the deposition of large amounts of sediments.

germinate *See* germination.

germination The process that a plant seed undergoes when it begins to grow.

giant molecular cloud A large mass of gas and dust found in the universe.

Giardia A specific type of disease-causing microorganism that is associated with human waste and is found in fresh surface water.

ginkgo A type of broad-leafed deciduous tree that dates back to the time of the dinosaurs.

gizzard The organ in a bird's digestive tract that helps to break down food by grinding it between stones.

glacial advance The forward movement of a glacier, caused when snow accumulation is greater than melting.

glacial deposits Unsorted sediments deposited by a melting glacier.

glacial front The face of a glacier where the ice breaks off and melts.

glacial lake A lake that is formed from glacial melt water.

glacial retreat The shrinking of a glacier caused when melting exceeds snow accumulation.

glacial sediments Unsorted, angular rock fragments deposited by a glacier.

glacial till An accumulation of glacial sediments.

glacial transport soils Soils that are formed from glacial sediments.

glacier A long-lasting, large mass of snow and ice that forms over land from the accumulation and compaction of snow that creeps down slope.

globular cluster A large grouping of stars that form portions of a galaxy.

glucose A common form of sugar that is produced by the process of photosynthesis.

Goldilocks syndrome A illustration of the relationship among the atmospheric carbon dioxide concentrations of Venus, Earth, and Mars and their surface temperatures that uses the Goldilocks fairy tale. Venus has too much carbon dioxide; therefore it is too hot. Mars has very little carbon dioxide and is too cold. The Earth is just right.

graded bedding A specific form of deposition of sediments that results from the reduction of the velocity of water at the mouth of a river that enters into a body of water. This causes larger particles to settle closest to the mouth and decrease in size as they move farther into the still water.

granite A type of intrusive, felsic igneous rock composed of coarse crystals that are primarily made up of quartz and feldspar minerals.

granule A small grain or particle.

graph A visual representation that shows the relationship between a specific set of numbers or data.

grassland A specific type of biome that experiences hot, dry summers and cold winters where grass is the main vegetation.

gravel A loose mixture of rock fragments.

gravity An attractive force between two bodies in the universe.

greenhouse effect A term that is used to describe short-wave radiation passing through the atmosphere and being absorbed by the Earth's surface, which then reradiates the energy in a long-wave form that is trapped by gases in the atmosphere and heats the planet. This process is the same method by which greenhouses are heated by the sun, whereby the glass of the greenhouse acts like the atmospheric gases.

greenhouse gases Specific gases in the atmosphere that trap long-wave radiation, such as water vapor, methane, and carbon dioxide.

Greenwich mean time The exact time of day on the prime meridian that is located in Greenwich, England, which is used to determine exact longitude locations on the Earth.

groundwater Naturally occurring freshwater that flows or is stored underground in rock or sediments.

groundwater flow The specific direction and movement of groundwater.

gulf stream A wind-driven, surface ocean current that brings warm water from the equator northward along the east coast of North America; first discovered by Benjamin Franklin.

gymnosperm The classification of a green plant that produces a seed without seed coat in a cone, such as a pine tree.

H

habitat The specific food, water, cover, and space requirements for a particular organism.

Hadley cell The large-scale convection cell that forms low pressure at the equator and high pressure at 30 degrees north and south of the equator, which produces the planetary scale winds; first proposed by George Hadley.

hail A ball-like type of precipitation that forms within a cumulonimbus cloud as a result of strong updrafts.

half-life The time it takes for half a mass of a radioactive isotope to decay into its daughter element.

halo Rings or arcs that encircle the Sun or Moon, caused by light passing through ice crystals in the atmosphere.

hardness A term used to describe a mineral's resistance to being scratched.

head waters The point of origin of a river.

heat The measurable or perceived effect of energy that is transferred between two objects that have different temperatures.

heat capacity The ability for a substance to absorb, contain, and release heat energy.

heat gradient The change in heat energy over a specific distance.

hemisphere One half of a sphere.

herbicides A group of chemicals that are used to kill or control the growth of undesired plants.

herbivores Animals that only eat plants.

heterotrophs Organisms that must consume other organisms to gain energy, such as humans.

high atmospheric pressure system An air mass that has relatively high atmospheric pressure, is cool and dry, and produces winds that spiral outward and clockwise in the Northern Hemisphere.

high clouds A classification of clouds that generally form at altitudes above 20,000 feet.

high-concentration deposit A mineral ore deposit that has a high concentration of the desired element; also known as a high-grade deposit.

high-pressure center An area on the Earth's surface that has relatively high atmospheric pressure compared with the surrounding air, is generally cooler, and is dry. Air circulates around high pressure outward and clockwise in the Northern Hemisphere.

high tide The time or times of day when the surface of the ocean is raised to its highest point by the gravitational attraction of the Sun and Moon.

hominid A term used to identify all members of the human family.

Homo sapiens The genus and species classification of human beings.

horizon The horizontal plane on the Earth where the sky meets the land.

horizontal sorting A specific type of sediment deposition that occurs in still water, which results in larger particles settling first, then progressively smaller particles settling on top of the larger ones.

hot house climate A term used to describe a period in the Earth's history when the average surface temperature on Earth was much warmer than today.

hot spot A term used to describe a specific point located near the middle of a tectonic plate that experiences volcanic activity.

HR diagram The Hertzprung-Russell diagram, which classifies stars by their color, temperature, and luminosity.

humid A term used describe a region that has high atmospheric moisture.

humidity A common term that refers to the amount of water vapor in the air.

humus The completely decomposed remains of organic debris, which is an important part of a soil.

hurricane A large-scale tropical cyclone with winds in excess of 74 miles per hour.

hydration To add water to something

hydraulic mining A method of mining minerals that uses high-pressure water.

hydrocarbon A chemical compound composed of hydrogen and carbon that is commonly associated with fuels.

hydrological cycle The circular pathway of water molecules as they move through the environment; also called the water cycle.

hydrophilic A term that describes a substance that attracts water molecules.

hydrosphere All the water on the Earth.

hydrothermal vents Ocean water that seeps through cracks in the sea floor and is superheated by magma close to the surface.

hypothesis An explanation based on a set of facts that can be tested.

I

ice The solid form of water that is less dense than liquid water which causes it to float.

ice age A specific period in the Earth's history when the average surface temperature was much lower than today, causing the formation of widespread glaciers.

ice cap A platelike or dome-shaped mass of ice and snow that covers an entire mountain and spreads outward in all directions

ice cores Long cylindrical sections of ice that are removed from glaciers by drilling and can be used to study the Earth's past climate.

ice field An extensive mass of ice that covers large areas of land, except high mountain peaks.

ice pellets A form of precipitation that consists of small frozen grains of ice.

ice sheet An extremely large and thick mass of glacial ice that covers an entire continent; also called a continental glacier

icebergs Large chunks of glacial ice that break off the leading edge of a glacier and float away into the sea.

igneous intrusion Magma that seeps into an existing rock mass to form new igneous rock and may metamorphasize surrounding rock by the heat of contact.

igneous mineral deposits Specific mineral deposits that exist within igneous rock formations.

igneous rocks A type of rock on the Earth that forms from solidifying magma or lava.

immune system The system in the body that is responsible for fighting off disease.

impact craters Large bowl-like depressions that are left on the surface of a celestial object as the result of an impact by another celestial object, usually an asteroid or comet.

impermeable Unable to pass through, such as certain rocks that do not allow water to pass through them.

inches of mercury A unit of measurement used to record atmospheric pressure.

index fossils A specific type of fossil organism that lived for a short time over a wide geographical area, which is used to identify specific rock formations and the period when they formed.

infectious agents Any substance or organism that transmits disease.

inference A conclusion based on a set of observed facts.

infertile soil A common term for a soil that does not support healthy crop growth.

infiltrate To enter into something, such as water entering into the ground

infiltration The process of infiltrating, or entering into something.

infrared radiation A form of long-wavelength electromagnetic radiation.

infrared satellite images Satellite imagery that is created by sensing the infrared energy given off by a substance, which allows images to be recorded at night without the presence of light.

ingestion To eat.

Inorganic Not containing carbon; also referring to something that is not or was not alive.

Insecta A taxonomical classification for a class of arthropods commonly called insects.

Insectivora A taxonomical classification for an order of mammals that includes insect-eating moles and shrews.

insolation The amount of incoming solar radiation on the Earth.

international date line A specific line of longitude located in the middle of the Pacific Ocean that is exactly 180 degrees east or west of the prime meridian.

intertidal zone An aquatic life zone that exists near the shoreline between the area of the highest and lowest tides.

intertropical convergence zone The zone surrounding the equator where warm rising air causes low pressure to form at the Earth's surface, resulting in the formation of planetary winds called the trade winds.

intrusive rock Igneous rock that is formed from magma seeping into an existing rock mass.

invertebrates A taxonomical term used to describe all organisms within the animal kingdom that have no backbone.

ionosphere A layer of the Earth's atmosphere where there exists a large number of ions and free electrons.

ion An atom or group of atoms that have an electric charge as a result of gaining or loosing electrons.

Iron Age A period in time when humans began to use iron to make tools, which began approximately 3000 years ago.

iron meteorites A type of meteorite that is composed mostly of iron compounds.

irregular galaxy A type galaxy that has no definite shape.

irrigation An artificial means of supplying water to plants.

isobars Isolines that connect points of equal atmospheric pressure on a weather map.

isostasy The theory that explains how the Earth's crust is in balance, causing the continents to float at different levels on the asthenosphere below, much like different-sized blocks of wood floating at different levels in water. The higher the continent above sea level, the lower it floats on the crust, and as material is eroded away from the continent, it tends to rise.

isothermal layer A layer in the atmosphere where temperature remains the same as the altitude increases.

isotherms Isolines that are used to connect points of equal temperature on a weather map.

isotopes Similar atoms that have the same number of protons but a different numbers of protons in the nucleus.

J

jawed fish A common name for the class Osteichthyes, also known as bony fish, which have jaws.

jet stream A region of high-velocity winds that exist high up in the atmosphere.

Jurassic period A period of the geologic time scale that began approximately 206 million years ago and ended 142 million years ago, marking the rise of the dinosaurs and the first appearance of birds.

K

katabatic wind A cold wind that blows down the side of a mountain.

kinetic energy The energy of motion or movement.

kingdom A broad, general category of living things.

Kuiper Belt An area located just outside of the orbit of the planet Pluto that may contain millions of comets.

L

La Niña The opposite effect of the El Niño southern oscillation, which occurs when the sea surface temperature along the equator off the western coast of South America is cooler than normal, resulting in widespread change in climate.

Lagomorpha A taxonomical classification for the order of mammals that has long back legs used for jumping, such as rabbits.

lahar A rapid flow of mud and debris formed from the rapid melting of snow and ice associated with a volcanic eruption.

lake A large, inland body of water.

land breeze A coastal wind system that occurs when high pressure forms over cool land and low pressure forms over warmer water, causing air to flow form the land toward the water.

landfills A term used to describe a place where large amounts of garbage are deposited.

lapse rate The rate at which temperature or moisture decrease with height.

latent heat Heat energy that is either absorbed or released during a phase change.

latent heat of condensation Heat energy that is released into the atmosphere by the condensation of water vapor.

lateral fault A large rupture in a rock mass that runs horizontally; also known as a strike-slip fault

lateral forces Forces that push in on something from both sides.

lateral moraines Glacial sediments that form along the sides of a glacier as it scrapes along rock.

latitude Parallel lines that run east and west across the Earth's surface, measuring location north or south of the equator.

lava Hot, molten volcanic rock that flows freely on the Earth's surface.

lava dome A domelike structure of cooled lava that is located within a larger, active volcanic crater.

lava vents A system of pipes or tunnels that funnel lava outward along the sides of a volcano.

leaching The movement of chemicals that are dissolved in water, from a higher layer of soil downward into a lower level of soil or groundwater.

leeward The side of something opposite toward which the wind is blowing

legumes A specific type of plant that houses symbiotic bacteria in its root system, which puts nitrogen into the soil. Common legumes include clover, peas, and alfalfa.

length A of measurement of horizontal distance.

lichen A symbiotic organism that consists of an alga and fungus that lives on rocks and trees.

light year A unit of measurement used in astronomy that records the distance that light travels in 1 year, based on the speed of light being 186,000 miles per second.

lightning A naturally occurring electrical discharge in the atmosphere usually associated with thunderstorms.

limestone A form of sedimentary rock composed of the mineral calcite (calcium carbonate).

limiting factor A specific nutrient that is lacking in an ecosystem and limits the growth of organisms.

limnetic zone The aquatic life zone that begins near the surface of a body of water and stretches down as far as light can penetrate and be used for photosynthesis.

liquid A state of matter between a solid and a gas where molecules are allowed to flow.

lithification The process of converting sediments into one solid mass of rock.

lithified To undergo the process of lithification.

lithosphere The solid outer layer of the Earth, which is composed of rock and soil.

littoral zone The aquatic life zone of body of water that is located along the shore, where light can reach the bottom.

loam Soil that contains specific portions of sand, silt, and clay.

Local Group A term used in astronomy to describe the group of galaxies that are close to our own galaxy, called the Milky Way.

loess A type of soil found in the American Midwest, that is formed from wind transported parent material.

long-period comets Comets that take 200 or more years to complete one orbit around the Sun.

longitude Coordinate lines used on the Earth's surface that run north and south from pole to pole and measure a location east or west of the prime meridian.

long-wave radiation A low-energy form of electromagnetic radiation, such as infrared or radio waves.

low atmospheric pressure system An air mass that has relatively low atmospheric pressure, is warm and moist, and produces winds that spiral inward and counterclockwise in the Northern Hemisphere.

low clouds A classification of clouds with bases that form lower than 6500 feet above the Earth's surface.

low-concentration deposit A mineral ore deposit that has a low concentration of the desired element, also known as a low-grade deposit.

low-pressure center An area on the Earth's surface that has relatively low atmospheric pressure compared with the surrounding air, is generally warmer, and is moist. Air circulates around low pressure inward and counterclockwise in the Northern Hemisphere.

low tide The time or times of day when the surface of the ocean is reduced to its lowest point by the gravitational attraction of the Sun and Moon.

luminescent The production of light without heat.

luminosity A measure of the rate at which stars radiate electromagnetic energy into space; commonly measured by comparing the luminosity of a star with that of the Sun's luminosity, which has a value of 1 Ls.

lunar eclipse The total or partial blocking of sunlight striking the Moon's surface by the Earth as it moves directly between the Moon and Sun.

lunar month The time it takes for the Moon to make one complete orbit around the Earth, which is approximately 29 days.

lungfish Specific fish species that have both lungs and gills, which allow them to breathe both underwater and on land.

luster A characteristic property of a mineral that describes how it reflects light from its surface; commonly classified as metallic or nonmetallic.

M

mafic rocks A specific class of igneous rocks that are generally dark and dense and contain iron and magnesium.

magma Hot, molten rock located within the Earth's crust.

magma chambers Tubes, tunnels, or large cavities in the Earth's crust through which magma travels or collects.

magnetic energy A form of energy that is associated with magnetic fields.

magnetic north The direction on the Earth toward which a magnetic needle points, which is slightly different than the geographical North Pole, which is located directly on the Earth's axis of rotation.

magnetosphere The term used to describe the magnetic field that surrounds the Earth and stretches out into space.

main sequence A classification of a star that is in the middle part of its life cycle and is actively undergoing nuclear fusion.

Mammalia A taxonomical classification for a class of animals commonly known as mammals.

mammals Organisms that produce milk for their young from specialized glands called mammaries.

mantle The extremely hot, dense inner layer of the Earth that makes up most of the planet's total volume.

mantle convection Large convection cells that are believed to exist in the Earth's upper mantle.

mares A Latin word for "seas," used to describe the flat, dark plains that cover the surface of the Moon.

marine A term that refers to anything associated with the oceans.

maritime polar air mass A large mass of cold, moist air that forms over the oceans near the poles.

maritime tropical air mass A large mass of warm, moist air that forms over the ocean near the equator.

marrow The soft, nutritious material found at the center of a bone.

Marsupials An order of mammals that raise their young in pouches attached to their bodies.

mass A measurement of the amount of matter a substance contains.

mass extinction The widespread disappearance of a great number of species on Earth in a short period.

mass wasting The rapid, down-slope movement of large masses of rock and soil.

matter Something that occupies space and has mass.

meandering The reoccurring S-shaped curves in of a river or stream.

mechanical energy A form of energy that involves physical movement by a mechanism.

mechanism A system of parts that operate or interact like a machine.

medial moraines Glacial sediments that are located near the middle of a glacier.

melting The change in phase from a solid to a liquid.

membrane A thin, flexible layer of tissue.

mesopause A transitional layer in the Earth's atmosphere that lies between the mesosphere and thermosphere, where temperature does not change with an increase in height.

mesoscale winds Medium-scale winds that affect areas over distances between 1 and 900 miles.

mesosphere A layer of the Earth's atmosphere between 30 and 60 miles above the Earth's surface where temperature decreases with height.

mesotrophic lake The classification for a middle-aged lake that is relatively clear, deep, and low in available nutrients.

Mesozoic era A specific division of the geologic time scale that existed between 251 and 65 million years ago, which includes the rise and fall of dinosaur species.

metallic mineral resources Specific mined mineral resources that are composed of metal elements.

metamorphic rocks A class of rocks that are formed when igneous or sedimentary rocks are changed into a new rock by exposure to intense heat and pressure.

meteor A small chunk of rock, no larger than a few feet in diameter, that is traveling through space and enters the Earth's atmosphere; commonly known as a shooting star.

meteor shower An event that describes a group of meteors entering and burning up in the Earth's atmosphere.

meteorites A meteor that does not burn up in the atmosphere, and strikes the Earth's surface.

meteoroids Small chunks or rock, no larger than a few feet in diameter, that travel through space.

meteorologist A scientist who studies all aspects of the Earth's atmosphere and weather.

meteorology The scientific discipline that studies all aspects of the Earth's atmosphere and weather.

methane A colorless, odorless gas that is flammable and is composed of one carbon atom and four hydrogen atoms (CH4).

metric system A specific system of measurement commonly used in science that uses the meter as a base for length, the kilogram as a base for mass, and the liter as a base for volume.

microscopic A term that describes something that is very small and cannot be seen with the naked eye.

mid-latitude cyclone A specific type of low-pressure weather system that develops over regions located in the middle latitudes.

mid-ocean ridges Divergent tectonic plate boundaries located on the ocean floor, where new crust is formed that pushes on the two plates, causing them to spread apart.

middle clouds Specific clouds that form between 6500 and 23,00 feet above the Earth's surface.

mildew A common term used to describe a type of fungus that grows in dark, damp areas.

Milky Way The name for the galaxy of stars in which our solar system is located.

millibar A unit of measurement commonly used in meteorology to record atmospheric pressure.

mineral deposit A relatively high concentration of a specific mineral located within the Earth's crust.

mineral ore The raw, unprocessed rock form of a desired mineral resource found in the Earth's crust.

mineral resources The valuable minerals that are located in specific locations in the Earth's crust and can be mined.

mineral A naturally occurring, inorganic, crystalline substance that has specific physical properties.

mining The process or specific technique of removing mineral resources from the Earth's crust

mixture A blend of different substances.

modified Mercalli scale A scale used to measure the intensity of an earthquake based on potential damage.

moist adiabatic lapse rate The rate of temperature decrease associated with an increase in height of moist air.

molds A type of fungi that grows on organic material, such as food.

molecules An arrangement of atoms in specific proportions that are bound together and have specific physical and chemical properties.

mollisol A specific order of soil with a thick, rich A-horizon, or topsoil, that typically forms in grassland biomes, such as the American Midwest.

Mollusca A taxonomical classification for a phylum that includes all mollusks, such as snails, squid, clams, and octopuses.

molten rock Rock that is in the liquid state of matter that only occurs at extremely high temperatures.

Monera A taxonomical classification for the kingdom of single-celled organisms that include bacteria.

monomineralic rock A specific type of rock that is made up from only one mineral, such as rock salt or limestone.

Monotremata A taxonomical classification for the order of mammals that lay eggs, such as the platypus.

Moon The name for any large celestial body that orbits around a planet.

Moon phases The series of different appearances of the Moon as observed from Earth, which results from the varying amount of light that illuminates the Moon at specific points in its orbit.

moraines Large amounts of unsorted glacial sediments that are deposited by a melting glacier.

mountain breeze A local wind system that occurs when air blows down the side of a mountain at night.

mouth The term used to describe the end point of a river where it empties into a lake or ocean.

mud slide The rapid down-slope movement of soil that is saturated with water.

multicelled An organism that is composed of many cells.

mushrooms A type of fungus that reproduces by forming a large fruiting body.

mutated To be altered by a change in the structure of an organism's genes.

mutualism A type of symbiotic relationship in which both organisms benefit.

mutualistic A type of relationship by which both parties benefit.

mycorrhizae Symbiotic fungi that are associated with the root systems of plants.

N

natural law A well-accepted fact that describes a process or processes that occur in nature.

neap tides The least amount of tidal activity that occurs when the Moon is at its first and third quarter positions in its orbit.

nekton A term used to describe aquatic organisms that can move freely by swimming, such as fish.

Nematoda A taxonomical classification for a phylum of organisms commonly known as roundworms.

Neogene period A division of the geologic time scale that began approximately 24 million years ago and ended 1.6 million years ago, which marked the appearance of grasses and grazing animals on the Earth.

Neolithic A term that means "new stone age" and is a division of time that began more than 53,000 years ago and ended approximately 5,500 years ago; it was marked by the widespread use of stone tools by human beings.

neritic zone A life zone in an aquatic ecosystem that stretches from the shoreline out to water that is approximately 600 feet deep.

neurotoxins A type of toxic chemical compound that affects the nervous system of animals.
neutron star A star that is extremely dense and mostly composed of neutrons.
neutrons A subatomic particle with a neutral electrical charge that is found in the nucleus of an atom.
New Moon A particular phase of the Moon in which it appears totally darkened and that occurs when it is aligned between the Sun and Earth. The New Moon phase also marks the starting point of the orbit of the Moon around the Earth.
nitrate A chemical compound composed of one atom of nitrogen and three atoms of oxygen (NO3-); a common plant fertilizer.
nitric acid A strong acid (HNO3) that is formed in the atmosphere when rain mixes with nitrogen compounds to create acid precipitation.
nitrification The process of converting atmospheric nitrogen (N2) to nitrogen compounds such as nitrate (NO3-) that are usable by plants; called nitrogen fixation.
nitrifying bacteria A specific type of bacteria that converts atmospheric nitrogen (N2) to plant-usable forms of nitrogen, such as nitrate (NO3-).
nitrogen cycle The movement and recycling of nitrogen compounds through the environment.
nitrogen-fixing bacteria *See* nitrifying bacteria.
nitrogen oxides Compounds that are composed of nitrogen and oxygen (NOx).
nitrous oxide A gas that contains two atoms of nitrogen and one atom of oxygen (N2O).
nocturnal A term used to describe an organism that is active during the night.
nonmetallic mineral resources Mineral resources that are not considered metals, such as limestone, rock salt, gypsum, diamonds, and marble.
nonvascular plants Plants that do not have an organized network of vascular bundles that delivers water and nutrients throughout the plant.
nor'easter A strong low-pressure storm system that affects the northeastern coast of North America, blowing in cold air and precipitation from the Northeast.
notocord A flexible rodlike structure that runs along the back of the embryos of all vertebrates.
nuclear fusion A nuclear reaction that is caused by combining, or fusing, two elements, which results in the creation of a great amount of energy.
nucleus The central core of an atom, which is composed of a specific number of protons and neutrons.
nutrient pollution A form of water pollution caused by an increase in nutrients in an aquatic ecosystem that leads to a rapid increase of aquatic plants and algae.

O

O-horizon The uppermost layer of a soil, which is composed of partially decayed organic material.
oblate A slightly flattened sphere.
observation The direct perception of something by use of one of the five human senses.
occluded front A frontal boundary that usually occurs when a rapidly moving cold air mass overtakes a warm air mass by wedging underneath it and lifting upward, creating very unstable weather.
ocean An extremely large body of saltwater that covers approximately 72% of the Earth's surface and divides the continents.
oceanic crust The portion of the Earth's crust that lies below the oceans; it is typically more dense than continental crust and is composed of basaltic rock.
oceanic zone A life zone in the open ocean that makes up the deep water environment that exists in water with a depth of 600 feet or greater.
oil A fossil fuel that is formed deep within the Earth's crust from the decayed remains of plankton that lived in the oceans millions of years ago.

oligotrophic lakes A classification of lake that has been recently formed from glacial melt waters, which has very clear, cold water and is very low in nutrients or aquatic life.

omnivores Organisms that eat both plants and animals, such as humans.

Oort cloud A hypothetical area that is located approximately 100,000 astronomical units from the orbit of the planet Pluto, where comets are believed to originate. It is named for the Dutch astronomer Jan Oort, who first proposed it.

open pit mining A method of mining that involves the extraction of minerals from the Earth by digging large open pits.

orbit The elliptical path that a celestial object takes as it revolves around another celestial object.

order A taxonomical classification for a group of animals that have similar characteristics.

Ordovician period A division of the geologic time scale that existed between 490 and 443 million years ago, which marked the rise of corals and coral reefs in the ocean.

organic A term that is used to describe any compound that contains carbon; also refers to something that is associated with a living thing.

organic material Material that is derived from a living thing.

original horizontality A natural law used in geology that assumes that all sedimentary rock layers were laid down in horizontal layers when they originally were formed.

orographical lifting A process that occurs when a large mass of air is forced to rise upward along the windward side of a mountain.

orographical precipitation A type of precipitation that occurs as a result of orographic lifting, which causes moist air to cool and condense when it is forced to rise up the windward side of a mountain.

oscilloscope A type of instrument used to display electrical motion in the form of a wave.

Osteichthyes A taxonomical classification used to describe a class of vertebrates commonly known as bony fish.

outgassing The process of expelling gas from a substance.

overburden A large mass of rock, soil, or ice that lies on top of something.

oxidation The process of adding oxygen to a chemical compound.

oxisol A highly weathered soil found in tropical rainforest biomes, which has a very thin A-horizon, or topsoil, and a deep B-horizon that contains a high concentration of iron oxide.

oxygen cycle The movement and recycling of oxygen through the environment.

ozone A colorless, gaseous compound composed of three atoms of oxygen (O3).

ozone hole An area over Antarctica that has a reduced level of ozone gas in the stratosphere.

ozone layer The specific area in the stratosphere, located at an altitude between 10 and 20 miles, that contains a high concentration of ozone gas.

P

P wave *See* compressional primary wave.

paleoclimatology The scientific discipline that studies the history of the Earth's climates.

Paleogene period A division of the geologic time scale that occurred between 65 and 24 million years ago and began with the extinction of the dinosaurs.

Paleolithic A term used to describe a period of time that ended approximately 100,000 years ago; also called the Old Stone Age.

Paleozoic era A large division of the geologic time scale that occurred between 544 and 251 million years ago, which began with the appearance of a variety of aquatic organisms and ended with the first appearance of the dinosaurs.

Pangea The name of a supercontinent that existed more than 250 million years ago, when all the present-day continents were joined together into one great landmass.

Parallel A term to describe two objects that are located in the same plane and are equally distant from each other

parasitism A symbiotic relationship between two organisms in which the host organism is harmed by a parasite.

parent element The specific element that is the starting point of the radioactive decay of a radioisotope.

parent material The term used to describe the specific rocks from which soil minerals come.

parent rocks The specific rocks from which the rock particles that make up sediments come.

partial eclipse A partial blocking of light caused by the movement of a celestial object in front of another celestial object.

pendulum A device that consists of a weight suspended from a fixed point, which is caused to swing back and forth under the influence of gravity.

penumbra The lighter area located next to the umbra, or darkened shadow, that occurs during an eclipse.

percent error A unit of measurement that mathematically describes the difference between a measured value and the accepted value (% error = difference between the measured value and the accepted value/the accepted value (100); also known as percent deviation.

perihelion The point in a planet's orbit around the Sun when it is closest to the Sun.

Perissodactyla A taxonomical classification used to describe a specific order of mammals that have hooves with an odd number of toes, such as horses, zebras, and rhinoceroses.

permafrost Permanently frozen soil.

permanent gases Gases that are in fixed amounts in the Earth's atmosphere.

Permian period A division of the geologic time scale that occurred between 251 and 290 million years ago; it was marked by the first appearance of mammal-like reptiles and the mass extinction of many marine species, including trilobites.

perpendicular Two objects or lines that are at a right angle to one another.

pH The unit of measurement used to measure the acidity or alkalinity of a solution.

Phanerozoic eon A large division of the geologic time scale that began 544 million years ago with the appearance of a variety of marine organisms and continues to this day.

phosphate A chemical compound composed of one atom of phosphorus and four atoms of oxygen (PO43-); an important plant fertilizer.

phosphorus cycle The movement and recycling of phosphorus through the environment.

photosphere The outer, visible layer of the Sun.

photosynthesis A chemical reaction that certain organisms utilize for energy, combining water, carbon dioxide, and solar energy to form carbohydrates, oxygen, and water.

photosynthetic A term to describe an organism that uses photosynthesis to derive energy.

phylum A taxonomical classification for a group of organisms that share common traits.

psychrometer A weather instrument used to measure the dry and wet bulb temperatures, which determines the dew point and relative humidity of the atmosphere.

physical weathering The process of breaking down rocks into smaller rock particles by a physical process, where no chemical changes take place.

physicist A scientist who studies physics, which is the study of matter and energy and their interactions

phytoplankton A type of plankton that utilizes photosynthesis to gain energy.

piedmont glacier A specific type of glacier that forms at the base of a mountain range where valley glaciers come together to form a large sheet of ice

piezoelectric The generation of an electrical current by crystals that are subjected to mechanical stress.

pioneer community A group of organisms that populate an area on the Earth where no organisms have lived before.

plane of the ecliptic *See* ecliptic.

planetary winds Large-scale winds that circulate air around the Earth, such as the trade winds or the westerlies.

planet A large celestial object that orbits around a star.

plankton Tiny, free-floating organisms that live in water.

plant uptake The process by which a plant absorbs water and nutrients into its root system.

Plantae The taxonomical classification for the kingdom of organisms that includes all plants.

plasma The fourth state of matter, which occurs when matter exists in the form of an ionized gas.

plate boundaries Specific areas on the Earth's crust where two or more tectonic plates interact with one another.

plate tectonics The theory that describes the Earth's crust as being divided into distinct plates that float on a semiliquid mantle and move relative to one another, which helps to explain the occurrence of earthquakes, volcanoes, mid-ocean ridges, mountain ranges, deep sea trenches, and deformed rock structures.

Platyhelminthes A taxonomical classification for a phylum of organisms commonly known as flatworms.

plesiosaurs Aquatic, reptilelike dinosaurs that lived during the Cretaceous period, which grew to more than 40 feet in length and used paddlelike feet to propel itself through the water.

point bar The point in a stream or river, located on the inside curve of a meander, where sediments are deposited.

polar jet stream An area of high-level winds located high in the atmosphere in the upper latitudes.

polar molecule A molecule that has a weak positive and negative electric charge.

polar regions The geographical areas that lie close to the poles of the planet.

Polaris A star that is located directly over the North Pole; also called the North Star.

pollen Tiny, dustlike male reproductive cells produced by a flower.

pollution An undesirable change in the quality of the environment that negatively affects the health of organisms living there.

polychlorinated biphenyls (PCBs) A class of human-created chemical compounds used in the production of electrical parts.

polymineralic rocks Rocks that are composed of two or more different minerals.

pond A very small, inland body of water.

population A group of the same species of organisms living in the same area.

pore spaces Tiny vessels in soil or rock that hold air and water.

pores Tiny openings in a substance.

Porifera A taxonomical classification for a specific phylum of animals commonly known as sponges.

porous A substance that is composed of many pores, which allow liquids and gases to pass through.

potential energy The stored energy, or capacity for something to perform work or cause change.

Precambrian A large unit of geologic time when not much life flourished on the Earth, which started with the formation of the planet and ended approximately 544 million years ago.

precession The wobbling motion of the axis of a rapidly rotating body, such as the Earth, which causes the tilt of the axis to change periodically.

precipitate The process of a solid forming out of a solution.

precipitation Liquid or solid water formed in clouds that falls to the surface of the Earth.

prefrontal fog A layer of fog close to the ground that forms in front of an approaching air mass.

pressure Force applied over a surface, which is measured in force per unit area.

pressure center A region on the Earth's surface where the atmospheric pressure is relatively low or high as compared with the surrounding air, around which air circulates.

pressure gradient The change in atmospheric pressure that occurs over a specific distance.

pressure gradient force The rate of atmospheric pressure decrease over a specific area, which forces air to move, forming wind.

pressure unloading The release of pressure off a rock mass, caused when overlying rock or ice is removed.

prevailing winds Dominant, planetary scale winds that move air along the surface of the Earth, such as the trade winds or the westerlies.

primary consumers A feeding classification for any organism that consumes plants; also called herbivores.

primary production The process by which plants utilize photosynthesis to convert solar energy into chemical energy that is stored in plant material.

primary succession The introduction of a community of organisms into an area where life has not existed before.

primary wave *See* compressional primary wave.

primate A taxonomical classification for an order of animals that includes monkeys, apes, lemurs, and humans.

prime meridian An imaginary line drawn on the Earth's surface that connects the North Pole to the South Pole, which runs through Greenwich, England, and represents zero degrees longitude.

primitive An early or original state of development.

principle of superposition A natural law used in geology that states that in undisturbed rock layers, the oldest layers are located at the bottom and the youngest are at the top.

principle of uniformity A natural law used in geology that states that the geologic processes that are currently shaping the Earth have been occurring all throughout the Earth's history.

procedure The series of steps taken to perform an experiment.

producers A feeding classification for all organisms that utilize photosynthesis or chemosynthesis as means to derive energy and create food.

productivity A term used to describe how much biomass is created by plants in a specific area.

profundal zone An aquatic life zone located in deep open water where it is too dark for photosynthesis too occur.

protein A nitrogen-containing chemical compound that is important for the growth and repair of an organism's tissues.

Proterozoic eon A large division of the geologic time scale that began approximately 2.5 million years ago and ended 544 million years ago with the appearance of many aquatic life forms on the Earth.

Protista A taxonomical classification for the kingdom of organisms commonly called protists, which include plankton, algae, and diatoms.

proto star A stage in the life cycle of a star when a stellar nebula has collapsed and nuclear fusion is beginning to occur, forming an embryonic star.

protons Subatomic particles with a positive electrical charge found in the nucleus of an atom.

pterosaur A type of long-winged dinosaur that could fly.

pulsars A type of neutron star that regularly emits periodic radio signals.

purpose The reason an experiment is performed.

pyroclastic Fragmented material ejected from a volcano, which includes ash, cinders, and volcanic rock.

pyroclastic flow The extremely hot gas, ash, and volcanic material that is ejected from a volcano during an eruption and rapidly moves downhill.

Q

quarks The theoretical basic particles that make up all subatomic particles, such as protons and neutrons.

quasars A very high energy celestial object believed to be a type of galaxy that is rapidly moving away from the center of the universe.

Quaternary period A division of the geologic time scale that began 1.6 million years ago with the appearance of human beings and includes the present day.

quiet eruption A type of volcanic eruption that releases lava in a gentle outpouring associated with a shield cone or fissure eruption.

R

radar A term that stands for Radio Detecting and Ranging, which describes a device that uses radio waves to bounce off an object and track its location, speed, and movement.

radiant energy Energy in the form of visible light.

radiation Waves or particles that are emitted from a substance; *see* electromagnetic radiation.

radiative cooling The cooling of an object as a result of its emitting electromagnetic radiation into the atmosphere or into space.

radioactive decay The breakdown of one element by the release of subatomic particles, which forms a new element.

radiometric dating A method of dating objects utilizing the known decay rate of certain radio isotopes.

rain The liquid form of precipitation.

recharge The process by which groundwater is replenished when surface water infiltrates into the ground.

recrystallization The rearrangement of crystals in a rock by exposure to extreme heat and pressure.

recycle To reuse something

red algae A harmful form of marine algae that produces neurotoxins; commonly called the red tide.

red dwarf star A classification for a dim star that is small, red, and has a relatively cool temperature.

red giants A classification for a star that is very bright, large, red, and has a relatively cool temperature.

red supergiant A classification of a star that is extremely bright and large, red, and has a medium temperature.

reflection The act of electromagnetic energy bouncing off a surface.

refraction The act of electromagnetic energy being altered in its direction of travel when it passes through certain substances

regeneration The process of reconstructing something.

relative humidity A unit of measurement that records the amount of moisture present in the air, usually expressed as a percentage.

replication The process of reproducing something.

reptiles A class of animals that reproduce by laying eggs and are cold blooded, such as lizards and snakes.

Reptilia A taxonomical classification for the class of animals commonly known as reptiles.

reradiate *See* reradiation.

reradiation The process by which an object takes in electromagnetic radiation and reemits it into the atmosphere.

reservoir A storage place for something.

residual heat Heat left over from something.

residual parent material Parent rock on which a soil has formed that has not been transported.

respiration The chemical process by which carbohydrates are broken down in the presence of oxygen to derive energy and produce carbon dioxide.

retractable claws A type of claw that can be pushed into and pulled out of an animal's foot, such as a cat's claw.

retreat To move back.

retrograde To move backward or in an opposite direction.

revolution The movement of an object in an orbit around another object.

revolutionize To change drastically.

rice paddies A type of agricultural field that is used to grow rice, which is periodically flooded with water.

Richter scale A scale that measures the energy released by an earthquake.

riffle The particular point in a flowing body of water that is relatively straight and shallow, where sediments are deposited, forming turbulent water.

rift valley A valley that forms along a divergent plate boundary, where two tectonic plates are spreading apart.

right ascension A coordinate used to measure the east-west position of a celestial body.

ring system An well-defined area of debris composed of rock and ice that is orbiting a planet and appears to form a ring or series of rings around the planet.

river A large flowing body of water

rock cycle A model of the processes that cause the formation, movement, and recycling of rock material on the Earth.

rock slide The rapid down-slope movement of rock.

rock The solid, crystalline substances that make up the Earth's crust, which are mostly composed of one or more mineral.

Rodentia A taxonomical classification for the order of mammals, commonly called rodents, that have continuously growing teeth, such as rats, mice, and squirrels.

rodents *See* Rodentia.

rotation The circular movement of a body around a central point called an axis.

runoff The rapid loss of soil, sediments, or other substances as a result of being washed away by rain or melting snow.

S

S wave (secondary wave) A seismic wave generated at the focus of an earthquake that travels in the form of a wave, is slower than a P wave, and can only pass through solids.

Saffir-Simpson scale A scale used to classify a hurricane by using its central pressure, wind speed, and damage.

salinity A measure of the mineral salt content of a solution.

sand A size classification for rock particles that are between 0.006 and 0.2 centimeters in diameter.

Santa Ana wind A warm, dry wind that blows off the high desert plateau of southern California.

saturated The point at which a substance can no longer absorb any more liquid.

sauropod A large type of long-necked, plant-eating dinosaur; commonly called the brontosaurus.

savanna A type of biome that experiences hot, dry summers and cool, rainy winters, where grass, shrubs, and small trees are the main type of vegetation. This biome is sometimes known as the tropical grassland.

scattering The random reflection and refraction of electromagnetic waves that occurs when it comes into contact with certain substances

scavenge To feed on dead or decaying organic material.

science The practice of observing, identifying, describing, and explaining natural phenomena.

scientific instruments Devices that are used to extend the human senses.

scientific method The specific set of procedures that scientists utilize to gain knowledge.

scientific notation A method of writing very large or very small numbers that uses exponents.

sea breeze A coastal wind that develops during the day when air blows from high pressure forming over cool water toward low pressure developing over the warmer land.

sea floor spreading The process of forming new oceanic crust at a mid-ocean ridge, which causes two tectonic plates to move, or spread apart from one another.

seawater Water that has a high concentration of minerals dissolved in it; also known as saltwater.

secondary consumers A classification of organisms that consume herbivores; also known as carnivores.

secondary succession The slow, gradual, and predictable replacement of one community of organisms by another in a specific area where life has already existed.

secondary wave *See* S wave.

sediment pollution A form of water pollution that is caused by the rapid introduction of sediments into an aquatic ecosystem as a result of runoff.

sedimentary mineral deposits Mineral resources that were deposited in the form of sediments.

sedimentary rocks A type of rock formed from rock particles that are compacted or cemented together to form one solid mass.

sedimentation The process of depositing sediments.

sediments Small fragments or particles of rock.

seepage The process of a liquid passing slowly through small openings in a substance.

segmented worms A specific phylum of worms that includes earthworms and leeches.

seismic waves Energy released by an earthquake that travels through the Earth in the form of waves.

seismograph A scientific instrument that is used to detect seismic waves generated by earthquakes.

semi-diurnal tides Two high and two low tides occurring each day.

shield cones The largest type of volcano, which is composed of piles of lava that form a steep, sloping, cone-shaped mountain.

short-period comets Comets that take less than 200 years to complete one orbit around the Sun.

short-wave radiation A high-energy form of electromagnetic energy that has a short wavelength.

siliceous ooze Mudlike sediments composed of silicates that form on the ocean floor.

silicate A chemical compound that is composed of atoms of silicon and oxygen.

silt A size classification for rock particles that are between 0.0004 and 0.006 centimeters in diameter.

Silurian period A division of the geologic time scale that existed between 443 and 418 million years ago, during which the first land plants and animals appeared on the Earth.

single-celled organisms Organisms that are composed of only one cell, such as bacteria.

Sirenia A taxonomical classification for an order of aquatic mammals that eat plants, such as the manatee.

slime molds A type of protist that grows in the damp organic material located on the forest floor.

slope An inclined surface that is at a particular angle to the horizon.

snowflakes A solid form of precipitation that has unique crystal shapes.

soil A mixture of minerals, organic material, air, and water that forms at the surface of the Earth.

soil horizon A well-defined layer of soil that has specific characteristics.

soil minerals The rock particles that are found in a soil.

soil moisture The amount of water that is present in a soil.

soil orders Specific classifications of unique soils that are found on the Earth.

soil profile The cross sectional view of a particular soil that shows all the soil's horizons.

solar cycle The time it takes the Sun to go from low sun spot activity to high sun spot activity and back again, which is approximately 11 years.

solar eclipse The total or partial blocking of the Sun as viewed from the Earth when the Moon passes in front of it.

solar flares Large flamelike emissions of hot plasma and radiation that leap off the surface of the Sun.

Solar Nebula The theoretical cloud of gas and dust that is believed to have formed the solar system.

solar radiation The electromagnetic radiation that is emitted from the Sun.

solar system The group of nine planets and other celestial bodies that orbit around a main sequence star called the Sun.

solar wind The stream of particles and electromagnetic radiation that is emitted from the Sun and travels out into space in all directions.

solid The state of matter in which atoms form a crystal structure and have the most restricted movement.

solidification The change in phase from a liquid to a solid; also called crystallization.

solution A mixture of two or more substances that have become dissolved.

solvent A substance that is capable of dissolving another substance.

south equatorial current A wind-driven, surface ocean current that moves water westward along the equator away from the west coast of South America.

southern oscillation The periodic change in the locations of low and high atmospheric pressure systems over the Pacific Ocean near the equator.

specialist populations Species of organisms that consume very specific types of food.

species A taxonomical classification used to describe a fundamental group of organisms that can interbreed.

specific gravity A unit of measurement used to describe the weight of a substance per unit volume.

specific heat A unit of measurement used to describe the ratio between the temperature change of a substance and its ability to absorb, hold, or release heat.

spectral class A classification used to describe a star's unique spectrum.

spherical A three-dimensional ball-like shape.

spiral galaxy A type of galaxy in which the stars are arranged in a spiral shape.

spores The reproductive cells of fungi.

spring tides The time of maximum tides, which occurs during the time of a Full or New Moon.

squall line A rapidly moving line of thunderstorms that is located ahead of a cold front.

standard atmosphere A theoretical state of the atmosphere that assumes it has equal density, temperature, and a pressure of 29.92 inches of mercury or 14.7 pounds per square inch.

standard system *See* metric system.

standing freshwater A still body of freshwater.

star A large, shining, spherical celestial object that is held together by its own gravity and is undergoing nuclear fusion.

starch A complex molecule that is made up of carbon, hydrogen, and oxygen and is produced by plants.

station model A coded symbol used on weather maps to display specific atmospheric variables.

stationary front A front that develops between two air masses that are not moving.

stegosaurus A type of dinosaur that had a series of large platelike scales running along its back and a tail that possessed four or five large spikes.

stellar nebula A large mass of collapsing gas and dust located in the universe that forms stars and planets.

stony meteorites A classification used to describe meteorites that are composed of silica.

stony-iron meteorites A classification used to describe meteorites that are composed of silica and iron.

storm surge The rise in water level that is caused by the strong winds associated with an approaching hurricane.

strata Horizontal layers of sedimentary rocks.

stratopause An isothermal layer in the atmosphere located between the stratosphere and the mesosphere.

stratosphere A layer in the Earth's atmosphere where the ozone layer exists and temperature increases with an increase in altitude.

streak The colored powder left behind after a mineral is rubbed against a surface.

stream erosion The movement of rock, soil, or sediments in a flowing body of water.

stream features The specific physical characteristics of a stream.

stream A small, flowing body of water.

strip mining A method of exposing and removing mineral resources from the ground that involves the digging up of rock and soil located above the mineral deposit.

stromatolites Large moundlike layers of sediments that form when cyanobacteria trap sand in warm, shallow ocean water.

subsoil The layer of soil, also known as the B-horizon, that lies directly below the topsoil.

subduction The movement of one tectonic plate underneath another at a convergent plate boundary.

subduction zone A narrow zone located at a convergent plate boundary where subduction is occurring, leading to the formation of deep ocean trenches, volcanoes, and earthquake activity.

sublimation The term to describe the phase change from a solid to a gas.

subtropical jet stream A narrow band of rapidly moving winds located in the upper atmosphere near the lower latitudes.

sulfur dioxide A gaseous chemical compound composed of one atom of sulfur and two atoms of oxygen (SO2).

sulfuric acid A strong acid (H2SO4) that forms in the atmosphere when sulfur dioxide gas reacts with atmospheric moisture.

summer solstice The time of the year when the Earth's tilted axis points the Northern Hemisphere toward the Sun, which usually occurs around June 21.

Sun The name of a medium-aged, main sequence star that lies at the center of the solar system.

sunspots Dark spots that appear on the surface of the Sun that are believed to be cooler areas on its surface.

supernova The violent explosion that is caused by a dying star when it blows off its atmosphere.

supercooled droplets Tiny drops of water that have temperatures below 32° Fahrenheit and exist in clouds.

surface mining A method of removing mineral resources from the Earth when they are located directly at the surface.

surface water Water that is located on the surface of the Earth.

surface wave A seismic wave formed from the interaction of other seismic waves at the Earth's surface caused by an earthquake, which causes the ground to move in a wavelike rolling pattern. This is the most destructive form of seismic wave; also known as a surface wave.

surface weather maps Maps that display weather conditions at the Earth's surface.

survey To determine the boundaries, elevations, distances, and other aspects of a specific portion of the Earth's surface.

suspension Free-moving, solid particles that are hanging in a liquid.

swamp A low region on land that is covered with shallow standing water.

symbiotic relationship A relationship between two different species of organisms in which one or both organisms benefit from the action of the other.

Syncline A U-shaped fold or depression in rock strata.

T

taiga *See* coniferous forest.

taxonomy A discipline of science that deals with the classification of organisms.

Technological Revolution A period of time that marked the beginning of the widespread use of technological processes in industry, which began near the middle of the nineteenth century. This period is often associated with the use of coal as fuel; also known as the Industrial Revolution.

technological system The specific process or series of processes that are associated with a specific technology.

technological systems model A model used to illustrate the steps in a technological process.

technology The application of human knowledge to solve problems or perform tasks.

tectonic forces The geologic processes and forces that are associated with plate tectonics.

tectonic plate A large portion of the Earth's crust that floats on the plasticlike upper mantle.

temperate forest A biome that exists in the middle latitudes, where the winters are very cold, the summers are hot, and the main type of vegetation is deciduous trees.

temperature The average amount of kinetic energy of the atoms and molecules in substance, which is commonly expressed as the degree of heat or cold measured by a thermometer.

temperature inversion The process by which a warm layer of air overlies a cold layer of air.

terminal moraines Unsorted glacial sediments that are located at the front or deposited near the front of a glacier.

terraces Steplike features located on a floodplain of a river or stream that mark the past location of the channel during a flood.

terrestrial A term that refers to land.

terrestrial ecosystems Ecosystems that are located on land.

terrestrial organisms Organisms that live on land.

terrestrial planet A planet that is mostly composed of rock.

terrestrial radiation Long-wavelength infrared radiation that is reradiated into the Earth's atmosphere from the land surface.

tertiary consumers A feeding classification used to describe carnivores that are at the top of a food chain.

Tertiary period A division of the geologic time scale that began approximately 65 million years ago with the extinction of the dinosaurs and ended 1.6 million years ago with the appearance of humans.

tetrahedron A four-sided geometric figure composed of four equilateral triangles.

texture The size and arrangements of crystals or sediments in a rock.

thawing The phase change that occurs when ice melts into liquid water.

theory A statement or statements used describe a phenomenon.

thermal energy Energy associated with heat.

thermal pollution A form of pollution that is associated with a change in temperature.

thermocline The layer of water located below the surface where the temperature drops rapidly.

thermohaline circulation The vertical distribution of water in the oceans that is caused by differences in the salinity and temperature of the water.

thermometer A scientific instrument used to measure temperature.

thermosphere A layer in the Earth's upper atmosphere where temperature increases with an increase in altitude.

Third Quarter Moon A particular phase of the Moon when it has completed three quarters of its orbit around the Earth and appears like a Half Moon when viewed from Earth.

thunderstorm A small-scale storm that is caused by the formation of a cumulonimbus cloud which produces strong winds, heavy rain or hail, lightning, and thunder.

tidal range The difference in height between the low and high tides.

tides The periodic rise and fall of the ocean surface that is caused by the gravitational attraction of the Sun and Moon.

tilling The process of turning over, or plowing, the soil.

tillites Sedimentary rocks made of sediments created by tilling.

tilted rock Disturbed rock strata that lies at an angle with the ground.

time A fundamental unit of measurement that records the specific interval that separates events.

topsoil The uppermost layer of soil that contains a high amount of organic material; also called the A-horizon.

topographic maps A two-dimensional map that displays the three-dimensional surface of the Earth by using contour lines that shows the elevation and shape of the land surface

tornado A rapidly rotating, funnel-shaped column of air located at the base of a cumulonimbus cloud that results in deadly high-velocity winds.

total eclipse The total blocking of light by one celestial body passing in front of another.

toxic heavy metals Naturally occurring, poisonous metal elements such as lead or mercury.

toxic inorganic chemicals Human-created chemical compounds that are poisonous to living organisms.

toxic organic compound A naturally occurring poisonous chemical compound.

trade winds Planetary scale winds in the Earth's atmosphere that form when areas of high atmospheric pressure located near 30 degrees north and south of the equator move air toward low pressure at the equator. The trade winds blow from the northeast toward the southwest in the Northern Hemisphere.

transform fault plate boundary A type of plate boundary where two tectonic plates slide along one another.

transmission The movement of electromagnetic radiation through a substance, without a change in its wavelength.

transpiration The movement of water through a soil, into a plant's body, and then back into the atmosphere when it evaporates off the leaf surface.

transported parent material Rock material that forms a soil that has been transported by wind, water, or a glacier.

transporting agent The means by which a substance is moved.

tree ring The ringlike growth of new wood in the trunk of tree that marks the occurrence of one growing season.

Triassic period A division of the geologic time scale that occurred between 251 and 206 million years ago, which began after the mass extinction of trilobite species and the appearance of the first dinosaurs.

tributaries Smaller flowing bodies of water that drain into a larger river system.

triceratops A type of dinosaur that had three horns located on its massive shield-like head.

trilobites A now extinct species of invertebrate aquatic animals that lived widely in the oceans of the Earth between 544 and 251 million years ago.

trophic level A term used to describe the feeding level classification of an organism in a food chain.

tropical A term that refers to something being located near the equator.

tropical depression A low atmospheric pressure system that forms in the tropics over the ocean, which consists of group of thunderstorms and cyclonic winds between 20 and 40 miles per hour.

tropical disturbance An organized group of thunderstorms with cyclonic winds that are less than 20 miles per hour.

tropical rainforest a type of biome that exists near the equator and experiences high temperatures and rainfall throughout the year, which supports rainforest vegetation.

tropical storm An organized group of thunderstorms associated with a strong low pressure system with cyclonic winds between 40 and 70 miles per hour.

tropics The geographic areas that lie close to the equator.

tropopause An isothermal layer in the atmosphere located between the troposphere and the stratosphere.

troposphere The lowest layer of the Earth's atmosphere that lies closest to the surface, where all weather takes place and the temperature decreases with an increase in altitude.

tundra A biome that is found in the higher latitudes where the temperature is below freezing for most of the year, and supports matlike vegetation.

typhoon A hurricane that forms and is located over the western Pacific Ocean.

Tyrannosaurus A type of large meat-eating dinosaur that walked upright on two legs.

U

ultraviolet radiation A specific high-energy form of electromagnetic radiation emitted from the Sun that can be harmful to living things.

umbra The area in shadow during an eclipse that is totally blocked from the light.

unconsolidated parent material Parent rock that forms a soil that has been broken apart and is no longer one mass.

underground mining A method of removing mineral resources from the Earth's crust that involves the digging of long, deep mine shafts to access the minerals.

unit A precise quantity that is used to describe a measurement.

universe The area in which all things exist.

updrafts Strong, vertical winds that move upward through a cloud.

uplift The process of lifting something upward.

upper air maps Weather maps that are used to display weather variables that exist in the upper atmosphere.

upwelling The uplift of cold ocean water from the bottom to the surface.

V

valley breeze A local wind that blows warm air up the sides of a mountain from the valley below.

valley glaciers Glaciers that are located in valleys between mountains.

vaporization The phase change that occurs when a liquid changes into a gas.

variable gases Gases in the Earth's atmosphere that change in their composition.

vascular plants A class of plants that have bundles of tiny tubes throughout their bodies that transport water and nutrients.

vector force A force that has both speed and direction.

velocity The speed at which something is moving.

vernal equinox The point in the Earth's orbit around the Sun, occurring on or around March 21, when the tilt of the Earth's axis is perpendicular to the plane of the Sun and the length of day light (12 hours) equals the amount of darkness (12 hours).

vertebrates A classification of organisms that have backbones.

visibility The greatest distance an observer can see through the atmosphere or through water.

visible light radiation A form of electromagnetic radiation that includes all the visible colors.

volcanic arc A chain of volcanic islands that forms near a convergent plate boundary located below the ocean.

volcanic ash Tiny particles of pyroclastic material produced by a volcanic eruption.

volcanic bomb A chunk of lava that is ejected into the sky by a volcanic eruption, which hardens into an aerodynamic shape before it strikes the ground.

volcanic eruption The often sudden and violent release of gas, dust, lava, and other pyroclastic material from a volcano.

volcanic rocks Igneous rocks that form from cooled lava produced by a volcano.

volcanic vents Cracks and tubes that carry lava that run along the side or below a volcano.

volcano An opening in the Earth's crust through which gas, dust, lava, and other pyroclastic materials flow to the surface.

volume The amount of space occupied by something, or how large an object is.

W

waning crescent A phase of the Moon that occurs near the end of its orbital period around the Earth, during which the only a crescent-shaped portion of the Moon is lit by the sun as viewed from Earth.

waning gibbous A phase of the Moon that occurs between the full and three quarter phases, during which three quarters of the Moon's surface is lit by the Sun as viewed from Earth.

warm front A front that develops when a warm air mass replaces a cold air mass.

wastewater Polluted water that is unfit for drinking or introduction into the environment because it has been used for some purpose and made unclean.

water molecule The chemical compound that is found in large quantities on the Earth which consists of two atoms of hydrogen and one atom of oxygen (H2O).

water table The uppermost level of the soil where the pore spaces are completely saturated by groundwater.

water vapor The gaseous form of water.

waterborne illness A type of disease or sickness that is caused by an organism that lives in water.

watershed The total land area that is drained by a particular river system.

waxing crescent A phase of the Moon that occurs near the beginning of its orbital period around the Earth, during which the only a crescent-shaped portion of the Moon is lit by the sun as viewed from Earth.

waxing gibbous A phase of the Moon that occurs between the first quarter and full Moon phases, during which three quarters of the Moon's surface is lit by the Sun as viewed from Earth.

weathered Something that has been exposed to the forces of weathering.

weathered mineral deposits Mineral resources that were formed as a result of the weathering of rocks.

weathering The physical or chemical processes by which large rocks are broken down into smaller rock particles.

weight The force that results from the gravitational attraction of the Earth on an object, which depends on its mass; commonly known as how heavy something is.

westerlies Planetary scale winds that occur in the middle latitudes, which move air from the southwest in the Northern Hemisphere and northwest in the Southern Hemisphere.

wet bulb temperature The lowest temperature that can be obtained by the evaporation of water into the air, usually measured using a psychrometer.

wetland An terrestrial ecosystem that is covered by shallow water for most of the year.

whirlwinds Small rotating winds that create weak cyclones, which kick up dust and debris.

white dwarf A classification of a star that is dimmer than the Sun, white, small, and extremely hot.

wilting point A term used to describe the state of soil moisture when a soil's pore spaces only contain a thin film of water, which is unavailable for use by plants, causing them to wilt.

wind The horizontal movement of air across the Earth's surface, from areas of high atmospheric pressure to areas of low atmospheric pressure.

wind-driven currents Surface ocean currents that are formed by planetary winds.

wind erosion The movement of soil and sediments by the force of wind.

wind sock An large cloth tube that is used to indicate the direction of the wind

wind vane A metal device used to indicate the direction of the wind.

windward The side of something that is facing away from the direction of the wind.

winter solstice The time of the year when the Earth's tilted axis points the Northern Hemisphere away from the Sun, which usually occurs on or around December 21.

Y

yeast A single-celled fungus that converts sugar into carbon dioxide and alcohol by the process of fermentation.

Z

zebra mussel A species of mollusk, named for its black and white shell, that is not native to North America and is currently overpopulating many fresh bodies of water in the northeastern United States.

zenith The point in the sky that is directly above the observer, or 90 degrees above the horizon.

zone of ablation The front of a glacier where ice is breaking off and melting.

zone of accumulation The area on a glacier where snowfall is building up, forming new glacial ice.

zone of flowage The area on a glacier where glacial ice is currently flowing.

zone of saturation The area of a soil where all the pores in a soil are filled with water.

zooplankton Tiny animals that float freely in water.

INDEX

Page numbers followed by a *t* or *f* indicate that the reference is specifically located in a table or figure, respectively. Page numbers in bold type indicate the location of a key term.

A
Abiotic, **500**
Ablation, 451
Abrasion, 188, 189*f*
Absorption, **406**
Abyssal zone, 506*f*
Acid deposition, **364**
  control of, 369–370
  effect on aquatic ecosystems, 364, 366, 367*f*, 368*f*
  effect on building materials, 369
  effect on humans, 367, 369
Acidic, **362**
Acid precipitation, 361, **362**–370
  anthropogenic gases and, 362–364, 365*f*
  control of, 369
  effect on aquatic ecosystems, 364, 366, 367*f*, 368*f*
  effect on building materials, 369
  effect on humans, 367, 369
  effect on terrestrial ecosystems, 366–367, 368*f*
  formation of, 362, 363*f*
  pH of, 362
  transport of pollutants in, 364, 366*f*
Acid shock, **366**
Acoustical physics, 37
Adhesion, **393**–394
Adiabatic cooling, **273**, 287
  cold fronts and, 314
Adirondack Mountains, acid precipitation in, 362*f*, 364, 368*f*
Aerobic bacteria, **463**
Aeronomer, 377
Aerovane, 310*f*, 311
African Rift Valley, 135
Agassiz, Alexander, 430
Agassiz, Jean, 453
Agnatha class, 580
Agricultural engineers, 14
Agricultural meteorologists, 293
Agricultural Revolution, 5, 7*f*
A-horizon, 206
Air, **242**. *See also* atmosphere.
  content in soil, 203–204
Air masses, **312**–314
  classification of, 312–314
  formation of, 312
  pressure systems and, 307
  source regions of, 312–314
Air pressure, 270–275
Air quality environmental technicians, 366
Albedo, **253**
Aldebaran, 77*f*, 78
Alfisol soil order, 207, 209*f*
Algae, **488**
  blue-green, 568
  brown, 570
  red, 570
  single-celled, 570
Algal bloom, **463**–464, 552
Alkalinity, **362**
Alligators, 583
Alluvial soils, **205**
Alpine glaciers, 452
Altocumulus clouds, 291, 292*f*
Altostratus clouds, 291, 292*f*
Aluminum, acid deposition and, 367
Alvarez, Luis, 230, 231
Alvarez, Walter, 231
Amber, 162, 166
American alligator, 583
Amoebae, 570
Amoebic dysentery, 467*t*

Amphibia class, 581, 582*f*, 583
Amphibians, 581, 582*f*, 583
Anadromous fish, 527
Anaerobic bacteria, 547
Analytical chemistry, 32
Andes mountains, 136
Andromeda galaxy, 106
Anemometer, 311
Angiosperms, **573**, 574*f*
Angle of insolation, **254**–256
Animalia kingdom, 568*f*, 574–579
Anions, 28
Annelida phylum, 577*f*, 578*f*
Anthracite coal, 174
Anthropogenic gases, 362–**364**, 365*f*
Anthropologists, 6
Anthropology, 6
Anticline, **127**
Anticyclone, 306
Aphelion, **52**, 53*f*
Aphotic zone, **416**, 417*f*
Apogee, 70
Apollo 65, 101
Apollo asteroids, 99
Appalachian mountain range, 136
Appalachian orogeny, 227
Apparent motion of celestial objects, 53–54
Aquaculture, 581
Aquatic ecosystems, 505, 506*f*, 511, **523**
  effect of acid deposition on, 364, 366, 367*f*, 368*f*
  marine ecosystems, 511, 523–529
Aquatic organisms, 500
Aquifers, 439*f*, 440*f*, 441
Arachnids, 579
Archaeologists, 6
Archaeopteryx, 229
Arctic air mass, 313–314
Arctic express, 314
Arctic tern, 501, 584
Area of convergence, 303
Area of discharge, 439
Area of divergence, 303
Arid climate, 518
Aristotle, 243
Arkansas River, 433
Arrhenius, Svante, 315
Artesian wells, **441**
Arthropoda phylum, 579–580
Arthropods, 579
Artiodactyla order, 586, 587*f*
Asteroid belt, **99**
Asteroids, **99**, 101
  Apollo, 99, 101
  Aten, 99
Asthenosphere, 48*f*, 118, **119**
  tectonic plates and, 132, 133*f*
Astronomers, 77
Astronomical unit (AU), 92, 93
Astrophysicists, 37
Aten asteroids, 99
Atlantic eel, 527
Atlantic Ocean, 390
  formation of, 228
Atmosphere, 48, **242**
  acid precipitation, 361, 362–370
  air masses, 312–314
  clouds, 281, 282–286
  composition of, 242–243
  fronts, 301, 314–319
  global climate change and, 345–356
  heating of, 256–258
  humidity, 281, 282–286
  insolation, 251, 252–258
  ozone depletion in, 375–382
  precipitation, 281, 292–295
  pressure, 263, 270–275
  storms, 325, 326–334
  structure of, 243–246
  temperature, 263, 264–269
  weather forecasting, 325, 334–339
  wind, 301, 302–312
Atmospheric convection cell, 304
Atmospheric moisture, **273**, 274*f*, 275, 282
  sources of, 282–283, 284*f*
Atmospheric pressure, 263, 270–275
  high and low, 271, 273, 274*f*
  measurement of, 270–271, 272*f*
  moisture and, 273, 274*f*, 275
Atmospheric scientist, 268
Atmospheric temperature, 263, 264–269
  distribution of heat on Earth, 264, 266–267
  greenhouse effect and, 268–269
  radiative cooling, 267–268
Atomic numbers, 26
Atoms, 26

Aurora borealis, 84
Australis borealis, 84
Australopithecus afarensis, 4*f*
Australopithecus africanus, 4*f*
Autotrophs, **536**, 537, 572
Autumnal equinox, 51, 52*f*
Aves class, 583–585
Axis, **50**–51, 52*f*
Azimuth, 54, 55*f*

B

Bacteria, 568–570
  aerobic, 463
  anaerobic, 547
  denitrifying, 550, 551*f*
  nitrogen cycling and, 547, 549, 550*f*
  nitrogen-fixing, 204
Bacterial dysentery, 467*t*
Barnard's star, 76, 77*f*
Barometer, **270**–271, 272*f*
Barometric pressure, 270–275
Basalt, **66**, 168, 169, 170*f*
Base 10 system, 18
Bathyal zone, 508*f*
Bats, 590, 591*f*
Bauxite, 179
Beaufort scale, 311–312
Becquerel, Antoine Henri, 217
Bedrock, 206*f*, 207
Benchmarks, 128–129
Benthic zone
  in lakes, 431, 432*f*
  in oceans, **416**–417, 530
Betelgeuse, 77*f*, 78, 79*f*
Bethe, Hans, 82
B-horizon, 206
Big bang, 104
Binder, 11
Bingham Canyon Copper Mine, 180
Binomial method, 568, 569
Biochemistry, 32
Biochemists, 550
Bioclastic texture, 169, 172*f*
Biogeochemical cycling, 535, **544**–552
  carbon cycling, 544–547, 548*f*
  nitrogen cycling, 547, 549–550, 551*f*
  oxygen cycling, 547, 549*f*
  phosphorous cycling, 550, 551*f*, 552
Biogeographers, 516
Biological anthropologists, 6
Biological characteristics, **568**
Biological diversity, **519**, 520*f*
Biological oceanographers, 528
Biological succession, 557, **558**–563
  pioneer communities, 558–559
  primary succession, 558, 559–560
  secondary succession, 560–563
Biologists, marine, 528
Biomass, **537**–538
Biomes, 511, **512**–523
  chaparral, 519, 521, 522*f*
  coniferous forests, 512, 513*f*, 514, 522*f*
  deserts, 518–519, 520*f*, 522*f*
  grasslands, 516, 517*f*, 522*f*
  mountains, 521, 522*f*
  savannas, 517–518, 522*f*
  temperate forests, 514, 515*f*, 516, 522*f*
  tropical rain forests, 519, 520*f*, 522*f*
  tundra, 512, 513*f*, 522*f*
Biometeorologists, 273
Biophysicists, 37, 537
Biosphere, 497
  biogeochemical cycling, 535, 544–552
  biological succession, 557, 558–563
  biomes, 511, 512–523
  ecosystems in, 499–505, 506*f*
  energy flow within, 535, 536–543
  taxonomy, 567, 568–591
Biotechnology, 15, 550, 571
Biotic, **500**
Biotite, 189
Bipedalism, 4
Birds, 583–585
Bituminous coal, 172
Bivalves, **577**–578, 579*f*
Bjerknes, Jacob, 477
Bjerknes, Wilhelm, 328
Black holes, **80**
Blue supergiants, 77*f*, **78**–79
Bohr, Niels, 26
Boiling point, **30**, **264**
Bony fish, 581, 582*f*
Boreal forests, 512, 513*f*, 514, 522*f*
Brahmagupta, 18
Brahmi numbers, 18
A Brief History of Time (1988), 80

Bristlecone pine, 572
Broad-leafed evergreens, 519
Bronze, 6
Bronze Age, 6, 9*f*
Brown algae, 570
Brown dwarf, 103
Buffer, **366**, 368*f*
Building materials, effects of acid deposition on, 369
Buoys, **482**
Burial, rock formation and, 176*f*

C

Calcite, 162*f*, 163*t*, 164, 165*t*, 166, 167, 189
  sedimentary rocks and, 170–171
Calcium carbonate, 546
Caldera, 151*f*, **152**
California current, **414**, 415*f*
Callisto, 96, 97*f*
Calories, 39
Calving, 452
Cambrian period, 220*f*, 221*f*, 224
Canadian Shield, 224
Canyons, submarine, 419, 421
Capillary action, **394**, 395*f*
Carbohydrates, 545
Carbonate rocks, 179
Carbonation, **189**, 190*f*
Carbon cycling, 544–547, 548*f*
Carbon dioxide
  in the atmosphere, 242*t*, 243
  carbon cycling and, 544–547
  climate change and, 351–352, 353, 355*f*
  coral bleaching and, 491
  effect on precipitation, 362
Carboniferous period, 220*f*, 221*f*, 227, 228*f*, 229*f*
  ice age in, 347
Carbon reservoir, **546**
Carnivora order, 586, 588*f*
Carnivores, **538**, 543*f*, 586, 588*f*
Carson, Rachel, 442
Cartilage, 580
Cartilaginous fish, 580–581
Cartographers, 57
Catadromous organisms, 527
Catastrophic events, **560**
Cations, 28
Caudal fin, 581
Cavendish, Henry, 282
Celestial objects, apparent motion of, 53–54
Celsius, Anders, 264
Celsius scale, 38*f*, 39, 264, 265*f*
Cementation, 169
Cenozoic era, 220*f*, 221*f*, 230, 232*f*, 233
  Quaternary period, 220*f*, 221*f*, 233
  Tertiary period, 220*f*, 221*f*, 230, 232*f*, 233
Cephalopods, 579
Ceres, 99
Cetacea order, 588, 589*f*
Channel, of a river, **435**, 437*f*
Chaparral, 519, 521, 522*f*
Charon, 99, 100*t*
Chemical engineers, 14, 15
Chemical limestone, 171, 172*f*
Chemical oceanographers, 391
Chemical weathering, **189**
  carbonation, 189, 191*f*
  hydration, 189
  oxidation, 189
Chemists, 32
Chemosynthesis, **536**
Chicxulub crater, 231
Chiron, 99
Chiroptera order, 588, 589*f*
Chloroflourocarbons (CFCs), 242*t*, 243, **354**, **378**–379, 380*f*
Cholera, 467*t*
Chondrichthyes class, 580
Choppers, 5*f*
Chordata phylum, 580–591
C-horizon, 206
Chromosphere of the Sun, 82*f*, 83
Cinder cone, 151*f*, 152
Cirrocumulus clouds, 291, 292*f*
Cirrostratus clouds, 291, 292*f*
Cirrus clouds, 291, 292*f*
Civil engineers, 14, 15
Clactonian flake, 5*f*
Clams, 578
Class
  Agnatha, 580
  Amphibia, 581, 582*f*, 583
  Aves, 583–585
  Chondrichthyes, 580–581
  Insecta, 580

Class—cont'd
  Mammalia, 585–591
  Osteichthyes, 581, 582*f*
  Reptila, 583
Classification of the living world. *See* taxonomy
Clastic rocks, 179
Clastic texture, 169, 172*f*
Claws
  nonretractable, 586
  retractable, 586, 588*f*
Clay, in soil, 204, 205*f*
Cleavage, mineral, 163–164, 165*t*
Climate, **345**, **476**
  arid, 518
  glaciers and, 455–456
  hot house, 346, 351
  ice age, 346, 347, 348*f*
  past, 346, 347*f*
Climate change, 345–356
  coral bleaching and, 490, 491*f*
  glaciations, 347, 348*f*
  Goldilocks syndrome and, 351–352, 353*f*
  hot house climates, 351
  human influence on, 352–356
  ice ages, 347, 348*f*
  Milankovitch cycles and, 347, 349–350
Climatologists, 256
Clouds, 281, **287**–291, 292*f*
  cold fronts and, 315*t*
  formation of, 287–289, 290*f*
  funnel, 329
  molecular, 104–105
  occluded fronts and, 317*t*
  types of, 290–291, 292*f*
  warm fronts and, 316*t*
Cnidaria phylum, 574, 576*f*, 577*f*
Coal, 166
  acid precipitation and, 364
  anthracite, 174
  bituminous, 172
Coal swamps, 227, 228*f*
Coastal wetlands, 523, 524*f*
Cobras, 583
Cohesion, **393**, 394
Cold-blooded vertebrates, 583
Cold front, **314**, 315*t*
Colorado River, 434*f*, 436
  pollution in, 470
Color, of a mineral, 162
Columbia River plateau, 150
Coma, of a comet, 102
Combustion, **363**, 547
Comets, **102**–103
  long-period, 103
  short-period, 103
Comma cloud, 319
Commensalism, **503**, 505
Community, 502*f*, **503**, 505, 527
Compaction, 170, 176*f*
Composite cone volcanoes, 151
Composite materials, 11
Composting, 203
Compounds, 31–32
Compression, 170
Compressional primary (P) wave, **145**, 146*f*, 147
Condensation, **30**, **287**, 404*f*, **405**
Condensation nuclei, **287**
Conduction, **37**, 257
Confined aquifer, **441**
Conglomerate, 170, 171*f*, 172*f*
Coniferous forests, 512, 513*f*, 514, 522*f*
Conifers, 227, 573, 574*f*
Conshelf program, 391
Constellations, 75
Consumers
  primary, 538, 543*f*
  secondary, 538, 539*f*, 543*f*
  tertiary, 538, 543*f*
Contact metamorphism, 174, 175*f*
Continental crust, **119**
Continental drift, 129–131, 153
Continental glaciers, 452, 453*f*
Continental polar air mass (cP), 312–313
Continental shelf, **419**, 421
Continental tropical air mass (cT), 313
Contour interval, **58**
Contour lines, **57**–58
Convection, **37**, **266–267**
  mantle, 136–137
Convection cells, **118**, 266*f*, **267**
Convection zone, of the Sun, 81, 82*f*
Convergence, **288**, 289*f*, 303
Convergent plate boundary, 134*f*, 135–136
Copernicus, Nicolas, 15, 67, 93

Coral bleaching, 487, **488**–492
  causes of, 489–491
  current research on, 491
  occurrence of, 489
Coral reefs, 346, **488**, 528
  bleaching of, 488–491
Coral reef scientists, 491
Corals, **488**
Core of the Sun, 81, 82*f*
Coriolis effect, **304**–306
Coriolis, Gustave-Gaspard, 305
Corona, 70, 82*f*, 83
Cousteau Foundation, 391
Cousteau, Jacques-Yves, 391
Crater Lake, 152*f*
Craters, impact, 66
Creep, 193
Cretaceous period, 220*f*, 221*f*, 229–230
  hot house climate in, 351
Crinoids, 227, 229*f*
Crows, 584
Crustaceans, 579, 580*f*
Crust of the Earth, 48, **118**–119
  continental, 119
  continental drift of, 129–131
  mantle convection and, 136–137
  movement of, 126–137
  oceanic, 118–119
  plate boundaries, 132, 134*f*, 135–136
  sea floor spreading and, 131–132, 133
  tectonic plates, 132, 134*f*
Crystalline, **162**, 169, 172*f*
Crystallization, **168**
Cultural anthropologists, 6
Cumulonimbus clouds, 290
  cold fronts and, 314, 315*t*
  precipitation and, 293*f*, 294
  thunderstorm development and, 326
Cumulus clouds, 290, 291*f*
  cold fronts and, 314, 315*t*
  thunderstorm development and, 326
  updrafts and, 292
Currents
  Gulf Stream, 338, 414, 415*f*
  ocean, 414, 415*f*, 416
  south equatorial, 476
Cut banks, 435
Cuvier, Georges, 219
Cyanobacteria, 569–570
Cycads, 229
Cyclones, 307, **318**–319, **330**

D

Dalton, John, 33
Dalton's law of partial pressure, 33
Dams, thermal pollution caused by, 469–470
Darwin, Charles, 507
Daughter element, 217
Davis, William Morris, 192
Deccan lava flows, 150
Deciduous forests, 514, 515*f*, 516, 522*f*
Declination, 54
Decomposers, **538**, 539*f*, 540, 543*f*
Deep pool, 435
Deficient, **66**
Deformed rock, 126–127, 128*f*
Delta, 433, 434*f*
Democritus, 26
Demos, 96
Deneb, 80
Denitrifying bacteria, 550, 551*f*
Dense soils, 204
Density, **18**–19, 20*t*
Depletion of the ozone layer, 375–382, **378**
Deposition, 187, **193**–194
  glacial till, 194
  graded bedding, 193, 194*f*
  horizontal sorting, 193–194
  moraines and, 194
De Revolutionibus (1543), 93
Derived units, 18
Deserts, 518–519, 520*f*, 522*f*
Detritivores, **539**, 540, 543*f*
Devonian period, 220*f*, 221*f*, 225, 226*f*
Dew point, **284**–285, 286*t*
  cloud formation and, 287
  cold fronts and, 315*t*
  occluded fronts and, 317*t*
  warm fronts and, 316*t*
Dew point depression, 285
D-horizon, 206
Dialogue Concerning the Two Chief World Systems (1632), 67
Diamonds, 163, 164*f*
Diatoms, 570
Dinosaurs, 583, 585
Discharge rate, **435**, 436*f*

Disease-causing agents, **466**–467
Disphotic zone, **416**, 417*f*
Displaced fossils, 128, 129*f*
Displaced rocks, 144
Dissolve, 396–**397**
Dissolved oxygen, **468**
Disturbed areas, **562**
Diurnal tides, 53
Divergent plate boundary, 132, 134*f*, 135
Dobson, Gordon, 379
Dobson ozone spectrometer, 380
Dobson unit, 378*f*, 379
Docile animals, 588
Drainage basins, 432, 433*f*
Dredging, 181
Drizzle, 293–294
Droughts, **516**
Dry adiabatic lapse rate, 288*f*
Dry bulb temperature, 284, 286*t*
Duration of insolation, **256**
Dust bowl, 192
Dust devils, 309
Dwarf ellipticals, 107
Dynamic equilibrium, **376**, **435**
Dynamic system, **557**
Dysentery
  amoebic, 467*t*
  bacterial, 467*t*

E

Earle, Sylvia, 490
Earth, 94, 100*t*, 389
  apparent motion of celestial objects, 53–54
  axis of, 50–51, 52*f*
  composition of, 48
  core of, 118
  crust of, 118–119
  distribution of heat on, 264, 266–267
  energy transfer in, 36–37
  freshwater, distribution on, 391–393
  geologic history of, 215, 216–234
  Goldilocks syndrome and, 352, 353*f*
  latitude and longitude on, 54–57
  mantle of, 118, 119*f*
  oceans on, 390–391, 413–421
  orbit of, 51–52, 53*f*
  place in universe, 104–109
  rotation of, 50–51
  shape of, 48–49
  size of, 49, 50*f*
  tides, 52–53, 54*f*
  topographical maps of, 57–58, 59*f*
Earthquakes, 143, **144**–148
  causes of, 144
  epicenter location, 145, 146*f*, 147
  measurement of, 147, 148*t*
  Mercalli scale, 147, 148*t*
  Richter scale, 147, 148*t*
  seismic waves of, 144–145, 146*f*
Earth's core, **118**
Earthworms, 204, 577, 578*f*
East Antarctic ice sheet, 453*f*
Eccentricity, **349**–350
Eclipse, **68**–70
  lunar, 68–69, 70
  solar, 68, 69–70
E. coli, 466
Ecological systems. *See* ecosystems.
Ecologists, 500, 501
Ecology, **499**, 500
Ecosphere, **504**
Ecosystems, **504**, 505*f*
  aquatic, 504, 506*f*, 507
  biogeochemical cycling, 535, 544–552
  communities in, 502*f*, 503, 507
  energy flow within, 535, 536–543
  habitats in, 500–501
  marine, 511, 523–529
  populations in, 501, 503
  terrestrial, 504, 505*f*, 507
Edentata order, 588, 589*f*
Einstein, Albert, 33
Ekman transport, 416*f*
Electrical engineers, 14, 15
Electromagnetic radiation, 34–35, 36*f*
Electromagnetic spectrum, **35**, 36*f*, 252*f*
Electrons, 26
Elementary particle physicists, 37
Elements, 26, 27*f*, 28
Ellipse, **52**
Elliptical galaxy, 107
El Niño southern oscillation (ENSO), 475–482, **479**
  coral bleaching and, 490
Energy, **16**, 17*t*, **33**–40
  chemosynthesis, 536

electromagnetic radiation, 34–35, 36*f*
flow through biosphere, 535, 536–543
food chains and, 540, 541*f*
food webs and, 540, 541*f*, 542*f*
heat and, 39
kinetic, 33, 34*t*, 264
law of the conservation of energy, 34
photosynthesis, 537
potential, 33–34
primary consumers and, 538, 543*f*
primary production, 537–538
secondary consumers and, 538, 539*f*, 543*f*
temperature and, 39
transfer of, 36–37
Energy pyramid, 540, 543
Engineers, 14, 15
marine, 481
water resource, 408
ENSO. *See* El Niño southern oscillation.
Enteritis, 467*t*
Entomologists, 577
Eon, **218**
Archean, 218–219, 220*f*, 221*f*, 222–223
Phanerozoic, 220*f*, 221*f*, 224–233
Proterozoic, 220*f*, 221*f*, 223–224
Epicenter of earthquakes, **145**, 146*f*, 147
Epochs, **233**
Equator, **54**, 55
Equilibrium, dynamic, 376
Equinoxes, 51
Eras, **224**
Cenozoic, 220*f*, 221*f*, 230, 232*f*, 233
Mesozoic, 220*f*, 221*f*, 228–230, 231*f*
Paleozoic, 220*f*, 221*f*, 224–228, 229*f*
Eratosthenes, 49, 50
Erosion, 187, **190**–193
agents of, 192–193
creep, 193
current research on, 192
mass wasting, 192–193
process of, 190–191
rock formation and, 176*f*
runoff, 191
soils and, 202, 203
stream, 191
wind, 192
Erratics, glacial, 452
Eruption, **149**
explosive, 150–152
fissure, 150
quiet, 149–150
Escherichia coli, 466
An Essay on Calcareous Manures (1832), 205
Estuarine zone, 506*f*
Eta Carinae, 79*f*
Euphotic zone, **416**, 417*f*, 506*f*
Europa, 96, 97, 98*f*
Eurypterids, 225
Eutrophication, **463**–464
Eutrophic lakes, 429*f*, **430**
Evaporate sedimentary rock, 171
Evaporation, 257, 283–284, **404**, 405*f*
sedimentary rock formation and, 171
Evapotranspiration, **283**, 404*f*, **406**–407, 408*f*
Evergreens
broad-leafed, 519
conifers, 227, 573, 574*f*
Evolution, 505
Exfoliation, 188
Exobiologists, 109
Exosphere, **45**–110, **245**
Earth, place in universe, 104–109
Earth, as a planet, 46–60
Moon, 64–71
Solar system, 91–103
stars, life of, 76–80
Sun, 81–85
Exotic species, 579
Experiments, **12**–14
The Exploration of the Colorado River (1875), 178
Explosive eruption volcanoes, 150–152
Extrusive rock, **168**, 170*f*
Eye of a hurricane, 331

F

The Face of the Earth (1909), 153
Fahrenheit, Daniel Gabriel, 264, 266
Fahrenheit scale, 38*f*, 39, 264, 265*f*
Farming, fish, 581
Faulted rock, 127, 128*f*
Faults, **127**, 216
Feathers, 584
Feldspar, 189
Felsic rocks, **169**, 170*f*
Fermentation, 571

Ferns, 573
Fiberglass, 11
Field capacity, 203
Filter feeders, 578
Fin, caudal, 581
Finger Lakes, 430
First Quarter Moon phase, 68, 69*f*
Fish
  anadromous, 527
  bony, 581, 582*f*
  cartilaginous, 580–581
Fish farming, 581
Fissure, 150
Fissure eruption, 150
Flatworms, 575, 576*f*, 578*f*
Floodplains, 434*f*, **435**–436, 437*f*
Flourite, 163, 164, 165*t*
Flowering plants, 573, 574*f*
Flowing springs, **408**
Flowing water, 431–438
Fluid physics, 37
Focus, **145**
Fog, **291**
  prefrontal, 315
Folded deformed rock, 126–127
Foliated rocks, **173**, 175*f*
Foliation, 173
Food chain, **540**, 541*f*
Food web, 540, 541*f*, 542*f*
Force
  Coriolis, 304–306
  pressure gradient, 302
  vector, 302
Foresters, 563
Forests
  coniferous, 512, 513*f*, 514, 522*f*
  deciduous, 514, 515*f*, 516, 522*f*
  temperate, 514, 515*f*, 516, 522*f*
  tropical rain, 519, 520*f*, 522*f*
Fossil fuels, **353**, 355*f*, **362**
  carbon cycling and, 546–547, 548*f*
  combustion, 362–364
Fossil limestone, 171, 172*f*
Fossils, 128, 129*f*, 169, 217
  index, 216*f*, 217
  past climates and, 346
Foucault, Jean, 51
Fractured minerals, 164, 165*t*
Franklin, Benjamin, 338
Freezing, 30
Freezing point, **264**
Freezing rain, 293*f*, 294
Freon, **378**
Freshwater, **391**
  distribution on Earth, 391–393
  groundwater, 427, 438–443
  surface water, 427, 428–438
Friction, **118**
Fronts, 301, **314**–319
  cold, 314, 315*t*
  mid-latitude cyclones and, 318–319
  occluded, 315–317
  stationary, 317, 318*f*
  warm, 314–315, 316*t*
Frost action, 188
Fruit, 573
Fujita scale, 330, 331*t*
Fujita, T. Theodore, 330
Full Moon, 68, 69*f*
Fungi kingdom, 568*f*, 571–572
Funnel clouds, 329
Fusion, nuclear, 76

G
Gabbro, 169, 170*f*
Gagnan, Emile, 391
Galaxy, **105**–107
  elliptical, 107
  irregular, 107, 108*f*
  spiral, 106
Galena, 162, 163*f*, 164*f*, 165*t*
Galilei, Galileo, 58, 67, 96
Game Management (1933), 519
Gamma rays, 35, 36*f*
Ganymede, 96
Garnet, 177, 178*t*
Gases
  greenhouse, 269
  permanent, 242–243
  trace, 243
  variable, 242*t*, 243
Gas giants, **92**, 96–99, 100*t*
Generalists, 501, 503
Geocentric model of the universe, 93
Geocentric theory, 67
Geochronology, 174

Geographers, 57
Geographic information systems specialists, 356
Geologic principles, 216–217
Geologic time scale, **216**, 218–233
  Archean eon, 218–219, 220*f*, 221*f*, 222–223
  Cenozoic era, 220*f*, 221*f*, 230, 232*f*, 233
  Mesozoic era, 220*f*, 221*f*, 228–230, 231*f*
  Paleozoic era, 220*f*, 221*f*, 224–228, 229*f*
  Phanerozoic eon, 220*f*, 221*f*, 224–233
  Proterozoic eon, 220*f*, 221*f*, 223–224
Geologists
  lunar, 68
  structural, 128
Geomorphologists, 190
Geophysicists, 119
Geosyncline, **129**
Germinate, **560**
Giant salamander, 581
Giant squid, 579
Giardia, 571
Giardiasis, 467*t*
Gizzard, **584**–585
Glacial advance, **454**
Glacial erratics, 452
Glacial front, **450**, 452
Glacial retreat, **454**
Glacial till, **194**, 205, 223, **451**
Glacial transport soils, 205
Glaciations, 347, 348*f*
  Milankovitch cycles and, 347, 349, 350
Glacier Bay National Park, 456*f*
Glaciers, **392**, 449, **450**–456
  anatomy of, 450–451
  biological succession and, 559
  current research on, 454
  deposition and, 194
  erosion and, 191
  global climate and, 455–456
  hydrologic cycle and, 406
  moraines and, 451–452
  movement of, 451
  types of, 452–454
Glaciologists, 452
Glands, mammary, 585
Gliese, 77*f*
Global climate change. *See* Climate change.
Globular cluster, 106
Glucose, 544–545
  photosynthesis and, 536, 544–545
Gneiss, 173*f*, 174, 175*f*
Gobi Desert, 518
Gold, 164
Goldilocks syndrome, 351–**352**, 353*f*
Gondwanaland, 153
Graded bedding, **193**, 194*f*
Gram, 17*t*, 18
Grand Canyon, rock strata of, 126*f*
Granite, 168, 169, 170*f*
  soil from, 205
Granules, 81, 83*f*
Grasslands, 516, 517*f*, 522*f*
Gravity, 52
  erosion and, 190, 192
Great white shark, 581
Greenhouse effect, **268**–269, 315
  climate change and, 351, 352*f*
Greenhouse gases, 243, **269**
  climate change and, 354, 356
Greenwich meridian, 56
Groundwater, **392**, 404*f*, **407**–408, 427, 438–443
  aquifers, 439*f*, 440*f*, 441
  flow of, 439, 441
  pollution of, 441–443
  recharge, 438–439
  water table, 438*f*, 439
Groundwater hydrologists, 443
Gulf Stream, 338, **414**, 415*f*
Gymnosperms, 572*f*, **573**, 574*f*

H

Habitats, **500**–501
Hadley cell, **304**, 305
Hadley, George, 305
Hail, 293*f*, 294–295
Half-life, **217**, 218*f*
Half Moon, 68
Halite, 162, 164, 165*t*, 166, 189
  sedimentary rock formation and, 171
Halley, Edmund, 257
Halley's comet, 103
Halo, 106
Hardness of a mineral, 163, 165*t*
Harrison, John, 57

Hawaiian Islands, formation of, 137
Hawking, Stephen, 80
Headwaters, 433, 434*f*
Heat, **39**
  distribution on Earth, 264, 266–267
  rock formation and, 176*f*
  specific, 39
Heat capacity, 39, **253**, **394**–395
Heat gradient, **264**
Heating the atmosphere, 256–258
Heat island effect, 328
Heat of vaporization, 30
Heliocentric theory, 67, 93
Hematite, 163*f*, 164, 165*t*
Hemispheres, 54
Hensen, Victor, 527
Hepatitis, 467*t*
Herbicides, **492**–493
Herbivores, **538**, 543*f*
Herpetologists, 577
Hertz, Heinrich, 328
Hertzsprung, Ejnar, 76
Hertzsprung-Russell (HR) diagram, 76, 77*f*
Hess, Harry, 133
Heterotrophs, **537**
High clouds, 291, 292*f*
High pressure, 273, 274*f*, 275
  polar highs, 303*f*, 304
  pressure center, 306–307
  pressure gradients and, 302
  subtropical highs, 304
Himalayan mountains, 135
Hindu-Arabic numeral system, 18
Holmes, Arthur, 219
Holocene period, hot house climate in, 351
Hominids, 6
Homo erectus, 4*f*
Homo habilis, 4
Homo sapiens, 4*f*, 5, **591**
Homo sapiens neanderthalensis, 4*f*
Horizontal sorting, **193**–194
Hornblende, 189
Horton, Robert, 405
Hot house climate, **346**, 351
Hot spot, 134*f*, **136**–137
Hubble constant, 105
Hubble, Edwin, 105
Hubble's law, 105
Hudson Canyon, 419, 421
Hudson River, 436
  pollution of, 466
Human origins, 4–6
Humid, **189**
Humidity, 281, **282**–286
  atmospheric moisture, 282
  dew point, 284–285, 286*t*
  relative, 283–284, 286*t*
  sources of atmospheric moisture, 282–283, 284*f*
Humus, **203**, 206*f*
Hurricanes, **330**–334
  Saffir-Simpson scale of, 333*t*
Hutton, James, 120, 216
Hydration, 189
Hydraulic mining, 180–181
Hydrocarbons, 227, **362**
  carbon cycling and, 546–547
Hydrogeologists, 443
Hydrolab project, 490
Hydrologic cycle, 282–283, **403**–408
  condensation, 404*f*, 405
  evaporation, 404, 405*f*
  evapotranspiration, 404*f*, 406–407, 408*f*
  groundwater and, 404*f*, 407–408
  infiltration, 404*f*, 407, 408*f*
  precipitation, 404*f*, 405–406
  runoff, 404*f*, 406, 407*f*, 408*f*
  surface water, 405
  water vapor, 404–405
Hydrologists, 397
  groundwater, 443
Hydrolysis, 189
Hydrophilic, **394**
Hydrosphere, **390**
  coral bleaching, 486–492
  El Niño southern oscillation (ENSO), 475–484
  glaciers, 449, 450–456
  groundwater, 427, 438–443
  hydrologic cycle, 403–408
  oceanography, 413–421
  pollution of, 461–470
  surface water, 427, 428–438
  water, on the Earth, 390–393
  water, properties of, 393–398

Hydrothermal vent communities, 528–529
  chemosynthesis and, 536
Hydrothermal vents, **418**–419
Hypothesis, **12**

I

Iapetus Ocean, 225
Ice, **396**
  hydrologic cycle and, 406
  properties of, 395–396
Ice age, **346**, 347, 348*f*
  Milankovitch cycles and, 347, 349, 350
Icebergs, **452**–453
Ice caps, 452, 454
Ice cores, **346**, 348*f*
Ice fields, 452, 454*f*
Ice pellets, 294
Ice sheets, 452, 453*f*
Icthyologists, 577
Igneous intrusions, 216
Igneous rocks, 168–169, 170*f*, 176*f*
  extrusive rocks, 168, 170*f*
  felsic rocks, 169, 170*f*
  identification of, 168–169, 170*f*
  intrusive rocks, 168, 170*f*
  mafic rocks, 169, 170*f*
Illness, waterborne, 466
Immune systems, **379**
Impact craters, **66**, 67*f*
Impact theory, 66
Impermeable, 439*f*, **441**
Index fossils, 216*f*, **217**
Indian Ocean, 390–391
Industrial Revolution, 8*f*, 9*t*
Infectious agents, **490**
Inference, **12**, 120
Infiltration, 404*f*, **407**, 408*f*
  of groundwater, 438, 439*f*
Information Age, 8*f*, 9
Infrared radiation, 268, 269
Infrared satellite images, **336**
Infrared waves, 36
Ingestion, 574
Inorganic, 162
Insecta class, 580
Insectivora order, 590
Insects, 580
Insolation, 251, **252**–258
  angle of, 254–256
  atmosphere and, 252–253
  duration of, 256
  Earth's surface and, 253–254
  heating the atmosphere, 256–258
  mesoscale winds and, 307–308
  Milankovitch cycles and, 349–350
  planetary winds and, 302
  radiation and, 252
International date line, 56
International system of measurement.
    See metric system.
Intertidal zones, 420*f*, **421**, **523**, 524*f*, 525
Intertropical convergence zone (ITCZ), 304
Intrusive rock, **168**, 170*f*
Inventions, 8, 10*t*
Invertebrate animals, **574**–580
Io, 96, 97
Ionosphere, 245–246
Ions, 28, **245**, 246
Iron Age, 6, 9*t*
Iron meteorites, 101
Irregular galaxy, 107, 108*f*
Irrigation, **408**
Isobars, **335**
Isostasy, **129**
Isothermal layer, **244**
Isotherms, **335**
Isotopes, 28

J

Jellyfish, 575
Jet stream, 309–**310**
Johanson, Don, 5
Jones, Ian, 417
Journey to the Center of the Earth (1864), 117
Juneau ice field, 452
Jupiter, 96–97, 100*t*
Jurassic period, 220*f*, 221*f*, 228–229

K

Kalstenius ice field, 454*f*
Katabatic wind, 309
Kelvin, Lord, 264
Kelvin scale, 38*f*, 39, 264, 265*f*
Keratin, 584
Kettle lakes, 453
Kevlar, 11

Kinetic energy, 16, 33, 34*f*, 264
  heat and, 39
  temperature and, 39
Kingdom
  Animalia, 568*f*, 574–591
  Fungi, 568*f*, 571–572
  Monera, 568–570
  Plantae, 568*f*, 572–574, 575*f*
  Protista, 568*f*, 570–571
Klepas, John, 491
Knot, 311
Koala bear, 586*f*
Komodo dragon, 583
Kuiper belt, 103

L
Lagomorpha order, 590
Lahars, **154**, 155*f*
Lake Baikal, 428
Lakes, 428–431, 432*f*
  eutrophic, 429*f*, 430
  kettle, 453
  life zones in, 430–431, 432*f*
  mesotrophic, 429*f*, 430
  oligotrophic, 428, 429*f*, 430
  oxbow, 434*f*, 436
  productivity of, 428, 429*f*, 430, 431*f*
Lake Superior, 405, 428
Lampreys, 580
Land breeze, 308–309
Landfills, **466**
La Niña, **480**–481
Lapse rate, **287**, 288*f*
Large Magellanic Cloud, 107, 108*f*
Latent heat, 257, 258
Latent heat of condensation, **39**, **257**–258
Latent heat of vaporization, 39
Lateral fault, 134*f*, 136
Lateral forces, **127**
Lateral moraines, 450*f*, **451**
Latitude, **55**, 56*f*
Lava, **149**
Lava domes, 151*f*, 152, 153*f*
Lava vents, **149**
Law of the conservation of energy, 34
Leaching, **202**
Leakey, Louis, 8
Leakey, Mary, 8
Leakey, Richard, 8
Leeward side of a mountain, 290*f*, 294*f*, 295
Legumes, 551, 552*f*
Length, **15**, 17*t*
Leopold, Aldo, 519
Levallois point, 5*f*
Lichen, 204, **503**, **558**–559
Lightning, **326**–327
Limestone, 177, 178*t*
  carbon cycling and, 546
  fossil, 171, 172*f*
  soil from, 205
Limiting factor, **552**
Limnetic zone, 431, 432*f*
Limnologist, 432
Linguists, 6
Linnaeus, Carolus, 219, 568, 569
Lions, 588*f*
Lithification, **169**
Lithosphere, 115, **118**, 119*f*
  deposition in, 187, 193–194
  earthquakes, 143, 144–148
  Earth's geologic history, 215, 216–234
  erosion in, 187, 190–193
  mineral resources, 161, 177–181
  minerals, 161, 162–167
  plate tectonics, 125–137
  rocks, 161, 167–177
  soil, 201, **202**–210
  volcanoes, 143, 149–155
  weathering in, 187, 188–190
Littoral zone, 431, 432*f*
Loam, **204**, 205*f*
  sandy, 204, 205*f*
Local Group, 106
Local winds, 309
Loess, 205
Longitude, **56**–57
Long-period comets, 102
Long-term weather forecasts, 336
Long-wave radiation, **252**
Low clouds, 290, 291*f*
Low pressure, 273, 274*f*, 275
  pressure center, 307
  pressure gradients and, 302
  subpolar lows, 304
Lucy, 5

Luminosity, **76**, 77*f*
Lunar eclipse, 68–**69**, 70
Lunar geologist, 68
Lunar month, **68**
Luster of a mineral, 162, 165*t*

M
Mafic rocks, **169**, 170*f*
Magma, 135
  metamorphic rock formation and, 174
Magma chambers, **149**
Magnetic north, **131**
Magnetite, 162, 164, 165*t*, 167
Magnetosphere, 48
Main sequence stage, 76
Malaria, 571
Mallett, Robert, 145
Mammalia class, 585–591
Mammalogists, 577
Mammals, 585–591
Mammary glands, 585
Manatees, 469, 588
Manhattan Project, 82, 231
Manta rays, 581
Mantle, 48, **118**, 119*f*
Mantle convection, **136–**137
Maps, topographical, 57–58, 59*f*
Mares, **66**, 67*f*
Marianis trench, 134*f*, 136
Marine biologists, 528
Marine ecosystems, 511, **523**–529
  benthic zone, 528
  coastal wetlands, 523, 524*f*
  neritic zone, 523, 524*f*, 525
  oceanic zone, 525, 526*f*, 527
Marine engineers, 482
Maritime polar air mass (mP), 313
Maritime tropical air mass (mT), 313
Mars, 94–96, 100*t*
  climate of, 352, 353*f*
Marsupials order, 585–586
Mass, **15**, 17*t*
Mass extinctions, 219, **224**, 227–228
Mass wasting, **192**–193
Matter, **26**–32
  atoms, 26
  compounds and molecules, 31–32
  elements, 26, 27*f*, 28
  particulate, 192
  states of, 29–31
Mauna Loa, carbon dioxide levels on, 355*f*
Maury, Matthew, 415
Maxima, 81
McGuire, Bill, 152
Meandering, **435**
Measurement
  metric system, 16, 17*t*, 18
  standard system of, 16, 17*t*
  units of, 16, 17*t*, 18
Mechanical engineers, 14
Mechanical weathering, 188–189
Medial moraines, 450*f*, **451**–452
Medieval warm period, 351
Melting point, 30
Melting, rock formation and, 176*f*
Meniscus, 395*f*
Mercalli scale, 147, 148*t*
Mercury, 92–93, 100*t*, 264
Meridians, 56
Mesopause, 245
Mesoscale winds, 307–309
Mesosphere, 245
  atmospheric pressure in, 271*f*
Mesotrophic lakes, 429*f*, **430**
Mesozoic era, 220*f*, 221*f*, 228–230, 231*f*
  Cretaceous period, 220*f*, 221*f*, 229–230
  Jurassic period, 220*f*, 221*f*, 228–229
  Triassic period, 220*f*, 221*f*, 228
Metal, heavy toxic, 364, 367
Metamorphic rocks, **173**–174, 175*f*, 176*f*
  identification of, 175*f*
  Meteorites, **101**, 102
  Meteoroids, **101**–102
Meteorologica, 243
Meteorologist
  agricultural, 293
  operational, 309
  physical, 246
  synoptic, 331
Meteors, **101**, 102
Meteor shower, **102**
Meter, 17*t*, 18
Methane, 547
  in the atmosphere, 242*t*, 243
  climate change and, 352–354, 356*f*

Metric system, 16, 17*t*, 18
Mica, 164, 165*t*, 167*t*
Microwaves, 35*f*, 36*f*
Mid-Atlantic Ridge, 131, 132*f*
  divergent plate boundaries and, 135
  magnetic reversals in, 133*f*
Middle clouds, 290–291, 292*f*
Mid-latitude cyclones, 318–319
Mid-ocean ridges, **131**
Milankovitch cycle, 347, 349–350
Milankovitch, Milutin, 349, 351
Mildew, 571
Milky Way, **106**, 107*f*
The Millennium Man, 5
Millibar, 271
Mineral deposits, 177, 178–179
Mineralologists, 166
Mineral ores, 177–178
Mineral resources, 161, **177**–181
  classification of, 177, 178*t*
  mineral deposits, 178–179
  mineral ores, 177–178
  mining techniques, 179–181
Minerals, 161, **162**–167
  common, 165*t*
  composition of, 166
  properties of, 162–166
  rock-forming, 167*t*
  rocks and, 166–167
  soil and, 202
Minima, 81
Mining techniques, 179–181
  dredging, 181
  hydraulic, 180–181
  open pit, 180
  strip, 180
  surface, 179–181
  underground, 179, 179*f*
Mississippi River, 433, 438
Mixture, 32
Mohs, Friedrich, 163
Mohs scale of mineral hardness, 163
Moist adiabatic lapse rate, 287, 288*f*
Moisture, atmospheric, 273, 274*f*, 275
Mojave Desert, 519
Moldboard plow, 8*f*
Molds, 571
  slime, 570–571
Molecular clouds, 104–105
Molecular physicists, 37
Molecules, 31–32
  polar, 393
  water, 393
Mollisol soil order, 207–208, 209*f*
Mollusca phylum, 576*f*, 577–580
Monera kingdom, 568–570
Monomineralic rocks, **166**
Monotremata order, 585
Monsoons, 476–477
Moon, 65–71, 100*t*
  composition of, 66–67
  eclipses, 68–70
  formation of, 66
  orbit, 68
  phases of, 68, 69*f*
  surface of, 67
Moon phases, **68**, 69*f*
Moons, Martian, **96**
Moraines, **194**, 450*f*, **451**–452
  lateral, 450*f*, 451
  medial, 450*f*, 451–452
  terminal, 450*f*, 451
Morrison Formation, 229
Mountains
  biome, 521, 522*f*
  cloud formation and, 288, 289*f*, 290*f*
  precipitation and, 294*f*, 295
Mount Pinatubo, 153, 155*t*
Mount St. Helens, 151*f*, 152, 153*f*, 154*t*
  biological succession and, 560–561
Mount Vesuvius, 143, 153
Mount Washington, 522
Mouth, of a river, 433, 434*f*
Murray, John, 103
Mushrooms, 571
Mussels, 577, 578–579
Mutated, **379**
Mutualism, **503**
Mycorrhizae, **571**–572

N

National Aeronautics and Space Administration (NASA), 570
  ozone monitoring of, 379
National Center for Atmospheric Research, 491

National Oceanographic and Atmospheric Administration (NOAA)
Equatorial Pacific Ocean Climate Studies program, 482
ozone monitoring of, 379
Natural selection, 504
Neap tide, **53**, 54*f*
Nebula
Solar, 81, 92
Stellar, 76
Nekton, 525, 527
Nematoda phylum, 575, 577, 578*f*
Nematodes, 575, 578*f*
Neolithic period, 6, 9*f*
Neptune, 98–99, 100*t*
Neritic zone, 523, 524*f*, 525
Neurotoxins, **570**
Neutrons, 26
Neutron stars, **79**
New Moon, 68, 69*f*
Newton, Sir Isaac, 52
Nichols, Mary, 8
Nile River, 406
Nimbostratus clouds, 290
precipitation and, 293*f*, 294
Nitric acid, **362**
Nitrification, 549
Nitrogen, in the atmosphere, 242
Nitrogen cycling, 547, 549–550, 551*f*
bacteria in, 569
Nitrogen-fixing bacteria, **204**
Nitrogen oxide, 550
emission of, 363–364, 365*f*
reduced visibility caused by, 369
Nitrous oxide, **354**
Nocturnal animals, **588**, 589*f*
Nonvascular plants, 573–574, 575*f*
Nor'easters, 309
North Atlantic deep water, 416
Northern Hemisphere, 54
Northern lights, 84
North Star, 49, 55, 78
Notocord, 580
Nuclear fusion, **76**
Nuclear physicists, 37
Nuclear power plants, thermal pollution from, 468–469
Nucleus, 568
of a comet, 102
of a spiral galaxy, 106
Numbers, 18
atomic, 26
Nutrient pollution, **463**–464

O
Oblate spheroid, 49
Observation, **12**, **15**
measurement and, 15–16
Obsidian, 168, 169, 170*f*
Occluded front, **315**–317
Oceanic crust, **118**–119
Oceanic zone, 525, 526*f*, 527
Oceanographers
biological, 528
chemical, 391
physical, 416
Oceans, 390–391, 413–421
continental shelves, 419, 421
current research on, 417
currents, 414, 415*f*, 416
formation of, 219
intertidal zones, 420*f*, 421
life zones in, 416–419
seawater, 414
thermohaline circulation, 416
Oglalla aquifer, 440*f*, 441
O-horizon, 206
Oil spills, 464, 465*f*
Oligotrophic lakes, **428**, 429*f*, 430
Olivine, 162, 165*t*, 167*t*, 190
Olympus Mons, 95
Omnivores, **538**, 543*f*
Oort clouds, **102**–103
Open pit mining, 180
Operational meteorologists, 309
Opposable thumb, 590
Optical physicists, 37
Orbit
of Earth, 51–52, 53*f*
of the Moon, 68
Order
Artiodactyla, 586, 587*f*
Carnivora, 587, 588*f*
Cetacea, 588, 589*f*
Chiroptera, 588, 589*f*

Order—cont'd
  Edentata, 588, 589*f*
  Insectivora, 590
  Lagomorpha, 590
  Marsupials, 585–586
  Monotremata, 585
  Perissodactyla, 586, 587*f*
  Primates, 590–591
  Rodentia, 589, 590*f*
  Sirenia, 588
Orders of soil, 207–208, 209*f*
Ordovician period, 220*f*, 221*f*, 224–225
  ice age in, 347
Ores, mineral, 177–178
Organic chemistry, 32
Organic compounds, toxic, 464–465
Organic layer of soil, 206
Organic material, **202**–203
Original horizontality, 126
The Origins of Continents and Oceans (1929), 131
Orion A., 104*f*
Ornithologists, 577
Orographic lifting, 288, 289*f*, 290*f*
Orographic precipitation, 294*f*, **295**
Orthoclase, 190
Osteichthyes class, 581, 582*f*
Outgassing, 219
Overburden, 170
Oxbow lakes, 434*f*, 436
Oxidation, **189**
Oxisol soil order, 208
Oxygen
  in the atmosphere, 242
  dissolved, 468
  ozone formation and, 376, 377*f*
  measurement of, 377–379
  photosynthesis and, 536
  in soil, 203–204
Oxygen cycling, 547, 549*f*
Oysters, 578
Ozone, 242*t*, 244, 252, **376**
Ozone depletion, 375–382
  effects of, 379
  measurement of, 377–379
  ozone hole, 379–380, 381*f*
  ozone layer, 376–377
  reducing, 380, 382
Ozone hole, **379**–380, 381*f*
Ozone layer, **244**–245, **376**–377

P

Pacific Ocean, 390
  monitoring, 482–483
Paleoclimatologists, 285
Paleoclimatology, **346**
Paleogene period, 230, 232*f*
Paleolithic period, 5, 9*f*
Paleontologists, 226*f*, 227
Paleozoic era, 220*f*, 221*f*, 224–228, 229*f*
  Cambrian period, 220*f*, 221*f*, 224
  Carboniferous period, 220*f*, 221*f*, 227, 228*f*, 229*f*
  Devonian period, 220*f*, 221*f*, 225, 226*f*
  Ordovician period, 220*f*, 221*f*, 224–225
  Permian period, 220*f*, 221*f*, 227–228
  Silurian period, 220*f*, 221*f*, 225
Pangea, 131, **227**, 229*f*
  breakup of, 228, 229
  formation of, 136
Parasitism, 503*f*, **504**
Parent element, 217
Parent material, **205**–206
  unconsolidated, 206
Parent rocks, 173
Parsec, 80
Particulate matter, 192
Past climates, 346, 347*f*
  glaciations, 347, 348*f*
  hot house climates, 351
  ice ages, 347, 348*f*
  Milankovitch cycles and, 347, 349–350
Pasteurization, 467
Pasteur, Louis, 467
Pauling, Linus, 536
PCBs, **466**
Pearls, 578
Penguins, 584
Penultimate Interglacial period, 351
Penumbra, **69**
Percent error, 20–21
Percolation, 404*f*, 408*f*
Perigee, 70
Perihelion, **52**, 53*f*
Perissodactyla order, 586, 587*f*
Permafrost, **512**

Permanent gases, **242**–243
Permian period, 220*f*, 221*f*, 227–228
  ice age in, 347
Perseid meteor shower, 102
Pesticides, pollution and, 465
Petrologists, 174
Phanerozoic eon
  Cenozoic era, 220*f*, 221*f*, 230, 232*f*, 233
  Mesozoic era, 220*f*, 221*f*, 228–230, 231*f*
  Paleozoic era, 220*f*, 221*f*, 224–228, 229*f*
Phase changes, 30
Phases of the Moon, 68, 69*f*
Phobos, 96
Phosphate, 177, 178*t*
Phosphorous cycling, 550, 551*f*, 552
Photosphere of the Sun, 81, 82*f*
Photosynthesis, 538
  carbon cycling and, 544, 545*f*, 548*f*
pH scale, 362
Phylum
  Annelida, 576*f*, 577, 578*f*
  Arthropoda, 579–580
  Chordata, 580–591
  Cnidaria, 575, 576*f*, 577*f*
  Mollusca, 576*f*, 577–579
  Nematoda, 575, 577, 578*f*
  Platyhelminthes, 575, 578*f*
  Porifera, 574–575
Physical anthropologists, 6
Physical chemistry, 32
Physical meteorologists, 246
Physical oceanographers, 416
Physical weathering, **188**–189
  abrasion, 188, 189*f*
  exfoliation, 188
  freezing and thawing, 188
  heating and cooling, 188
  pressure unloading, 188–189
Physicists, 37
Phytoplankton, **525**, 526*f*
Piedmont glaciers, 452
Piezoelectric minerals, 164
Pinchot, Gifford, 560
Pine, bristlecone, 572
Pioneer community, **558**–559
Placoderm, 226*f*
Planck constant, 36
Planck, Max, 36
Plane of the ecliptic, 51
Planetary scientists, 94
Planetary winds, **302**–306
Planets, **92**–99, 100*t*
  gaseous, 92, 96–99, 100*t*
  terrestrial, 48, 92–96, 100*t*
Plankton, 525, 526*f*, 527, 570
Plantae kingdom, 568*f*, 572–574, 575*f*
Plants
  flowering, 573, 574*f*
  nonvascular, 573–574, 575*f*
  seed, 572–573
  seedless, 573
  vascular, 572–573
Plant uptake, 202
Plasma, 30–31
Plate tectonics, 125–137
  continental drift, 129–131
  evidence of crustal movement, 126–129
  mantle convection, 136–137
  plate boundaries, **132**, 134*f*, 135–136
  sea floor spreading, 131–132, 133
  tectonic plates, 132, 134*f*
Platyhelminthes phylum, 575, 578*f*
Platypus, 585
Plesiosaur, 229
Plow, moldboard, 7*f*
Pluto, 99, 100*t*
Point bar, 435
Polar front jet stream, 310*f*
Polar highs, 303*f*, 304
Polaris, 49, 55, 77*f*, 78
Polar molecule, **393**
Pollen, **346**
Pollution, **461**
  disease-causing agents, 466–467
  groundwater, 441–443
  hydrosphere, 461–470
  nutrient, 463–464
  sediment, 462–463
  thermal, 468–470
  toxic inorganic compounds, 465–466
  toxic organic compounds, 464–465
  transport of acid-causing, 364, 366*f*
Polychlorinated biphenyls (PCBs), **466**
Polymineralic rocks, **166**
Population, **501**, 503
Pore spaces in soil, 203

Porifera phylum, 574–575
Portuguese man-of-war, 575
Potential energy, 33–34
Powell, John Wesley, 178
Precambrian time, 220*f*, 221*f*, **224**
Precession, 349, 350*f*
Precipitate rocks, 179
Precipitation, 281, **292**–295, 404*f*, **405**–406
  acid, 361, 362–370
  cold fronts and, 315*t*
  formation of, 292, 293*f*
  geologic, **170**–171
  occluded fronts and, 317*t*
  orographic, 294*f*, 295
  pH of, 362
  types of, 292–295
  warm fronts and, 316*t*
Prefrontal fog, 315
Pressure, 30
  atmospheric, 263, 270–275
  cold fronts and, 315*t*
  density and, 19
  hurricanes and, 333*t*
  occluded fronts and, 317*t*
  rock formation and, 176*f*
  warm fronts and, 316*t*
  weather maps of, 335, 336*f*, 337*f*
Pressure center, **306**, 307
Pressure gradient, **302**
Pressure system, 306–307
Pressure unloading, 188–189
Prevailing winds, 302–306
Primary consumers, 538, 543*f*
  energy pyramid and, 540, 543*f*
Primary production, **537**–538
Primary succession, **558**, 559–560
Primates, 590–591
Prime meridian, 56
Principle of superposition, **216**
Principle of uniformity, **216**
Procyon B, 77*f*, 78
Producers, 537, 543*f*
  energy pyramid and, 540, 543*f*
Production, primary, 537–538, 540, 543*f*
Productivity of lakes, 428, 429*f*, 430, 431*f*
Profundal zone, 431, 432*f*
Prominences, 82*f*
Prospero, Joseph, 192
Proterozoic eon, ice age in, 347
Protista kingdom, 568*f*, 570–571
Protons, 26
Proto star stage, 76
Protozoologists, 577
Psychrometer, 284–285
Pterosaur, 228
Ptolemy, 93
Pulsars, **79**
Pumice, 168, 169, 170*f*
P-wave, **145**, 146*f*, 147
Pyroclastic flow, **153**
Pyroclastic material, 151
Pyroxene, 162, 165*t*, 167*t*, 190

Q
Quantum theory, 36
Quarks, 26
Quartz, 163*t*, 164, 165*t*, 167*t*, 190
Quartzite, 174, 175*f*
Quasars, 107–109, **108**
Quaternary period, 220*f*, 221*f*, 233
Quiet eruption volcanoes, 149–150

R
Racloir, 5*f*
Radar, **336**
Radiation, **37**
  electromagnetic, 34–35, 36*f*
  infrared, 268, 269
  long-wave, 252
  short-wave, 252
  terrestrial, 257
  ultraviolet, 376, 377, 380, 381*f*
Radiative cooling, **267**–268
Radioactive decay, **118**
Radio communications, ionosphere and, 246
Radiometric dating, **217**–218
Radio waves, 35, 36*f*
Rain, 293*f*, 294
  acid. *See* acid precipitation.
  freezing, 293*f*, 294
Rapid mass wasting, 192–193
Rattlesnakes, 583
Rays, 581
Recharge areas, 438, 439*f*
Recrystallization, 173

Red algae, 570
Red dwarf stars, **76**, 77*f*
Red giants, 77*f*, **78**
Red shift, 104
Red supergiants, 78
Red tide, 570
Reflection, 35, 37*f*
Refraction, 35, 37*f*
Regeneration, **376**
Regional winds, 307–309
Reinforcer, 11
Relative humidity, 283–284, 286*t*
Remora, 503, 505
Reptila class, 583
Reptiles, 583
Reradiate, **253**, 256–257
Residual heat, **118**
Residual soils, 205
Respiration, **545**, 548*f*
Retractable claws, 586, 588*f*
Retrograde motion, 94
Revelle, Roger, 548
Revolution, **51**
Rhyolite, 169*f*, 170*f*
Rice paddies, **354**
Richter scale, 147, 148*t*
Riffle, 435
Rift valleys, **135**
Rigel, 77*f*, 78–79
Right ascension, 54
Rivers, 431–438
  delta of, 433, 434*f*
  flood plains of, 435–436, 437*f*
  headwaters of, 433, 434*f*
  life cycle of, 436, 437*f*, 438
  stream features, 434–435, 436*f*
  watersheds and, 432, 433*f*
Rock-forming minerals, 167*t*
Rocks, 161, 167–177
  cycle of, 175–176
  deformed, 126–127, 128*f*
  displaced, 144
  fossils in, 128, 129*f*
  igneous, 168–169, 170*f*, 176*f*
  metamorphic, 173–174, 175*f*, 176*f*
  minerals and, 166–167
  sedimentary, 169–172, 176*f*
  soils derived from, 205
  strata, 126–127, 128*f*
  volcanic, 66
Rock salt, sedimentary rock formation and, 171
Rocky intertidal zones, 421
Rodentia order, 589, 590*f*
Rodents, 589, 590*f*
Roman numeral system, 18
Ross, Sir John, 397
Roundworms, 574, 576*f*, 577
Rotation, **51**
  of Earth, 50–51
Ruffin, Edmund, 205
Runoff, **191**, 404*f*, **406**, 407*f*, 408*f*, 432
  nutrient pollution and, 463*f*
  sediment pollution and, 462
Russell, Henry, 76
Rust, 189

S
Saffir-Simpson scale, 333*t*
Sagan, Carl, 108
Sahara Desert, 519
Salamander, giant, 581
Salinity, **414**
Salmon, 527
Saltwater, 391, 392*f*
San Andreas fault, 134*f*, 136
Sandstone, 170, 171*f*, 172*f*
  soil from, 205
Sandy intertidal zones, 421
Sandy loam, 204, 205*f*
Santa Ana winds, 309
Satellites
  earthquakes and, 147
  weather forecasting and, 336
Saturation, 283–284
  of soil, 203
  zone of, 438–439
Saturn, 97–98, 100*t*
Sauropods, 229
Savannas, 516–518, 522*f*
Scallops, 577
Scattering, 35, 37*f*
Schistosomiasis, 467*t*
Science, **11**–15
  birth of, 11–12
  foundations of, 1–43

Science—cont'd
  scientific measurement, 15–21
  scientific method, 12, 13*t*
  technology and, 14–15
Scientific experiments, 12–14
Scientific measurement, 15–21
  density, 18–19, 20*t*
  observations and, 15–16
  percent error, 20–21
  scientific notation, 19–20
  units of measurement, 16, 17*t*, 18
Scientific method, **12**, 13*t*
Scientific notation, 19–20
Sea breeze, 307–308
Seawater, **391**, 392*f*, **414**
Seaweed, 570
Secondary consumers, 538, 539*f*, 543*f*
  energy pyramid and, 540, 543*f*
Secondary succession, **560**–563
Secondary (S) wave, 145, 146*f*, 147
Sedimentary rocks, **169**–172, 176*f*
  evaporation and, 171
  identification of, 172*f*
Sedimentation, 176*f*
Sedimentologists, 179
Sediment pollution, **462**–463
Sediments, 188, 190
  deposition of, 193
  glacial, 191
  transport of, 191
Seed coat, 573
Seedless plants, 573
Seed plants, 572–573
Seepage, of groundwater, 438
Segmented worms, 577
Seismic waves, **144**–145, 146*f*
Seismography, **144**–145, 146*f*
Seismologists, 147
Semi-diurnal tides, 53
Serengeti Plains, 518
Shale, 170, 172*f*
  soil from, 205
Sharks, 580–581
Shelter, 500
Shield cones, 149, 150*f*, 151*f*
Shoemaker-Levy 9 comet, 103
Shooting star, 101
Short-period comets, 103
Short-term weather forecasts, 336
Short-wave radiation, **252**
Silicate, **66**, 166
Siliceous ooze, 570
Siltstone, soil from, 205
Silurian period, 220*f*, 222*f*, 225
  ice age in, 347
Single-celled algae, 570
Sink, 39
Sirenia order, 588
Slate, 174, 175*f*
Slime molds, 570–571
Slope, of a river, 433, 434*f*
Sloth, tree, 588, 589*f*
Smith, Robert Angus, 363
Smith, William, 173
Snowflakes, 293*f*, 294
Soil, 190, 201, **202**–210
  air in, 203–204
  alluvial, 205
  classification of, 207–208, 209*f*
  layers of, 206–207
  minerals in, 202
  organic material in, 202–203
  organisms in, 204
  parent material, 205–206
  structure, 204, 205*f*
  water in, 203
Soil conservation technicians, 204
Soil horizons, **206**–207
Soil moisture, **203**
Soil profile, **206**
Solar astronomers, 83
Solar corona, 70
Solar cycle, 81–82
Solar eclipse, 68, **69**–70
Solar flares, **81**, 82, 83*f*, 84
Solar and Heliospheric Observatory (SOHO), 84
Solar Nebula, 81, 92
Solar radiation, **245**–246
Solar system, 91–103
  asteroids in, 99, 101
  formation of, 92
  gaseous planets, 92, 96–99, 100*t*
  location in the Milky Way, 107*f*
  meteoroids in, 101–102
  terrestrial planets, 92–96, 100*t*

Solar wind, **83**–84
Solidification, rock formation and, 176*f*
Solution, 32
Solvent, **396**–397
  universal, 396–397
Sonoran Desert, 519
Source, 39
Source regions of air masses, 312–314
South equatorial current, **476**, 477
Southern Cross, 55
Southern Hemisphere, 54
Southern oscillation, **476**–479
Space, habitats and, 501
Space weather, 83
Species, **501**
  exotic, 579
Specific gravity, 164
Specific heat, 39
Spectral class, **76**
Speed, 18
Spiral galaxy, 106
Spoilage, 572
Sponges, 574–575, 576*f*
Spores, 571, **573**
Springs, flowing, 408
Spring tides, **53**, 54*f*
Squall line, 314
Squid, 579
Standard system of measurement, 16, 17*t*
Standing water, 428–431, 432*f*
Starches, 536, 540, 545
Starfish, 580
Stars, 75–85, **76**
  black holes and, 80
  classification of, 76, 77*f*
  dwarf, 76, 77*f*, 78
  giants, 77*f*, 78
  neutron, 79
  shooting, 101
  supergiants, 77*f*, 78–79
Stationary fronts, **317**, 318*f*
Station model, **335**, 336*f*
Stegosaurus, 229
Stellar nebula, **76**, 105
Stingrays, 581
Stonehenge, 12*f*
Stony-iron meteorites, 101
Stony meteorites, 101
Storms, 325, 326–334
  hurricanes, 330–334
  thunderstorms, 326–328
Storm surge, **331**–332, 333*t*
Strata, **126**–127, 128*f*, 169, 171*f*
Stratocumulus clouds, 291*f*
Stratopause, 245
Stratosphere, **244**–245, **376**
  atmospheric pressure in, 271*f*
  ozone in, 376, 377–379
Stratus clouds, 290
Streak of a mineral, 162
Streak plate, 162
Stream erosion, **191**
Strip mining, 180
Stromatolites, **219**, 222–223, 571
Structural engineers, 14, 15
Structural geologists, 128
Structure of a soil, 204, 205*f*
Subduction, **136**
Subduction zones, 136
Sublimation, **30**, **283**, 284*f*
Submarine canyons, 419, 421
Subpolar lows, 304
Subsoil, **206**
Subtropical highs, 304
Subtropical jet stream, 310
Suess, Eduard, 153
Sulfur, 162, 165*t*
Sulfur dioxide, **362**
  climate change and, 350
  emissions of, 364, 365*f*
  reduced visibility caused by, 369
Sulfuric acid, **362**
Summer solstice, 51, 52*f*
Sun, 75, 81–85
  composition of, 81–84
  electromagnetic energy of, 36, 37
  formation of, 81
  insolation, 251, 252–258
  life cycle of, 84–85
  photosynthesis and, 536, 537, 543*f*
Sunquakes, 84
Sunspot cycle, 81–82
Sunspots, **81**–82, 83*f*
Supernova, 79
Supervolcanoes, 152
Surface mining, 179–181

Surface water, 392, **405**, 427, 428–438
  lakes, 428–431, 432*f*
  rivers, 431–438
  watersheds, 432, 433*f*
Surface wave, **145**
Surface weather maps, 335, 337*f*
Surveyors, 19
Suspension, **191**
S-wave, **145**, 146*f*, 147
Symbiotic relationships, **488**, **503**
Syncline, **127**
Synoptic meteorologists, 331
Synoptic weather maps, 335, 336*f*, 337*f*

T

Taconic Mountains, 225
Taiga, 512, 513*f*, 514, 522*f*
Tail of a comet, 102
Talc, 163, 164, 165*t*, 167
Talus, 188*f*
TAO array, 483
Taxonomy, 567, **568**–591
  invertebrate animals, 574–580
  Kingdom Animalia, 568*f*, 574–591
  Kingdom Fungi, 568*f*, 571–572
  Kingdom Monera, 568–570
  Kingdom Plantae, 568*f*, 572–574, 575*f*
  Kingdom Protista, 568*f*, 570–571
  vertebrate animals, 580–591
Technological Revolution, 7, 9*f*, 10*t*
Technological systems model, **9**, 10*f*
Technology, 4–11, **6**
  Agricultural Revolution, 6, 9*f*
  Bronze Age, 6–7, 9*f*
  composite, 11
  human origins and, 4–6
  Industrial Revolution, 7–9, 9*f*, 10*t*
  Iron Age, 6–7, 8*f*, 9*f*
  science and, 14–15
  Stone Age, 6, 9*f*
  technological systems, 9, 10*f*
  timeline of, 9*f*
Tectonic plates, 132, 134*f*
Television waves, 35*f*
Temperate forests, 514, 515*f*, 516, 522*f*
Temperature, 17*t*, **39**, **264**
  atmospheric, 263, 264–269
  atmospheric moisture and, 282
  Celsius scale, 264, 265*f*
  climate change and, 345–356
  cold fronts and, 315*t*
  density and, 19, 20*t*
  dew point, 284–285, 286*t*
  dry bulb, 284, 286*t*
  Fahrenheit scale, 264, 265*f*
  Kelvin scale, 264, 265*f*
  occluded fronts and, 317*t*
  scales of, 38*f*, 39
  warm fronts and, 316*t*
  weather maps of, 335, 336*f*
  wet bulb, 284–285, 286*t*
Temperature inversion, **244**, 245, 252
Tentacles, 579
Terminal moraines, 450*f*, **451**
Tern, arctic, 501, 584
Terraces, 435, 437*f*
Terrestrial ecosystems, 504*f*, 505, 511, **512**
  chaparral, 520, 521, 522*f*
  coniferous forests, 512, 513*f*, 514, 522*f*
  deserts, 518–519, 520*f*, 522*f*
  effects of acid precipitation on, 366–367, 368*f*
  grasslands, 516, 517*f*, 522*f*
  mountains, 521, 522*f*
  savannas, 516–518, 522*f*
  temperate forests, 514, 515*f*, 516, 522*f*
  tropical rain forests, 519, 520*f*, 522*f*
  tundra, 512, 513*f*, 522*f*
Terrestrial organisms, 500
Terrestrial planet, **48**, **92**–96, 100*t*
Terrestrial radiation, **257**
Tertiary consumers, 538, 543*f*
Tertiary period, 220*f*, 221*f*, 230, 232*f*, 233
  ice age in, 347
Tetrahedron, 166
Texture
  bioclastic, 169, 172*f*
  clastic, 169, 172*f*
  crystalline, 169, 172*f*
  vesicular, 168, 170*f*
Thales of Miletus, 12
Theory, **12**
  unifying, 125
Theory of natural selection, 504
Thermal pollution, **468**–470
Thermocline, **417**–418, 431

Thermohaline circulation, **416**
Thermometer, **264**
Thermosphere, 245
  atmospheric pressure in, 271*f*
Third Quarter Moon Phase, 68, 69*f*
Thompson, Lonnie, 454
Thumb, opposable, 590
Thunder, 327
Thunderstorms, **326**–328
  current research on, 328
  dangers of, 327–328
  life cycle of, 326, 328
  lightning from, 326–327
Tidal range, 53
Tidal waves, 144
Tides, 52–53, 54*f*
Tilling, **204**
Tillites, 223
Tilted rock, 127
Time, **16**
Titan, 98
Topographical maps, 57–58, 59*f*
Topography, cloud formation and, 288, 289*f*
Topsoil, 206
Tornadoes, **328**–330*f*, 331*t*
  Fujita scale of, 330, 331*t*
Torricelli, Evangelista, 271, 272
Toxic heavy metals, **364**, 367
Toxic inorganic compounds, 465–466
Toxic organic compounds, 464–465
Trace gases, 243
Trade winds, 302, 303*f*, 305, **476**
Transform fault plate boundary, 134*f*, 136
Transmission, 35, 37*f*
Transpiration, 404*f*, 406–407, 408*f*
Transported parent material, 205–206
Transporting agent, 190
Tree rings, climate change and, **346**, 347*f*
Tree sloth, 588, 589*f*
Triassic period, 220*f*, 221*f*, 228
Tributaries, 432, 434*f*
Trichinosis, 575, 577
Triton, 99
Trophic level, 541, 543*f*
Tropical Atmosphere Ocean (TAO) array, 482–483
Tropical depression, **331**
Tropical disturbance, 330–**331**
Tropical rain forests, 519, 520*f*, 522*f*
Tropical storm, **331**
Tropics, **519**
Tropopause, 244, 245*f*
  jet stream and, 310
Troposphere, **243**–244, 245*f*
  atmospheric pressure in, 271*f*
  convection in, 267
Tsunamis, 144
Tundra, **512**, 513*f*, 522*f*
Tunguska blast, 101
Turtles, 583
Tyndall, John, 289
Typhoid fever, 467*t*
Typhoons, 332*f*, **334**

U

Ultraviolet radiation, **244**, **376**
  coral bleaching and, 492
  ozone depletion and, 377, 380, 381*f*
Ultraviolet waves, 35*f*, 36*f*
Umbra, **69**
Unconsolidated parent material, 206
Underground mining, 179, 179*f*
Uniformitarianism, 120
Unifying theory, 125
Universal law of gravitation, 52
Universal solvent, 396–397
Universe
  age of, 104
  Earth's place in, 104–109
  galaxies in, 105–107, 108*f*
  life in, 109
  quasars in, 107–109
  size of, 104
Updrafts, **292**
Uplift, **287**–289
  rock formation and, 176*f*
Upper air maps, 335
Upwelling, **414**, 416, 476, 477*f*
Uranus, 98, 99*f*, 100*t*
Urban heat islands, 328
U.S. Forest Service, 560

V

Valley glaciers, 452, 455*f*
Vaporization, **30**
Variable gases, 242*t*, **243**
Vascular plants, 572–573

Vector force, 302
Vent, 149
Venus, 93–94, 100*t*
climate of, 352, 353*f*
Vernal equinox, 51, 52*f*
Vertebrate animals, 580–591
Vesicular texture, 168, 170*f*
Visibility
cold fronts and, 315*t*
occluded fronts and, 317*t*
warm fronts and, 316*t*
Volcanic arc, **136**
Volcanic ash, 153
Volcanic bomb, 152
Volcanic crater, 151*f*, 152
Volcanic eruptions, acid precipitation and, 362, 363*f*
Volcanic glass, 168, 170*f*
Volcanic rocks, **66**
Volcanoes, 143, **149**–155
active, 154*t*
carbon cycling and, 546
climate change and, 350
explosive eruption, 150–152
formation of, 149
hazards of, 153–154
quiet eruption, 149–150
Volcanologists, 151
Volume, 17*t*, 18

W

Walker circulation, 479
Walker, Sir Gilbert, 476–477, 478*f*, 479
Waning Crescent phase, 68, 69*f*
Waning Gibbous phase, 68, 69*f*
Warm front, **314**–315, 316*t*
Wastewater, **467**, **491**
Wastewater treatment operators, 468
Water, 32
adhesion and, 393–394
cohesion and, 393, 394
content in soil, 203
erosion and, 190–191
flowing, 431–438
freshwater, 391–393
groundwater, 392, 404*f*, 407–408, 427, 438–443
habitats and, 500
heat capacity of, 394–395
hydrologic cycle, 403–408
ice, properties of, 395–396
properties of, 393–398
seawater, 391, 392*f*, 414
as a solvent, 396–398
standing, 428–431, 432*f*
surface, 392, 405, 427, 428–438
Waterborne illnesses, **466**
Water resource engineers, 408
Watersheds, **432**, 433*f*
Waterspouts, 330
Water table, 438*f*, **439**
Water-transported soils, 205
Water vapor, 242*t*, **243**, **282**, **396**, **404**
in the atmosphere, 282–286
Waves
compressional primary (P), 145, 146*f*, 147
secondary (S), 145, 146*f*, 147
surface, 145
tidal, 144
Waxing Crescent phase, 68, 69*f*
Waxing Gibbous phase, 68, 69*f*
Weather forecasting, 325, 334–339
current research on, 338
data collection, 334, 335*f*
forecasts, 335–336
radar and, 336
satellites and, 336
synoptic weather maps, 335, 336*f*, 337*f*
Weathering, 187, 188–190
chemical, 189
physical, 188–189
rate of, 189–190
rock formation and, 176*f*
Weather observers, 316
Weather reporters, 338
Wegener, Alfred, 130, 131
Weight, **15**–16
Wells, artesian, 441
Werner, Abraham, 175
Westerlies, 302–303
Wet bulb temperature, 284–285, 286*t*
Wetlands, coastal, 523, 524*f*
Whale shark, 581
Whirlwinds, 309
White button mushrooms, 571
White dwarf stars, **76**, 77*f*, 78

Wien's law, 257
Wilting point, 203
Wind, 301, **302**–312
  cold fronts and, 315*t*
  hurricanes and, 333*t*
  jet stream, 309–310
  local, 309
  measurement of, 311–312
  mesoscale, 307–309
  occluded fronts and, 317*t*
  planetary, 302–306
  pressure gradient and, 302
  pressure systems and, 306–307
  tornadoes and, 329
  trade, 476
  warm fronts and, 316*t*
Wind-driven current, **414**
Wind erosion, 192
Wind sock, 311
Wind-transported soils, 205
Wind vane, 311
Windward side of a mountain, 290*f*, 294*f*, 295
Winter solstice, 51, 52*f*
World biomes. *See* biomes.
Worms
  earthworms, 577, 578*f*
  flatworms, 575, 576*f*, 578*f*
  roundworms, 575, 576*f*, 577
  segmented, 577

X

X rays, 35*f*, 36*f*

Y

Yeasts, 571

Z

Zebra mussels, **578**–579
Zenith, **54**, 55*f*
Zone of ablation, 450*f*, **451**
Zone of accumulation, **450**
Zone of flowage, **450**
Zone of saturation, 438–439
Zoologists, 577
Zooplankton, **525**, 526*f*